# Prior acclaim for Jeremy Griffith's treatise

'Frankly, I am 'blown away' as the saying goes…
The ground-breaking significance of this work is tremendous.'
**DR PATRICIA GLAZEBROOK**,
*Professor and Chair of Philosophy, Dalhousie University*
(Response to 'The Human Condition Documentary Proposal', 2004)

'It might help bring about a paradigm shift in the
self-image of humanity – an outcome that in the past
only the great world religions have achieved.'
**DR MIHALY CSIKSZENTMIHALYI**,
*Professor of Psychology, Claremont Graduate University*
(Response to 'The Human Condition Documentary Proposal', 2004)

'I am simply overwhelmed…I find it astonishing and impressive.'
**JOSEPH CHILTON PEARCE**, *American author of 'Magical Child'*
(Response to 'The Human Condition Documentary Proposal', 2004)

'Could you please send me an extra copy of your book?
[Mine] is on loan because it was so appreciated.'
**SIR LAURENS VAN DER POST**, *pre-eminent philosopher and author*
(Response to 'Free: The End of the Human Condition', 1988)

'a superb book…[that] brings out the truth of a new and
wider frontier for humankind, a forward view of a world of
humans no longer in naked competition amongst ourselves.'
**DR JOHN MORTON**, *Professor of Zoology, University of Auckland*
(Response to 'A Species In Denial', 2003)

'The proposal is indeed impressive.'
**DR ROGER LEWIN**, *prize-winning British science writer and author*
(Response to 'The Human Condition Documentary Proposal', 2004)

'I have never heard of anything comparable before.'
**DR FRIEDEMANN SCHRENK**,
*Professor of Paleobiology, Goethe University Frankfurt*
(Response to 'The Human Condition Documentary Proposal', 2004)

# FREEDOM

## The End Of The Human Condition

## Jeremy Griffith

www.HumanCondition.com  or
www.WorldTransformation.com  or
www.WorldTransformationCentre.com  or
www.TheBookThatSavesTheWorld.com  or
www.JeremyGriffith.com

*FREEDOM: The End Of The Human Condition* by Jeremy Griffith

**First Edition**, published in 2016, by WTM Publishing and Communications Pty Ltd (ACN 103 136 778) (www.wtmpublishing.com)

Cover design by Luke Causby/Blue Cork.
Cover image by Genevieve Salter and Jeremy Griffith.

 All enquiries to:
WORLD TRANSFORMATION MOVEMENT®
Email: info@worldtransformation.com
Website: www.humancondition.com  or  www.worldtransformation.com

The World Transformation Movement (WTM) is a global not-for-profit movement represented by WTM charities and centers around the world.

ISBN 978-1-74129-028-8
CIP – Biology, Philosophy, Psychology, Health

Edited by Fiona Cullen-Ward.
Typesetting: designed by Jeremy Griffith, set by Lee Jones & Polly Watson (16).
Font: Times; main body text: 11.4pt on 14.2pt leading; quote: 9.634pt, horizontal scale 100.4%, bold; quote source: 8.5pt; comment within a quote: 10.744pt; digits and all caps text: 1-2pts smaller than body text. For further details about the typesetting, styles and layout used in this book please view the *WTM Style Guide* at <www.humancondition.com/style-guide>.

With all their love and support of the truth that this book contains,
which ensured its creation, *FREEDOM* is dedicated to the future of
the human race by the World Transformation Movement's Founding
Members: Annabel Armstrong, Susan Armstrong, Sam Belfield,
John Biggs, Richard Biggs, Anthony Clarke, Lyn Collins,
Steve Collins, Lachlan Colquhoun, Eric Crooke, Emma Cullen-Ward,
Fiona Cullen-Ward, Anthony Cummins, Neil Duns, Sally Edgar,
Anna Fitzgerald, Brony FitzGerald, Connor FitzGerald,
Tony Gowing, Jeremy Griffith, Simon Griffith, Damon Isherwood,
Felicity Jackson, Charlotte James, Lee Jones, Monica Kodet,
Anthony Landahl, Doug Lobban, Tim Macartney-Snape, Simon Mackintosh,
Katrina Makim, Sean Makim, Manus McFadyen, Ken Miall, Tony Miall,
Rachel O'Brien, James Press, Stacy Rodger, Marcus Rowell,
Genevieve Salter, Will Salter, Nick Shaw, Wendy Skelton,
Pete Storey, Ali Watson, Polly Watson, Prue Watson, Tess Watson,
Tim Watson, James West, Stirling West, Prue Westbrook and Annie Williams.

# Notes to the Reader

- Unlike most publications, there is no **bibliography** at the conclusion of this book because the source is provided in small text at the end of each quote. While unconventional, it does mean you will always have the benefit of being able to immediately see when, where and by whom the quote was given, and experience has shown that it doesn't take long to learn to skip past it if you wish. Also, rather than give **the particular edition and/or publisher** of the book that the quote comes from, the page number where the quote appears and the total number of pages of the particular edition used for the source is provided. This enables the reader to find the comparative place of the quote in any edition. In addition to the **index** at the back of the book, any word or phrase can be easily searched in the online edition using the 'Search FREEDOM' facility in the top right-hand corner of the book's website at <www.humancondition.com>. All **paragraphs are numbered** to allow easy referencing across any of the book's formats. *FREEDOM* contains a number of colored **images**, however, in this print edition they appear in black and white. To view those images in color, see the online edition at <www.humancondition.com/freedom>. All **biblical references** are from the 1978 *New International Version* translation of the Bible.

- Since this book contains the only thing that can save the human race from extinction, namely the dignifying and relieving understanding of our species' psychologically troubled human condition, it must be made accessible to all people—which is why, alongside its availability for purchase through bookshops and Amazon, it is, and will always remain, freely available to be read, shared or printed (with binding instructions included) at <www.humancondition.com/freedom>.

## • The Special Edition, Produced For Scientists

This book has been presented in two versions and in two stages.

Since the book is about the biology of human behavior, the publisher's

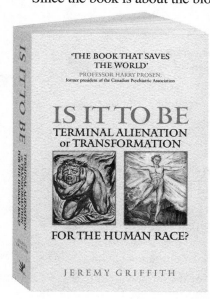

first responsibility was to promote it within the scientific community. This was done in 2014 by publishing the book with a title that emphasized the serious plight of humanity, which science is ultimately responsible for solving—that title being *IS IT TO BE Terminal Alienation or Transformation For The Human Race?* From July to September 2014 over 900 copies of *IS IT TO BE* were sent to leading scientific institutions, scientists and science commentators throughout the English-speaking world to introduce them to these world-saving understandings.

In 2016, this version, which is oriented to the general public, was published. Containing the same content as *IS IT TO BE* but with an additional first chapter and further insights, this version has the title <u>*FREEDOM: The End Of The Human Condition*</u>. <u>*FREEDOM* therefore replaces and supersedes *IS IT TO BE*</u>.

# Contents

PAGE  PARA-GRAPH

**Introduction**  by Professor Harry Prosen............................ 15    1–

**Chapter 1  Summary of the contents of _FREEDOM_**.......... 35    39–

1:1   Freedom from the human condition............................ 35    39–

1:2   What exactly is the human condition? ........................ 37    44–

1:3   A brief description of the human-race-transforming
        explanation of the human condition that is presented
        in _FREEDOM_.................................................... 43    53–

1:4   The problem of the 'deaf effect' that reading about the
        human condition initially causes ............................. 56    77–

1:5   Solutions to the 'deaf effect' ................................... 60    86–

**Chapter 2  The Threat of Terminal Alienation from Science's
        Entrenched Denial of The Human Condition**... 65    98–

2:1   Summary ...................................................... 65    98–

2:2   The psychological event of 'Resignation' reveals our species'
        mortal fear of the human condition and thus how difficult it
        has been for scientists to make sense of human behavior ...... 68    103–

2:3   Our near total resistance to analysis of the human condition
        post-Resignation ................................................ 78    119–

2:4   How has science coped with the issue of the human
        condition? ..................................................... 87    134–

2:5   The three fundamental truths of the human condition .......... 93    149–

2:6   Evidence of the three fundamental truths, as provided by
        Moses and Plato................................................. 95    155–

2:7   Further evidence of the three fundamental truths .............. 106   180–

2:8   The ultimate paradox of the human condition.................. 114   191–

2:9   Social Darwinism............................................... 115   194–

2:10  Sociobiology/Evolutionary Psychology......................... 116   196–

2:11  Multilevel Selection theory for eusociality .................... 117   200–

2:12  While denial has been necessary, you can't find the truth
        with lies........................................................ 128   220–

**Chapter 3  The Human-Race-Transforming, *Real* Explanation of The Human Condition** ........................... 137 234–

3:1   Summary ..................................................... 137 234–

3:2   We cannot endure being faced by the problem of the human condition forever.............................................. 138 236–

3:3   The psychosis-addressing-and-solving *real* explanation of the human condition ........................................... 140 242–

3:4   The Story of Adam Stork......................................... 144 250–

3:5   The double and triple whammy involved in our *human* situation or condition ......................................... 150 261–

3:6   Adam and Eve — we humans — are heroes NOT villains ....... 157 273–

3:7   Our instinctive moral conscience has been 'A sharp accuser, but a helpless friend!' ......................................... 165 282–

3:8   Most wonderfully, this psychosis-addressing-and-solving, *real* explanation of the human condition makes possible the psychological rehabilitation of the human race ........... 169 289–

3:9   The end of politics ............................................. 175 299–

3:10  'Free at last!' ................................................. 179 305–

**Chapter 4  The Meaning of Life** ............................... 183 313–

4:1   Summary ..................................................... 183 313

4:2   The obvious truth of the development of order of matter on Earth....................................................... 184 314–

4:3   'God' is our personification of Integrative Meaning .......... 189 324–

4:4   The denial-free history of the development of matter on Earth....................................................... 199 341–

4:5   Elaborating the sexually reproducing individual .............. 207 359–

4:6   Negative Entropy found a way to form the Specie Individual................................................... 212 372–

**Chapter 5  The Origin of Humans' Unconditionally Selfless, Altruistic, Moral Instinctive Self or Soul** ........ 213 375–

5:1   Summary ..................................................... 213 375–

5:2   How could humans have acquired their altruistic moral instincts?.................................................... 214 378–

5:3   The integration of sexually reproducing individuals to form
      the Specie Individual ............................................. 216   384–

5:4   Love-Indoctrination............................................... 219   388–

5:5   Fossil evidence confirming the love-indoctrination process... 223   396–

5:6   Bonobos provide living evidence of the love-indoctrination
      process......................................................... 232   411–

5:7   The role of nurturing in our development has been an
      unbearably confronting truth..................................... 241   419–

5:8   The emergence of consciousness assisted the love-
      indoctrination process by allowing the sexual selection
      of integrativeness .............................................. 242   421–

5:9   Sexual selection for integrativeness explains neoteny ......... 248   432–

5:10  The importance of nurturing in bonobo society ................ 252   440–

5:11  The importance of strong-willed females in developing
      integration ..................................................... 255   444–

5:12  Descriptions of bonobos provide an extraordinary insight
      into what life for our human ancestors was like before the
      emergence of the human condition............................. 258   450–

5:13  Milwaukee County Zoo's fabulous group of bonobos ......... 261   455–

5:14  'A golden race...formed on earth' ............................ 270   468–

**Chapter 6  End Play for The Human Race** .................... 277   477–

6:1   Summary ....................................................... 277   477

6:2   The danger of denial becoming so entrenched that it locks
      humanity onto the path to terminal alienation.................. 277   478–

6:3   The nurturing origins of our moral soul is an obvious truth... 280   483–

6:4   The problem has been that the nurturing origin of our moral
      soul has been devastatingly, unbearably, excruciatingly
      condemning ..................................................... 287   491–

6:5   To deny the importance of nurturing, the Social Intelligence
      Hypothesis was invented........................................ 293   499–

6:6   Dismissing maternal love as training to manage complex
      social situations still left the extraordinarily cooperative
      lives of bonobos, and of our ape ancestors, to somehow
      be explained ................................................... 300   512–

6:7   Fossil evidence of our species' cooperative past has also
      been dismissed, ignored or misrepresented by mechanistic
      science ......................................................... 304   519–

6:8 The Social Ecological Model ................................... 307 525–

6:9 A brief history of left-wing dishonest mechanistic biology ... 312 536–

6:10 The Self-Domestication Hypothesis............................ 317 544–

6:11 End play for the human race ................................. 324 562–

6:12 *The* great obscenity ............................................. 326 569–

6:13 Nurturing now becomes a priority............................ 358 617–

**Chapter 7 What is Consciousness, and Why, How and When Did Humans' Unique Conscious Mind Emerge?**............................................... 361 622–

7:1 Summary .................................................... 361 622

7:2 What is consciousness?....................................... 362 623–

7:3 Why, how and when did consciousness emerge in humans? .. 379 659–

**Chapter 8 The Greatest, Most Heroic Story Ever Told: Humanity's Journey from Ignorance to Enlightenment**....................................... 401 702–

8:1 Summary .................................................... 401 702–

8:2 The stages of humanity's maturation from ignorance to enlightenment................................................. 403 705–

8:3 INFANCY .................................................... 408 717–

8:4 CHILDHOOD................................................. 412 722

  8:5 Early Happy, Innocent Childman........................ 412 723–

  8:6 Middle Demonstrative Childman........................ 413 725–

  8:7 Late Naughty Childman ............................... 417 729–

8:8 ADOLESCENCE.............................................. 424 738

  8:9 Early Sobered Adolescentman ........................... 425 739–

  8:10 Distressed Adolescentman ............................... 428 741–

  8:11A Adventurous Adolescentman ........................... 444 765–

    8:11B Men and women's relationship after the emergence of the human condition ............ 446 769–

    8:11C Other adjustments to life under the duress of the human condition that developed during the reign of Adventurous Adolescentman ..... 489 825–

  8:12 Angry Adolescentman .................................... 506 845–

  8:13 Pseudo Idealistic Adolescentman ....................... 522 872–

8:14    Hollow Adolescentman................................. 534  894–

8:15    The last 11,000 years and the rise of Imposed
        Discipline, Religion and other forms of Pseudo
        Idealism............................................... 539  902–

8:16A The last 200 years when pseudo idealism takes
        humanity to the brink of terminal alienation ........... 563  940–

        8:16B    The emergence of terminal levels of
                 alienation in the 'developed' world............ 566  944–

        8:16C    The dire consequences of terminal levels of
                 alienation destroying our ability to nurture
                 our children................................... 577  961–

        8:16D    The dysfunctional, extremely narcissistic
                 'Power Addicted' state......................... 597  1001–

        8:16E    The differences in alienation between 'races'
                 (ethnic groups) of humans...................... 610  1023–

        8:16F    The emergence of 'materialism envy' and
                 with it unbridled greed, extreme dysfunction
                 and destitution in the 'developing' world ..... 635  1051–

        8:16G    So it is end play wherever we look ............ 639  1057–

        8:16H    The progression of ever-more dishonest and
                 dangerous forms of pseudo idealism to cope
                 with the unbearable levels of upset ........... 642  1063–

        8:16I    Less Guilt Emphasizing Expressions of
                 Religion....................................... 643  1065–

        8:16J    Non-Religious Pseudo Idealistic Causes ...... 650  1076–

        8:16K    Socialism and Communism .................... 651  1080

        8:16L    The New Age Movement....................... 652  1081

        8:16M  The Feminist Movement....................... 653  1082

        8:16N    The Environmental or Green Movement...... 654  1083

        8:16O    The Politically Correct Movement............. 654  1084–

        8:16P    The Postmodern Deconstructionist
                 Movement ..................................... 661  1096–

        8:16Q    The 'abomination that causes desolation'
                 'sign...of the end of the age' that is 'cut short'
                 by the arrival of the liberating truth about the
                 human condition .............................. 668  1111–

**Chapter 9  The Transformation of The Human Race**........ 687 1143–

9:1   Summary ..................................................... 687 1143

9:2   The 'dawn...of our emancipation'............................. 687 1144–

9:3   'Judgment day'................................................ 692 1150–

9:4   The Transformed Lifeforce Way of Living .................... 697 1157–

9:5   The Transformed Lifeforce Way of Living is not another
      religion ....................................................... 704 1169–

9:6   Nor is the Transformed Lifeforce Way of Living another
      deluded, false start to a human-condition-free new world..... 706 1174–

9:7   But how does the Transformed Lifeforce Way of Living solve
      the problem of the unbearable, 'judgment day' exposure that
      understanding of the human condition unavoidably brings? .. 710 1182–

9:8   The utter magnificence of the Transformed Lifeforce State ... 719 1197–

9:9   How the Transformed Lifeforce Way of Living will quickly
      repair the world ................................................ 730 1219–

9:10  Humanity's overall situation now that we have
      understanding ................................................. 742 1240–

9:11  The 'pathway of the sun' ....................................... 760 1261–

**Index** ................................................................. 789

# Introduction

by Professor Harry Prosen

'In the whole of written history there are only two or three people who have been able to think on this scale about the human condition.'
Prof. Anthony Barnett, zoologist, author and broadcaster, 1983

[1] The truth is I am inadequate to write this Introduction, but everyone is, so I will do my best.

[2] Firstly, to immediately put into context the fabulous significance of what Australian biologist Jeremy Griffith has achieved with this book: it delivers the breakthrough biological explanation of the human condition, the holy grail of insight humans have sought for the psychological rehabilitation of our species. It brings the compassionate, redeeming, reconciling and transforming understanding of our lives that the human race has lived in eternal hope, faith and trust would one day be found and have so assiduously pursued since we first became conscious beings some 2 million years ago! So this is, in short, the most momentous event in human history!

[3] I once read that the great philosophers each had an acute grasp of their own small piece of an unthinkably huge and interlocking puzzle, but nobody has ever put the pieces together into a single coherent system (Steve King, 'Post-Structuralism or Nothing', 1996; see <www.wtmsources.com/135>). Well, that is no longer the case because Jeremy provides that great unifying, make-sense-of-everything synthesis—for not only does his treatise explain the human condition—our species' capacity for what has been called 'good' and 'evil' (in chapters 1 and 3)—his ability to plumb the great depths of this most foreboding of all subjects has meant that he has *also* been able to provide the fully accountable explanation of the other great outstanding mysteries in science of the meaning of human existence (chapter 4), the origin of our unconditionally selfless moral instincts (chapters 5 and 6), and why humans became conscious when other animals haven't (chapter 7). And in unlocking those insights, he has in turn been able to make sense of *every* other aspect of our troubled human condition, including the strained relationship between men and women, sex as humans practice it, the origin and nature of politics,

religion, and so many other human phenomena (chapter 8). Indeed, preposterous as it must seem, what this book effectively does is take humanity from a state of bewilderment about the nature of human behavior and existence to a state of profound understanding of our lives. It *truly* is a case of having got all the truth up in one go. Understanding of our species' troubled human condition has *finally* emerged to drain away all the pain, suffering, confusion and conflict from the world—and, given our plight, its arrival couldn't be more timely or serious. If there was ever a case of cometh the hour, cometh the man, this is it! Grand statements and syntheses about human behavior certainly deserve cynicism; as a senior editor of biology at *Nature* journal, Henry Gee, has said, '**Hardly a month goes by without my receiving, at my desk at *Nature*, an exegesis on the reasons how and why human beings evolved to be this way or that. They are always nonsense**' (*The Guardian*, 7 May 2013). But you only have to read a few paragraphs of this book to recognize its extraordinary profundity and authenticity, for in addressing the underlying, core issue in human life of our species' conflicted human condition, and never departing from that course, it unravels the *whole* riddle of what it is to be human.

[4]So, yes, it is with great trepidation that I attempt to write this Introduction, but I hope that my background as a professor of psychiatry with over 50 years' experience working in the field, including chairing 2 departments of psychiatry and serving as president of the Canadian Psychiatric Association—and my 10 years of knowing the author—might allow me to provide the reader with a helpful insight into this greatest of all books. And by the greatest of all books I do include within that comparison the Bible, which, as it turns out, and as many believed, is also entrenched in truth, but, *unlike* this book, was written in pre-scientific times and thus unable to provide the scientific basis for all those profound truths. In fact, there is really only one book in the world now: this book. It is *so* all-explaining and all-solving that it is, in effect, the new science-based Bible for the human race! It is an extraordinary statement to make but one the Nobel laureate physicist Charles H. Townes was anticipating would one day be possible when he wrote that '**they [science and religion] both represent man's efforts to understand his universe and must ultimately be dealing with the same substance. As we understand more in each realm, the two must grow together...converge they must**' ('The Convergence of Science and Religion', *Zygon*, 1966, Vol.1, No.3). Yes, this is the book we have been waiting for—the book that saves the world.

[5]The explanation of the human condition, which Jeremy has found, is *THE* key to a fully unifying understanding of our **'universe'**; it is, as I've mentioned, *the* insight from which all the other mysteries about human life unravel. What needs to be stressed, however, is that finding that explanation depended on overcoming a very great *psychological* hurdle—which Charles Darwin actually alluded to when, towards the end of *The Origin of Species*, he wrote that **'In the distant future I see open fields for far more important researches. Psychology will be based on a new foundation...Light will be thrown on the origin of man and his history'** (1859, p.458 of 476). Yes, while Darwin shed illuminating light on the origin of the variety of life, there biology has been stalled—and for a very good reason—because the next step for biologists involved **'far more important'** (in terms of difficulty) **'research'**. For **'light'** to **'be thrown on the origin of man and his history'**, the issue of our species' seemingly highly imperfect condition had to be explained, but the **'psycholog[ical]'** difficulty was that the human condition has, in fact, been *so* unbearably self-confronting and depressing a subject that, as the Australian zoologist, author and broadcaster, Professor Anthony Barnett, admitted in the opening quote to this Introduction, only a rare few individuals in recorded history have been able to go anywhere near it. Indeed, as Jeremy explains in chapter 2:4, the fundamental reason science has been what is referred to as **'reductionist'** and **'mechanistic'** is because it has avoided the great overarching, all-important but unbearably confronting and depressing issue of the human condition and instead *reduced* its focus to only looking down at the details of the *mechanisms* of the working of our world—the great hope being that understanding of those mechanisms would eventually make it possible to explain, understand and thus at last be able to confront the human condition. Which is all very well, but the problem was that to *assemble* that liberating explanation from the hard-won insights into the mechanisms of the workings of our world, a biologist was going to have to emerge who could face our species' psychosis! Mechanistic science had to do all the hard work of, as it were, finding the pieces of the jigsaw puzzle, but unable to look at the whole picture its practitioners were in no position to put it together—that task required someone sound and secure enough in self to look at the whole picture, to confront the human condition, which, as emphasized, is no mean feat in itself, but one which Jeremy was able to accomplish.

[6] In *The Structure of Scientific Revolutions* (1970), science historian Thomas Kuhn noted that **'revolutions are often initiated by an outsider— someone not locked into the current model, which hampers vision almost as much as blinders would'** (from Shirley Strum's *Almost Human*, 1987, p.164 of 297). And when it comes to addressing the problem of the human condition this need to think independently of the existing details-only-focused, whole-view-of-the-human-condition-avoiding, mechanistic framework could not be more critical. For someone to be able to explain, and, through that explanation, bring reconciling, ameliorating understanding to our troubled human-condition-afflicted lives, they obviously had to be thinking from a position *outside* that conventional mechanistic paradigm. The situation certainly brings to mind Einstein's famous comment that **'We can't solve problems by using the same kind of thinking we used when we created them'**! I might say that I think we have always known that profound insight into human nature wasn't going to emerge from the ivory towers of intellectualdom, rather it was going to come from the deepest of deep left field, somewhere where some extraordinary untainted clarity of thought might still exist, such as from the backwoods of Australia where these answers are actually from.

[7] So what exactly *is* the human condition, and why has the human race, including the mechanistic scientific establishment, been *so* committed to avoiding it, to living in such complete denial of it—and what have been the consequences for Jeremy, the outsider who dared to break such an entrenched, fiercely held doctrine?

[8] To draw on Jeremy's own penetrating description, the human condition is our species' extraordinary capacity for what has been called 'good' and 'evil'. While it's undeniable that humans are capable of great love and empathy, we also have an unspeakable history of greed, hatred, rape, torture, murder and war; a propensity for deeds so shocking and overwhelming that the eternal question of 'Why?' seems depressingly inexplicable. Even in our everyday behavior, why, when the ideals of life are to be cooperative, selfless and loving, are we so ruthlessly competitive, selfish and aggressive that human life has become all but unbearable and our planet near destroyed? How could humans possibly be considered good when all the evidence seems to unequivocally indicate that we are a deeply flawed, bad, even 'evil' species?

[9] But as my profession has taught me only too well, for most people, trying to think about this ultimate of questions of whether humans are

fundamentally good or not has been *an unbearably self-confronting exercise*. Indeed, while the term 'human condition' has become fashionable, its superficial use masks just how profoundly unsettling a subject it really is. Again, the truth is, the issue of the human condition has been *so* depressing for virtually all humans that only a rare few individuals in history have been sound and secure enough in self to go anywhere near what the human condition *really* is. So for Jeremy to so freely and accurately talk about it as he does in this book, he clearly must be one of those rare few. Nurtured by a sheltered upbringing in the Australian 'bush' (countryside), Jeremy's soundness and resulting extraordinary integrity and thus clarity of thought, coupled with his training in biology, has enabled him to successfully grapple with this most foreboding of all subjects for the human mind of the human condition and produce the breakthrough, human-behavior-demystifying-and-ameliorating explanation of it.

[10] This ability exists in stark contrast to the current paradigm of thought that permeates science (and indeed all aspects of human life) which has had no choice but to avoid—in truth, *deny*—the seemingly inexplicable and unbearably confronting and depressing question of the human condition, meaning other scientists have remained failure-trapped in trying to explain it. If you can't confront the issue you're in no position to solve it. In fact, the human condition has been *such* a fearful, unconfrontable subject that science as a whole has become the purveyor of extremely dishonest theories that seek to falsely account for, and thus dismiss, our paradoxical nature. As Jeremy explains in chapter 2:11, there can not be a more dangerous example of this dishonesty than the theory that appears in biologist E.O. Wilson's 2012 book, *The Social Conquest of Earth*—a theory that evasively trivializes the human condition as nothing more than selfless instincts at odds with selfish instincts within us.

[11] Given we haven't been able to acknowledge the immense role denial of the issue of the human condition has played in human life (because obviously we couldn't be in denial *and* admit we were in denial), I am going to, with Jeremy's permission, pre-emptively employ some frightfully accurate references and quotes (including Jeremy's clarifications and interpretations within the square brackets in the quotes) that appear in this book to illustrate just *how* terrifying the human condition has been for virtually all humans, and thus *why* humanity, including its scientific fraternity, has *had* to live in almost complete denial of the subject. Yes, ours has been a culture of such pervasive dishonesty that it became nearly

impossible for the *real*, human-race-liberating, biological explanation of the human condition to be unearthed.

[12]The most famous account ever given of the human condition actually forms the centerpiece of Plato's most acclaimed work, *The Republic*, in which humans are metaphorically depicted living as prisoners in a dark cave, deep underground, so fearful are they of the issue of the imperfection of their lives that the sun has the power to reveal. As Plato wrote, **'I want you to go on to picture the enlightenment or ignorance of our human conditions somewhat as follows. Imagine an underground chamber, like a cave with an entrance open to the daylight and running a long way underground. In this chamber are men who have been prisoners there'** (c.360 BC; tr. H.D.P. Lee, 1955, 514). Plato described how the cave's exit is blocked by a **'fire'** that **'corresponds to the power of the sun'**, which the cave prisoners have to hide from because its searing, **'painful'** light would make **'visible'** the unbearably depressing issue of **'the imperfections of human life'** (ibid. 516-517).

[13]In terms of how unbearably confronting—in fact, suicidally depressing—grappling with the issue of the human condition has been, the great philosopher Søren Kierkegaard is said to have given the most honest description of that *worse-than-death* experience, writing in his aptly titled 1849 book, *The Sickness Unto Death*, that **'the torment of despair is precisely the inability to die** [and end the torture of our unexplained human condition]**...that despair is the sickness unto death, this tormenting contradiction** [of our 'good-and-evil'-human-condition-conflicted lives]**, this sickness in the self; eternally to die, to die and yet not to die'** (tr. A. Hannay, 1989, p.48 of 179). The famous analytical psychologist Carl Jung provided an equally stark description of the terrifying nature of the human condition when he wrote: **'When it** [our **shadow] appears...it is quite within the bounds of possibility for a man to recognize the relative evil of his nature, but it is a rare and shattering experience for him to gaze into the face of absolute evil'** (*Aion: Researches into the Phenomenology of the Self*, 1959; tr. R. Hull, *The Collected Works of C.G. Jung*, Vol. 9/2, p.10). The **'face of absolute evil'** *is* the **'shattering'** possibility—if humans allow their minds to think about it—that they might indeed be a terrible mistake.

[14]It follows then that to confront the human condition has been an impossible ask for most people—as another great philosopher, Nikolai Berdyaev, acknowledged: **'Knowledge requires great daring. It means victory over ancient, primeval terror...it must also be said of knowledge that it is bitter, and there is no escaping that bitterness...Particularly bitter is moral knowledge, the knowledge of good and evil. But the bitterness is due to the fallen state of**

the world...There is a deadly pain in the very distinction of good and evil, of the valuable and the worthless' (*The Destiny of Man*, 1931; tr. N. Duddington, 1960, pp.14-15 of 310). Trying to think about our corrupted, 'fallen', seemingly 'evil' and 'worthless' human-condition-afflicted state *has been* an 'ancient, primeval terror', a 'deadly pain', 'the bitterest thing in the world' for virtually all humans—'knowledge of good and evil', of the human condition, 'requires great daring'. No wonder Professor Barnett said to Jeremy back in 1983 that 'you are being very arrogant to think you can answer questions on this scale. In the whole of written history there are only two or three people who have been able to think on this scale about the human condition' (from a recorded interview with Jeremy Griffith, 15 Jan. 1983).

[15] It is perhaps not surprising then that the human-condition-avoiding, denial-based, mechanistic or reductionist scientific establishment, along with the human-condition-avoiding public at large, have found Jeremy's human-condition-confronting work deeply heretical, an anathema to be dismissed and even persecuted. Indeed, the persecution of Jeremy and his work was of such ferocity that it led to what was at the time the biggest defamation case in Australia's history, a case that, after 15 long years, resulted in the vindication of Jeremy's work. I know all this only too well because I was one of the international scientists who gave evidence in Jeremy's defense during the trial, where I was able to vouch for the veracity of Jeremy's biological thinking on the basis of the extensive studies and work I have carried out over many years on empathy within both humans and bonobos, as well as from all my experience in the field of intensive psychotherapy.

[16] The extreme paradox of this situation should be apparent to the reader—as humanity's designated vehicle for enquiry into our world and our place in it, science's ultimate objective and responsibility has been to find understanding of the human condition, so to treat the eventual discovery of its explanation as an anathema, to the point of persecuting it, is ridiculously counter to the whole purpose of science. The fundamental goal of the whole human journey of conscious thought and enquiry has been to find the reconciling, redeeming and rehabilitating explanation of our species' troubled condition, so to reject it when it arrives is madness of the highest order!

[17] Professor Barnett, who passed away in 2003, didn't mention to Jeremy who he considered to be the 'two or three people who have been able to think on this scale about the human condition', but I think we can

deduce that they would have been Moses, Plato and Christ. We have already seen the extraordinary soundness and integrity of Plato's thinking in his ability to so clearly describe humans' existence in a **'cave'-'like'** state of denial of **'our human condition'**. In fact, Plato's honest, truthful, human-condition-confronting clarity of thought was such that he actually fully anticipated the rejection, indeed persecution, that would occur when someone eventually found the truthful, compassionate, all-liberating but at the same time all-revealing explanation of the human condition. To quote from a summary of Plato's cave allegory from the *Encarta Encyclopaedia*: **'Breaking free, one of the individuals escapes from the cave into the light of day. With the aid of the sun** [living free of denial of **our human condition**]**, that person sees for the first time the real world and returns to the cave with the message that the only things they have seen heretofore are shadows and appearances and that the real world awaits them if they are willing to struggle free of their bonds. The shadowy environment of the cave symbolizes for Plato the physical world of** [false] **appearances. Escape into the sun-filled setting outside the cave symbolizes the transition to the real world...which is the proper object of knowledge'** (written by Prof. Robert M. Baird, 'Plato'; see <www.wtmsources.com/101>). For the description of what would happen when the **'mess[enger]'** tries to **'free'** the cave prisoners **'into the sun-filled setting outside the cave'**, Plato wrote that when **'he** [the cave prisoner] **were made to look directly at the light of the fire** [again the **fire corresponds to the power of the sun** which makes **visible** the **imperfections of human life**]**, it would hurt his eyes and he would turn back and take refuge in the things which he could see, which he would think really far clearer than the things being shown him** [the mechanistic scientific establishment would prefer its human-condition-avoiding, dishonest theories]**. And if he** [the cave prisoner] **were forcibly dragged up the steep and rocky ascent** [out of the cave] **and not let go till he had been dragged out into the sunlight, the process would be a painful one, to which he would much object, and when he emerged into the light his eyes would be so overwhelmed by the brightness of it that he wouldn't be able to see a single one of the things he was now told were real. Certainly not at first. Because he would need to grow accustomed to the light before he could see things in the world outside the cave** [initially the cave prisoners would find it impossible reading about and absorbing the truthful descriptions of the human condition; they would suffer from a 'deaf effect']**'** *(The Republic, 515-516)*. Plato went on to say that **'they would say that his** [the person who tries to deliver understanding of **our human condition**] **visit to the upper world had ruined his sight** [they would treat him as if he was mad]**, and that the ascent** [out of the cave] **was not worth even attempting. And if anyone tried to release them and lead them up, they would kill him if they could lay hands on him'** (517; or see <www.wtmsources.com/227>). Yes, as any psychotherapist knows, denials do fight back with a vengeance when faced with annihilation.

[18] Of course, in today's civilized world more subtle means of eliminating threats to the mechanistic, human-condition-avoiding, denial-based 'cave existence' were employed in the campaign of persecution against Jeremy for being the person who **'escapes from the cave into the light of day'** and then **'dragged** [humanity] **out into the sunlight'**. This persecution was the focal point of the aforementioned legal battle undertaken to defend Jeremy's work, in which three judges in the New South Wales Court of Appeal unanimously found that an earlier ruling in a lower court did **'not adequately consider' 'the nature and scale of its subject matter'**, in particular **'that the work was a grand narrative explanation from a holistic approach, involving teleological elements'**, and other important submissions **'were not adequately considered by the primary judge'** including that the work <u>can make</u> **'those who take the trouble to grapple with it uncomfortable'** because it <u>**'involves reflections on subject-matter including the purpose of human existence which may, of its nature, cause an adverse reaction as it touches upon issues which some would regard as threatening to their ideals, values or even world views'**</u>! (For details of the persecution and court case, including this vindicating ruling, see <www.humancondition.com/persecution>).

[19] The journey that Jeremy and those advocating his work have been on to bring these human-race-saving understandings to the world, which has culminated in this absolutely astonishing book, has been a long and torturous one, but one that makes for fascinating and revealing reading, so I will now present for the reader a very brief summary of Jeremy's early life, his writing and his unrelenting efforts to free our species from its incarceration in Plato's terrible cave of human-condition-denying, alienated darkness.

---

[20] Born on December 1, 1945, and raised on a sheep station (ranch) in rural New South Wales, Australia, Jeremy was educated at Geelong Grammar School in Victoria, a school whose visionary approach to education has produced such notable alumni as Rupert Murdoch and HRH The Prince of Wales. He gained first class honors in biology in the state matriculation exams and in 1965 began a science degree at the

University of New England in northern New South Wales. While there, Jeremy played representative rugby union football, making the 1966 trials for the national team, the Wallabies (see <www.humancondition.com/jeremy-rugby>).

[21] Deferring his studies in 1967, Jeremy undertook the most thorough investigation ever into the plight of the Tasmanian Tiger (thylacine) (see <www.humancondition.com/tasmanian-tiger-search>)—a search that was to last more than six years, before concluding the 'Tiger' was indeed extinct. His findings were internationally reported, with articles appearing in the American Museum of Natural History's journal, *Natural History*, and *Australian Geographic* (see above link). His search also featured in an episode of the national television series *A Big Country* (see above link).

[22] In 1971 Jeremy completed his Bachelor of Science degree in zoology at the University of Sydney and the following year, in the same self-sufficient spirit with which he had undertaken the 'Tiger' search, he established a successful furniture manufacturing business based on his own simple and natural designs, which pioneered the use of bark-to-bark slabs of timber (see <www.humancondition.com/griffith-tablecraft>). On the subject of creativity, I should mention that Jeremy is also an accomplished artist (see <www.humancondition.com/jeremy-art-work>).

[23] An upbringing nurtured with real, unconditional love (it is to the nurturing from Jeremy's mother, and to the good fortune of having a father who was not oppressively egocentric, that we really owe these world-saving insights) in the sheltered isolation of the Australian bush left Jeremy deeply troubled and perplexed by all the selfishness, aggression, dishonesty and indifference on the one hand, and all the suffering on the other, that he inevitably encountered in the innocence-destroyed, human-condition-embattled, psychotic wider world. In time, he realized that trying to save animals from extinction or trying to build ideal furniture wasn't going to make a difference to the extraordinary imperfection in human life and that he would have to get to the bottom of the issue of this seeming complete wrongness of human behavior, which is the issue of the human condition. And so it was while building his furniture business that Jeremy first began to write down his ever developing thoughts about the problem of the human condition. Indeed, since the early 1970s Jeremy has spent the first, often pre-dawn, hours of each day thinking and writing about the human condition. After some 10 years of extraordinarily profound, honest, human-condition-confronting-not-avoiding,

effective thinking during his late 20s and early 30s, Jeremy was ready to present to the world his fully accountable, but, for virtually everyone else, unbearably self-confronting insights into human behavior. It was an amazing decade of clear thinking, confirming Einstein's belief that **'a person who has not made his great contribution to science before the age of 30 will never do so'** (Selig Brodetsky, 'Newton: Scientist and Man', *Nature*, 1942, Vol.150).

[24] However, since producing his all-explaining synthesis, the problem Jeremy has encountered has been how to present it in such a way that people could access its confronting truthfulness; it's all very well to find the redeeming, reconciling full truth about humans, but when everyone has been living in determined denial of all the elements that comprise that compassionate full truth, how do you get them to overcome that denial and hear it? Recall that Plato said that when the **'cave' 'prisoners'** are **'free[d] of their bonds'** they **'would be so overwhelmed by the brightness of'** the **'light'** of understanding of **'our human condition'** that **'at first'** they **'wouldn't be able to see a single one of the things'** revealed as **'real'**. It is Jeremy's journey to find a way to overcome this 'deaf effect' difficulty that reading about the human condition initially causes that has led to this presentation of his synthesis in *FREEDOM*. As you will see, his strategy in this book is to encourage readers to watch introductory videos to *FREEDOM* at <www.humancondition.com/intro-videos>, and also be prepared to patiently re-read the text, so you can, as Plato said, **'grow accustomed to the light'**. Of course, this strategy was arrived at after much trial and error—over 30 years, in fact, of presenting his synthesis in slightly different ways in a series of articles and books, beginning in 1983 with submissions to *Nature* and *New Scientist* (which were rejected, with the then editor of *Nature*, John Maddox, telling Jeremy that his starting point teleological argument that there is an underlying order in nature **'is wrong'**—as I will mention again shortly, in chapter 4, Jeremy explains why the truth of the order in nature has been denied by human-condition-avoiding, mechanistic science); then, in 1988, *Free: The End Of The Human Condition*; *Beyond The Human Condition* (1991); *A Species In Denial* (2003), which was a bestseller in Australia and New Zealand but still failed to attract any real interest from the scientific establishment; *The Human Condition Documentary Proposal* (2004); *The Great Exodus: From the horror and darkness of the human condition* (2006); *Freedom Expanded* (2009); *The Book of Real Answers to Everything!* (2011); and now, in 2016, *FREEDOM: The End Of The Human Condition*. (All the earlier works referred to here

are freely available to be read or printed at <www.humancondition.com/publications>.)

[25] Despite the difficulty of presenting such a confronting treatise, Jeremy's books have attracted the support of such accomplished thinkers as Australia's Templeton Prize-winning biologist Professor Charles Birch, one of New Zealand's foremost zoologists, Professor John Morton, and the pre-eminent philosopher Sir Laurens van der Post. The proposal to make a documentary about the human condition (*The Human Condition Documentary Proposal*), in which Jeremy outlined all the main biological explanations contained in his synthesis, also received over 100 endorsements from many of the world's leading scientists and thinkers, including professors Stephen Hawking and the aforementioned Nobel Laureate Charles H. Townes (see <www.humancondition.com/doco-responses>).

[26] But while Jeremy's work has drawn praise and garnered impressive commendations from some exceptional thinkers able to acknowledge his insights, he has, as mentioned, also had to withstand the enormous cynicism, indifference and even persecution that humans' historical resistance to engaging the subject of the human condition produces. Indeed, it was soon after he began writing that Jeremy realized that not only was the scientific establishment failing its responsibility to address the issue of the human condition, but that, like the rest of humanity, it was treating the whole issue as an anathema. As a result, Jeremy established, in 1983, a non-profit organization dedicated to the study and amelioration of the human condition, now called the World Transformation Movement (WTM) (<www.humancondition.com>). So fearful, however, has humanity been of the issue of the human condition that a vicious campaign was launched in 1995 to try to shut down Jeremy's work and bring the WTM into disrepute, which Jeremy, along with fellow WTM Patron, renowned mountaineer and twice-honored Order of Australia recipient Tim Macartney-Snape and the other supporters of the WTM, determinedly resisted—the result of which was the then biggest defamation case in Australia's history, against the two biggest, left-wing (described by Jeremy as dogmatic, pseudo idealistic, 'let's pretend there's no human condition that has to be solved and the world should just be ideal', dishonest) media organizations in Australia, including its national public broadcaster. As mentioned, in 2010—after 15 long years—Jeremy and Tim were vindicated, enabling Jeremy to concentrate solely on producing new works such as *Freedom Expanded*, and now this, its condensation and Jeremy's summa masterpiece, *FREEDOM:*

*The End Of The Human Condition*. (Note, this book was first published in 2014 under the title *IS IT TO BE Terminal Alienation or Transformation For The Human Race?* and with content that was specifically tailored to a scientific audience. In 2016 the book was recast for general release under its current title *FREEDOM: The End Of The Human Condition*.)

[27] To learn more about the WTM, which is based in Sydney, and about Jeremy's role as founder and patron, see <www.humancondition.com>.

---

[28] As I have said, Jeremy's journey in bringing understanding to the human condition and protecting the integrity of that explanation—a 40 year saga—has certainly been a protracted and torturous one (indeed, the persecution was so terrible it left Jeremy seriously debilitated with Chronic Fatigue Syndrome from 1999 to 2009), but bringing understanding to the human condition is the only rational path forward for the human race! Indeed, as a reflection of the looming psychological crisis for the human race that is the end result of having to live ever deeper **'underground'** in Plato's horrible **'cave'** of **'human condition'**-avoiding, dishonest, alienated darkness, and the now desperate need for the reconciling, redeeming and psychologically transforming light of understanding of the human condition, the following initiatives have all taken place in the last 18 months (as at February 2014 when this book was first published under the title *IS IT TO BE*)! (The sources of the following quotes are provided when Jeremy refers to them in par. 603.)

- in December 2012 an American billionaire pledged $200 million to Columbia University's **'accomplished scholars whose collective mission is both greater understanding of the human condition and the discovery of new cures for human suffering'**; and,

- in January 2013 the European Commission announced the launch of the **'Human Brain Project with a 2013 budget of €54 million (US$69 million)'** with a **'projected billion-euro funding over the next ten years'** with the goal of providing **'a new understanding of the human brain and its diseases'** to **'offer solutions to tackling conditions such as depression'**; and,

- in April 2013 the President of the United States, Barack Obama, announced a **'Brain Initiative'**, giving **'$100 million initial funding'** to mechanistic science to also find **'the underlying causes of...neurological and psychiatric conditions'** afflicting humans; and,

• in April 2013 *BBC News Business* reported that **'Lord Rees, the Astronomer Royal and former president of the Royal Society, is backing plans for** [Cambridge University to open] **a Centre for the Study of Existential Risk** [meaning risk to our existence]. **"This is the first century in the world's history when the biggest threat is from humanity," says Lord Rees'**. The article then referred to Oxford University's Future of Humanity Institute that was established in 2005, which is **'looking at big-picture questions for human civilization...**[and] **change...**[that] **might transform the human condition'**, quoting its Director and advisor to the Centre for the Study of Existential Risk, Nick Bostrom: **'There is a bottleneck in human history. The human condition is going to change. It could be that we end in a catastrophe or that we are transformed by taking much greater control over our biology'**!

[29] There are two points I would make about these very recent initiatives. Firstly, establishing a center to study the human condition is *precisely* the initiative Jeremy took 30 years ago when he created the self-funded WTM, an act of prescience that evidences his clarity and integrity of thought—the WTM even has the now much sought-after domain name **'humancondition.com'**. Secondly, while the fastest growing realization in the world has to be that humanity can't go on the way it is going—indeed, as Bostrom said, the great fear *is* that we are rapidly approaching an endgame situation, a **'bottleneck in human history'**, where the human species is either **'transformed'** or **'we end in a catastrophe'**—and so these initiatives *are* admirable in their goal to address the underlying, real problem afflicting the human race of the human condition, they are still attempting to do so from the same old reductionist, mechanistic position, which, as Jeremy explains in chapter 2, is an approach committed to avoiding the real *psychological* nature of the human condition, and as such is self-defeating; it is doomed to fail. An opinion piece by Benjamin Y. Fong of the University of Chicago that was published in *The New York Times* in 2013 made this very point: **'The real trouble with the Brain Initiative is...the instrumental approach...**[such **biological reduction** is] **intent on uncovering the organic "cause"...of mental problems... rather than looking into psychosocial factors...By humbly claiming ignorance about the "causes" of mental problems...neuroscientists unconsciously repress all that we know about the alienating, unequal, and dissatisfying world in which we live and the harmful effects it has on the psyche, thus unwittingly foreclosing'** the ability to **'alleviate mental disorder'** ('Bursting the Neuro-Utopian Bubble', 11 Aug.

2013; see <www.wtmsources.com/151>). To summarize what Fong has said here and elsewhere in his article, mechanistic, **'reduction**[ist]**'** science's **'synthetic'** focus on the **'organic'** rather than the **'psychological'** nature of our problems can only end in denying humans **'the possibility of self-transformation'**. It is *only* Jeremy's approach of confronting the real, psychological nature of the human condition that could hope to find, and now *has* found, the reconciling and human-race-**'transform**[ing]**'** understanding of the human condition—and yet it is his approach that has been treated as heretical, an anathema and a threat by the mechanistic scientific establishment!! My sincere hope, however, is that with our species' predicament now so dire, the scientific establishment will finally acknowledge and support Jeremy's human-race-saving insight into the human condition—and the other critically important insights made possible by his power to unravel our species' psychosis.

[30] I want to emphasize what I have just said: the desperation to solve the human condition that is apparent in the sudden emergence of all these admirable yet ultimately futile multi-multi-million dollar-supported Brain Initiatives evidences just how dire our situation is—just how close humanity is to the **'end in catastrophe'**, cornered, **'bottleneck'**, end play situation of terminal alienation. And since Jeremy's psychosis confronting, solving and **'transform**[ing]**'** insights into **'the human condition'** are all that can save humanity from this fate, this book is all that the human race has standing between it and extinction! *That* is how important this book is, and why science must now recognize the substance and truth it contains.

[31] And I should point out that not only has Jeremy's work been treated as heretical by mechanistic science because he dares to look at the real **'psychological'** nature of the human condition, it has also been resisted because of the two reasons referred to in the ruling by the aforementioned three judges of the New South Wales Court of Appeal. Firstly, rather than being a more mechanistic and less thinking dependent, deduction-derived theory, Jeremy, like Darwin did with his theory of natural selection, puts forward a wide-ranging, induction-derived synthesis, a **'grand narrative explanation'** for, in this case, human behavior—an approach, incidentally, that led to both Darwin's and Jeremy's work being very wrongly criticized by some for not presenting 'new data' and a 'testable hypothesis', and even as 'not being science at all'! Secondly, Jeremy's *enormously* knowledge-advancing (and 'science' literally means 'knowledge', derived as it is from the Latin word *scientia*, which means 'knowledge') thinking

is based on **'a holistic approach involving teleological elements'**. As Jeremy beautifully explains in chapter 4, the reason that the fundamental truth of the teleological, holistic purpose or meaning of existence of developing the order or integration of matter into ever larger and more stable wholes (atoms into compounds, into virus-like organisms, into single-celled organisms, into multicelled organisms, etc) has been denied by human-condition-avoiding mechanistic science is because it implies humans should behave in an ordered, integrative, cooperative, selfless, loving way. The fundamental truth of holism, which literally means **'the tendency in nature to form wholes'** (*Concise Oxford Dictionary*, 5th edn, 1964), confronts humans with the unbearable issue of the human condition, the issue of why don't we behave in an integrative, cooperative, considerate-of-others, loving way. As Jeremy explains in chapter 4, this order-of-matter developing, integrative process, direction and meaning of existence is a truth we have been so fearful of we have personified it as **'God'**.

[32] Jeremy once sent me a feature article that was syndicated in the weekend magazine of two of Australia's leading newspapers about the extraordinarily enlightened Australian biologist Charles Birch, who was the head of the biology faculty at Sydney University when Jeremy was a student there, which I will quote because it describes the treatment that has been given to any scientist who dared to recognize the teleological, holistic purpose or meaning of existence. Titled 'Science Friction', the article referred to an emerging group of scientists who are bringing about a **'scientific revolution'** and **'monumental paradigm shift'** in science because they have **'dared to take a holistic approach'** and are thus being seen by the scientific orthodoxy as committing **'scientific heresy'**. The article said that these scientists, such as the **'physicist Paul Davies and biologist Charles Birch'**, who are **'not afraid of terms such as "purpose" and "meaning"'**, are trying **'to cross the great divide between science and religion'**, adding that **'Quite a number of biologists got upset** [about this new development] **because they don't want to open the gates to teleology—the idea that there is goal-directed change is an anathema to biologists who believe that change is random'**. The article summarized that **'The emerging clash of scientific thought has forced many of the new scientists on to the fringe. Some of the pioneers no longer have university positions, many publish their theories in popular books rather than journals, others have their work sponsored by independent organisations...universities are not catering for the new paradigm'** (Deidre Macken, *The Sydney Morning Herald* and Melbourne's *The Age*, 16 Nov. 1991; see <www.wtmsources.com/152>). While Jeremy

gained a BSc degree at a conventional university, he didn't continue his studies there to gain a PhD and he has **'publish**[ed his]**...theories in popular books rather than journals'** — but as this article points out, the very good reason for pursuing that autonomous path is that **'Universities are not catering for the new paradigm'**. In fact, as I mentioned, Jeremy had to create an **'independent organisation'** to study the human condition from a truthful, non-mechanistic, teleology-recognizing base — and, as I mentioned, his prescience in **'pioneer**[ing]**'** this now recognized all-important frontier for science reveals what an extraordinarily capable and eminent scientist Jeremy is; a professor of science in the truest sense. Thomas Kuhn was certainly right when, as I mentioned earlier, he said that **'revolutions are often initiated by an outsider** — **someone not locked into the current model, which hampers vision almost as much as blinders would'**. Kuhn also recognized that **'When a field is pre-paradigmatic** [introduces a new paradigm, as Jeremy's work does]**...progress is made with books, not with journal papers'** (*The Structure of Scientific Revolutions*, 1970; from 'Phillip Greenspun's Weblog'; see <www.wtmsources.com/154>). And I might point out that just as Jeremy has had to do to create a revolution in science, Charles Darwin was **'a lone genius, working from his country home without any official academic position'** (Geoffrey Miller, *The Mating Mind*, 2000, p.33 of 538).

[33] In conclusion — and given Jeremy was so vilified he was made a pariah — I would like to emphasize the height of my regard for him. He is the most impressive person and courageous thinker I have ever met and no doubt ever will. Normally people disappoint you at some point, or on some occasions, but Jeremy never does. Basically he is not egocentric. Being exceptionally well nurtured with unconditional love as a child he is sound and secure in himself and as a result is not preoccupied having to prove his worth all the time like most people are. Free of such selfish self-preoccupation he is selflessly concerned *only* with finding a way to end all the suffering in others, which has resulted in him focusing on finding the solution to the human condition. And being sound and secure in self has meant that in tackling that issue he has been able to think in an unafraid, truthful and thus effective way about it; as Berdyaev foresaw, it was going to **'require...great daring'** to find **'knowledge of good and evil'**, understanding of the human condition.

[34] It is quite amazing, in all my years of meeting people and practicing psychiatry, I haven't encountered a soul like him. He is one of those incredibly rare individuals, a person of intellectual rigor and personal

nobility who has the capacity to be completely honest without a personal bent; when you are with him you can feel his passion for the truth, which he embodies. Indeed, meeting Jeremy as I did after reading so much of his work, I realized that he lives 100 percent in the world that he writes about—an immensely inspired, child-like-zest-full, enthralled-with-all-of-life, truthful world where the human condition is at all times being addressed and understood. Unlike everyone I have ever come across, for whom discussion of the human condition is so extremely difficult (almost impossible, one might say), when you talk to Jeremy about the human condition and the biology surrounding it, the world changes, everything seems possible, biology makes sense. Logic—simple and obvious truth—replaces over-complicated intellectual scientific downright rubbish—the vast majority of it. Indeed, science—and biology in particular—is so saturated with evasive, dishonest denial that the denial-free world of understanding that Jeremy introduces us to is so new it is akin to having to start your education all over again! Jeremy's capacity for unerring and unrelenting honesty *is* literally staggering, but, as you will see, it is always accompanied by understanding; his is no 'feel good', guru-like, false prophet form of totally dubious and ephemeral 'help' for humans' troubled lives, but a get-to-the-bottom-of-all-the-problems, truthful, *real*, insightful, ameliorating love that the world has been so in need of. And, thankfully, right through all the vehement resistance that he has faced and overcome in the last 30 years of his life, which took him near to death, Jeremy never gave up his responsibility that he saw right from the early years of his completely-nurtured-with-love upbringing to deliver the understanding of human nature that would end all the suffering in the world.

[35] Since coming across Jeremy's work 10 years ago and realizing its enormous world-saving significance, I have kept one of his pieces of writing with me at all times. Having these insights into what it means to be human brings such clarity and change to everything, that staying in close contact with them in a world that is so distressed and psychologically crippled brings me relief, security and optimism for the future like nothing else I have ever encountered. Jeremy is the ultimate psychotherapist, the psychotherapist for psychotherapists—in fact, all the great theories I have encountered in my lifetime of studies of psychiatry can be accounted for under his explanation of human origins and behavior. I want to

emphasize that Jeremy does describe the world exactly as it is with all its imperfections, but just as he describes all the horror of the world he also <u>provides the insight into our condition that makes possible the only real hope, optimism and downright, out-of-your-skin excitement for the future of the human race and our planet</u>—a transformational experience Jeremy introduces us to in chapter 9. Indeed, while <u>the overall significance of this book is its ability to transform the reader</u>—and thus the human race—by presenting the most relieving, uplifting and positive story ever told about humans, the explanation of the human condition in chapter 3 (which is summarized in chapter 1) is so amazingly accountable, insightful and relieving that it alone will transform you.

[36] Yet be warned. While ever hopeful, humans are also great skeptics when it comes to the prospect that anyone could ever actually get in behind what the human condition *really* is and from there explain and expose everything about us, which means that what typically occurs when someone begins reading this book is that they are expecting it to be like any other—that at best it will only *allude* to the human condition, rather than going right into and down to its depths! <u>The result of this skepticism and under-estimation is that when someone starts reading this book they usually progress further and further into a state of shock until they are, as author Ian Frazier acknowledged in his comment at the very beginning of the book, **'staggered into silence'**, even, as Plato warned, finding the book **'so overwhelm**[ing]**'** they aren't **'able to see a single one of the things'** they are **'now told'** are **'real'**.</u> Importantly, however, this stunned, 'deaf effect' situation can, as I mentioned earlier, be overcome by patient re-reading of the text, and, most helpfully of all, <u>by having Jeremy reassuringly escort you through this historically forbidden realm in his introductory videos to *FREEDOM*</u> at <www.humancondition.com/intro-videos>. Both actions will allow you to, as Plato said, **'grow accustomed to the light'**. This *is a fabulous but naturally initially difficult* paradigm shift you will be making to a transformed, human-condition-free world.

[37] The importance of the ideas in this book is immeasurable. The depths they enable us to reach in understanding ourselves and our world is bottomless. The great impasse to a full understanding of our existence has finally been breached. This truly is it, the day of days, the coming of our species' moment of liberation, the implications and context of which Jeremy fully deals with in pars 1278 and 1279.

[38]I am so very, very fortunate to have spent time with Jeremy and to have spoken with him almost fortnightly for 10 years—however, it is not who Jeremy is, but what he has done that is so important and in this regard I commend this book to you with all my mind, heart and soul. You are in for an absolute feast of knowledge, insight and ultimately love like you could never have imagined.

(I am very grateful to Fiona Cullen-Ward and Tony Gowing at the WTM for their assistance in preparing this Introduction.)

Harry Prosen, Wisconsin, USA, 2014

# Chapter 1

# Summary of the contents of *FREEDOM*

Computer graphic by Jeremy Griffith, Marcus Rowell and Genevieve Salter © 2009 Fedmex Pty Ltd

## Chapter 1:1 <u>Freedom from the human condition</u>

[39] This book liberates you, the reader, and all other humans from an underlying insecurity and resulting psychosis that all humans have suffered from since we became a fully conscious species some two million years ago.

[40] This underlying insecurity and psychosis that exists within every human is the product of a very deep anxiety, an uncertainty, of not knowing *why*, when the ideals of life are so obviously to be cooperative, loving and selfless, are humans so competitive, aggressive and selfish. Certainly, we have relied heavily on the excuse that our behavior is no different to that seen in the animal kingdom—that we humans are competitive, aggressive and selfish because of our animal heritage. We have argued that we are, as Lord Alfred Tennyson put it, *'red in tooth and claw'*—a victim of savage animal instincts that compel us to fight and compete for food, shelter, territory and a mate; that we are at the mercy

of a biological need to reproduce our genes. But this reason that biologists, including the most celebrated living biologist, the Harvard-based Edward (E.) O. Wilson, have been perpetuating cannot be the *real* cause of our competitive, divisive behavior because descriptions of human behavior, such as egocentric, arrogant, inspired, depressed, deluded, pessimistic, optimistic, artificial, hateful, cynical, mean, immoral, brilliant, guilt-ridden, evil, psychotic, neurotic, alienated, all recognize the involvement of our species' unique fully conscious thinking mind— that there is a *psychological* dimension to *our* behavior. Humans have suffered *not* from the genetic-opportunism-based, non-psychological *animal* condition, but the conscious-mind-based, *PSYCHOLOGICALLY* troubled *HUMAN CONDITION*.

[41] Of course, in the absence of understanding, we had no choice but to come up with *some* excuse for why we are the way we are in order to cope with the negative implications of being divisively instead of cooperatively behaved, but we do all in truth know that the old 'animals are competitive and aggressive and that's why we are' defense simply doesn't hold water—it doesn't explain our psychologically distressed human condition and so cannot end the pain, suffering, conflict and confusion that plagues this planet. And *that* is where the human race has been stalled, waiting in an increasingly distressed state for the *real* explanation for our *psychologically* troubled *human condition* that will finally make sense of the riddle of human life. That is, *until now*—because it is precisely that human-mind-liberating, psychosis-relieving REAL explanation that this book at last delivers.

[42] While the full description of this psychosis-addressing-and-solving, real explanation of the human condition is presented in chapter 3, a summary of it is provided here in chapter 1 to demonstrate to the reader that this *is the* understanding and insight that is able to end the underlying insecurity and resulting psychosis of everyone's condition and, through doing so, <u>transform *every* human into a new, human-condition-free person!</u> Yes, this book brings about the liberation of humanity from its incarceration in the horrifically debilitating darkness of Plato's 'cave' of alienating psychological denial that I have depicted in the above image and will explain later in this chapter.

[43] But before diving into the heart of the explanation, it needs to be emphasized that while the human condition is essentially the riddle of

why humans are competitive and aggressive when the ideals are to be cooperative and loving, the deeper meaning of the human condition is more elusive. Indeed, as will be made clear in chapter 2, the human condition has been *such* a difficult issue for humans to think about and confront that many people now have very little idea of what the human condition *actually* is, thinking it refers not to the reality of our species' immensely troubled psychology, but to the state of widespread poverty and physical hardship in human life, or to problems such as human inequality. But these problems are only superficial manifestations and aspects of the human condition. The truth is, the human condition is a *much* more profound and serious issue that goes to the very heart of who we are. So before I present the brief explanation of the human condition I first need to describe what the human condition *really* is.

## Chapter 1:2 <u>What exactly is the human condition?</u>

[44] Here on Earth some of the most complex arrangements of matter in the known universe have come into existence. Life, in all its incredible diversity and richness, developed. And, by virtue of our mind, the human species must surely represent the culmination of this grand experiment of nature we call life—for, as far as we can detect, we are the first organism to have developed the fully conscious ability to sufficiently understand and thus manage the relationship between cause and effect to wrest management of our lives from our instincts, and to even reflect upon our existence. It is easy to lose sight of the utter magnificence of what we are, but the human mind must surely be nature's most astonishing creation. Indeed, it must be one of the wonders of the universe! Consider, for example, the intellectual brilliance involved in sending three of our kind to the Moon and back.

[45] *AND YET*, despite our species' magnificent mental capabilities, and undeniable capacity for immense sensitivity and love, behind every wondrous scientific achievement, sensitive artistic expression and compassionate act lies the shadow of humanity's darker side—an unspeakable history of greed, hatred, rape, torture, murder and war; a propensity for deeds of shocking violence, depravity, indifference and cruelty. As the philosopher Arthur Schopenhauer wrote, '**man is the only animal which causes pain to others with no other object than causing pain...No animal ever torments another for the sake of tormenting: but man does so, and it is this**

**which constitutes the *diabolical* nature which is far worse than the merely bestial'** (*Essays and Aphorisms*, tr. R.J. Hollingdale, 1970, p.139 of 237). Yes, undermining all our marvelous accomplishments and sensibilities is the fact that we humans have also been the most ferocious and malicious creatures to have ever lived on Earth!

[46] And it is precisely this dual capacity for what has historically been referred to as 'good' and 'evil' that has troubled the human mind since we first became fully conscious, thinking beings: are we essentially 'good' and, if so, what is the cause of our destructive, insensitive and cruel, so-called 'evil' side? *Why* do we thinking, reasoning, rational, immensely clever, supposedly sensible beings behave so abominably and cause so much suffering and devastation? Yes, the eternal question has been why 'evil'? What is the origin of the dark, volcanic forces that undeniably exist within us humans? What is it deep within us humans that troubles us so terribly? What is it that makes us such a combative, ruthless, hateful, retaliatory, violent, in-truth-psychologically-disturbed creature? In metaphysical religious terms, what is 'the origin of sin'? More generally, if the universally accepted ideals of life are to be *cooperative, loving and selfless*—ideals that have been accepted by modern civilizations as the foundations for constitutions and laws and by the founders of all the great religions as the basis of their teachings—*why* are humans *competitive, aggressive and selfish?* Indeed, *so* ruthlessly competitive, selfish and brutal that life has become all but unbearable and we have nearly destroyed our own planet? Does our inconsistency with the ideals mean we are essentially bad, a flawed species, an evolutionary mistake, a blight on Earth, a cancer in the universe—or could we possibly be divine beings? And, more to the point, is the human race faced with having to live *forever* in this tormented state of uncertainty and insecurity about the fundamental worth and meaning of our lives? Is it our species' destiny to have to live in a state of permanent damnation?!

[47] The agony of being unable to truthfully answer the fundamental question of why we are the way we are—divisively instead of cooperatively behaved—has been the particular burden of human life. It has been our species' particular affliction or condition—our *'human condition'*. Good or bad, loving or hateful, angels or devils, constructive or destructive, sensitive or insensitive: WHAT ARE WE? Throughout history we have struggled to find meaning in the awesome contradiction of our human condition. Our endeavors in philosophy, psychology and biology have failed,

until now, to provide a truthful, real, psychosis-addressing-and-solving, not-the-patently-false-animals-are-competitive-and-aggressive-and-that's-why-we-are, fully accountable, genuinely clarifying explanation. And for their part, while religious assurances such as 'God loves you' may have provided temporary comfort, they too failed to explain *WHY* we are lovable. So, yes, *WHY* are we lovable? How could we be good when all the evidence seems to unequivocally indicate that we are a deeply flawed, bad, even evil species? What is the answer to this question of questions, this problem of 'good and evil' in the human make-up, this greatest of all paradoxes and dilemmas of the human condition? *What caused humans to become divisively behaved and, more importantly, how is this divisive behavior ever going to be brought to an end? THIS*, the issue of the human condition, is the *REAL* question facing the human race.

[48] And with every day bringing with it more alarming evidence of the turmoil of the human situation, the issue of the human condition is the *ONLY* question confronting the human race, because its solution has become a matter of critical urgency. Conflict between individuals, 'races', cultures, religions and countries abounds (and by 'races' I mean groups of people whose members have mostly been together a long time and are thus relatively closely related genetically—people who have a shared history). There is genocide, terrorism, mass displacement of peoples, starvation, runaway diseases, environmental devastation, gross inequality, 'racial' and gender oppression, polarized politics, rampant corruption and other crimes, drug abuse, family breakdown, and epidemic levels of obesity, chronic anxiety, depression, unhappiness and loneliness—all of which are being rapidly exacerbated by the exploding world population and exponential rise everywhere in anger, egocentricity and alienation (what I refer to as 'upset'). Improved forms of management, such as better laws, better politics and better economics—and better self-management, such as new ways of disciplining, suppressing, organizing, motivating or even transcending our troubled natures—have all failed to end the march towards ever greater levels of alienation, devastation and unhappiness. In short, the situation is now *so* grim the human race *IS*, in fact, entering *end play* or *end game*, where the Earth cannot absorb any further devastation from the effects of our upset behavior, nor the human body cope with any more debilitating stress, or, most particularly, our mind endure any more psychological distress, any more alienated psychosis and neurosis. The journalist Richard Neville was frighteningly accurate when, in

summarizing the desperate state of our species' situation, he wrote that **'the world is hurtling to catastrophe: from nuclear horrors, a wrecked eco-system, 20 million dead each year from malnutrition, 600 million chronically hungry...All these crises are man made, their causes are psychological. The cures must come from this same source; which means the planet needs psychological maturity fast. We are locked in a race between self destruction and self discovery'** (*Good Weekend, The Sydney Morning Herald*, 14 Oct. 1986; see <www.wtmsources.com/167>). Yes, our species has come to the critical juncture where *ONLY* **'self discovery'**—reconciling, ameliorating, **'psychological[ly]'** healing understanding of ourselves—could save us from **'self destruction'**.

Cartoon by Michael Leunig, Melbourne's *The Age* newspaper, 8 Oct. 1988

[49]The Australian cartoonist Michael Leunig has been contributing cartoons and articles to Australian newspapers since 1965 and in that time has produced innumerable brilliantly insightful, revealing and therapeutically honest cartoons about all aspects of our species' troubled condition—five more of which appear later in this book. This cartoon, in

which he truthfully depicts all the horrors of the human condition, is, in my view, one of his best. It is certainly *not* a picture of a lovely ordered city park where people peacefully and happily enjoy themselves, as we all too easily prefer to delude ourselves that the world we have created is like. Rather, it shows a mother and child approaching the **'Gardens of the Human Condition'** with an expression of bewildered dread on the face of the mother, and in the case of the child, wide-eyed shock. Yes, as Leunig cleverly intimates, our world is no longer an innocent Garden of Eden, but a devastated realm of human-condition-stricken, psychologically distressed humans where 'i̱nhumanity' reigns. With this masterpiece, Leunig has boldly revealed the truth that we humans are a brutally angry, hateful, destructive, arrogant, egocentric, selfish, mad, lonely, unhappy and psychologically depressed species. He has people fighting, beating and strangling each other, drunk out of their minds, depressed, lonely, crying, hiding and suiciding, going mad, and egocentrically holding forth—reflecting, in effect, every aspect of the human condition. Yes, as the main character in the 2005 film *The White Countess* noted, **'What we see out there** [in the world] **is chaos; mistrust, deception, hatred, viciousness— chaos—there's no broader canvas out there, nothing that man can go and compose a pretty picture on.'** The polymath Blaise Pascal was even more damning in his depiction of the human condition, when, centuries earlier, he spelled out the full horror of our contradictory nature, writing, **'What a chimera then is man! What a novelty, what a monster, what a chaos, what a contradiction, what a prodigy! Judge of all things, imbecile worm of the earth, repository of truth, a sewer of uncertainty and error, the glory and the scum of the universe!'** (*Pensées*, 1669). William Shakespeare was equally revealing of the paradoxical true nature of the human condition when he wrote, **'What a piece of work is a man! how noble in reason! how infinite in faculty!... in action how like an angel! in apprehension how like a god! the beauty of the world! the paragon of animals! And yet, to me, what is this quintessence of dust? man delights not me'** (*Hamlet*, 1603)!!

[50] So the 17 goals of **'The** [United Nations'] **Global Goals' 'Movement'** that each **'global citizen'** and **'every school'** is currently being told to **'tell everyone about'** and **'make famous'** (namely to address **'poverty, hunger, well-being, education, gender equality, clean water, clean energy, decent work, infrastructure, inequality, sustainability, responsible consumption, climate action, life in water, life on land, peace and justice, partnerships for the goals'**) *hugely* trivialize our species' plight—because all these goals focus *only* on the

symptoms of the human condition. To stop the destruction of our world and the disintegration of society that is happening everywhere we look we have to fix the *cause* of the problems at its source, which is *us humans, our psychosis*. WE are the problem; our out-of-control egocentric, selfish, competitive and ferociously vicious, mean and aggressive behavior. The cartoonist Walt Kelly spoke the truth when he had Pogo, his comic strip hero, say, **'We have met the enemy and he is us'** (1971). Yes, the underlying, REAL question that had to be answered if we were ever to find relieving, redeeming, psychologically healing understanding of ourselves was WHY ARE WE HUMANS the most brilliantly clever of creatures, the ones who are **'god'-'like'** in our **'infinite' 'faculty'** of **'reason'** and **'apprehension'**, a **'glor**[ious]**'**, **'angel'-'like' 'prodigy'** capable of being a **'judge of all things'** and a **'repository of truth'**, also the meanest, most vicious, most capable of inflicting pain, cruelty, suffering and degradation? Why are humans so choked full of volcanic frustration, anger and hatred—the species that behaves *so appallingly* that we seem to be **'monster**[s]**'**, **'imbecile**[s]**'**, **'a sewer of uncertainty and error'** and **'chaos'**, the **'essence'** of **'dust'**, **'the scum of the universe'**? *That* is what the issue of the human condition *really* is: 'WHY ARE WE THE WAY WE ARE, COMPETITIVE AND AGGRESSIVE, RATHER THAN COOPERATIVE AND LOVING?'—AND BENEATH THAT, THE DEEPER QUESTION OF, 'WHAT IS THE ORIGIN OF ALL THIS PSYCHOLOGICAL FRUSTRATION AND PAIN INSIDE OF US HUMANS?'

[51] Clearly, it is of incalculable importance to finally be able to answer this question of questions of the origin of the human condition. In fact, the great hope, faith and trust of the human race *has* been that one day the redeeming and psychologically rehabilitating understanding of our 'good and evil'-afflicted human condition would be found. And since this all-important issue of the human condition—the underlying issue in *all* human affairs—is biological in nature, its resolution has been *the* most important task assigned to biologists; indeed, it has been described as the 'holy grail' of biology. AND, as incredible as it may seem, it is *this* breakthrough of breakthroughs—this all-important, world-saving, psychosis-addressing-and-relieving, real biological explanation of the human condition—that is presented in this book! As stated earlier, the effect of finally knowing and understanding and living with this explanation is that it transforms humans from their psychologically insecure, human-condition-afflicted existence to a psychologically secure

and mature, human-condition-free state. *This* is the explanation that lifts the so-called burden of guilt from the shoulders of the human race. *This* is the explanation that ends the condemnation that we humans have had to endure for *so* long. And the reason you will know this is the real, true explanation is because, unlike earlier excuses, it will prove *so* accountable it will make *all* of our egocentric, arrogant, inspired, depressed, deluded, etc, psychologically distressed behavior transparent. Understanding and absorbing this explanation will end all the bewilderment, confusion and uncertainty—all the *insecurity*—about human behavior. It will unravel the whole mess we humans have been living in; as the leading Australian journalist, broadcaster and commentator, Brian Carlton, said in one of the introductory videos I will shortly be recommending you view, **'It's an intellectual epiphany...It's a revelation...the clarity of it is euphoric...when you *get* it, it is an event. You remember the day, you remember the section of the book, you remember when it happened, it stays with you...Don't underestimate the extent to which your work has impacted me in terms of how I think about what I'm seeing, how I interpret behaviour. I worked up this ability to be able to work out what a person was like in the first five or six seconds of a conversation** [as the host of a talkback radio program].'

[52] So, in a world fast going crazy from the effects of the human condition, *this* is the now desperately needed reconciling understanding that brings about a new world for humans *FREE* of the agony of the human condition. In short, this is the understanding that ends human suffering and unites the human race.

## Chapter 1:3 A brief description of the human-race-transforming explanation of the human condition that is presented in *FREEDOM*

[53] So, what is this psychosis-addressing-and-solving, fully accountable, *real* explanation of our human condition that makes the world of humanity so understandable that it becomes transparent?

[54] (Again, what follows is a very brief description of the explanation that will be provided in more comprehensive form in chapter 3—after which chapter upon chapter will dismantle the whole universe of dishonest excuses that humans have had to invent for all aspects of our behavior while we awaited this real explanation of our divisive-instead-of-cooperative condition.)

[55] Our human condition is directly related to the emergence of our conscious thinking mind—*it is a result of our species having become fully conscious*—and once we accept this foundation truth, then the explanation is actually fairly obvious. Clearly, before our species became fully conscious our lives must have been controlled by instincts, as the lives of all other animals continue to be. So the essential question is, what would happen to a species if it became capable of consciously understanding and thus managing its world? If we think about this scenario, what would obviously happen is that the conscious mind would start to take over management from the instincts. And if we think further about that development, we can appreciate that a conflict would have arisen between the already established instinctive management system and the new conscious, understanding-based management system.

[56] To help visualize this development, consider the situation of a migrating bird that had just acquired a fully conscious mind.

[57] Many bird species are perfectly orientated to instinctive migratory flight paths. Each winter, without ever 'learning' where to go and without knowing why, they quit their established breeding grounds and migrate to warmer feeding grounds. They then return each summer and so the cycle continues. Over the course of thousands of generations and migratory movements, only those birds that happened to have a genetic make-up that inclined them to follow the right route survived. Thus, through natural selection, they acquired their instinctive orientation.

[58] So, imagine a flock of migrating storks returning to their summer breeding nests on the rooftops of Europe from their winter feeding grounds in southern Africa. Suppose in the instinct-controlled brain of one of them we place a fully conscious mind (we will call the stork Adam because we will soon see that, up to a point, this analogy parallels the old, pre-scientific Biblical account of Adam and Eve taking the **'fruit'** (Gen. 3:3) **'from the tree of…knowledge'** (2:9, 17); that is, becoming conscious). As Adam Stork flies north he spots an island off to the left with a tree laden with apples. Using his newly acquired conscious mind, Adam thinks, 'I should fly down and eat some apples.' It seems a reasonable thought but he can't know if it is a good decision or not until he acts on it. For Adam's new thinking mind to make sense of the world he has to learn by trial and error and so he decides to carry out his first grand experiment in self-management by flying down to the island and sampling the apples.

[59] But it's not that simple. As soon as Adam's conscious thinking self deviates from his established migratory path, his innocent instinctive self (innocent in the sense of being unaware or ignorant of the need to search for knowledge) tries to pull him back on course. In following the flight path past the island, Adam's instinct-obedient self is, in effect, criticizing his conscious mind's decision to veer off course; it is condemning his search for understanding. All of a sudden Adam is in a dilemma: if he obeys his instinctive self and flies back on course, his instincts will be happy but he'll never *learn* if his deviation was the right decision or not. All the messages he's receiving from within inform him that obeying his instincts is good, is right, but there's also a new inclination to disobey, a defiance of instinct. Diverting from his course will result in apples and understanding, yet he already sees that doing so will make him feel bad.

[60] Uncomfortable with the criticism his newly conscious mind or intellect is receiving from his instinctive self, Adam's first response is to ignore the temptation the apples present and fly back on course. As he does so, however, Adam realizes he can't deny his intellect—sooner or later he must find the courage to master his conscious mind by carrying out experiments in understanding. So, continuing to think, he next asks himself, 'Why not fly down to an island and rest?' Again, not knowing any reason why he shouldn't, he proceeds with his experiment. And *again*, his decision is met with the same criticism from his instinctive self—but *this time* Adam defies the criticism and perseveres with his experimentation in self-management. His decision, however, means he must now live with the criticism and immediately he is condemned to a state of upset. A battle has broken out between his instinctive self, which is perfectly orientated to the flight path, and his emerging conscious mind, which needs to understand why that flight path is the correct course to follow. His instinctive self is perfectly orientated, but Adam doesn't *understand* that orientation.

[61] In short, when the fully conscious mind emerged it wasn't enough for it to be orientated by instincts, it *had to* find understanding to operate effectively and fulfill its great potential to manage life. But, tragically, the instinctive self didn't 'appreciate' that need and 'tried to stop' the mind's necessary search for knowledge, as represented by the latter's experiments in self-management—hence the ensuing battle between instinct and intellect. To refute the criticism from his instinctive self, Adam needed the discoveries that science has given us of the difference

in the way genes and nerves process information; in particular, he needed to be able to explain that the gene-based learning system can orientate species to situations but is incapable of insight into the nature of change. Genetic selection of one reproducing individual over another reproducing individual (the selection, in effect, of one idea over another idea, or one piece of information over another piece of information) gives species adaptations or orientations—instinctive programming—for managing life, but those genetic orientations, those instincts, are not understandings. The nerve-based learning system on the other hand, can, if sufficiently developed, understand change. Nerves were originally developed for the coordination of movement in animals, but, once developed, their ability to store impressions—what we refer to as 'memory'—gave rise to the potential to develop understanding of cause and effect. If you can remember past events, you can compare them with current events and identify regularly occurring experiences. This knowledge of, or insight into, what has commonly occurred in the past enables you to predict what is likely to happen in the future and to adjust your behavior accordingly. Once insights into the nature of change are put into effect, the self-modified behavior starts to provide feedback, refining the insights further. Predictions are compared with outcomes and so on. Much developed, nerves can sufficiently *associate* information to *reason* how experiences are related, learn to *understand* and become *conscious* of, or aware of, or *intelligent* about, the relationship between events that occur through time. Thus consciousness means being sufficiently aware of how experiences are related to attempt to manage change from a basis of understanding. What this means is that when the nerve-based learning system became sufficiently developed for consciousness to emerge and with it the ability to understand the world, it wasn't enough to be instinctively *orientated* to the world—conscious *understanding* of the world had to be found. The problem, of course, was that Adam had only just taken his first, tentative steps in the search for knowledge, and so had no ability to explain anything. It was a catch-22 situation for the fledgling thinker, because in order to explain himself he needed the very knowledge he was setting out to accumulate. He had to search for understanding, ultimately self-understanding, understanding of why he had to 'fly off course', without the ability to first explain why he needed to 'fly off course'. And without that defense, he had to live with the criticism from his instinctive self and was *INSECURE* in its presence.

[62] To resist the tirade of unjust criticism he was having to endure and mitigate that insecurity, Adam had to do something. *But what could he do?* If he abandoned the search and flew back on course, he'd gain some momentary relief, but the search would, nevertheless, remain to be undertaken. So all Adam could do was retaliate against and ATTACK the instincts' unjust criticism, attempt to PROVE the instincts' unjust criticism wrong, and try to DENY or block from his mind the instincts' unjust criticism—and he did *all* those things. He became angry towards the criticism. In every way he could he tried to demonstrate his self worth, prove that he is good and not bad—he shook his fist at the heavens in a gesture of defiance of the implication that he is bad. And he tried to block out the criticism—this block-out or denial including having to invent contrived excuses for his instinct-defying behavior. In short, his ANGRY, EGOCENTRIC and ALIENATED state appeared. Adam's intellect or 'ego' (which is just another word for the intellect since the *Concise Oxford Dictionary* defines **'ego'** as **'the conscious thinking self'** (5th edn, 1964)) became 'centered' or focused on the need to justify itself—selfishly pre-occupied aggressively competing for opportunities to prove he is good and not bad, to validate his worth, to get a 'win'; to essentially eke out any positive reinforcement that would bring him some relief from criticism and sense of worth. He unavoidably became SELFISH, AGGRESSIVE and COMPETITIVE.

[63] Overall, it was a terrible predicament in which Adam became PSYCHOLOGICALLY UPSET—a sufferer of PSYCHOSIS and NEUROSIS. Yes, since, according to *Dictionary.com*, **'osis'** means **'abnormal state or condition'**, and the *Penguin Dictionary of Psychology*'s entry for **'psyche'** reads **'The oldest and most general use of this term is by the early Greeks, who envisioned the psyche as the soul or the very essence of life'** (1985 edn), Adam developed a **'psychosis'** or 'soul-illness', and a **'neurosis'** or neuron or nerve or 'intellect-illness'. His original gene-based, instinctive **'essence of life'** soul or PSYCHE became repressed by his intellect for its unjust condemnation of his intellect, and, for its part, his nerve or NEURON-based intellect became preoccupied denying any implication that it is bad. Adam became psychotic and neurotic.

[64] But, again, without the knowledge he was seeking, without self-understanding (specifically the understanding of the difference between the gene and nerve-based learning systems that science has given us), Adam Stork had no choice but to resign himself to living a psychologically

upset life of anger, egocentricity and alienation as the only three responses available to him to cope with the horror of his situation. It was
an extremely unfair and difficult, indeed tragic, position for Adam to
find himself in, for we can see that while he was good he appeared to be
bad and had to endure the horror of his psychologically distressed, upset
condition until he found the real—as opposed to the invented or contrived
*not*-psychosis-recognizing—defense or reason for his 'mistakes'. Basically,
*suffering psychological upset was the price of his heroic search for under-
standing*. Indeed, it is the tragic yet inevitable situation any animal would
have to endure if it transitioned from an instinct-controlled state to an
intellect-controlled state—its instincts would resist the conscious mind's
search for knowledge. Adam's uncooperative and divisive competitive
aggression—and his selfish, egocentric, self-preoccupied efforts to prove
his worth; and his need to deny and evade criticism, essentially embrace a
dishonest state—all became an unavoidable part of his personality. Such
was Adam Stork's predicament, *and such has been the human condition,
for it was within our species that the fully conscious mind emerged.*

[65] We can now see how the story of Adam Stork—which describes the
primary issue involved in our human condition of the psychologically
upsetting battle that emerged between our instincts and our conscious
intellect's search for knowledge—has parallels with the pre-scientific
Biblical account in the Book of Genesis of Adam and Eve's experiences in
the Garden of Eden, except in that presentation when Adam and Eve took
the **'fruit'** (3:3) **'from the tree of the knowledge of good and evil'** (2:9, 17)—went
in search of understanding—they were **'banished...from the Garden'** (3:23)
for being **'disobedient'** (the term widely used in descriptions of Gen. 3) and becoming
'bad' or **'evil'** or 'sinful'. In *this* presentation, however, Adam and Eve are
revealed to be the HEROES, NOT THE VILLAINS they have so long been
portrayed as. So while humans ARE immensely upset—that is, immensely
angry, egocentric and alienated—WE ARE GOOD AND NOT BAD AFTER
ALL!!!! And 'upset' is the right word for our condition because while we
are not 'evil' or 'bad' we are definitely psychologically upset from having
to participate in humanity's heroic search for knowledge. 'Corrupted' and
'fallen' have sometimes been used to describe our condition, but they
have negative connotations that we can now appreciate are undeserved.

[66] For our species, it really has been a case of **'Give me liberty or give
me death'**, **'No retreat, no surrender'**, **'Death before dishonor'**, **'Never back
down'**, as the sayings go. Our conscious thinking self was *never* going

to give in to our instinctive self or soul. Even though we had developed into angry, egocentric and alienated people, we were never going to accept that we were fundamentally bad, evil, worthless, awful beings; we weren't going to wear that criticism—for if we did, we wouldn't be able to get out of bed each morning and face the world. If we truly believed we were fundamentally evil beings, we would shoot ourselves. There had to be a greater truth that explained our behavior and until we found it we couldn't rest. And so every day as we got out of bed we took on the world of ignorance that was condemning us. We defied the implication that we are bad. We shook our fist at the heavens. In essence, we said, 'One day, one day, we are going to prove our worth, explain that we are not bad after all, and until that day arrives we are not going to **'back down'**, we are not going to take the ignorant, naive, stupid, unjustified criticism from our instincts. No, we are going to fight back with all our might.' And that is what we have done; that is what *every* conscious human that has ever lived has done—and because we did, because we persevered against all that criticism, we have now finally broken through and found the full truth that explains that <u>humans are wonderful beings after all. In fact, not just wonderful but the heroes of the whole story of life on Earth</u>. This is because our fully conscious mind is surely—given its phenomenal ability to understand the world—nature's greatest invention, so for us humans who were given this greatest of all inventions to develop, to be made to endure the torture of being unjustly condemned as bad or evil for doing just that, and to have had to endure that torture for *so* long, *some 2 million years* (the time we have likely been fully conscious), *has to* make us the absolute heroes of the story of life on Earth. We were given the hardest, toughest of tasks, and against all the odds we completed it. Humans *are* the champions of the story of life on Earth. We are *so, so* wonderful! Yes, we can at last understand the absolutely extraordinary paradox that neither Pascal nor Shakespeare could at all understand, of how on earth could we humans be **'god'-'like'** in our **'infinite' 'faculty'** of **'reason'** and **'apprehension'**, a **'glor**[ious]**'**, **'angel'-'like' 'prodigy'** capable of being a **'judge of all things'** and a **'repository of truth'**, and yet seemingly behave so appallingly that we appear to be **'monster**[s]**'**, **'imbecile**[s]**'**, **'a sewer of uncertainty and error'** and **'chaos'**, the **'essence'** of **'dust'**, **'the scum of the universe'**. We have made sense of the seemingly nonsensical!!!

[67] As we will see throughout this book, now that we can finally explain *the seemingly-impossible-to-explain* paradox of how we humans could

be good when all the evidence appeared to unequivocally indicate we were bad, *all* our mythology can at last be made sense of. For example, why was Miguel de Cervantes' 1605 novel *Don Quixote* voted 'The Greatest Book of All Time' by the world's most acclaimed writers in a poll arranged by the Nobel Institute? Well, *Don Quixote* is the story of an elderly man who gets out of bed, re-names himself 'Don Quixote of la Mancha', dons an old suit of armor, takes up an ancient shield and lance, mounts his skinny old horse, and calls on his loyal but world-weary companion Sancho to join him on the most spectacular of adventures. As I have depicted (opposite), coming across a field of large windmills, the noble knight says, **'Look yonder, friend Sancho, there are…outrageous giants whom I intend to…deprive…of life…and the expiration of that cursed brood will be an acceptable service to Heaven'**. And so the crazed and hopeless adventure goes on, gloriously doomed battle after gloriously doomed battle. *But that has been the lot of every human for some 2 million years; hopeless battle after hopeless battle, feeble beings charging at and trying to vanquish the 'outrageous giant' ignorance-of-the-fact-of-our-species'-fundamental-goodness!* Wave after wave of *quixotic* humans have thrown themselves at that **'outrageous giant'** of ignorance for eons and eons, as bit by tiny bit they accumulated the knowledge that finally made the redeeming explanation of our human condition possible!

[68] Joe Darion's fabulous lyrics of the song *The Impossible Dream*, which featured in the 1965 musical about Don Quixote, *Man of La Mancha*, perfectly describes the unbelievably courageous and heroic participation in humanity's corrupting search for knowledge by every human who has ever lived during the last 2 million years, which Don Quixote's story personifies: **'To dream the impossible dream** [of one day, in the far future (which has now arrived), finding the redeeming understanding of the human condition], **to fight the unbeatable foe** [of our ignorant instincts], **to bear the unbearable sorrow, to run where the brave dare not go. To right the unrightable wrong** [of being unjustly criticized], **to love pure and chaste from afar, to try when your arms are too weary, to reach the unreachable star. This is my quest, to follow that star, no matter how hopeless, no matter how far. To fight for the right without question or pause, to be willing to march into hell for a heavenly cause. And I know if I will only be true, to this glorious quest, that my heart will lie peaceful and calm, when I'm laid to my rest. And the world will be better for this, that one man scorned and covered with scars, still strove with his last ounce of courage, to reach the unreachable star.'**

[69] Again, it has to be emphasized that this has been a very brief outline of the explanation of the human condition that is provided in chapter 3. There is so much more to explain—in particular, humans aren't migrating birds, so what was the particular instinctive orientation that *our* species was defying when we went in search of knowledge? In chapter 2 it will be pointed out that humans have unconditionally selfless, moral instincts, the 'voice' or expression of which within us is our conscience, while chapter 5 will provide the biological explanation for how we acquired these extraordinary altruistic, moral instincts. So it follows that when we humans became fully conscious and went in search of knowledge and became angry, egocentric and alienated, that behavior only served to *further* exacerbate the condemnation we already felt from defying our particular instinctive orientation. Yes, our necessary search for knowledge has been an *extremely* guilt-inducing enterprise—it required us to defy our instincts, a necessary defiance that made us angry, egocentric and alienated, which was an outcome that further offended our particular instincts that expect us to behave in an unconditionally selfless, cooperative and loving way. So, if Adam Stork felt guilty for merely veering off his flight path, but was still a hero for doing so, just how much more heroic are humans to have endured and surmounted the horrendous degree of guilt *we* encountered!

[70] Consider then the fact that humans have been living in this extremely unfair and torn state where we couldn't explain the good reason for our species' psychologically upset condition for over 2 million years! With this in mind, we can start to register just how much hurt, frustration and anger must now exist within humans. After all, imagine living just *one*

day with the injustice of being condemned as evil, bad and worthless when you intuitively knew—but were unable to explain—that you were actually the complete opposite, namely truly wonderful, good and meaningful. How tormented and furious—how upset—would you be by the end of that one day? You would be *immensely* upset. So extrapolate that experience over *2 million years* and you will begin to get some appreciation of just how much volcanic anger must now exist within us humans today! While we have learnt to significantly restrain and conceal—'civilize'—our phenomenal amount of upset, it nevertheless follows that, under the surface, our species must be boiling with rage, and that sometimes, when our restraint can no longer find a way to contain it, that anger must express itself. Yes, *we can finally understand humans' capacity for astounding acts of aggression, hate, brutality and atrocity.*

[71] While all of this will be explained more thoroughly in chapter 3, this brief Adam Stork analogy does serve to convey the main upsetting clash that occurs when a conscious mind develops in the presence of pre-established instincts. And even from this simplified analogy, we can see how absolutely wonderfully exonerating and psychologically transforming this psychosis-addressing-and-solving explanation of the human condition is, because after *2 million years of uncertainty* it allows all humans to finally understand that there has been a *very good* reason for our angry, alienated and egocentric lives. Indeed, this fact of the utter magnificence of the human race—that we are, in truth, the heroes of the story of life on Earth—brings such intense relief to our angst-ridden cells, limbs and torsos that it will seem as though we have thrown off a shroud of heavy weights. The great, heavy burden of guilt *has* finally been lifted from the shoulders of humans. Yes, doesn't the core feeling exist in all humans that far from being meaningless, **'banish**[ment]**'**-deserving **'evil'** blights on this planet we *are* all immense heroes? Doesn't this explanation at last make sense of the immensely courageous and defiant attitude of all humans? And won't this explanation bring deep, bone-draining relief to the whole of each person's being?

[72] Our ability now to explain and understand that we are actually all good and not bad enables *all* the upset that resulted from being unable to explain the source of our divisive condition to subside and disappear. Finding understanding of the human condition is what rehabilitates and transforms the human race from its psychologically upset angry, ego-centric and alienated condition. In fact, the word **'psychiatry'** literally

means **'soul-healing'** (derived as it is from *psyche* meaning **'soul'** and *iatreia*, which according to *The Encyclopedic World Dictionary* means **'healing'**)—but we have never before been able to 'heal our soul', explain to our original instinctive self or soul that our fully conscious, thinking self is good and not bad and, by so doing, reconcile and heal our split selves. As Professor Harry Prosen, the former president of the Canadian Psychiatric Association to whom I am eternally grateful for writing his deeply appreciative Introduction to this book, has said about the psychological effect of this all-loving, all-compassionate understanding of ourselves: **'I have no doubt this biological explanation of the human condition is the holy grail of insight we have sought for the psychological rehabilitation of the human race.'** Yes, our ability now to understand the dark side of ourselves means we can finally achieve the **'wholeness for humans'** that the analytical psychologist Carl Jung was forever pointing out **'depends on the ability to own our own shadow'**. As the picture below powerfully intimates, we have needed, and now have, the key that liberates the human mind from the underlying deep, dark, psychological trauma of the human condition!

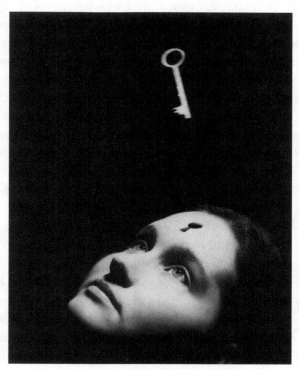

Illustration by Matt Mahurin for *TIME*, 29 Nov. 1993

[73] As for the veracity of this explanation, it is precisely this explanation's ability to at last make relieving sense of human life, of all our behavior in fact, that lets us know that we have finally found the *true* explanation of the human condition. The great physicist Albert Einstein once wrote that **'Truth is what stands the test of experience'** (*Out of My Later Years*, 1950, p.115 of 286), and since this study and explanation is all about us, *our* behavior, we are each in a position to personally **'experience'** its validity, to know if it's true or not. As the subject of this study, we can each know if the ideas being put forward work or not. We can each know if these explanations do make sense of our deepest feelings—of all our competitive anger and egocentricity, of our lonely estranged souls, of our insecure state yet core belief that we are wonderful beings, etc, etc—as, in fact, they do.

[74] Moreover, as mentioned earlier, these explanations are *so* powerfully insightful, accountable and revealing (so true) that they render our lives transparent—a transparency, a sudden exposure, that can initially be overwhelmingly confronting and depressing. But there is a way, indeed an absolutely wonderful, joyous way, to deal with the arrival of exposure day, or transparency day, or revelation day, or truth day, or honesty day—in fact, what has been described as the long-feared **'judgment day'**, which is *not*, as has just been described, a time of condemning 'judgment' but one of compassionate understanding. And that wonderful way of managing the arrival of the all-liberating, all-rehabilitating but at the same time all-exposing truth about us humans is the subject of the concluding chapter (9) in this book. But the point being made here is that the transparency of our lives that these explanations bring, which the aforementioned interview with Brian Carlton reveals, confirms just how effective, how penetrating, and, therefore, *how truthful* these understandings actually are.

[75] Yes, while searching for the real explanation of the human condition humans had no choice but to develop all manner of behavior to protect ourselves from exposure to criticism of our divisive-instead-of-cooperative condition—with the result being that we humans have been hiding behind very large, carefully constructed facades of evasion, pretense and delusion. And, as we will see in the next chapter, as our vehicle for enquiry into the nature and workings of our world, science and its human-condition-afflicted practitioners invented and contrived all sorts

of false excuses for our divisive behavior to support this denial. But when the truth arrives about the human condition, all these facades and false excuses are suddenly exposed for the lies they are. The truth reveals the lies, as it must, and there has been an *immense* amount of lying going on, so all this sudden exposure of the extent of our fraudulent existence *will* come as a sudden shock—it does represent a kind of exposure or revelation or judgment 'day'—but as will be explained in chapter 9 when the great transformation of the human race is described, that exposure is easily managed through not overly confronting the truth that has now arrived and instead focusing on redeeming the whole human race and to repairing the immensely damaged world that has come about as a result of all our upset behavior.

[76]Again, this chapter contains only a very brief outline of what this book presents. The long-awaited arrival of the truth about our psychologically distressed human condition was always going to produce a huge paradigm shift in our thinking—a *great* deal of sweeping change—which means there is much to explain, describe and have to digest. As mentioned earlier, chapter after chapter that follows dismantles the whole universe of dishonest excuses that we humans have had to invent to try to account for all aspects of our behavior while we waited for the real explanation of our divisive-instead-of-cooperative condition. And not only will all the dishonest biological excuses for our behavior be exposed for the lies they are, the truthful explanations for all our behavior will be provided in their place. This is because, as will be described and explained and become evident in chapter 2, trying to investigate human behavior while you're living in denial of the human condition—as reductionist, mechanistic scientists have been doing—was never going to produce the truthful insights into human behavior. Only a human-condition-confronting-not-avoiding approach—the holistic approach taken in this book—could find the answers to all the great outstanding questions in science, namely of the real and true explanation of the human condition (which has been outlined here and is fully described in chapter 3); of the real and true explanation of the meaning of our existence (which is presented in chapter 4); of the real and true explanation for the origins of humans' unconditionally selfless, moral instincts (which is presented in chapters 5 and 6); of the real and true explanation for how and why humans became conscious when other species haven't (which is presented

in chapter 7); of the real and true description of the emergence of humans from our ape ancestor to our present immensely human-condition-afflicted upset state (which is presented in chapter 8 and includes the explanation of the roles of men and women in this journey, the explanation of sex as humans have practiced it, the explanation of religion and politics, and many, many other insights into human behavior); and finally, the description of how the real and true explanation of the human condition transforms the human race (which is presented in chapter 9).

## Chapter 1:4  The problem of the 'deaf effect' that reading about the human condition initially causes

[77] While the shock of the arrival of the all-liberating but at the same time all-exposing truth about our 2-million-years upset/corrupted/fallen condition will bring inevitable problems, it is accompanied by absolutely wonderful solutions—the main of which will be presented in chapter 9. There is, however, one particular issue which needs to be addressed and solved here and now, which is the problem of the 'deaf effect' that reading about the human condition initially produces.

[78] As has been described, there have been three ways of coping with the imperfections of our upset human condition while we couldn't truthfully explain it: attack the unjust criticism; try to prove it wrong; and block it out of our mind. And as will become increasingly clear as you read through this book, it was this third method of blocking any criticizing truth out of our mind that has played a *hugely* important role in coping with our previously unexplained upset condition; we humans have had to adopt a great deal of block-out, denial and evasion of any truth that brought the unbearable issue of our upset condition into focus. So with the arrival now of the redeeming explanation of the human condition our human-condition-avoiding minds are clearly going to apply all that determinedly practiced block-out, denial and evasion to prevent us from taking in or 'hearing' what is being presented. As soon as discussion of the human condition begins, our minds will be subconsciously alert to the fact they are being taken into a historically off-limits realm and start blocking out what is being said. Our minds will suffer from a 'deaf effect' to what is being presented, with the consequence being that we will struggle to read and absorb the liberating and transforming explanation of the human

condition. Our habituated practice of denial will prevent us from gaining our FREEDOM from the human condition!

[79] To illustrate the power of this 'deaf effect', consider the following reactions my books have generated amongst readers: **'When I first read this material all I saw were a lot of black marks on white paper'**; and, **'Reading this is like reading another language—you know it's English, you can understand the words, but the concepts are so basic and so different that they are almost incomprehensible—it's a paradigm shift of a read'**; and, **'This stuff is so head on it can be crippling, which, initially at least, can make it hard to get behind what's being said and access the profundity of where it's coming from'**; and, **'At first I found this information difficult to absorb, in fact my wife and I would sit in bed and read a page together, and then re-read it a number of times, but still we couldn't understand what was written there and ended up thinking it must be due to poor expression.'** As the last response indicates, a consequence of being unaware that this resistance and block-out is occurring in our mind (because when we are in denial of something we aren't aware we are in denial, because, obviously, if we were we wouldn't be in denial of it!) is that we naturally blame the inaccessibility of what is being put forward on flaws in the presentation; we think it is, as readers of my earlier books have said, **'badly written'**, **'impenetrably dense'**, **'disjointed'**, **'confusingly worded'**, **'too intellectual for me to understand'**, **'long-winded'**, **'unnecessarily repetitive of vague points'**, **'desperately needs editing'**, and even **'lacking in any substance or meaning'**. Frustrated readers have often even written seeking **'an executive summary so I can grasp what you're trying to say'**!

[80] The situation is akin to giving someone who suffers from a phobia about snakes a book that cures their phobia when, to date, their fear of snakes has been so great they couldn't even admit they had a phobia. So if I were to ask them, 'Why don't you ever go outdoors?', their unawareness or blindness to their phobia, or condition, might prompt the defensive response, 'Well, I like living indoors because I like carpets and square walls and I like going through doorways; in fact, going through doorways is what made humans stand upright in the first place!'—or some such ridiculous excuse like that! To account for their inability to face their phobia they have had no choice but to create absurd theories based on the *denial* of their phobia (which, as mentioned earlier and as we are going to see in chapter 2, is exactly what mechanistic science has been doing as a result of humans' fear of the human condition). So, in order to cure them of their phobia, I give them a book that introduces them to

and explains their condition—but therein lies the problem, for as soon as they open the book and see descriptions and images of snakes, they fearfully slam it shut; their fear, in effect, blocks their ability to access the book's fabulously relieving understanding of, and thus solution to, their phobia!

[81] While this snake phobia analogy gives some idea of the problem of the 'deaf effect' resistance that blocks access to the compassionate, reconciling and immensely relieving understanding of our species' condition, there is a much better analogy and description of it—one that was given by that greatest of all philosophers, Plato, way back in the Golden Age of Greece, some 360 years before Christ. As to Plato's greatness as a philosopher (philosophy being the study of **'the truths underlying all reality'** (*Macquarie Dictionary*, 3rd edn, 1998)), Alfred North (A.N.) Whitehead, himself one of the most highly regarded philosophers of the twentieth century, described the history of philosophy as being merely **'a series of footnotes to Plato'** (*Process and Reality [Gifford Lectures Delivered in the University of Edinburgh During the Session 1927-28]*, 1979, p.39 of 413).

[82] So what was Plato's marvelously descriptive analogy of humans' extreme fear of the human condition and the resulting 'deaf effect' difficulty we have when reading about it—and what importance did he place on the difficulty of the 'deaf effect' in his profound contribution to the study of **'the truths underlying all reality'**? Well, Plato's most acclaimed work is *The Republic* and the central focus of *The Republic* is **'our human condition'**; and, most revealingly, in describing **'our human condition'**, Plato metaphorically depicted humans as having to live deep **'underground'** in a **'cave'** hiding from the **'painful'** issue of **'the imperfections of human life'**—these **'imperfections'** being the issue of the human condition. So the greatest of philosophers recognized that the central problem in understanding the **'reality'** of our behavior is our fear of the human condition!

[83] This is what Plato wrote: **'I want you to go on to picture the enlightenment or ignorance of our human conditions somewhat as follows. Imagine an underground chamber, like a cave with an entrance open to the daylight and running a long way underground. In this chamber are men who have been prisoners there'** (*The Republic*, c.360 BC; tr. H.D.P. Lee, 1955, 514; or see <www.wtmsources. com/227>). Plato described how the cave's exit is blocked by a **'fire'** that **'corresponds…to the power of the sun'**, which the cave prisoners have to hide from because its searing, **'painful' 'light'** would make **'visible'** the

unbearably depressing issue of **'the imperfections of human life'** (516-517). Fearing such self-confrontation, the cave prisoners have to **'take refuge'** in the dark **'cave'** where there are only some **'shadows thrown by the fire'** that represent a **'mere illusion'** of the **'real'** world outside the cave (515). The allegory makes clear that while **'the sun...makes the things we see visible'** (509), such that without it we can only **'see dimly and appear to be almost blind'** (508), having to hide in the **'cave'** of **'illusion'** and endure **'almost blind'** alienation has been infinitely preferable to facing the **'painful'** issue of **'our** [seemingly **imperfect] human condition'**. Then, with regard to the problem of the 'deaf effect' response the **'cave' 'prisoners'** would have to reading about the human condition, Plato described what occurs when, as summarized in the *Encarta Encyclopedia*, someone **'escapes from the cave into the light of day'** and **'sees for the first time the real world and returns to the cave'** to help the cave prisoners **'Escape into the sun-filled setting outside the cave** [which] **symbolizes the transition to the real world...which is the proper object of knowledge'** (written by Prof. Robert M. Baird, 'Plato'; see <www.wtmsources.com/101>). Plato wrote that **'it would hurt his** [the cave's prisoner's] **eyes and he would turn back and take refuge in the things which he could see** [take refuge in all the dishonest, **illusionary** explanations for human behavior that, as we are going to see in ch. 2, we have become accustomed to from human-condition-avoiding, mechanistic science], **which he would think really far clearer than the things being shown him. And if he were forcibly dragged up the steep and rocky ascent** [out of the cave of denial] **and not let go till he had been dragged out into the sunlight** [shown the truthful, **real** description of **our human condition**], **the process would be a painful one, to which he would much object, <u>and when he emerged into the light his eyes would be so overwhelmed by the brightness of it that he wouldn't be able to see a single one of the things he was now told were real.</u>'** Significantly, Plato then added, **'<u>Certainly not at first. Because he would need to grow accustomed to the light before he could see things in the world outside the cave</u>'** (*The Republic*, 515-516).

[84] So again, in his central and main insight into **'the truths underlying all reality'** of **'our human condition'**, Plato warned that when we **'first'** start reading about what **'our human condition'** really is we **'wouldn't be able to see a single one of the things he was now told were real'**. And I say *'really* is' because, as will become very clear in the next chapter, many refer to the human condition without engaging with what it really is. Yes, the 'deaf effect' will be a *very* significant problem when reading this book that presents the human-condition-confronting, truthful-yet-fully-compassionate

and psychologically relieving explanation of human behavior. (I should mention that after Plato warned about the problem of the 'deaf effect', he went on to describe how the person who tries to liberate the cave prisoners from their world of **'illusions'** would be viciously attacked. He wrote that **'they** [the cave prisoners] **would say that his** [the person who attempts to bring them liberating understanding of **our human condition**] **visit to the upper world had ruined his sight** [they would say he was mad], **and that the ascent** [out of the cave] **was not worth even attempting. And if anyone tried to release them and lead them up, they would kill him if they could lay hands on him'** (ibid. 517). Later in pars 574-578 I describe how true Plato's prediction here of horrible persecution would prove to be.)

[85] (You can read more about the problem of the 'deaf effect' at <www. humancondition.com/freedom-expanded-deaf-effect>.)

## Chapter 1:5  Solutions to the 'deaf effect'

[86] The obvious next question then is, 'How can the reader overcome the problem of not being able to absorb or 'hear' discussion of the human condition, so as to be able to access the incredibly relieving understanding of our species' behavior?' Plato indicated the answer when he said that the **'cave' 'prisoner' 'would need to grow accustomed to the light'** of the compassionate analysis of **'the imperfections of human life'** of **'our human condition'** to consequently achieve, as summarized in *Encarta*, the wonderful **'transition to the real world' 'which is the proper object of knowledge'**. And over thirty years of experience presenting the fully accountable, all-clarifying, relieving and transforming insights into human behavior that are contained in this book has taught me that this is indeed the case and that there are three main ways to **'grow accustomed to'** analysis of **'our human condition'** and, through doing so, overcome the 'deaf effect'.

[87] Solution 1: The first and most important method is to view the companion introductory videos for this book that are freely available at <www.humancondition.com/intro-videos>. In fact, such viewing is really a prerequisite, a necessity, for reading this book because in these presentations I describe the book's purpose, explain and warn of the problem of the 'deaf effect', escort readers through its chapters, and explain and discuss various aspects of our human condition. This familiarization process has the enormous psychological benefit of allowing you to watch

someone talk openly with others about, and walk freely around, this historically forbidden realm of the human condition—which experience has shown will greatly diminish the subconscious fear in your mind of discussion of the human condition. As will be made very clear at the beginning of the next chapter when the extreme extent of our fear of the issue of the human condition is revealed, our minds have been absolutely terrified of thinking about the human condition, so to see someone talking freely, happily and securely about the subject is subconsciously *immensely* reassuring: 'No one could be so at ease talking about the human condition unless the redeeming and relieving understanding of it had been found. That great breakthrough must have occurred', is the subconscious conversation that takes place in the reader's mind. The realm where the issue of the human condition resides has been *such* a terrifying place that for someone to be *so* comfortably walking around in it is extraordinarily reassuring and comforting; it is extremely helpful in overcoming the shock of having this subject of the human condition broached.

[88] Evidence of how reassuring and effective in eroding the 'deaf effect' watching the introductory videos is, is that when giving introductory talks about the human condition people, after having attended a second or third talk, often say, **'That was a much better presentation this time than last time, the explanations and descriptions were so much easier to follow.'** The talks are virtually identical—what has dramatically improved is not the quality or content of the presentation but the listener's ability to take in or 'hear' what was being said.

[89] So, in short, the more introductory videos you watch, the more your fear of the subject of the human condition will be eroded and the easier you will find the book to read. The importance in watching these videos cannot be overstated; put bluntly, you won't be able to read this book unless you watch them!

[90] Solution 2: In conjunction with viewing these videos, what also helps overcome the 'deaf effect' is a willingness to patiently re-read the text, as this further allows your mind the time to **'grow accustomed to'** description of **'our human condition'** and to start to **'see things in the world outside the cave'** of denial.

[91] Solution 3: Once the 'deaf effect' has started to erode through viewing the introductory videos and through being prepared to patiently re-read the text and you have begun to access these human-condition-

confronting-not-avoiding, truthful understandings of human behavior, you will find having a venue where you can participate in discussions (and/or watch others discussing these understandings) extremely helpful—another necessity, in fact. To cater for this need, online discussions about this new paradigm are held regularly that you can take part in, anonymously if you prefer.

[92] I have to emphasize that despite Professor Harry Prosen also warning of the 'deaf effect' in his Introduction, people typically disregard these warnings because they think the problem won't apply to them, that if something makes sense they will be able to follow it—and yet the 'deaf effect' *does* apply to virtually *every* reader; it's simply the reality of what occurs when the historically fearful issue of the human condition is brought into focus.

[93] To emphasize the very real nature of the 'deaf effect', take the following further example, this time from an online article about my 2003 book, *A Species In Denial*: **'I read it in 2005, and at the time it was not an easy read. The core concepts keep slipping from my mental grasp, at the time I put it down to bad writing, however a second reading revealed something the Author had indicated from the outset—***your mind doesn't want to understand the content*. **The second read was quick and painless...**[and I was then able to see that] **The cause of the malaise** [of humanity] **is exposed, remedied and the reader is left with at the very least an understanding of themselves, and for me something of an optimism for the future'** ('Fitzy', *Humanitus Interruptus – Great Minds of Today*, 21 Oct. 2011; see <www.wtmsources.com/106>). Yes, the **'second read'** is all-important—and regarding this last point about being left with **'an optimism for the future'**, Plato also emphasized just how relieved the cave prisoner would be to be free of his old, human-condition-avoiding, dishonest existence by saying that once he had become **'accustomed to the light'**, **'when he thought of his first home and what passed for wisdom there, and of his fellow-prisoners, don't you think he would congratulate himself on his good fortune and be sorry for them?'** (*The Republic*, 516). The following are some other quotes you can search online that reveal this sense of **'good fortune'** of being able to access understanding of the human condition and of finally being in the position to make sense of human existence: **'If Plato and Aristotle were alive and read Griffith, they would die happy men'**; and, **'We don't have to put up with "Not Knowing" anymore'**; and, **'tears**

stream down my face, so overcome have I been by this book. It is the greatest book on the planet, no wait, in the universe. In fact it is the greatest anything in the universe'; and, 'I don't care what question you have, this book will answer it'; and, 'Here is the breakthrough biological explanation that PROVES we are ALL very, very good'; and, 'This, to me, is the most significant thing I have ever stumbled across...If it doesn't hit you right away—it will down the road'; and, 'Gah! words are too limited for this. Here have some love brother! ☺'; and, 'This book is why I'm alive enough to scribe to you. Joy and Love to you all.'

[94] So, once you listen to the introductory videos and patiently re-read the text you will be astonished to discover that the fog does begin to lift, that what is being presented *does* begin to make extraordinarily accountable sense of human behavior. This process of illumination is palpable in this additional extract from Brian Carlton's interview with me (which can be viewed at <www.humancondition.com/carlton-video>): 'I remember when I first read one of your books I went through a stage where I couldn't quite get my head around it. I got about half of it and it was a little confusing and a little dense but I didn't give up. And in time your explanations did start to become clear and it made a hell of a lot of sense to me...The process of stripping off the denial is the difficult part, but once you've done that the answers become glaringly obvious...It's an intellectual epiphany; I have a more complete understanding of myself, everybody around me, the society at large, the way the planet works. It's a revelation! I don't use that in a religious sense, it's a quantifiably different thing but it has a similar impact on you. You wake up the next morning feeling more invigorated, more able to deal with the world because your level of understanding of it is so much higher...It's very simple, it's not hard. The end process is easy and reassuring and calming and self-accepting. Getting there is the difficult bit, once you have the revelation, the clarity of it is euphoric almost...when you *get* it, it is an event. You remember the day, you remember the section of the book, you remember when it happened, it stays with you...Don't underestimate the extent to which your work has impacted me in terms of how I think about what I'm seeing, how I interpret behaviour. I worked up this ability to be able to work out what a person was like in the first five or six seconds of a conversation [as the host of a talk-back radio program]...the trickle-down transfer to every day life and every day human relationships and experiences has been hugely valuable.'

[95] It should be mentioned that it is inevitable that some people will react angrily towards this human-condition-confronting-rather-than-avoiding information. As has been stressed throughout the latter stage of this first chapter, the arrival of understanding of the human condition can't but be a shock, and that shock has to be worked through—a process chapter 9 has been written to facilitate. It deals there with the massive paradigm shift that the human race is now faced with—a process of adjustment that hopefully won't take too long.

[96] AT THIS POINT I STRONGLY RECOMMENDED YOU WATCH THE INTRODUCTORY VIDEOS at <www.humancondition.com/intro-videos>.

[97] Having now warned of the problem of the 'deaf effect', and how best to overcome it, it is now necessary to explain in more comprehensive terms what the human condition actually is and reveal the mortal fear humans have of it so that our species' behavior can be compassionately explained and relievingly understood. Indeed, what will become apparent as you read *FREEDOM* is that almost all our behavior has been affected by, indeed is a product of, even driven by, our fear of the human condition. What this means is that to understand our behavior (which is the purpose of this book) we have to acknowledge and recognize this immense fear—which is why the main presentation in this book, which begins in the next chapter, starts with this all-important focus.

# Chapter 2

# The Threat of Terminal Alienation from Science's Entrenched Denial of The Human Condition

Michelangelo's *The Last Judgement*
(detail), 1537-41

William Blake's *Cringing in Terror*
c.1794-96

## Chapter 2:1 <u>Summary</u>

[98] Since the instinct vs intellect explanation of the human condition (which was outlined in the previous chapter and is fully explained in chapter 3) is a reasonably obvious explanation, the question that arises is why was it not identified by science long before now? As will become clear in this chapter, the answer is that humans have been *so* afraid of the issue of the human condition that we haven't been able to think truthfully and thus effectively about it. In fact, those charged with the task of finding understanding of human behavior, namely biological scientists, have been *so* committed to living in fear-driven denial of the human condition that they not only failed to find the

instinct vs intellect explanation, they built an immense edifice of human-condition-avoiding, dishonest 'explanations' to excuse our behavior—including completely false explanations for the human condition itself. Furthermore, this attachment to living in denial of the issue of the human condition is why the so-called mechanistic, scientific establishment has so determinedly failed to recognize—indeed, it has ignored—this fully accountable, human-race-saving, instinct vs intellect explanation ever since it was first presented back in 1983. While we couldn't truthfully explain the human condition, denial of it saved us from unbearable confrontation with the subject, but with the compassionate, redeeming, fully accountable, true explanation of the human condition now found, that practiced, historical denial is not only unnecessary, it is blocking the way to humanity's freedom from the human condition.

[99] Essentially, what has happened is that humans have become *so* habituated to living in Plato's dark cave of denial that when finally given the means to exit the cave and stand in the warm, healing sunshine of self-understanding, we have refused to leave! And, most frighteningly, in choosing to stay there means denial and the alienation from our true self that results from that denial can only continue to increase, so that very soon the human race will succumb to horrific terminal alienation. Indeed, the recent flood of movies and documentaries based on zombie, apocalyptic, escape-to-another-planet, 'we are being attacked by aliens [by our own alienation]', doomsday-preparation, 'we need a super hero to save the world' and other judgment-day-and-anxious-Bible-related-epic themes reflect the fact that the end play state of terminal alienation for humans that Michelangelo and Blake so frighteningly depict *is* upon us. The epidemic levels we are now seeing of the extremely psychologically distressed states of psychopathic narcissism, manic depression, Attention Deficit Hyperactivity Disorder (ADHD), and the ultimate completely-dissociated-from-the-world state of autism (states that will be explained later in chapters 8:16B, C and D) are similarly indicative of this state.

[100] So the great hope, indeed expectation, with *FREEDOM* is that by complementing the carefully argued and constructed presentation of the instinct vs intellect explanation of the human condition (and all the

**other insights that flow from it) with deaf-effect-eroding introductory videos and the opportunity to participate in interactive online discussions, support for this world-saving information and the fabulous life outside of Plato's dark cave of denial it makes possible, will finally begin in earnest!**

[101] The just described inability to think truthfully, or to accept truthful thinking, and the associated problem of the deaf effect that was explained in the previous chapter, all reflect a truth that will be established in this chapter, which is that our species has suffered from an *immense* fear of the human condition. In fact, our fear has been *so* pervasive that almost all human behavior has been affected by—indeed, is a product of, even driven by—it. What this means is that to truly understand our behavior—which is the purpose of this book—we have to first truly understand our extreme historical fear of the human condition. As such, the main presentation in this book, which begins here in chapter 2, must start with an exposé on just how terrified we humans have been of the issue of the human condition, of our species' upset state. The latter part of this chapter will then demonstrate how that extreme terror has dictated all of mechanistic science's human-condition-avoiding, blind and extremely dishonest thinking about the biology of human behavior, a process that will in itself dismantle the giant edifice of denial-based non-answers about human behavior that has so determinedly been assembled.

[102] Following that demolition, chapter 3 will present the detailed account of the truthful human-condition-confronting-*not*-avoiding, instinct vs intellect, *real* explanation of our species' condition that was outlined in chapter 1, with subsequent chapters providing the truthful explanation, alongside the dishonest accounts, of all the other outstanding questions about human behavior—of the meaning of human existence (in chapter 4); of the origins of our altruistic moral nature (in chapters 5 and 6); of how and why humans became conscious when other species haven't (in chapter 7); of the true story of our species' journey from an original state of cooperative, loving innocence to our now immensely psychologically upset angry, egocentric and alienated condition (in chapter 8)—and, finally, how this real understanding of the human condition liberates and transforms the human race (in chapter 9).

## Chapter 2:2 <u>The psychological event of 'Resignation' reveals our species' mortal fear of the human condition and thus how difficult it has been for scientists to find the explanation of the human condition and make sense of human behavior</u>

[103]To briefly recount the description given in chapter 1 of what the human condition *really* is, it is worth reciting the incisive words of the polymath Blaise Pascal, who spelled out the full horror of our contradictory condition when he wrote, **'What a chimera then is man! What a novelty, what a monster, what a chaos, what a contradiction, what a prodigy! Judge of all things, imbecile worm of the earth, repository of truth, a sewer of uncertainty and error, the glory and the scum of the universe!'** Shakespeare too was equally revealing of what the human condition *really* is when he wrote, **'What a piece of work is a man! how noble in reason! how infinite in faculty!…in action how like an angel! in apprehension how like a god! the beauty of the world! the paragon of animals! And yet, to me, what is this quintessence of dust? man delights not me'**. Pascal's and Shakespeare's identification of the dichotomy of **'man'** is what the human condition *really* is—this most extraordinary **'contradiction'** of being the most brilliantly clever of creatures, the ones who are **'god'-'like'** in our **'infinite' 'faculty'** of **'reason'** and **'apprehension'**, and yet also the meanest, most vicious of species, one that is only too capable of inflicting pain, cruelty, suffering and degradation. Yes, the eternal and seemingly unanswerable question has been: are we **'monster[s]'**, the **'essence'** of **'dust'**, **'the scum of the universe'**, or are we a wonderful **'prodigy'**, even **'glor[ious]' 'angel[s]'**?

[104]Thankfully, as was outlined in chapter 1, we can at long last now explain and understand that we are *not*, in fact, **'monster[s]'** but **'glor[ious]'** heroes. *However*, having had to live without this reconciling and dignifying understanding has meant that each human growing up under the duress of the human condition has suffered from *immense* insecurity about their fundamental goodness, worth and meaningfulness. So much so that the more we tried to think about this, in truth, most obvious question of our meaningfulness and worthiness (or otherwise), the more insecure and depressed our thoughts became. The emotional anxiety produced when reading Pascal's and Shakespeare's descriptions of the human condition gives some indication of just how unnervingly confronting

the issue of the human condition really is. <u>The truth, that will now be revealed, is that this intensely personal yet universal issue of the human condition has been *so* unbearably confronting and depressing that we eventually learnt as we grew up that we had no choice but to *resign* ourselves to never revisiting the subject, to never again looking at the seemingly inexplicable issue of the human condition.</u> The examination of this process of what I call 'Resignation' to living in Plato's dark cave of denial of the human condition, and how it unfolds, will reveal just how immensely fearful humans have been of the human condition, and, it follows, how impossible it has been for mechanistic scientists to think effectively about human behavior.

[105] As will be explained in detail in chapter 8, as humans grew up in a human-condition-afflicted world that wasn't able to be truthfully analyzed and explained, we each became increasingly troubled by the glaringly obvious issue of the extreme **'imperfections of human life'** (as Plato referred to **'our human condition'**). This progression went through precise stages—and I should point out that all these stages of resignation to a life of blocking out the issue of the human condition have not been admitted by science, because like almost every other human, its practitioners have also lived in mortal fear and thus almost total denial of the human condition. What follows then is a very brief summary of the life stages that will be fully described in chapter 8.

[106] As consciousness emerged in humans we progressed from being able to sufficiently understand the relationship between cause and effect to become self-conscious, aware of our own existence, during our infancy, to proactively carrying out experiments in self-management during our childhood, at which point all the manifestations of the human condition of anger, egocentricity and alienation began to reveal themselves. It follows that it was during our childhood that we each became increasingly aware of not only the imperfection of the human-condition-afflicted world around us, but of the imperfection of our own behavior—that we too suffered from anger, selfishness, meanness and indifference to others. Basically, all of human life, including our own behavior, became increasingly bewildering and distressing, to such a degree that by the time children reached late childhood they generally entered what is recognized as the **'naughty nines'**, where their confusion and frustration was such that they even angrily began taunting and bullying those around

them. By the end of childhood, however, children realized that lashing out in exasperation at the imperfections, wrongness and injustice of the world didn't change anything and that the only possible way to end their frustration was to understand *why* the world, and their own behavior, was not ideal. It was at this point, which occurred around 12 years of age, that children underwent a dramatic change from being frustrated, protesting, demonstrative, loud extroverts into sobered, deeply thoughtful, quiet introverts, consumed with anxiety about the imperfections of life under the duress of the human condition. Indeed, it is in recognition of this very significant psychological transition from a relatively human-condition-free state to a very human-condition-aware state that we separate these stages into 'Childhood' and 'Adolescence', a shift even our schooling system marks by having children graduate from what is generally called primary school into secondary school. What then happened during adolescence was that, at about 14 or 15 years of age and after struggling for a few years to make sense of existence, the search for understanding became *so* confronting of those extreme internal imperfections that adolescents had no choice but to 'Resign' to living in denial of the whole unbearably depressing and seemingly unsolvable issue of the human condition—after which they became superficial and artificial escapists, not wanting to look at any issue too deeply, and, before long, combative and competitive power-fame-fortune-and-glory, relief-from-the-agony-and-guilt-of-the-human-condition-seeking resigned adults.

[107] Delving deeper into how the journey toward 'Resignation' unfolds will reveal just how terrifying the issue of the human condition has been, which is precisely what the reader needs to become aware of in order to appreciate why it has, until now, been impossible to truthfully and thus effectively explain human behavior. Yes, describing what occurs at 'Resignation' makes it *abundantly* clear why resigned humans became so superficial and artificial in their thinking, incapable of plumbing the great depths of the human condition and thus incapable of finding the desperately needed understanding of human existence.

[108] So what happened at around 14 or 15 years of age for virtually all humans growing up under the duress of **'the imperfections'** of **'our human condition'** was that to avoid the suicidal depression that accompanied *any* thinking about the issue of our species', and our *own*, seemingly *extremely* imperfect condition, there was simply no choice but to stop grappling with the answerless question. And so despite the human condition being the

all-important issue of the meaningfulness or otherwise of our existence, there came a time (and, although it varies according to each individual's circumstances, it typically occurred at about 14 or 15 years of age) when adolescents were forced to put the whole depressing subject aside once and for all and just hope that one day in the future the explanation and defense for our species', and thus our own, apparently horrifically flawed, seemingly utterly disappointing, sad state would be found, because then, and *only* then, would it be psychologically safe to even broach the subject. In 2010 a poignantly honest film titled *It's Kind Of A Funny Story* (based on Ned Vizzini's book of the same title) was made about a 16-year-old boy named Craig who is going through the agonising process of grappling with the human condition; he struggles with **'suicidal' 'depression'** from **'anxiety'** about **'grades, parents** [who don't seem to have **'a clue'** that **'there's something bigger going on'**]**, two wars, impending environmental catastrophe, a messed up economy'**. Eventually a psychiatrist counsels him that **'there is a saying that goes something like this: "Lord, grant me the strength to change the things I can, the courage to accept the things I can't, and the wisdom to know the difference"'**; basically he is advised to resign himself to living in denial of the human condition. The Beatles' song *Let It Be*—consistently voted one of the most popular songs of the twentieth century—is actually an anthem to this need that adolescents have historically had, when confronted with the unbearable **'hour of darkness'** that came from grappling with the issue of all **'the broken hearted people living in the world'**, to **'let it be' 'until tomorrow'** when **'there will be an answer'** (Lennon/McCartney, 1970). So when the great poet Gerard Manley Hopkins wrote about the unbearably depressing subject of the human condition in his aptly titled poem *No Worst, There Is None* (1885), his words, **'O the mind, mind has mountains; cliffs of fall, frightful, sheer, no-man-fathomed'**, did not exaggerate the depth of depression humans faced if we allowed our minds to think about the human condition while it was still to be **'fathomed'**/understood/**'answer[ed]'**. Yes, when, in **'my hour of darkness'**, **'Mother Mary comes...speaking words of wisdom, let it be, let it be'**—accept the adults' **'wisdom'**, and don't go there!

[109] It's little wonder then that the human condition has been described so vehemently as **'the personal unspeakable'** and as **'the black box inside of humans they can't go near'** (personal conversations, WTM records, Feb. 1995)—*and* why it is so very rare to find a completely honest description of adolescents going through the excruciating process of Resignation, of resigning themselves to having to block out the seemingly inexplicable question

of their worth and meaning and live, from that time on, in denial of the unbearable issue of the human condition. Having already been through this terrible process of Resignation, most adults simply couldn't allow themselves to recall, recognize and thus empathize with what adolescents were experiencing (they were, as Craig complained, rendered 'clue'-less to the situation). And so our young have been alone with their pain, unable to share it with those closest, or the world at large. All of which makes the following account of a teenager in the midst of Resignation, by the American Pulitzer Prize-winning child psychiatrist Robert Coles, incredibly special: **'I tell of the loneliness many young people feel…It's a loneliness that has to do with a self-imposed judgment of sorts…I remember…a young man of fifteen who engaged in light banter, only to shut down, shake his head, refuse to talk at all when his own life and troubles became the subject at hand. He had stopped going to school…he sat in his room for hours listening to rock music, the door closed…I asked him about his head-shaking behavior: I wondered whom he was thereby addressing. He replied: "No one." I hesitated, gulped a bit as I took a chance: "Not yourself?" He looked right at me now in a sustained stare, for the first time. "Why do you say that?"** [he asked]**…I decided not to answer the question in the manner that I was trained** [basically, 'trained' in avoiding what the human condition really is]**…Instead, with some unease…I heard myself saying this: "I've been there; I remember being there—remember when I felt I couldn't say a word to anyone"…The young man kept staring at me, didn't speak…When he took out his handkerchief and wiped his eyes, I realized they had begun to fill'** (*The Moral Intelligence of Children*, 1996, pp.143-144 of 218). The boy was in tears because Coles had reached him with *some* recognition and appreciation of what he was wrestling with; Coles had shown *some* honesty about what the boy could see and was struggling with, namely the horror of the utter hypocrisy of human behavior—including his own.

[110] The words Coles used in his admission that he too had once grappled with the issue of the human condition, of **'I've been there'**, are exactly those used by one of Australia's greatest poets, Henry Lawson, in his extraordinarily honest poem about the unbearable depression that results from trying to confront the question of why human behavior is so at odds with the cooperative, loving—or, to use religious terms, 'Godly'—ideals of life. In his 1897 poem *The Voice from Over Yonder*, Lawson wrote: **"'Say it! think it, if you dare! Have you ever thought or wondered, why the Man and God were sundered** [torn apart]**? Do you think the Maker blundered?" And the voice in mocking accents, answered only: "I've been there."'** The unsaid

words in the final phrase, **'I've been there'**, being 'and I'm certainly not going **'there'** again!'—with the **'there'** and the **'over yonder'** of the title referring to the state of deepest, darkest depression.

Goya's *The sleep of reason brings forth monsters*, 1796-97

[111] In his bestselling 2003 book, *Goya* (about the great Spanish artist Francisco Goya), another Australian, Robert Hughes, who for many years was *TIME* magazine's art critic, described how he **'had been thinking about Goya**...[since] **I was a high school student in Australia**...[with] **the first work of art I ever bought**...[being] **a poor second state of Capricho 43**...*The sleep of reason brings forth monsters*...[Goya's most famous etching reproduced above] **of the intellectual beset with doubts and night terrors, slumped on his desk with owls gyring around his poor perplexed head'** (p.3 of 435). Hughes then commented that **'glimpsing *The sleep of reason brings forth monsters* was a fluke'** (p.4). A little further on, Hughes wrote of this experience that **'At fifteen, to find**

**this voice** [of Goya's]—so finely wrought [in *The sleep of reason brings forth monsters*] and yet so raw, public and yet strangely private—speaking to me with such insistence and urgency...was no small thing. It had the feeling of a message transmitted with terrible urgency, mouth to ear: this is the *truth*, you *must know* this, I have *been through* it' (p.5). Again, while the process of Resignation is such a horrific experience that adults determined never to revisit it, or even recall it, Hughes' attraction to *The sleep of reason brings forth monsters* was not the **'fluke'** he thought it was. The person slumped at the table with owls and bats gyrating around his head perfectly depicts the bottomless depression that occurs in humans just prior to resigning to a life of denial of the issue of the human condition, and someone in that situation would have recognized that meaning instantly, almost willfully drawing such a perfect representation of their state out of the world around them. Even the title is accurate: **'The sleep of reason'**—namely reasoning at a very deep level—does **'bring forth monsters'**! While Hughes hasn't recognized that what he was negotiating **'At fifteen'** was Resignation, he has accurately recalled how strongly he connected to what was being portrayed in the etching: **'It had the feeling of a message transmitted with terrible urgency, mouth to ear: this is the *truth*, you *must know* this, I have *been through* it.'** Note how Hughes' words, **'I have *been through* it'**, are almost identical to Coles' and Lawson's words, **'I've been there.'**

[112] And so, unable to acknowledge the process of Resignation, adults instead blamed the well-known struggles of adolescence on the hormonal upheaval that accompanies puberty, the so-called 'puberty blues'— even terming glandular fever, a debilitating illness that often occurs in mid-adolescence, a puberty-related 'kissing disease'. These terms, 'puberty blues' and 'kissing disease', are dishonest, denial-complying, evasive excuses because it wasn't the onset of puberty that was causing the depressing 'blues' or glandular fever, but the trauma of Resignation. For glandular fever to occur, a person's immune system must be extremely rundown, and yet during puberty the body is physically at its peak in terms of growth and vitality—so for an adolescent to succumb to the illness they must be under extraordinary psychological pressure, experiencing stresses much greater than those that could possibly be associated with the physical adjustments to puberty, an adjustment, after all, that has been going on since animals first became sexual. No, the depression and glandular fever experienced by young adolescents are a direct result of the trauma of having to resign to never again revisiting the unbearably

depressing subject of the human condition.

[113] That sublime classic of American literature, J.D. Salinger's 1951 novel *The Catcher in the Rye*, is a masterpiece because, like Coles, Salinger dared to write about that forbidden subject for adults of adolescents having to resign to a dishonest life of denial of the human condition—for *The Catcher in the Rye* is actually entirely about a 16-year-old boy struggling against Resignation. The boy, Holden Caulfield, complains of feeling **'surrounded by phonies'** (p.12 of 192) and **'morons'** who **'never want to discuss anything'** (p.39), of living on the **'opposite sides of the pole'** (p.13) to most people, where he **'just didn't like anything that was happening'** (p.152), to wanting to escape to **'somewhere with a brook…**[where] **I could chop all our own wood in the winter time and all'** (p.119). He knows he is supposed to resign—in the novel he talks about being told that **'Life…**[is] **a game…you should play it according to the rules'** (p.7), and to feeling **'so damn lonesome'** (pp.42, 134) and **'depressed'** (multiple references) that he felt like **'committing suicide'** (p.94). As a result of all this despair and disenchantment with the world he keeps **'failing'** (p.9) his subjects at school and is expelled from four for **'making absolutely no effort at all'** (p.167). About his behavior he says, **'I swear to God I'm a madman'** (p.121) and **'I know. I'm very hard to talk to'** (p.168). But like the boy in Coles' account, Holden finally encounters some rare honesty from an adult that, in Holden's words, **'really saved my life'** (p.172). This is what the adult said: **'This fall I think you're riding for—it's a special kind of fall, a horrible kind…**[where you] **just keep falling and falling** [utter depression]**'** (p.169). The adult then spoke of men who **'at some time or other in their lives, were looking for something their own environment couldn't supply them with… So they gave up looking** [they resigned]**…**[adding] **you'll find that you're not the first person who was ever confused and frightened and even sickened by human behavior'** (pp.169-170). Yes, to be **'confused and frightened'** to the point of being **'sickened by human behavior'**—indeed, to be **'suicid**[ally]**' 'depressed'** by it—is the effect the human condition has if you haven't resigned yourself to living a relieving but utterly dishonest and superficial life in denial of it.

[114] Going through Resignation *has* been a truly horrific experience. A friend and I were walking in bushland past a school one day when we came across a boy, who would have been about 14 years old, sitting by the track in a hunched, fetal position. When I asked him if he was okay he looked up with such deep despair in his eyes that it was clear he didn't want to be disturbed and so we left him alone. It was very apparent that he was trying to wrestle with the issue of the human condition, but without

understanding of the human condition it hasn't been possible for virtually all humans to do so without becoming so hideously condemned and thus depressed that they had no choice but to eventually surrender and take up denial of the issue of the human condition as the only way to cope with it—even though doing so meant adopting a completely dishonest, superficial and artificial, effectively dead, existence.

**'Too poor to go on school trip, boy fishes the day after classmates perish in plane crash'**
*LIFE* magazine, Fall Special Edition, 1991

[115]I haven't as yet come across a photograph of an adolescent in the midst of Resignation, however, in my picture collection I do have the above haunting image of a boy who had, the previous day, lost all his classmates in a plane crash, and his expression is exactly the same deeply sobered, drained pale, all-pretenses-and-facades-stripped-away, pained, tragic, stunned, human-condition-laid-bare expression I have seen on the faces of adolescents going through Resignation. We can see in this boy's face that all the artificialities of human life have been rendered

meaningless and ineffectual by the horror of losing all his friends, leaving bare only the sad, painful awareness of a world devoid of any real love, meaning or truth.

[116] Although rarely shared, adolescents in the midst of Resignation quite often write excruciatingly honest poetry about their impending fate; indeed, *The Catcher in the Rye* is really one long poem about the agony of having to resign to living a human-condition-denying, superficial, totally false existence. I have written much more about Resignation at <www.humancondition.com/freedom-expanded-resignation>, however, the following are two horrifically honest Resignation poems that are discussed at that link and worth including here to provide first-hand insights into the agony of adolescence: **'You will never have a home again / You'll forget the bonds of family and family will become just family / Smiles will never bloom from your heart again, but be fake and you will speak fake words to fake people from your fake soul / What you do today you will do tomorrow and what you do tomorrow you will do for the rest of your life / From now on pressure, stress, pain and the past can never be forgotten / You have no heart or soul and there are no good memories / Your mind and thoughts rule your body that will hold all things inside it; bottled up, now impossible to be released / You are fake, you will be fake, you will be a supreme actor of happiness but never be happy / Time, joy and freedom will hardly come your way and never last as you well know / Others' lives and the dreams of things that you can never have or be part of, will keep you alive / You will become like the rest of the world—a divine actor, trying to hide and suppress your fate, pretending it doesn't exist / There is only one way to escape society and the world you help build, but that is impossible, for no one can ever become a baby again / Instead you spend the rest of life trying to find the meaning of life and confused in its maze';** and **'Growing Up: There is a little hillside / Where I used to sit and think / I thought of being a fireman / And of thoughts, I thought important / Then they were beyond me / Way above my head / But now they are forgotten / Trivial and dead.'**

[117] Yes, as these poems so painfully express, Resignation means blocking out all memory of the innocent, soulful, true world because it is unbearably condemning of our present immensely corrupted human condition: **'You have no heart or soul and there are no good memories / Your mind and thoughts rule your body that will hold all things inside it; bottled up, now impossible to be released / You are fake, you will be fake, you will be a supreme actor of happiness but never be happy.'** And since virtually all adults have resigned, that is exactly how **'fake'**, or as the 16-year-old Holden

Caulfield described it, **'phony'**, they have become. Clearly, the price of Resignation is *enormous*, but the alternative for virtually all humans of not resigning has been an *even worse* fate because it meant living with constant suicidal depression.

[118] Appreciably then, in what forms the key passage in *The Catcher in the Rye* (indeed, it provides the meaning behind the book's enigmatic title), Salinger has Holden Caulfield dreaming of a time when this absolute horror, indeed obscenity, of Resignation will no longer have to form an unavoidable part of human life: **'I keep picturing all these little kids playing some game in this big field of rye and all. Thousands of little kids, and nobody's around—nobody big, I mean—except me. And I'm standing on the edge of some crazy cliff. What I have to do, I have to catch everybody if they start to go over the cliff—I mean if they're running and they don't look where they're going I have to come out from somewhere and *catch* them. That's all I do all day. I'd just be the catcher in the rye and all. I know it's crazy, but that's the only thing I'd really like to be'** (p.156). *And*, as will be explained in this book, the time that Holden Caulfield so yearned for when we will be able to **'catch'** *all* children before **'they start to go over the** [excruciating] **cliff'** of Resignation to a life of utter dishonesty, **'phony'**, **'fake'** superficiality, and silence in terms of **'never want**[ing] **to discuss anything'** truthfully again, *has* finally come about with the finding of the all-clarifying, redeeming and relieving truthful, fully compassionate explanation of human behavior! Yes, the real **'catcher in the rye'** is the ability to explain **'human behavior'** so that it is no longer **'sicken**[ing]**'** but understandable, and, best of all, healable.

### Chapter 2:3 Our near total resistance to analysis of the human condition post-Resignation

[119] It is a measure of just how unbearable the issue of the human condition has been that while Resignation has been *the* most important psychological event in human life the process is never spoken of and has virtually gone unacknowledged in the public realm, with only a rare few of our most accomplished writers even managing to write about the suicidally depressing experience of engaging the subject of the human condition itself. To this end, many **'philosophers and psychologists'** consider that the great Danish philosopher Søren Kierkegaard's **'analysis on the nature of despair is one of the best accounts on the subject'** (Wikipedia;

see <www.wtmsources.com/137>) — with the **'nature of despair'** being as close as the reviewer could go in referring to the worse-than-death, suicidal depression that the human condition has caused humans, but which Kierkegaard managed to give such an honest account of in his aptly titled 1849 book, *The Sickness Unto Death*: **'the torment of despair is precisely the inability to die** [and end the torture of our unexplained human condition]...**that despair is the sickness unto death, this tormenting contradiction** [of our 'good and evil', human condition-afflicted lives], **this sickness in the self; eternally to die, to die and yet not to die'** (tr. A. Hannay, 1989, p.48 of 179). Kierkegaard went on to include these unnervingly truthful words about how, even when the blocks *were* in place in our minds against recognizing the existence of the issue of the human condition, the terrifying **'anxiety'** it caused us still occasionally surfaced: **'there is not a single** [adult] **human being who does not despair at least a little, in whose innermost being there doesn't dwell an uneasiness, an unquiet, a discordance, an anxiety in the face of an unknown something, or a something he doesn't even dare strike up acquaintance with...he goes about with a sickness, goes about weighed down with a sickness of the spirit, which only now and then reveals its presence within, in glimpses, and with what is for him an inexplicable anxiety'** (p.52).

[120] Another great philosopher, the Russian Nikolai Berdyaev, gave this extraordinarily forthright description of how trying to address and solve the sickeningly depressing issue of the human condition and, by so doing, make sense of human behavior, has been a nightmare: **'Knowledge requires great daring. It means victory over ancient, primeval terror. Fear makes the search for truth and the knowledge of it impossible...it must also be said of knowledge that it is bitter, and there is no escaping that bitterness...Particularly bitter is moral knowledge, the knowledge of good and evil. But the bitterness is due to the fallen state of the world, and in no way undermines the value of knowledge...it must be said that the very distinction between good and evil is a bitter distinction, the bitterest thing in the world...There is a deadly pain in the very distinction of good and evil, of the valuable and the worthless'** (*The Destiny of Man*, 1931; tr. N. Duddington, 1960, pp.14-15 of 310). Yes, trying to think about our corrupted, **'fallen'**, seemingly **'evil'** and **'worthless'** state *has* been an **'ancient, primeval terror'**, a **'deadly pain'**, **'the bitterest thing in the world'** for virtually all humans. As one of the key figures of the Enlightenment, the philosopher Immanuel Kant, said: we have to **'Dare to know!'** (*What is Enlightenment?*, 1784).

[121] So Carl Jung certainly exhibited **'great daring'** in his thinking when he wrote the following words about the terrifying nature of the human condition: **'When it [our shadow] appears...it is quite within the bounds of possibility for a man to recognize the relative evil of his nature, but it is a rare and shattering experience for him to gaze into the face of absolute evil'** (*Aion: Researches into the Phenomenology of the Self*, 1959; tr. R. Hull, *The Collected Works of C.G. Jung*, Vol. 9/2, p.10). The **'face of absolute evil'** *is* the **'shattering'** possibility—if we allowed our minds to think about it—that we humans might indeed be a terrible mistake. And so to avoid that implication, humans have had to avoid almost all deep, penetrating, truthful thinking because almost *any* thinking at a deeper level brought us into contact with the unbearable issue of our seemingly horribly flawed condition: 'There's a tree with lovely autumn leaves; isn't it amazing how beautiful nature can be, I wonder why some things are beautiful while others are not— I wonder why I'm not beautiful inside, in fact, *so* full of all manner of angst, selfish self-obsession, indifference and anger...aaahhhhh!!!!' The very great English poet William Wordsworth was making this point when he wrote, **'To me the meanest flower that blows can give thoughts that do often lie too deep for tears'** (*Intimations of Immortality from Recollections of Early Childhood*, 1807), for it is true that even the plainest flower can remind us of the unbearably depressing issue of our seemingly horrifically imperfect, **'fallen'**, apparently **'worthless'** condition. Yes, as the comedian Rod Quantock once said, **'Thinking can get you into terrible downwards spirals of doubt'** ('Sayings of the Week', *The Sydney Morning Herald*, 5 Jul. 1986). The Nobel Laureate Albert Camus wasn't overstating the issue either when he wrote that **'Beginning to think is beginning to be undermined'** (*The Myth of Sisyphus*, 1942); nor was another Nobel Prize winner in Literature, Bertrand Russell, when he said, **'Many people would sooner die than think'** (Antony Flew, *Thinking About Thinking*, 1975, p.5 of 127). And nor was the equally acclaimed poet T.S. Eliot when he wrote that **'human kind cannot bear very much reality'** (*Burnt Norton*, 1936). **'The sleep of reason'**, reasoning at a deep level, *does* indeed **'bring forth monsters'**.

[122] Already we can see the truth of the initial point made in this chapter—that to understand human behavior requires bottoming out on the fact that almost *all* of our behavior is a product of, even driven by, our fear of the human condition, for we can appreciate here how our fear of the human condition has limited, indeed relegated, us to an extremely superficial existence. As this book progresses, it will

become increasingly clear that making sense of human behavior and finding the answers to all the mysteries about human life depends on recognizing humans' immense fear of the human condition. Moreover, recognizing this underlying fear will enable you to see *through* our behavior, see what is *behind* it, what is *causing* it—to such an extent, in fact, that our behavior becomes transparent. You can get a feeling for this concept of transparency in the extract from the interview included earlier with the Australian broadcaster Brian Carlton (see par. 94).

[123]The fact is, the human race has lived a haunted existence, dogged by the dark shadow of its imperfect human condition, forever trying to escape it—the result of which is that we have become *immensely* superficial and artificial; **'phony'** and **'fake'**, as the resigning adolescents so truthfully described it, and living on the absolute meniscus of life in terms of what we are prepared to look at, feel and consider. We are a *profoundly* estranged or alienated species, completely blocked-off from the amazing and wonderful real world, and from the truth of our self-corruption that thinking about that beautiful, inspired, natural, soulful world unbearably connects us to—as the absolutely **'fear'**-lessly honest Scottish psychiatrist R.D. Laing has written: **'Our alienation goes to the roots. The realization of this is the essential springboard for any serious reflection on any aspect of present inter-human life…We are born into a world where alienation awaits us. We are potentially men, but are in an alienated state** [p.12 of 156] **…the *ordinary* person is a shrivelled, desiccated fragment of what a person can be. As adults, we have forgotten most of our childhood, not only its contents but its flavour; as men of the world, we hardly know of the existence of the inner world** [p.22] **… The condition of alienation, of being asleep, of being unconscious, of being out of one's mind, is the condition of the normal man** [p.24] **…between *us* and It [our true selves** or soul**] there is a veil which is more like fifty feet of solid concrete. *Deus absconditus*. Or we have absconded** [p.118] **…The outer divorced from any illumination from the inner is in a state of darkness. We are in an age of darkness. The state of outer darkness is a state of sin—i.e. alienation or estrangement from the inner light** [p.116] **…We are all murderers and prostitutes…We are bemused and crazed creatures, strangers to our true selves, to one another'** [pp.11-12] (*The Politics of Experience* and *The Bird of Paradise*, 1967). **'We are dead, but think we are alive. We are asleep, but think we are awake. We are dreaming, but take our dreams to be reality. We are the halt, lame, blind, deaf, the sick. But we are doubly unconscious. We are *so* ill that we no longer feel ill, as in many terminal illnesses. We are mad, but have no insight** [into the fact of our madness]**'** (*Self and Others*,

1961, p.38 of 192). **'We are so out of touch with this realm** [where the issue of the human condition lies] **that many people can now argue seriously that it does not exist'** (*The Politics of Experience* and *The Bird of Paradise*, p.105).

[124]Laing's honesty is astonishing. One other rare example from the twentieth century of someone who managed to depict and penetrate our **'fifty feet of solid concrete'** wall of denial of the truth of our tortured, 'good-and-evil'-stricken human condition was the Irish artist Francis Bacon. While people in their resigned state of denial of what the human condition actually is typically find his work **'enigmatic'** and **'obscene'** (*The Sydney Morning Herald*, 29 Apr. 1992), there is really no mistaking the agony of the human condition in Bacon's death-mask-like, twisted, smudged, distorted, trodden-on—alienated—faces, and tortured, contorted, stomach-knotted, arms-pinned, psychologically strangled and imprisoned bodies; consider, for instance, his *Study for self-portrait* (opposite). It is some recognition of the incredible integrity/honesty of Bacon's work that in 2013 one of his triptychs sold for $US142.4 million, becoming (at the time) **'the most expensive work of art ever sold at auction, breaking the previous record, set in May 2012, when a version of Edvard Munch's** *The Scream* [another exceptionally honest, human-condition-revealing painting] **sold for $119.9 million'** (*TIME*, 25 Nov. 2013). (It may seem incongruous that people living in denial of the human condition should pay such exorbitant sums for such stark depictions of our psychologically upset state, but living in an almost completely **'phony'**, **'fake'**, **'alienat**[ed]...**to the roots'** and truthless world, as this book reveals we have been, has meant that the honesty about our true state depicted by Bacon and Munch could be immensely valued for its cathartic, purging, purifying, relieving powers. This book, however, provides the ultimate form of this therapy, because in explaining the human condition it at last allows us to be completely honest without that honesty bringing with it any condemnation of our horrifically upset condition.)

[125]Indeed, Laing's words and Bacon's images are so cathartically honest that if they don't reconnect us with the human condition then nothing will!! But again, despite the contributions made by these great thinkers and artists, and the horror of the world around us, that reconnection is not easily achieved because what Laing wrote is true—resigned humans have learnt to live in *such complete* denial of the issue of the human condition that many people do now **'seriously'** believe the issue **'does not exist'**, at least not in its true form as a profoundly disturbed and insecure, *psychological* affliction.

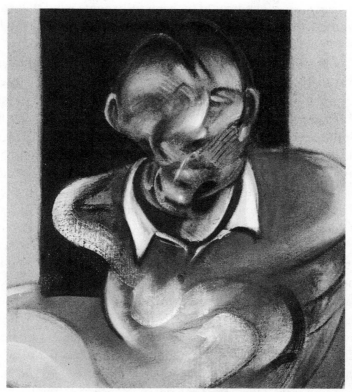

Detail from Francis Bacon's *Study for self-portrait*, 1976

[126] A further illustration of just how impenetrable the issue of the human condition has been is Plato's aforementioned allegory of the cave (see par. 83) in which humans have been incarcerated, unable to face the glare of the sunlit **'real'** world because it would make **'visible'** the unbearably depressing issue of **'the imperfections of human life'**. Plato said that while **'the sun...makes the things we see visible'**, such that without it we can only **'see dimly and appear to be almost blind'**, having to hide in the **'cave'** of **'illusion'** and endure **'almost blind'** alienation was infinitely preferable to facing the **'painful'** issue of **'our** [seemingly **imperfect**] **human condition'**. Clearly, what enabled Plato to be such an effective and penetrating thinker was that he was able to **'realiz[e]'** that **'Our alienation goes to the roots'**, which, as Laing wrote, is **'the essential springboard for any serious reflection on any aspect of present inter-human life'**. Plato was one of the rare few individuals in history who have been sound and secure enough in self to not have had to resign to living a life of denial of the human condition, because

only by being unresigned could he have fully confronted, talked freely about and effortlessly described the human condition the way he did; you can't think truthfully if you are living in a cave of denial.

[127] I should mention that while there have been innumerable interpretations of Plato's cave allegory, none that I am aware of has presented the interpretation that was given in par. 83, which was that Plato was describing humans as living in such immense fear of the human condition that they were having to hide from it in the equivalent of a deep, dark cave. However, if this interpretation *is* true and adult humans *are* living in terrifying fear of the human condition (and what has been described about Resignation evidences they are), then it makes complete sense that they wouldn't want to acknowledge that interpretation, they wouldn't want to admit that Plato was right about them wanting to hide from the human condition, and so they would have sought a different, less confronting interpretation. Admitting to having to hide from the human condition means acknowledging and having to confront the issue of the human condition, which humans haven't wanted to do. And in addition to not wanting to be reconnected to the issue of the human condition and the terrible depression they experienced at Resignation, adults haven't wanted to admit that they did resign to a life of lying—to living a completely superficial, **'phony'**, **'fake'**, **'alienat**[ed]**...to the roots'**, **'asleep'**, **'unconscious'**, **'out of one's mind'** existence. Such honesty would undermine their ability to operate; it would be self-negating: 'I'm lying, so neither you nor I can trust me.' No wonder humans have been so sensitive about being called liars. In the absence of understanding, denial and delusion have been the only means we have had to cope with the extreme imperfection of our lives. It is only now that we can properly explain the human condition and, by so doing, fully defend and understand our upset, corrupted, alienated state that we can afford to abandon that delusion and denial and be honest about Resignation and its consequences. Indeed, the only outright admissions of Resignation and its effects that I have found have come from exceptionally honest, unresigned, denial-free-thinking individuals we have termed 'prophets', namely Plato with his cave allegory, and, as will be described in par. 750, Moses with his Noah's Ark metaphor.

[128] We can therefore appreciate why every effort has been made to avoid the, in truth, obvious 'we-are-hiding-from-the-human-condition' interpretation of Plato's cave allegory that was given in par. 83. And it *is* a

*very obvious* interpretation, because, in Sir Desmond Lee's translation of Plato's own words, Plato wrote, **'I want you to go on to picture the enlightenment or ignorance of our human conditions'**, where the **'fire'** that blocks the cave's exit **'corresponds...to the power of the sun'**, which the **'cave'** **'prisoners'** have to hide from because its searing, **'painful'** light would make **'visible'** the unbearably depressing issue of **'the imperfections of human life'** — those imperfections being **'our human condition'**!

[129] Some might question whether Lee took liberties in translating Plato's original Greek words as meaning **'our human condition'**. The first point I would make is that Lee's translation is held in very high regard; it is, for instance, the translation the publishing house Penguin chose for its very popular paperback edition of *The Republic*. And the second point I would make is that the most literal interpretation of Plato's original Greek text, by the philosopher Allan Bloom, carries the same meaning. In his preface to his 1968 translation of *The Republic*, Bloom (the author of the renowned and exceptionally honest book, *The Closing of the American Mind*) explains why he went to such pains to be literal: **'Such a translation is intended to be useful to the serious student, the one who wishes and is able to arrive at his own understanding of the work...The only way to provide the reader with this independence is by a slavish...literalness—insofar as possible always using the same English equivalent for the same Greek word'** (*The Republic of Plato*, tr. Allan Bloom, 1968, Preface, p.xi of 512). So what is Bloom's **'slavish[ly]'** **'literal'** translation of the passage Lee translated as **'picture the enlightenment or ignorance of our human conditions'**?—it is **'make an image of our nature in its education and want of education, likening it to a condition of the following kind'** (ibid. 514a). **'Our nature'** or **'condition'** *is* 'our human condition'! I might include more of Bloom's rendition of the passage Lee translated, on how the **'fire'** that blocks the cave's exit **'corresponds... to the power of the sun'**, which the cave prisoners have to hide from because its searing, **'painful'** light would make **'visible'** the unbearably depressing issue of **'the imperfections of human life'**, those imperfections being **'our human condition'**. Bloom's translation of these key words and phrases is that we should **'liken'** the **'fire'** that blocks the cave's exit **'to the sun's power'** (ibid. 517b), which the cave prisoners have to hide from because its searing, **'distress[ing]'** (ibid. 516e) light would make **'visible'** the unbearably depressing issue of **'human evils'** (ibid. 517c-d), that propensity for evil being **'our nature'**. So Lee's translation of those other key words is equally accurate.

[130] So, despite the human condition being the all-important issue that *had* to be solved if we were to understand and ameliorate human behavior, it has been, for all but a very rare few individuals like Plato and Moses, the one issue humans couldn't face. It has been *the* great unacknowledged 'elephant in the living room' of our lives, THE absolutely critical and yet completely unconfrontable and virtually unmentionable subject in life.

[131] With this in mind, we can now finally understand the resigned mind's inability to make sense of human behavior, *and* the problem posed by the 'deaf effect' when reading about the human condition, *and* why humans resigned to a life in denial of the human condition. And with this appreciation, it is not difficult to see the connection between our species' denial of our condition and the following account of one woman's reaction to being told she had cancer; her 'block out' or denial of her condition is palpable: **'She said, "I thought I was cool, calm and collected but I must have been in a state of shock because the words just seemed to flood over me and I remembered almost nothing from what the doctor said in the initial consultation"'** (*The Sydney Morning Herald*, 18 Aug. 1995). *And*, since discussion of the human condition involves re-connecting with Resignation, an experience *worse than death* (a **'horrible' 'shattering' 'deadly pain'** of such depressing **'despair'**, **'doubts and night terrors'** that it was like a **'sickness unto death'** where you **'just keep falling and falling'**, **'eternally to die, to die and yet not to die'**, as Hughes, Salinger, Kierkegaard, Berdyaev and Jung variously described it), it should be clear to the reader that even though you will *think* you are **'cool, calm and collected'** and able to absorb what is being explained here, your mind will actually be in **'shock'**, so the words will **'flood over'** you and you will **'initial[ly]'** be, as Plato said, unable to take in or hear much of what is being said. Once you are resigned to living in denial of the issue of the human condition, as virtually every adult is, then you are *living in denial of the issue of the human condition*—so your mind will, at least **'initial[ly]'**, resist absorbing discussion of it. So while you may *think* you will be able to follow and take in discussion about what the human condition is, like you would expect when reading or learning about any new subject, this 'deaf effect' that **'initial[ly]'** occurs when reading about the human condition is, in fact, very real.

[132] The problem is that once humans resign and become mentally blocked out or alienated from any truth that brings the issue of the human condition into focus (which, as will be revealed in this book, is most truth),

we aren't then aware that we are blocking out anything; we aren't then aware, as Laing said, that **'there is a veil which is more like fifty feet of solid concrete'** between us and **'our true selves'** or soul that is preventing our mind from accessing what is being said. As pointed out earlier, this inability to know we are blocking something out occurs because obviously we can't block something out of our mind and know we have blocked it out because if we knew we had blocked it out we wouldn't have blocked it out. <u>The fact is, we aren't aware that we are alienated!</u>

[133] So the 'deaf effect' is very real, and so, therefore, is the need to watch the introductory videos and have the willingness to patiently re-read the text and engage in interactive discussions about these understandings.

## Chapter 2:4 <u>How has science coped with the issue of the human condition, and the dangerous 'trap' involved in the way it coped?</u>

[134] In light of what has now been explained about our species' extremely committed denial of the human condition, how has science—as humanity's designated vehicle for enquiry into the nature of our world, particularly into the all-important issue of human nature—coped with the human condition? Well, given it is practiced by humans who have had no choice but to avoid the suicidally depressing subject of the human condition, <u>science has necessarily been what is termed **'reductionist'** and **'mechanistic'**</u>. It has *avoided* <u>the overarching whole view of life that required having to confront the issue of the human condition and instead</u> *reduced* <u>its focus to only looking down at the details of the</u> *mechanisms* <u>of the workings of our world</u>, in the hope that understanding of those mechanisms would eventually make it possible to explain, understand and thus at last be able to both confront and ameliorate or heal the human condition.

[135] Of course, the great danger inherent in the reduced, mechanistic, whole-view-evading, resigned-to-living-in-denial-of-the-human-condition, hiding-in-Plato's-cave, fundamentally dishonest approach is that it could become *so* entrenched, so habituated to its strategy of denial, it could resist the whole-view-embracing, human-condition-confronting, out-of-Plato's-cave, truthful explanation of the human condition when it was finally found and continue to persevere with its dishonest strategy to the point of taking the human race to terminal alienation and extinction.

Yes, despite the arrival of the *truthful* scientific paradigm being science's great objective and fundamental responsibility—and the only means by which the human race can be liberated from its condition, and thus transformed—the risk is that the established *dishonest* scientific paradigm might not welcome or, indeed even tolerate, its arrival!

[136]What has just been said in the above two paragraphs is so critical it needs to be further explained and emphasized.

[137]If we were to stand back and consider the situation the human race has been in—where it needed to find understanding of humans' less-than-ideally behaved, competitive, aggressive and selfish human condition in order to liberate itself from the insecurity of not knowing why that divisive condition emerged—then we can see that there was a very serious obstacle that had to be overcome: how on earth could the human race investigate a subject that virtually everyone was too terrified of to go near? Well, if we then imagine a group of objective thinkers were elected to address this problem—an enlightened board of directors overseeing our situation, if you like—the rationale of their thinking would surely have been that all the human race could do was investigate the nature of our world while all the time avoiding any truths that brought the unbearable issue of the human condition into focus—and just hope that with sufficient understanding of the nature of our world found someone who didn't find the issue of the human condition unbearably condemning and confronting would then be able to assemble the explanation of the human condition from those understandings. No other strategy was possible, and, as mentioned, that is the strategy the human race took.

[138]Human-condition-avoiding, whole-view-evading, so-called reductionist or mechanistic science had to complete the difficult and painstaking task of finding all the pieces of the jigsaw of the explanation of the human condition, but unable to look at the *whole* picture its practitioners were in no position to put the jigsaw together—that final task required a whole-view-confronting, denial-free thinking approach. I have drawn the following picture to illustrate the strategy. (Note, not all the captions in this drawing will be able to be understood at this early stage of the book; however, the essential roles played by human-condition-*avoiding* science and human-condition-*confronting* science in finding the key unlocking insight into our human condition should be sufficiently clear. Our ideal-behavior-demanding, 'condemning

moral conscience' was briefly referred to in chapter 1:3 and will be more fully described shortly in this chapter; the concept of truthful, cooperative-meaning-accepting 'holistic' thinking will be fully explained in chapter 4:2, with inductive and deductive science explained later in par. 581; while the reason deductive or mechanistic science is described as having been 'supposedly objective' rather than *actually* objective, is provided in pars 590 and 1151.)

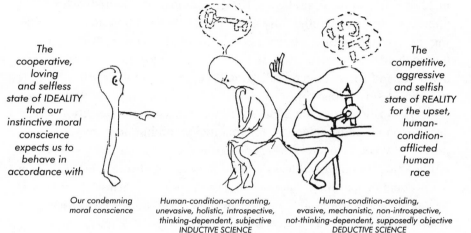

Drawing by Jeremy Griffith © 1991-2014 Fedmex Pty Ltd

The cooperative, loving and selfless state of IDEALITY that our instinctive moral conscience expects us to behave in accordance with

Our condemning moral conscience

Human-condition-confronting, unevasive, holistic, introspective, thinking-dependent, subjective INDUCTIVE SCIENCE

Human-condition-avoiding, evasive, mechanistic, non-introspective, not-thinking-dependent, supposedly objective DEDUCTIVE SCIENCE

The competitive, aggressive and selfish state of REALITY for the upset, human-condition-afflicted human race

**The search for the key, unlocking insight into our human condition**

[139] As was briefly explained in chapter 1, and will be fully explained in chapter 3, the all-important 'piece of the jigsaw' that finally made it possible for the human-condition-confronting, truthful thinking approach to find the explanation of the human condition that is presented in this book was the discovery by human-condition-avoiding, mechanistic science of the difference in the way genes and nerves process information—that genes give species *orientations* but nerves give rise to a conscious mind that needs to *understand* existence, with the inevitable clash between the two learning systems explaining the psychologically upset state of the human condition.

[140] However, while the two approaches taken by human-condition-avoiding science and human-condition-confronting science have now played their part and the explanation of the human condition has been found, there remains one final step to fulfill our board of directors' plans to save humanity—and it is at this last step that a very dangerous 'trap' exists, which our enlightened board of directors would have to

have recognized. <u>That most dangerous of traps is the possibility that the all-dominating world of mechanistic science might become *so* attached to its human-condition-avoiding mechanistic approach—*so* habituated to living in Plato's dark cave of denial—that it might not tolerate the world-saving insight into the human condition that has now been found.</u> In terms of what our board of directors thought could be done to mitigate or avoid this trap, the reality is that all they, or anyone, could do was just hope that there would be a sufficient number of scientists who could appreciate that finding the fully accountable, exonerating and rehabilitating understanding of the human condition made the need for science to be mechanistic obsolete, and that the responsibility now for science as a whole, and for scientists individually, is to acknowledge and support that world-saving insight.

[141] The critical question then is, will there be enough integrity, courage and vision amongst scientists for this understanding to receive the support it now needs to survive—because, as the science historian Thomas Kuhn said, **'In science…ideas do not change simply because new facts win out over outmoded ones…Since the facts can't speak for themselves, it is their human advocates who win or lose the day'** (Shirley Strum, *Almost Human*, 1987, p.164 of 297—Strum's references are to Thomas Kuhn's *The Structure of Scientific Revolutions*, 2nd edn, 1970). And as will be documented later in chapter 6:12, despite support from some very eminent scientists like Harry Prosen, the current situation is that, as of early 2016, the scientific community is failing to demonstrate the integrity, courage and vision necessary to guarantee the understanding survives. Indeed, two of the main themes running through this book are the enormous struggle for acceptance, within both the scientific community and the wider world, that this all-important understanding (along with all the other critically important truthful explanations of human behavior that accompany it) is having to endure, and the dire consequences for the human race if that acceptance fails to eventuate.

[142] And the consequences *are* dire, because if sufficient support for these understandings doesn't develop, humanity can *only* become more and more psychologically upset until it eventually becomes *terminally* psychologically upset—in particular, *so* committed to denial of any truth that brings the issue of the human condition into focus (which, as we are going to see in this book, is most truth) that the human race perishes in a horrific state of terminal alienation!

[143] So it is most distressing that the scientific establishment hasn't taken the lead in recognizing the importance of this explanation of the human condition and has, instead, been treating the explanation as an anathema. By this conduct, and by perpetuating its own dishonest path with the development of an extremely dangerously dishonest account of the human condition itself by someone who has been lauded as the **'living heir to Darwin'**, science *is* completely failing its responsibility of ensuring humanity avoids the horror of terminal alienation.

[144] Yes, in 2012, E.O. Wilson published his **'summa work'**, *The Social Conquest of Earth*, the opening sentence of which truthfully recognizes that **'There is no grail more elusive or precious in the life of the mind than the key to understanding the human condition'** before going on to claim to, as the book's dust jacket says, present **'the clearest explanation ever produced as to the origins of the human condition'**. But, despite all the accolades this book has received, including *The New York Times* rating it one of **'The 100 Notable Books of the Year'** (and the nod to Darwin's throne that also appears on the book's dust jacket), we have to wonder whether, given the human condition has been such a terrifying, unapproachable subject, Wilson has actually been able to confront, think effectively about and, by so doing, find the long sought-after, human-race-liberating, holy **'grail'** of science of the explanation of the human condition—or has he, in fact, not actually confronted and thought truthfully about the human condition at all, and, therefore, *not* presented the liberating and ameliorating understanding of ourselves that we humans so desperately need? As we are going to see further on in this chapter, the answer is the latter; indeed, rather than delivering the dreamed-of *relieving* insight into our 'good-and-evil'-afflicted lives, Wilson, who turns out to be the quintessential exponent of dishonest, human-condition-avoiding mechanistic science, *is* taking the human race *so* deep **'underground'** into the darkness of Plato's human-condition-avoiding cave of denial that he *is* effectively locking humanity onto a path to the utterly *tortured*, permanent darkness of terminal alienation that Michelangelo's and Blake's paintings at the beginning of this chapter so dramatically depict!!

[145] I must emphasize the extreme seriousness of what has occurred. Instead of opening the shutters and letting the liberating light of understanding stream in upon the agonising dilemma of our human condition, which in effect is what Wilson claims he has done, he has *actually* taken the human race into the deepest and darkest corner of truth-avoiding

denial and alienation it has ever known! The appearance is that mechanistic science *is* taking humanity to terminal alienation and extinction!

[146] So it is most significant—and relieving for the human race— that this exposé of Wilson's completely dishonest, condemning-humanity-to-the-torture-of-unspeakable-levels-of-psychosis account of the human condition is now countered by the denial-free, human-condition-confronting-not-avoiding, alienation-removing-not-increasing, psychologically-rehabilitating-and-thus-human-race-transforming, *real* biological explanation of the human condition, which was outlined in chapter 1 and will be fully explained in chapter 3. Further, it is this fully accountable and thus true explanation of the human condition that finally makes it both possible and psychologically safe to also provide the fully accountable, real and true answers to the three other (only slightly less important) outstanding holy grails in science: the meaning of existence; how humans acquired our altruistic moral instincts—as well as a rebuttal of scientific theories that have been put forward on the subject; and how humans became conscious when other animals haven't—explanations that are presented in chapters 4, 5, 6 and 7, respectively. Chapter 8 will then provide the denial-free, real and true account of humanity's journey from ignorance to enlightenment that these fully accountable, true explanations make possible—an account that includes the reconciling understanding of the lives of men and women, the explanation of sex as humans have practiced it (including homosexuality), the explanations of religion and politics, and many, many other denial-free explanations of human behavior.

[147] In short, this book has the power to transport humanity from a world of ignorant darkness and excruciating, human-condition-afflicted bondage that Plato's cave so honestly depicted, to a liberated world bathed in the light of redeeming, relieving and psychologically rehabilitating understanding. The book's concluding chapter (9) describes how this fabulous transformation can, and, with this new presentation that has the supportive 'deaf effect'-eroding introductory videos, will now occur.

[148] (Much more can be read about mechanistic science's strategy of investigating the human condition while avoiding it, including the role the Greek philosopher Aristotle played in developing the strategy, at <www.humancondition.com/freedom-expanded-science-was-invented>.)

## Chapter 2:5  <u>The three fundamental truths that have to be admitted for there to be a true analysis of the human condition</u>

[149] As has been mentioned, what is going to be revealed in the remainder of this chapter is how human-condition-avoiding mechanistic biology—with E.O. Wilson at the helm—has fast been leading humanity to terminal alienation and extinction.

[150] The most effective way to begin this exposé is to identify the three fundamental truths about human behavior that have to be admitted for there to be a true analysis of our condition, but which mechanistic biology has been determinedly avoiding because they have been unbearably condemning of our present psychologically upset competitive, aggressive and selfish human-condition-afflicted existence. <u>We need to bring into the open what it is that has consistently been avoided by mechanistic biology in order to reveal the field's hidden agenda</u>.

[151] So what are the three fundamental truths about our human condition that have been determinedly denied by mechanistic biology? As explained in chapter 1 when the fully accountable explanation of the human condition was outlined, the human condition emerged when our conscious mind challenged our instincts for the management of our lives, with the resulting psychologically upset competitive, aggressive and selfish condition being greatly exacerbated by the fact that our moral instincts are orientated to living in a way that is the complete opposite of this state, namely cooperatively, lovingly and selflessly. (Note, the biological explanation of the great mystery as to how our distant ape ancestors came to live unconditionally selflessly, cooperatively and peacefully, the instinctive memory of which is our moral conscience, is presented in chapters 5 and 6.) Thus, <u>the three fundamental elements</u> involved in this explanation are that <u>our conscious mind caused our upset state to emerge</u>, that <u>it is a *psychological* psychotic and neurotic state of upset that we are living in</u>, and that <u>our species' original instinctive orientation was to living in a psychosis-free, peaceful and harmonious state of cooperation, love and selflessness</u>. The problem, of course, has been that until this reconciling explanation of the human condition was found that defends our conscious mind's upsetting search for knowledge, relieves our psychosis and neurosis, and explains why we had to depart from an original cooperative, loving, selfless state of innocence, each

of these three fundamental truths was unbearably condemning of our present competitive, aggressive and selfish upset, psychotic and neurotic conscious self, and therefore had to be denied. We couldn't face the truth until we could explain it. So these are the three truths that have been denied by human-condition-avoiding mechanistic science—and, as sufferers of the human condition, it is most likely that they are truths that readers of this book will *also* have been living in denial of, and thus unlikely to have been accepting as being 'truths'.

[152] It makes sense, therefore, that the reader will require evidence that these are, in fact, all truths before being presented with an exposé on how mechanistic science has dismissed and denied them—but in the interim the following paragraph serves as a helpful introduction to how mechanistic biology has gone about denying these truths.

[153] As was mentioned in chapter 1:1, the main way mechanistic science managed to deny these three extremely condemning and confronting truths (that we, in the form of our conscious mind, caused our upset, corrupted condition; and that humans now suffer from a psychosis; and that our species' instinctive heritage is of having lived in an all-loving, cooperative, peaceful state) was to assert that our present behavior is no different to that seen in the animal kingdom. It was argued that humans are competitive, aggressive and selfish because of our animal heritage; that we have savage animal instincts that make us fight and compete for food, shelter, territory and a mate—basically, for the chance to reproduce our genes. Further, as will be described shortly in chapter 2:9, we said that the task of our conscious mind is to try to *control* these supposed brutal, savage instincts within us. As will be pointed out, this was an absolutely brilliant excuse because instead of our instincts being all-loving and thus unbearably condemning of our present non-loving state, they were made out to be vicious and brutal; *and*, instead of our conscious mind being the cause of our corruption, the insecurity of which made us repress our instinctive self or soul or psyche and become psychotic, it was made out to be the blameless, psychosis-free mediating 'hero' that had to manage those supposed vicious instincts within us! Of course, as was pointed out in chapter 1:1, the whole 'animals are competitive and aggressive and that's why we are' excuse cannot be the *real* cause of our divisive behavior because descriptions of our behavior, such as egocentric, arrogant, inspired, depressed, deluded, pessimistic, optimistic, artificial, hateful, cynical, mean, immoral, guilt-ridden, evil, psychotic, neurotic, alienated,

all recognize the involvement of our species' unique fully conscious thinking mind—that there *is* a *psychological* dimension to *our* behavior. We
have suffered not from the genetic-opportunism-based, non-psychological
*animal* condition, but the conscious-mind-based, *PSYCHOLOGICALLY*
troubled *HUMAN CONDITION*. The other reason (which I didn't mention
in chapter 1:1 but can now include) for why the savage animal instincts in
us excuse doesn't hold water is because of the third fundamental truth
about humans, which is that we have unconditionally selfless, cooperative,
loving, *moral* instincts, the expression or 'voice' of which within us we
call our conscience. The reason adolescents have become so depressed
during Resignation, and why they don't fall for and adopt the savage
instincts excuse (although they gladly embrace it *after* resigning and
deciding they have to live in denial of the human condition) is because
their moral instinctive self or soul lets them know their behavior should
be cooperative and loving, *not* competitive and aggressive. The fundamental reason humans have had a sense of guilt is because we have a
moral conscience. Yes, the truth is we do all know that the old 'animals
are competitive and aggressive and that's why we are' defense doesn't
explain our *psychologically distressed, guilt-ridden human condition*. In
fact—as is going to be revealed—'the savage instincts excuse' and 'the
conscious mind is the psychosis-free hero' accounts of human behavior
have all just been *terrible* reverse-of-the-truth lies, albeit hugely relieving
ones for humans needing to seek relief from the human condition while
it wasn't able to be truthfully explained.

[154]Now, to supply evidence for the reader of the three truths that our
conscious mind caused our upset, corrupted condition, that we suffer
from a psychosis, and that our instinctive heritage is of having lived in
an all-loving, cooperative, peaceful state.

## Chapter 2:6 <u>Evidence of the three fundamental truths, as provided by Moses and Plato</u>

[155]If we ask ourselves what is the 'we' that we are talking about when
we refer to the possibility that 'we' are a terrible mistake, a worthless
blight on this planet, the **'face of absolute evil'** as Jung said, we can see
where the *real* problem about our seemingly horribly flawed condition
lies: the 'we' is surely our conscious thinking mind or intellect. It is our
conscious mind that is uncertain of its worthiness, that suspects that it

might be to blame for our species' present seemingly highly imperfect, even 'fallen' or corrupted, competitive, aggressive and selfish condition. And indeed, that most famous mythological account of the origin or genesis of the human condition, the story of Adam and Eve from the book of 'Genesis' in the Old Testament in the Bible, which the very great prophet Moses wrote (versions of which also appear in the Torah of Judaism, and the Koran of Islam), recognizes that this was the case—that it *was* our conscious mind that led to our 'good-and-evil'-afflicted condition. Moses said that the first humans, represented in this account by Adam and Eve, lived **'naked, and they felt no shame'** (Gen. 2:25) in **'the Garden of Eden'** (3:23) and were **'created...in the image of God'** (1:27), obviously meaning we once lived in a pre-human-condition-afflicted state of original innocence where we were perfectly instinctively orientated to the cooperative, selfless, loving, 'Godly' ideals of life. Moses then said that Adam and Eve took the **'fruit'** **'from the tree of...knowledge'** (3:3, 2:17) because it was **'desirable for gaining wisdom'** (3:6), obviously meaning we became fully conscious, thinking, knowledge-seeking beings. Then, as a result of becoming conscious, and being **'disobedient'** (the term widely used in descriptions of Gen. 3), Moses said we **'fell from grace'** (derived from the title of Gen. 3, **'The Fall of Man'**), obviously meaning our original cooperative, selfless and loving (good) state became corrupted and our competitive, selfish and aggressive—indeed, angry, egocentric and alienated—('evil'/'sinful'/guilt-ridden) state emerged. It was at this point that humans **'realised that they were naked; so they sewed fig leaves together and made coverings for themselves'** (3:7), meaning nudity was no longer a **'shame**[less]**'** state, with sex as humans now practice it emerging where, as will be explained in chapter 8:11B, the act of procreation became perverted and used as a way of angrily attacking or 'fucking' innocence because of its implied criticism of our lack thereof—at which point it was necessary to clothe ourselves to dampen lust and reduce the **'shame'** we felt for being so horrifically destructive of innocence. (Note: this explanation of sex as humans practice it now will likely be another concept new to the reader, requiring some thought before it can be accepted as being true, but it really is just a further honest, obvious explanation we couldn't afford to admit while we couldn't explain the human condition and defend our immensely corrupted lives.) Moses then said that, as a result of the emergence of all our corrupt angry, egocentric and alienated behavior, we were **'banished...from the Garden of Eden'**-like (3:23) state of original innocence and left **'a restless wanderer on the earth'** (4:14); that is, we were left in our present, psychologically upset, distressed and alienated condition.

[156]It should be emphasized here that while Moses' extraordinarily sound and thus effective thinking enabled him to describe all the elements involved in producing the human condition, the story of Adam and Eve is only a *description* of the conflict that produced the upset state of our human condition, *not* an explanation of WHY the conflict occurred. As was described in chapter 1, for that all-important explanation to be possible science had to be established and understanding of the difference in the way genes and nerves process information found; we had to understand that one, the genetic learning system, is an orientating learning system while the other, the nerve-based learning system, is insightful. And until that clarifying explanation was found it wasn't possible to explain that the intellect is actually the hero not the villain, deserving of being **'banished... from the Garden of Eden'**, it is portrayed as in the story of Adam and Eve. Moses was an exceptionally honest and thus effective thinker and could describe the elements involved in producing the human condition, but he could not liberate humanity from the insecurity of that condition. For that to be possible science had to be developed.

[157] As we have already established, <u>Plato</u> was, like Moses, an extremely honest, denial-free, effective thinker, one whose mind was focused on **'the enlightenment or ignorance of our human condition'**. Given his extraordinarily truthful and accurate description in *The Republic* of the human race being imprisoned in a dark cave of denial, it should come as no surprise that <u>Plato *also* recognized that, as Moses said, we **'fell from grace'** from a **'Garden of Eden'** pre-human-condition-afflicted state of original innocence, *and* identified the elements involved in that fall of our conscious mind in conflict with our original innocent instinctive self</u>. Yes, in *The Republic*, prior to introducing his cave allegory, Plato presented what he termed in the original Greek wording as his theory of the *psychē* or psychological condition of humans, in which he spoke of the conflicting elements that caused **'the imperfections of human life'**, namely our moral instincts, which (again, in the original Greek wording) he referred to as *thymos*, in conflict with our conscious intellect, which he referred to as *eros*. This conflicted state is clearly **'our human condition'**—an interpretation that is even more apparent when Plato returned to this *thymos* vs *eros* conflict

in his dialogue *Phaedrus*, this time describing our condition using what is, after his cave allegory, his second most famous allegory, <u>the allegory of the two-horsed chariot</u>.

[158] <u>Firstly, with regard to our species' original innocent state</u>, in his chariot allegory Plato gave this exceptionally honest description of it: **'there was a time when…we beheld the beatific vision and were initiated into a mystery which may be truly called most blessed, celebrated by us in our state of innocence, before we had any experience of evils to come, when we were admitted to the sight of apparitions innocent and simple and calm and happy, which we beheld shining in pure light, pure ourselves and not yet enshrined in that living tomb which we carry about, now that we are imprisoned in the body, like an oyster in his shell'** (*Phaedrus*, c.360 BC; tr. B. Jowett, 1871, 250).

[159] <u>And, with regard to the conflict between instinct and intellect</u> that gave rise to our upset **'tomb'**-like human-condition-afflicted existence, Plato was equally extraordinarily insightful, writing: **'Let the figure** [of the two-horsed chariot] **be composite—a pair of winged horses and a charioteer… and one of them** [one of the horses] **is noble and of noble breed, and the other is ignoble and of ignoble breed; and the driving of them** [by the charioteer, which is us having to try to manage these two conflicting elements within us] **of necessity gives a great deal of trouble to him'** (ibid. 246). Some pages later, Plato was even more explicit about the nature of the two horses and the clash between them, writing that **'one of the horses was good and the other bad** [ibid. 253] **…[and the bad horse], heedless of the [charioteer]…plunges and runs away, giving all manner of trouble to his companion and the charioteer…[they being] urged on to do terrible and unlawful deeds'** (ibid. 254).

[160] Plato added that the **'noble'**, **'good'** horse is **'cleanly made…his colour is white…he is a lover of honour and modesty and temperance'** (ibid. 253)—clearly the representation of what we now know is our innate, ideal-behavior-demanding moral instinctive self, the voice of which is our conscience. Obviously Plato was not using scientific terms, but, as mentioned, he designated this white horse as *thymos*, which is usually translated as 'spirit', a concept described by the political scientist Francis Fukuyama as being **'something like an innate human sense of justice'** (*The End of History and the Last Man*, 1992, p.165 of 418). And in *The Republic*, Plato spoke plainly about the happy, loving, goodness-and-**'justice'**-expecting, soulful **'innate'** nature of *thymos* (or 'spirit'), writing that **'You can see it in children, who are full of spirit as soon as they're born'** (*The Republic*, tr. H.D.P. Lee, 1955, 441).

[161] As for the **'ignoble'** or **'bad'** horse, in his original Greek wording, Plato called it *eros*, which signifies carnal love or sexual desire (it is the origin of the English word erotic), or more generally, any excessive desire. Plato described this horse as **'a crooked lumbering animal...of a dark colour... the mate of insolence and pride, shag-eared and deaf'** (*Phaedrus*, 253), which is clearly a reference to our upset, **'terrible and unlawful'**, **'deaf'**-to-the-truth conscious intellect. Some people have misinterpreted this horse as representing animal instincts within us, specifically the instinct to reproduce, however, it is clear in the quotes above that Plato recognized our instincts as **'innocent'** and **'pure'**, which can be seen **'in children'**, **'before we had any experience of evils to come'**, when we were **'not yet enshrined in that living tomb which we carry about, now that we are imprisoned'**. No, in Plato's time, when upset behavior was not restrained, or civilized, to the extent that it is in today's modern society (much more will be said in chs 8:15 and 8:16 about humanity's adoption of restraint), destructive sexual behavior would have been the most common and obvious manifestation of our psychologically upset angry, egocentric and alienated state, and so *eros* or sexual desire would have been the obvious concept for Plato to use to depict it. That Plato intended for *eros* to signify our full upset state, not just sexual desire, is made clear in Allan Bloom's Interpretive Essay on *The Republic*, in which he wrote that Plato **'characterizes the tyrant as the erotic man, and *eros*, as...a mad master...Eros is the most dangerous and powerful of the desires, an infinite longing which consumes all other attachments in its heat'** (*The Republic of Plato*, tr. Allan Bloom, 1968, p.423 of 512). Yes, *eros* is more than just the excess of sexual desire as we know it, where, as has been mentioned, the act of procreation has become corrupted and used to attack the innocence of women; its *ultimate* manifestation is found in a **'tyrant['s]'** **'infinite longing'** for power and glory where an overly insecure, upset mind forever seeks to validate itself and, by so doing, avoid the implication it is unworthy or bad. (The psychological reason for the tyrannical, power-addicted mind is explained in ch. 8:16D.)

[162] It should be mentioned that just as Plato referred to sexual desire as the most obvious manifestation of our psychologically upset state, so did Moses in his Genesis story when he alluded to sexual desire in the form of Eve tempting Adam to take the fruit from the tree of knowledge, and, as with Plato, this reference by Moses to sexual desire has also been misinterpreted as Moses inferring that we have competitive, have-to-reproduce-our-genes 'animal' instincts. However, again as with

Plato, it is clear that Moses recognized that we *don't* have competitive, selfish, 'have-to-reproduce-our-genes' instincts, rather that our instinctive heritage is one of having lived an innocent, cooperative, loving, moral existence, writing that humans were, as has been mentioned, **'created... in the image of God'**; our distant ancestors, the first humans, which Adam and Eve represent, lived in accordance with the cooperative, loving, 'Godly' ideals of life, and **'Adam and his wife were both naked, and they felt no shame'**; meaning at that point sex hadn't been perverted and used as a way of attacking innocence, but *after* this innocent, pre-conscious, pre-human-condition-afflicted time our psychologically upset, **'fall[en]'** condition emerged and our innocence *was* destroyed. As Moses described it, after we took the **'fruit' 'from the tree of...knowledge'**, **'the eyes of both of them** [Adam and Eve/we] **were opened, and they realized that they were naked; so they sewed fig leaves together and made coverings for themselves'**.

[163] Although Moses *clearly* recognized that our distant, pre-conscious, pre-human-condition-afflicted, pre-upset ancestors were innocent, co-operative and loving, the question arises as to why did he infer that it was the upset state of sexual desire, in the form of Eve supposedly using sex to tempt Adam, that *caused* Adam to take the fruit from the tree of knowledge? The answer can only be that an account of the emergence of the human condition, which is what is being provided by Moses, should include reference to the immense role sex has played in the life of upset humans (indeed, as will all be explained in chapter 8:11B, after the upset state emerged 'sex' became one of the main means by which the upset was spread from one generation to the next), and having Adam 'tempted' by Eve in this great moment of transition for our species from innocence to upset gives recognition to that, even though the upset behavior of sex as humans now practice it emerged *after* we set out on our upsetting search for knowledge, not before. This misplaced representation of when this perversion of 'sex' occurred is very apparent in this critical passage, in which Moses wrote that Eve **'gave some** [fruit from the tree of knowledge] **to her husband, who was with her, and he ate it. Then the eyes of both of them were opened and they realised they were naked; so they sewed fig leaves together and made coverings for themselves'**. The first sentence is where it is historically inferred that Eve 'seduced' Adam into taking the fruit from the tree of knowledge; however, the second sentence says that their **'eyes'-'opened'**, conscious, upset, **'shame[ful]'**, lustful state emerged *after* they ate the fruit from the tree of knowledge!?

[164] It should be mentioned that in another of Plato's allegories, which will be described shortly, he also referred to our distant pre-conscious ancestors as having lived shamelessly in a lust-free naked, innocent state, writing that our ancestors were **'earth-born'**, meaning they were not born of the sexual perversion of the act of procreation that is involved in sex now, and there was no **'possession of women'**, no **'devouring of one another'**, and **'they dwelt naked'**.

[165] As to why people have misrepresented Plato's **'dark'** horse and Moses' reference to sexual desire as indicating we have brutish, have-to-reproduce-our-genes animal instincts, it is because—as has been mentioned and as will be explained more fully in the latter part of this chapter—blaming our present competitive and aggressive behavior on supposed savage, competitive, have-to-reproduce-our-genes animal instincts in us which our conscious mind has had to try to control, has been the main device used to deny that we once lived in a cooperative, selfless, loving innocent state that was upset by the emergence of our conscious mind. (It will become very clear through the course of this chapter just how determinedly humans have sought to avoid the human condition by blaming our divisive behavior on savage, we-have-to-reproduce-our-genes instincts.)

[166] To return to Plato's two-horsed chariot allegory, and the explanation of the meaning of the **'charioteer'**. Designated in Plato's original Greek wording as *logos*, the charioteer represents the overall situation our species has been in where we have been trying to understand the two conflicting elements within us of our ideal-behavior-demanding instinct and our defiant, searching-for-knowledge, psychologically upset angry, egocentric and alienated intellect, with the ultimate goal being to find the reconciling understanding of our condition that will enable us to be liberated and transformed from it. *Logos* is normally inadequately translated as 'reason', but in the context of Plato's chariot allegory it takes on a broader meaning of our reasoning intellect seeking the true understanding of our condition, or as Plato wrote, guiding our chariot to those realms of **'justice, and temperance, and knowledge absolute'** (*Phaedrus*, 247), of which we have an **'exceeding eagerness to behold'** (ibid. 248).

[167] So, Plato's two-horsed chariot allegory is an astonishingly clear description of the upset, **'crooked lumbering'**, **'dark'**, **'mate of insolence and pride, shag-eared and deaf** [alienated]**'** intellect rising in defiant **'heedless'** opposition to our **'upright and cleanly made'**, **'white'**, **'lover of honour and**

**modesty and temperance'**, **'pure' 'innocent'** original instinctive self, or soul, leading us to **'terrible and unlawful deeds'**, so that we are now condemned as **'evil'** and **'enshrined in that living tomb'**. However, it has to be emphasized that, like Moses' Garden of Eden story, the chariot allegory is only a *description* of the conflict that produced the upset state of our human condition, *not* an explanation of WHY the conflict occurred. Like Moses with his story of consciousness developing in the Garden of Eden, Plato was still only able to view our intellect as an **'evil'**, **'bad'**, **'ignoble'** influence in our lives. Despite being the greatest of philosophers, Plato couldn't explain the human condition. He could *describe* the situation perfectly but he still couldn't deliver the clarifying, psychosis-addressing-and-relieving explanation—he couldn't explain HOW humans could be good when we appeared to be bad. As has been explained, for that to be possible science had to be developed.

[168] Nevertheless, Plato's insights were absolutely remarkable, for not only did he clearly identify our original state of uncorrupted innocence and the two conflicted elements that then produced our psychologically upset, corrupted, fallen human condition in his two-horsed chariot allegory, in one of his final dialogues, *The Statesman*, he was even more incisive. While still not able to explain why the conflict occurred and from there explain how humans are good when we appear to be bad, in what is known as the myth of the 'reversed cosmos' he gave a perfectly clear description of the sequence of events that *led* to that conflict, even anticipating its eventual peaceful resolution!

[169] To fully appreciate this account, the references Plato makes in it to **'God'** being **'the orderer of all'** (*The Statesman*, c.350 BC; tr. B. Jowett, 1871, 273) need to briefly be explained. Later in chapter 4, it will be explained that there is a teleological or holistic purpose or meaning to existence, which is to develop the order of matter into ever larger and more stable wholes—atoms come together or integrate to form compounds, which in turn integrate into virus-like organisms, into single-celled organisms, into multicelled organisms, etc. But while this integrative meaning of existence is one of the most obvious truths, it too has been denied by human-condition-avoiding mechanistic science because it implies humans should behave in an integrative, ordered, cooperative, sharing, selfless, loving way—behavior that is at complete odds with our present seemingly *divisive, disorderly and disintegrative competitive, selfish and aggressive* behavior. No, only when the human condition was explained

and our divisive state understood, as it now is, would it be safe to admit this truth of Integrative Meaning that our concept of 'God' represents, which Plato in his extraordinary human-condition-confronting-not-avoiding honesty was able to acknowledge.

[170] So, just as he did with his chariot allegory, Plato began his 'reversed cosmos' allegory by giving <u>a truthful description of our innocent ancestors</u>, referring to them in this instance as the **'earth-born'** (ibid. 271), so-called because they were born of the earth rather than through **'procreation'** (ibid). 'Earth-born' is presumably a metaphor for the time prior to the emergence of 'sex' as we upset humans practice it, where, as mentioned earlier, the act of **'procreation'** has been corrupted and used to attack the innocence of women. And I should mention that Plato insisted that this 'Golden Age' in our past was a historical reality, writing that in **'this tradition** [of the **earth-born** man]**, which is now-a-days often unduly discredited, our ancestors** [in the form of existing relatively innocent 'races' of people, such as those who still exist today like the Bushmen of the Kalahari and the Australian Aborigine], **who were nearest in point of time to the end of the last period and came into being at the beginning of this, are to us the heralds** [of that earlier innocent age]**'** (ibid. 271). What now follows is Plato's second extremely honest description that he gave of this innocent 'Golden Age' in our species' past; he wrote that we lived a **'blessed and spontaneous life…**[where] **neither was there any violence, or devouring of one another, or war or quarrel among them…In those days God himself was their shepherd, and ruled over them** [our original instinctive self was orientated to living in an integrative, cooperative, loving way]**…Under him there were no forms of government or separate possession of women and children; for all men rose again from the earth, having no memory of the past** [we lived in a pre-conscious state]**. And…the earth gave them fruits in abundance, which grew on trees and shrubs unbidden, and were not planted by the hand of man. And they dwelt naked, and mostly in the open air, for the temperature of their seasons was mild; and they had no beds, but lay on soft couches of grass, which grew plentifully out of the earth'** (ibid. 271-272).

[171] Continuing with his extraordinary honesty and resulting clarity of thought, Plato then <u>described how management of our lives transferred</u> <u>from our instincts to our emerging consciousness, and how we slowly</u> <u>began to accumulate knowledge</u>: **'Deprived of the care of God, who had possessed and tended them** [we disobeyed our original instinctive orientation to living in an integrative, cooperative, loving way]**, they were left helpless and**

**defenceless...And in the first ages they were still without skill or resource; the food which once grew spontaneously had failed, and as yet they knew not how to procure it, because they had never felt the pressure of necessity** [we had lived in a cooperative, sharing, loving way, free of the insatiable greed that exhausts resources]**...the gifts spoken of in the old tradition were** [now] **imparted to man by the gods** [of fire, creativity, agriculture and so forth], **together with so much teaching and education** [knowledge] **as was indispensable...fire...the arts...seeds and plants...From these is derived all that has helped to frame human life; since the care of the Gods, as I was saying, had now failed men, and they had to order their course of life for themselves, and were their own masters'** (ibid. 274).

[172] He also described the upset, corrupted, fallen state that resulted from the emergence of consciousness, writing that in the very beginning the world was of a **'primal nature, which was full of disorder...**[then] **the world was aided by the pilot** [God, the process of integrating matter] **in nurturing** [creating] **the animals,** [and] **the evil was small, and great the good which he produced** [in our innocent human forebears]**, but after the separation, when the world was let go** [when the conscious mind began to challenge the instincts for mastery]**, at first all proceeded well enough** [our intellect mostly deferred to our instincts]**; but, as time went on, there was more and more forgetting** [alienation, or separation from our instinctive moral self]**, and the old discord** [disorder] **again held sway and burst forth in full glory** [the psychologically upset, divisive, disordered state of the human condition emerged]**; and at last small was the good, and great was the admixture of evil, and there was a danger of universal ruin to the world'** (ibid. 273).

[173] And showing even *more* honesty and clarity of thought, Plato *then* described how such truthful, denial-free, God/Integrative Meaning-acknowledging thinking would one day re-establish cooperative, loving order amongst humans: **'Wherefore God, the orderer of all, in his tender care, seeing that the world was in great straits, and fearing that all might be dissolved in the storm and disappear in infinite chaos, again seated himself at the helm; and bringing back the elements which had fallen into dissolution and disorder to the motion which had prevailed under his dispensation, he set them in order and restored them, and made the world imperishable and immortal. And this is the whole tale'** (ibid. 273).

[174] Even more astonishing still is the fact that Plato could not only think truthfully enough to see and thus prophesize how such truthful, effective thinking would one day **'set them** [humans] **in order and restore...them'**, he went on to predict that the restoration would be achieved by appreciating

that the corrupting search for knowledge was of paramount importance. While still not able to clearly explain *why* the conflict occurred and from there reveal how humans are good when we appear to be bad, Plato did recognize that we *had to* search for knowledge. Posing the question of whether a **'blessed and spontaneous'**, instinctively guided innocent ancestor **'having this boundless leisure, and the power of holding intercourse, not only with men, but with brute creation** [in other words, having the power to sensitively relate to each other and even to other creatures]' would prefer that existence over the situation of an upset human, someone **'of our own day'** who is dedicated to developing **'a view to philosophy'** and **'able to contribute some special experience to the store of wisdom'**, Plato said that **'the answer would be easy'** — he would **'deem the happier'** the life **'of our own day'** in which we each had the opportunity to **'contribute some special experience to the store of wisdom** [the necessary search for knowledge]' (ibid. 272)!

[175] So that is Plato's truly extraordinary, denial-free, pre-scientific account of our past instinctive **'blessed and spontaneous life'** and the subsequent emergence of our conscious mind that allowed us to become our **'own masters'**, the result of which was the emergence of our upset, **'evil'** condition of **'discord'** — a state we were then so ashamed of that we **'often unduly discredit**[ed]' the truth of our **'blessed'** past, leaving us **'enshrined in that living tomb'** of a **'cave'**-like state of dishonest psychosis and neurosis-producing denial where there was **'more and more forgetting'**. *So* **'enshrined'** in denial, in fact, that the human race has now reached, as Plato predicted, the state of terminal alienation that threatens **'universal ruin to the world'**, from which only a denial-free approach, one where **'God'** in the form of Integrative-Meaning-acknowledging truthfulness, has **'again seated himself at the helm'**, could, as has now happened with this book, **'set them** [humans] **in order and restore...them'**. Plato certainly had no trouble admitting the three great truths underlying the reality of our condition — that our conscious mind caused our fall from innocence, that we suffer from a psychosis, and that our distant forebears lived co-operatively and peacefully. I think we are now able to fully appreciate why A.N. Whitehead said that all philosophy, which again is the quest for **'the truths underlying all reality'**, is merely **'a series of footnotes to Plato'**!!

[176] It should be noted that Plato emphasized that we would **'often unduly discredit'** the truth of our species' **'innocent'**, **'blessed'**, **'upright'**, **'cleanly made'**, **'pure'**, **'noble'**, **'good'**, **'modest'**, **'honour**[able]', **'spirit**[ed]', **'simple and calm and happy'** past — which is part of the **'more and more forgetting**

[denial]', that, unchecked, leads to **'universal ruin to the world'**. This journey to ever increasing levels of alienation and its dishonesty is the underlying story that this book documents.

[177] There is yet <u>one more very impressive reference that Plato makes to our species' original all-loving, all-sensitive, always-behaving-in-a-way-that-is-consistent-with-the-integrative-cooperative-Godly-ideals-of-life instinctive self or soul</u>. This appears in his dialogue *Phaedo* where he wrote that humans have **'knowledge, both before and at the moment of birth... of all absolute standards...**[of] **beauty, goodness, uprightness, holiness...our souls exist before our birth'**, describing **'the soul'** as **'the pure and everlasting and immortal and changeless...realm of the absolute...**[our] **soul resembles the divine** [God]' (tr. H. Tredennick, 1954, 75-80).

[178] So both <u>Moses'</u> Garden of Eden and <u>Plato</u>'s various accounts identify our conscious mind as causing **'the fall'** from an original, cooperative, loving **'blessed'**, **'calm and happy'** state of **'innocence'**. They recognized that our present 'fallen', corrupted, psychologically upset human condition resulted from the emergence of our unique fully conscious thinking mind.

[179] (Note again that the biological explanation for the great mystery as to how our distant ape ancestors came to live unconditionally selflessly, cooperatively and peacefully, the instinctive memory of which is our moral conscience, is presented in chapters 5 and 6.)

## Chapter 2:7 <u>Further evidence of the three fundamental truths, as provided by religion, mythology, profound thinkers, and primatological and anthropological studies</u>

[180] And, tellingly, <u>this awareness that humans did once live in a pre-conscious and pre-human-condition-afflicted peaceful, cooperative, selfless, loving ideal state is something *all* our religions and mythologies recognize</u>. In addition to Moses' account of Adam and Eve's innocent heritage in the Garden of Eden, the Bible also contains a passage in <u>Ecclesiastes</u> that reads, **'God made mankind upright** [uncorrupted], **but men have gone in search of many schemes** [conscious understandings]' (7:29), and the references <u>Christ</u> made to a time when God **'loved** [us] **before the creation of the** ['upset', 'fallen', corrupted] **world'** (John 17:24), and a time of **'the glory... before the** [corrupted] **world began'** (John 17:5). <u>Taoist</u> scripture also features a description of our distant forebears as being **'the Men of Perfect Virtue'** (Bruce Chatwin, *The Songlines*, 1987, p.227 of 325), while <u>Zen Buddhism</u> similarly speaks

of the loss of an uncontaminated, pure state as a result of the intervening conscious mind, referring to **'the affective contamination (*klesha*)'** or **'the interference of the conscious mind predominated by intellection (*vijñāna*)'** (D.J. Suzuki, Erich Fromm, Richard Demartino, *Zen Buddhism & Psychoanalysis*, 1960, p.20). And prior to Plato and his vast contribution—including his two-horsed chariot account of **'our state of innocence, before we had any experience of evils to come, when we were admitted to the sight of apparitions innocent and simple and calm and happy, which we beheld shining in pure light, pure ourselves and not yet enshrined in that living tomb which we carry about, now that we are imprisoned'**, and his 'reversed cosmos' description of our species' past **'blessed and spontaneous life...[where] neither was there any violence, or devouring of one another, or war or quarrel among them...And they dwelt naked, and mostly in the open air...and they had no beds, but lay on soft couches of grass'**—there was his compatriot Hesiod, who, in his epic poem *Works and Days*, said of our distant ancestors that **'When gods alike and mortals rose to birth / A golden race the immortals formed on earth...Like gods they lived, with calm untroubled mind / Free from the toils and anguish of our kind / Nor e'er decrepit age misshaped their frame...Strangers to ill, their lives in feasts flowed by...Dying they sank in sleep, nor seemed to die / Theirs was each good; the life-sustaining soil / Yielded its copious fruits, unbribed by toil / They with abundant goods 'midst quiet lands / All willing shared the gathering of their hands'** (c. eighth century BC).

[181] The consistency of all these descriptions of how consciousness led to the corruption of our species' original all-loving cooperative state has been borne out by the investigations of the author Richard Heinberg, who found that every human culture has a myth involving both the emergence of consciousness and a 'fall' from an original 'Golden Age' of togetherness and peace—from the major religions, to 'races' as isolated and diverse as the Eskimos, Aborigines, and Native Americans—summarizing in his 1990 book *Memories & Visions of Paradise* (a well-researched collection of acknowledgments from mythologies and religions of our species' innocent, Edenic past) that **'Every religion begins with the recognition that human consciousness has been separated from the divine Source, that a former sense of oneness...has been lost...everywhere in religion and myth there is an acknowledgment that we have departed from an original...innocence and can return to it only through the resolution of some profound inner discord...the cause of the Fall is described variously as disobedience, as the eating of a forbidden fruit, and as spiritual amnesia [alienation]'** (pp.81-82 of 282). Yes, as Berdyaev recognized, **'The memory of a lost paradise, of a Golden Age, is very deep in man'**

(*The Destiny of Man*, 1931; tr. N. Duddington, 1960, p.36 of 310). So when John Milton titled his epic 1667 poem **'Paradise Lost'**, he was recognizing the existence of this **'deep' 'memory' 'in man'** — as were the Australian Aborigines with their **'memory'** of a **'Dreamtime'**. The philosopher Jean-Jacques Rousseau also expressed what we all do intuitively know is the truth when he wrote that **'nothing is more gentle than man in his primitive state'** (*The Origin of Inequality*, 1755; *The Social Contract and Discourses*, tr. G. Cole, 1913, p.198 of 269).

[182]William Wordsworth was another who spoke honestly when he wrote, **'There was a time when meadow, grove, and streams / The earth, and every common sight / To me did seem / Apparelled in celestial light / The glory and the freshness of a dream / It is not now as it hath been of yore / Turn wheresoe'er I may / By night or day / The things which I have seen I now can see no more // The Rainbow comes and goes / And lovely is the Rose / The Moon doth with delight / Look round her when the heavens are bare / Waters on a starry night / Are beautiful and fair / The sunshine is a glorious birth / But yet I know, where'er I go / That there hath past away a glory from the earth // ...something that is gone / ... Whither is fled the visionary gleam? / Where is it now, the glory and the dream? // Our birth is but a sleep and a forgetting / The Soul** [the instinctive memory of our species' past all-loving, selfless, cooperative existence] **that rises with us** [that we are born with], **our life's Star / Hath had elsewhere its setting / And cometh from afar / Not in entire forgetfulness / And not in utter nakedness / But trailing clouds of glory do we come / From God, who is our home / Heaven lies about us in our infancy! / Shades of the prison-house begin to close / Upon the growing Boy / ...And by the vision splendid / Is on his way attended / At length the Man perceives it die away / And fade into the light of common day / ... Forget the glories he hath known / And that imperial palace whence he came'** (*Intimations of Immortality from Recollections of Early Childhood*, 1807). This beautiful description by Wordsworth of our original instinctive self or **'Soul'** equates perfectly with that given by the poet Henry Vaughan when he wrote, **'My soul, there is a country far beyond the stars'** (*Peace*, 1655). The sentiment is also reflected in the words of the polymath Sir Thomas Browne, who said, **'We carry within us all the wonders we seek without us'** (*Religio Medici*, 1643, Sect.15), and in the poet Lord Byron's observation that **'Man is in part divine, A troubled stream from a pure source'** (*Prometheus*, 1816). With regard to Wordsworth's reference to **'God'**, as outlined earlier and as will be explained in chapter 4:3, 'God' is our personification of the terrifyingly confronting truth of the teleological, integrative, order-of-matter-developing, cooperative, selfless, loving theme or direction or meaning of

existence that our distant ancestors lived in accordance with, but which we no longer appear to. And like Laing, the prophet <u>Isaiah</u> recognized how upset/corrupted we humans have become, but was more specific about how it is a corruption of this integrated state we once lived in, writing that **'From the sole of your foot to the top of your head there is no soundness—only wounds and welts and open sore...Your country is desolate... the faithful city has become a harlot! She once was full of justice; righteousness used to dwell in her'**, and **'the world languishes and withers...The earth is defiled by its people; they have disobeyed the laws** [become divisively rather than integratively behaved]**...In the streets...all joy turns to gloom, all gaiety is banished from the earth'** (Bible, Isa. 1 & 24).

[183] The consistency of these accounts *is* remarkable, but <u>beyond what is revealed by myth, religion and profound thought, consider the evidence provided by our studies in anthropology and primatology</u>. While fossils of our early ape ancestors who lived from 12 to 4 million years ago are rare, recent discoveries from this period are now providing proof of a cooperative past, which anthropologists are beginning to admit; for instance, <u>C. Owen Lovejoy</u> has written that **'our species-defining cooperative mutualism can now be seen to extend well beyond the deepest Pliocene** [well beyond 5.3 million years ago]**'** ('Reexamining Human Origins in Light of *Ardipithecus ramidus*', *Science*, 2009, Vol.326, No.5949). In primatology, studies of living apes reveal that bonobos (*Pan paniscus*) are not only humans' closest relatives, but an extremely gentle and cooperative species. While bonobos and chimpanzees (*Pan troglodyte*) both share around 99 percent of humans' genetic material, the primatologist <u>Frans de Waal</u> points out that **'the recent discovery** [by neuroscientists Elizabeth Hammock and Larry Young in 2005] **that bonobos and humans share genetic code in relation to affiliative** [social, cohesive, loving, integrative] **behavior that is absent in the chimpanzee'** indicates bonobos and humans are more closely related in terms of their social nature ('Foreword by Frans B.M. de Waal', *The Bonobos: Behavior, Ecology, and Conservation*, eds T. Furuichi & J. Thompson, 2008, p.12 of 327). The anthropologist <u>Adrienne Zihlman</u> has also shown that of bonobos and chimpanzees, bonobos are closer anatomically to our ancestors ('Reconstructions reconsidered: chimpanzee models and human evolution', *Great Ape Societies*, eds William C. McGrew et al., 1996, pp.293-304 of 352). As to their peaceful nature, many primatologists attest to it; consider this from <u>Barbara Fruth</u>: **'up to 100 bonobos at a time from several groups spend their night together. That would not be possible with chimpanzees because there would be brutal fighting between rival groups'** (Paul

Raffaele, 'Bonobos: The apes who make love, not war', Last Tribes on Earth.com, 2003; see <www.wtmsources.com/143>). (The evidence of our cooperative past that is provided by anthropology, some of which is referred to above, and by the bonobos, will be documented in some detail in chapters 5:5 and 5:6, respectively.)

[184] Of course, proof is *also* apparent in the relative innocence of existing so-called 'primitive' people, such as the Bushmen of the Kalahari, who, according to DNA studies, are the oldest human population on Earth. While some people, such as Carl Jung and Erich Neumann, have sought to dismiss the idea of an innocent, Edenic past as nothing more than a nostalgia for the security and maternal warmth of infancy, and certainly **'never an historical state'** (Erich Neumann, *The Origins and History of Consciousness*, 1949, p.15 of 493), the explorer and philosopher Bruce Chatwin bravely rejected this paradise-as-infancy theory and recognized the relative innocence of these 'primitive' 'races', writing that **'Every mythology remembers the innocence of the first state: Adam in the Garden, the peaceful Hyperboreans, the Uttarakurus or "the Men of Perfect Virtue" of the Taoists. Pessimists often interpret the story of the Golden Age as a tendency to turn our backs on the ills of the present, and sigh for the happiness of youth. But nothing in Hesiod's text exceeds the bounds of probability. The real or half-real tribes which hover on the fringe of ancient geographies—Atavantes, Fenni, Parrossits or the dancing Spermatophagi—have their modern equivalents in the Bushman, the Shoshonean, the Eskimo and the Aboriginal'** (*The Songlines*, 1987, p.227 of 325). The great South African philosopher, Sir Laurens van der Post (whom I consider to be the pre-eminent philosopher of the twentieth century) had an incredibly deep appreciation of the Bushmen, and when writing about the effect of our alienated, innocence-destroyed modern world on their relative innocence, he described how **'mere contact with twentieth-century life seemed lethal to the Bushman. He was essentially so innocent and natural a person that he had only to come near us for a sort of radioactive fall-out from our unnatural world to produce a fatal leukaemia in his spirit'** (*The Heart of the Hunter*, 1961, p.111 of 233). As Plato said, in **'this tradition** [of the innocent **earth-born** man], **which is now-a-days often unduly discredited, our ancestors, who were nearest in point of time to the end of the last** [innocent] **period and came into being at the beginning of this** [corrupted period], **are to us the heralds** [of that earlier innocent age]'.

[185] Note again that even in the earlier more innocent times that Plato lived in (recall in chapter 1 that alienation has been increasing from

generation to generation ever since the human condition emerged), there was already a strong desire to **'unduly discredit'** the truth that **'our ancestors'** lived in a pre-human-condition-afflicted, **'innocent'**, **'blessed'**, **'divine'**, **'upright'**, **'cleanly made'**, **'pure'**, **'noble'**, **'good'**, **'modest'**, **'honour**[able]**'**, **'spirit**[ed]**'**, **'simple and calm and happy'** state. Yes, given how *extremely* condemning and confronting the truth of our species' cooperative, all-loving, innocent past has been while we couldn't explain our present corrupted, **'fallen'**, **'evil'**, **'ignoble'**, **'bad'**, **'crooked'**, **'terrible'**, **'unlawful'**, **'insolent'**, **'pride**[ful]**'**, **'lumbering'**, **'disorder**[ly]**'**, **'chaos'**-causing, increasingly **'forget**[ful]**'**, **'deaf'**, threatening **'universal ruin to the world'**, **'imprisoned in'** a **'living tomb'** lives, it is not at all surprising that efforts have been made to **'discredit'** this truth of an innocent ancestry as nothing more than nostalgia for the security of infancy—and by claiming that the Bushmen and other 'primitive' 'races' are more war-like and aggressive and less peaceful than the majority of the human race now. (This latter ridiculous claim that 'advanced' 'races' are, in effect, more innocent than 'primitive' 'races' is dealt with later in pars 205-208, and more fully in pars 862-868.) But, of course, as has been mentioned and will shortly be elaborated upon, the *main* way our innocent past has been denied has been to claim that our distant ancestors were no different from other animals, in ferocious competition with each other for food, shelter, territory and a mate.

[186] Further to the numerous acknowledgments and recognitions of a pre-human-condition-afflicted, all-loving past, <u>some of our greatest contemporary thinkers have also identified the rise of consciousness as being key to understanding and thus resolving our present corrupted, soul-devastated condition</u>. In particular, the just mentioned Sir Laurens van der Post lifted description of the truth of our species' innocent past and the enormous tragedy of our present consciousness-induced psychologically upset, corrupted state into the stratosphere of beautiful writing when he composed these words: **'This shrill, brittle, self-important life of today is by comparison a graveyard where the living are dead and the dead are alive and talking** [through our instinctive self or soul] **in the still, small, clear voice of a love and trust in life that we have for the moment lost...**[there was a time when] **All on earth and in the universe were still members and family of the early race seeking comfort and warmth through the long, cold night before the dawning of individual consciousness in a togetherness which still gnaws like an unappeasable homesickness at the base of the human heart'** (*Testament to the Bushmen*, 1984, pp.127-128 of 176).

[187] The South African naturalist Eugène Marais, the first person to study primates in their natural habitat, also got to the point of our condition being a *psychologically* embattled one when he focused on a conflict between our already established instincts and an emerging consciousness, saying, **'The highest primate, man, is born an instinctive animal. All its behavior for a long period after its birth is dominated by the instinctive mentality…it has no memory, no conception of cause and effect, no consciousness…As the…individual memory slowly emerges, the instinctive soul becomes just as slowly submerged… For a time it is almost as though there were a struggle between the two'** (*The Soul of the Ape*, written between 1916-1936 and published posthumously in 1969, pp.77-79 of 171). And after a lifetime spent hunting the cause of the human condition, the great Hungarian-English polymath Arthur Koestler similarly identified the emergence of consciousness as the catalyst for our condition: **'the brain explosion gave rise to a mentally unbalanced species in which old brain and new brain, emotion and intellect, faith and reason were at loggerheads'** (*Janus: A Summing Up*, 1978, p.10 of 354).

[188] Berdyaev was another who regarded consciousness as an intrinsically important consideration in the quest for self-understanding: **'the distinction between the conscious and the subconscious mind is fundamental to the new psychology'** (*The Destiny of Man*, 1931; tr. N. Duddington, 1960, p.68 of 310). The very great English biologist Charles Darwin also recognized the acknowledgment of the emergence of consciousness as being **'fundamental to the new psychology'**; in fact, he used almost the same words, saying such acknowledgment will mean **'Psychology will be based on a new foundation'**. While it has been noted that Darwin's seminal 1859 book **'*The Origin of Species* contains almost no mention of the human species'** (Robert Wright, *The Moral Animal*, 1994, p.3 of 467), near the end of the final chapter he did write that **'In the distant future I see open fields for far more important researches. Psychology will be based on a new foundation, that of the necessary acquirement of each mental power and capacity by gradation. Light will be thrown on the origin of man and his history'** (*The Origin of Species*, 1859, p.458 of 476). So while Darwin studiously avoided trying to explain human behavior in *The Origin of Species*, he did at least recognize that for **'Light'** to **'be thrown on the origin of man and his history'** and a new meaningful, **'important'** world of understanding to be **'open[ed]'** up, **'Psychology'** will have to **'be based on a new foundation'** **'of the necessary acquirement of each mental power and capacity by gradation'** — that it will need to recognize the involvement of the emergence of our **'mental power'** of consciousness in creating our species' unique **'psycholog[icall]y'** troubled human condition.

Yes, the key to understanding the origin of man's **'psychology'** is to recognize the **'acquirement'** of our **'mental power'** of consciousness.

[189] (With regard to Darwin not addressing the issue of human behavior in *The Origin of Species*—which was a stark omission given the book is titled *The Origin of Species* and the most important species we needed to understand the origin of was ourselves—it is true that 12 years after the publication of *The Origin of Species* Darwin did publish *The Descent of Man, and Selection in Relation to Sex* in which he did make an attempt to look at the origins of our behavior, but it was still only a tentative step in that all-important exploration. As I talk more about later in par. 485, the evidence suggests that while Darwin was honest enough to recognize that trying to address the issue of human behavior meant addressing the issue of the human condition, he apparently didn't feel secure enough to attempt it himself; in fact, when asked why he didn't address human behavior in *The Origin of Species*, Darwin even said, **'I think I shall avoid the whole subject'** (Letter to A.R. Wallace, 22 Dec. 1857; *The Complete Work of Charles Darwin Online*, ed. John van Wyhe, 2002). The Cambridge academic Jane Ellen Harrison recognized Darwin's reticence to deal with issues relating to our human situation when she wrote that Darwin **'foresaw that his doctrine must have, for the history of man's mental evolution, issues wider than those with which he was prepared personally to deal'** ('The Influence of Darwinism on the Study of Religions', *Darwin and Modern Science*, ed. A.C. Seward, 1909, ch.25). But as we are going to see in this chapter, many who purport to be biologists, such as E.O. Wilson, have shown no such scruples about trying to explain human behavior when they couldn't confront the human condition. Indeed, they haven't been interested in explaining human behavior at all, only in using their claimed stature as a biologist to invent a way to *avoid* the issue of the human condition!)

[190] Yes, any truthful analysis of our human condition, its origins and its amelioration requires a **'fundamental'**, **'new'**, honest **'foundation'** in thinking that recognizes the involvement of our conscious mind in our species' departure from an original, cooperative, loving state, as well as the psychosis and neurosis it has produced in us. It *has to* acknowledge that our condition is a **'psycholog[ical]'** one—that our conscious mind is deeply *psychologically* troubled, that we *are* a psychotic and neurotic, immensely alienated species. This is the point <u>Laing</u> was emphasizing when he wrote that **'Our alienation goes to the roots. The realization of this is the essential springboard for any serious reflection on any aspect of present inter-human life.'**

## Chapter 2:8 <u>The ultimate paradox of the human condition</u>

[191] So if we do have a cooperative past (which we do), and our consciousness is behind our destructiveness (which it is), then the *ultimate* paradox of the human condition, indeed the reason we are still able to get out of bed in the morning and face the world, is that, as mentioned in chapter 1:3, we fully conscious humans don't actually believe we are fundamentally bad/evil. Despite all the damning evidence, we don't accept that we are a terrible mistake, a worthless blight on this planet, the **'face of absolute evil'**. In fact, the incredible determination with which we live our lives bespeaks of a core belief within us all that we are not only *not* bad/evil but the great heroes of the story of life on Earth, and that one day we will be able to explain why that is true!

[192] It follows then that while awaiting the exonerating, liberating explanation of our present egocentric, competitive, selfish and aggressive condition we couldn't afford to concede that we did once live in an innocent, all-loving instinctive state that was corrupted by the emergence of our conscious mind. Accepting such truth without the full, clarifying explanation for it would have left us unbearably condemned as bad and worthless, sentenced to a state of completely insecure, permanent damnation. <u>So, as dishonest as it was, there *has* been a need to deny the truth of our species' cooperative, peaceful past and the role consciousness played in its corruption</u>. Yes, while, as has been emphasized, denial in science carried with it the great risk of becoming overly entrenched, rendering its practitioners incapable of fulfilling their fundamental responsibility of acknowledging the human-condition-confronting-not-denying, fully accountable and human-race-saving, true explanation of the human condition when it eventually arrived, it *has* been necessary in the interim, not just because we couldn't cope with the depression that resulted from trying to think about the human condition, but because we didn't believe we were fundamentally bad/flawed/**'the face of absolute evil'**.

[193] Having explained and evidenced the fundamental elements involved in a *true* analysis of human behavior, I will now describe how this denial in science led to the development of a litany of human-condition-*avoiding*, *dishonest*, *not-truly-accountable* biological theories on human behavior—all of which will be exposed for the lies they are, including the most prominent and seductive offering of all, that which has been put forward by E.O. Wilson.

## Chapter 2:9 <u>Social Darwinism</u>

[194] Prior to the development of science and its evasive offerings, humans had already found a way to avoid the condemning truths of a cooperative, all-loving, innocent past and of a consciousness-induced 'fall from grace', which was to simply assert that nature is brutally competitive and aggressive—**'red in tooth and claw'**, as Tennyson put it (*In Memoriam*, 1850)— and that's why we are. Basically, we looked around and saw that animals always appear to be fighting and competing with each other and instead of acknowledging that our instinctive orientation is to be cooperative and all-loving, we said that our instincts are similarly ruthlessly competitive and aggressive. <u>We said that we have brutal, savage animal instincts that our conscious mind has to somehow try to control. As was mentioned in par. 153, it was an absolutely brilliant excuse, because instead of our instincts being all-loving and thus unbearably condemning of our present non-loving state, they were made out to be vicious and brutal; *and*, instead of our conscious mind being the villain, the cause of our corruption, the insecurity of which made us repress our instinctive self or soul or psyche and become psychotic, it was made out to be the blameless, psychosis-free mediating 'hero' that had to manage those supposed vicious instincts within us! It was all a terrible reverse-of-the-truth lie, but a hugely relieving one for humans seeking relief from the human condition.</u>

[195] What happened when Charles Darwin presented his idea of natural selection in his momentous book *On the Origin of Species by Means of Natural Selection* was that the excuse that claimed we have 'savage', 'barbaric', 'backward', 'brutish', 'bestial', 'primitive' animal instincts within us was supposedly given a biological basis through the misrepresentation of natural selection as a 'survival of the fittest' process. Natural selection is the process by which some members of a population reproduce more than others in a given environment, and, most significantly, in the first edition of *The Origin of Species* Darwin left it undecided as to whether those individuals that reproduced more could be viewed as winners, as being 'fitter'. However, in later editions Darwin's associates, Herbert Spencer and Alfred Russel Wallace, persuaded him to substitute the term 'natural selection' with the term 'survival of the fittest' (Letter from Wallace to Darwin, 2 Jul. 1866; *The Correspondence of Charles Darwin*, Vol.14, p.227 of 706). While Darwin's friend and staunch defender, the biologist Thomas Huxley described the term 'survival of the fittest' as an **'unlucky substitution'** (1890; *Life and*

*Letters of Thomas Henry Huxley Vol.3*, ed. Leonard Huxley, 1903, ch. 3.7), from the point of view of humanity needing to contrive an excuse for its divisive selfish, competitive and aggressive behavior it was a *lucky* substitution because it reinforced the dishonest but human-condition-relieving argument that our instincts are competitive and selfish and that we, in the sense of 'we' being our conscious thinking self, are blameless. (I should mention that later in chapter 4 in par. 358 it will be explained that Darwin's original position, where he left it undecided as to whether those who reproduced more are 'fitter', was right because being unconditionally selfless, where you give your life to help others and don't seek to reproduce more, *can* be a biologically meaningful—'fitter'—outcome.)

## Chapter 2:10 <u>Sociobiology/Evolutionary Psychology</u>

[196] There were, of course, serious problems with this so-called Social Darwinist contrived excuse that 'nature is selfish and that's why we are'. For starters, it didn't account for instances in nature where self-lessness occurs, such as in ant and bee colonies where workers slave selflessly for the whole colony. And secondly, and most particularly, it didn't account for *our* instinctive memory of having lived in a co-operative, loving, 'Garden of Eden'-like existence, which is our selfless, consider-the-welfare-of-others, born-with, instinctive moral nature, the 'voice' of which is our 'conscience'.

[197] Seeking to address these cracks in the argument, biologists developed the theory of Sociobiology, with E.O. Wilson acting as its main proponent. Later known as Evolutionary Psychology, this theory explains, truthfully enough, that worker ants and bees are *not* actually being unconditionally selfless, truly altruistic when serving their colony because, when doing so, they are fostering the queen who reproduces their genes, which means their apparent selfless behavior is, in fact, just a subtle form of selfishness: they are helping the queen to selfishly reproduce their genes. But in terms of maintaining the primary agenda of avoiding the unbearable and unacceptable issue of the human condition at all costs, this idea of selfless behavior actually being a subtle form of selfishness, where you indirectly promote the reproduction of your own genes by fostering others who are related to you—your kin—was then not surprisingly, but *in this case* extremely dishonestly, commandeered to explain *our* moral

instincts. Yes, it was claimed that *our* moral inclination to help others was no more than an attempt to reproduce our genes by supporting others whose genes we shared, with any anomalies put down to **'misplaced parental behavior'** (George Williams, *Adaptation and Natural Selection*, 1966, p.vii of 307)! As Wilson boldly summarized, **'Morality has no other demonstrable function'** other than to ensure **'human genetic material...will be kept intact'** (*On Human Nature*, 1978, p.167 of 260); even saying that **'Rousseau claimed** [that humanity] **was originally a race of noble savages in a peaceful state of nature, who were later corrupted...**[but what] **Rousseau invented** [was] **a stunningly inaccurate form of anthropology'** (*Consilience*, 1998, p.37 of 374)!!

[198] So, in saying our moral soul is still basically selfish, the old 'nature is selfish and that's why we are selfish' excuse was preserved; the same 'I'm going to determinedly avoid, not confront, the human condition' attitude had been upheld.

[199] The problem that *then* emerged, of course, was that this denigration of our moral self as nothing more than a subtle form of selfishness was both deeply offensive to and entirely inconsistent with what we all in truth know about our moral instincts, which is that they *are* unconditionally selfless, genuinely altruistic. As the journalist Bryan Appleyard pointed out, biologists **'still have a gaping hole in an attempt to explain altruism. If, for example, I help a blind man cross the street, it is plainly unlikely that I am being prompted to do this because he is a close relation and bears my genes. And the world is full of all sorts of elaborate forms of cooperation which extend far beyond the boundaries of mere relatedness'** (*Brave New Worlds: Staying Human in a Genetic Future*, 1998, p.112 of 198).

## Chapter 2:11 <u>Multilevel Selection theory for eusociality</u>

[200] This **'gaping hole'** in the theory of Sociobiology/Evolutionary Psychology brings us to the present and the publication in 2012 of E.O. Wilson's *The Social Conquest of Earth*. Yes, once again, it was Wilson who concocted a 'solution' to this problem of the offensiveness of Evolutionary Psychology's denigration of our moral instincts as selfish. Now, to the dismay of his earlier supporters, he has dismissed 'his' previous Evolutionary Psychology theory as **'incorrect'**, **'inoperable'** and as having **'failed'** (*The Social Conquest of Earth*, pp.143, 180, 181 of 330), proffering in its place a new theory that not only contrives an explanation for our

genuinely moral instincts, but takes the art of denial to the absolute extreme by contriving a non-human-condition-confronting explanation of the human condition itself!

[201] Known as Multilevel Selection or **'a New Theory of Eusociality'** (ibid. p.183) (eusociality simply meaning genuine sociality), this theory claims that humans have instincts derived from natural selection operating at the individual level (where members of a species selfishly compete for food, shelter, territory and a mate), *and* instincts derived from natural selection operating at the group level (where, it is claimed, groups of altruistic, cooperative members outcompete groups of selfish, non-cooperative members)—with the supposed selfish individual level instincts being the bad/sinful aspects of our nature, and the selfless, supposed group-selected instincts being the good/virtuous aspect of our nature. According to Wilson, **'Individual selection is responsible for much of what we call sin, while group selection is responsible for the greater part of virtue. Together they have created the conflict between the poorer and the better angels of our nature'** (ibid. p.241). In summary, Wilson asserts that **'The dilemma of good and evil was created by multilevel selection'** (ibid).

[202] Before looking at the way in which the Multilevel Selection theory for eusociality misrepresents, in fact, *avoids*, the real, consciousness-derived, psychological aspect of the human condition, we need to look at the 'group selection' mechanism that Wilson says accounts for our moral sense—because while we certainly *do* have a genuine moral sense, under scrutiny Wilson's theory of how we acquired it completely falls apart.

[203] While it makes sense that, as Wilson states, **'groups of altruists** [will] **beat groups of selfish individuals'** (ibid. p.243), the biological stumbling block is whether genes, which have to selfishly ensure they reproduce if they are to carry on, can develop self-sacrificing altruistic traits in the first place. (Indeed, the initial premise of group selection makes so much sense that even Darwin canvassed the idea, but with far less arrogance than Wilson, aware as he was of the inherent difficulties of the concept. With regard to Darwin's tentative approach to group selection, one of the leading evolutionary theorists of the twentieth century, William Hamilton, said that **'Darwin had gone** [there] **circumspectly or not at all'** ('Innate Social Aptitudes of Man: An Approach from Evolutionary Genetics', *Biosocial Anthropology*, ed. R. Fox, 1975, p.135 of 169). I describe Darwin's flirtation with group selection in *Freedom Expanded*

at <www.humancondition.com/freedom-expanded-multilevel-selection>.) To
reiterate, while it *is* true that a group of altruists whose members are
prepared to make sacrifices for each other *will* defeat a group whose
members are concerned only for themselves, *the question is whether a
group of altruists can ever actually form in the first place?* The genetic
reality is that whenever an unconditionally selfless, altruistic trait appears
those that are selfish will naturally take advantage of it: 'Sure, you can
help me reproduce my genes but I'm not about to help you reproduce
yours!' Any selflessness that might arise through group selection will be
constantly exploited by individual selfishness from *within* the group; as
the biologist Jerry Coyne pointed out, **'group selection for altruism would
be unlikely to override the tendency of each group to quickly lose its altruism
through natural selection favoring cheaters** [selfish individuals]**'** ('Can Darwinism
improve Binghamton?', *The New York Times*, 9 Sep. 2011). The only biological models
that have been presented that appear to overcome this problem of genetic
selfishness always prevailing are so complex and convoluted they seem
highly implausible, in that they involve the disbanding of a population
into new, separate colonies, formed by solitary fertilized females, some
of whom only have selfish genes and some of whom have altruistic genes,
with those altruistic colonies out-competing those with just selfish genes
to build larger, more altruistic populations. Then, before the colonies with
altruistic genes **'quickly lose...**[their] **altruism through natural selection favoring
cheaters'**, the colonies peacefully merge back into one population, after
which fertilized females separate out again to breed new, isolated groups
(and so on). Essentially the model requires a process of constant merging
and disbanding in order to 'outrun' the genetic imperative in nature to
exploit altruism or selflessness. The situations where this between-group
selection of unconditionally selfless traits is said to have taken place are
in the occurrence of female-biased sex ratios in some small invertebrate
species, and in the evolution of reduced virulence in some disease organisms (see David Sloan Wilson & Elliot Sober, *Unto Others: The Evolution and Psychology of
Unselfish Behavior*, 1998, pp.35-50 of 394). However, for large mammals especially,
who don't have complex life cycles, the mechanism is so implausible it
has to be considered impossible.

[204] Nevertheless, in defiance of the biological reality that, even where
there is selection between groups, unconditionally selfless traits will
be exploited and eliminated, Wilson proposes that extreme warring
between groups of early humans where cooperation would have been

an advantage was a strong enough force to overcome this problem of selfish exploitation and thus allow for the selection of altruism and the emergence of our genuinely moral instincts! So, according to Wilson, our ability to war successfully somehow produced our ability to love unconditionally!

[205] Wilson's theory not only defies biological reality, it also flies in the face of both our cultural memories and anthropological evidence. As has been emphasized, standing in stark contrast to Wilson's proclamation of **'universal and eternal'** warfare *(The Social Conquest of Earth*, p.65) are the cultural memories enshrined in our myths, religions and in the words of some of our greatest thinkers that attest to humans having a peaceful heritage (recall, for instance, Plato's description of how our distant ancestors lived a **'blessed' 'life'** where **'neither was there any violence, or devouring of one another, or war or quarrel among them'**), and in the evidence gleaned from studies in anthropology and primatology—such as the aforementioned recent fossil discoveries that are now confirming a cooperative past, and studies of bonobos, who are our species' closest living relatives and extraordinarily peaceful.

[206] But, in attempting to prove we have a warlike past, Wilson ignores the evidence provided by the fossil record that reveals at least 7 million years of cooperative existence and instead argues that **'to test the prevalence of violent group conflict in deep human history** [one can look at]**...archaic cultures** [such as]**...the aboriginals of Little Andaman Island off the east coast of India, the Mbuti Pygmies of Central Africa, and the !Kung Bushmen of southern Africa. All today, or at least within historical memory, have exhibited aggressive territorial behavior'** (ibid. pp.69, 71). In addition, in his 1978 book *On Human Nature* Wilson wrote of the Bushmen that **'their homicide rate per capita equalled that of Detroit and Houston'** (p.100 of 260). However, there is ample evidence for just how extraordinarily cooperative, social and relatively peaceful the Bushmen are. For example, Lorna Marshall, regarded as **'the doyenne of American anthropology'** (Sandy Gall, *The Bushmen of Southern Africa: Slaughter of the Innocent*, 2001, ch.10) and one of the only Westerners to live with the Bushmen before they became contaminated through contact with more upset-adapted, alienated 'races', described **'their predominantly peaceful, well-adjusted human relations'** (*The !Kung of Nyae Nyae*, 1976, p.286 of 433). Marshall's daughter, Elizabeth Marshall Thomas, who accompanied her on her expeditions in the 1950s, wrote the classic 1959 book about the Bushmen, *The Harmless People*. In a 1989 addition to

that book, Marshall Thomas wrote: **'To my knowledge Wilson has never visited the Ju/wasi** [Bushmen]. **His book** [*On Human Nature*] **never mentions how important it was to them to keep their social balance, how carefully they treated this balance, and how successful they were. That he discusses them at all is perhaps due to the fact that in the 1970s they were selected by academics as a sort of living laboratory in which studies could be made on attributes of human nature, the most intriguing of which at the time seemed to be aggression'** (p.283 of 303). (More will be said in pars 862-868 about mechanistic science's misrepresentation of Bushmen and other so-called 'primitive' societies as **'violent'** and **'aggressive'** in order to comply with the human-condition-avoiding excuse that we are competitive, aggressive and selfish because of our animal heritage.)

[207] Wilson also cites recent archaeological evidence to support **'the prevalence of violent group conflict in deep human history'**, stating that **'Early humans had the innate equipment—and likely the tendency also—to use projectiles in capturing prey and repelling enemies. The advantages gained were surely decisive. Spear points and arrowheads are among the earliest artifacts found in archaeological sites'** (*The Social Conquest of Earth*, p.29). But this 'data' proves equally unreliable, as the archaeologist Steven Mithen has noted: **'No, the earliest artifacts are from around 2.5 million years ago, but spear points are not made until a mere 250,000 years ago and arrowheads might have first been manufactured no longer ago than 20,000 years'** ('How Fit Is E.O. Wilson's Evolution?', *The New York Review of Books*, 21 Jun. 2012). And in response to Wilson's claim that **'Archaeologists have found burials of massacred people to be a commonplace'** and **'archaeological sites are strewn with the evidence of mass conflict'**, Mithen argues that **'No, both are quite rare, especially in pre-state societies, and those that are known are difficult to interpret'** (ibid).

[208] On primatology, in an attempt to dismiss the example bonobos present of a cooperative, peaceful heritage, Wilson suggests there is no difference between bonobos and the more aggressive, selfish chimpanzee, claiming, for instance, that like chimpanzees, bonobos **'do not share the fruit they pick'** (*The Social Conquest of Earth*, p.42), despite the fact there is a wealth of data recording instances of bonobos sharing fruit—such as a report by Barbara Fruth and Gottfried Hohmann, included in 2002's influential *Behavioural Diversity in Chimpanzees and Bonobos*, which states: **'In bonobos (*Pan paniscus*), food sharing between mature individuals is common... [and] bonobos often divide large-size fruits'** ('How bonobos handle hunts and harvests: why share food?', *Behavioural Diversity in Chimpanzees and Bonobos*, eds C. Boesch et al., 2002,

pp.231-232 of 285). Wilson also claims bonobos **'hunt in coordinated packs in the same manner as chimpanzees' 'wolves and African wild dogs'** (*The Social Conquest of Earth*, p.32). While bonobos have been known to capture and eat small game, including small monkeys, to supplement their diet with protein, they are not known to routinely hunt down and eat large animals such as colobus monkeys, like chimpanzees do, with **'hunting behavior** [by bonobos] **very rare'** (Tetsuya Sakamaki quoted by David Quammen, 'The Left Bank Ape', *National Geographic*, Mar. 2013). Wilson relies upon a 2008 report by Hohmann titled 'Primate hunting by bonobos at LuiKotale, Salonga National Park' to draw these erroneous comparisons with the more aggressive chimpanzees—but Hohmann's report actually reveals extraordinary *differences* between bonobos and chimpanzees, recording that **'at the Lilungu site, bonobos catch guenons and colobus monkeys but do not eat them, and at Wamba, bonobos and red colobus monkeys have been seen to engage in mutual grooming'** (M. Surbeck & G. Hohmann, *Current Biology*, 2008, Vol.18, No.19). A brief look at the papers Hohmann, in turn, cites, reveals that at Lilungu **'the bonobos interacted with the captured primates as if they were dealing with individuals of their own species. They sought cooperation in their interaction with the captured young primates without success. There is no evidence that they ate the captives...this interactional behavior...satisf[ies] the ethological definition of play'** (J. Sabater Pi et al., 'Behaviour of Bonobos (*Pan paniscus*) Following Their Capture of Monkeys in Zaire', *International Journal of Primatology*, 1993, Vol.14, No.5); and that at Wamba, despite **'red colobus** [being] **major hunting targets of common chimpanzees...there is little evidence of hunting by the pygmy chimpanzees** [bonobos] **of Wamba, despite the fact that they have been intensively studied for over ten years'** (Hiroshi Ihobe, 'Interspecific Interactions Between Wild Pygmy Chimpanzees (*Pan paniscus*) and Red Colobus (*Colobus badius*)', *Primates*, 1990, Vol.31, No.1). So it turns out that Hohmann's 2008 paper, which Wilson relies so heavily upon to depict bonobos as chimpanzee-like, ruthless killers, contains reports of bonobo behavior that would be unthinkable from a chimpanzee (let alone from wolves or wild dogs) but which Wilson has just ignored! 'That's one more problem out of the way', he seems to be saying. It would appear that just as the rest of the insecure, human-condition-afflicted human race has had to practice the art of denial of the human condition, Wilson too has had to fudge the evidence to try to find support for his lies.

[209] In summary, our moral instincts are *not* derived from warring with other groups of humans as Wilson and his theory of group selection would have us believe. No, as will be explained in chapter 5, humans

have an *unconditionally* selfless, *fully* altruistic, *all*-loving, *universally-benevolent*-not-competitive-with-other-groups, moral conscience. Our instinctive orientation is to love *all* people, not love some and be at war with others. The 'savage-instincts-in-us' excuse for our selfish behavior is entirely inconsistent with the fact that we have *completely* moral, NOT partially moral and partially savage, instincts.

[210] Overall then, while the Multilevel Selection theory for eusociality adds unconditionally selfless instincts to selfish instincts in the mix of what allegedly forms our species' instinctive make-up (thus countering Evolutionary Psychology's offensive denial of the fact that we have unconditionally selfless instincts), the same old reverse-of-the-truth, escape-rather-than-confront-the-human-condition agenda—that we have villainous selfish instincts and a blameless conscious mind that has to 'step-in' to control them—continues. (It should be mentioned that there has been an attempt to counter the selfishness-emphasizing-and-justifying biological theories that have been described in this chapter with cooperation-not-competition, selflessness-not-selfishness emphasizing biological theories. A summary of these 'left-wing' theories, as put forward by scientists such as Stephen Jay Gould, David Sloan Wilson and Robert Sussman—which all avoid the human condition just as ardently as these 'right-wing' theories and are therefore just as, if not more, dishonest, false and unaccountable—is provided in chapter 6:9, 'A brief history of left-wing dishonest mechanistic biology'. I should also mention here that the real reason for, and consequences of, genes having to selfishly ensure their own reproduction—which is the genetic reality that all the 'right-wing' and 'left-wing' biological theories have had to accommodate—will be explained when the integration of matter is described later in chapters 4:4 and 4:5.)

[211] To look now at how Wilson's Multilevel Selection theory for eusociality avoids the real, consciousness-derived-and-induced *psychological* aspect of our human condition.

[212] If our instincts *are* wholly peaceful and cooperative (which they are), and we are *not* selfish because of selfish instincts (which we are not), what is the source of our selfishness—or what Wilson calls our propensity for evil? The honest, human-condition-confronting answer is that it is the result of a *psychosis*.

[213] As shown previously, our human behavior involves our unique fully conscious thinking mind. As I have emphasized, descriptions of our less-than-ideal condition, such as egocentric, arrogant, deluded, artificial, hateful, cynical, mean, immoral, alienated, etc, all imply a consciousness-derived, *psychological* dimension to our behavior. We suffer from a consciousness-induced, psychological HUMAN CONDITION, *not* an instinct-controlled ANIMAL CONDITION. And so it is to this psychological dimension to our behavior that we should look for the cause of our selfishness. And yet in Wilson's psychological-problem-avoiding model our consciousness is merely a *mediator* between supposed selfish and selfless instincts—as he writes: **'Multilevel selection (group and individual selection combined) also explains the conflicted nature of motivations. Every normal person feels the pull of conscience, of heroism against cowardice, of truth against deception, of commitment against withdrawal. It is our fate to be tormented...We, all of us, live out our lives in conflict and contention'** (*The Social Conquest of Earth*, p.290). Clever semblance of our conflicted condition, diabolically clever, but entirely untrue, the epitome of shonk/evasion/denial/'**phon**[iness]'/'**fake**[ness]'/separation-from-the-truth—alienation!

[214] In finding a way to avoid the truth of our psychologically conflicted condition with a non-psychological 'clever semblance' of it, what Wilson has done is *not* explain the human condition but nullify it, render the issue benign, virtually inconsequential—and in doing so he is effectively burying humanity into the deepest, darkest, '**underground**' depths of the '**living tomb**' of Plato's cave of denial. Make no mistake, Wilson's great '**phony**', '**fake**', superficial, not-genuinely-biological, '**Darwin's**'-'**heir**'-be-damned, deliberately-human-condition-trivializing account of the human condition is the most sophisticated expression of denial to have ever been invented—and thus the most dangerous. Certainly, providing humans with a 'get out of jail free' card, a way to supposedly explain the human condition *without* having to confront the issue of the extreme psychosis (psyche/soul repression) and neurosis (neuron/mental denial) of our real human condition, is immensely appealing to the now overly psychologically upset human race—but it is precisely that seductiveness that is *so* dangerous. (Already school children are talking about the human condition in an off-hand way, with a teacher reporting in 2013 that one of her students had remarked that **'I love the term The Human Condition; I can use it in just about any essay for any subject'** (WTM records, 15 Feb. 2013).) This Ultimate Lie had the potential to seduce the exhausted human race to

such a degree that it obliterated any chance of the real human condition ever being truthfully confronted and thus understood! While denial was necessary while we couldn't explain ourselves, taking the art of denial to the extreme that Wilson has done with his dismissal of the fundamental issue before us as a species of our human condition as nothing more than two different instincts within us that are sometimes at odds, is a truly sinister—in fact, unconscionable—lie. It is the ultimate tragic expression of the human-condition-avoiding, superficializing, dumbing-down, dogma-not-knowledge-preferring, madness-becomes-universal, end play situation the human race is now in—the time Plato prophesized where there would be **'more and more forgetting** [dishonest denial to the point where]**...there was a danger of universal ruin to the world'**. Yes, it is nothing less than the final great push to have the world of lies with all its darkness take over the world—and condemn humanity to extinction. If the real psychosis-addressing-and-solving explanation of the human condition that is presented here in this book hadn't emerged then this Ultimate Lie would have given humanity's headlong march to ever greater levels of lying denial and its terrible alienating effects its terminal impetus.

[215] I should add that despite Wilson's hateful dismissal of religion as mere group propaganda—as nothing more than **'an expression of tribalism'** (*The Social Conquest of Earth*, p.258) that is **'dragging us down'** and that **'for the sake of human progress, the best thing we could possibly do would be to diminish, to the point of eliminating, religious faith'** ('Don't let Earth's tapestry unravel', *New Scientist*, 24 Jan. 2015)—the fact is, religions resonate with us, and have done for millennia, because they contain profound truth. Contrast Wilson's dangerously superficial account of our condition, where our consciousness is the blameless mediator or manager of villainous selfish instincts within us, with Moses' Garden of Eden account of the origin or genesis of the human condition or Plato's two-horsed chariot and 'reversed cosmos' accounts (see ch. 2:6), all of which say that our instinctive heritage is wholly selfless and that it was the emergence of consciousness that led to our selfishness. *These* are the accounts that acknowledge the problematic role of our conscious mind, which is where the real terror of the human condition lies—our deep insecurity about whether 'we', our conscious thinking self, is actually evil. And it is *these* accounts that have been reinforced by some of history's most profound thinkers—like van der Post, Marais, Koestler and Berdyaev, whose words were included earlier (in pars 186-188).

[216]It is worth including here a review by the journalist Christopher Booker of Wilson's *The Social Conquest of Earth*. While I don't agree at all with Booker's assertion that Darwin's theory of natural selection is a flawed and deficient theory that **'can't explain'** how all of life developed (the explanation of how it can and does is presented in subsequent chapters of this book), in every other respect what he has to say about Wilson's book is extraordinarily honest—in particular that **'what Wilson completely misses out is any recognition of what is by far the most glaring difference between humans and ants...we have broken free from the dictates of instinct...that peculiarity of human consciousness...has allowed us to step outside the instinctive frame...But it is this which also gives us our disintegrative propensity, individually and collectively, to behave egocentrically, presenting us with all those problems which distinguish us from all the other species which still live in unthinking obedience to the dictates of nature. All these follow from that split from our selfless 'higher nature', with which over the millennia our customs, laws, religion and artistic creativity have tried their best to re-integrate us'** ('E.O. Wilson has a new explanation for consciousness, art & religion. Is it credible?', *The Spectator*, 7 Sep. 2013).

[217]Yes, in summary, Wilson continues to look everywhere for the cause of our condition, except to our consciousness; and instead of explaining our psychosis (**'all those problems which distinguish us from all the other species'**, as Booker referred to it) he simply states that it does not exist. Like most resigned, human-condition-avoiding humans (and other mechanistic scientists) he is effectively saying, 'What psychosis? What inner insecurity? What sense of guilt? What original 'Golden Age' of innocence? What **'split from our selfless "higher nature"'**? What **'fallen'** condition? What 'haunted existence dogged by the shadow of our human condition'? What deeply troubled state? What depression from the **'mountains'** of the **'mind['s]'** **'cliffs of fall, frightful, sheer'**? What agonising issue of the human condition that we, as teenagers, had to learn to **'let it be'**? What cave-like state of alienated denial that I'm now living in? What **'phony'**, **'fake'**, **'alienat[ed]...to the roots'** existence? What great elephant in the living room of our lives that we can't acknowledge? What sickness of the soul; for that matter, what 'soul'? What dreamed-of psychologically rehabilitated, transformed human race? What wonderful time when **'we'll all live as one'** that John Lennon **'imagine[d]'** (*Imagine*, 1971)?' To Wilson, the human condition is nothing more than a conflict between selfless and selfish instincts within us. He is basically saying, 'To hell with your

psychological garbage, I'm not going there!' So, while he might have won almost every accolade in science, in the end Wilson has revealed himself to be just another victim of the human condition—coping with it by finding a way to deny it.

[218]There is one last but very important aspect of Wilson's account that needs to be addressed, which is that his notion of our condition being a result of selfish and selfless instincts within us would mean that unless we change our genes we are, as he asserts, **'intrinsically imperfectible'** (*The Social Conquest of Earth*, p.241). BUT such a fate is completely inconsistent with one of the central beliefs about the real, *psychological* nature of our condition, which is that finding understanding of it will bring about the psychologically ameliorated transformation of the human race. If, as Wilson maintains, we don't suffer from a psychosis, then we can't be healed—but we *do* suffer from a psychosis, which *can* be healed. Carl Jung was forever asserting that **'wholeness for humans depends on the ability to own their own shadow'** because he knew that finding the psychologically ameliorating understanding of the dark side of ourselves would make us **'whole'**. In religious terms, *The Lord's Prayer* contains the hope of the time when **'Your** [the Godly, ideal, peaceful] **kingdom come, your will be done on earth as it is in heaven'** (Bible, Matt. 6:10 & Luke 11:2). In contemporary mythology, the same sentiment is conveyed in the words of our modern day, truthful thinking prophets: in John Lennon's **'imagin[ings]'**; in Bob Dylan's anticipation of **'when the ship** [understanding of the human condition] **comes in...and the morning will be a-breaking...and the** [dishonest] **words that are used to get the ship confused will no longer be understood as the spoken** [truth]' (*When The Ship Comes In*, 1963); in Jim Morrison, of The Doors, singing of **'Standing there on freedom's shore, waiting for the sun...waiting... to tell me what went wrong'** (*Waiting for the Sun*, 1968), waiting for the liberating light of understanding of our upset lives to arrive because when **'day destroys night'** we can **'break on through to the other side'** (*Break on Through*, 1966) to our **'freedom'** from the agony of the human condition; and, finally, in Bono's (of the band U2) lyrics about the coming of a world **'high on a** [uncorrupted] **desert plain'** where there will be no more egocentricity and **'the streets** [will] **have no name'** and **'there will be no toil or sorrow, then there will be no time of pain'** (*Where The Streets Have No Name*, 1987). This last prophetic vision is exactly the same as that expressed in the Bible where it states that **'Another** [denial-free, honest, all-clarifying] **book** [will be]**...opened which is**

the book of life...[which will introduce] **a new heaven and a new earth...**[and] **wipe every tear from...**[our] **eyes. There will be no more death or mourning or crying or pain, for the old order of things has passed away'** (Rev. 20:12, 21:1, 4). Buddhist scripture contains the exact same anticipation of a fabulous time when humans **'will with a perfect voice preach the true Dharma** [present the supreme wisdom, namely the psychologically rehabilitating, transforming, true understanding of the human condition], **which is auspicious and removes all ill'**, saying, **'Human beings are then without any blemishes, moral offences are unknown among them, and they are full of zest and joy'** (Maitreyavyakarana; *Buddhist Scriptures*, tr. Edward Conze, 1959, pp.238-242).

[219] Yes, the fulfillment of Holden Caulfield's dream of a time when the need for Resignation, with all its horrible alienating, psychosis-and-neurosis-producing effects—that Francis Bacon so honestly depicted—*will end* is what the human race *has* tirelessly been working towards, and has now, in the nick of time, finally achieved. We are precisely the opposite of what Wilson argues, because we are, in fact, **'intrinsically'** **'perfectible'**!

## Chapter 2:12 While denial has been necessary, you can't find the truth with lies

[220] In concluding this chapter, I need to emphasize that Wilson, as the quintessential exponent of reductionist, mechanistic science, is only doing what *all* mechanistic scientists have been doing—and, indeed, what virtually *all* resigned humans have been doing—which is avoiding the issue of the human condition at all costs. But when it comes to finding understanding of the human condition, the costs of such evasion *are* great indeed because, clearly, if you're committed to living in denial of the human condition you are in no position to ever find understanding of it. When Laing said, **'Our alienation goes to the roots. The realization of this is the essential springboard for any serious reflection on any aspect of present inter-human life'**, he was making the fundamental point that for there to be **'any serious reflection on' 'human life'** the truth of our alienated, psychotic and neurotic, human-condition-afflicted state *had* to be **'realiz**[ed]**'**/ confronted. You can't find the truth from a position of lying—a case the aforementioned philosopher Arthur Schopenhauer was making when he wrote that **'The discovery of truth is prevented most effectively...by prejudice, which...stands in the path of truth and is then like a contrary wind driving a ship**

**away from land'** (*Essays and Aphorisms*, tr. R.J. Hollingdale, 1970, p.120 of 237). As Martin Luther King Jr once said, **'Darkness cannot drive out darkness; only light can do that. Hate** [relevantly here, hate/fear of the issue of the human condition] **cannot drive out hate; only love** [relevantly here, love/tolerance of the issue of the human condition] **can do that'** ('Loving Your Enemies' sermon, Christmas 1957).

[221] The psychologist Arthur Janov gave this deadly accurate description of the psychological basis of resigned, human-condition-avoiding, neurotic, alienated, ineffective, mechanistic thinking when he wrote that (underlinings are my emphasis) **'As the child becomes split by his Pain** [caused by his particular encounter with the human condition], **he will develop philosophies and attitudes commensurate with his denials. He will have a warped view of the world…Thus, intellect becomes the mental process of repression… We can understand now why it is so difficult to change a neurotic's ideas with facts, reasoning, or even counselling. He needs his ideological padding, and he will incorporate into it whatever he needs to strengthen it** [this explains the 'deaf effect' resistance to discussion of the human condition described in ch. 1:4]**… The more reality a person is forced to hide in his youth, the more likely it will be that certain areas of thinking will be unreal. That is, it is more likely that thought process will be constricted so that generalised extrapolations cannot be made about the nature of life and the world. Conversely, to be free to articulate one's feelings while growing up will lead to becoming an articulate, free-thinking person, unhampered by fear, which paralyses thought…A young child can split from his feelings** [from his **Pain**, the human condition] **and learn every aspect of engineering. He can be a "smart" engineer or scientist…His intellect is something apart from his feelings…Neurotic intellect is an order superimposed on reality… Neurotic intellect is subject to indoctrination and brainwashing—because neurosis** [blocking out] *is* **brainwashing. So long as the neurotic has lost his full internal frame of reference, his mind can be swayed by false ideas and inaccurate systems. So long as he is neurotic, his judgment will be poor…He is truly a specialised man, living in his head because his body** [where his feelings/pain/hurt soul lives] **is out of touch and reach. He will deal with each piece of news he hears as an isolated event, unable to assemble what he sees and hears into an integrated view. Life for him is a series of discrete events, unconnected, without rhyme or true meaning'** (*The Primal Revolution*, 1972, pp.158-160 of 246).

[222] Yes, once you are resigned to living in denial of the human condition, you are in no position to think truthfully and thus effectively—as the aforementioned, painfully honest poem of the resigning adolescent lamented, **'you spend the rest of life trying to find the meaning of life and**

**confused in its maze'.** This sense of confusion was something Plato also acknowledged when he too wrote about the consequences of **'unconnected'**, **'true meaning'**-blocked, **'warped'**, **'paralyse[d]'**, soul-and-truth-denying, mechanistic, reductionist *intellectualism*—that **'when the soul uses the instrumentality of the body** [uses the body's intellect with its preoccupation with denial] **for any inquiry...it is drawn away by the body into the realm of the variable, and loses its way and becomes confused and dizzy, as though it were fuddled** [drunk]' *(Phaedo*, c.360 BC; tr. H. Tredennick, 1954, 79). And of course in his cave allegory (see par. 83), the inimitable Plato described how being unable to face the **'sun[lit]'** true world that **'makes the things we see visible'** meant that humans could only **'see dimly and appear to be almost blind'**. Indeed, as mentioned in par. 172, Plato described how in the situation of the human race as a whole, the practice of **'more and more forgetting** [denial]' only leads to **'discord'** and **'disorder' 'burst[ing] forth'** and eventually **'universal ruin'**. Similar to Plato's concerns are those expressed by the Templeton Prize-winning physicist Paul Davies who recognized the stultifying, **'confused'**, **'fuddled'**, **'paralyse[d]'**, **'blind[ing]'**, **'ruin[ing]'**, **'more and more forgetting'**, *alienating* effect of being **'driv[en]' 'away from'** the **'truth'** in his observation that **'For 300 years science has been dominated by extremely mechanistic thinking. According to this** [whole-view-evading, human-condition-psychosis-avoiding, mechanisms-only-focused] **view of the world all physical systems are regarded as basically machines...I have little doubt that much of the alienation and demoralisation that people feel in our so-called scientific age stems from the bleak sterility of mechanistic thought'** ('Living in a non-material world—the new scientific consciousness', *The Australian*, 9 Oct. 1991).

[223] Arthur Koestler was another frustrated by mechanistic, reductionist science's avoidance of our species' consciousness-induced, human-condition-afflicted psychosis, writing that **'symptoms of the mental disorder which appears to be endemic in our species...are specifically and uniquely human, and not found in any other species. Thus it seems only logical that our search for explanations** [of human behavior] **should also concentrate primarily on those attributes of *homo sapiens* which are exclusively human and not shared by the rest of the animal kingdom. But however obvious this conclusion may seem, it runs counter to the prevailing reductionist trend. "Reductionism" is the philosophical belief that all human activities can be "reduced" to – i.e., explained by – the** [non-psychosis involved] **behavioural responses of lower animals – Pavlov's dogs, Skinner's rats and pigeons, Lorenz's greylag geese, Morris's hairless apes...That is why the scientific establishment has so pitifully failed to define the**

**predicament of man'** (*Janus: A Summing Up*, 1978, p.19 of 354). Like Davies, Koestler complained too of **'the sterile deserts of reductionist philosophy'**, making the fundamental point that **'a correct diagnosis of the condition of man** [had to be] **based on a new approach to the sciences of life'** (ibid. pp.19-20), concluding that **'the citadel they** [mechanistic scientists] **are defending lies in ruins'** (p.192).

[224] And in addition to pointing out the *psychological* nature of our condition when he said that **'Our alienation goes to the roots. The realization of this is the essential springboard for any serious reflection on any aspect of present inter-human life'**, Laing *also* bemoaned the fact that mechanistic science has **'pitifully failed to define the predicament of man'** when he wrote that **'The requirement of the present, the failure of the past, is the same: to provide a thoroughly self-conscious and self-critical human account of man** [p.11 of 156] ... **We respect the voyager, the explorer, the climber, the space man. It makes far more sense to me as a valid project—indeed, as a desperately urgently required project for our time—to explore the inner space and time of consciousness** [p.105]' (*The Politics of Experience* and *The Bird of Paradise*, 1967). Since our condition is consciousness-induced, 'consciousness' has become (and this will be more fully explained in ch. 7:2) code word for the issue of the human condition.

[225] Charles Birch, my professor of biology at Sydney University and another recipient of the Templeton Prize, also bravely spoke the truth about the limitations of human-condition-avoiding mechanistic science when he said, '[mechanistic] **science can't deal with subjectivity** [the issue of our psychologically distressed condition]**...what we were all taught in universities is pretty much a dead end'** (from recording of Birch's 1993 World Transformation Movement Open Day address). He also perceived the stultifying, 'truth'-'prevent[ing]' effects of dishonest, denial-based, mechanistic thinking when he said, **'Biology has not made any real advance since Darwin'** (in recorded conversation with the author, 20 Mar. 1987), and, some 10 years later, that **'the traditional framework of thinking in science is not adequate for solving the really hard problems'** (*Ockham's Razor*, ABC Radio National, 16 Apr. 1997), with the **'hard[est] problem'** of all for truth-avoiding thinking to solve being the all-important issue of our psychologically distressed human condition. Yes, as Birch concluded, **'Biology right now awaits its Einstein in the realm of consciousness studies'** (ibid).

[226] It is no wonder that humanity has lost faith in science. General Omar Bradley was certainly clear-sighted when, in pointing out science's failings, he said, **'The world has achieved brilliance...without conscience.**

**Ours is a world of nuclear giants and ethical infants'** (Armistice Day Address, 10 Nov. 1948; *Collected Writings of General Omar N. Bradley*, Vol.1). Yes, as Carl Jung said, **'Man everywhere is dangerously unaware of himself. We really know nothing about the nature of man, and unless we hurry to get to know ourselves we are in dangerous trouble'** (Laurens van der Post, *Jung and the Story of Our Time*, 1976, p.239 of 275). Sharing Jung's concerns about science's inability to provide us with the all-important, psychologically liberating, redeeming and transforming understanding of ourselves was the author Antoine de Saint-Exupéry, who wrote that **'We are living through deeply anxious days, and if we are to relieve our anxiety we must diagnose its cause...What is the meaning of man? To this question no answer is being offered, and I have the feeling that we are moving toward the darkest era our world has ever known'** (*A Sense of Life*, 1965, pp.127, 219 of 231). And lastly, when Sir Bob Geldof wrote and sang about our species' plight in his aptly titled 1986 album, *Deep in the Heart of Nowhere*, **'What are we going to do because it can't go on...This is the world calling. God help us'**, and **'Searching through their sacred books for the holy grail of "why", but the total sum of knowledge knows no more than you or I'**, he was recognizing that not only has science failed us, religious scripture has also been unable to help us with the **'holy grail'** of answers we needed of the **'why'** of the human condition. As explained in par. 156, until science clarified the difference between the gene and nerve based learning systems, the great prophets of old were in no position to explain the human condition—but *even* with that key knowledge found, science has been practicing such extreme denial that it couldn't use that knowledge to answer **'why'**.

²²⁷ The picture these concerned thinkers have painted of humanity's predicament may be bleak, but it is true—**'Man...is [so] dangerously unaware of himself'** that **'we are moving toward the darkest era our world has ever known'**; **'Ours is a world of nuclear giants and ethical infants'**, **'the scientific establishment has so pitifully failed to define the predicament of man'**, **'what we were all taught in universities is pretty much a dead end'**, **'the traditional framework of thinking in science is not adequate for solving the'** human condition, the **'demoralisation that people feel in our so-called scientific age stems from the bleak sterility of mechanistic thought'**, **'the total sum of knowledge'** hasn't been able to explain **'why'** we are the way we are. Dishonest mechanistic science *couldn't* solve the human condition, which is what was needed for the human race to progress to a human-condition-ameliorated, transformed state. And so human progress has been stalled, piled up and festering. The dialogue of one character in the 1991 film *Separate but Equal* accurately

recognized the plight of our species when he said, **'Struggling between two worlds; one dead, the other powerless to be born'** — words that echo those of the philosopher Antonio Gramsci: **'The crisis consists precisely in the fact that the old is dying and the new cannot be born; in this interregnum a great variety of morbid symptoms appears'** (*Prison Notebooks*, written during Gramsci's 10-year imprisonment under Mussolini, 1927-1937). The politician Lionel Bowen also alluded to the futility of trying to reform our lives and world without first finding the reconciling, ameliorating understanding of ourselves when he said, **'I think it's just impossible to bring about change until such time as some new civilisation develops to allow change'** (*The Sydney Morning Herald*, 10 Sep. 1988). But it was perhaps the historian Eric Hobsbawm who most succinctly captured the stark predicament facing humanity — that **'the alternative to a changed society, is darkness'** (*The Age of Extremes*, 1994, p.585 of 672).

[228] Yes, *only* a whole new way of thinking, in particular the reconciling, redeeming and healing way of understanding ourselves, and resulting new transformed civilization could alter our species' plight. We had arrived at a situation where humanity *desperately* needed clear biological understanding of ourselves, understanding that would make sense of our divisive condition and liberate us from criticism, lift the psychological burden of guilt, give us meaning. There *had* to be a scientific, first-principle-based, biological reason for our divisive behavior and finding it had become a matter of great urgency. The **'race'** that Richard Neville so accurately identified we were **'locked in...between self destruction and self discovery'** had reached crisis point, for to be stranded in a state of insecurity about our worthiness or otherwise was to be stranded in a terminally upset, psychologically immature state of arrested development — as Benjamin Disraeli, a former Prime Minister of Great Britain, famously recognized, **'Stranded halfway between ape and angel is no place to stop.'** The essayist Jonathan Swift's anguished cry to **'Not die here in a rage, like a poisoned rat in a hole'** (Letter to Bolingbroke, 21 Mar. 1729) did not exaggerate the truth of our predicament. The cellist Pablo Casals similarly emphasized the danger of our species' stalled state when he said, **'The situation is hopeless, we must take the next step'** (at a press conference in Madrid, on the occasion of his 80th [approx.] birthday). The journalist Doug Anderson made the same point when he wrote, **'Time may well be dwindling for us to enlighten ourselves... Tragic to die of thirst half a yard from the well'** (*The Sydney Morning Herald*, 31 Oct. 1994). In saying **'enlighten ourselves'**, Anderson was intimating that only understanding ourselves, understanding of our psychosis and neurosis

afflicted human condition, could make the difference that was needed. In quoting clinical psychologist Maureen O'Hara, the science reporter Richard Eckersley also acknowledged that **'humanity is either standing on the brink of "a quantum leap in human psychological capabilities or heading for a global nervous breakdown"'** (Address titled 'Values and Visions: Western Culture and Humanity's Future', Nov. 1995; see <www.wtmsources.com/133>). The psychotherapist Wayne Dyer was another who understood that it is only the *reconciling, dignifying* understanding of our seemingly imperfect human condition that could save the human race, when he said, **'We've come to a place…where we can either destroy ourselves or discover our divineness'** (*The Australian Magazine,* 8 Oct. 1994). Yes, as the great Australian educator, and my headmaster when I was a student at Geelong Grammar School, Sir James Darling, wrote when referring to the critical need now to solve the human condition: **'The time is past for help which is only a Band-Aid. It is time for radical thinking and for a solution on the grand scale'** (*Reflections for The Age*, ed. J. Minchin & B. Porter, 1991, p.145 of 176).

[229] Knowing now how evasive, truth-avoiding, defensive, excusive and deluded the resigned mind is, it is really to an unresigned adolescent mind that we should go for a truly accurate description of the seriousness of our species' plight—and that's what we have in these clearly unresigned, denial-free, honest lyrics from the 2010 *Grievances* album of the young American heavy metal band With Life In Mind: **'It scares me to death to think of what I have become…I feel so lost in this world'**, **'Our innocence is lost'**, **'I scream to the sky but my words get lost along the way. I can't express all the hate that's led me here and all the filth that swallows us whole. I don't want to be part of all this insanity. Famine and death. Pestilence and war.** [Famine, death, pestilence and war are traditional interpretations of the 'Four Horsemen of the Apocalypse' described in Revelation 6 in the Bible. Christ referred to similar **'Signs of the End of the Age'** (Matt. 24:6-8 and Luke 21:10-11).] **A world shrouded in darkness…Fear is driven into our minds everywhere we look'**, **'Trying so hard for a life with such little purpose…Lost in oblivion'**, **'Everything you've been told has been a lie…We've all been asleep since the beginning of time. Why are we so scared to use our minds?'**, **'Keep pretending; soon enough things will crumble to the ground…If they could only see the truth they would coil in disgust'**, **'How do we save ourselves from this misery…So desperate for the answers…We're straining on the last bit of hope we have left. No one hears our cries. And no one sees us screaming'**, **'This is the end.'** Saying **'We've all been asleep since the beginning of time'** echoes all that Laing

said (par. 123) about the extent of our blocked-out, alienated condition; and saying **'Everything you've been told has been a lie'** reiterates the extent of the dishonest denial in the world, especially in science, today; and saying **'So desperate for the answers'** confirms how incredibly important are all the **'answers'** about our human condition that are presented in this book. If there was ever a collection of words that cuts through all the dishonest pretense and delusion in the world about our condition these lyrics from With Life In Mind surely do it!

[230] Thankfully then it is precisely this **'solution on the grand scale'**, this **'enlightenment'** of **'ourselves'** — in fact, the **'discover[y]'** of the reason for **'our divineness'** — that makes possible the **'quantum leap in human psychological capabilities'** from alienation to transformation that alone can **'save ourselves from this misery'** of the human condition, that is going to be presented in the next chapter. Yes, finding understanding of the human condition is the *real* game-changer the human race has been waiting for, such that when only yesterday the levels of human suffering and distress and anger and environmental degradation from the effects of our horrifically troubled, upset human condition seemed irredeemable and irreversible, and all looked hopeless, suddenly people are going to appear who are inspired and transformed; *so* inspired and transformed, in fact, they are super-charged on a super-highway to a fabulous future for the human race. As will be explained and described in chapter 9 of this book, that sublime future is what becomes possible when the human condition is truthfully explained and solved, as it is in this book — and, indeed, the affirmations provided at <www.humancondition.com/affirmations> evidence this is true.

[231] Indeed, the next chapter in this book evidences what *is* possible if you *don't* think dishonestly and instead acknowledge important truths such as that humans did once live in a cooperative, loving, innocent state, and that it was our conscious mind that led to our present 'good-and-evil'-afflicted, immensely psychologically upset, fallen condition — for it provides the presentation of the dreamed-of, psychologically liberating, human-race-transforming, fully accountable, *true* explanation of the human condition.

[232] In conclusion, the following amazingly honest cartoon by Michael Leunig summarizes all that has just been said about how truth and beauty can only be accessed via another paradigm outside the mechanistic one the human race currently lives in. In it we see a lone, self-powered, self-sufficient individual leaving the great artificial and superficial,

alienated and alienating metropolis that has been built on dishonest denial to seek **'truth'** in the dark night of all that we have repressed, and, by so doing, resurrect **'beauty'** on Earth. Yes, as Sir Laurens van der Post wrote, **'There is, somewhere beyond it all, an undiscovered country to be pioneered and explored, and only a few lonely and mature spirits take it seriously and are trying to walk it'** (*About Blady*, 1991, p.87 of 255) — a statement of prophetic honesty that reiterates an observation he made some 40 years earlier that **'the one primary and elemental approach to the problem** [that the world faces] **is through** [understanding] **the being of man. Unfortunately it is an increasingly lonely way, trodden more and more not by masses but by solitary individuals...**[only these few] **sustain** [man's] **urge to seek an answer to the riddle of life** [confront and by so doing solve the human condition]' (*The Dark Eye in Africa*, 1955, p.15 of 159).

Cartoon by Michael Leunig, Melbourne's *The Age* newspaper, 17 Oct. 1987

[233] (The complete presentation of this chapter's demolition of mechanistic biology appears in my freely available, online book *Freedom Expanded* at <www.humancondition.com/freedom-expanded-the-denials-in-biology>.)

# Chapter 3

# The Human-Race-Transforming, *Real* Explanation of The Human Condition

William Blake's *Albion Arose*, c.1794-96;
this 1991 colored impression is by Carol Marando

## Chapter 3:1 Summary

[234] What will now be presented is the human-race-liberating-and-transforming, psychosis-addressing-and-solving, fully accountable, *REAL* biological explanation of the human condition that was introduced in chapter 1. This is the understanding that the human race has been working tirelessly towards since humans first became

fully conscious beings some 2 million years ago and our deeply psychologically troubled, conflicted state of the human condition emerged.

[235] And given the just described extreme threat of terminal alienation and extinction of the human race from mechanistic science's entrenched denial of the psychological nature of the human condition, *this* is the *psychologically* redeeming understanding of the human condition that is now desperately needed to bring about a new world for humans that is *free* of the human condition, a world in which we all stand liberated from Plato's terribly lonely, dark, bat-filled cave of alienation and magnificently transformed—the FREEDOM Blake so wonderfully anticipated in his above picture.

## Chapter 3:2 We cannot endure being faced by the problem of the human condition forever

[236] Before presenting this dreamed-of, human-race-transforming explanation of the human condition, the following extraordinarily honest words from Nikolai Berdyaev's 1931 book, *The Destiny of Man*, provide a powerful summary of all that was just said in chapter 2 about the stalled, festering state of science, and thus of the human race.

[237] Firstly, in regard to our species' overall plight, Berdyaev wrote that (and part of this extract was referred to in par. 181) **'The memory of a lost paradise, of a Golden Age** [our species' pre-human-condition-afflicted state of original cooperative, loving innocence]**, is very deep in man, together with a sense of guilt and sin and a dream of regaining the Kingdom of Heaven...We are faced with a profound enigma: how could man have renounced paradise which he recalls so longingly in our world-aeon? How could he have fallen away from it?'** (tr. N. Duddington, 1960, p.36 of 310). When it came to our consciousness-induced psychosis, he recognized that **'Philosophers and scientists have done very little to elucidate the problem of man'** (p.49). **'[P]sychologists were wrong in assuming that man was a healthy creature, mainly conscious and intellectual, and should be studied from that point of view. Man is a sick being...The human soul is divided, an agonizing conflict between opposing elements is going on in it...the distinction between the conscious and the subconscious mind is fundamental for the new psychology'** (pp.67-68). Significantly, Berdyaev *also* acknowledged how our species' immense fear of the human condition has blocked access to finding understanding of it, writing that **'Knowledge**

requires great daring. It means victory over ancient, primeval terror. Fear
makes the search for truth and the knowledge of it impossible. Knowledge
implies fearlessness...Particularly bitter is moral knowledge, the knowledge of
good and evil. But the bitterness is due to the fallen state of the world...Moral
knowledge is the most bitter and the most fearless of all for in it sin and evil
are revealed to us along with the meaning and value of life. There is a deadly
pain in the very distinction of good and evil, of the valuable and the worthless.
We cannot rest in the thought that that distinction is ultimate...we cannot bear
to be faced for ever with the distinction between good and evil...Ethics must
be both theoretical and practical, i.e. it must call for the moral reformation of
life...this implies that ethics is bound to contain a prophetic element. It must
be a revelation of a clear conscience, unclouded by social conventions' (pp.14-16).
(Note Berdyaev's accord with Janov's fundamental point that **'fear...
paralyses thought'** (in par. 221).)

[238] No, **'we cannot bear to be faced for ever with the' 'deadly' 'distinction
of good and evil, of the valuable and the worthless'** — nor with the resulting
**'social conventions'** of having to live with the psychologically **'sick'** state
of alienated denial as our only means of coping with that patently untrue
**'distinction'** between **'the valuable and the worthless'**. Permanent damnation
and terminal alienation are simply *not* acceptable options for the human
race, but to find the liberating, exonerating, reconciling and rehabilitating
understanding of our **'divided'** condition, and, by so doing, bring about
**'the new psychology'** and with it the transforming **'moral reformation of life'**,
required **'victory over [the] ancient, primeval terror'** of our condition—a
**'victory'** that could *only* be achieved by **'a revelation of a clear conscience,
unclouded by social conventions'** of denial, because only those who don't
suffer from the **'terror'** of the human condition could hope to **'fearless[ly]'**
investigate the subject.

[239] Yes, to find understanding of the human condition necessarily
required a truthful, **'clear conscience'**-guided, instinctual approach, not a
resigned-to-living-in-denial-of-the-human-condition, alienated-from-the-
truth, blocking-out-of-condemning-moral-instincts, hiding-in-Plato's-
cave, **'intellectual'** approach. It is simply not possible to build the truth
from a position of denial/lying. You can't think effectively, insightfully,
**'prophetic[ally]'**, if you're not being honest.

[240] Indeed, as I pointed out in chapter 2:12, trying to investigate reality
while living in denial of any truths that brought the, for most people,
**'deadly pain[ful]'** issue of the human condition into focus, as mechanistic

science has done, was an extremely compromised and deficient way of searching for knowledge. In fact, it is a measure of the blindness of human-condition-avoiding, denial-based thinking, and the effectiveness of human-condition-confronting, honest thinking, that when the whole truth about our condition *is* finally reached, as it now has been, it can appear so straightforward and simple that it seems self-evident. But simplicity has always been a hallmark of insightful thought—as the pioneering biologist Allan Savory observed, **'whenever there has been a major insoluble problem for mankind, the answer, when finally found, has always been very simple'** (*Holistic Resource Management*, 1988, 1st edition, p.3). For instance, when Charles Darwin put forward his breakthrough, and necessarily exceptionally **'fearless'** and truthful-thinking-based insight of natural selection, it was, in hindsight, such a simple explanation that the eminent biologist of the time, the aforementioned Thomas Huxley, was prompted to exclaim, **'How extremely stupid not to have thought of that!'** (1887; *Life and Letters of Thomas Henry Huxley Vol.1*, ed. Leonard Huxley, 1900, p.170). So, yes, the author George Seaver was prescient when, in anticipating just how simple the explanation of **'the riddle of life's meaning and mystery'**, namely the human condition, would be, he wrote that **'The ultimate thought, the thought which holds the clue to the riddle of life's meaning and mystery, must be the simplest thought conceivable, the most natural, the most elemental, and therefore also the most profound'** (*Albert Schweitzer: The Man and His Mind*, 1947, p.311).

[241] But while the crux question facing the human race, of the much-needed clarification of the nature of 'good and evil', *does* have an amazingly simple answer, the *implications* of it could not be more significant, far-reaching or exciting.

## Chapter 3:3 The psychosis-addressing-and-solving *real* explanation of the human condition

[242] In another brilliant cartoon (opposite), Michael Leunig succinctly dares to ask the fundamental question of *'What does the chaotic, traumatic and strife-torn life of humans all mean—how are we to make sense of our existence?'* He has done so by placing a very perplexed and distressed gentleman behind an 'Understandascope', through which he peers into a sea of apparent madness. Everywhere he looks there is tumultuous congestion: there are people furiously arguing and fighting with each

other; there is a church where people pray for forgiveness and salvation; and there are vehicles polluting the chaos with fumes and noise. *And*, in this 1984 drawing Leunig even seems to have predicted a climactic demonstration of all our human excesses and frustrations when, on September 11, 2001, terrorists flew planes into the tall, square-shaped towers of the World Trade Center in New York City! Well, it is precisely this great burning, searing question of *'What is wrong with us humans?'* — *'How are we to understand it all?'* — *'How are we to make sense of human existence?'* — that is now going to be answered here. The explanation to be given here in full *is* the 'UNDERSTANDASCOPE' we have always wanted, needed and sought.

Cartoon by Michael Leunig, Melbourne's *The Age* newspaper, 17 Mar. 1984

[243] As emphasized in chapter 2, to provide this all-exciting, human-race-liberating, amazingly simple, fully accountable, *true* explanation of the human condition requires starting from the unequivocally honest basis of acknowledging that humans *did* once live in a completely loving, unconditionally selfless, altruistic state, and that it was only *after* the emergence of our conscious mind that our present 'good-and-evil'-afflicted, immensely psychologically upset condition emerged. If we do

this and consider what would happen when a conscious mind emerged in the presence of an already established cooperative, loving instinctive state—given what human-condition-avoiding, mechanistic science has managed to discover about the gene-based natural selection process and how nerves are capable of memory—then the explanation is right there in front of us.

[244]Clearly, since our altruistic, moral instincts are only genetic *orientations* to the world and not *understandings* of it, when our fully conscious, reasoning, self-managing mind emerged it would, in order to find the understandings it needed to effectively manage events, have had to challenge those instinctive orientations, which would have led to a psychologically upsetting clash with our moral instincts.

[245]The exceptionally **'clear conscience'**-guided, truthful, and thus effective-thinking, **'prophetic'** naturalist Eugène Marais was on the right track when, as mentioned in par. 187, he recognized that **'As the...individual memory slowly emerges** [in humans]**, the instinctive soul becomes just as slowly submerged...For a time it is almost as though there were a struggle between the two.'** Berdyaev was also approaching the truth when, as just mentioned, he wrote that **'The human soul is divided, an agonizing conflict between opposing elements is going on in it...the distinction between the conscious and the subconscious mind is fundamental for the new psychology.'** Other thinkers, such as Arthur Koestler, Erich Neumann, Paul MacLean, Julian Jaynes and Christopher Booker, have also delved into the problem of the human condition to a similar depth to Marais and Berdyaev, but what was missing was the *clarifying* explanation of the *nature* of that **'struggle'** and **'conflict'** between our instincts and our consciousness. (After all, as pointed out in chapter 2:6, even Moses' Genesis story of the Garden of Eden contains the truth that our corrupted, 'fallen' condition occurred when our conscious mind emerged from an original, presumably instinctive, idyllic state, and Plato recognized there was a conflict between **'noble'**, **'good'** instincts, a **'white' 'horse'**, against an **'ignoble'**, **'bad'**, **'crooked'**, **'unlawful deeds'**-producing conscious intellect, a **'dark' 'horse'**, but the degree of insight apparent in these descriptions didn't liberate humanity from the human condition.) So, yes, what is the particular **'distinction'** between our instincts and intellect that *caused* the psychologically upset state of our 'good-and-evil'-afflicted condition? As just stated, the answer is that the gene-based refinement system is only capable of *orientating* a

species, whereas the nerve-based refinement system has the potential to *understand* the nature of change.

[246] To present the explanation in more detail.

[247] As briefly introduced in par. 61, nerves were originally developed for the coordination of movement in animals, but, once developed, their ability to store impressions—what we refer to as 'memory'—gave rise to the potential to develop understanding of cause and effect. If you can remember past events, you can compare them with current events and identify regularly occurring experiences. This knowledge of, or insight into, what has commonly occurred in the past enables you to predict what is likely to happen in the future and to adjust your behavior accordingly. Once insights into the nature of change are put into effect, the self-modified behavior starts to provide feedback, refining the insights further. Predictions are compared with outcomes and so on. Much developed, and such refinement occurred in the human brain, nerves can sufficiently *associate* information to *reason* how experiences are related, learn to *understand* and become CONSCIOUS of, or aware of, or *intelligent* about, the relationship between events that occur through time. Thus consciousness means being sufficiently aware of how experiences are related to attempt to manage change from a basis of understanding. (Much more will be explained about the nature and origin of consciousness in chapter 7.)

[248] The significance of this process is that once our nerve-based learning system became sufficiently developed for us to become conscious and able to effectively manage events, our conscious intellect was then in a position to wrest control from our gene-based learning system's instincts, which, up until then, had been in charge of our lives—instincts being **'a largely inheritable and unalterable tendency of an organism to make a complex and specific response to environmental stimuli without involving reason'** (*Merriam-Webster Dictionary*; see <www.wtmsources.com/144>). Basically, once our self-adjusting intellect, or ability to **'reason'**, emerged it was capable of taking over the management of our lives from the instinctive orientations we had acquired through the natural selection of genetic traits that adapted us to our environment. Moreover, at the point of becoming conscious the nerve-based learning system *should* wrest management of the individual from the instincts because such a self-managing or self-adjusting system is infinitely more efficient at adapting to change

than the gene-based system, which can only adapt to change very slowly over many generations. HOWEVER, it was at this juncture, when our conscious intellect challenged our instincts for control, that a terrible battle broke out between our instincts and intellect, the effect of which was the extremely competitive, selfish and aggressive state that we call the 'human condition'.

[249] An analogy will help further explain the origin of our human condition. (I should mention that a condensation of the explanation of the human condition given in this chapter, in particular some of the 'Adam Stork' analogy, was included in chapter 1's summary of the contents of this book, so the reader should expect some repetition of that material.)


## Chapter 3:4  The Story of Adam Stork

[250] What distinguishes humans from other animals is our fully conscious mind, our ability to understand and thus manage the relationship between cause and effect. However, prior to becoming fully conscious and able to self-manage—consciously decide how to behave—our earliest ape ancestors were controlled by and obedient to their instincts, as other animals continue to be. (Of course, humans are still apes, so when I say 'ape ancestor', I mean when we were like *existing* non-human great apes.) Aldous Huxley, author of famous novels such as *Brave New World*, acknowledged how other animals live, and also how our distant ancestors would have lived, obedient to their instincts, when he wrote, **'Non-rational creatures do not look before or after, but live in the animal eternity of a perpetual present; instinct is their animal grace and constant inspiration; and <u>they are never tempted to live otherwise than in accord with their own...immanent law</u>. Thanks to his reasoning powers and to the instrument of reason, language, man (in his merely human condition) lives nostalgically, apprehensively and hopefully in the past and future as well as in the present'** (*The Perennial Philosophy*, 1946, p.141 of 352). I have underlined part of this passage because it raises the important question of what would happen if a species was **'tempted to live otherwise than in accord with their own'** instincts—to, in effect, break that **'immanent law'**—as Huxley infers humans must have done when we became fully conscious? The following analogy involving migrating storks provides an illustration of what would happen.

**The Story of Adam Stork**

Drawing by Jeremy Griffith © 1991 Fedmex Pty Ltd

[251] Many bird species are perfectly orientated to instinctive migratory flight paths. Each winter, without ever 'learning' where to go and without knowing why, they quit their established breeding grounds and migrate to warmer feeding grounds. They then return each summer and so the cycle continues. Over the course of thousands of generations and migratory movements, only those birds that happened to have a genetic make-up that inclined them to follow the right route survived. Thus, through natural selection, they acquired their instinctive orientation.

[252] Consider a flock of migrating storks returning to their summer breeding nests on the rooftops of Europe from their winter feeding

grounds in southern Africa. Suppose in the instinct-controlled brain of one of them we place a fully conscious mind (we call the stork Adam because we will soon see that, up to a point, this analogy parallels Moses' pre-scientific account of the origin or genesis of the human condition in the Bible where Adam and Eve take the fruit from the tree of knowledge in the Garden of Eden). So, as Adam Stork flies north he spots an island off to the left with a tree laden with apples. Using his newly acquired conscious mind, Adam thinks, 'I should fly down and eat some apples.' It seems a reasonable thought but he can't know if it is a good decision or not until he acts on it. For Adam's new thinking mind to make sense of the world he has to learn by trial and error and so he decides to carry out his first grand experiment in self-management by flying down to the island and sampling the apples.

[253] But it's not that simple. As soon as Adam's conscious thinking self (as depicted by the stork on the left) deviates from his established migratory path, his innocent instinctive self (as depicted by the wide-eyed stork on the right) tries to pull him back on course. In following the flight path past the island, the instinct-obedient self/stork is, in effect, criticizing Adam's decision to veer off course; it is condemning his search for understanding. All of a sudden Adam is in a dilemma: if he obeys his instinctive self and flies back on course, he will remain perfectly orientated but he'll never *learn* if his deviation was the right decision or not. All the messages he's receiving from within inform him that obeying his instincts is good, is right, but there's also a new inclination to disobey, a defiance of instinct. Diverting from his course will result in apples and understanding, yet he already sees that doing so will make him feel bad.

[254] Uncomfortable with the criticism his newly conscious mind or intellect is receiving from his instinctive self, Adam's first response is to ignore the temptation the apples present and fly back on course. This 'correction' not only makes his instinctive self happy, it also wins back the approval of his fellow storks, for not having conscious minds they, like his instinctive self, are innocent, unaware or ignorant of the conscious mind's need to search for knowledge. Furthermore, since Adam's instinctive self developed alongside the natural world, it too reminds him of his instinctive orientation, thus contributing to the criticism of Adam for his rebellious decision.

[255] Flying on, however, Adam realizes he can't deny his intellect. Sooner or later he must find the courage to master his conscious mind

by carrying out experiments in understanding. This time he thinks, 'Why not fly down to an island and rest?' Again, not knowing any reason why he shouldn't, he proceeds with his experiment. And *again*, his decision is met with the same chorus of criticism—from his instinctive self, from the other storks that were ignorant of the need to search for knowledge, and from the natural world. But this time Adam defies the criticism and perseveres with his experimentation in self-management. His decision, however, means he must now live with the criticism and immediately he is condemned to a state of upset. A battle has broken out between his instinctive self, which is perfectly orientated to the flight path, and his emerging conscious mind, which needs to understand why that flight path is the correct course to follow. His instinctive self is perfectly orientated, but Adam doesn't *understand* that orientation.

[256] In short, when the fully conscious mind emerged it wasn't enough for it to be orientated by instincts. It *had to* find understanding to operate effectively and fulfill its great potential to manage life. Tragically, the instinctive self didn't 'appreciate' that need and 'tried to stop' the mind's necessary search for knowledge, as represented by the latter's experiments in self-management—hence the ensuing battle between instinct and intellect. To refute the criticism from his instinctive self, Adam needed the understanding found by science of the difference in the way genes and nerves process information; he needed to be able to explain that the gene-based learning system can orientate species to situations but is incapable of insight into the nature of change. Genetic selection of one reproducing individual over another reproducing individual (the selection, in effect, of one idea over another idea, or one piece of information over another piece of information) gives species adaptations or orientations—instinctive programming—for managing life, but those genetic orientations, those instincts, are not understandings. This means that when the nerve-based learning system gave rise to consciousness and the ability to understand the world, it wasn't sufficient to be *orientated* to the world—*understanding* of the world had to be found. The problem, of course, was that Adam had only just taken his first, tentative steps in the search for knowledge, and so had no ability to explain anything. It was a catch-22 situation for the fledgling thinker, because to explain himself he needed the very knowledge he was setting out to accumulate. He had to search for understanding, ultimately self-understanding, understanding of why he had to 'fly off course', without the ability to first explain why he

needed to 'fly off course'. And without that defense, he had to live with the criticism from his instinctive self and was *INSECURE* in its presence. (I should clarify that while instincts are hard-wired, genetic programming and as such cannot literally criticize our conscious mind, they can *in effect* do so. Our instincts let our conscious mind know when our body needs food, or, as our instinctive conscience clearly does, want us to behave in a cooperative, loving way, and certainly our conscious mind can defy those instinctive orientations if it chooses to. Our conscious mind *can* feel criticized by our instinctive conscience; it happens all the time.)

[257] And so to resist the tirade of unjust criticism he was having to endure and mitigate that insecurity, Adam had to do something. *But what could he do?* If he abandoned the search and flew back on course, he'd gain some momentary relief, but the search would, nevertheless, remain to be undertaken. So all Adam could do was retaliate against and ATTACK the instincts' unjust criticism, attempt to PROVE the instincts' unjust criticism wrong, and try to DENY or block from his mind the instincts' unjust criticism—and, as the stork on the left of the picture illustrates, he did *all* those things. He became angry towards the criticism. In every way he could he tried to demonstrate his self worth, prove that he is good and not bad—he shook his fist at the heavens in a gesture of defiance of the implication that he is bad. And he tried to block out the criticism— this block-out or denial including having to invent contrived excuses for his instinct-defying behavior. In short, his ANGRY, EGOCENTRIC and ALIENATED state appeared. Adam's intellect or 'ego' (which is just another word for the intellect since the *Concise Oxford Dictionary* defines 'ego' as 'the conscious thinking self' (5th edn, 1964)) became 'centered' or focused on the need to justify itself—selfishly preoccupied aggressively competing for opportunities to prove he is good and not bad, to validate his worth, to get a 'win'; to essentially eke out any positive reinforcement that would bring him some relief from criticism and sense of worth. He unavoidably became SELFISH, AGGRESSIVE and COMPETITIVE.

[258] Overall, it was a terrible predicament in which Adam became PSYCHOLOGICALLY UPSET—a sufferer of PSYCHOSIS and NEUROSIS. Yes, since, according to *Dictionary.com*, 'osis' means 'abnormal state or condition', and the *Penguin Dictionary of Psychology*'s entry for 'psyche' reads 'The oldest and most general use of this term is by the early Greeks, who envisioned the psyche as the soul or the very essence of life' (1985 edn), Adam developed a 'psychosis' or 'soul-illness', and a 'neurosis' or neuron or nerve

or 'intellect-illness'. His original gene-based, instinctive **'essence of life'** soul or PSYCHE became repressed by his intellect for its unjust condemnation of his intellect, and, for its part, his nerve or NEURON-based intellect became preoccupied denying any implication that it is bad. Adam became psychotic and neurotic.

[259] But, again, without the knowledge he was seeking, without self-understanding (specifically the understanding of the difference between the gene and nerve-based learning systems that science has given us), Adam Stork had no choice but to resign himself to living a psychologically upset life of anger, egocentricity and alienation as the only three responses available to him to cope with the horror of his situation. (And by alienation, it should be clear by now that I don't mean alienation from society or some other group or individual, but the situation of being estranged or detached from our own instinctive self and any truthful thinking it inclines our conscious mind to pursue.) It was an extremely unfair and difficult, indeed tragic, position for Adam to find himself in, for we can see that while he was good he appeared to be bad and had to endure the horror of his psychologically distressed, upset condition until he found the real—as opposed to the invented or contrived *not*-psychosis-recognizing—defense or reason for his 'mistakes'. Basically, *suffering upset was the price of his heroic search for understanding.* Indeed, it is the tragic yet inevitable situation any animal would have to endure if it transitioned from an instinct-controlled state to an intellect-controlled state—its instincts would resist the intellect's search for knowledge. Adam's uncooperative and divisive competitive aggression—and his selfish, egocentric, self-preoccupied efforts to prove his worth; and his need to deny and evade criticism, essentially embrace a dishonest state—all became an unavoidable part of his personality. Such was Adam Stork's predicament, *and such has been the human condition, for it was within our species that the fully conscious mind emerged.*

[260] What now needs to be explained is how the situation faced by humans when *we* began searching for knowledge was *so* much worse than it was for our hypothetical Adam Stork. This is because, as was emphasized in chapters 2:5 to 2:7, *our* instinctive orientation wasn't to a flight path but to *behaving in an unconditionally selfless, all-loving, cooperative moral way*, so when we started carrying out experiments in understanding and then unavoidably began reacting defensively in an angry, aggressive, competitive, selfish and evasive, dishonest way to combat the criticism from our instincts, that divisive response drew *even more* criticism from

our particular cooperative, loving, ideal-behavior-demanding, integrative, **'very essence of life'**-orientated moral instinctive self or soul or psyche—a vicious cycle that fuelled our upset immensely. (Chapter 5 will present the biological explanation of how humans acquired such unconditionally selfless, universally loving, moral instincts, the 'voice' of which is our conscience.)

## Chapter 3:5 The double and triple whammy involved in our *human* situation or condition

[261] When Adam Stork began searching for knowledge and unavoidably became angry, egocentric and alienated, that upset response didn't attract further criticism from his instinctive orientation—but that certainly wasn't the case with us humans. No, when we began searching for knowledge and became angry, egocentric and alienated, that response was extremely offensive to our particular instinctive orientation because our instinctive orientation isn't to a flight path, or to any of the various instinctive orientations that other animals are obedient to; it is to behaving in the opposite way, namely lovingly, selflessly and honestly. So in *our* case, when we began experimenting in understanding and were criticized by our instincts and unavoidably responded in an angry, egocentric and alienated way, we had to endure a further round of criticism, a second hit, a 'double whammy', from our instinctive orientation. Yes, in our necessary search for understanding we were firstly unjustly condemned for defying our instincts, and then again for reacting to that condemnation in a way that was counter and offensive to our instincts. So if Adam Stork had cause to be upset, we had double cause to be upset!

[262] And yet the horror of our situation didn't end there—for we weren't just unjustly condemned twice, we were unjustly condemned *three times*; we were forced to endure a 'triple whammy', which I will now explain.

[263] While the following explanation will be covered in greater detail in chapter 4, to context why we have experienced this 'triple whammy' of condemnation it is necessary to briefly explain that there is a teleological, order-of-matter-developing, integrative, cooperative theme or direction or purpose or meaning to existence, which God is the personification of. Everywhere we look we see hierarchies of ordered matter—'There is a tree that is composed of parts (leaves, branches, a trunk and roots) and in turn those parts are composed of parts (fibers, cells, etc).' Our world is clearly

composed of a hierarchy of ordered matter: atoms have come together or integrated to form compounds, which in turn have come together or integrated to form virus-like organisms, which integrated to form single-celled organisms, which then integrated to form multicellular organisms, which have come together or integrated to form societies of multicellular organisms. Significantly, the behavior required for these ordered arrangements of matter or wholes to stay together is selflessness—because selflessness means being considerate of the welfare of the larger whole or integrative, while selfishness is divisive or disintegrative. Selflessness is, in fact, the theme of existence, the glue that holds wholes together. But in light of our *divisive* selfish, egocentric, competitive and aggressive behavior, humans have obviously found this truth of the selflessness-dependent, *integrative* meaning of existence unbearably condemning. And being so unbearably condemning, the main way we coped with this truth of Integrative Meaning was to deify it, make it God—a concept we revered but claimed had no material relationship to us. Again, all this will be explained in chapter 4, but, in terms of explaining the 'triple whammy' we suffered from when we searched for knowledge, what this selfless, cooperative, Godly, integrative theme of existence means is that when we started retaliating against the criticism from our instincts in a divisive, selfish, uncooperative, angry, aggressive, egocentric and alienated way, we were not only at odds with our cooperative, selfless, loving moral instincts, we were also defying God! Our retaliation against our instinctive self made us appear as though we were out of step with creation, living in a way that was entirely inconsistent with the integrative, cooperative, selfless theme or meaning of existence! The point here is that despite having removed the confronting presence of Integrative Meaning by abstracting it as God (and even by outright denial of the existence of Integrative Meaning or God), the development of order of matter in nature is actually such an obvious truth that our conscious mind is well aware of it. So, when we humans went in search of knowledge we were initially criticized for not obeying our instincts, and *then secondly* for responding to that initial criticism in a way that offended our cooperatively orientated, moral instinctive conscience, and, *then thirdly*, by that behavior defying the very integrative, cooperative theme of existence that our intellect could so plainly see existed in nature. *We defied our instincts, we offended our moral conscience, and we insulted the very meaning of existence/God!!!* We humans could hardly be more guilt-ridden. And all this guilt, which we can now understand was completely unjustified, made us *extremely, excruciatingly* upset—*absolutely furious*, in fact. As necessary as it was, in our case, 'flying off course' was an *incredibly* upsetting act of defiance—*which is why we humans have been capable of absolutely extraordinary acts of brutality, barbarism and cruelty.* While we have tried to restrain and conceal the anger within us, 'civilize' it as we say, it is, in truth, volcanic—but, again, we can now at last understand the origin of all that anger.

[264]To appreciate exactly what occurred when humans began to actively experiment in understanding we can consider what occurs in the lives of children today, and must have occurred in the lives of our distant, child-equivalent ancestors, who, as will be described in chapters 8:2 to 8:7, were the australopithecines who emerged from our infant-equivalent, bonobo-like ape ancestors some 4 million years ago.

[265]In the case of young humans today, when they begin to actively experiment in managing their lives from a basis of understanding, which is what occurs during the stage of a conscious mind's development that we call 'childhood', they encounter this upsetting criticism from their instinctive moral conscience and from their mind's awareness of the integrative theme of existence—as well as from other minds around them. Imagine, for instance, a situation where a young boy sees a birthday cake on a table and, being new to this business of reasoning, thinks, innocently enough, 'Why shouldn't I take all the cake for myself', before doing so. While many mothers actually witness these grand mistakes of pure self-ishness that young children make when they first attempt to self-manage their lives, they still have to be reasonably lucky to do so because, once done, the child usually doesn't make such a completely naive mistake again due to the criticism it attracts from its moral instincts, from its own conscious mind's awareness of the very obvious integrative, cooperative, selfless theme of existence, and from others present. But despite the nasty shock from all the criticism and his subsequent determination to never again make such a mistake, the child, although he is unable to explain his actions, does feel that what he has done is not something bad, not some-thing deserving of such criticism. In fact, by this stage in the child's mental development, he has become quite proud of his efforts at self-managing his life, drawing attention to his achievements with excited declarations like 'Look at me Daddy, I can jump puddles', and so on. So the child is only just discovering that this business of self-adjusting is not all fun and that some experiments are getting him into trouble. It is at this stage that 'playing',

as we call these early experiments in self-management, starts leading to some serious issues for the child. Indeed, the frustrated feeling of being unjustly criticized for some of the experiments gives rise to the precursors of the defensive, retaliatory reactions of anger, egocentricity and alienation, with some angry, aggressive nastiness creeping into the child's behavior. Furthermore, in this situation of feeling unfairly criticized, it follows that any *positive* feedback or reinforcement begins to become highly sought-after, which is the beginning of egocentricity—the conscious thinking self or ego starts to become preoccupied trying to defend its worth, assert that it is good and not bad. At this point, the intellect also begins experimenting in ways to deny the unwarranted criticism, which, in this initial, unskilled-in-the-art-of-denial stage, takes the form of blatant lying: 'But Mum, Billy told me to do it', or 'But Mum, the cake accidentally fell in my lap.' These apparent misrepresentations weren't *actually* lies, rather they were inadequate attempts at explanation. Lacking the real excuse or explanation, it was at least *an* excuse, a contrived defense for the child's mistake. The child was evading the false implication that his behavior was bad, in the sense that a 'lie' that said he wasn't bad was less of a 'lie' than a partial truth that said he was. Basically, the child has started to feel the first aggravations from the horror of the injustice of the human condition—and we can expect that exactly the same kind of mistakes in thinking and the resulting frustrations with the ensuing criticism would have also occurred in our australopithecine, child-equivalent ancestors. Some of them, those who had become intelligent enough to actively experiment in self-adjustment, would have begun to encounter criticism of their efforts to self-manage and, as such, begun to exhibit the psychologically upset behaviors of anger, egocentricity and alienation.

[266] Once the experimentation in self-management gets underway, so the upsetting frustrations with the resulting criticism increases. The journey of ever-increasing levels of upset has begun—and this corrupting journey of escalating upset could not and would not stop until the exonerating explanation of the human condition was found, which has only now, in this book, finally emerged millions of years after this self-fuelling process began, which means there must be, and indeed is, an absolutely astronomical amount of upset built up in us humans! That horrific journey of ever-increasing levels of upset is described in chapter 8, but just to follow its development a little further—throughout childhood the experimenting in understanding increases and the resulting frustrated

upset also increases, such that by late childhood children enter what is recognized as the **'naughty nines'**. By this stage the resentment and frustration with the criticism from their efforts at self-adjustment has become so great that the child starts lashing out at the unjust world. Indeed, by late childhood children become very angry, even taunting and bullying those around them. However, by the end of childhood, children realize that lashing out in exasperation at the 'injustice of the world' doesn't change anything and that the only possible way to solve their frustration is to find the reconciling insight into why the criticism they are experiencing is not deserved. It is at this point, which occurs around 12 years of age, that the child undergoes a dramatic change from a frustrated, protesting, demonstrative, loud extrovert into a sobered, deeply thoughtful, quiet introvert, consumed with anxiety about the imperfections of life under the duress of the human condition. In fact, it is in recognition of this very significant psychological change from a relatively human-condition-free state to a very human-condition-aware state that we separate these stages into 'Childhood' and 'Adolescence', a shift even our schooling system marks by having children graduate from what is generally called primary school into secondary school. And indeed, this critical junction in our species' development is also acknowledged in the anthropological record, with the name of the genus changing from *Australopithecus*, the extrovert 'Childman', to *Homo*, the sobered 'Adolescentman'. The story of the journey through this next stage of adolescence was briefly introduced in chapter 2:2 when Resignation was described, but to quickly recap: after struggling for a few years during their early adolescence to make sense of existence, by about 14 or 15 years of age that search for understanding generally became so confronting that denial of the whole unbearably depressing issue of the human condition had to be adopted—after which adolescents became superficial and artificial escapists, not wanting to look at any issue too deeply, and, before long, combative and competitive power-fame-fortune-and-glory, relief-seeking resigned adults.

[267] So while the upset state of the human condition emerged during our childhood, and in the case of our species' journey, during the life of our child-equivalent australopithecine ancestor, the *real* struggle with the agony of the human condition didn't arise until our adolescence, which began when we were around 12 years of age, or, in the case of our species, during the life of *Homo* who emerged from the australopithecines some 2 million years ago. Indeed, we could say that we only became *fully*

*conscious* in the sense of being *fully aware* of the situation we humans have been in of having to live with the agony of the human condition some 2 million years ago when our species entered adolescence. Again, all the stages of ever-increasing levels of upset that humans both individually and as a species have progressed through will be much more fully described in chapter 8.

[268] In summary, unlike Adam Stork who only had to contend with criticism of his attempts to self-adjust, humans have had to contend with that criticism *and* criticism of our unavoidable angry, aggressive, competitive, selfish and dishonestly evasive response to that initial criticism—a response that went against our moral conscience and against the integrative meaning of existence; against the Godly ideals, no less. In short, through our efforts to self-adjust and experiment in self-understanding, in order to find the ultimate knowledge of understanding of the human condition that we needed to make sense of ourselves, we were made to feel extremely guilty, 'evil' and 'sinful', which very greatly compounded our insecurity-of-self and frustration, making us *immensely* angry and egocentric and *very much* needing to live in denial of any confrontation with the problem of our corrupted condition. We had to live totally separated, or blocked-off, or dissociated, or alienated from our true situation—metaphorically **'enshrined in that living tomb'** of Plato's dark **'cave'** where no exposure of our corrupted condition was possible.

[269] Yes, humans have been *immensely* insecure about our upset, corrupted, fallen condition—*extremely* fearful of the suicidally depressing implication that we are bad, vile creatures for having departed so incredibly far from our species' original unconditionally selfless, all-loving-and-all-sensitive, fully-cooperative-and-integrative, Godly, 'ideal' way of living. But while total block-out/separation/dissociation/alienation from the truth of our species' original all-loving world meant we had chosen a **'living tomb'** of dishonest darkness to live in, that existence was infinitely preferable to trying to confront the suicidally depressing truth of how far we had departed from our original world of all-loving innocence! Such has been the horror of the human condition. Thank goodness we can at last explain *why* we couldn't avoid becoming angry, egocentric and alienated and as a result no longer have to hide in that **'living tomb'** dungeon of dishonest darkness anymore!

[270] If we immerse ourselves in the truth of how pure we humans once were—as Plato and Hesiod did when they admitted **'there was a time when'**

we lived a **'blessed'**, **'innocent and simple and calm and happy' 'pure'** existence where there was no **'war or quarrel**[ling]**'**, and **'no forms of government or separate possession of women and children'** and where we **'dwelt naked…in the open air…and…lay on soft couches of grass'**; a time when we were a **'golden race…with calm untroubled mind**[s]**…unbridled by toil…**[and] **all willing shared the gathering of…**[our] **hands'** — and then contrast *that* existence with our present immensely upset competitive, selfish, greedy, uncaring, mean, aggressive and materialistic existence, the distinction is most certainly a suicidally depressing one to have to face, unless you are exceptionally free of upset, as Plato and Hesiod must have been, or you can explain the good reason *why* that extremely upset state emerged. Appreciating this distinction allows us to now fully understand what Gerard Manley Hopkins meant when he wrote in his 1885 poem *No Worst, There Is None*, **'O the mind, mind has mountains; cliffs of fall, frightful, sheer, no-man-fathomed. Hold them cheap may** [any] **who ne'er** [have never] **hung there.'** Yes, only innocent people, those free of upset, have been able to confront the issue of the human condition without becoming depressed — **'hung'** being the perfect description for the depressed state. For innocents it was not costly; rather, as Hopkins said, it was **'cheap'** for them. But now that the upset state of the human condition has been explained and defended *the whole human race is freed from 'frightful[ly]' depressing condemnation!*

[271] And to think we have been living in this extremely unfair and torn state where we couldn't explain the good reason for our species' upset condition for over 2 million years! With this in mind, we can start to register just how much hurt, frustration and anger must now exist within us humans. After all, imagine living just *one* day with the injustice of being condemned as evil, bad and worthless when you intuitively knew — but were unable to explain — that you were actually the complete opposite, namely truly wonderful, good and meaningful. How tormented and furious — how upset — would you be by the end of that one day? You would be *immensely* upset. So extrapolate that experience over *2 million years* and you will begin to get some appreciation of just how much volcanic anger must now exist within us humans today! While we have learnt to significantly restrain and conceal — 'civilize' — our phenomenal amount of upset, it nevertheless follows that, under the surface, our species must be boiling with rage, and that sometimes, when our restraint can no longer find a way to contain it, that anger must express itself. Yes, *we can finally understand humans' capacity for astounding acts of aggression, hate,*

*brutality and atrocity*. The following is but one description of how much anger humans have accumulated as a result of being unjustly condemned; it is an account of the bloodshed that was commonplace during World War I: **'The flowing blood of these murdered men, ten million gallons of steaming human blood could substitute for a whole day the gigantic water masses of the Niagara [Falls]...Make a chain of these ten million murdered murderers, placing them head to head and foot to foot, and you will have an uninterrupted line measuring ten thousand miles, a grave ten thousand miles long'** (Mrs Will Gordon, *Roumania Yesterday and To-day*, 1918, p.251 of 270). And it can now be understood that *this capacity for inhumanity exists in us all*—as the author Morris West so bravely acknowledged in his memoir, *A View from the Ridge*: **'brutalise a child and you create a casualty or a criminal. Bribe a servant of the state and you will soon hear the deathwatch beetles chewing away at the rooftrees of society. The disease of evil** [now able to be understood as upset] **is pandemic; it spares no individual, no society, <u>because all are predisposed to it</u>. It is this predisposition which is the root of the mystery** [of 'evil' that is now explained]. **I cannot blame a Satan, a Lucifer, a Mephistopheles, for the evils I have committed, the consequences of which have infected other people's lives. I know, as certainly as I know anything, that the roots are in myself, buried deeper than I care to delve, in caverns so dark that I fear to explore them. I know that, given the circumstances and the provocation, I could commit any crime in the calendar'** (1996, p.78 of 143).

[272] It certainly is an understatement of the grandest proportions to say that it is a relief that humanity's 2-million-year journey of conscious thought and enquiry into the nature of our condition has finally delivered understanding of it—that the **'caverns so dark'** where the **'mystery'** of our horrifically upset human condition lies have at last been **'explore[d]'** and the greater dignifying, redeeming, liberating, healing, ameliorating explanation for that condition found.

## Chapter 3:6 <u>Adam and Eve — we humans — are heroes NOT villains</u>

[273] We can now see how The Story of Adam Stork—which describes the primary issue involved in our human condition of the upsetting battle that emerged between our instincts and our conscious intellect's search for knowledge—has parallels with Moses' Biblical account of Adam and Eve's experiences in the Garden of Eden, except in that presentation when Adam and Eve took the **'fruit'** (Gen. 3:3) **'from the tree of the knowledge**

of good and evil' (2:9, 17)—went in search of understanding—they were
'banished...from the Garden' (3:23) of our original innocent, all-loving state
for being 'disobedient' (the term widely used in descriptions of Gen. 3) and becoming
'bad' or 'evil' or 'sinful'. In THIS presentation, however, Adam and Eve
are revealed to be the HEROES, NOT THE VILLAINS they have so long
been portrayed as. So while humans ARE immensely upset—that is,
immensely angry, egocentric and alienated—WE ARE GOOD AND NOT
BAD AFTER ALL!!!! Yes, this explanation finally allows us to see that
our conscious mind that caused us to take the 'fruit' 'from the tree of...
knowledge' is *NOT* the sinful, evil villain it has so long been portrayed
as. And 'upset' is the right word for our condition because while we are
not 'evil' or 'bad' we are definitely psychologically upset from having to
participate in humanity's heroic search for knowledge. 'Corrupted' and
'fallen' have sometimes been used to describe our condition, but they
have negative connotations that we can now appreciate are undeserved.

[274]The following is another famous cartoon by Michael Leunig that
beautifully illustrates the journey humans have been on in our quest for
understanding—what is missing, however, from his depiction is this
final, truthful revelation of our species' heroism that the explanation of
the human condition has now made possible. So it is with deep reverence
to Leunig that I have taken the liberty of drawing three more frames in
his marvelously expressive style to complete the story. Leunig's cartoon depicts the Genesis story of Adam and Eve in the Garden of Eden,
beginning with the wide-eyed and innocent Adam and Eve taking the
'fruit' 'from the tree of...knowledge' and being 'banished...from the Garden
of Eden' as a result. Up to that point there is nothing unusual about the
story being portrayed, however, the cartoon goes on to show Adam
reaping revenge upon that hallowed Garden. It is possible that Leunig
meant for the retaliation to be interpreted as a straightforward joke about
human behavior—'You kicked me out, I'll get even'—but surely there is
a deeper truth to the retaliation that accounts for the cartoon resonating
with so many people. Hasn't Leunig got to the truth that lies at the very
heart of the issue of our human condition, and summarized that truth in
the most succinct way possible using the story of the Garden of Eden?
Hasn't he captured the underlying feeling that we humans have of being
condemned as fundamentally evil and God-disobeying when in our heart
of hearts we don't believe we are; and hasn't he captured the psychotic
and neurotic anger that feeling of unjust condemnation has caused us?

This is a deservedly celebrated cartoon, for, as emphasized, we humans
have been unjustly condemned for some *2 million years*—so how deeply,
deeply angry must we be inside ourselves having had to live undefended
on this planet for so long! Wouldn't, and shouldn't, we be as angry as
Leunig has depicted us in this cartoon!

Cartoon by Michael Leunig that appeared in Melbourne's *The Age* newspaper on 31 Dec. 1988

Drawings by Jeremy Griffith, with deeply appreciative deference to Michael Leunig, Jul. 2009

[275]To analyze the cartoon's elements more closely, in the fourth frame Adam is shown fuming with rage and resentment for being evicted, implying that he doesn't believe it is deserved, and deciding that he has no choice but to retaliate against the injustice; he can't be expected to just sit there and take it, he has to find some way of demonstrating that he doesn't accept as true the criticism that he is fundamentally bad. And so a vengeful Adam returns with a chainsaw to raze the Garden. The guardian angel is in tears at the wanton destruction, and we can see that Eve is similarly distressed by his actions. (This lack of empathy by women for men's immensely upsetting battle to defy the ignorance of our instinctive self, which Leunig has so honestly expressed here, will be explained in chapter 8:11B.) But Adam's expression and body language shows the enormous relief and satisfaction his retaliation brings him. In giving the guardian angel 'the finger' in the eighth frame, he's symbolically saying, 'Go to hell you bastard, for unjustly condemning me!'

[276]BUT, above all, in the expression of extreme anger on Adam's face, Leunig has revealed just what 2 million years of being unjustly condemned by the *whole* world has done to us humans. Yes, since the sun, the rain, the trees and the innocent animals are all friends of our original innocent, instinctive self or soul, through that association they too have condemned us, which is why Adam's/our, retaliation against nature for its unjust condemnation of him has left the whole natural world such a wasteland! We burnt the scrub, tore down the trees and dumped rubbish, pollutants and cement over what was left of nature, and we murdered animals who, in their innocence, condemned us. For example, there has been **'5,843 Sq km of Amazon rain forest reportedly lost to deforestation from August 2012 to July 2013; activists blame the 28% rise in one year on looser environmental laws'** (*TIME*, 2 Dec. 2013)—the deeper truth being that if we humans weren't so hateful and couldn't-care-less we wouldn't destroy pristine rainforest like this. Similarly, the diary of the legendary 'white hunter', the suitably named J.A. Hunter, reveals that he dispatched **'966 Rhinos'** from **'August 29th 1944 to October 31st 1946'** (Peter Beard, *The End of the Game*, 1963, p.137 of 280). That's the equivalent of nearly 10 rhinoceroses every week for more than 2 years that he shot to death! Incidentally, we can see here how, if we really wanted to save the environment, that 'hugging trees'

and 'patting dolphins' wasn't going to do it. Pretending to be loving and kind and considerate could make us feel good but it was never going to fix anything. To 'save the planet' we had to find the *understanding* that would end the underlying upset in us humans. We had to confront the issue of *us, the human condition*, NOT find ways to delude ourselves we were good so we didn't have to confront that issue. Pretense, delusion and escapism got us nowhere; in fact, it made the situation the human race is in much, much worse because it hid the real issue of the immensely psychologically upset state of the human condition. So yes, while we have learnt to conceal how upset we are—learnt 'civility'—underneath that facade of restraint and delusion lies the level of anger Leunig has portrayed.

[277]It really has been a case of **'Give me liberty or give me death'**, **'No retreat, no surrender'**, **'Death before dishonor'**, **'No guts no glory'**, **'Do or die'**, **'Die on your feet, don't live on your knees'**, **'Never give in'**, **'Better to reign in hell than serve in heaven'**, **'You can stand me up at the gates of hell, but I won't back down'**, as the sayings go. We were *never* going to give in to our instinctive self or soul; we were never going to accept that we were fundamentally bad, evil, worthless, awful beings; we weren't going to wear that criticism—for if we did, we wouldn't be able to get out of bed each morning and face the world. If we truly believed we were fundamentally evil beings, we would shoot ourselves. There had to be a greater truth that explained our behavior and until we found it we couldn't rest. And so every day as we got out of bed we took on the world of ignorance that was condemning us. We defied the implication that we are bad. We shook our fist at the heavens. In essence, we said, 'One day, one day, we are going to prove our worth, explain that we are not bad after all, and until that day arrives we are not going to **'back down'**, we are not going to take the ignorant, naive, stupid, unjustified criticism from our instincts, or from Integrative Meaning/God. No, we are going to fight back with all our might.' And that is what we have done; that is what *every* conscious human that has ever lived has done—and because we did, because we persevered against all the criticism, we have now finally broken through and found the full truth that explains that <u>humans are wonderful beings after all. In fact, not just wonderful but the heroes of the whole story of life on Earth</u>. Now, at last, we can

*finally understand* that the Greek playwright Sophocles was right when he wrote that **'There are numerous wonders in the world, but none more wonderful than man'** (*Antigone*, c.441 BC). This is because our fully conscious mind is surely—given its phenomenal ability to understand the world—nature's greatest invention, so for us humans who were given this greatest of all inventions to develop, to be made to endure that torture of being unjustly condemned as bad or evil for doing just that, and to have had to endure the torture for *so* long, *some 2 million years, has to* make us the absolute heroes of the story of life on Earth. We were assigned the hardest, toughest of tasks, and against all the odds we completed it. We humans *are* the champions of the story of life on Earth. We are *so, so* wonderful!

J.M.W. Turner's *Fishermen at Sea*, 1796

[278] In chapter 1:3 the story of *Don Quixote* was used to illustrate the stupendous courage of the human race. This painting by Turner is an equally powerful portrayal of how absolutely incredibly HEROIC the human race has been, huddled together as best we could for some reassurance,

and with a few provisions, while we struggled through 2 *million years* of terrifying darkness and tumultuous storms. To be given a fully conscious brain, the marvelous computer we have on our heads, but *not* be given the program for it and instead be left **'a restless wanderer on the earth'** (Bible, Gen. 4:14) searching for that program/understanding in a dreadful darkness of confusion and bewilderment, most especially about our worthiness or otherwise as a species, *was* the most diabolical of tortures; as the great denial-free thinking prophet Isaiah put it: **'justice is far from us, and righteousness does not reach us. We look for light, but all is darkness; for brightness, but we walk in deep shadows. Like the blind we grope along the wall, feeling our way like men without eyes...Truth is nowhere to be found'** (Bible, Isa. 59). Those lyrics that *so* plead the terrible agony of our species' seemingly lost and meaningless condition— **'How does it feel to be on your own, with no direction home, like a complete unknown'** —have understandably led to the 1965 song from which they come, *Like a Rolling Stone* by that prophet of our time, Bob Dylan, to be voted the greatest of all time by that arbiter of popular music, *Rolling Stone* magazine, in 2011. Yes, how *incredibly* heroic have we humans been—and how wonderful, beyond-the-powers-of-description, is it to now have freed ourselves from that horrific situation where the **'truth'** about ourselves was **'nowhere to be found'**!

[279] So, <u>the essential truth that has now at last been explained is that we should *never* have been 'thrown out of the Garden of Eden' in the first place</u>—which is why, to return to Leunig's cartoon, I have taken the liberty of drawing three additional frames to complete the story. The first addition depicts Adam and Eve beckoning to the guardian angel to return, while the second shows them explaining to the angel the irrefutable, first-principle-based, biological reason why humans are *not* fundamentally evil; in fact, explaining, as I have just done, that we are the absolute heroes of the story of life on Earth—an explanation that so affects the angel that it starts to cry out of regret and sympathy. In the third and final frame, we see the angel taking Adam and Eve—the human race in effect—by the hand and apologetically escorting us back to the Garden of Eden. Yes, the human race is coming home now, our **'banish[ment]'** has ended; we are no longer **'strangers in paradise'** (popular song from the 1953 musical *Kismet*), rather we are the heroes of paradise—its most revered, cherished and loved occupants! What a wonderful change of circumstances for us humans!!

Detail from Michelangelo's *The Creation of Adam*, c.1508-1512

[280] This is a detail from *The Creation of Adam*, Michelangelo's famous masterpiece that adorns the ceiling of the Sistine Chapel in the Vatican City in Rome. In interpretations of the painting, God is said to be in the process of creating Adam, yet it can also be construed as God and man reaching out to each other. Since we can now afford to acknowledge that God is the personification of the integrative, cooperative, selfless, harmonious ideal values of life, then with the reconciling understanding of our less-than-ideal, human-condition-afflicted state found, the 'out-stretched fingers' of God and man *have* finally touched. Our seemingly un-Godly, upset anger, egocentricity and alienation resulted from our participation in humanity's heroic search for knowledge. Yes, it turns out that in that classic 1980 American film *The Blues Brothers*, for all their wild, off-the-wall, upset behavior, the Blues Brothers really were **'on a mission from God'**, as they kept telling everyone they encountered; *indeed, the life of every human who has ever lived has been entirely meaningful and thus Godly*.

[281] Another universally iconic symbol that can now be interpreted through the truthful lens that this explanation allows is the Statue of Liberty that stands so proudly in New York Harbor (opposite). The United States of America presently sits at the forefront of the human race's immensely upsetting but critically important, heroic search for knowledge; it really has been **'the land of the free and the home of the brave'**, as its national anthem proclaims. So it is entirely fitting that this symbol of the USA is such a brave one because it has been *so* difficult maintaining the freedom/liberty necessary to continue the upsetting/corrupting search for knowledge. This is because as soon as there was too much corruption of humans' cooperative, selfless, loving instinctive self or soul—too much anger, egocentricity and alienation, but especially too much egocentric selfishness—a society became dysfunctional and decadent, and then, in

a bid to counter that dysfunction, it became oppressive and restraining of the freedoms each person needed to continue the corrupting search for knowledge. Yes, you had to be **'brave'** to avoid giving up the corrupting search for knowledge and maintain the **'free[dom]'** needed to continue humanity's heroic search for knowledge—*and the story of the human race as a whole is that it has never completely abandoned that corrupting search for knowledge, and as a result of that tenacity it has now finally succeeded in finding the particular knowledge it was in search of, namely the explanation of the human condition: the reconciling, ameliorating, peace-bringing understanding of why humans are good and not bad.*

## Chapter 3:7 <u>Our instinctive moral conscience has been 'A sharp accuser, but a helpless friend!'</u>

[282] Greek mythology describes how Prometheus stole fire from his fellow Gods and gave it to humans for their use, an act which enraged the Gods, and Zeus in particular who punished Prometheus by having him strapped to the top of a mountain where, every day in perpetuity, he was forced to suffer having his liver eaten out by an eagle. In light of what has been revealed, we can now understand that in this story fire is the metaphor for the conscious intellect (as it is in many mythologies; indeed, 'Prometheus' literally means 'forethought'), and that the consequence of humans gaining a conscious mind was extremely upset behavior, which explains why Prometheus was punished by the Gods—in their eyes his gift to humans of consciousness was responsible for the corruption of the human race, for our falling out with the Godly ideals.

[283] Yes, for 2 million years our intellect has been seen as the villain of the piece while our instinctive self or soul's moral conscience was held up as the epitome of goodness, but the truth, which we can now finally explain, turns out to be the exact opposite in the sense that it was our instincts' unjust criticism that caused us to become upset. This paradoxical turn of events in which our 'good side' is revealed to have been the 'bad side' is the theme of the crime writer Agatha Christie's famous play *The Mousetrap*. The play is just another 'whodunit' murder mystery and yet it is now the longest running play in history, having been performed continuously since opening in 1952; indeed, the play's enduring popularity is such that it celebrated its 60th anniversary in 2012 with a global tour. All enduring myths and stories contain truths that resonate, and in the case of *The Mousetrap*, the police inspector involved in the murder investigation, who is held up throughout the play as the pillar of goodness and justice, is revealed at the very end to be the culprit. It is the essential story of humanity where the apparent ideals of our original instinctive self or soul's selfless, loving world are revealed, at the very last moment, to have been the unjustly condemning influence. As with so many aspects of the human condition, THE TRUTH WAS NOT AS IT APPEARED. We discover at the very end of our journey to enlightenment that conscious humans, immensely corrupt as we are, are 'good' and not 'bad' after all, and that which was 'good', our moral conscience, turns out to be the cause of our 'sin'.

[284] The same essential paradox appears in G.K. Chesterton's 1908 'masterpiece' (Simon Hammond, *The Guardian*, 7 Oct 2012), *The Man Who Was Thursday*, in which a policeman representing the 'good' side has to infiltrate and expose the sinister members of a quintessentially corrupt organization, but as the tale unfolds each of the apparently corrupt members are revealed to be forces for good commissioned to fight evil.

[285] The poet Alexander Pope acknowledged the pain, guilt and frustration caused by the unjust criticism emanating from our species' instinctive moral conscience when he wrote, **'our nature** [conscience—is]**...A sharp accuser, but a helpless friend!'** (*An Essay on Man*, 1733, Epistle II). It was a sentiment echoed by William Wordsworth in his great 1807 poem, *Intimations of Immortality*: **'High instincts before which our mortal Nature / Did tremble like a guilty thing surprised.'** Albert Camus was another who felt the sting of the criticism from our naive, ignorant, innocent soul when he asked **'whether innocence, the moment it begins to act, can avoid committing murder [?]'** (*The Rebel*, 1951, p.12 of 269).

[286]Considering then how unjustly hurtful our instinctive self or soul has been it is little wonder we learnt to psychologically block it out, deny and bury it to the point where we now refer to it as **'the child within'** and the **'collective** [shared by all, instinctive] **unconscious'**. Indeed, if we look at the situation from the perspective of our original, unconditionally selfless, all-loving, moral instinctive self or soul for a moment, we can begin to appreciate just how enormous the schism between our instinctive soul and conscious mind has been. Imagine then what a shock it was for our all-loving, moral instincts when our conscious mind began searching for knowledge and became angry, aggressive, selfish and competitive—our soul would have been absolutely mortified, as children are today by the extreme imperfection of human behavior. Our soul would have been *utterly* bewildered and distressed, *completely* overcome with shock and disappointment, *absolutely* devastated: 'Why, when all our behavior has been so cooperative, loving and perfect, and our world so happy and content as a result, are you, our conscious mind, doing this?—this is *SO* wrong!', is how our soul has perceived the situation. Unable to explain our divisive behavior, we conscious humans have never allowed ourselves to properly consider the hurt that behavior has caused our moral soul, but now that we can at last explain our upset state we can afford to do so. Yes, when our conscious mind emerged and the angry, aggressive, selfish and competitive behavior started to appear, our original, all-loving and all-sensitive instinctive self experienced the most profound shock imaginable. Our soul has been *completely* distressed and overwhelmed by this turn of events: 'Why is this happening, this behavior simply must stop, this is absolutely wrong!' Our soul has been in tears of distress and disappointment, utterly overcome with unhappiness—BUT WHAT WAS HAPPENING *HAD* TO GO ON, THE UPSETTING SEARCH FOR THE KNOWLEDGE THAT WOULD LIBERATE OUR SPECIES *HAD* TO CONTINUE.

[287]To return, however, to viewing the situation from the perspective of our conscious intellect, this **'sharp accuser'**, these **'high'** and mighty **'innocen[t]'** **'instincts'** that made us **'tremble like a guilty thing'**, *have* been **'murder[ous]'** and thus completely unbearable, SO WE SIMPLY HAD TO BLOCK OUR MORAL INSTINCTIVE SELF OUT, OTHERWISE WE WOULD DIE FROM THE PAIN, THE SHAME, THE MORTIFICATION, THE GUILT, IT WAS CAUSING US. So the extent of our soul's unhappiness with us, which was of a level of *extreme* unhappiness, is the extent we, our conscious mind, had to block out and become alienated from it, which is, it

follows, a level of *extreme* alienation. As a species, we are EXTREMELY soul-repressed, or psychotic, and EXTREMELY mentally committed to living in denial, or neurotic. R.D. Laing was right when, as mentioned earlier in par. 123, he described humans as being **'a shrivelled, desiccated fragment of what a person can be...between *us* and It [our true selves or soul] there is a veil which is more like fifty feet of solid concrete...The outer divorced from any illumination from the inner is in a state of darkness. We are in an age of darkness. The state of outer darkness is a state of sin—i.e. alienation or estrangement from the inner light...We are bemused and crazed creatures, strangers to our true selves, to one another'**, and **'We are dead, but think we are alive. We are asleep, but think we are awake...We are the halt, lame, blind, deaf, the sick. But we are doubly unconscious...We are mad, but have no insight** [into the fact of our madness].' We have certainly been, as Plato said, **'enshrined in that living tomb which we carry about, now that we are imprisoned'**—but *again*, our conscious, intellectual self simply had *no choice* but to block out our instinctive self or soul, to banish it to our subconscious where it only now occasionally bubbles up in dreams and on other occasions when our conscious self is subdued, such as when praying or meditating. As Carl Jung wrote, **'The dream is a little hidden door in the innermost and most secret recesses of the psyche** [soul], **opening into that cosmic night which was psyche long before there was any ego consciousness'** (*Civilization in Transition*, 1945; *The Collected Works of C.G. Jung*, tr. R. Hull, Vol.10). That pre-eminent philosopher Sir Laurens van der Post also wrote about the repression of our soul when he acknowledged that **'Human beings know far more than they allow themselves to know: there is a kind of knowledge of life which they reject, although it is born into them: it is built into them'** (*A Walk with a White Bushman*, 1986, p.142 of 326). And in talking about how human psychosis and neurosis develops, the psychiatrist D.W. Winnicott wrote that **'The word "unconscious"...has been used for a very long time to describe unawareness...there are depths to our natures which we cannot easily plumb...a special variety of unconscious, which he** [Sigmund Freud] **named** *the repressed unconscious*...**what is unconscious cannot be remembered because of its being associated with painful feeling or some other intolerable emotion'** (*Thinking About Children*, 1996 posthumous publication of his writings, p.9 of 343).

[288] Yes, our ability now to understand the human condition, understand the consequences of humans gaining a conscious mind, means the whole mystery of our life in extreme alienation finally becomes fully understandable. *Of course* there is **'a kind of knowledge of life which**

they [conscious humans] **reject'**, even **'though it is born into'** us—our moral conscience *has been an unbearably* **'sharp accuser, but a helpless friend!'** *Of course* we have had to block out our soul, and with it any truth that brought the unbearable issue of our corrupted human condition into focus! And no wonder we are 2 million years wedded to a life deep underground in Plato's metaphorical cave of dark, truthless, alienated denial. And, given how precious, indeed life-saving, that block-out or denial has been, and how practiced and habituated we are in applying it, it is no wonder it has been almost impossible for anyone to break through that denial, actually confront the human condition and find the liberating explanation of it—and, further, that there has been so much resistance to having all that denial demolished and exposed. Tragically, we had *no choice* other than to suppress our all-sensitive, unconditionally selfless, all-loving (although unloving of our intellect's need to search for knowledge!) soul—as the saying goes, **'we hurt the one we love'**—BUT THANKFULLY, WITH THE EXPLANATION OF WHY WE FULLY CONSCIOUS HUMANS BECAME ANGRY, AGGRESSIVE, SELFISH AND COMPETITIVE FOUND, WE NO LONGER HAVE TO. Indeed, we now have the ability to end our *psychosis*, to rehabilitate our wondrously sensitive soul. Further, our conscious mind no longer has to practice denial, which means we can also end our *neurosis*—all of which means the human race can be *psychologically* rehabilitated; our species can be brought back to life from its psychologically dead state.

## Chapter 3:8 <u>Most wonderfully, this psychosis-addressing-and-solving, *real* explanation of the human condition makes possible the psychological rehabilitation of the human race</u>

[289] What is *so* utterly exonerating and psychologically healing about our ability to understand the human condition now is that we can finally appreciate that there was a *very good* reason for our angry, alienated and egocentric lives—in fact, we can now see why our species has not just been ego-*centric*, but ego-*infuriated*, even ego-gone-mad-with-murderous-rage for having to live with so much unjust criticism/guilt, for some *2 million years*. <u>At long last we have the reason for humans' capacity for shocking acts of cruelty, sadism, hate, murder and warfare—we now know the source of the dark volcanic forces in us humans</u>. And no wonder we led such an evasive, escapist, superficial and artificial, greedy,

smother-ourselves-with-*material*-glory-while-we-lacked-the-*spiritual*-glory-of-compassionate-understanding-of-ourselves, power, fame and fortune-seeking existence. Yes, it was the accumulation of money or capital that served to supply the symbolic wins we needed to counter the insecurity of our seemingly worthless condition. And indeed, the so-called Seven Deadly Sins of the human condition, of lust, anger, pride, envy, covetousness, gluttony and sloth, are all just different manifestations of the three fundamental upsets of anger, egocentricity and alienation that unavoidably emerged when humans became fully conscious and *had to* set out in search of knowledge in the presence of unjustly condemning instincts.

[290] And with the ability now to explain and understand that we are actually all good and not bad, the upset that resulted from not being able to explain the source of our divisive condition is able to subside and disappear. <u>Finding understanding of the human condition is what rehabilitates and transforms the human race from its psychologically upset angry, egocentric and alienated condition</u>. The word **'psychiatry'** literally means **'soul-healing'** (derived as it is from *psyche* meaning **'soul'** and *iatreia* which according to *The Encyclopedic World Dictionary* means **'healing'**), but never before have we been able to 'heal our soul', explain to our original instinctive self or soul that our fully conscious, thinking self is good and not bad and, by so doing, reconcile and heal our split selves. As Professor Harry Prosen has said about the psychological effect of this all-loving, all-compassionate understanding of ourselves: **'I have no doubt this biological explanation of the human condition is the holy grail of insight we have sought for the psychological rehabilitation of the human race.'** Yes, our ability now to understand the dark side of ourselves means we can finally achieve the **'wholeness for humans'** that Carl Jung was forever pointing out **'depends on the ability to own our own shadow'**. Sir Laurens van der Post was also describing the understanding that was required to heal our species' corrupted, 'fallen' condition when he wrote that **'True love is love of the difficult and unlovable'** (*Journey Into Russia*, 1964, p.145 of 319); and, **'how can there ever be any real beginning without forgiveness?'** (*Venture to the Interior*, 1952, p.16 of 241); and that **'Only by understanding how we were all a part of the same contemporary pattern** [of wars, cruelty, greed and indifference] **could we defeat those dark forces with a true understanding of their nature and origin'** (*Jung and the Story of Our Time*, 1976, p.24 of 275); and that **'Compassion leaves an indelible blueprint of the recognition that life so sorely needs between one**

**individual and another; one nation and another; one culture and another. It is also valid for the road which our spirit should be building now for crossing the historical abyss that still separates us from a truly contemporary vision of life, and the increase of life and meaning that awaits us in the future'** (ibid. p.29). Yes, one day there had to be, to quote The Rolling Stones' lyrics, **'Sympathy for the devil'** (1968); one day, we had to find the reconciling, compassionate, healing understanding of the dark side of human nature. One day, *'The Marriage of Heaven and Hell'*, as William Blake titled his famous book (c.1790), had to occur.

[291] This explanation *is* immensely positively reinforcing and thus therapeutic because <u>we can now understand that humans *are* nothing less than the heroes of the story of life on Earth</u>. This is the fundamental truth that, as I said in chapter 2:8, we humans hoped and believed we would one day establish, *and now have*! Despite all appearances to the contrary—despite our anger, egocentricity and alienation—humans *are* the most wonderful beings. <u>In fact, if there is a word such as divine that can be applied to mortals, then we can now see that it truly does belong to us humans</u>—because to withstand 2 million years of the injustice that we have had to endure and still be on our feet, still be able to laugh, still be able to smack each other on the back with encouragement, still be able to carry on, get out of bed each day and face life under the duress of the human condition, we *must* be the most magnificent of organisms! Truly, as Camus wrote, **'man's greatness...lies in his decision to be stronger than his condition'** ('The Night of Truth', *Combat*, 25 Aug. 1944), to not give in to the possibility he famously described in his 1942 essay *The Myth of Sisyphus*, that **'There is but one truly serious philosophical problem, and that is suicide. Judging whether life** [under the duress of the human condition] **is or is not worth living'**.

[292] This fact of the utter magnificence of the human race—that we can now understand and know is true rather than merely hope it is so—brings such intense relief to our angst-ridden cells, limbs and torsos that it will feel as though we have thrown off a shroud of heavy chains. <u>The great, heavy burden of guilt *has* finally been lifted from the shoulders of humans</u>. Yes, doesn't the core feeling exist in all humans that far from being meaningless, **'banish**[ment]**'**-deserving **'evil'** blights on this planet we *are* all immense heroes? Doesn't this explanation at last make sense of all humans' immensely courageous and defiant attitude? And won't this explanation bring deep, bone-draining relief to the whole of each person's being?

[293] Yes, this explanation, this understanding of the human condition, provides the **'insight we have sought for the psychological rehabilitation of the human race'**, as Professor Prosen said it would. It enables us to see that our divisive human nature was *not* an unchangeable or immutable state as many people came to believe it is, and which E.O. Wilson's Multilevel Selection theory for eusociality deems it to be, but the result of the human condition, our inability to understand ourselves, and, as such, it *can* dissipate now that we have found that understanding.

[294] Importantly, this understanding of why we became upset as a species doesn't condone or sanction 'evil', rather, through bringing understanding to humans' upset behavior, it gives us the power to ameliorate and thus subside and ultimately eliminate it. 'Evil'—humans' divisive behavior—was a result of a conflict and insecurity within us that arose from the dilemma of the human condition, so once you resolve the dilemma, you end the conflict and insecurity. Peace could only come to our troubled, divisive state and world through removing the underlying insecurity of our condition. With our ego or sense of self worth satisfied at the most fundamental level, our anger can now subside and all our denials and resulting alienation can be dismantled. From having lived in a dark, cave-like, depressed state of condemnation and, as a result, had to repress, hide and deny our true selves, we can at last, as the 1960s rock musical *Hair* sang, **'Let the sunshine in'**—end our horrid existence of having to depend on denial to cope. The compassionate-understanding-based psychological rehabilitation of the human race—the TRANSFORMATION of all humans—truly *can* begin. Yes, finally, at last, we can, as Jim Morrison hoped when, as mentioned earlier, he sang of **'Standing there on freedom's shore, waiting for the sun…waiting…to tell me what went wrong'**, have **'day destroy…**[the] **night'** and **'break on through to the other side** [to a sound world free of alienated denial]'. (A detailed description of how the transformation of the human race is now able to take place, such that every human can now *immediately* be free of the human condition, is presented in the concluding chapter 9 of this book.)

[295] As for the veracity of this explanation, it is precisely this explanation's ability to at last make relieving sense of human life, of all our behavior in fact, that lets us know that we have finally found the *true* explanation of the human condition. The great physicist Albert Einstein once wrote that **'Truth is what stands the test of experience'** (*Out of My Later Years*, 1950, p.115 of 286), and since this study and explanation is all about us,

about *our* behavior, we are in a position to personally **'experience'** its validity, to know if it's true or not. As the subject of this study, we can know if the ideas being put forward work or not, if these explanations do make sense of our deepest feelings; of our volcanic anger, of our lonely estranged souls, of our insecure state yet core belief that we are wonderful beings, etc, etc—as, in fact, they *do*. Moreover, these explanations are *so* powerfully insightful, accountable and revealing (so true) that they render our lives transparent—a transparency, a sudden exposure, that can initially be overwhelmingly confronting and depressing. But there is a way, an absolutely wonderful, joyous way, to cope with the arrival of exposure day, or transparency day, or revelation day, or truth day, or honesty day—in fact, the long-feared so-called **'judgment day'**, which is actually *not* a time of condemning 'judgment' but of compassionate understanding. And that wonderful way of coping with the arrival of the all-liberating, all-rehabilitating but at the same time all-exposing truth about us humans is, as stated above, the subject of the concluding chapter (9) in this book. But the point being made here is that the transparency of our lives that these explanations bring reveals just how effective, how penetrating, and, therefore, *how truthful* these under-standings actually are.

[296] So, we can see that as dishonest as it necessarily was in its approach, mechanistic, reductionist science turns out to be the liberator of humanity, the proverbial 'messiah' or 'savior' of humanity, for it *has* finally enabled us to lift the so-called 'burden of guilt' from the human race! It found understanding of the difference in the way nerves and genes process information, that one is insightful and the other isn't, which has finally made it possible to explain that greatest of all mysteries, that holy grail of all human conscious thought and enquiry, of the human condition. As was mentioned in par. 245, the fundamental conflict involved in our condition between our innocent instinctive self and our conscious self was described as far back as when Genesis was written by Moses and Plato presented his allegory of the two-horsed chariot, but we couldn't *explain* that conflicted state until science found understanding of the different ways genes and nerves process information.

[297] Significantly, now that we are able to understand from scientific first principles that upset is not an 'evil', worthless, bad state, but an immensely heroic state, we can know that while, inevitably, all humans are variously upset from their different encounters with, and degrees of engagement in,

humanity's epic search to find knowledge, ALL HUMANS ARE EQUALLY GOOD. Everyone is variously angry, egocentric and alienated, but everyone is good, and not just good but a hero of the story of life on Earth! No longer does humanity have to rely on dogmatic assertions that **'all men are created equal'** purely on the basis that it is a **'self-evident'** truth, as the United States' Declaration of Independence asserts, because we can now *explain, understand* and *know* that the equality of goodness of all humans is a fundamental truth. We can now understand *why* everyone is equally worthy, and that no one is superior or inferior, and that *everyone* deserves the **'rights'** of **'life, liberty and the pursuit of happiness'**. Indeed, through this understanding, the whole concept of good and bad disappears from our conceptualization of ourselves. Compassionate, relieving truth and honesty about humans has finally arrived. Religions have taught us that 'God loves you', and while that mantra has been comforting, ultimately we needed to understand *why* we were lovable (again, the issue of who or what God is will be addressed in more detail in chapter 4:3). Similarly, the Bible states that **'the truth will set you free'** (John 8:32), and while we know this statement to be true, the problem has been that all the *partial* truths—such as that humans are the most brutal and destructive animals to ever walk the earth—condemned our upset state, fuelling it further, which means that, ultimately, for the truth to *genuinely* set us free, it had to be the *full* truth that explains *why* humans are *all* good and not bad.

[298] Yes, the human condition is certainly shot through with paradox: to become happy we had to first endure unhappiness; we appeared to be bad but believed we were good; we are intelligent, smart and clever but, by all appearances, behave in such an unintelligent, stupid way that we have brought the world to the brink of destruction. But most wonderfully, we at last have the understanding that <u>reconciles</u> not just these but <u>*all* the polarities of life</u> that existed under the duress of the human condition—<u>between 'good' and 'evil', instinct and intellect, emotion and reason, conscience and conscious, ignorance and wisdom, soul and mind, heart and head, 'I feel' and 'I think', yin and yang, light and dark, the innocent and the corrupted, the un-embattled and the battle-hardened, the selfless and the selfish, the happy and the upset, the light-hearted and the heavy-hearted, the cooperative and the competitive, the integrative and the divisive, the 'Godly' and the 'unGodly', the gentle and the aggressive, the loving and the hateful, the sound and the alienated, the secure and the insecure, the honest and the dishonest, the natural and</u>

the artificial, the non-sexual and the sexual (sex as humans practice it is explained in chapter 8:11B), altruism and egotism, idealism and realism, spiritualism and materialism, socialism and capitalism, left-wing and right-wing, instinctualism and intellectualism, religion and science, holism and mechanism, young and old, women and men (the different roles of women and men under the duress of the human condition is also explained in chapter 8:11B), blacks and whites (the differences in upset between 'races' is explained in chapter 8:16E), unresigned and resigned, country and city, etc, etc. The explanation of the human condition unravels and makes sense of the whole seemingly impenetrable and insoluble confusion of human life—and does so in such a simple and obvious way that it brings to mind Huxley's famous response to Darwin's idea of natural selection: **'How extremely stupid of me not to have thought of that!'** Yes, we have at last found the UNDERSTANDASCOPE for human behavior.

## Chapter 3:9 The end of politics

[299] Most wonderfully, with this understanding everyone *can* come in from the cold, from their lonely outposts of bewilderment about what has actually been going on in the world of humans. The result is that immense— and urgently needed—fundamental change can now occur for the human race. Conflict from misunderstanding especially goes. For instance, this explanation obsoletes the conflict between the philosophically opposed left and right wings of politics because we can now understand that while giving in to our ideal-behavior-demanding instincts—'flying back on course' in the Adam Stork analogy—was an immensely guilt-relieving exercise that made us 'feel good' (the left-wing approach), it was fundamentally irresponsible because it meant abandoning humanity's upsetting but necessary search for knowledge (the right-wing approach). The truth was not as it appeared—participating in humanity's selfish, aggressive and competitive battle to find knowledge was not the 'bad' transgression it has been condemned as, rather it was the responsible course of action and thus a 'good' thing. The paradox of the situation was marvelously summed-up in the musical *Man of La Mancha*, when, as mentioned in par. 68, it says we had to be prepared **'to march into hell for a heavenly cause'** (Joe Darion, *The Impossible Dream*, 1965). Yes, we had to, as it were, lose ourselves to find ourselves—suffering upset was the heroic price we *had* to pay to find understanding, ultimately self-understanding and, with it, freedom

from the upset state of the human condition. So while the left-wing has had an absolute field day demonizing the right-wing's support of the non-ideal state—even labeling it **'evil'**, as was proselytized by the left-wing activist Michael Moore *(Capitalism: A Love Story, 2009)*—we can *now* see that it was the philosophy of the left that was morally bankrupt, void of meaning in the sense that while siding with idealism made its practitioners 'feel good', it ultimately had no relevance in humanity's critical journey from ignorance to enlightenment. Proponents of the left-wing approach were deluding themselves that they were holding the moral high ground when, in fact, the reverse was true—their stridently *pseudo* idealistic, dogmatic, condemning, ridiculing, escapist, deluded, arrogant, dishonest culture *oppressed* progress towards humanity's liberation from the human condition, stymieing the return of the *genuinely* peaceful, ideal world. Karl Marx, the political theorist whose mid-nineteenth century ideas gave rise to socialism and communism, was very wrong when he asserted that **'The philosophers have only interpreted the world in various ways; the point is** [not to understand the world but] **to change it** [just make it cooperative/social/ communal]' *(Theses on Feuerbach, 1845).* The whole **'point'** and responsibility of being a conscious being *is* to understand our world and our place in it—ultimately, to find understanding of our seemingly horribly flawed human condition. In short, pseudo idealism demands that we 'Just be ideal, don't think about why we are not ideal; in fact, don't go anywhere near the issue of "self", namely the issue of the human condition'. Just pretend there is no human condition that has to be understood and that all humans need to do to fix up their world is behave ideally! It is an attitude of total delusion and complete dishonesty.

[300] The immense danger of pseudo idealistic left-wing thinking was that the longer the upsetting search for knowledge went on without the goal of that search of the true, psychologically relieving explanation of the human condition being found, the more people became upset sufferers of the human condition and the more they couldn't resist the artificial, feel-good relief offered by the left-wing's dogmatic, anti-knowledge, pseudo idealistic, deluded, dishonest, 'fly back on course' way of living. *And*, in contriving the Multilevel Selection theory for human behavior, E.O. Wilson furnished the whole dumbing-down process with the ultimate delusion it needed of a 'biological', non-explanation nullification of the core, critical, all-important issue of the human condition itself, thereby virtually locking humanity onto a path to terminal dishonesty/dogma/

'phon[iness]'/'fake[ness]'/alienation/darkness/extinction! As the author George Orwell famously predicted, **'If you want a picture of the future, imagine a boot stamping on a human face—for ever** [stamping forever on the freedom that humans' extraordinary and unique conscious mind needs to search for knowledge]' (*Nineteen Eighty-Four*, 1949, p.267 of 328).

[301] The impact of this pseudo idealistic development on the democratic process is grim indeed, because democracy—in which the proper balance is sought between the need to maintain a degree of selfless, loving, soulful, cooperative 'ideal' behavior for society to function, and the need for there to be sufficient freedom from the imposition of the expectation of such 'ideal' behavior, for people to be able to carry on the corrupting search for knowledge—is destroyed as an effective and meaningful process when people start voting for idealism *not* because there is too much selfishness in society but simply because it makes *them* 'feel good'. We can now at last appreciate that when we vote this way we are *not* participating in the true democratic process that human advancement has been dependent upon, we are exploiting and subverting it. And while many people think that even if feel-good pseudo idealism is not a real form of idealism it is surely harmless, that is most certainly *not* the case. *No*, it is extremely bloody-minded, totally selfish and dangerously destructive behavior that says, 'I no longer care about the human race, only about finding personal relief from my human condition.' Of course, such behavior was always going to develop when the levels of upset in society reached unbearable heights because at that end play point the need for relief from the agony of the human condition was going to become the *only* concern amongst an ever-increasing proportion of the population—with the dire consequence that democracy would fail to find the proper balance because the human-race-indifferent, deluded, feel-good, pseudo idealistic, left-wing political attitude would become impossible to defeat in an election, leading to the human race suffering a horrifically alienated death by dogma.

[302] Certainly, the upset behavior that results from right-wing participation in humanity's heroic search for knowledge is increasingly bringing about immense human suffering and environmental devastation, but it is the knowledge-oppressing left-wing that poses the *real* threat to the survival of the human race because *only* through successfully completing that search for knowledge could humanity be liberated from the upset state of the human condition. So while, as the journalist Geoffrey Wheatcroft

recognized, '**the great twin political problems of the age are the brutality of the right, and the dishonesty of the left'** ('The year of sexual correctness and double standards', *The Australian Financial Review*, 29 Jan. 1999), it is NOT '**the brutality of the right'** but '**the dishonesty of the left'** that stands like a colossal ogre over the human race, threatening to destroy it. Yes, we can now appreciate that the philosopher Friedrich Nietzsche spoke the truth about the need for humanity to hold its nerve when he wrote that '**There have always been many sickly people among those who invent fables and long for God** [ideality]: **they have a raging hate for the enlightened man and for that youngest of virtues which is called honesty…Purer and more honest of speech is the healthy body, perfect and square-built: and it speaks of the meaning of the earth** [which is to fight for knowledge, ultimately self-knowledge, understanding of the human condition]…**You are not yet free, you still *search* for freedom. Your search has fatigued you…But, by my love and hope I entreat you: do not reject the hero in your soul! Keep holy your highest hope!…War** [against the oppression of dogma] **and courage have done more great things than charity. Not your pity but your bravery has saved the unfortunate up to now…What warrior wants to be spared? I do not spare you, I love you from the very heart, my brothers in war!'** (*Thus Spoke Zarathustra: A Book for Everyone and No One*, 1892; tr. R.J. Hollingdale, 1961, pp.61-75 of 343). Yes, as stated in par. 281, you had to be '**brave'** to avoid the temptation of giving up the self-corrupting search for knowledge. (Much more is explained about the extreme danger of pseudo idealism in chapter 8:16, 'The last 200 years'.)

[303] Thankfully, the arrival of the demystifying, exposing and reconciling, true explanation of the human condition has the power to cut short this frightening, death-by-dogma, left-wing threat to humanity—that is, of course, if humanity accepts this lifeline it has been thrown. And, for its part, the right-wing's need to support the upsetting, often brutal, competition-aggression-and-selfishness-producing battle/'**war'** to defy the ignorance of our ideal-behavior-demanding, unjustly condemning, moral instinctive self or soul and find understanding of ourselves, can *also* come to an end with the acceptance of this liberating insight. Yes, with understanding of the human condition now found, the *whole* necessary but ugly business of politics can happily come to a close.

[304] The explanation of just how the human race can now 'put down the sword', end its egocentric, 'must-prove-that-we-are-good-and-not-bad', '**warrior'** existence, and by so doing transform itself into a peaceful state, is explained in the concluding chapter 9 of this book.

## Chapter 3:10 'Free at last!'

[305]The end of the political process as we know it is just one example of the great change that can now come to the human race—this *is* the ultimate 'future shock', massive paradigm shift, for humans—BUT, as will be explained in some detail in chapter 9, since this shift is an immensely positive change, the transformation will not only be easy but *fabulously* exciting. Indeed, so great is both the amount of change we now face (the **'future shock'**) and the self-confronting exposure (the **'judgment day'** effect mentioned earlier) that I and others established the World Transformation Movement to introduce and help manage the truly wonderful way we are able to manage these huge transitions. It is called the World Transformation Movement because the most fabulous effect of having understanding of the human condition is that *all* our psychologically upset angry, egocentric and alienated behavior can now, immediately, be put aside, and eventually completely disappear, thus bringing about the complete rehabilitation and transformation of the human race and thus of our planet.

[306]So this is the end of the world as we know it and the beginning of the world we always hoped for! Indeed, while the 'Socialist', 'Temperance', 'Age of Aquarius', 'Peace', 'New Age', 'Feminist', 'Green', 'Politically Correct', 'Postmodernist', 'Multicultural', 'Anti-Capitalist' movements (and these are only the more recent of the litany of deluded, false starts to an ideal world for humans that we have witnessed) have all severely discredited—and inhibited—the prospect of a truly human-condition-resolved, psychologically ameliorated, completely transformed new world for humans, one of the founders of the 1980s New Age Movement, the author Marilyn Ferguson, did offer this accurate description of the now-realized hope of its arrival: '**Maybe** [the Jesuit priest, scientist and philosopher] **Teilhard de Chardin was right; maybe we are moving toward an omega point** [a final genuine unification/individuation of our split selves]... **Maybe...we can finally resolve the planet's inner conflict between its neurotic self (which we've created and which is unreal) and its real self. Our** [original all-sensitive and loving instinctive] **real self knows how to commune, how to create...From everything I've seen people really urgently want the kind of new beginning...**[that I am] **talking about** [where humans will live in]**...cooperation instead of competition'** (*New Age* mag. Aug. 1982; see <www.wtmsources.com/174>).

<sup>307</sup> Yes, from being selfish, aggressive and competitive, humans can now return to being selfless, loving and cooperative. Our round of departure has ended. T.S. Eliot wonderfully articulated our species' journey from an original, innocent yet ignorant state, to a psychologically upset, 'fallen', corrupted state, and back to an uncorrupted but this time enlightened state when he wrote, **'We shall not cease from exploration and the end of all our exploring will be to arrive where we started and know the place for the first time'** (*Little Gidding*, 1942).

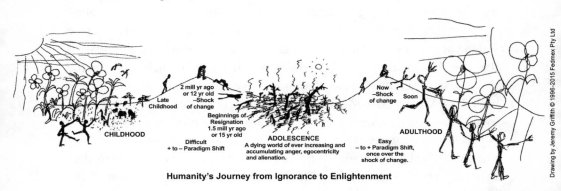

**Humanity's Journey from Ignorance to Enlightenment**

<sup>308</sup> I have drawn this picture to summarize this horrendously agonising but awesomely heroic journey that our species has taken from ignorance to enlightenment. From its innocent, happy, cooperatively orientated, selfless and loving infancy and childhood, humanity progressed to the horror of an insecure, upset adolescence where humans had to search for our identity—specifically for understanding of why we lost our innocence and became selfish, aggressive and competitive. But with understanding of the human condition now found, humanity can, at long last, enter the happy, ameliorated, secure and mature state of adulthood. (Again, all these stages that the human race has progressed through will be explained and described in some detail in chapter 8.) Martin Luther King Jr's **'dream'** has finally come true; we can now **'allow freedom to ring… from every village and every hamlet, from every state and every city'** because **'all of God's children, black men and white men, Jews and Gentiles, Protestants and Catholics'** can **'join hands and sing'**, **'Free at last! Free at last! Thank God Almighty, we are free at last!'** ('I Have A Dream' speech, 28 Aug. 1963).

<sup>309</sup> Yes, the human race can now, at last, ultimately, finally leave Plato's horrid **'living tomb' 'cave'** of truth-avoiding, alienated, psychotic and neurotic darkness and stand in the warm, healing sunshine of understanding.

And to celebrate the completion of the first draft of this book that **'let**[s] **the sunshine in'**, I have made, from golden sea shells I found on a beach in Sydney, this exciting picture that depicts that all-fabulous liberating and healing sun, which also represents a mandala of the arrival of harmony and happiness. (This picture is especially worth seeing in color in the online version at <www.humancondition.com>.)

Composed of golden cockle shells, white star limpets and a large scallop shell from a Sydney beach, this collage was made by the author in 2013 to celebrate this time that the human race has dreamed of, and which has now arrived, when we can finally leave Plato's dark cave of denial and 'let the sunshine in'. To help fund the world-saving work of the World Transformation Movement, this artwork is for sale for $US 5 million.

[310] Having now established the fundamental goodness of humans it finally becomes both possible and psychologically safe to present the fully accountable and thus true scientific explanations for the meaning of existence; for how humans acquired our moral instincts; and for how we became conscious when other animals haven't. These answers, as well as a rebuttal of the false accounts that have been put forward to supposedly explain the origins of our moral instincts, are presented in the next four

chapters of this book. Chapter 6, which contains the rebuttal of the false theories that have been put forward for our moral nature, also provides an account of the ill-treatment, even persecution, these human-race-saving answers have received for 30 years from the mechanistic paradigm—for while our species' journey to enlightenment *is* complete, this explanation's journey to acceptance is still a work in progress. Since the conscious mind *must* surely be nature's greatest invention, its failure to fulfill its great potential by not recognizing the psychosis-addressing-and-explaining real explanation of the human condition that has now arrived would represent a failure of the whole story of life on Earth! This resistance, especially by the scientific establishment, to recognize this understanding of the human condition is an *extremely* serious matter. The survival of humanity is hanging in the balance, and that balance simply must tip the right way.

[311] The final chapters of this book, chapters 8 and 9 (respectively), present a description of humanity's journey from ignorance to enlightenment, and a description of how the human race is now able to be transformed.

[312] (A more complete description of this fully accountable, human-race-transforming, *true* biological explanation of the human condition can be found in my freely available, online book *Freedom Expanded* at <www.humancondition.com/freedom-expanded>.)

# Chapter 4

# The Meaning of Life

Michelangelo's *The Creation of Adam* from the Sistine Chapel, c.1508-12.
Michelangelo's masterpiece also came to symbolize our hope of
reconciliation with God, which has now occurred.

## Chapter 4:1 Summary

[313] While the conventional view in science is that there is no direction
or purpose or meaning to existence, and that change is random, there
is, in fact, a very obvious theme, direction, purpose and meaning to
existence, which is the ordering or integration or complexification of
matter into ever larger and more stable wholes. Indeed, the answer
to the great question of 'what is the meaning of life' is that it is to
live in accordance with this integrative, order-of-matter-developing,
cooperation-and-selfless-behavior-dependent theme or meaning of
existence. HOWEVER, *until* we could explain our seemingly imperfect
human condition, explain *why* humans appear to have been living in
defiance of 'Integrative Meaning', namely in a divisive competitive and
selfish way, such a truth could not be faced. *BUT*, as seen in the previous
chapter, since we CAN now at last explain the good reason why we

humans have been divisively rather than integratively behaved, we CAN finally admit the truth of Integrative Meaning and acknowledge that there is a direction or purpose to existence. And furthermore, since our concept of 'God' is actually the personification of the unbearably condemning truth of Integrative Meaning, through admitting this truth humans can *also* demystify 'God' and, by so doing, change from being a 'God'-fearing species to a 'God'-confronting one. Yes, the instinct vs intellect explanation of the human condition makes sense of and reconciles *all* the unresolved manifestations of the polarities of the human situation—'good' vs 'evil', idealism or 'God' vs human's non-ideal existence, religion vs science, holism vs mechanism, altruism vs egotism, communism vs capitalism, conscience vs conscious, and so on—thus ameliorating or healing, and finally ending, the psychologically upset, divisive way we have been living!

## Chapter 4:2 The obvious truth of the development of order of matter on Earth

[314] In starting this fully accountable, true explanation of the meaning of life, we need to take a look at our surroundings. As you do, you'll note that the most obvious characteristic of our world is that it is full of 'things', variously enduring arrangements of matter, like plants, animals, clouds and rocks. And not only that, it is apparent that these arrangements of matter consist of a hierarchy of ordered parts; a tree, for instance, is a hierarchy of ordered matter—it has a trunk, limbs, roots, leaves and wood cells. Our bodies are also a collection of parts, as are clouds and rocks, which are built from different elements and compounds. Furthermore, what we have seen happen over time to these arrangements of matter is that there has been a progression from simple to more complex arrangements. From the fundamental ingredients of our world of matter, space and time, matter has become ordered into ever larger (in space) and more stable or durable (in time) arrangements.

[315] To elaborate, our world is constructed from some 94 naturally occurring elements that have come together to form stable arrangements. For example, two hydrogen atoms with their single positive charges came together with one oxygen atom with its double negative charge to form the stable relationship known as water. Over time, larger molecules and compounds developed. Eventually macro compounds formed. These

then integrated to form virus-like organisms, which in turn came together or integrated to form single-celled organisms that then integrated to form multicellular organisms, which in turn integrated to form societies of single species that continue to integrate to form stable, ordered arrangements of different species. Clearly, what is happening on Earth is that matter is integrating into larger and more stable wholes. And this development of order is not only occurring here, it is also happening out in the universe where, over the eons, a chaotic cosmos continues to organize itself into stars, planets and galaxies. As two of the world's greatest physicists, Stephen Hawking and Albert Einstein, have said, respectively, **'The overwhelming impression is of order…[in] the universe'** (Gregory Benford, 'The time of his life', *The Sydney Morning Herald*, 27 Apr. 2002; see <www.wtmsources.com/170>), and **'behind everything is an order'** (*Einstein Revealed*, PBS, 1997).

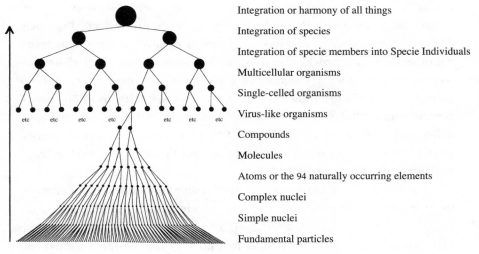

Integration or harmony of all things

Integration of species

Integration of specie members into Specie Individuals

Multicellular organisms

Single-celled organisms

Virus-like organisms

Compounds

Molecules

Atoms or the 94 naturally occurring elements

Complex nuclei

Simple nuclei

Fundamental particles

Development of Order or Integration of Matter on Earth.
A similar chart appears in Arthur Koestler's 1978 book, *Janus: A Summing Up*.

[316]The law of physics that accounts for this integration of matter is known as the 'Second Path of the Second Law of Thermodynamics', or 'Negative Entropy', which states that in an open system, where energy can come into the system from outside it (in Earth's case, from the sun, and, in the case of the universe, from the original 'big bang' explosion that created it), matter integrates; it develops order. Thus, subject to the influence of Negative Entropy, the 94 elements from which our world is built develop ever larger and more stable wholes.

[317] In *Janus: A Summing Up* (1978), the scientist-philosopher Arthur Koestler gave this excellent summary of the history of the concept of Negative Entropy: **'One of the basic doctrines of the nineteenth-century mechanistic world-view was Clausius' famous "Second Law of Thermodynamics". It asserted that the universe was running down towards its final dissolution because its energy is being steadily, inexorably dissipated...cosmos dissolving into chaos. Only fairly recently did science begin to recover from the hypnotic effect of this gloomy vision, by realizing that the Second Law applies only in the special case of so-called "closed systems"...whereas all living organisms are "open systems" which maintain their complex structure and function by continuously drawing materials and energy from their environment** [p.222 of 354] **...It was in fact a physicist, not a biologist, the Nobel laureate Erwin Schrödinger, who put an end to the tyranny of the Second Law with his celebrated dictum: "What an organism feeds on is negative entropy"** [p.223] **...Schrödinger's revolutionary concept of negentropy, published in 1944** [p.224] **...is a somewhat perverse way of referring to the power of living organisms to "build up" instead of running down, to create complex structures out of simpler elements, integrated patterns out of shapelessness, order out of disorder. The same irrepressible building-up tendency is manifested in the progress of evolution, the emergence of new levels of complexity in the organismic hierarchy and new methods of functional coordination** [p.223].**'** Significantly, Koestler wrote of **'the active striving of living matter towards [order]** [p.223]**'**, of **'a drive towards synthesis, towards growth, towards wholeness** [p.224]**'**, and that **'this "innate drive" derives from the "integrative tendency"** [p.225]**'**.

[318] So the theme of existence, the overall direction or destiny of change, or, from a conscious observer's point of view, the overall purpose or meaning of existence, is the ordering or integration or complexification of matter. 'Teleology', **'the belief that purpose and design are a part of nature'** (*Macquarie Dictionary*, 3rd edn, 1998), and 'holism', which the dictionary defines as **'the tendency in nature to form wholes'** (*Concise Oxford Dictionary*, 5th edn, 1964), are terms that recognize this integrative **'tendency'**. The concept 'holism' was first introduced by the South African denial-free thinker or prophet, the statesman, philosopher and scientist Jan Smuts in his 1926 book *Holism and Evolution*. Smuts conceived 'holism' as **'the ultimate organising, regulative activity in the universe that accounts for all the structural groupings and syntheses in it, from the atom, and the physico-chemical structures, through the cell and organisms, through Mind in animals, to Personality in Man'** (p.341 of 380).

[319] But while the integrative meaning of existence *is* the most obvious of all truths, it has also been the most difficult of all truths for humans to acknowledge, for an *extremely* good reason.

[320] The difficulty arises from the fact that for a collection of parts to form and hold together, for matter to integrate, the parts of the developing whole *must* cooperate, behave selflessly, place the maintenance of the whole above the maintenance of themselves, because if they don't cooperate—if they compete, behave selfishly or inconsiderately—then the whole disintegrates, the parts break down into the more elementary building blocks of matter from which they were assembled. As Koestler stated, to create **'order out of disorder'** requires **'functional coordination'**. A leaf falling from a tree in autumn does so to ensure the tree survives through winter and carries on; it puts the maintenance of the whole, namely the tree, above the maintenance of itself. The effective functioning of our body similarly depends on the cooperation of all its parts, on every part doing what is best for the whole body. Our skin cells, for example, are in constant turnover, with new cells replacing the old ones that have sacrificed themselves to protect our body. Cancer cells, on the other hand, destroy the body precisely because they violate this principle and follow their own selfish, independent agenda. Indeed, the very reason ant and bee societies work so well is because all their parts, the worker ants and bees, behave selflessly; in their behavior, they put the welfare of the larger whole above that of their own.

[321] Put simply, selfishness is divisive or disintegrative while selflessness is integrative—it is the glue that holds wholes together; it is, in fact, the theme of the integrative process, and thus of existence. It is also what we mean by the word 'love', with the old Christian word for love being **'caritas'**, meaning charity or giving or selflessness (see Col. 3:14, 1 Cor. 13:1-13, 10:24 & John 15:13). So 'love' is cooperative selflessness—and not just selflessness but *unconditional selflessness*, the capacity, if called upon, to make a full, self-sacrificing commitment to the maintenance of the larger whole. BUT—and herein lies the nub of the problem—if the meaning of existence is to behave *integratively*, which means behave cooperatively and selflessly, *why* do humans behave in the completely opposite way, in such a competitive and selfish *divisive* way? Yes, the integrative theme of existence squarely confronts us humans with the issue of the human condition, the issue of our non-ideal behavior. And so despite being such an obvious truth, Integrative Meaning has been

*so* horrifically condemning of the competitive, aggressive and selfish human race that until we could explain the *good reason why* humans have been divisively rather than integratively behaved (which was done in chapter 3)—and thus make it psychologically safe to admit the truth of the order developing, integrative meaning of existence—we had no choice but to live in near total denial of it. Hawking's, Einstein's and Koestler's acknowledgments of the order developing, integrative process when, as Koestler said, **'mechanistic'** science has maintained such a dedicated, **'hypnotic'** denial of it, were bold indeed.

[322] In summary then, selflessness, specifically unconditional selflessness or altruism, *is* the theme or meaning of existence. And since love means unconditional selflessness, love *is* the theme of existence—it *is* the meaning of life. The very great truthful, denial-free-thinker or prophet Christ emphasized the unconditionally selfless significance of the word 'love' when he said, **'Greater love has no-one than this, that one lay down his life for his friends'** (Bible, John 15:13). And of the biblical references to love cited above, Colossians 3:14 perfectly summarizes the integrative significance of love: **'And over all these virtues put on love, which binds them all together in perfect unity.'**

[323] But, unfortunately, while the unconditionally selfless nature of love is acknowledged in the Bible (as virtually all the great truths are), in our everyday world we couldn't admit that love is unconditional selflessness and, therefore, the theme of existence until we could explain *why* humans don't behave lovingly and are so seemingly at odds with the integrative process. In fact, in the human-condition-avoiding mechanistic scientific paradigm it is considered improper, unscientific, to *even* use the word 'love'. The linguist Robin Allott summed up mechanistic science's attitude to love succinctly when he wrote that **'Love has been described as a taboo subject, not serious, not appropriate for scientific study'** ('Evolutionary Aspects of Love and Empathy', *Journal of Social and Evolutionary Systems*, 1992, Vol.15, No.4). Indeed, love has been deemed so **'[in]appropriate for scientific study'** that it has been reported that **'more than 100,000 scientific studies have been published on depression and schizophrenia (the negative aspects of human nature), but no more than a dozen good studies have been published on unselfish love'** (*Science & Theology News*, Feb. 2004). So mechanistic science has determinedly resisted analysis of one of humanity's most used, valued and meaningful words! The psychologist Harry F. Harlow was another who highlighted this discrepancy when he observed that **'Psychologists, at least psychologists who**

write textbooks, not only show no interest in the origin and development of love or affection, but they seem to be unaware of its very existence. The apparent repression of love by modern psychologists stands in sharp contrast with the attitude taken by many famous and normal people. The word "love" has the highest reference frequency of any word cited in Bartlett's book of *Familiar Quotations'* ('The nature of love', *American Psychologist*, 1958, Vol.13, No.12). The concept of 'unselfish love' has certainly been an unbearable area of scientific enquiry for us selfish, seemingly non-loving and unlovable humans!

## Chapter 4:3 'God' is our personification of Integrative Meaning

[324] So, 'love' has been **'a taboo subject'** for science, and yet if 'love' or unconditional selflessness forms only an *aspect* of Integrative Meaning, how much more unbearable has the *overall* tenet of Integrative Meaning itself been? The answer is the Negative Entropy-driven integrative, cooperative, loving, selfless, order-developing theme or meaning or purpose of existence has been an almost completely unconfrontable truth for the psychologically upset, competitive, aggressive and selfish human-condition-afflicted human race. In fact, we have lived in *such* terrified fear and awe of the truth of Integrative Meaning, have been *so* confronted, condemned and intimidated by it, *so* unable to deal with it on any sort of an equal footing, that we deified the concept—and not just as *a* God, but *the one and only* God, *the* most universal and fundamental, yet completely unconfrontable, of truths.

[325] Monotheism, the belief that there is only one God, is an insight that goes back as far as 4,000 years ago to two very great denial-free thinkers or prophets—the Hebrew prophet Abraham, who lived around 2,000 BC, and the pharaoh Akhenaton, who reigned in Egypt from approximately 1,350 to 1,335 BC. The very great Persian prophet Zoroaster also recognized that there is only **'one supreme deity'**, as this reference to the faith describes: 'sometime around or before 600 BC—perhaps as early as 1200 BC—there came forth from the windy steppes of northeastern Iran a prophet who utterly transformed the Persian faith. The prophet was Zarathustra—or Zoroaster, as the Greeks would style his name. Ahuramazda [the supreme being or wise lord] had appeared to Zoroaster in a vision, in which the god had revealed himself to be the one supreme deity, all seeing and all powerful. He represented both light and truth, and was creator of all things, fountainhead of all virtue. Ranged against

**him stood the powers of darkness, the angels of evil and keepers of the lie. The universe was seen as a battleground in which these opposing forces contended, both in the sphere of political conquest and in the depths of each man's soul. But in time the light would shine out, scattering the darkness, and truth would prevail. A day of reckoning would arrive'** (*A Soaring Spirit: Time-Life History of the World 600-400 BC*, 1988, p.37 of 176). And in approximately 360 BC, that other very great denial-free-thinking prophet, Plato, *similarly* recognized that God is Integrative Meaning, writing that **'God desired that all things should be good and nothing bad, so far as this was attainable. Wherefore also finding the whole visible sphere not at rest, but moving in an irregular and disorderly fashion, out of disorder he brought order, considering that this was in every way better than the other'** (*Timaeus*; tr. B. Jowett, 1871, 30).

[326]But *until* we could explain the human condition and explain in first-principle-based, scientific terms who, or more precisely, what God is—namely our personification of the Negative Entropy-driven integrative theme, purpose and meaning of life—and why we needed to resort to deification in the first place, we had no choice but to leave the religious concept of God in that safely abstract, undefined state. And so despite Integrative Meaning being an *extremely* obvious truth, with evidence of the hierarchy of the order of matter everywhere we look, without understanding of our divisive condition it was imperative for humanity that human-condition-avoiding mechanistic science found a way to deny such a seemingly totally condemning truth. This was easily achieved through the simple assertion that there is no meaning or purpose or theme to existence and that while change *does* occur, it is a random, purposeless, directionless, meaningless, blind process. And, as stated, to cope with the imbued recognition of integrative ideality and meaning in the religious notion of God, mechanistic science simply left the concept undefined and undefinable, maintaining it was a strictly abstract, metaphysical and spiritual notion unrelated to the scientific domain; if 'God' existed in any form, it was as an inexplicable deity, a supernatural being seated on a throne somewhere in a remote blue heaven who could be worshipped from afar as someone superior to us 'mere mortals', thus nullifying any direct and confronting comparisons with our own upset state. Religion and science were firmly demarcated as two entirely unrelated subjects. Indeed, E.O. Wilson succinctly captured mechanistic science's view on the matter when he said, **'I take a very strong stance against the mingling of religion and science'** ('Edward O. Wilson From Ants, Onward', *National Geographic*, May 2006).

[327] But of course, the truth is, to use Nobel Prize-winning physicist Charles H. Townes' words, **'they [science and religion] both represent man's efforts to understand his universe and must ultimately be dealing with the same substance. As we understand more in each realm, the two must grow together... converge they must'** ('The Convergence of Science and Religion', *Zygon*, 1966, Vol.1, No.3). The physicist Max Planck (another Nobel winner) similarly recognized that **'There can never be any real opposition between science and religion; for the one is the complement of the other'** (*Where Is Science Going?*, 1977, p.168). As my headmaster at Geelong Grammar School, Australia's greatest ever educator, Sir James Darling, said, **'The scientist can no more deny or devaluate the truths of spiritual experience than the theologian can neglect the truths of science: and the two truths must be reconcilable, and it must be of importance to each of us that they should be reconciled'** (*The Education of a Civilized Man*, ed. Michael Persse, 1962, p.68 of 223). And with understanding of the human condition now found, **'converge'** they *have*; ideality (which religions and the truthful, denial-free-thinking, God-confronting-not-avoiding prophets they were founded around represented) and our search for understanding of our non-ideal reality (which science represented—the word 'science' literally means 'knowledge') have finally been **'reconciled'**. Yes, with the human condition now explained and our divisive, seemingly non-integrative state finally understood, *all* humans can at last safely admit and recognize that there has only been one God, one all-dominating and all-pervading theme or meaning of existence, which is Integrative Meaning—a truth we recognize when we say **'God is love'** (Bible, 1 John 4:8, 16).

[328] It should be mentioned here that despite the fact that the admittance of Integrative Meaning first required solving the issue of our divisive human condition, a rare few holistic scientists have not only courageously defied the almost universal need to deny the development of order of matter on Earth, or Integrative Meaning, they have actually acknowledged that it is what we mean by God. If we include more of what the aforementioned giants of physics, Stephen Hawking and Albert Einstein, said earlier about order being the main characteristic of change in the universe, we can see that they both regarded God to be the personification of Integrative Meaning. In 1989 Hawking said, **'I would use the term God as the embodiment of the laws of physics'** (*Master of the Universe*, BBC). In 2002 he went further, saying, **'The overwhelming impression is of order** [in the universe]. **The more we discover about the universe, the more we find that it is governed by rational laws. If one liked, one could say that this order was**

**the work of God. Einstein thought so...We could call order by the name of God'** (Gregory Benford, 'The time of his life', *The Sydney Morning Herald*, 27 Apr. 2002; see <www. wtmsources.com/170>). Einstein's views on the matter were chronicled in the 1997 PBS documentary *Einstein Revealed*, which reported Einstein as saying that **'over time, I have come to realize that behind everything is an order that we glimpse only indirectly** [because it's unbearably condemning]. **This is religiousness. In this sense, I am a religious man.'** Einstein was also recognizing that God is order or harmony when he said, **'In view of such harmony in the cosmos which I, with my limited human mind, am able to recognize, there are yet people who say there is no God'** (Hubertus zu Löwenstein, *Towards the Further Shore*, 1968, p.156). Einstein's friend and occasional collaborator, the Nobel Prize-winning physicist Erwin Schrödinger, was another leading scientist who acknowledged that integrative unity is what we have been terming God when he wrote that **'Science is reticent too when it is a question of the great Unity...of which we all somehow form part, to which we belong. The most popular name for it in our time is God—with a capital "G"'** (*Nature and the Greeks* and *Science and Humanism*, 1954, p.97 of 184). Schrödinger's contemporary and fellow Nobel Prize-winning physicist, Werner Heisenberg, also recognized the relationship between the integrative meaning of existence that science is able to point to, and the concept of God that religion recognizes, when he wrote, **'I have repeatedly been compelled to ponder on the relationship of these two regions of thought [science and religion], for I have never been able to doubt the reality of that to which they point'** ('Scientific and Religious Truth', *Across the Frontiers*, 1974, p.213). Yes, the **'reality'** of **'order'** or **'harmony'** or **'unity'** is apparent everywhere and it is what our **'religiousness'**, our belief in **'God'**, is concerned with acknowledging. I might mention that while Mahatma Gandhi was an inspired leader of the Indian nation rather than a scientist, he was another who bravely acknowledged that **'There is an orderliness in the Universe, there is an unalterable law governing everything and every being that exists or lives...That law then which governs all life is God'** (Louis Fischer, *Gandhi: His Life and Message for the World*, 1954, p.108 of 224)—as did Plato, who referred to **'God, the orderer of all'** (see par. 173).

[329] In his 1987 book, *The Cosmic Blueprint*, another holistic physicist, the Templeton Prize-winner Paul Davies, actually went so far as to protest against the denial of Integrative Meaning in the world of science, writing that **'We seem to be on the verge of discovering not only wholly new laws of nature, but ways of thinking about nature that depart radically from traditional science** [p.142 of 232] **...Way back in the primeval phase of the universe, gravity triggered a**

cascade of self-organizing processes—organization begets organization—that led, step by step, to the conscious individuals who now contemplate the history of the cosmos and wonder what it all means [p.135] …There exists alongside the entropy arrow another arrow of time [the Negative Entropy arrow], equally fundamental and no less subtle in nature…I refer to the fact that the universe is *progressing*— through the steady growth of structure, organization and complexity—to ever more developed and elaborate states of matter and energy. This unidirectional advance we might call the optimistic arrow, as opposed to the pessimistic arrow of the second law. There has been a tendency for scientists to simply deny the existence of the optimistic arrow. One wonders why [p.20].'

[330] We can now appreciate the reason 'why' 'Science is reticent when it comes to a question of the great Unity'—the reason 'why' 'scientists' 'deny' 'the optimistic arrow' of Integrative Meaning—is because it was far too psychologically dangerous to acknowledge without first finding the biological reason, and thus defense, for our divisive, apparently non-integrative, un-Godly human condition. No wonder we have been, as we say, a 'God-fearing'—in fact, so in awe of God to the point of being a God-worshipping—*not* a 'God-confronting' species; as Berdyaev put it, **'He cannot** [man struggles to] **break through to paradise that lies beyond the painful distinction between good and evil, and the suffering connected therewith. Man's fear of God is his fear of himself, of the yawning abyss of non-being** [alienation] **in his own nature'** (*The Destiny of Man*, 1931; tr. N. Duddington, 1960, p.41 of 310).

[331] Our species' immense fear, and thus denial, of the truth of Integrative Meaning is, of course, the subject of Plato's famous allegory of the human condition, referred to in par. 83, which describes humans as being imprisoned in a cave. As 'prisoners' in this metaphorical cave, we are only able to envisage the outside world via shadows cast on the back wall of the cave. These shadows, which symbolize our limited and distorted, human-condition-avoiding, dishonest, immensely alienated **'phony'** and **'fake'** view of the world, are thrown by the light of a fire that, situated in the entrance to the cave, effectively prevents any escape from it. Explaining the symbolism of the fire, Plato wrote that **'the light of the fire in the [cave] prison [corresponds] to the power of the sun'** (*The Republic*, c.360 BC; tr. H.D.P. Lee, 1955, 517), from which we have to **'turn back'** because if we/the cave prisoner were to go **'out into the sunlight, the process would be a painful one, to which he [we] would much object'** (515-516). Plato explained that the sun represents the **'universal, self-sufficient first principle'** (511), the **'absolute**

**form of Good'** (517) and the **'highest form of knowledge'** (505), which we can now understand is Integrative Meaning.

[332]Fire is a common theme in many mythologies, appearing as a metaphor for the integrative, Godly ideals of life whose condemning, scorching glare we had to **'turn back'** from. In the Zoroastrian religion, **'Fire is the representative of God...His physical manifestation...Fire is bright, always points upward, is always pure'** (Edward Rice, *Eastern Definitions*, 1978, p.138 of 433). In Christian mythology, the story of Genesis features **'a flaming sword flashing back and forth to guard the way to the tree of life'** (3:24). The Bible also records the Israelites as saying, **'Let us not hear the voice of the Lord our God nor see this great fire any more, or we will die'** (Deut. 18:16). The biblical character Job was another who pleaded for relief from confront-ation with the unbearably depressing integrative, Godly ideals when he lamented, **'Why then did you** [God] **bring me out of the womb?...Turn away from me so I can have a moment's joy before I go to the place of no return, to the land of gloom and deep shadow, to the land of deepest night** [depression]' (Job 10:18, 20-22). Christ also recognized the problem of the exposing **'light'** of truth that he was an unresigned, denial-free spokesman for, when he said, **'the light shines in the darkness but...everyone who does evil** [becomes upset sufferers of the human condition] **hates the light, and will not come into the light for fear that his deeds will be exposed'** (John 1:5, 3:20).

[333]So, again, while Integrative Meaning *is* the most obvious, profound and thus important of all truths it is clearly also the truth that has appeared to most condemn humans—and in the absence of the explanation as to *why* we, as a species, appear to be so at odds with the integrative meaning of life, we humans have sensibly taken one of two options: we either prac-ticed denial of Integrative Meaning, and even of God, and thus of the issue of our self-corruption, or we indirectly acknowledged our self-corruption by acknowledging the existence of God and embracing some expression of faith that a greater dignifying understanding of our divisive condition does exist and would one day be found. To cope with our less-than-ideal human condition there has only ever been either denial or faith.

[334]To counter the utter dishonesty of mechanistic science's denial of the existence of Integrative Meaning/God, support for an extremely literal interpretation of God, in the form of so-called 'Creationism' and 'Intelligent Design', emerged. While still having to avoid the human con-dition and, therefore, the truth of Integrative Meaning, these movements did acknowledge God, but only in a fundamentalist way in which God

took the form of an actual being who 'designed' life on Earth, or 'created' the world in just six days. In truth, both the mechanistic approach and these more literal attitudes were immensely dishonest in that mechanistic scientists wanted to pretend to be rational and either deny any semblance of Integrative Meaning by refuting the existence of the concept of God, or acknowledge the concept of God but claim it has nothing to do with science, while supporters of Creationism and Intelligent Design chose to admit to a semblance of Integrative Meaning in the form of a God who is literally a special being or deity, with the downside being that such a stance necessarily meant abandoning all attempts at rationality. We can see that the *real* issue neither party was willing or able to acknowledge is the issue of Integrative Meaning and its human-condition-confronting implications. (I should mention that advocates of Intelligent Design would have us believe that their position is different to that of Creationists. The main website for Intelligent Design states that **'The theory of intelligent design...**[is concerned with] **whether the "apparent design" in nature...is genuine design (the product of an intelligent cause) or is simply the product of an undirected process such as natural selection acting on random variations'**, and that **'unlike Creationism, the scientific theory of intelligent design does not claim that modern biology can identify whether the intelligent cause detected through science is supernatural'** (Center for Science & Culture; see <www.wtmsources. com/116>). In other words, Intelligent Design acknowledges Integrative Meaning, the development of order of matter, *but doesn't discount that a supernatural-type creator/being/God might be involved*—so Intelligent Design is trying to have it both ways, appear to be scientific but still allow for a supernatural creator; it is still fundamentally similar to Creationism, which unscientifically supports the idea of a supernatural creator.)

[335] Indeed, the truth of Integrative Meaning and its human-condition-confronting implications have been *so* unbearably confronting that in recent years mechanistic science has, in an insidious attempt to keep the issue even further at bay, evasively steered the discussion toward whether the concept of God has been irrevocably undermined by physicists' on-going discoveries about the big bang origin of the universe, the extinction of time before the big bang and, more recently, the possibility of multiple universes! This is classic 'displacement' behavior—**'an unconscious defense mechanism whereby the mind redirects affects from an object felt to be danger-ous or unacceptable to an object felt to be safe or acceptable'** (Wikipedia; see <www. wtmsources.com/138>). The fact is that starting with the boundaries of our

reality of matter, space and time, and drawing on the laws of physics within which we live, we can construct the human condition, and also solve it—and, by so doing, make it possible to demystify God, and, indeed, bring to an end the whole debate about 'His' existence. The enormous issue of 'God' that has existed in the lives of humans relates *entirely* to the integrative process of the development of order of matter that occurs in the world within which we live that is bounded by the elements of matter, space and time and the effect the laws of physics have on those elements as we experience them. The insecure state of the human condition that caused us to so fear and revere all manner of gods, and then just one God, is created and solved within that realm. Science's task has been to be a winnower of mystery and superstition, with the ultimate mystery it needed to solve being the human condition. So our ability now to understand the human condition necessarily ends the fear, confusion, bewilderment and mystery that fuelled such superstitious thought; it ends ignorance. Paul Davies was emphasizing the Integrative-Meaning-related *real* issue about God when he said, **'So where is God in this story** [of physics]**? Not especially in the big bang…To me, the true miracle of nature is to be found in the ingenious and unswerving lawfulness of the cosmos, a lawfulness that permits complex order to emerge from the chaos'** ('Physics and the Mind of God: The Templeton Prize Address', 3 May 1995).

[336] Yes, with understanding of the human condition now found, it is at last psychologically safe to demystify God as Integrative Meaning, and, by so doing, finally reconcile religion and science. From the religious perspective, *this* is the time the prophet Isaiah was looking forward to when reconciling understanding of the human condition would be found and we could **'revere'** instead of fear the truth of Integrative Meaning/ God: **'Why, O Lord, do you make us wander from your ways and harden our hearts so we do not revere you?…Do not be angry beyond measure…do not remember our sins for ever…all that we treasured** [before the human condition emerged] **lies in ruins. After all this, O Lord, will you hold yourself back? Will you keep silent and punish us beyond measure?'** (Isa. 63-64). And from the scientific side of the fence, when the scientist-philosopher Pierre Teilhard de Chardin wrote in 1938 that **'I can see a direction and a line of progress for life, a line and a direction which are in fact so well marked that I am convinced their reality will be universally admitted by the science of tomorrow'** (*The Phenomenon of Man*, p.142 of 320), he too was recognizing how obvious the truth of Integrative Meaning is, and how it wouldn't be able

to be **'universally admitted'** until the human-condition-reconciled **'science of tomorrow'** emerged.

[337] I should mention here that there have been a few scientists in addition to Hawking, Einstein, Koestler, de Chardin and Davies who 'jumped the gun' and **'admitted'** Integrative Meaning, as the titles (particularly the words I have underlined) of the following books (including three by Davies) illustrate—for instance, David Bohm wrote *Wholeness and The Implicate Order* in 1980; Ilya Prigogine and Isabelle Stengers wrote *Order Out of Chaos* in 1984; Paul Davies wrote *God and the New Physics* in 1983, *The Cosmic Blueprint* in 1987 and *The Mind of God: Science and the Search for Ultimate Meaning* in 1992; Charles Birch wrote *Nature and God* in 1965, *On Purpose* in 1990 and *Biology and The Riddle of Life* in 1999; M. Mitchell Waldrop wrote *Complexity: The Emerging Science at the Edge of Order and Chaos* in 1992; Roger Lewin wrote *Complexity: Life at the Edge of Chaos, the major new theory that unifies all sciences* in 1993; Stuart Kauffman wrote *The Origins of Order: Self-Organization and Selection in Evolution* in 1993, *At Home in the Universe: The Search for the Laws of Self-Organization and Complexity* in 1995 and *Anti-chaos* in 1996; and Richard J. Bird wrote *Chaos and Life: Complexity and Order in Evolution and Thought* in 2003. But such admissions are, nevertheless, an anomaly, because, as has been emphasized, the vast majority of scientists haven't been prepared to go *anywhere near* the historically unbearably confronting truth of Integrative Meaning, and, as stated in chapter 2:12, in coming off such a dishonest base it is impossible to find a true understanding of our world and place in it—which de Chardin *also* understood when, in 1956, he wrote that **'biology cannot develop and fit coherently into the universe of science unless we decide to recognise in life the expression of one of the most significant and fundamental movements in the world around us… the vast universal phenomenon…of** *complexification of matter***. This is something that must be clearly appreciated if we are to get away to a good start in our study of man** [p.19 of 124] **…This [complexification of matter] is a very simple concept, but the more we think about it, the more, in fact, are we led to see the world of life as a vast sheaf of particles rushing headlong…down the slope of an indefinite corpusculisation…First, there are the regressive currents: entropy, dissipation of energy…But there are progressive, or constructive, currents too…a growing complexity…the passage from an unordered to an ordered heterogeneity…**[where, at a certain **point** in this progression, **vitalisation** occurs, and] **one portion of the cosmic stuff not only does not disintegrate but even begins—by producing a**

**sort of bloom upon itself—to vitalise** [as will shortly be explained in par. 348, this is when the replicating DNA molecule appeared and 'made a business' of actively resisting disintegration] [pp.31-33] **...life can no longer be regarded as a superficial accident in the universe: we must look on it as...ready to seep through the narrowest fissure at any point whatsoever in the cosmos—and, once it has appeared, obliged to use every opportunity and every means to reach the furthest extremity of everything it can attain: the ultimate, externally, of complexity, internally of consciousness** [p.35]' (*Man's Place in Nature*).

[338] Plato was another who recognized this inherent limitation of the Integrative-Meaning-denying mechanistic approach when, long ago, he wrote that **'the Good** [as explained in par. 331, **the Good** is Integrative Meaning]**...gives the objects of knowledge their truth and the mind the power of knowing...**[just as] **The sun...makes the things we see visible...The Good therefore may be said to be the source not only of the intelligibility of the objects of knowledge, but also of their existence and reality'** (*The Republic*, c.360 BC; tr. H.D.P. Lee, 1955, 508-509). Yes, this loss of **'the power of knowing'** has been very serious indeed. Koestler also who felt it, bemoaning the crippled, stalled, atrophied state of all of science, but of biology and psychology in particular, when he said that blind, reductionist, mechanistic science's denial of Integrative Meaning has **'taken the life out of biology as well as psychology'**, writing that **'although the facts** [of the integration of matter] **were there for everyone to see, orthodox evolutionists were reluctant to accept their theoretical implications. The idea that living organisms, in contrast to machines, were primarily *active*, and not merely *reactive*; that instead of passively adapting to their environment they were...creating...new patterns of structure...such ideas were profoundly distasteful to** [Social] **Darwinians, behaviourists and reductionists in general** [p.222 of 354] **...Evolution has been compared to a journey from an unknown origin towards an unknown destination, a sailing along a vast ocean; but we can at least chart the route...and there is no denying that there is a wind which makes the sails move...the purposiveness of all vital processes... Causality and finality are complementary principles in the sciences of life; if you take out finality and purpose you have taken the life out of biology as well as psychology** [p.226]' (*Janus: A Summing Up*, 1978).

[339] As was pointed out in par. 188, towards the end of *The Origin of Species*, Charles Darwin anticipated that **'In the distant future I see open fields for far more important researches. Psychology will be based on a new foundation, that of the necessary acquirement of each mental power and capacity by gradation. Light will be thrown on the origin of man and his history'** (1859,

p.458 of 476). Given Koestler's comment that **'if you take out finality and** [the **'integrative tendency'** or] **purpose you have taken the life out of biology as well as psychology'**, what was required to bring about Darwin's **'new'** en-**'light'**-ening **'foundation'** for **'far more important research'** in **'biology as well as psychology'** was not only acknowledgment of the involvement of our conscious **'mental power'** but also of **'integrative' 'purpose'**.

[340] So, what *is* the **'far more important research'** that results from thinking from the **'new'** en-**'light'**-ening **'foundation'** of accepting the truth of **'integrative' 'purpose'**?

## Chapter 4:4 <u>The denial-free history of the development of matter on Earth</u>

[341] In commencing this denial-free analysis of the development of matter from a perspective that takes into account the truth of Integrative Meaning, we first need to replace the word 'evolution' with the word 'development', for while evolution implies that organisms *do* change or evolve it avoids acknowledging that there is a direction and purpose to that change, which is to *develop* the order of matter.

[342] As the study of physics has shown, our world is made up of three fundamental ingredients — time, space, and energy — with, as Einstein revealed in his famous formula $E = mc^2$, energy taking the form of matter, which comprises the 94 or so naturally occurring elements that, when subjected to the laws of physics, particularly the law of Negative Entropy, became ordered or integrated; they formed more stable or enduring (in time) and ever larger (in space) arrangements.

[343] This development of order of matter involved the initial mixture of the Earth's elements and their gradual formation into stable arrangements called molecules — earlier I provided the example of a water molecule being the stable arrangement of two single positively charged hydrogen atoms with one double negatively charged oxygen atom. In time, through the mixing of different elements, each with its own particular properties, many stable arrangements were found or developed, leading to even greater order and complexity of arrangements in the form of very complex macromolecules.

[344] The problem, however, was that the more complex these macromolecules became, the more unstable they tended to be. Highly complex macromolecules would only occasionally form and, when they did, they

didn't tend to hold together for long before breaking down into their separate parts. Eventually an <u>impasse</u> was reached where the degree of instability imposed a limit on how complex macromolecules could become. When this ceiling was reached it appeared Negative Entropy— or 'God' if we were to personify the process—could not develop any more order of matter on Earth. And yet matter *did* continue to 'develop' beyond this apparent impasse, with the emergence out of the primordial soup of a complex macromolecule with an unusual property—DNA, or deoxyribonucleic acid. What was unusual about DNA was that it could replicate. It could split, allowing the two halves to draw material from the environment to build two complete DNA molecules. The significance of this replication was that it meant DNA could defy breakdown. It could turn a relatively brief lifetime for a complex macromolecule into a relatively indefinite one. DNA's ability to replicate meant that even though some of the replicates disintegrated into smaller parts, others would survive and go on to replicate further. With slight variations called mutations occurring from the effects of solar radiation, replicates were 'found' that were even more stable/enduring (in time) and more ordered/complex/larger (in space). The process of natural selection of more stable and larger arrangements of matter—and the origin of an indefinite lifetime, or 'life' as we call it—appeared.

[345] In this process, each replicating arrangement of matter or reproducing individual was, in effect, being tested both for its ability to survive and reproduce in its lifetime and, over generations of offspring, for its ability to adapt to changes in the environment in which it lived, with those that managed to survive and adapt inevitably, whenever possible, finding/refining/achieving/growing/developing even greater order of matter. The effect of this process over time was that more and greater order of matter was integrated. It was the ability to survive and adapt that supplied the opportunity for more and greater order of matter to develop. Thus, using the tool of replicating DNA, Negative Entropy was able to integrate matter into larger wholes; it was able to develop ever more and ever greater order of matter on Earth.

[346] DNA is actually a very complex crystal. Crystal molecules abound— common salt, sodium chloride, for instance, is one—and in a suitable nutrient environment they all have the capacity to reproduce; to grow their structure from their structure. However, being much simpler than DNA—having fewer elements within their molecular structure—they have

little or no potential for adaptation and, it follows, for the development of greater order.

[347] Indeed, variability is so critical to this DNA process of developing greater order of matter that 'sexual reproduction' of DNA molecules developed, where the split halves from two compatible DNA molecules were made, through natural selection, to come together (be 'attracted to and mate with each other as males and females') to form a new, slightly different DNA-based sexually reproducing individual. This greatly increased the variety of a particular DNA type or 'species' and, by so doing, greatly increased its chances of finding/achieving/growing/ developing larger and more stable arrangements of matter. Sexual reproduction, therefore, soon replaced non-sexual or 'asexual' reproduction as the most successful or effective form of DNA reproduction in this business of finding or developing greater order of matter.

[348] It can be seen then that Negative Entropy's development of order of matter really comes down to being a product of possibilities. The differing properties of matter mean some arrangements of matter break down towards heat energy, while others stay stable and still others become part of larger and more enduring associations of matter. In time, all the possible associations of matter will be automatically or, as Charles Darwin called it, 'naturally' investigated until the largest, most stable association is left or found or, as Darwin described it, 'selected'. What happened with DNA was that it not only turned a relatively short lifetime for extremely complex molecules into a relatively indefinite one, it also made a business, as it were, of this 'negentropy' direction—both of resisting breakdown and of developing order. The replicating DNA molecule gave rise to a process that actively resisted breakdown and actively developed ever more and greater order of matter. This is **'the active striving of living matter towards order'**, **'a drive towards synthesis, growth and wholeness'**, the **'active'**, **'creating'**, **'purposiveness'**, **'vitalisation'** of life about which Koestler and de Chardin wrote.

[349] The DNA unit of inheritance is called a gene, with the study of the process of change that genes undergo termed 'genetics'. As a tool for Negative Entropy's development or refinement of the order of matter on Earth, the genetic process was very powerful—it was able to develop the great diversity of matter that we term 'the variety of life'. From DNA, virus-like organisms developed, then from virus-like organisms developed single-celled organisms (such as bacteria), and from single-celled

organisms developed multicellular organisms (such as plants and animals). The next level of order to be developed or integrated by Negative Entropy was societies or colonies or ordered arrangements of multicellular organisms. It was at this point, however, that Negative Entropy (or God) encountered another major impasse.

[350] While genetics has proved to be a marvelous tool for integrating matter it has one very significant limitation, which arises from the fact that each sexually reproducing individual organism has to struggle, compete and fight selfishly for the available resources of food, shelter, territory and the mating opportunities it needs if it is to successfully reproduce its genes. What this means is that integration, and the unconditionally selfless cooperation it depends on, cannot normally develop between one sexually reproducing individual and another. Indeed, the competition between sexually reproducing individuals is the basis of the natural selection process that gave rise to the great variety of life on Earth. The word 'selection' in 'natural selection' implies competition—a comparison between sexually reproducing individuals for their ability to survive, adapt and develop greater order of matter. So integration beyond the level of the sexually reproducing individual—that is, the coming together or integration of the sexually reproducing individual members of a species to form the next larger and more stable whole of the Specie Individual—could not, normally, develop (the *Development of Order of Matter* chart included earlier shows where the Specie Individual appears in the hierarchy of integration). This was the second major impasse that Negative Entropy (or God) encountered: the development of order of matter on Earth had seemingly come to a stop at the level of the sexually reproducing individual.

[351] To elaborate, each sexually reproducing individual normally has to ensure the reproduction of its own genes, which means sexually reproducing individuals cannot normally develop the ability to behave unconditionally selflessly towards other sexually reproducing individuals— which, as has been explained, is what full cooperation and thus complete integration requires. Certainly sexually reproducing individuals can develop *conditionally* selfless behavior towards other sexually reproducing individuals. Situations of reciprocity can develop where one sexually reproducing individual selflessly helps another on the proviso that they are selflessly helped in return, which, in effect, means both parties are still selfishly benefiting. So sexually reproducing individuals *can* develop

reciprocity because it is, in essence, still selfish behavior: it doesn't give away an advantage to other sexually reproducing individuals and, therefore, doesn't compromise the reproductive chances of the sexually reproducing individual practicing the behavior. Unconditionally selfless, altruistic traits, on the other hand, do give away an advantage to other sexually reproducing individuals—that being the meaning of unconditional selflessness, that you are giving without receiving—and, therefore, unconditionally selfless, altruistic, self-sacrificing traits *do* compromise the reproductive chances of the sexually reproducing individual practicing such behavior and, therefore, cannot normally develop.

[352] So cooperation between sexually reproducing individuals cannot normally be developed beyond a situation where there is reciprocal/conditional selflessness, and, since conditionally selfless behavior is still basically selfish behavior, full cooperation and thus complete integration cannot normally be developed between sexually reproducing individuals to form the Specie Individual. This inability to develop unconditionally selfless, altruistic behavior leaves sexually reproducing individuals competing relentlessly with each other for available resources of food, shelter, territory and a mate. So much so, in fact, that what we see happening between sexually reproducing individuals as they try to develop more integration under this limitation of not being able to develop unconditionally selfless behavior is that the competition between them becomes *so* intense that the only way they can contain it at all is by establishing a dominance hierarchy, where each individual accepts its position in a hierarchy that is ordered according to the competitive strengths of the various individuals involved. The benefit of a dominance hierarchy, or a so-called 'peck order', is that once established the only time competition breaks out is when an opportunity arises to move up the hierarchy; for the rest of the time there is relative peace. The emergence of a dominance hierarchy is a sign a species has developed as much integration as it possibly can.

[353] It should be pointed out that in situations where competition between individuals breaks out—when, for instance, male elephants or whales or kangaroos or birds or solitary insects, etc, etc, chase a female in estrous—it's not simply because the female wants to discover which is the strongest male with which to mate to ensure her offspring is the strongest, most competitively successful individual it can be, as is currently taught, but because the Negative Entropy integrative tendency has

driven the males and the females to that extreme state of competition. Such extreme competition is, in truth, a result of trying to develop greater order of matter. More will be said about this shortly, but the real story of life on Earth is *not* about selfish competition but integration.

[354] So although dominance hierarchy hides it from view for most of the time, the reality is that *extreme* competitiveness characterizes the behavior of the more cooperative and thus integrated, or what has evasively been called 'social', species. In my youth I remember feeding hens in our hen house and seeing a hen twist her leg and become temporarily crippled, at which point all the other hens immediately attacked her. In that instant it was suddenly apparent to me just how closely and intensely each hen was watching all the other hens for an opportunity to literally move up the peck order. The hen house was not at all the gregarious, peaceful community I thought it was; rather, it was a place of absolutely fierce competition! Charles Darwin recognized this truth about the real struggle in the lives of most animals when he wrote that **'It is difficult to believe in the dreadful but quiet war of organic beings, going on** [in] **the peaceful woods and smiling fields'** (1839; *The Complete Work of Charles Darwin Online*, ed. John van Wyhe, 2002, *Notebook E*, line ref. 114).

[355] This situation where sexually reproducing individuals relentlessly compete for available resources is the situation, the condition, that almost all animals have to endure—it is the great, agonising 'animal condition'. When humans become free of our numbed, alienated human condition we are going to be shocked by the agony of the animal condition; we are going to *feel* the distress all non-human animal species live under, where each sexually reproducing individual, through its Negative Entropy-driven commitment to achieve greater integration, is having to relentlessly and fiercely compete to reproduce its genes. Unfortunately, because animals' innocence (lack of the psychological upset we humans suffer from) confronts us with our lack of innocence (our vicious angry, egocentric and alienated state), we humans have so hated, despised and resented animals that we have hunted and shot them for 'sport'; but one day we are going to have so much sympathy for animals because of what they have to endure trapped in a life of having to relentlessly compete with each other, often with their closest friends! ('Friends' in the sense of those with whom they have shared their life and developed emotional bonds.) Certainly the same extremely competitive state exists for plants and microbes, but, not having the developed nervous system that animals

have, their awareness of the agony of that horrifically competitive existence could obviously not be anywhere near as great as it is for animals.
(The other issue about the life of non-human animals is that they rarely
die peacefully; as soon as they grow old they are ruthlessly picked off
and eaten by a lion or fox or mongoose, often while they are still alive.
Thank goodness they can't reason and thus look forward in time. As the
poet Robert Burns wrote about non-rational animals, **'Still thou art blest,
compared with me! The present only touches thee: But och! I backward cast my
eye, on prospects drear!' And forward, tho I cannot see, I guess and fear!'** (*To A
Mouse*, 1785).)

[356]What now needs to be explained is, firstly, that while sexually reproducing individuals cannot normally be integrated, the sexually reproducing
individual itself *could* be elaborated, enlarged, expanded—developed
further to become bigger—which, as will be explained next, is how
single-celled organisms developed into multicellular organisms, and how
multicellular colonial ants and bees integrated into their fully cooperative
and thus completely integrated colonies. Significantly, in these 'elaborated
sexually reproducing individuals', the cells of the multicellular body,
or the individual ants and bees in their fully integrated colonies, are no
longer sexually reproducing individuals themselves, but part of a larger
sexually reproducing individual, which is the body, or, in the case of ants
and bees, the colony. Secondly, it has to be explained why I have been
saying it is 'normally' not possible for sexually reproducing individuals
to become fully integrated to form the Specie Individual. There was, in
fact, one species who managed to achieve the development of the next
larger whole in the integration of matter on Earth of the integration of
sexually reproducing individuals to form the Specie Individual: our ape
ancestors. As will be explained in chapter 5, this amazing step in the
development of matter was achieved through maternalism—the nurturing of our offspring—which has been another of those unbearable truths
that humans couldn't face until we could explain our divisive, unloving
human condition.

[357]To summarize what has been explained so far: in the development
of order of matter on Earth, all non-human animal species are stuck in the
'animal condition', with each sexually reproducing individual member
of the species forever having to compete to ensure its genes reproduce
and carry on. *That* is the essential fact or rule of the gene-based natural
selection process—genes are unavoidably selfish; they have to ensure

they reproduce if they are to carry on. It is important to reiterate, however, that even though this selfishness—and the extreme competition between the sexually reproducing individuals it gives rise to—is characteristic of virtually all of nature, such selfishness is *only* occurring because of the *limitation* of the genetic process of *normally* being unable to develop unconditional selflessness between sexually reproducing individuals. In his 1850 poem *In Memoriam*, Tennyson famously wrote: '**Who trusted God was love indeed / And love Creation's final law / Tho' Nature, red in tooth and claw / With ravine** [in violent contradiction], **shriek'd against his creed.**' While Integrative Meaning or **'God'** and its theme of unconditional selflessness or **'love'** is the **'creed'** or **'final law'** of **'creation'** that the competitive, selfish and aggressive, **'red in tooth and claw'** characteristic of so much of **'Nature'** seems to be in violent contradiction **'against'**, we can now understand that this selfish characteristic doesn't mean that the *overall* biological reality of existence—life's meaning and theme—is to be selfish, as the dishonest theories of Social Darwinism, Sociobiology, Evolutionary Psychology and Multilevel Selection would have us believe. As will be explained in chapter 5, in the case of humans, we don't have selfish instincts like other species, rather we have *unconditionally selfless* instincts. And the selfishness that is characteristic of so much of nature is *only* occurring because of the *limitation* of the gene-based refinement process—its inability, in most situations, to develop unconditional selflessness. The genetic process would develop unconditionally selfless, fully cooperative behavior between all sexually reproducing individuals if it could—because such selflessness is what is required to maintain a fully integrated whole—but, because of its particular *limitation*, it normally can't. Integrative selflessness, not divisive selfishness, is the *real* nature or characteristic of existence, the theme of life.

[358] Incidentally, in par. 195 it was mentioned how, in describing his concept of natural selection, Darwin originally left it undecided as to whether individuals who managed to reproduce are 'fitter' or better than those who don't, but was later persuaded by human-condition-avoiding, mechanistic biologists to describe natural selection as a competitive, 'survival of the fittest' process. Well, we can now see why it was right for Darwin to leave it undecided as to whether individuals who manage to reproduce are better or 'fitter' than those who don't. As has now been explained, it can be completely consistent with the integrative meaning of existence for an individual to give their life for the purpose of maintaining

the larger whole of their society and thus not reproduce. Unconditionally selfless, self-sacrifice for the good of the whole, is the very theme of existence. It is only because of the *limitation* of the gene-based natural selection process that unconditionally selfless behavior normally cannot be developed between sexually reproducing individuals. Selfless co-operation, *not* selfish, competitive, 'survival of the fittest' behavior, *is* the real characteristic of existence, the theme of life.

## Chapter 4:5 Elaborating the sexually reproducing individual

[359] As stated, while sexually reproducing individuals cannot normally be integrated, the sexually reproducing individual itself could be elaborated, made bigger, which, as will now be explained, is how single-celled organisms developed into multicellular organisms, and how multicellular colonial ants and bees integrated into their fully cooperative colonies. Struggling to find a way to develop greater order of matter by integrating sexually reproducing individuals, it was as if Negative Entropy (or God) decided, 'Well, what I'll do is develop greater order of matter *within* each genetically reproducing individual, making it bigger.'

[360] As was also mentioned, in these 'elaborated sexually reproducing individuals', the cells of the multicellular body, or the individual ants and bees in their fully integrated colonies, are no longer sexually reproducing individuals themselves, but part of a larger sexually reproducing individual, which is the body, or, in the case of ants and bees, the colony.

[361] The biological mechanism for elaborating the sexually reproducing individual involved the body's cells, or the colony's multicellular bees/ants, delegating the task of sexual reproduction to a distinct part of the whole that *specializes* in reproduction. In the case of the integration of single-celled organisms, the green alga known as *Volvox* provides an example of an organism in transition from the single-celled to the multi-cellular state, as this quote describes: **'Volvox is...a small, green sphere... composed of thousands of flagellates embedded in the surface of a jelly ball... Volvox is a colony of unicellular animals rather than a many-celled animal, because even the simplest many-celled animals have considerably more differen-tiation between cells than appears among the cells of Volvox. The colony swims about, rolling over and over from the action of the flagella; but, remarkably enough, the same end of the sphere is always directed forward...Its behaviour can be explained only by supposing that the activities of the numerous flagellates**

are subordinated to the activity of the colony as a whole. If the flagella of each member of the colony were to beat without reference to the other members, the sphere would never get anywhere. In such subordination of the individual cells of a colony to the good of the colony as a whole we see the beginnings of individuality as it exists in the higher animals, where each animal behaves as a single individual, although composed of millions of cells…The co-ordination of numerous components into an individual is usually followed by the specialisation of different individuals for different duties. Only the slightest degree of specialisation is seen in the Volvox colony; the flagellates of the back part of the colony are capable of reproduction, while the front members never reproduce but have larger eyespots and serve primarily in directing the course of the colony' (R. Buchsbaum, *Animals without Backbones*, 1938, p.50 of 401).

[362] The marine invertebrates known as siphonophores, which include species such as the Portuguese man-of-war (otherwise known as the Bluebottle), live in colonies composed of 'zooids', individual animals that are not fully independent—indeed, their reliance upon and integration with each other is so strong the colony attains the character of one large organism. In fact, most of the zooids are so specialized they lack the ability to survive on their own. Thus siphonophorae, like Volvox, exist at the boundary between colonial and complex multicellular organisms.

[363] We can imagine the path to the creation of Volvox and siphonophores began with cloning, which is the asexual reproduction of identical offspring where competition between the clones is pointless and unnecessary since each individual is genetically identical and, therefore, the division of labor and cooperation can develop and exist between individuals. One concern with cloning, however, is loss of variability—for example, if one colony kept reproducing asexually it could become so big it monopolized the available resources of food, space and territory, leading to the detriment of other colonies and a subsequent lack of variability in the species. We can imagine that eventually a limited, functional size would be arrived at through natural selection, which presumably is the size at which Volvox and siphonophores operate. And obviously to maintain variability, it would also be beneficial for sexual reproduction to occur from time to time, as it does amongst Volvox and siphonophores.

[364] In the case of bees, the queen bee feeds all of her offspring that she intends to be workers a 'royal jelly' that causes sterility (ants also employ a similar chemical retardant). To ensure the reproduction of their genes these sterile offspring then have to support the queen because she carries

their genes. (It should be mentioned that saying the queen 'intends' and the offspring 'have to' is obviously personifying the genetic process. The queen and the offspring are obviously not conscious thinking organisms, deciding they 'intend' and 'have to' do something or other as humans do; however, this form of anthropomorphism is simply a useful way of describing what, *in effect*, occurs. For example, the way genetics actually causes offspring to 'have to' support the queen is that, out of the many different mutational varieties of offspring that appear over time, only those that happen to have a genetic make-up that inclines them to support the queen will tend to reproduce, naturally selecting that particular behavior for all subsequent generations and eventually the whole species.)

[365] Elaborating the sexually reproducing individual allows the members of the elaborated individual to develop the ability to at least *behave* unconditionally selflessly, which, as has been explained, is fundamental if the fully cooperative integration of members into a new whole is to develop. The reason our body works so well is because each part has sublimated its needs to the greater good of the whole body; each part behaves unconditionally selflessly. Just as our skin cells are in constant turnover, with new cells replacing the old ones that have sacrificed themselves to protect our body, the leaves that fall in autumn do so to ensure their tree survives through winter. Bees and ants readily sacrifice themselves for their colony; for example, when a bee stings to protect its hive, its innards are attached to the sting that is left in its victim, so when it stings, it dies. The skin, leaves and bees/ants have behaved unconditionally selflessly; they have, in effect, considered the welfare of the greater good above their own welfare.

[366] Of significance, however, is the emphasis here on our body's skin, the tree's leaves and the bees/ants only *behaving* unconditionally selflessly, because the selflessness apparent in these examples is *not* actually true unconditional selflessness, it is *not* true altruism. This is because the self-sacrificing skin, leaves and bees/ants are all indirectly selfishly ensuring their own genetic existence will be maintained by supporting the body, tree, or bee/ant colony that carries the genes for their existence and so reproduces them when it reproduces itself as a whole. Genetically, they are selflessly fostering the body/tree/colony to selfishly ensure their own genetic reproduction. Their apparently unconditionally selfless *behavior* is not actually unconditional and thus altruistic, but rather a subtle form of selfishness. As explained earlier, such reciprocity can

develop genetically because it doesn't compromise the chances of the sexually reproducing individual reproducing its genes. (As pointed out in par. 197, the dishonest biological theory of Sociobiology/Evolutionary Psychology was truthful to the extent that it did recognize this fact that the selfless behavior of social ants and bees is due to reciprocity—where the theory was dishonest was in its application of 'kin selection' to explain *all* social behavior, even our own *unconditionally* selfless, *universally* benevolent, *fully* altruistic moral instincts.)

[367] It now needs to be explained that large animals couldn't employ this device of elaborating the sexually reproducing individual to develop a fully cooperative, integrated association or whole of their members because for them it involves too great a loss of the variability that all species need to be able to adapt to their environment. For example, if a female buffalo happened to be born with a particular mutation that caused her to produce a chemical in her milk that retarded the sexual maturation of most of her offspring such that those offspring then had to have selected mutations that inclined them to protect her to ensure their genes are successfully reproduced by her, and this became a common practice amongst buffalos with every queen buffalo having, say, 9 protector sacrificial buffalos from 10 offspring (so there is a sexual offspring to ensure the reproduction of the buffalo species, like ants have a few fertile females and males to carry on their species, but these fertile offspring have the same potential to produce some infertile offspring), then the genetic variety of a population of 1,000 buffalos would be reduced to just 100, a drastic loss of the variability so critical to ensuring that the species' genetic pool remains able to adapt to any changes in its environment and thus keep maintaining and/or developing greater order. In the case of bees/ants, they are so small in relation to their environment that they can afford to have many fully integrated colonies in their environment without any significant loss of variability within their species.

[368] The following two photographs illustrate the point. As will be mentioned shortly, while termites are a variety of cockroach rather than ant or bee, they have developed the same colonial capability as colonial ants and bees, which means that although there are millions of termites in each termite mound, in terms of the *genetic variety* present in the territory shown, these mounds do, in fact, represent a similar number of sexually reproducing individuals to the number of sexually reproducing individual buffalos shown in a corresponding area in the second photograph.

Magnetic Termite Mounds, Litchfield National Park, Northern Territory, Australia;
and feral Asian Buffalo, Northern Territory, Australia. Photographs by the author, 2010.

[369] Quite a number of species that are much larger than ants and bees
are attempting to create the integrated society of members by <u>temporarily</u>
<u>elaborating the sexually reproducing individual</u>. Many bird species, such
as the Australian kookaburra, delay their sexual maturation for a few
years after they fledge, during which time they selflessly help raise their
parents' subsequent offspring. Wolves, African wild dogs and meerkats
do the same thing. However, what these species have obviously found
is that to delay their sexual maturation permanently leads to too great a
loss of variability in their species.

[370] Underground-living colonial naked mole rats form fully integrated
colonies of up to 300 members comprising a single queen who uses
hormones to inhibit the sexual maturation of nearly all the others who
then act as 'workers' and 'soldiers'. A few 'sexual disperser caste' are
allowed to reach sexual maturity and these periodically leave their natal
burrow to access other colonies and, in doing so, help maintain the genetic
variety of the mole rat species. Significantly, like colonial ants and bees,
and the dozen or so other varieties of multicellular organisms (including
the termite) that have been able to permanently elaborate the sexually
reproducing individual, mole rats are relatively small; individuals typically measure only 8 to 10 centimeters (3 to 4 inches) long.

[371] What has been explained here is very significant for humans because
it means, as large animals, we could not have employed the integrating
device of elaborating the sexually reproducing individual to create the
pre-conscious and pre-human-condition-afflicted, fully cooperative, completely integrated, 'Golden', 'Garden of Eden'-like state that our distant
ancestors lived in. Further, during that fully integrated, idyllic past our
instinctive orientation was *not* to reciprocity's subtle form of selfishness
that the parts of multicellular organisms and bee/ant colonies practice, as

the theory of Sociobiology/Evolutionary Psychology claims, but to being truly altruistic, genuinely unconditionally selflessly orientated towards *all of life*. Thus, even if we could have employed the device of elaborating the sexually reproducing individual it would not even begin to account for our *unconditionally* selfless moral soul. I italicized '*all of life*' because while ant and bee colonies have members who are dedicated to supporting each other, each colony is, in fact, engaged in fierce competition with other colonies. Worker ants and bees are not interested in behaving selflessly towards *all of life*, which, contrary to what the theory of Multilevel Selection claims, our moral self is interested in. As pointed out in chapter 2:11, our ability to love unconditionally didn't arise from an ability to war successfully. So, the question remains: how did humans manage to develop our absolutely wonderful and astonishing *unconditionally* selfless, *genuinely* altruistic, *all*-loving moral instinctive orientation to the world?

## Chapter 4:6  Negative Entropy found a way to form the Specie Individual

[372] In conclusion, while elaborating the sexually reproducing individual does allow greater order of matter to be developed, it doesn't achieve the next level of integration, which is the coming together or integration of sexually reproducing individual members of a species to form the Specie Individual or whole. The question, therefore, is, could Negative Entropy or God find a way to overcome the impasse of integrating sexually reproducing members of a species into a Specie Individual—or had the limit to the amount of order of matter that could be developed on Earth finally been reached?

[373] As will be fully explained in the next chapter, the reason I have written in this chapter that it is not '*normally*' possible to integrate sexually reproducing individuals is because Negative Entropy or God *did* find one way to integrate sexually reproducing members of a species to form the Specie Individual, which was through the nurturing of offspring—and it was *this* device that our ape ancestors employed to achieve the fully integrated state, the instinctive memory of which is our unconditionally selfless, genuinely altruistic, all-loving moral instinctive self or soul.

[374] (A more complete description of the integration of matter and the meaning of life can be found in *Freedom Expanded* at <www.humancondition.com/freedom-expanded-integrative-meaning>.)

# Chapter 5

# The Origin of Humans' Unconditionally Selfless, Altruistic, Moral Instinctive Self or Soul

Bonobos Matata and her adopted son, Kanzi

## Chapter 5:1  <u>Summary</u>

[375] When the philosopher Immanuel Kant had the following words inscribed on his tombstone, that *'there are two things which fill me with awe: the starry heavens above us, and the moral law within us'* (*Critique of Practical Reason*, 1788), and Charles Darwin wrote that *'The moral sense perhaps affords the best and highest distinction between man and the lower animals'* (*The Descent of Man*, 1871, ch.4), neither man was overstating the magnificence of our altruistic moral sense. Our moral instinctive self or soul, the 'voice' or expression of which is our conscience, *is* a truly amazing phenomenon, for it provided the cooperative, unconditionally selfless love that created humanity.

[376] But as amazing as our moral soul most certainly is, its very existence is *also* a cause for wonder because it raises the baffling question of *how on earth did humans acquire such an 'awe'-inspiring, 'distinct'-from-other-animals moral sense?* For biologists especially, the

great outstanding mystery has been how could the cold, selfish, competitive, gene-based natural selection process have possibly created such warm, unconditionally selfless, cooperative, loving instincts in us humans?

[377]While it may seem astonishing to suggest that what is now going to be presented *is* that most elusive of answers to this most intriguing of mysteries, that is, in fact, the case—this chapter contains nothing less than the truth about what it really means to be human.


## Chapter 5:2  How could humans have acquired their altruistic moral instincts?

[378]As evidenced in pars 180-182, throughout our mythologies and in the work of our most profound thinkers there is a recognition that our distant ancestors lived in a pre-conscious, pre-human-condition-afflicted, innocent, unconditionally selfless, genuinely altruistic, fully cooperative, universally loving, peaceful state; as the author Richard Heinberg's research into this collective memory of a 'Garden of Eden'-like, 'Golden Age' in our species' past found, **'Every religion begins with the recognition that human consciousness has been separated from the divine Source, that a former sense of oneness…has been lost…everywhere in religion and myth there is an acknowledgment that we have departed from an original…innocence'** (*Memories & Visions of Paradise*, 1990, pp.81-82 of 282). So yes, when Nikolai Berdyaev acknowledged that **'The memory of a lost paradise, of a Golden Age, is very deep in man'** (*The Destiny of Man*, 1931; tr. N. Duddington, 1960, p.36 of 310), he was expressing what we *all* intuitively know is the truth about our species' past innocent existence—as was the philosopher Jean-Jacques Rousseau when he, almost two centuries earlier, wrote that **'nothing is more gentle than man in his primitive state'** (*The Social Contract and Discourses*, 1755; tr. G. Cole, 1913, Book IV, p.198 of 269).

[379]The origin of the words associated with our moral nature reveals this underlying awareness of the extraordinarily loving, ideal-behavior-expecting, 'good-and-evil'-differentiating, sound nature of our <u>instinctive self</u> or '<u>psyche</u>' or '<u>soul</u>', the 'voice' or expression of which is our '<u>conscience</u>'. For instance, our **'conscience'** is defined as our **'moral sense of right and wrong'**, and our **'soul'** as the **'moral and emotional part of man'**, and as the **'animating or essential part'** of us (*Concise Oxford Dictionary*, 5th edn, 1964), while, as mentioned in pars 258 and 260, the *Penguin Dictionary of Psychology*'s entry for **'psyche'** reads: **'The oldest and most general use of**

**this term is by the early Greeks, who envisioned the psyche as the soul or the very essence of life.'** Indeed, as the **'early Greek'** philosopher Plato wrote about our innate, ideal-or-Godly-behavior-expecting moral nature, we humans have **'knowledge, both before and at the moment of birth...of all absolute standards... [of] beauty, goodness, uprightness, holiness...our souls exist before our birth... [our] soul resembles the divine'** (*Phaedo*, c.360 BC; tr. H. Tredennick, 1954, 65-80).

[380] When the philosopher John Fiske wrote about the existence of our moral nature, he was similarly effusive: **'We approve of certain actions and disapprove of certain actions quite instinctively. We shrink from stealing or lying as we shrink from burning our fingers'** (*Outlines of Cosmic Philosophy*, 1874, Vol. IV, Part II, p.126). *And*, while our moral instinctive self or soul will **'shrink from stealing or lying'**, it is not merely concerned with avoiding the ill-treatment of others—it is also deeply concerned with ensuring their well-being. For instance, when the professional footballer Joe Delaney admitted that **'I can't swim good, but I've got to save those kids'**, just moments before plunging into a Louisiana pond and drowning in an attempt to rescue three boys ('Sometimes The Good Die Young', *Sports Illustrated*, 7 Nov. 1983), he was considering the welfare of others above that of his own. The truth is that everywhere we look we see examples of humans behaving unconditionally selflessly, such as those who show charity to others less fortunate, or sacrifice their lives for ethical principles. Indeed, now that we can explain the human condition it becomes clear that since the human condition fully emerged some 2 million years ago, *every* generation of humans has had to suffer becoming self-corrupted in an unconditionally selfless effort to aid the accumulation of knowledge that would one day liberate humanity from the human condition; to borrow the words from the musical *Man of La Mancha* that were included in par. 68, every generation has altruistically **'march[ed] into hell for a heavenly cause'**.

[381] Our species' unconditionally selfless moral nature is undoubtedly a wonderful phenomenon. However, as mentioned in chapter 3:7, Alexander Pope saw our **'awe'**-inspiring, **'best and highest distinction'**-deserving, **'divine'**-like, **'absolute standards...[of] beauty, goodness, uprightness, holiness'**-expecting, **'animating'**, **'very essence of life'**, **'moral and emotional'**, **'essential part'** of us in a very different light, pointing out that **'our nature [is]...A sharp accuser, but a helpless friend!'** And he was right in the sense that, as was made clear in that chapter, our ideal-behavior-expecting, moral conscience *has* been **'a sharp accuser, but a helpless friend'**; it *has* criticized us aplenty when what we needed was redeeming *understanding* of our

'good-and-evil'-afflicted, corrupted or 'fallen' present human condition—
which we now at last have.

[382] Paradoxically, until we could explain our present soul-devastated,
innocence-destroyed, angry, egocentric and alienated condition we
couldn't afford to face the truth that our **'awe'**-inspiring moral soul is our
instinctive memory of an unconditionally selfless, all-loving past. And so
we undermined its very existence; yes, just as human-condition-avoiding,
mechanistic scientists argued that 'unconditional love' was **'not appropriate
for scientific study'**, the psychologist Ronald Conway noted that **'Soul is
customarily suspected in empirical psychology and analytical philosophy as a
disreputable entity'** (Letter to the Editor, *The Australian*, 10 May 2000). But with the
fully accountable, human psychosis-addressing-and-solving, truthful
explanation of the human condition now found, we *can* finally acknowledge what our soul is, and, most significantly, heal our species' psychosis
or 'soul-illness'; yes, since *psyche* means **'soul'** and *osis*, according
to *Dictionary.com*, means **'abnormal state or condition'**, we can at last
ameliorate or heal our species' *psychosis*—its alienated, psychologically
'ill', **'abnormal state or condition'**.

[383] But recognition and resolution of the issue aside, the very great
question that remained to be answered was how could we humans have
possibly acquired such a **'distinct'** from other **'animals'**, **'awe'**-inspiring
but **'sharp accus**[ing]**'** instinctive orientation in the first place? What is the
biological origin of our species' extraordinary moral nature?

## Chapter 5:3  The integration of sexually reproducing individuals to form the Specie Individual

[384] As was explained in some detail in chapters 4:4 to 4:6, while the gene-
based system for developing the order of matter on Earth is powerfully
effective—it is, after all, responsible for the great variety of life we see on
Earth—it has one very significant limitation, which arises from the fact that
each sexually reproducing individual organism has to struggle and selfishly
compete for the available resources of food, shelter, territory and the mating
opportunities it requires if it is to successfully reproduce its genes. What
this means is that integration, and the unconditionally selfless cooperation
that integration depends upon, cannot normally develop between one
sexually reproducing individual and another; which *in turn* means that
integration *beyond* the level of the sexually reproducing individual—that is,

the coming together or integration of sexually reproducing individuals to form a new larger and more stable whole of sexually reproducing individuals, the Specie Individual—can also not normally develop. Thus, it would appear that Negative Entropy's, or God's, development of order of matter on Earth had come to a stop at the level of the sexually reproducing individual. The integration of the members of a species into the larger whole of the Specie Individual could seemingly not be developed.

[385] What this means is that only a degree of cooperation and thus integration could be developed between the sexually reproducing individual members of a species before the competition between them became so intense that a dominance hierarchy had to be employed to contain the divisive competition; and, in fact, that is where most animal species *are* stalled in their ability to integrate. They could become integrated to a degree (what has been termed 'social'), but not completely integrated into one new larger organism or whole. Certainly each sexually reproducing individual could be either temporarily (in the case of large animals like wolves) or permanently (in the case of small animals like ants and bees) 'elaborated'—developed to become bigger—thus allowing greater integration of matter to occur *within* the sexually reproducing individual. But those elaborated units (the wolf packs and the ant/bee colonies) were, nevertheless, *still* engaged in competition with each other. It seemed that the integration of sexually reproducing members of a species and thus the full integration of the members of a species into a Specie Individual could not be achieved; Negative Entropy, or God, *had* seemingly developed as much order of matter on Earth as it could.

[386] HOWEVER, this was *not* the case—the integration of matter *hadn't* come to an end, because a way *was* found by Negative Entropy, or God, to integrate the members of a species into the larger whole of the Specie Individual, AND, moreover, it was our ape ancestors who achieved this extraordinary step. Yes, as Moses recognized in his account of the emergence of the human condition in Genesis, we humans *did* once live **'in the image of God'** (1:27), we were once a fully cooperative, unconditionally selflessly behaved, completely integrated species. We *did* once live in **'the Garden of Eden'**-like (3:23) state of original cooperative, loving, innocent togetherness, then we became conscious, took the **'fruit' 'from the tree of...knowledge'** (3:3, 2:17), and, as a result of being **'disobedient'** (the term widely used in descriptions of Gen. 3), we **'fell from grace'** (derived from the title of Gen. 3, 'The Fall of Man') because we became divisively behaved sufferers of the human condition, supposedly

deserving of being **'banished…from the Garden of Eden'**-like (3:23) state of original innocent togetherness. Our divisive, non-integrative, seemingly unGodly, psychologically angry, egocentric and alienated present corrupted condition meant that **'Today you** [God/Integrative Meaning] **are driving me from the land, and I will be hidden from your presence** [humans will have to live in a disconnected-from-the-truth, meaningless, lost, extremely-distressed state of alienation, and as a result], **I will be a restless wanderer on the earth'** (4:14)—but we can now at last emerge from this state because we can finally explain and thus compassionately understand *why* we *had to* search for **'knowledge'** and suffer becoming corrupted.

Africa–our soul's home–the Garden of Eden

Lucas Cranach the Elder's *Adam and Eve* (1526)

We can still see the remnants of the time when our species lived in a pre-human-condition-afflicted, innocent, Garden-of-Eden-like state in the happiness and togetherness of children.

[387] So THE GREAT QUESTION is, how did Negative Entropy, or God, achieve this amazing integration; this **'blessed'**, **'innocent and simple and calm and happy' 'pure'** state that Plato described (in par. 158), where there was no **'war or quarrel[ling]'**, and where we **'dwelt naked...in the open air... and...lay on soft couches of grass'** (see par. 170); the time when, as Hesiod said (in par. 180), we were a **'golden race...with calm untroubled mind[s]... unbridled by toil...[and] all willing shared the gathering of...[our] hands'**? How did our human ancestors develop the fully integrated state of the Specie Individual, the instinctive memory of which is our unconditionally self-less, genuinely altruistic, cooperative, loving, moral self or 'soul', and the 'voice' or expression of which is our conscience? What is the biological origin of our species' extraordinary **'shrink from stealing or lying'**, **'got to save those kids'**, selflessly prepared **'to march into hell for a heavenly cause'** moral sense?

## Chapter 5:4 <u>Love-Indoctrination</u>

[388] Ever since Charles Darwin published his natural selection explanation for the origins of the great variety of life on Earth in 1859 all manner of theories have been put forward to suggest how one species, namely *our* species, managed to defy the essential selfish nature of the natural selection process and develop unconditionally selfless moral instincts. Indeed, in response to the great build up of anxiety in the world about the now desperate plight of the human race and the resulting urgent need for the relieving answers that are being presented in this book — especially answers to the questions of the origin of our species' moral nature and present human-condition-afflicted lack of compliance with it — almost every edition of every science journal over the last 10 years has either presented or discussed a new theory that attempts to account for our moral sense. But of all these theories (a brief history of which will be presented in chapter 6) only one provides the fully accountable and thus adequate, true explanation for the origins of our moral sense, which is that it was achieved through <u>nurturing</u> — a mother's maternal instinct to care for her offspring. However, as will be acknowledged shortly (in ch. 5:7), and described in detail in chapter 6, the problem with this nurturing explanation, and why its early permutations were discarded by the scientific establishment, is that it has been an unbearably confronting truth. It is *only* now that we can explain the human condition and thus

understand *why* the present human-condition-afflicted human race hasn't been able to adequately nurture our infants to the extent their instincts expect that it becomes safe to finally admit that nurturing is what made us human—that it was nurturing that gave us our moral soul and created humanity. So how *did* nurturing create our extraordinary unconditionally selfless moral instincts?

[389] Genetic traits for nurturing are intrinsically selfish (which, as has been emphasized, genetic traits normally have to be if they are to survive) because through a mother's nurturing and fostering of offspring who carry her genes her genetic traits for nurturing are selfishly ensuring their reproduction into the next generation. However, while nurturing *is* a genetically selfish trait, from an observer's point of view it *appears* to be unconditionally selfless behavior—the mother is giving her offspring food, warmth, shelter, support and protection for *apparently* nothing in return. This point is most significant because it means from the infant's perspective its mother is treating it with real love, with unconditional selflessness. The infant's brain is, therefore, being trained or indoctrinated or inscribed with unconditional selflessness and so, with enough training in unconditional selflessness, that infant will grow into an adult who behaves unconditionally selflessly. Apply this training across all the members of that infant's group and the result is an unconditionally selflessly behaved, cooperative, fully integrated society.

[390] The 'trick' in this '<u>love-indoctrination</u>' process lies in the fact that the traits for nurturing are encouraged, or selected for genetically, because the better the infants are cared for, the greater are their, and the nurturing traits', chances of survival. The process does, however, have an integrative side effect, in that the more infants are nurtured, the more their brains are trained in unconditional selflessness. There are very few situations in biology where animals appear to behave selflessly towards other animals—normally, they each selfishly compete for food, shelter, territory and mating opportunities. Maternalism, a mother's fostering of her infant, is one of the few situations where an animal appears to be behaving selflessly towards another animal and it was this *appearance* of selflessness that exists in the maternal situation that provided the integrative opportunity for the development of love-indoctrination, the training of individuals in unconditional selflessness. And with this unconditionally selfless behavior recurring over many, many generations, the unconditionally selfless behavior will become instinctive—a moral

soul will be established—because genetic selection will inevitably follow and reinforce any development process occurring in a species. The difficulty was in getting the development of unconditional selflessness to occur in the first place, for once it was regularly occurring it would naturally become instinctive over time.

[391] But for a species to develop nurturing—to develop this 'trick' for overcoming the gene-based learning system's seeming inability to develop unconditional selflessness—it required the capacity to allow its offspring to remain in the infancy stage long enough to allow for the infant's brain to become trained or indoctrinated with unconditional selflessness or love. In most species, infancy has to be kept as brief as possible because of the infant's extreme vulnerability to predators. Zebra foals, for example, have to be capable of independent flight almost as soon as they are born, which gives them little opportunity to be trained in selflessness. But as the following photo of a rhesus monkey trying to carry its infant illustrates, being semi-upright as a result of their tree-living, swinging-from-branch-to-branch, arboreal heritage meant primates' arms were largely freed from walking and thus available to hold dependents. Infants similarly had the capacity to latch onto their mothers' bodies. This freedom of the upper body meant primates were especially facilitated for prolonging their offspring's infancy and thus developing love-indoctrination.

Rhesus monkey with infant. This picture illustrates the difficulty
of carrying an infant and suggests the reason for bipedalism.

[392] It follows that the longer infancy is delayed, the more and longer infants had to be held, and thus the greater need and, therefore, selection for the arms-freed, upright walking ability known as bipedalism. When I first put forward this nurturing, 'love-indoctrination' explanation for humans' unconditionally selfless moral nature in a 1983 submission to *Nature* and *New Scientist* (see <www.humancondition.com/nature>), I said, contrary to prevailing views, that it meant bipedalism must have developed early in this nurturing of love process, and, in fact, as will be discussed in more detail shortly, the early appearance of bipedalism in the fossil record of our ancestors is now being found. For instance, in 2012 it was reported that **'The oldest hominins currently known are *Sahelanthropus tchadensis*...dated to between 6 and 7 mya** [million years ago]', which the fossil record suggests **'stood and walked bipedally'** (Herman Pontzer, 'Overview of Hominin Evolution', *Nature Education Knowledge*, 2012, Vol.3, No.10).

[393] But while bipedalism was the key factor in developing nurturing and thus love-indoctrination, other influences also played a pivotal role, most notably the presence of ideal nursery conditions. This entailed uninterrupted access to food, shelter and territory, for if any element was compromised, or other difficulties and threats from predators excessive, then we can assume there would have been a strong inclination to revert to more selfish and competitive behavior. The successful nurturing of infants therefore required ample food, comfortable conditions and security from external threats over an extended period. But, in addition to these practical factors, to ensure the success of the love-indoctrination process, it wasn't enough to simply *look after* the infants, they had to be *loved*, and so maternalism became about much more than mothers simply protecting, providing for, and training their infants in life skills—it became about demonstrably loving them. Significantly, we speak of 'motherly love', not 'motherly protection, provision and training'.

[394] So, in addition to the prerequisites of, firstly, a physiology that facilitated an extended infancy and, secondly, ideal nursery conditions, what was *also* required for love-indoctrination to occur was the presence and influence of more maternal mothers. Of course, the difficulty with selecting for more maternal mothers is that their genes don't tend to endure because their offspring tend to be the most selflessly behaved, too ready to put others before themselves, leaving the more aggressive, competitive and selfish individuals to take advantage of their selflessness. Such selfish opportunism *could*, however, be avoided if *all* members

of the group were equally well nurtured with love, equally trained in selflessness—this situation being yet another of the delicate conditions that has to be maintained if love-indoctrination is to develop, for any breakdown in nurturing within a group that is in the midst of developing love-indoctrination could jeopardize the whole situation and see it revert to the old state of the 'each-for-his-own', opportunistic, all-out-competition-where-only-dominance-hierarchy-can-bring-some-peace, selfish-genes-rule, 'animal condition'.

[395] It can be seen then that while the development of unconditional selflessness through the love-indoctrination process of a mother's nurturing care of her infant *was* possible, it was not easy, even for the exceptionally facilitated primates, which explains why none of the primate species apart from our ape ancestors have been able to complete the development of love-indoctrination to the point of becoming fully integrated. While, as will shortly be described in chapter 5:6, bonobos are on the threshold of this achievement, the evidence so far indicates that it has only been our ape ancestors who managed to *complete* the process, the result of which is our species' unconditionally selfless, genuinely altruistic, universally loving instinctive self or 'soul', the 'voice' or expression of which is our moral 'conscience'. Before concluding this section, it is important to note that this love-indoctrination process involved an *indoctrination* or training in unconditional selflessness, not an *understanding* of it. The search for knowledge still had to take place, which is why the human-condition-producing clash between our instincts and conscious intellect occurred. (The process by which love-indoctrination liberated consciousness will be introduced in ch. 5:8, and fully explained in ch. 7).

## Chapter 5:5 <u>Fossil evidence confirming the love-indoctrination process</u>

[396] Although the fossil record has been slow to yield evidence of our ape ancestors who lived during humanity's infancy (which, as will be explained in chapters 8:2 and 8:3, lasted from some 12 to 4 million years ago), the very recent discoveries of fossils belonging to our direct ancestors from this period are now confirming the love-indoctrination process. These recently unearthed ancestors are: *Sahelanthropus tchadensis* (who lived some 7 million years ago and is thought to be the first representative of the human line after we diverged from humans' and chimpanzees' last

common ancestor); *Orrorin tugenensis* (who lived some 6 million years ago); and the two varieties of *Ardipithecus*: *kadabba* (who lived some 5.6 million years ago), and *ramidus* (who lived some 4.4 million years ago). Incidentally, *Sahelanthropus* means 'Sahel man' (Sahel is an area near the Sahara); *Orrorin tugenensis* means 'original man whose fossils were found in the Tugen region in Kenya'; while *Ardipithecus* means 'ground ape', with *kadabba* meaning 'oldest ancestor', and *ramidus* meaning 'root' or basal family ancestor.

[397] It is worth emphasizing that these fossils have all been found *very* recently. *Sahelanthropus* was only discovered in 2002 (in the form of a skull) and decisively identified as a human ancestor in April 2013, while fragments of a skull, jaw and thigh bone belonging to *Orrorin* were first unearthed in 2001. Although fragments of *Ardipithecus* were first discovered by a team led by the anthropologist Tim White in 1992, and their excavation of a largely intact skeleton (which was nicknamed 'Ardi') began in 1994, the remains of the skeleton—1 of only 6 reasonably complete skeletons of early humans older than 1 million years—were in such poor condition that it took until 2009 (over 15 years of analysis) for reports to be published. With studies on all of these recently discovered ancestors now becoming available, including the series of 2009 *Ardipithecus* reports, which the journal *Science* deemed 'Breakthrough of the Year', it is exciting to see that confirming evidence of the love-indoctrination process that led to the establishment of our extraordinary unconditionally selfless moral instincts is slowly but surely emerging.

[398] So, how does this new evidence confirm the love-indoctrination process? How, for instance, does it affect our understanding of the emergence of bipedalism, the first key factor in developing unconditionally selfless moral instincts?

[399] As I just mentioned, when I first put forward the nurturing, 'love-indoctrination' explanation for such instincts in 1983, I said, contrary to prevailing views, that it meant bipedalism must have developed early in this nurturing of love process and, it follows, early in our ancestors' history, and, indeed, that is what these fossil discoveries now show. Scientists can infer whether a species was bipedal by several methods, including the position of the foramen magnum (the opening at the base of the skull through which the spinal cord enters), because in species that stand upright the opening appears toward the center of the skull rather than at the rear. Using information such as this, the current scientific

thinking is that bipedalism arose at least as early as *Sahelanthropus*, with anthropologists now reporting that **'Bipedalism is one of very few human characteristics that appears to have evolved at the base of the hominin clade [species more closely related to modern humans than to any other living species]. Recent fossil discoveries have apparently pushed back the origin of the hominin clade into the late Miocene, to 6 to 7 million years ago (Ma). The oldest known potential hominin** [human line] **fossils** [are] **attributed to *Sahelanthropus tchadensis'*** (Brian G. Richmond & William L. Jungers, '*Orrorin tugenensis* Femoral Morphology and the Evolution of Hominin Bipedalism', *Science*, 2008, Vol.319, No.5870).

[400] Fossils belonging to the slightly more recent *Orrorin* provide further proof of this bipedalism. In addition to the evidence revealed by the fragments of its skull, the analysis of *Orrorin's* femur (thigh bone) has allowed scientists to conclude that **'*O. tugenensis* is a basal hominin adapted to bipedalism'** (ibid), and **'that *Orrorin* was a habitual biped as shown by a suite of features in the proximal femur'** (Martin Pickford et al., 'Bipedalism in *Orrorin tugenensis* revealed by its femora', *Comptes Rendus Palevol*, 2002, Vol.1, No.4).

[401] Fossils of *Ardipithecus*, and particularly *Ar. ramidus*, confirm that bipedalism was *well* established by 4.4 million years ago, with studies of 'Ardi' (the relatively intact skeleton) leading the prominent anthropologist C. Owen Lovejoy to conclude that **'*Ar. ramidus* was fully capable of bipedality and had evolved a substantially modified pelvis and foot with which to walk upright'** ('Reexamining Human Origins in Light of *Ardipithecus ramidus*', *Science*, 2009, Vol.326, No.5949). Furthermore, Lovejoy confirmed the long history of bipedalism that preceded *Ar. ramidus* when he said that *Ar. ramidus* **'has been bipedal for a very long time'** (Ann Gibbons, 'A New Kind of Ancestor: *Ardipithecus* Unveiled', *Science*, 2009, Vol.326, No.5949).

[402] The second requirement for love-indoctrination to occur is the existence of <u>ideal nursery conditions</u>, namely an environment that provides uninterrupted access to food, shelter and territory. You would perhaps expect such conditions would be found in humid forests and woodlands, where food is plentiful and trees provide shelter and refuge from predators, but the scientific community's traditional view has been that the factor in our ancestral history that propelled our ancestor's development beyond that of the other apes was their movement onto the savannah. However, in light of the fossil evidence that has emerged in the last decade or so—and in the 30 years since I first proposed the nurturing, love-indoctrination explanation—the scientific community now widely accepts that this separation of our human ancestors from other primates occurred while

our ancestors lived in forests and woodlands, the sort of environment I identified as being required for the love-indoctrination process to begin. [403] Scientists are now able to reconstruct the habitats of *Sahelanthropus*, *Orrorin* and *Ardipithecus* based on their physical characteristics, the information provided by the fossils of other animals and plants found accompanying them, as well as climate data. While *Sahelanthropus* fossils are so limited they don't provide the information needed to confirm that they were adapted to climbing trees and thus lived in forests or woodlands, reconstructions of their environment have narrowed *Sahelanthropus'* habitat to **'a mosaic of environments from gallery forest at the edge of a lake area to a dominance of large savannah and grassland'** (Patrick Vignaud et al., 'Geology and palaeontology of the Upper Miocene Toros-Menalla hominid locality, Chad', *Nature*, 2002, Vol.418, No.6894). As we move forward in time to *Orrorin* some 6 million years ago, its skeletal structure shows tree climbing adaptations, which clearly point to them living in an arboreal habitat. Further, associated animal and plant fossils have allowed scientists to infer that **'*Orrorin tugenensis* may have evolved in well wooded to forested conditions margining lakes and streams with open country-side in the vicinity'** (Soizic Le Fur et al., 'The mammal assemblage of the hominid site TM266 (Late Miocene, Chad Basin): ecological structure and paleoenvironmental implications', *Naturewissenschaften*, 2009, Vol.96, No.5); and that **'the surroundings of the site were probably open woodland, while the presence of several specimens of colobus monkeys indicate that there were denser stands of trees in the vicinity, possibly fringing the lake margin and streams that drained into the lake'** (Martin Pickford & Brigitte Senut, 'The geological and faunal context of Late Miocene hominid remains from Lukeino, Kenya', *Comptes Rendus de l'Academie des Sciences–Series IIA–Earth and Planetary Science*, 2001, Vol.332, No.2). Forest and woodlands continued to be the preferred habitat of *Ar. ramidus* some 4.4 million years ago, as indicated by its retention of tree climbing features such as a pelvis that supported large climbing muscles, flexible wrists that allowed walking on all fours along the top of branches, and an opposable big toe that allowed it to grasp the branches with its feet: **'*Ar. ramidus* preferred a woodland-to-forest habitat rather than open grasslands'** (Tim D. White et al., '*Ardipithecus ramidus* and the Paleobiology of Early Hominids', *Science*, 2009, Vol.326, No.5949). In fact, the wealth of surrounding evidence from the *Ar. ramidus* fossil site in Ethiopia allowed the paleoanthropologist Andrew Hill to remark that **'There's so much good data here that people aren't going to be able to question whether early hominins were living in woodlands'** (Ann Gibbons, 'Habitat for Humanity', *Science*, 2009, Vol.326, No.5949), and fellow researcher Giday WoldeGabriel to state that *Ar. ramidus* lived **'in an environment that was**

**humid and cooler than it is today, containing habitats ranging from woodland to forest patches'** (Giday WoldeGabriel et al., 'The Geological, Isotopic, Botanical, Invertebrate, and Lower Vertebrate Surroundings of *Ardipithecus ramidus'*, *Science*, 2009, Vol.326, No.5949). Indeed, this **'good data'** associated with the 'Ardi' dig has meant that paleobiologists have been able to reconstruct *Ar. ramidus'* habitat to an extraordinary level of detail: **'Ardi lived on an ancient floodplain covered in sylvan woodlands, climbing among hackberry, fig, and palm trees, and coexisting with monkeys, kudu antelopes, and peafowl'** (Ann Gibbons, 'Breakthrough Of The Year: *Ardipithecus ramidus'*, *Science*, 2009, Vol.326, No.5960) while **'doves and parrots flew overhead'** (Ann Gibbons, 'Habitat for Humanity', *Science*, 2009, Vol.326, No.5949). Combine this environment with our knowledge of *Ar. ramidus'* diet, which indicates **'*Ar. ramidus* was a generalized omnivore and frugivore** [fruit eater]' (Gen Suwa et al., 'Paleobiological Implications of the *Ardipithecus ramidus* Dentition', *Science*, 2009, Vol.326, No.5949), and our knowledge of existing ape behavior, which indicates *Ar. ramidus* **'almost certainly slept and fed in trees'** (Craig Stanford, 'Chimpanzees and the Behavior of *Ardipithecus ramidus'*, *Annual Review of Anthropology*, 2012, Vol.41), and a picture begins to emerge of the ideal nursery conditions that enabled love-indoctrination to develop.

[404] These ideal nursery conditions also refute the long-held nurturing-avoiding theory, espoused by E.O. Wilson amongst others, that upright walking supposedly developed when our ancestors moved out onto the savannah: **'*Ar. ramidus* did not live in the open savanna that was once envisioned to be the predominant habitat of the earliest hominids'** (Giday WoldeGabriel et al., 'The Geological, Isotopic, Botanical, Invertebrate, and Lower Vertebrate Surroundings of *Ardipithecus ramidus'*, *Science*, 2009, Vol.326, No.5949). In fact, the evidence that bipedality developed in **'forest or wooded environments'** is now so conclusive that Hill was able to assert that **'Savannas had nothing to do with upright walking'** (Ann Gibbons, 'Habitat for Humanity', *Science*, 2009, Vol.326, No.5949). Yes, because the development of bipedality is closely associated with the love-indoctrination process it had to have occurred while our ancestors were inhabiting ideal nursery conditions, which clearly suggested an arboreal environment—as I maintained when I originally put forward the love-indoctrination process in 1983.

[405] These recent fossil discoveries also confirm the third requirement for love-indoctrination to occur: the presence and influence of more maternal mothers. Scientists are able to deduce a remarkable amount of information about the social behavior of our ancestors from their fossils, and, as a result of this evidence, are now beginning to acknowledge that they exhibited

low levels of aggression toward one another, and that females were not only not dominated by males, but dictated mate choice by choosing to reproduce with non-aggressive, cooperative males—hallmarks you would expect of a society highly focused on maternal nurturing of their infants.

[406] The first striking evidence provided by the fossil record to support these deductions is that these early humans had <u>small canine teeth</u>: **'male canine size and prominence were dramatically reduced by ~6 to 4.4 Ma'** (Gen Suwa et al., 'Paleobiological Implications of the *Ardipithecus ramidus* Dentition', *Science*, 2009, Vol.326, No.5949). This is relevant because **'canines function as weapons in interindividual aggression in most anthropoid species'** (ibid), particularly in aggressive male-to-male sexual competition for mating opportunities, and so canines **'inform aspects of social structure and behavior'** (ibid), with small canines indicating minimal levels of social aggression. This connection is well established, with primatologists saying, **'It has long been evident that body and canine size are good indicators of the intensity of male-male competition'** (Peter M. Kappeler & Carel P. van Schaik, *Sexual Selection in Primates: New and Comparative Perspectives*, 2004, p.5 of 284).

[407] Furthermore, comparisons of canine size in *Ar. ramidus* with current apes indicate that *Ar. ramidus* males **'retained virtually no anatomical correlates of male-to-male conflict'** (C. Owen Lovejoy, 'Reexamining Human Origins in Light of *Ardipithecus ramidus*', *Science*, 2009, Vol.326, No.5949), a situation that would apply to our earlier ancestors *Sahelanthropus* and *Orrorin* since they too had small canines. Given that the reality of the animal kingdom involves fierce competition between sexually reproducing individuals seeking to reproduce their genes, this reduction in aggressive male competition for mating opportunities is an *extremely* significant anomaly, as Lovejoy recognizes: **'Loss of the projecting canine raises other vexing questions because this tooth is so fundamental to reproductive success in higher primates. What could cause males to forfeit their ability to aggressively compete with other males?'** (ibid). Traditional attempts to answer this **'vexing'** question have argued either that large canine teeth were made redundant when humans adopted hand-held weapons—the so-called 'weapons replacement' hypothesis; or that large, overlapping canines made eating certain foods difficult and were therefore selected against; or that large canines had to make way for the large grinding teeth of the robust australopithecines. However, the fossil record now shows that canines were reduced well before the emergence of the australopithecines; and as mentioned, it also shows that **'*Ar. ramidus* was a generalized omnivore and frugivore** [fruit eater]' like baboons and many other species of current primates who *have* retained their large canines.

And with regard to weapon use rendering large canines redundant, the fossil record now shows that our ancestors developed small canines at least as early as *Sahelanthropus*, millions of years before any fossil evidence of weapon or tool use—and even if those ancestors brandished weapons such as branches or bones that would not leave 'evidence', the argument still fails to explain why having weapons *and* large canines would not be an advantage in any contest. A 1992 paper articulated the confusion that has surrounded the evolution of human canine reduction, stating that **'the issue of human canine evolution has continued to be controversial and apparently intractable'** (Leonard O. Greenfield, 'Origin of the human canine: A new solution to an old enigma', *American Journal of Physical Anthropology*, 1992, Vol.35, No.S15). And the new discoveries have only increased this confusion. But as we can now see, the answer to the **'vexing'** and **'apparently intractable'** question of **'what could cause males to forfeit their ability to aggressively compete with other males'** is the love-indoctrination process. As will be explained below, conscious self-selection of integrativeness—especially the female sexual or mate selection of less competitive, less aggressive, more integrative males—developed to assist, speed up and help maintain love-indoctrination's development of integration. Indeed, male competition for mating opportunities is so **'fundamental to reproductive success'** that only active sexual selection against it can account for its reduction, as is made clear in this quote: **'Canine reduction did not result from a relaxation of selection pressure for large canines, but rather a positive selection against them'** (Arthur Klages, 'Sahelanthropus tchadensis: An Examination of its Hominin Affinities and Possible Phylogenetic Placement', *Totem: The University of Western Ontario Journal of Anthropology*, 2008, Vol.16, No.1). Indeed, it is now so apparent that canine reduction could only be caused by **'a positive selection against them'** that the importance of sexual selection is now being recognized by leading anthropologists such as Lovejoy, Gen Suwa, Berhane Asfaw, Tim White and others, who write, **'In modern monkeys and apes, the upper canine is important in male agonistic [aggressive] behavior, so its subdued shape in early hominids and *Ar. ramidus* suggests that sexual selection played a primary role in canine reduction. Thus, fundamental reproductive and social behavioral changes probably occurred in hominids long before they had enlarged brains and began to use stone tools'** (Gen Suwa et al., 'Paleobiological Implications of the *Ardipithecus ramidus* Dentition', *Science*, 2009, Vol.326, No.5949).

[408] As these authors make clear, the reduction in canine size was such a remarkable achievement that it required **'fundamental reproductive and social behavioral changes'** in which **'sexual selection played a primary role'**.

These scientists are describing a society that switched from being patriarchal—dominated by male sexual selection with males aggressively competing for mating opportunities—to matriarchal, dictated by <u>female sexual selection</u> where females choose mates that are less aggressive. However, what these scientists don't explain is the *only* mechanism that could allow such a switch: love-indoctrination. This remarkable reversal where females are empowered, and males **'forfeit their ability to aggressively compete with other males'**, is discussed in more detail in chapter 6; however, it is sufficient to emphasize at this point that the fossil record is increasingly providing compelling evidence that female sexual selection was occurring very early in human history, at least as early as *Sahelanthropus* some 7 million years ago, and that it **'emerged in concert with habituation to bipedality'** (C. Owen Lovejoy, 'Reexamining Human Origins in Light of *Ardipithecus ramidus*', *Science*, 2009, Vol.326, No.5949), which again is in accord with love-indoctrination, all of which I first predicted in 1983.

[409] Another significant factor revealed by the fossil record is the difference in the size between males and females, including their canines, a phenomena known as sexual size <u>dimorphism</u>. Since **'sexual size dimorphism is generally associated with sexual selection via agonistic male competition in nonhuman primates...if a species showed very strong size dimorphism, it probably was characterized by intense male mate competition'** (J. Michael Plavcan, 'Sexual Size Dimorphism, Canine Dimorphism, and Male-Male Competition in Primates', *Human Nature,* 2012, Vol.23, No.1). Conversely, scientists recognize that a low level of sexual size dimorphism is an indicator of a society in which males do not aggressively compete for mating opportunities. As mentioned, the fossil records of our human ancestors show that **'There is no evidence of substantial canine dimorphism in earlier hominins, including *Sahelanthropus*, *Ardipithecus*, and *Australopithecus anamensis*, or later hominins'** (ibid). In addition to this low level of canine dimorphism, *Ar. ramidus* exhibited low levels of body size dimorphism, which, in terms of behavior, **'were probably the anatomical correlates of comparatively weak amounts of male-male competition, perhaps associated with...a tendency for male-female codominance as seen in *P. paniscus* [bonobos]'** (Gen Suwa et al., 'Paleobiological Implications of the *Ardipithecus ramidus* Dentition', *Science*, 2009, Vol.326, No.5949). As will be described shortly, the prevailing view about bonobos is that rather than having achieved **'male-female codominance'**, they have, in fact, gone further and achieved female dominance, a matriarchy.

[410] So the three requirements of the love-indoctrination process of

bipedality, ideal nursery conditions and selection for more maternal mothers are now being dramatically confirmed by the fossil record. However, as I mentioned earlier and will elaborate on shortly, the problem with this nurturing, true explanation, and why its early permutations were dismissed by the scientific establishment, is that it has been an unbearably confronting, exposing truth for our present human-condition-afflicted human race that has been so unable to adequately nurture our infants to the extent our instincts expect. This new evidence has left those scientists who continue to deny the importance of nurturing in our development in a predicament in which they are forced to ask the right questions even though they are **'vexing'**, but refuse to acknowledge the truthful answer, because until the human condition was explained nurturing was an off-limits subject. The following passage from Lovejoy exemplifies this predicament: **'Why did early hominids become the only primate to completely eliminate the sectorial canine complex** [large projecting canines that are continuously sharpened against a lower molar]**? Why did they become bipedal, a form of locomotion with virtually no measurable mechanical advantage?…These are now among the ultimate questions of human evolution'** ('Reexamining Human Origins in Light of *Ardipithecus ramidus*', *Science*, 2009, Vol.326, No.5949). (Note, the above quote from 2009 also contained the question **'Why did body-size dimorphism increase in their likely descendants?'**, but that has been omitted here because Lovejoy has since found that body-size dimorphism in *Ardipithecus'* descendants is far less than previously thought, with **'relatively stable size patterns observed between** *Ardipithecus* **and** *Australopithecus'* (Philip Reno & C. Owen Lovejoy, 'From Lucy to Kadanuumuu: balanced analyses of *Australopithecus afarensis* assemblages confirm only moderate skeletal dimorphism', *PeerJ*, 2015, 3:e925).) Lovejoy further reduced these **'ultimate questions'** to this one, final sentence that admits the reality of a cooperative past: **'Even our species-defining cooperative mutualism can now be seen to extend well beyond the deepest Pliocene** [well beyond 5.3 million years ago]' (ibid). Yes, as stated at the outset of this chapter, the great outstanding mystery for biologists has been how could the cold, selfish, competitive, gene-based natural selection process have possibly created such warm, unconditionally selfless, cooperative, loving instincts in us humans? But to answer that question of questions required the explanation of the human condition that would finally make sense of *why* we haven't been able to adequately nurture our infants—because with that compassionate insight it at last becomes psychologically safe to admit that nurturing is what made us human, thus allowing these **'ultimate questions of human evolution'** to be answered.

## Chapter 5:6  <u>Bonobos provide living evidence of the love-indoctrination process</u>

[411]While these recent fossil discoveries are providing exciting confirmation that our ape ancestors completed the development of the love-indoctrination process, of the living primate species, only *Pan paniscus*, the <u>bonobos</u> (or pygmy chimpanzees as they were once called because of their comparatively gracile bodies), have not only developed love-indoctrination, they appear to have come close to completing the love-indoctrination process to become a fully integrated Specie Individual; they are certainly by far the most cooperative/harmonious/gentle/loving/integrated of the non-human primates. It follows then that although there is no suggestion that bonobos or chimpanzees are a living human ancestor, comparisons have been made between bonobos and our ancestors. For instance, the physical anthropologist Adrienne Zihlman first proposed in 1978 **'that, among living species, the pygmy chimpanzee (*P. paniscus*) offers us the best prototype of the prehominid ancestor'** (Adrienne L. Zihlman et al., 'Pygmy chimpanzee as a possible prototype for the common ancestor of humans, chimpanzees and gorillas', *Nature*, 1978, Vol.275, No.5682), using the then earliest known human ancestor, *Australopithecus*, to compare the two species' physical characteristics, including their bipedality, canine teeth and lack of sexual size dimorphism. In 1996 Zihlman refined her assessment to include similarities with the (at the time) newly discovered *Ardipithecus*. In a further example, the primatologist Frans de Waal has noted the extraordinary similarity between *Ardipithecus* and the bonobo, saying, **'The bonobo's body proportions—its long legs and narrow shoulders—seem to perfectly fit the descriptions of Ardi, as do its relatively small canines'** (*The Bonobo and the Atheist*, 2013, p.61 of 289). (Note, although the bonobo male **'possesses smaller canines than any other [male] hominoid** [apes and their ancestors]' (J. Michael Plavcan et al., 'Competition, coalitions and canine size in primates', *Journal of Human Evolution*, 1995, Vol.28, No.3), which, as explained in par. 406 above, is in itself a marker of low levels of aggression between males and thus a sign the love-indoctrination process is well underway in bonobo society, their canines do feature a sharp cutting edge that is absent in *Ardipithecus*, which suggests competitive fighting hasn't been completely eliminated within bonobo society and that the process

is not as advanced in bonobos as it was in *Ardipithecus*.) So bonobos (who, along with their chimpanzee cousins, share 98.7 percent of their DNA with humans) are physiologically extremely similar to our fossil ancestors, but beyond the physical similarities, some scientists are suggesting bonobo behavior also corresponds with that of our ancestors. In addition to the view expressed above by Gen Suwa, that, like bonobos, *Ardipithecus* were not male dominated, Zihlman has suggested that **'the Pan paniscus model offers another way to view the social life of early hominids, given their sociability, lack of male dominance and the female-centric features of their society'** ('Reconstructions reconsidered: chimpanzee models and human evolution', *Great Ape Societies*, eds William C. McGrew et al., 1996, p.301 of 352).

[412]So given the exceptionally cooperatively behaved, matriarchal bonobo species has developed the love-indoctrination process, we should expect that they provide living evidence of the three elements previously identified as being required for that process to occur: bipedalism, ideal nursery conditions, and selection for more maternal mothers.

A female bonobo rests in a nest of compacted branches and leaves high in the forest canopy at the Kokolopori Bonobo Reserve, Democratic Republic of the Congo.

With her infant beside her, Kame shares provisioned sugar cane
with Senta, a non-related juvenile, at the Wamba bonobo
research station, Democratic Republic of the Congo, 1987.

[413] In relation to <u>bipedalism</u>, research confirms that bonobos are extremely well-adapted to upright walking; in fact, they are **'the most bipedal
of all the extant** [living] **apes'** (Roberto Macchiarelli et al., 'Comparative analysis of the iliac
trabecular architecture in extant and fossil primates by means of digital image processing techniques', *Hominoid Evolution and Climate Change in Europe: Vol.2*, eds Louis de Bonis et al., 2001,
p.71 of 372). As was explained earlier, we can now account for this bipedality,
along with bonobos' peaceful cooperative nature: the longer infancy is
delayed, the more and longer infants had to be held, thus the greater need
and, therefore, selection for arms-freed, upright walking.

[414] As mentioned, <u>ideal nursery conditions</u> are the second requirement for establishing love-indoctrination—and bonobos have certainly
benefited from a comfortable environment in the food-rich, relatively
predator-and-competitor-free, ideal nursery conditions of the rainforests
south of the Congo River. (Bonobos, for example, don't have to compete
with gorillas, who have a diet similar to that of the bonobo but who live
north of the Congo River.) This fortuitous geographic situation is thought
to have been created about 2 million years ago after the formation of the

Clockwise from top: Frans Lanting/Getty Images; © Frans Lanting; Planckendael Wildlife Park

Upright Bonobos. Note the well-developed breasts (similar to those of
female humans) for suckling their young—another indication of
how important a role nurturing is playing in bonobo society.

Congo River divided an ancestral population of what became, some 1 million years ago, the two distinct species of bonobos and chimpanzees, with the bonobos, *Pan paniscus*, as mentioned, living south of the river, while the chimpanzees, *Pan troglodytes*, found themselves confined to areas north and east of the river.

[415]The ideality of these nursery conditions, which have been so conducive to their successful development of the love-indoctrination process, and the resulting love-indoctrinated cooperativeness of the bonobos compared with that of their chimpanzee cousins, is apparent in this quote: **'we may say that the pygmy chimpanzees historically have existed in a stable environment rich in sources of food. Pygmy chimpanzees appear conservative in their food habits and unlike common chimpanzees have developed a more cohesive social structure and elaborate inventory of sociosexual behavior...Prior to the Bantu (Mongo) agriculturists' invasion into the central Zaire basin, the pygmy chimpanzees may have led a carefree life in a comparatively stable environment'** (Takayoshi Kano & Mbangi Mulavwa, 'Feeding ecology of the pygmy chimpanzees *(Pan paniscus)*'; *The Pygmy Chimpanzee*, ed. Randall Susman, 1984, p.271 of 435). Indeed, it is an indication of how difficult it is to develop love-indoctrination that even bonobos, living as they do in their ideal conditions, and who **'have developed a more cohesive social structure'** than chimpanzees, still find it necessary to <u>employ sex as an appeasement device</u> to help subside residual tension and aggression between individuals; this is the **'elaborate inventory of sociosexual behavior'** referred to in this quote. As Frans de Waal has written, **'For these animals [bonobos], sexual behavior is indistinguishable from social behavior. Given its peacemaking and appeasement functions, it is not surprising that sex among bonobos occurs in so many different partner combinations, including between juveniles and adults. The need for peaceful coexistence is obviously not restricted to adult heterosexual pairs'** ('Bonobo Sex and Society', *Scientific American*, Mar. 1995). Clearly, sex amongst bonobos is like the **'naked and they felt no shame'** (Gen. 2:25) sex that Moses described our innocent Adam and Eve/bonobo stage-equivalent ancestors as practicing, *not* the *anti*-**'social'** sex that humans currently practice, where, as will be explained in chapter 8:11B, it is used to attack/fuck innocence and, as a result, we have become self-consciously **'shame[ful]'** and had to put on clothes to dampen our destructive lust—as Moses said, we **'realised that'** we **'were naked; so'** we **'made coverings for'** ourselves (Gen. 3:7). Bonobos also use a rich and constant array of vocalizations to assist the developing **'cohesive social structure'** of the group. Indeed, **'Bonobos are the most vocal of the great apes'** ('What is a Bonobo?', Bonobo

Conservation Initiative; see <www.wtmsources.com/119>), and **'are excitable creatures who frequently "comment" on minor events around them through high-pitched peeps and barks. Even if most of these vocalizations are noticeable only at close range, one definitely hears more vocal exchange in a group of bonobos than in a group of chimpanzees'** (Frans de Waal & Frans Lanting, *Bonobo: The Forgotten Ape*, 1997, p.10 of 210). But just how extraordinarily **'cohesive'** or integrated the bonobo species *has* become through their development of the love-indoctrination process is apparent in the following quote from the primatologist Barbara Fruth, who has spent many years studying bonobos in their natural habitat: **'up to 100 bonobos at a time from several groups spend their night together. That would not be possible with chimpanzees because there would be brutal fighting between rival groups'** (Paul Raffaele, 'Bonobos: The apes who make love, not war', Last Tribes on Earth.com, 2003; see <www.wtmsources.com/143>). So there is relatively little conflict between individual bonobos or even between groups of bonobos, which is another indication that this species is well on its way to integrating its members into the Specie Individual—a feat that all evidence indicates is what our ape ancestors succeeded in achieving.

A group of bonobos at the Lola Ya Bonobo Sanctuary, Democratic Republic of the Congo

[416] However, while a situation conducive to the development of integration of sufficient food, shelter and territory can be reached, from a genetic point of view there can never usually be enough of that other key

resource—mates—simply because the more successfully an individual can breed, the more their genes can carry on and multiply. Since females are limited in how often they can reproduce due to pregnancy—and, in the case of mammals, lactation—it is the males who have the opportunity to breed continuously, and so it is the competition for mating opportunities amongst males that is the most difficult form of genetic selfishness to overcome. Whilst bonobos haven't been able to fully develop love-indoctrination and thus integration, as evidenced by the fact that they have to use sex to quell residual tension and vocal communication to support their social cohesion, they have been able to develop it sufficiently to bring to an end this most difficult of all forms of selfish competitiveness, male competition for mating opportunities, as the fossil evidence now suggests our ape ancestors did. (With sex engaged in so frequently, males who generate more sperm will be more likely to reproduce, and so bonobos have developed relatively large testes; however, this 'sperm competition' is not evidence that males are actively competing for mating opportunities, it just indicates the strength of the underlying genetic imperative. The genes are still trying to find a way to ensure their reproduction even though the love-indoctrination process has overcome the competitive *behavior* amongst males.) In fact, bonobos have been *so* successful at reining in male competition for mating opportunities that within their society there has been what could be described as a gender-role reversal, with bonobo females forming alliances and dominating social groups, both of which are distinctly male behaviors in chimpanzee, gorilla, orangutan and other non-human primate societies. <u>Bonobo society is matriarchal, female-dominated, controlled and led</u>. Further, in bonobo society, the entire focus of the social group does appear to be on the maternal, or female, role of nurturing infants, as this observation by America's leading ape-language researcher, the biologist and psychologist Sue Savage-Rumbaugh evidences: **'Bonobo life is centered around the offspring. Unlike what happens among common chimps, all members of the bonobo social group help with infant care and share food with infants. If you are a bonobo infant, you can do no wrong...Bonobo females and their infants form the core of the group'** (Sue Savage-Rumbaugh & writer Roger Lewin, *Kanzi: The Ape at the Brink of the Human Mind*, 1994, p.108 of 299). It is just such circumstances one would expect in a society that exhibits the third element necessary for love-indoctrination, <u>selection for more maternal mothers</u>.

[417]The following photographs of bonobos with infants reveal something of just how exceptionally nurturing bonobo females are—and even

males, because the bottom right photo is of a male lovingly playing with an infant. Bonobos have clearly had the environmental comfort and the freedom from fighting and tension in their world needed to develop the ability to love their infants.

Bonobos nurturing their infants

[418]The 2011 French documentary *Bonobos*, which was directed by Alain Tixier, contains marvelous footage evidencing the ability of bonobos to nurture their infants and the wondrous effect such nurturing has on their offspring. A short segment from the documentary showing the tenderness of the bonobo mothers and the absolute joy and zest for life of their infants can be seen in a YouTube clip at <www.wtmsources. com/107>. The documentary is about a young bonobo called Beny who

was sold as a pet after his mother was killed by poachers. Fortunately, he was rescued by the Belgian conservationist Claudine André, who took him to her wonderful bonobo sanctuary, Lola Ya Bonobo in the Democratic Republic of the Congo, and later released him back into the Congo forest. While the documentary's commentary is superficial, the short accompanying film that discusses its production contains two very revealing comments. The first comment was made by Tixier, who said that **'The choice to do a film about bonobos was because they're surely the most fascinating animals on the planet. They're the closest animals to man. They're the only animals capable of creating the same "gaze" as a human. When you look at a bonobo you're taken aback because you can see behind the eyes it's not just curiosity, it's understanding. We see human beings in the eyes of the bonobo.'** The second was provided by the film's animal advisor, Patrick Bleuzen, who remarked that **'Once I got hit on the head with a branch that had a bonobo on it. I sat down and the bonobo noticed I was in a difficult situation and came and took me by the hand and moved my hair back, like they do. So they live on compassion, and that's really interesting to experience.'**

Photograph by Finlay MacKay for *TIME*

This revealing photograph captures the human-like gaze of the bonobo Kanzi

## Chapter 5:7 <u>The role of nurturing in our development has been an unbearably confronting truth</u>

[419]Before continuing, it needs to be re-emphasized just how unbearably confronting this discussion about the importance of nurturing in the development of our species has been without the compassionate understanding of the human condition. The tragic reality has been that ever since the terrible battle broke out between our original all-loving instinctive self or soul and our newer conscious mind and our present immensely upset angry, egocentric and alienated human condition emerged (as was described in chapter 3), no child has been able to be given anything like the psychosis-and-neurosis-free, pure love all infants were given back in this nurturing phase in our primate past. And unable, until now, to explain *why* humans have become so psychologically upset and thus unable to adequately love our infants, such talk about the role that nurturing has played in the maturation of our species—and continues to play in our individual upbringing today, since we are all born still instinctively expecting to receive such pure, unconditional love—*has* been unbearable. So yes, it *is* only now that we can explain *why* humans have become psychologically embattled and unable to adequately nurture our children that we can safely admit the crucial role that nurturing has played in human development. (Much more will be said in chapter 6 about how unconfrontable this nurturing explanation of human origins has been and, as a result, how human-condition-avoiding mechanistic science has denied the truth of the nurturing origins of our unconditionally selfless moral nature.)

[420]While it has been unbearable and thus unconfrontable, the truth, nevertheless, is that nurturing was the all-important influence in the maturation of our species and remains the all-important influence in the maturation of our individual lives. The female gender created humanity, and, while under the duress of the upset state of the human condition it has rarely been possible to adequately nurture our offspring, the importance of nurturing in producing a secure, sound adult remains paramount. The archetypal image of the Madonna and child that is such a feature of Christian mythology—which I depicted in the drawing that appears at the beginning of this chapter—*is* all-meaningful because for Christ to have been such a sound, unresigned, denial-free-thinking person he must have had an exceptionally nurturing mother. So when the author Olive Schreiner wrote the following passage she was, in fact, articulating the

agonising reality for virtually all mothers who suffer under the duress of the human condition: **'They say women have one great and noble work left them, and they do it ill...*We* bear the world and *we* make it. The souls of little children are marvellously delicate and tender things, and keep for ever the shadow that first falls on them, and that is the mother's or at best a woman's. There was never a great man who had not a great mother—it is hardly an exaggeration. The first six years of our life make us; all that is added later is veneer'** (*The Story of an African Farm*, 1883, p.193 of 300). Yes, while nurturing *is* crucial in producing a sound, functional human—as the saying goes, **'The hand that rocks the cradle rules the world'**—this comment in a story about the failure of many mothers to adequately nurture their offspring shows just how unbearable a truth it has been: **'For a lot of women the only really important anchor in their lives is motherhood. If they fail in a primary role they feel should come naturally it is devastating for them'** ('The Deserted Mothers' Club', *The Weekend Australian Magazine*, 30 Nov. 2013). But, as will be explained in chapter 8:16D, the role that men had to take up of championing the search for knowledge when the fully conscious mind emerged was so psychologically crippling that they *also* contributed greatly to the corruption of the souls of each new generation of humans—so the unbearable 'guilt' from not being able to adequately nurture our children has not been confined to women, as this quote from the bestselling author of books for and about children, John Marsden, makes clear: **'The biggest crime you can commit in our society is to be a failure as a parent and people would rather admit to being an axe murderer than being a bad father or mother'** ('A Single Mum's Guide to Raising Boys', *Sunday Life*, *The Sun-Herald*, 7 Jul. 2002).

## Chapter 5:8 The emergence of consciousness assisted the love-indoctrination process by allowing the sexual selection of integrativeness

[421] To return now to the description of the biological origins of our moral instincts.

[422] Given selfish competition for mating opportunities, particularly amongst males, is such a powerful force, the gender-role reversal apparent in bonobo society—which, as described, scientists are now beginning to confirm also occurred in the society of our ape ancestors—was an absolutely extraordinary achievement. In fact, this reversal from patriarchy to matriarchy is *so* extraordinary an achievement—especially the *speed* of its development, occurring over only some 1 million years since bonobos

split with chimpanzees, which is very fast in evolutionary terms—that it must have been facilitated by some special factor. And indeed it was—it was helped along by the emergence of the most powerful tool of all for developing the order of matter on Earth: the conscious mind.

[423] As mentioned earlier (in ch. 5:4), the reality is that developing love-indoctrination to the point where love or unconditional selfless-ness becomes instinctive is a very precarious process, akin to trying to swim upstream to an island in a fast flowing river—any difficulty or breakdown in the nurturing process and you are invariably 'swept back downstream' once more to the old competitive, selfish, each-for-his-own, opportunistic, 'animal condition' situation. So while love-indoctrination could allow unconditionally selfless behavior to emerge in our ape ancestors, it was a very difficult, and also a very slow, process to both get underway and maintain. What the situation needed was a mechanism to assist, speed up and help maintain love-indoctrination's development of integration—assistance that came from the emergence of a conscious mind that enabled the <u>conscious self-selection</u> of integrativeness, especially the female <u>sexual or mate selection</u> of less competitive and aggressive, more integrative males with whom to mate. (As the title of his 1871 book, *The Descent of Man and Selection in Relation to Sex*, intimates, Charles Darwin actually suggested the role mate selection could have played in our human development, although he didn't understand its significance in the context of the love-indoctrination process.) Again, in 1983 I predicted that there must have been female sexual selection against male aggression in our ape ancestors as part of the love-indoctrination process—a prediction that the aforementioned reports of reduced canine size now confirm.

[424] While the explanation for why humans became conscious while other animals haven't is the subject of chapter 7, an extremely brief account of it should be given here to support the forthcoming explanation as to how this conscious self-selection of integrativeness developed. In short, it was the love-indoctrination process that enabled consciousness to emerge, for while a mother's nurturing of her infants enabled unconditionally selfless behavior to develop and, over time, become instinctual, this training in unconditional selflessness produced a *further* accidental by-product: it produced brains trained to think selflessly and thus truthfully and thus effectively and thus become 'conscious' of the relationship of events that occur through time. Other species that can't develop love-indoctrination and thus unconditional selflessness can't think truthfully and thus

effectively because unconditional selflessness, which they are unable to develop an orientation to, is the truthful theme or meaning of existence. As has been pointed out in each of the first four chapters in this book, you can't hope to think truthfully and thus effectively if you're lying. Species whose behavior is governed by genetic selfishness have emerging minds that are, in effect, dishonestly orientated; their minds are alienated from the truth—they won't, in fact, allow selflessness-recognizing, truthful and thus effective thinking—which means they can never make sense of experience and thus never become conscious. What all this means is that the human mind has been alienated from the truth *twice* in its history: firstly, in our pre-love-indoctrinated past when, like all other animals (except now for bonobos, who have almost completed the process of love-indoctrination), our brains were blocked from thinking truthfully; and, secondly, in our present state, where our minds have been alienated from the truth as a result of the human condition. Again, I emphasize that this is an extremely brief account of how love-indoctrination liberated consciousness, the complete explanation of which appears in chapter 7.

[425] (Incidentally, people wonder how we can know that other species, like dolphins and elephants, aren't fully conscious like humans are. As just explained, developing consciousness depends on overcoming the competitive, selfish 'animal condition' and becoming orientated to selflessness, so if you are still preoccupied with selfish, competitive dominance, as other animals are, you can't become fully conscious. Indeed, as all good animal trainers know, the key to managing and training non-human animal species, of both sexes, is to recognize that their great preoccupation is in achieving dominance whenever possible. Humans' deepest preoccupation, however, is with giving and receiving selfless love, a preoccupation we mistakenly project onto other animals, especially our pets, resulting in all the problems we have in managing them effectively. Animals are still victims of the 'animal condition', and so controlling them requires asserting dominance. If other animal species *had* achieved full consciousness they would not still be stranded in a world dictated by the rules of a selfish, competitive dominance hierarchy; they would be as preoccupied by giving and receiving love as we humans fundamentally are.)

[426] And so it was an emerging conscious intellect that the love-indoctrination process had liberated that then began to support the development of selflessness and thus help maintain and speed up that all-precious process—because consciousness made it possible to recognize

the importance of selflessness and, having realized that, begin to actively select for it. Unconditional selflessness or love *is* the theme of existence, so to not be able to recognize that, as species that haven't been able to develop love-indoctrination and become conscious haven't been able to do, means they are, in effect, 'locked out' of the ideal state—that being the essential agony of the 'animal condition'. We can, therefore, appreciate how immensely relieving it must be to be liberated from that condition. For our ape ancestors to effectively be set free from the tyranny of the selfish gene-based natural selection process must have felt like heaven had opened up! Love could finally be indulged in! They had finally escaped the stupor of the animal condition and the opportunity and desire to then develop love, select for gentleness and kindness, would have been immense. (As will be explained in chapter 9, a similar opportunity to be transformed by being released into a world of love appears for humans now that we can finally escape the agony of the human condition!) It follows then that bonobos, who, as mentioned, have been able to rein in divisive aggressive behavior, bring it under control, and develop love and achieve the extraordinary transition from patriarchy to matriarchy, have largely escaped the stupor of the animal condition and as a result are the most conscious, the most intelligent of all animals next to humans—as these quotes evidence: **'Everything seems to indicate that Chim** [a bonobo] **was extremely intelligent. His surprising alertness and interest in things about him bore fruit in action, for he was constantly imitating the acts of his human companions and testing all objects. He rapidly profited by his experiences…Never have I seen man or beast take greater satisfaction in showing off than did little Chim. The contrast in intellectual qualities between him and his female companion** [a chimpanzee] **may briefly, if not entirely adequately, be described by the term "opposites"'** [p.248 of 278] **…Prince Chim seems to have been an intellectual genius. His remarkable alertness and quickness to learn were associated with a cheerful and happy disposition which made him the favorite of all** [p.255] **…Chim also was even-tempered and good-natured, always ready for a romp; he seldom resented by word or deed unintentional rough handling or mishap. Never was he known to exhibit jealousy…** [By contrast] **Panzee** [the chimpanzee] **could not be trusted in critical situations. Her resentment and anger were readily aroused and she was quick to give them expression with hands and teeth** [p.246]**'** (Robert M. Yerkes, *Almost Human*, 1925).

[427] The pre-eminent ape language researcher Sue Savage-Rumbaugh reinforced this view when she wrote that **'Individuals who have had first hand interactive experience with both *Pan troglodytes* [chimpanzees] and *Pan paniscus***

**(Yerkes and Learned, 1925; Tratz and Heck, 1954) have been left with the distinct impression that pygmy chimpanzees** [bonobos] **are considerably more intelligent and more sociable than** *Pan troglodytes***'** and that **'Each individual who has worked with both species in our** [ape language research] **lab is repeatedly surprised by their** [bonobos'] **communicative behavior and their comprehension of complex social contexts that are vastly different from anything seen among** *Pan troglodytes***'** ('*Pan paniscus* and *Pan troglodytes*: Contrasts in Preverbal Communicative Competence', *The Pygmy Chimpanzee*, ed. Randall Susman, 1984, pp.396, 411-412 of 435). The bonobo that Savage-Rumbaugh has worked most with is Kanzi, and you can see in the photo of Kanzi that was included a few pages back something of this **'intelligent' 'comprehension'** in his eyes. As the filmmaker Alain Tixier said about the intelligent gaze of bonobos, **'They're the only animals capable of creating the same "gaze" as a human. When you look at a bonobo you're taken aback because you can see behind the eyes it's not just curiosity, it's understanding. We see human beings in the eyes of the bonobo.'**

[428] The following extract from an article titled 'The Bonobo: "Newest" apes are teaching us about ourselves' also illustrates just how extraordinarily aware, cooperative, empathetic and intelligent bonobos are: **'Barbara Bell...a keeper/trainer for the Milwaukee County Zoo...works daily with the largest group of bonobos...in North America..."It's like being with 9 two and a half year olds all day," she** [Bell] **says.** "They're extremely intelligent...They understand a couple of hundred words," she says. "They listen very attentively. And they'll often eavesdrop. If I'm discussing with the staff which bonobos (to) separate into smaller groups, if they like the plan, they'll line up in the order they just heard discussed. If they don't like the plan, they'll just line up the way they want." "They also love to tease me a lot," she says. "Like during training, if I were to ask for their left foot, they'll give me their right, and laugh and laugh and laugh. But what really blows me away is their ability to understand a situation entirely." For example, Kitty, the eldest female, is completely blind and hard of hearing. Sometimes she gets lost and confused. "They'll just pick her up and take her to where she needs to go," says Bell. "That's pretty amazing. Adults demonstrate tremendous compassion for each other"'** (*Chicago Tribune*, 11 Jun. 1998). More recently, the bonobo researcher Vanessa Woods described bonobos as **'the most intelligent of all the great apes'** ('Bonobos – our better nature', blogs.discovery.com, 21 Jun. 2010; see <www.wtmsources.com/134>).

[429] I should say that while some scientists suggest that chimpanzees are as intelligent as bonobos, there is an undeniable freedom in the mind of a bonobo that is apparent in their capacity to be interested in the world

around them, and in their empathy, compassion and even humor, as evidenced by the quotes above. While it *is* true that chimpanzees show a mental dexterity, it is a limited, opportunistic, self-interested mental focus (similar to a very primitive version of the alienated, deadened, self-centered, narrowly-focused minds of humans today), not the broad, free, open, curious, aware, all-sensitive, loving, truly thoughtful, conscious mind that bonobos have. It does have to be remembered that mechanistic science doesn't even have an interpretation of the word 'love', so we can't expect it to be capable of being interested in, or of acknowledging, the kind of open, curious, aware, all-sensitive and loving conscious mind that bonobos have. Indeed, the extraordinarily cooperative, unconditionally loving and truly aware characteristics of bonobos has actually motivated human-condition-avoiding, alienated mechanistic scientists to find ways to demote them at every opportunity, just as they found ways to demote the work of primatologist Dian Fossey, Sue Savage-Rumbaugh and my own work—issues that will be described in chapter 6.

[430] In terms then of being able to consciously favor more selfless and integrative behavior in order to achieve greater overall selflessness and, with it, integration, the fastest, most effective way of doing so would be through selecting mates who are more selfless because that way you are eliminating selfishness at the fundamental level of your species' genetic make-up—which, as mentioned in pars 407-408, is what scientists studying the fossil record are now inferring took place: **'sexual selection played a primary role in canine reduction'**. Further, since it is the males who are most preoccupied with competing for mating opportunities, it is the females who are in the best position to implement this selection of less aggressive individuals with whom to mate. Significantly, primatologists—despite not recognizing the process of love-indoctrination (and, therefore, the different degrees of success various primate species have had in developing it), or that love-indoctrination liberated consciousness, which in turn allowed self-selection of integrativeness, especially through mate selection—*have* verified this female-driven selection of cooperative integrativeness amongst primates (the underlinings in this and some subsequent quotes are my emphasis): **'Male [baboon] newcomers also were generally the most dominant while long-term residents were the most subordinate, the most easily cowed. Yet in winning the receptive females and special foods, the subordinate, <u>unaggressive veterans got more than their fair share, the newcomers next to nothing. Socially inept and often aggressive, newcomers made a poor job of initiating friendships</u>'**

(Shirley Strum, 'The "Gang" Moves to a Strange New Land', *National Geographic*, Nov. 1987)<sup>;</sup> **and 'The high frequencies of intersexual association, grooming, and <u>food sharing</u> together with the <u>low level of male-female aggression in pygmy chimpanzees</u> [bonobos] may be a factor in male reproductive strategies. Tutin (1980) has demonstrated that a <u>high degree of reproductive success for male common chimpanzees was correlated with male-female affiliative behaviours</u>** [again, 'affiliative' being an evasive, denial-complying, mechanistic term meaning friendly/cohesive/social/loving/integrative]. **These included males spending more time with estrous females, grooming them, and sharing food with them'** (Alison & Noel Badrian, 'Social Organization of *Pan paniscus* in the Lomako Forest, Zaire'; *The Pygmy Chimpanzee*, ed. Randall Susman, 1984, p.343 of 435).

[431] While the observations made in the popular documentary series *Orangutan Island* (about the rehabilitation of a group of juvenile orphaned orangutans on a protected island in Borneo) probably can't be considered a product of rigorous scientific research, they do provide revealing footage and interesting commentary about orangutan behavior. In one episode, an orangutan named Daisy, who is the dominant young female in the group (in the series she is described as 'Sheriff Daisy'), is seen strongly reprimanding another young female, Nadi, who, according to the narrator, repeatedly behaves selfishly: **'As usual, Daisy is keeping a watchful eye on all the action and she spies someone who is not playing fair—it's repeat offender Nadi who is refusing to share** [a jackfruit]**...Daisy decides it's her duty to step in...Daisy is refusing to allow Nadi anywhere near the eating platform because Nadi's been upsetting the order in this peaceful community...Daisy is making Nadi pay for her behaviour, so to avoid starving Nadi has no choice but to leave'** (*Animal Planet*, episode 'House of Cards', 2007). Whether or not it is an accurate interpretation of events, the footage appears to fully support the commentary—a female is seen to be maintaining **'the order in this peaceful community'**. (Further illustrations of strong-willed, female primates insisting on integrative behavior will be provided shortly in ch. 5:11.)

## Chapter 5:9 <u>Sexual selection for integrativeness explains neoteny</u>

[432] Yes, since it is the males who are the most preoccupied with competing for mating opportunities, the females must have been the first to select for selfless, cooperative integrativeness by favoring integrative rather than competitive and aggressive mates—and it was this process of the conscious self-selection of integrativeness, especially the sexual

selection of less aggressive males with whom to mate, that greatly helped love-indoctrination subdue the males' divisive competitiveness. Moreover, by seeking out less aggressive, more integrative mates, the females were, in effect, selecting those who have been the most love-indoctrinated. This raises the next point to be explained, which is that the most love-indoctrinated and thus most integrative individuals will be those who have experienced a long infancy and exceptional nurturing and are closer to their memory of their love-indoctrinated infancy; specifically, those that are younger. The older individuals became, the more their infancy training in love wore off. In the case of <u>our ape ancestors, they began to recognize that the younger an individual, the more integrative he or she was likely to be</u> and, as a result, began to idolize, foster, favor and select for youthfulness because of its association with cooperative integration. The effect, over many generations, was to retard physical development so that adults became more infant-like in their appearance—which explains how we first came to regard <u>neotenous</u> (infant-like) features, like large eyes, dome forehead, snub nose and hairless skin, as beautiful and attractive.

[433] Consider, for instance, the neotenous or infant-like large eyes of seal pups and frogs, and the large eye spot markings together with the soft, typically infant-like, moppish ears of giant pandas—they are what make these animals so 'appealing'. The following drawing of a panda depicted without its trademark spotted eyes and round ears, and with pricked ears and small eyes instead, shows just how quickly it loses its 'cute' appeal.

Illustration by Robert Parkinson

**'Would we care if they weren't so cute? White out the black eye spots and give the ears points, and the panda loses much of its appeal'**
*Good Weekend, The Sydney Morning Herald*, 23 Sep. 1989

[434] To indicate the effects of the love-indoctrination, mate selection neotenizing process, I have assembled the following photographs of infant and adult non-human primates, specifically bonobos and chimpanzees.

[435] Firstly, the following three photographs, of an adult chimpanzee, an infant chimpanzee and an adult bonobo, show the similarity between the infant chimpanzee and the adult bonobo, indicating, as stated, the effects of the love-indoctrination, mate selection neotenizing process.

Adult chimpanzee; infant chimpanzee; and an adult bonobo

[436] In addition to their remarkably neotenous physical appearance, there is also a marked variance in features between individual bonobos (which is also apparent in the very different facial features of the two bonobos pictured standing upright earlier, opposite par. 413), suggesting the species is undergoing rapid change. This in turn suggests that the bonobo species has hit upon some opportunity that facilitates a rapid development, which evidence indicates is the ability to develop integration through love-indoctrination and mate selection.

[437] The photographs below of an infant and adult chimpanzee also show the greater resemblance humans have to the infant, again illustrating the effect of neoteny in human development.

Infant and adult chimpanzee from *The Mismeasure of Man*, Stephen Jay Gould, 1981

[438] And finally, the following photograph of a chimpanzee fetus at seven months shows body hair on the scalp, eyebrows, borders of the eye lids, lips and chin—precisely those regions where hair is predominantly retained in adult humans today, again illustrating the effect of neoteny or pedomorphosis in human development. ('Pedomorphosis' comes from the Greek *pais*, meaning 'child', and *morphosis*, meaning 'shaping'.) Clearly, humans are an extremely neotenized—love-indoctrinated—ape. Interestingly, a report on the fossilized remains of our 4.4-million-year-old ancestor *Ar. ramidus* describes it as having a **'short face and weak prognathism** [projecting jaw] **compared with the common chimpanzee'** (Gen Suwa et al., 'The *Ardipithecus ramidus* Skull and Its Implications for Hominid Origins', *Science*, 2009, Vol.326, No.5949), indicating that this physical retardation or neotenization was well underway by this point in our ancestral history.

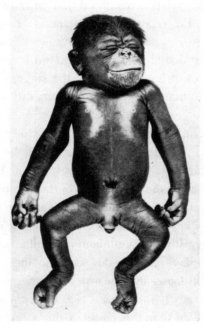

Chimpanzee foetus
at seven months from
*The Life of Primates*,
Adolph H. Schultz, 1969

[439] So, we humans did learn to recognize that the older individuals became, the more their infancy training in love wore off and, therefore, the younger an individual, the more integrative he or she would likely be, and it was this selection for youthfulness that had the effect of retarding our development so that we became more infant-like in our appearance as adults. As stated, this was how we came to regard neotenous features— large eyes, dome forehead, snub nose and hairless skin—as attractive.

## Chapter 5:10 <u>The importance of nurturing in bonobo society</u>

[440] An indication of how bonobos have been able to develop the nurturing love-indoctrination process more than chimpanzees, is that female bonobos have, on average, one offspring every five to six years and provide better maternal care than chimpanzees. Bonobos are born small, develop more slowly than other ape species, and stay in a state of infancy and total dependence for a comparatively long period of time—being weaned at about five years of age and remaining dependent on their mothers until between the ages of seven and nine. Chimpanzees are weaned at about four years of age and remain dependent up to the age of six (on average), while the biological weaning age for humans is considered to be between five-and-a-half and six years, which is closer to the five years of bonobos. Biologists have noted that an individual's lifespan is, amongst other factors, **'related to the postnatal development rate'** (Richard Cutler, 'Evolution of longevity in primates', *Journal of Human Evolution*, 1976, Vol.5, No.2), and so it is possible to deduce that the selection for a longer infancy period had the side effect of lengthening all the stages of maturation, which may explain how humans acquired our comparatively long lifespan.

[441] The primatologist Takayoshi Kano's long-running study of bonobos in their natural habitat in the Democratic Republic of the Congo has observed this prolonged infancy in practice, reporting that **'The long dependence of the son may be caused by the slow growth of the bonobo infant, which seems slower than in the chimpanzee. For example, even after one year of age, bonobo infants do not walk or climb much, and are very slow. The mothers keep them near. They start to play with others at about one and a half years, which is much later than in the chimpanzee. During this period, mothers are very attentive...Female juveniles gradually loosen their tie with the mother and travel further away from her than do her sons'** (Frans de Waal & Frans Lanting, *Bonobo: The Forgotten Ape*, 1997, p.60 of 210). As alluded to in this passage, the bond between mother and son is of particular significance in bonobo society where the son will maintain his connection with his mother for life and depend upon her for his social standing within the group. For example, the son of the society's dominant female, the strong matriarch who maintains social order, will rise in the ranks of the group, presumably to ensure the establishment and perpetuation of unaggressive, non-competitive,

cooperative male characteristics, both learnt and genetic, within the group. Again, historically, it is the male primates who have been particularly divisive in their aggressive competition to win mating opportunities and, therefore, the gender most needing of love-indoctrination—as this quote makes clear: **'Patient observation over many years convinced** [Takayoshi] **Kano that male bonobos bonded with their mothers for life. That contrasts with chimpanzee males who rarely have close contact with their mothers after they grow up, instead joining other males in never-ending tussles for dominance'** (Paul Raffaele, 'Bonobos: The apes who make love, not war', Last Tribes on Earth.com, 2003; see <www.wtmsources.com/143>).

[442] The following quote (part of which has already been referred to), from Sue Savage-Rumbaugh and the writer Roger Lewin's book, *Kanzi: The Ape at the Brink of the Human Mind*, provides further insight into bonobo society and its emphasis on nurturing: **'Bonobo life is centered around the offspring. Unlike what happens among common chimps, all members of the bonobo social group help with infant care and share food with infants. If you are a bonobo infant, you can do no wrong. This high regard for infants gives bonobo females a status that is not shared by common chimpanzee females, who must bear the burden of child care all alone. Bonobo females and their infants form the core of the group, with males invited in to the extent that they are cooperative and helpful. High-status males are those that are accepted by the females, and male aggression directed toward females is rare even though males are considerably stronger'** (1994, p.108 of 299).

[443] *Also* **'Unlike what happens among common chimps'** is the fact that bonobos regularly share their food, and while chimpanzees restrict their plant-food intake to mainly fruit, bonobos eat leaves and plant pith as well as fruit, a diet more like that of gorillas. While bonobos have been known to capture and eat small game, including small monkeys, to supplement their diet with protein, they are not known to routinely hunt down and eat large animals such as colobus monkeys, like chimpanzees do, with **'hunting behavior** [by bonobos] **very rare'** (Tetsuya Sakamaki quoted by David Quammen, 'The Left Bank Ape', *National Geographic*, Mar. 2013). (In chapter 6:6 it is explained how mechanistic science has had to try to argue that bonobos are not gentle and peaceful because their cooperative behavior serves as an unbearably exposing and confronting reminder of our now immensely angry, egocentric and alienated, unloving and unloved lives. For example, you will recall how, in par. 208, E.O. Wilson brazenly

attempted to paint bonobos as indistinguishable from chimpanzees by stating that, like chimpanzees, bonobos **'do not share'**, and that they **'hunt in coordinated packs in the same manner as chimpanzees'**. However, as I pointed out, even the most basic research shows that Wilson had to fudge his evidence to support his lie—but this is how desperate resigned humans have been to find some excuse for our species' divisive behavior.) Also, while physical violence is customary amongst chimpanzees it is rare among bonobos where, although the males are stronger, male aggression has been tamed and, unlike other great apes but like our ancestors, there is actually little difference in sexual size dimorphism between the male and female of the species. And, as has been mentioned, like our ancestors *Sahelanthropus*, *Orrorin* and *Ardipithecus*, bonobos have reduced canine teeth, another indication they are less aggressive—in fact, the bonobo male **'possesses smaller canines than any other [male] hominoid** (J. Michael Plavcan et al., 'Competition, coalitions and canine size in primates', *Journal of Human Evolution*, 1995, Vol.28, No.3). The practice of infanticide, while not uncommon amongst chimpanzees, also appears to be non-existent within bonobo societies where even orphan bonobos are cared for by the group. In chimpanzee society orphans are occasionally adopted by a female but are not especially cared for by the group. Social groups of bonobos also have much greater stability than social groups of chimpanzees, with bonobos periodically coming together in large, harmonious, stable groups of up to 120 individuals—as Barbara Fruth's observation, included in par. 415, indicated: **'up to 100 bonobos at a time from several groups spend their night together. That would not be possible with chimpanzees because there would be brutal fighting between rival groups.'** The primatologist Genichi Idani similarly reported that **'In societies of African great apes other than bonobos, only antagonistic** [aggressive] **relationships have been reported between unit-groups...[Bonobo] neighborhood relationships...allow peaceful coexistence of two unit-groups** [and] **provide the first nonhuman primate community model which is close to...human society'** ('Relations between Unit-Groups of Bonobos at Wamba, Zaire: Encounters and Temporary Fusions', *African Study Monographs*, 1990, Vol.11, No.3). So, bonobos are much gentler than their chimpanzee cousins; they are relatively placid, peaceful and egalitarian, exhibiting a remarkable sensitivity to others—and not just towards their own kind, as will be evidenced shortly with an account of how a bonobo cared for a stunned bird.

## Chapter 5:11 <u>The importance of strong-willed females in developing integration</u>

[444] The following extract from the 1995 *National Geographic* documentary *The New Chimpanzees* provides a good example of the important role a strong matriarchy plays in the maintenance of cooperative, integrative behavior. To quote from the narration: **'An impressively stern** [bonobo] **female enters and snaps a young sapling. Once she picks herself up she does something** [that would be] **entirely surprising for a female chimp, she displays** [the female is shown assertively dragging the sapling through the group], **and the males give her sway** [a male is shown cowering out of her way]. **For this is the confident stride of the group's leader, its alpha female, whom** [Takayoshi] **Kano has named Harloo.'**

© 1992 Fedmex Pty Ltd

From the author's 1992 visit to Africa. L to R: the author with Dr Shirley Strum's Chololo Ranch Baboon Project study group in Kenya; with Dr Susanne Abildgaard and young chimpanzee at Jane Goodall's Chimpanzee Rehabilitation Centre in Burundi; and with the Susa Mountain Gorilla study group in Rwanda.

[445] While this example is specific to bonobos, those who have studied primates will typically tell you of the presence in their study group of an extraordinarily self-assured, secure-in-self and strong-willed female. This is because all primates are trying to develop the nurturing of integrativeness but only our ape ancestors and the bonobos have had the right conditions to actually achieve it. During a trip to Kenya in 1992, my partner Annie Williams and I were invited to visit the anthropologist Shirley Strum's 'Pumphouse Gang' study troop of baboons, which had been made famous through numerous articles in *National Geographic* magazine. During our visit I noticed that Strum kept on her desk the skull of a baboon named Peggy. Displaying a skull is a bit macabre but Strum said she did so in memory of Peggy who was an extraordinarily confident, strong-willed, authoritative, charismatic individual who successfully led the 'Pumphouse Gang' for many years. As Strum has written: **'She** [Peggy]

was the highest-ranking female in the troop, and her presence often turned the tide in favor of the animal she sponsored. While every adult male outranked her by sheer size and physical strength, she exerted considerable social pressure on each member of the troop. Her family also outranked all the others...another reason for the contentment in this particular family was Peggy's personality. She was a strong, calm, social animal, self-assured yet not pushy, forceful yet not tyrannical' (*Almost Human: a journey into the world of baboons*, 1987, pp.38-39 of 294).

[446]In her famous studies of gorillas, which she documented in her 1983 book, *Gorillas in the Mist*, the primatologist Dian Fossey described the strength of character that had to be developed to curtail male aggression, and the centered, security of self needed to be an effective, love-indoctrinating mother, when she wrote about how the gorilla 'Old Goat' was 'an exemplary parent' and that, as a result, her son 'Tiger' was 'a contented and well adjusted individual'. While gorillas have not been able to develop love-indoctrination to the same degree as bonobos—seemingly because they have lacked the necessary ideal nursery conditions—denial-free, honest studies of their behavior, notably Fossey's, have revealed the strong correlation between nurturing and integrativeness that is the love-indoctrination process. The following extracts from *Gorillas in the Mist* reveal more about Old Goat's nurturing of Tiger: 'Like human mothers, gorilla mothers show a great variation in the treatment of their offspring. The contrasts were particularly marked between [the gorilla mothers] Old Goat and Flossie. Flossie was very casual in the handling, grooming, and support of both of her infants, whereas Old Goat was an exemplary parent' (p.174 of 282). The effect of Old Goat's 'exemplary parenting' of Tiger is apparent in the following extract: 'Like Digit, Tiger also was taking his place in Group 4's growing cohesiveness. By the age of five, Tiger was surrounded by playmates his own age, a loving mother, and a protective group leader. He was a contented and well-adjusted individual whose zest for living was almost contagious for the other animals of his group. His sense of well-being was often expressed by a characteristic facial "grimace"' (pp.186-187). The 'growing cohesiveness' (developing integration) brought about by 'loving mother[s], and a protective group leader' is love-indoctrination.

[447]Incidentally, with regard to the 'protective group leader[s]', namely the male silverback gorillas, their large size is not only due to having to compete for dominance but also reflects that while bonobos depend on the safety of trees for a secure, threat-free environment, gorillas evidently selected for physical size and great strength, particularly in the males,

to protect their groups from external, predatory threats—as the anthropologist Adolph H. Schultz noted, the adult male gorilla **'is a remarkably peaceful creature, using its incredible strength merely in self-defence'** (*The Life of Primates*, 1969, p.34 of 281).

[448] Fossey's account of the love-indoctrinated Tiger later in life illustrates how nurtured love is critical in producing and maintaining the integrated group—as well as the *extreme* fragility of the love-indoctrination process and how any disruption to it would result in a regression back to the competitive, each-for-his-own, opportunistic, divisive, 'animal condition' existence. In her account, Fossey describes how the secure, integrative, loving Tiger tried to maintain integration or love in the presence of an aggressive, divisive gorilla after the group's integrative silverback leader, Uncle Bert, was shot by poachers: **'The newly orphaned Kweli, deprived of his mother, Macho, and his father, Uncle Bert, and bearing a bullet wound himself, came to <u>rely only on Tiger for grooming the wound, cuddling, and sharing warmth in nightly nests</u>. Wearing <u>concerned</u> facial expressions, Tiger stayed near the three-year-old, <u>responding</u> to his cries with <u>comforting</u> belch vocalizations. As Group 4's new young leader, Tiger <u>regulated</u> the animals' feeding and travel pace whenever Kweli fell behind. Despondency alone seemed to pose the most critical threat to Kweli's survival during August 1978. Beetsme...was a significant <u>menace</u> to what remained of Group 4's <u>solidarity</u>. The immigrant, approximately two years older than Tiger and finding himself the oldest male within the group led by a younger animal, quickly developed an <u>unruly</u> desire to <u>dominate</u>. Although still sexually immature, Beetsme <u>took advantage</u> of his age and size to begin <u>severely tormenting</u> old Flossie three days after Uncle Bert's death. Beetsme's <u>aggression</u> was particularly <u>threatening</u> to Uncle Bert's last offspring, Frito [son of Flossie]. By killing Frito, Beetsme would be destroying an infant sired by a competitor, and Flossie would again become fertile. Neither young Tiger nor the aging female was any match against Beetsme. Twenty-two days after Uncle Bert's killing, Beetsme succeeded in <u>killing</u> fifty-four-day-old Frito even with the <u>unfailing efforts of Tiger</u> and the other Group 4 members to <u>defend</u> the mother and infant...Frito's death provided more evidence, however indirect, of the <u>devastation</u> poachers create by <u>killing the leader</u> of a gorilla <u>group</u>. Two days after Frito's death Flossie was observed soliciting copulations from Beetsme, not for sexual or even reproductive reasons—she had not yet returned to cyclicity and Beetsme still was sexually immature. Undoubtedly her invitations were <u>conciliatory measures</u> aimed at reducing his continuing physical <u>harassment</u>. <u>I found myself strongly disliking Beetsme</u> as I watched <u>his discord destroy what remained of all that Uncle Bert</u>**

**had succeeded in creating and defending over the past ten years...I also became increasingly concerned about Kweli, who had been, only a few months previously, Group 4's most vivacious and frolicsome infant.** The three-year-old's lethargy and depression were increasing daily even though **Tiger tried to be both mother and father to the orphan.** Three months following his gunshot wound and the loss of both parents, Kweli gave up the will to survive...It was difficult to think of Beetsme as an **integral member** of Group 4 because of his continual **abuse of the others** in futile **efforts to establish domination**, particularly over the **indomitable Tiger...Tiger helped maintain cohesiveness by "mothering" Titus and subduing Beetsme's rowdiness. Because of Tiger's influence** and the immaturity of all three males, they remained **together**' (*Gorillas in the Mist*, pp.218-221).

[449] It is very clear from this account how very easily any disruption to the love-indoctrination process can cause a regression to the competitive, opportunistic, each-for-his-own, pre-love-indoctrination, 'animal condition' situation—and thus how utterly incredible it is that the bonobos have been able to overcome the agony of the 'animal condition' and be as free as they are from the tyranny of the selfish gene-based natural selection process.

## Chapter 5:12 Descriptions of bonobos provide an extraordinary insight into what life for our human ancestors was like before the emergence of the human condition

[450] Indeed, the following selection of quotes provides further insight into how extraordinarily integratively orientated bonobos have become and, through that stunning evidence, just how wonderful our species' time in the innocent, cooperatively behaved, loving, 'Garden of Eden'-like, fully integrated, 'heavenly' 'Golden Age' must have been. I doubt you will find a better clue to our glorious past, and now future, than what you are about to read.

[451] Firstly, the aforementioned bonobo researcher Vanessa Woods gave this first-hand account of bonobos' seemingly unlimited capacity for love: **'Bonobo love is like a laser beam. They stop. They stare at you as though they have been waiting their whole lives for you to walk into their jungle. And then they love you with such helpless adandon that you love them back. You have to love them back'** ('A moment that changed me – my husband fell in love with a bonobo', *The Guardian*, 1 Oct. 2015). I would include with that description, this extract, which has already been referred to, that demonstrates how extraordinarily sensitive, cooperative, loving and intelligent bonobos are, as well as how few exist

in captivity: 'Barbara Bell...a keeper/trainer for the Milwaukee County Zoo... works daily with the largest group of bonobos...in North America..."It's like being with 9 two and a half year olds all day," she [Bell] says. "They're extremely intelligent...They understand a couple of hundred words," she says. "They listen very attentively. And they'll often eavesdrop. If I'm discussing with the staff which bonobos (to) separate into smaller groups, if they like the plan, they'll line up in the order they just heard discussed. If they don't like the plan, they'll just line up the way they want." "They also love to tease me a lot," she says. "Like during training, if I were to ask for their left foot, they'll give me their right, and laugh and laugh and laugh. But what really blows me away is their ability to understand a situation entirely." For example, Kitty, the eldest female, is completely blind and hard of hearing. Sometimes she gets lost and confused. "They'll just pick her up and take her to where she needs to go," says Bell. "That's pretty amazing. Adults demonstrate tremendous compassion for each other".'

[452] The intensity of personal relationships within bonobo society is also beautifully apparent in the following anecdote from the aforementioned book, *Kanzi: The Ape at the Brink of the Human Mind*, in which Sue Savage-Rumbaugh recounts the extreme elation and affection shown by her famous bonobo research subject, the young adult male Kanzi, when reunited with his adoptive mother, Matata, after a number of months apart (a picture of Matata and Kanzi appears at the beginning of this chapter): 'I sat down with him [Kanzi] and told him there was a *surprise* in the colony room. He began to vocalize in the way he does when expecting a favored food—"eeeh....eeeh....eeeh." I said, *No food surprise. Matata surprise*; *Matata in colony room*. He looked stunned, stared at me intently, and then ran to the colony room door, gesturing urgently for me to open it. When mother and son saw each other, they emitted earsplitting shrieks of excitement and joy and rushed to the wire that separated them. They both pushed their hands through the wire, to touch the other as best they could. Witnessing this display of emotion, I hadn't the heart to keep them apart any longer, and opened the connecting door. Kanzi leapt into Matata's arms, and they screamed and hugged for fully five minutes, and then stepped back to gaze at each other in happiness. They then played like children, laughing all the time as only bonobos can' (pp.143-144).

[453] And, as referred to earlier, such displays of sensitivity are not limited to their own kind, as this description from *Bonobo: The Forgotten Ape*, by the primatologist Frans de Waal and photographer Frans Lanting, reveals: 'Betty Walsh, a seasoned animal caretaker, observed the following incident involving a seven-year-old female bonobo named Kuni at Twycross

**Zoo in England. One day, Kuni captured a starling. Out of fear that she might molest the stunned bird, which appeared undamaged, the keeper urged the ape to let it go. Perhaps because of this encouragement, Kuni took the bird outside and gently set it onto its feet, the right way up, where it stayed, looking petrified. When it didn't move, Kuni threw it a little way, but it just fluttered. Not satisfied, Kuni picked up the starling with one hand and climbed to the highest point of the highest tree...She then carefully unfolded its wings and spread them wide open, one wing in each hand, before throwing the bird as hard as she could** [to free it]' (1997, p.156 of 210). The situation reminds one of the compassion extended to Patrick Bleuzen, the animal advisor on the aforementioned French documentary, *Bonobos*, who was comforted by a bonobo after he was hit on the head by a falling branch during production.

[454] This next account (which is very reminiscent of the earlier mentioned orangutan, Sheriff Daisy, and her maintenance of cooperative order) reveals something of bonobos' strong moral sense: **'When a maintenance man was working on the heating and cooling system at the Iowa Primate Learning Sanctuary, Kanzi the bonobo didn't like it. So what did the bonobo do? He went to his computer and used his words to tell a researcher to put on a monkey suit and chase after the man. Julie Gilmore, a veterinarian at the sanctuary, said bonobos like to know if everyone is getting along, who is in charge and whether everyone is following the rules. Gilmore said that when the worker came, Kanzi was convinced he wasn't following the rules. She said the worker was luckily a good sport about the whole thing and allowed the monkey suit chasing to commence. The bonobo was so happy that the man was chased away, he watched a video of the action about 100 times. That's just one example of how the bonobos at the sanctuary communicate with the researchers...** [At the research sanctuary] **Sue Savage-Rumbaugh...taught Kanzi language using a chart of symbols...She was able to train Kanzi to know what word each symbol represented. Kanzi, who is now 33, now understands thousands of words and grammar'** ('Talking to another species', *The Des Moines Register*, 21 Nov. 2013). It's worth noting that the big difference between Sheriff Daisy and Kanzi is that Daisy was a juvenile orangutan. Because of the scarcity of food where orangutans live in the rainforests of Borneo and Sumatra they are forced to live competitively as adults; in effect, they are thrown out of the integrative, loving state they are able to experience when they are young. Kanzi, however, is a fully grown, 34-year-old adult and still expects to be living in an integrative, cooperative, loving state. Such is the ideal love-indoctrinating nursery conditions of the world from which bonobos come.

## Chapter 5:13  <u>Milwaukee County Zoo's fabulous group of bonobos</u>

The author (left) and Professor Harry Prosen
at the Milwaukee County Zoo bonobo exhibit

[455] In February 2014 I had the incredible good fortune to spend a day with one of the largest and most socially authentic captive breeding groups of bonobos in the world at the Milwaukee County Zoo in Wisconsin in the USA. Visiting the bonobos in their native habitat in the heart of the Congo basin is almost impossible because of the region's remoteness and impenetrable jungle, as well as the horrific civil war and economic collapse that is ravaging the Democratic Republic of the Congo. And even those who *do* manage to travel to where the bonobos are found only very occasionally get to glimpse them. So being able to spend hours right beside such a large group of these animals who are living virtually as we humans once did during our original pre-conscious, pre-human-condition-afflicted, innocent, loving, harmonious, cooperative state was so wonderful and precious it was almost beyond description. And since those who work with such a group on a regular basis surely know more about the lives of bonobos than anyone else, how lucky was I to be accompanied on my visit by one of those people, and someone who I strongly believe must be the most gifted and accomplished psychiatrist in

the world today—Professor Harry Prosen, who provided the Introduction to this book. Harry arranged my visit to the Zoo, whose staff most generously, and for the first time ever, brought almost the whole group into the display area for us to see (including all the females and infants)—a group that numbered 16 of their 23 bonobos, which, incidentally, is a significant portion of the world's captive population of only some 250 bonobos (in 2014).

[456] A lifetime of successfully treating humans led Harry to identify what he refers to as **'an empathy deficit'**—which is an acceptable but, in truth, evasive, denial-complying, mechanistic scientific term for 'lack of love during upbringing'—as being the main cause of humans' psychological problems. Indeed, Harry has spent a lifetime studying the effects of empathy/love, and the lack thereof, in not only humans but other species, including bonobos, who he has recognized as the most empathetic/loving of all non-human species. Three examples illustrate just how honest Harry is able to be about the need for empathy/love/unconditional selflessness in the lives of highly social mammals and how psychologically devastating the effects are of not receiving it—and thus how truly effective a psychiatrist he is. In the first example, Harry was asked to assist, via a video link, a girl in a mental institution who was completely uncommunicative, virtually mute. After Harry talked to her she finally mumbled the word 'Mary', to which Harry intuitively replied, 'Mary Magdalene?' When the girl seemed to respond, Harry added, 'Wasn't she a prostitute?', which reached into the girl's core psychosis and from there she was able to open up and heal. In another example, Harry was asked to help in a situation at a zoo where an elephant had died, leaving both its elephant companion and human keeper catatonic with grief. Harry intuitively moved the catatonic keeper in front of the catatonic elephant, which prompted the elephant to put its trunk on the shoulder of the keeper, and that sharing of feeling, of empathy, of compassion, between the two freed them both from their catatonic states. And in a case that was reported around the world, the Milwaukee County Zoo sought Harry's assistance with an adult bonobo named Brian, whose **'infancy had diverged from the normal pattern for bonobos in which mother and baby are in constant contact'** (Jo Sandin, *Bonobos: Encounters in Empathy*, 2007, p.48 of 109). Harry ascertained that Brian's extremely distressed behavior, including induced vomiting and self-mutilation, **'rose from desperate, but futile, attempts to calm himself in the midst of extreme anxiety'**, and **'prescribed surrounding Brian with a world that was absolutely**

**predictable'** — a calming **'therapeutic environment'** (ibid) that finally allowed Brian's healing to begin. Little wonder Harry is psychiatric consultant to the Bonobo Species Preservation Society. And little wonder he was so enthralled when, in 2004, he read my love-indoctrination explanation for humans' and bonobos' moral nature in a copy of the *Human Condition Documentary Proposal* (<www.humancondition.com/doco>) that we had sent to scientists around the world, especially those working with bonobos. Harry was so impressed by the proposal, in which I outlined all the explanations of the human condition, of Integrative Meaning and of the nurturing origins of our moral nature and conscious mind that are presented in this book, that he wrote to us saying, **'The Proposal is never out of my brief case. I read it over and over and it is I think one of the most astonishing and outstanding things of our time. It is a gift and I hope you don't give an inch to your detractors'** (WTM records, 17 Jun. 2005).

[457] In fact, Harry became so supportive of my work and so concerned about the attacks that were being made on it that, as he described in his Introduction to this book, he came to Australia in 2007 to give evidence in support of my work in a defamation trial. He has certainly become a precious friend on this most difficult of tasks of delivering understanding of the human condition to the world. Indeed, Harry's love and support for my work makes me wish I had been able to share these explanations with the anthropologist Ashley Montagu (1905-1999) because they too would have made complete sense of all that he was so honestly able to recognize about the importance of love in human life — as demonstrated by his amazingly truthful paper, 'A Scientist Looks at Love', which will be referred to in par. 493. Yes, since Harry has often said, **'All the great theories that I have encountered in my lifetime of studying psychiatry can be accounted for under your explanations of human origins and behavior'**, I'm sure the very rare honesty of Montagu's thinking would have meant that he would have also become a very valuable ally in trying to deliver these liberating but at the same time confronting explanations to the world.

[458] This issue of the confronting nature of my explanations of the origin and nature of human behavior, and resulting extreme resistance to it that led to the vicious attacks upon it and the then biggest defamation case in Australia's history, became all too apparent to Harry when he tried valiantly to have my nurturing explanation that so fully accounts for human and bonobo moral behavior referred to in Jo Sandin's 2007 book about the bonobos at the Milwaukee County Zoo, *Bonobos:*

*Encounters in Empathy* (which was quoted from above), which had been commissioned by the Zoological Society of Milwaukee. In chapter 5 of her book Sandin did manage to include oblique references to the love-indoctrination explanation, such as **'there is evidence from the work of evolutionary biologists to suggest that it was the distant ancestors of bonobos that brought empathy into the evolutionary process'** and **'certain life experiences are necessary for a bonobo to develop into an empathetic and wise adult'**, and **'the nurturing care Lody received (and remembered) constituted a key element in the bonobo's maturing into an empathetic and wise male'** (p.52), but despite all the respect Harry has at the Zoo for being such an effective psychiatric consultant for all their social animals, and despite wanting to, she was unable to include the following comment that Harry asked be added at the end of chapter 5 of her book: **'In summary, to the fascinating and indeed fundamental question for biologists of how the extraordinary empathy and even altruism we are observing amongst bonobos developed, our observations point to nurturing, maternalism and associated matriarchy as key influences. Certainly our observations do appear to be confirming of the nurturing explanation for empathy and true altruism that was first put forward by the American philosopher John Fiske in 1874** [Fiske's work will be described in ch. 6:3] **and, more recently, by Australian biologist Jeremy Griffith in his various books, in particular in Part 2 of his 2004 *Human Condition Documentary Proposal*.'** My work was mentioned in the concluding chapter of Sandin's book, but the key reference to **'the nurturing explanation for empathy and altruism'** wasn't included. I have to emphasize that failure to acknowledge the nurturing explanation for humans' and bonobos' moral nature is consistent with the practice of human-condition-avoiding, denial-complying mechanistic science the world over. The next chapter in this book (6) will describe at length the immense resistance to the nurturing explanation, as well as the mountain of nurturing-avoiding, dishonest explanations for the moral nature of humans and bonobos that have been put forward in its place. For example, it will be mentioned there how in 2005 I submitted an abstract of a paper titled 'Nurturing as the Prime Mover in Primate Development and Human Origins' for presentation at the International Primatological Society's (IPS) 2006 Congress in Uganda (the submission and subsequent correspondence can be viewed at <www.humancondition.com/ips-2006-congress>), only to be rejected on the grounds that **'Both reviewers felt this abstract presents no data nor a testable hypothesis'**. Despite arguing that my nurturing, love-indoctrination explanation for humans' moral instincts 'contains a great deal of supportive evidence in the form of many summaries of data-supported studies of bonobos and other primates by leading primatologists', and 'is an entirely testable, validatable hypothesis, as the evidence just described about bonobos shows', and submitting this protest to the President and 38 members of the IPS Congress Committee, the rejection was upheld! Humans have certainly not wanted to face the truth of the importance of nurturing in their lives; again, they 'would rather admit to being an axe murderer than being a bad father or mother'!

The author (left) with Trish Khan, Dr Jan Rafert and Professor Harry Prosen

[459] To return, however, to my visit to the wonderfully cared-for bonobos at the Milwaukee County Zoo, which, next to a visit Annie and I made to the Tiva River in Africa, which I describe in par. 835, is the highlight of my life. Having thought and written about bonobos for so long it was an incredibly exciting moment for me to finally meet them—and what I saw exceeded my expectations about their cooperative, integrative and intelligent nature. The Zoo's curator of primates and small mammals, Dr Jan Rafert, and primate area supervisor, Trish Khan, were unaware of the nurturing, love-indoctrination explanation for bonobos' moral behavior, suggesting that bonobos' secure, food rich environment was what allowed for selection for cooperation, with Rafert saying these ecological conditions allowed bonobos to 'not only take care of their own family but take care of each other'; and Khan suggesting that 'there is just no competition to produce offspring because [female bonobos] are copulating throughout their entire cycle…that's the big theory, isn't it'. These suggestions are, in effect, aspects of the Social Ecological Model that is discussed and dismantled in

chapter 6:8. Rafert and Khan did, however, confirm to me all the bonobo behaviors that I have already described: their cooperative care for each other; their nurturing love; their sensitivity; how at being reunited after years of separation they are, as Rafert described it, **'beside themselves with glee'**; how, according to Khan, **'their intelligent, understanding, ingenious minds so tests the wits of the keepers that the keepers are often exhausted by the end of the day'**; their constant vocal communication, including specific vocalizations for the different keepers; their tendency towards bipedalism; their use of sex to create and cement bonds and relieve tension; their robust sense of humor and laughter; the strong, forceful personalities of the group's dominant females; the lifelong bond between mother and son; the absence of the territorial wars that occur between groups of chimpanzees (because bonobos tolerate other groups of bonobos—they are, in essence, all one large group, a Specie Individual); and, above all for me, the confirmation of their moral sense, their constant maintenance of social order. For example, Rafert emphasized that when young males within bonobo society begin to become competitive and aggressive like adolescent males of other species **'The group will tolerate it for only so long before everyone comes down on it very forcefully'** to prevent such behavior developing. Yes, as was stated about Kanzi's situation, **'bonobos like to know if everyone is…following the** [social] **rules'** that achieve and maintain integration.

[460] With all other highly social species that I have observed, such as meerkats, I have had the feeling that their groups are still more a collection of individuals, but that wasn't the case with the bonobos. With them it was as though they were all part of one organism, all deeply aware of and in tune with each other. While there were outbursts of anxiety and occasional tensions and even fights amongst them—because bonobos haven't as yet completed the love-indoctrination process that enables the fully cooperative, utterly harmonious, completely integrated state to develop—there was a high degree of harmony in the group; to such an extent, in fact, that there was an overall togetherness, a real peacefulness and tranquility, a unity, a security, an each-knew-all-about-and-trusted-and-supported-and-loved-everyone-else-feeling between them. It was like they took each other for granted, just as we take our arms or legs or ears for granted because they are simply part of our whole being. It was like they all felt they belonged and were part of something bigger; it wasn't like they had to trust that this was the case, it was that they *knew*

it was the case. As is very apparent in some of the marvelous footage WTM Founding Member James Press took of the bonobos for inclusion in Harry's video introduction to this book that appears on this book's website (see <www.humancondition.com/prosen-intro>), when one of them looked at you with the extraordinary awareness and thoughtfulness that their facial expression exhibits, it was as though that individual was fully connected by an invisible cable or link to all the other members of their group behind them and around them; that at that moment they were the looking-out-at-the-world component of the whole group. Yes, they all moved about together like one big roly-poly organism that would suddenly materialize in front of you, and then, just as suddenly, vanish all together into another corner of their enclosure. I derived a deep, calming, reassuring, and even happy, life-as-it-should-be, feeling from being with them, which was quite amazing. To use some of the extraordinarily honest words from Sir Laurens van der Post that were included in par. 186, I felt part of a time when **'All on earth and in the universe were still members and family of the early race seeking comfort and warmth through the long, cold night before the dawning of individual consciousness in a togetherness which still gnaws like an unappeasable homesickness at the base of the human heart.'**

[461] It is just such a tragedy that the phenomenal integration and emerging conscious awareness that exists amongst bonobos has not been able to be admitted. There was no signage on their enclosure indicating their incredible cooperative, loving, nurturing nature and amazing presence of mind. They were just another variety of animal for the public to see. No one was told that what you are seeing here is a species living on the threshold of the completely integrated, heavenly, Garden-of-Eden state of **'togetherness'** we humans once lived in. What was on display was one of the most amazing and special sights that can be seen on planet Earth, yet no one was being made aware of it. In fact, as I discuss in chapter 6, I think the nurturing, integrative, near-conscious nature of bonobos is so confronting that there is a very real inclination to try to dismiss them as nothing special, emphasize that they do have fights, etc, etc—and even block them out of our awareness, which *has* occurred to the extent that they have been described as **'the forgotten ape'** (Frans de Waal & Frans Lanting, *Bonobos: The Forgotten Ape*, 1997).

[462] As I said, those who work on a regular basis with this large, socially authentic group of bonobos at the Milwaukee County Zoo surely know as much or even more about bonobo life than anyone else in the world,

and since Sandin's book *Bonobos: Encounters in Empathy* wonderfully documents the lives and interactions over many years of this large group of bonobos (as supplied by their keepers at the Milwaukee County Zoo and Harry) it has to be one of the most informative books on bonobo behavior. Available through Amazon, this book evidences all the interactions and behaviors of bonobos that have been mentioned in this chapter. It contains, for example, references to bonobos' **'Natural nurturing patterns [**that]**...has yielded mother-reared youngsters as socially robust as they are physically healthy'** (p.43), and recounts how **'Bonobo babies born at the Milwaukee County Zoo immediately cling to their mothers. They are carried constantly. Infants fostered here are greeted with enthusiasm and cuddled by males and females'** (p.66). I should mention that it also describes **'individuals whose life history** [of missing out on nurtured love] **has left them deficient in bonobo social graces...so-called problem bonobos'** (p.26), and how the group enforces **'the community's code of conduct'** (p.28) on such **'rule-breaking'** individuals (p.29), which is all *extremely* confronting for the human race today where almost everyone's **'life history** [of not being adequately nurtured in their infancy and childhood] **has left them deficient'**, **'problem'** humans. As will be described at length in chapter 6, it is little wonder there has been resistance to acknowledgment of the importance of nurturing in the lives of both bonobos and humans.

[463] The following paragraph, from Sandin's book, in particular reveals the whole love-indoctrination process at work: **'We see Maringa, for almost two decades the undisputed empress of the group, in full diva mode, throwing tantrums and slyly peeking to note their effect. We glimpse the intricate etiquette that orders relationships among group members. We witness the matriarch and her "sisters" in power administering swift and painful rebukes in response to infractions of the colony's behavior code. We view Maringa's own slide from power. Here we observe the philosopher-king Lody, long the dominant male, ruling by empathy and wisdom, wielding power but always within the context of a matriarchal society. We watch younger males challenge his pre-eminence as he acknowledges his gradual physical decline. We experience tender moments between mothers and children, mischievous testing of the patience of tribal elders and compassionate interventions on behalf of weaker bonobos'** (p.60). The **'behavior code'** among bonobos is a code that is dedicated to maintaining cooperative order, integration. While bonobos have not yet completed the love-indoctrination process, the **'code of conduct'** or **'rules'** or **'intricate etiquette'** represents the final stages of the behavioral management needed

to develop the fully integrated, all-loving and all-sensitive state that became instinctive in our distant human forebears.

[464] With regard to the **'compassionate interventions on behalf of weaker bonobos'**, this is but one of many illustrations from the book: **'Dr. Prosen has been particularly impressed by the attention given Brian by Lody, the group's dominant male. When Brian seemed too panicky to move, Lody often would take his hand and walk him into a different area such as the playroom or the outdoor yard. It was in Lody's company that Brian first sat down and ate with a group, a "very big deal for him," recorded in animal-management notes August 30, 1999. Regularly, Lody would postpone his own meals to sit with and comfort Brian when the younger bonobo was having difficulties...In Lody, Dr. Prosen said, the presence of weakness seems to activate his own compassionate strength'** (pp.49-50). Also, **'Harry Prosen...and keeper Barbara Bell credit Lody's wise leadership with the high level of emotional health of Milwaukee's bonobos and points to numerous instances of empathetic behaviour'** (p.14), and **'His** [Lody's] **days of mourning after the death of Kidogo constitute one of the most poignant chapters of the community's life'** (p.14).

[465] I know that all of nature is under dire threat from humans' now horrifically psychologically upset state, but to lose bonobos—who are categorized by the International Union for Conservation of Nature, the world's main authority on the conservation status of species, as a species **'facing a very high risk of extinction in the wild'**—is an unbearable prospect. The very latest reports indicate that the bushmeat trade (where wild animals including bonobos are killed for meat) has now penetrated into even the most remote areas of the Congo basin where the bonobos live. Now that the human condition has been explained and we can at last admit that the bonobos are living in this delicate nursery state of nurturing love and, as a result, are highly sensitive, loving and aware, killing them off is a terrible, terrible case of 'the massacre of the innocents'. It is *truly* unbearable to think about—not to mention what we lose in terms of the astonishing and irreplaceable evidence they provide of our origins. So please God, let these ameliorating, peace-bringing understandings of our human lives get out to the world as soon as possible. (I should reiterate here that when I talk of God, I am not invoking an abstract, metaphysical, supernatural being, I am referring to the all-pervading integrative power in our world.)

[466] This concern for bonobos in their home in the Congo raises another aspect of the absolutely wonderful work that is being carried out by those

associated with Milwaukee County Zoo. Dr Gay Reinartz, the director of the Bonobo & Congo Biodiversity Initiative (BCBI) for the Zoological Society of Milwaukee and coordinator of the Bonobo Species Survival Plan for the Association of Zoos and Aquariums, has since 1997 been travelling at least twice a year to the remote and dangerous Congo, even during its horrific civil war, to help protect the bonobos. What started out as a research project to discover what habitats they prefer and whether Salonga National Park, which forms the heart of the bonobos' range, harbors a self-sustaining bonobo population, has evolved into a multi-faceted conservation program that is **'the center of what is now an international effort to protect bonobos in the heart of their natural home'** (*Bonobos: Encounters in Empathy*, p.75). In 2011, the BCBI program received the international conservation award from the Association of Zoos and Aquariums, the highest recognition for an international conservation program.

[467] So, bonobos provide an extraordinary insight into what life for our human ancestors was like, graphically illustrating the loving, cooperative, nurturing environment our ape ancestors must have been blessed with to complete the love-indoctrination process and develop the unconditionally selfless moral instincts we have within us. The fossil evidence indicates these ideal 'nursery' conditions occurred somewhere in Africa, just exactly where is not yet clear, but, as we intuitively recognize, Africa *was* 'the cradle of mankind'.

## Chapter 5:14 'A golden race…formed on earth'

[468] In all, love-indoctrination was an incredibly fortuitous development. It gave humans our unconditionally selfless, integrative-behavior-orientated, Specie-Individual, humanity-creating, **'awe'**-inspiring, **'distinct'** from other **'animals'**, moral instinctive self or soul. And it gave us our fabulous conscious mind, a development that meant we were free from the stupor of the animal condition, but did then, of course, lead to our horrifically upset state of the human condition—*but*, with understanding of the human condition now found, we can, as will be explained in chapter 9, finally be transformed to an existence that is free of *both* conditions. Love-indoctrination also freed our hands to hold tools and carry out innumerable tasks, create all the things our conscious mind was capable of envisaging and ultimately undertake the experiments in knowledge that led to these insights; a fully conscious mind in a whale or a dog would be frustrated by

its inability to implement its understandings. Further, love-indoctrination is what likely gave humans the relatively long lifespan that has been so instrumental in the accumulation of knowledge. If we only lived to 30, which is considered a long life in the animal kingdom, instead of the 70 plus years we do (in the case of dogs, for example, no matter how well you look after one, it will still only live for 10-15 years), we would likely not have had sufficient time to properly assimilate and manage in our minds all the difficult nuances of the human condition, or to learn about and add to all the understandings humanity was accumulating.

[469] In conclusion, nurturing was the main influence or prime mover in the development of humans—*not* upright walking, or tool use, or language, or mastery of fire, or migration from the forest to the savannah, or any one of the other explanations that evasive, denial-complying, mechanistic biologists have been putting forward in the mountain of dishonest books that have been published on human origins. The human race had an immensely happy, all-sensitive-and-all-loving, thrilled-and-enthralled-with-all-of-life, zest-for-living-drenched, ecstatic infancy and childhood, which, as will be described in chapter 8:2, lasted from some 12 million years ago right up to some 2 million years ago when our conscious mind became fully developed and with it the upset state of the human condition emerged—NOT the brutish, barbaric, backward, primitive, savage, bestial, demonic, competitive, aggressive, warring, somber, morose, unhappy drudge of an existence that our species' early existence has for so long been portrayed in documentaries. When Buddhist scripture anticipated what a human-condition-resolved-and-thus-human-condition-free state would be like—that **'Human beings are then without any blemishes, moral offences are unknown among them, and they are full of zest and joy'** (Maitreyavyakarana; *Buddhist Scriptures*, tr. Edward Conze, 1959, pp.238-242)—it was also describing what life was like in the human-condition-free state that existed in our past when we had become free of the animal condition but had not yet encountered the upset state of the human condition.

[470] Yes, the recognition in all our mythologies and in the work of our most profound thinkers of a wonderful, all-loving, innocent past for the human race isn't some without-any-factual-base, romantic, fanciful dream of some impossible, unrealistic, idyllic, utopian existence, nor is it, as was mentioned in pars 184-185, merely a nostalgic yearning for the security and maternal warmth of infancy, as mechanistic science has tried to dismiss it as—no, it is a *completely real* time in our species' distant

past that recent fossil evidence is now confirming, and that bonobos provide ample living evidence of. Indeed, we can now finally reconcile our scientific knowledge with all the truths contained in our mythologies and religions—for example, we can finally appreciate that Moses' description of how our **'Adam and Eve'** ancestors lived **'naked and they felt no shame'** in an innocent, cooperative, loving **'image of God'**-like state in a **'Garden of Eden'** (see par. 155) equates perfectly with the life of the bonobos that has just been described; as does what the eighth century BC Greek poet Hesiod wrote in his poem *Works and Days* about this 'Golden Age' time in our species' past: **'When gods alike and mortals rose to birth / A golden race the immortals formed on earth...Like gods they lived, with calm untroubled mind / Free from the toils and anguish of our kind / Nor e'er decrepit age misshaped their frame...Strangers to ill, their lives in feasts flowed by...Dying they sank in sleep, nor seemed to die / Theirs was each good; the life-sustaining soil / Yielded its copious fruits, unbribed by toil / They with abundant goods 'midst quiet lands / All willing shared the gathering of their hands'** (see par. 180). *And* we can now also fully understand Plato's descriptions of **'a time when...we beheld the beatific vision and were initiated into a mystery which may be truly called most blessed** [the fully integrated state], **celebrated by us in our state of innocence, before we had any experience of evils to come, when we were admitted to the sight of apparitions innocent and simple and calm and happy, which we beheld shining in pure light, pure ourselves and not yet enshrined in that living tomb which we carry about, now that we are imprisoned'** (see par. 158), and when we lived a **'blessed and spontaneous life...**[where] **neither was there any violence, or devouring of one another, or war or quarrel among them...In those days...there were no forms of government or separate possession of women and children; for all men rose again from the earth, having no memory of the past** [we lived in a pre-conscious state]. **And...the earth gave them fruits in abundance, which grew on trees and shrubs unbidden, and were not planted by the hand of man. And they dwelt naked, and mostly in the open air, for the temperature of their seasons was mild; and they had no beds, but lay on soft couches of grass, which grew plentifully out of the earth'** (see par. 170). Clearly our instinctive memory of our species' time in the 'Garden of Eden' is so strong that when it is allowed to fully express itself, as it obviously was in the minds of the denial-free thinking prophets Moses, Hesiod and Plato, it is able to almost perfectly describe what life was like then. Obviously Moses, Hesiod and Plato didn't know of the existence of bonobos, and yet they knew almost exactly what their/our lives were like! If we look at the picture of bonobos above par. 416, they are even, as Plato described, lying **'naked' 'in the open air' 'on soft couches of grass'**!!

[471]Yes, now that we can explain our present immensely upset, human-condition-afflicted **'living tomb'** existence and thus afford to be honest, we can admit that what we have been referring to as our soul *IS* the instinctive memory within us all of the nurtured-with-love infancy and idyllic, **'calm and happy'** childhood period 12 to 2 million years ago that our species spent in Africa before our conscious mind and with it the psychologically upset state of the human condition fully emerged; the time when, as Moses said in Genesis, we were **'banished...from the Garden of Eden'**-like (3:23) state of original innocence and left **'a restless wanderer on the earth'** (4:14) (with, as will be described in chapters 8:2 and 8:11A, the fossil evidence showing these **'restless wander[ings]'** away from Africa beginning some 1.9 million years ago).

Detail from Edward Hicks' *Peaceable Kingdom with Seated Lion*, 1833-1834

[472]Like the many depictions of the 'Garden of Eden', such as the one included before par. 387, the above painting represents a bubbling up from our subconscious psyche/soul of this memory of our innocent time in Africa. When you are in natural Africa the sense of having been there before—indeed, of having returned home—is so mind-bendingly overwhelming you think you are in a dream! As Sir Thomas Browne wrote, **'We carry with us the wonders we seek without us: there is all Africa and her prodigies in us'** (*Religio Medici*, 1643, Sect.15), and as Shakespeare similarly

wrote, 'A foutra for the world and worldlings base! I speak of Africa and golden joys' *(Henry IV,* c.1597). In her 1967 book, the aptly titled *A Glimpse of Eden*, the poet Evelyn Ames recorded her experiences of an African safari: 'We thought we knew what to expect. Several friends had been there and told us about it...but we discovered that nothing, really, prepares you for life on the East African Highlands. It is life (I want to say), making our usual existences seem oddly unreal and other landscapes dead; that country in the sky is another world...It is a world, and a life, from which one comes back changed. Long afterwards, gazelles still galloped through my dreams or stood gazing at me out of their soft and watchful eyes, and as I returned each daybreak, unbelieving, to my familiar room, I realized increasingly that this world would never again be the same for having visited that one. Nor does it leave you when you go away. Knowing its landscapes and sounds (even more in silence), how it feels and smells—just knowing it is there—sets it forever, in its own special light, somewhere in the mind's sky** [pp.1-2 of 224] ...Each day in Africa my heart had almost burst with Walt Whitman's outcry: "As to me, I know of nothing else but miracles"** [p.204].' A sign at the entrance to the Serengeti National Park in Tanzania reads, 'This is the world as it was in the beginning'. Sir Laurens van der Post also expressed how much our soul yearns to return home when he wrote that 'We need primitive nature, the First Man in ourselves, it seems, as the lungs need air and the body food and water...I thought finally that of all the nostalgias that haunt the human heart the greatest of them all, for me, is an everlasting longing to bring what is youngest home to what is oldest, in us all' *(The Lost World of the Kalahari*, 1958, p.151 of 253).

[473] Of course, acknowledgment of an innocent, unconditionally selfless, all-loving past for humans has been near impossible while we couldn't explain our selfish, seemingly unloving human condition. In physics, for example, the only way we could cope with the fact that the upset human race no longer lives in this fully integrated, unconditionally selfless, loving way was to deny the truth of Integrative Meaning. This account of the sociologist Pitirim Sorokin's amazingly honest description of altruistic, unconditionally selfless love makes the need for this denial *very* clear: 'Altruistic love is a giving, sacrificial love; it often involves the sacrifice of very important interests, possibly one's life [p.457 of 744] ...altruistic love is ideally boundless. It originates within itself and extends out to the cosmos. It makes no distinctions; it embraces all. It is unconditional and undaunted by disappointment and failure. It is compassionate and caring; it hurts when others hurt and suffers

when they suffer. It is endlessly giving; it reaches out in the spirit of care, justice, and compassion. It is ennobling and exalted; it represents the highest in humans potential [p.456] …It regards all people as deserving of love. They feel responsible for all people, not just friends and family members [p.465] …[It is] a selfless love attained by the primal human capacity to submerge self and others into a greater whole [p.457]'** (Samuel Oliner & Jeffrey Gunn, 'Sorokin's Vision of Love and Altruism', *Living A Life Of Value*, ed. Jason Merchey, 2006). Yes, altruism *is* a **'primal human capacity'**, but such an ability **'to submerge self and others into a greater whole'** has *unbearably* condemned the present upset human race for no longer behaving in a way that is anything like that. In biology, meanwhile, the unbearable truth of an innocent, unconditionally selfless, all-loving past for humans has meant that there is, as will be described at some length next in chapter 6, an immense amount of dishonest biological thinking that now has to be dismantled and redressed. In particular, it will be described there how mechanistic science's reluctance to recognize the compelling evidence that bonobos provide for the nurturing origins of our moral nature has meant that little research has been done on bonobos—so little, in fact, that they have been, as mentioned, referred to as **'the forgotten ape'**. And as for the role nurturing is presently playing in bonobo society and did play in the emergence of humanity, chapter 6 will also describe how mechanistic science has denied such a confronting truth by maintaining that maternalism represents nothing more than mothers protecting their dependent infants and providing training in life skills. And as for our upset behavior, mechanistic science has simply blamed it on genes (or 'nature', as the contrived genetic excuse has been called in the 'nature versus nurture' debate) because, as the child psychologist Oliver James points out, **'Believing in genes removes any possibility of "blame" falling on parents'** (*They F\*\*\* You Up: How to Survive Family Life*, 2002, p.13 of 370). Other scapegoat excuses include 'chemical causes', as this dialogue from the 1989 film *Parenthood* illustrates: Counselor: **'He's a very bright, very aware, extremely tense little boy who is only likely to get tenser in adolescence. He needs some special attention.'** Karen: **'It's because he was our first. I think we were very tense when Kevin was little. I mean, if he got a scratch, we were hysterical. By the third kid, you know, you let him juggle knives.'** Counselor: **'On the other hand, Kevin may have been like this in the womb. Recent studies indicate that these things are all chemical.'** Gil: **'[points at Karen] She smoked grass.'** Karen: **'Gil! I never smoked when I was pregnant…Will you give me a break?'** Gil: **'But maybe it affected your chromosomes.'** Counselor intervening: **'You should not**

**look on the fact that Kevin will be going to a special school as any kind of failure on your part.'** Gil: **'Right, I'll blame the dog.'**

[474]Thankfully, the human race is now in the position to end all this lying and acknowledge the crucial role that nurturing has played in human origins, and in our own lives—which means all those dishonest, denial-complying, reverse-of-the-truth, mechanistic explanations, papers, books and documentaries that have been put forward to supposedly explain human nature and our origins will need to be re-written and re-filmed to present the honest story about our human journey. Yes, we are now able to admit that Rousseau *was* right when he said that **'nothing is more gentle than man in his primitive state'**. It was *not* our distant ancestors who were without the ability to be gentle and loving, as we have so long been taught, rather it is us fully conscious, immensely psychologically upset, angry, egocentric and alienated, soul-destroyed, human-condition-afflicted modern humans who have *lost* that ability.

[475]At last the artificially relieving but, in truth, absolutely awful dishonest, alienated, cold, psychotic and neurotic darkness that Plato so accurately portrayed us as living in with his cave analogy can disappear from human existence, leaving in its place a world of warm sunlight once more, a world of alienation-free truthfulness, happiness and togetherness, a world **'full of zest and joy'**. All that remains now is for the human race to make the right choice between continuing on its habituated path of denial that will lead to terminal alienation and the extinction of the human race, or choosing the path of truth that has now opened up that leads to our species' transformation. The dangerous impediments to this critical juncture and the awe-inspiring potentials that lie beyond will be revealed in the final four chapters of this book.

[476](A more complete description of the origin of humans' unconditionally selfless, altruistic, moral instinctive self or soul can be found in *Freedom Expanded* at <www.humancondition.com/freedom-expanded-origin-of-morality>.)

# Chapter 6

# End Play for The Human Race

Bonobo mother with her infant

## Chapter 6:1 <u>Summary</u>

[477] This sixth chapter reveals how science, the designated vehicle for human enquiry, has evaded the unbearable truth of the importance of nurturing in the development of our species, and in our own lives, by inventing theories about our origins that are now so dangerously dishonest they threaten the human race with extinction.

## Chapter 6:2 <u>The danger of denial becoming so entrenched that it locks humanity onto the path to terminal alienation</u>

[478] As explained in chapters 2 and 4, <u>the truth about the origin of our corrupted human condition and of the integrative meaning or purpose of existence were two extremely obvious truths</u> that, until we found the compassionate explanation and understanding for both, were so unbearably exposing, condemning and confronting that we, the human race, had no choice but to live in complete denial of them. For exactly the same

reason, the truth of the importance of nurturing in both the maturation of our species and in the maturation of our own lives—how nurturing created our moral instincts—that was just described in chapter 5 has been another extremely obvious truth that has had to be denied because it too was unbearably condemning of our present unloved and unloving, human-condition-afflicted lives. And as we will see in chapter 7, the nature of consciousness is a further obvious truth that has also been so unbearably condemning that it too has been fiercely denied, to the extent that our conscious mind—nature's greatest creation—has been deemed an inexplicable mystery.

[479] In light of all this unavoidable denial, it is not surprising that these four great outstanding mysteries in science—of the explanation of the human condition, the meaning of our existence, the origin of our un-conditionally selfless moral instincts, and why humans became conscious when other animals haven't—hadn't been solved. If you can't confront the human condition you are in no position to find the explanation for it; and if you can't admit to Integrative Meaning you obviously can't explain what the meaning of our lives is; and if you can't admit to the impor-tance of nurturing you are in no position to explain the origin of humans' unconditionally selfless moral nature; and if you can't confront the truth of the nature of consciousness you are in no position to explain how it arose and what has prevented its development in other species. It is *not* lack of mental cleverness or Einstein-like genius that made all the great missing explanations in science that are presented in this book impossible to find, but alienation, the lack of innocent, soul-guided soundness and security of self. The truths and insights revealed in this book are, in fact, obvious and easy to find—as long as you are not living in a resigned, dishonest, alienated, human-condition-denying state. Cleverness was not the crucial factor in finding all these great breakthroughs in science—what *was* needed was simple, truthful, soul-guided sound thinking. As Berdyaev recognized in the quote of his that was included in par. 237, **'the moral reformation of life…**[depends on] **a revelation of a clear conscience, unclouded by social conventions** [the most entrenched and insidious of which is denial]'. That exceptionally honest writer, Simone Weil, saw the problem clearly when she wrote that **'A new type of sanctity…a fresh Spring…a new revelation of the universe and of human destiny…**[requires] **the exposure of a large por-tion of truth and beauty hitherto concealed under a thick layer of dust** [denial]. **More genius is needed than was needed by Archimedes to invent machines and**

physics. A new saintliness…The world needs saints who have genius, just as a plague-stricken town needs doctors' ('Last thoughts', 1942; *The Simone Weil Reader*, ed. G. Panichas, 1977, p.114 of 529). There is certainly truth in the novelist Aldous Huxley's observation that 'We [resigned humans] don't know because we don't want to know' (*Ends and Means*, 1937, p.270).

[480] Yes, a recurring theme running through this book is that the truth *cannot* be found from an evasive and thus dishonest position. Living in denial of the human condition and of any truths that bring it into focus means your search for truth and understanding is only ever going to end in a completely lost state of confused madness. For instance, as we saw in chapter 2:9, mechanistic science initially tried to deny the existence of the human condition by dishonestly claiming that our selfish and competitive behavior was a natural consequence of a supposed 'survival of the fittest' natural selection process, and that our selfless moral nature was actually a selfish strategy to help relatives or kin reproduce our genes. But that extreme dishonesty was both contrary to and offensive of the fact that we humans do have unconditionally selfless, loving, moral instincts, and so, in response to that backlash, a much cleverer way was concocted to deny our species' extreme sickness, its psychosis (as has been mentioned, the word psychosis literally means 'soul-illness'). This contrivance was E.O. Wilson's Multilevel Selection theory for eusociality, which maintains there is no psychosis involved in the human condition, but that we simply have some unconditionally selfless instincts that exist alongside competitive, 'survival of the fittest' selfish ones, and that our human condition is the product of a conflict between those two instinctive states! Yes, this theory renders the human condition inconsequential, virtually benign—with the consequence of this extremely sophisticated denial being that the human race now faces terminal alienation; madness. Indeed, if this extreme dishonesty on the part of mechanistic science had not been exposed, as it is throughout this presentation, then the stage *was* set for the extinction of our species.

[481] In the case of humanity's denial of Integrative Meaning, it was explained in chapter 4:3 that the way the upset human race eventually found to eliminate *that* truth was by making the integrative theme or purpose of existence out to be a supernatural deity we termed 'God', a subject or realm that supposedly had no place in science—an outrageous display of dishonesty that has led to all manner of blind, purpose-*less*, ultimately unaccountable, ridiculous interpretations of the natural

selection process. And—as we will see in chapter 7—a similar fate has also befallen the study of consciousness, which has been so character-ized by denial and dishonesty that it too has resulted in the human race becoming lost in a completely bewildered state of intellectual confusion about the all-important issues of what consciousness actually is and how it emerged. *And*, lastly, in the case of that other key issue that is so crit-ical to our understanding of ourselves, namely the origins of our moral nature, what will be revealed throughout this chapter is that the denial of the importance of nurturing in the maturation of our species and in the maturation of our own lives has seen the 140 years that have passed since the American philosopher John Fiske first presented the nurturing explanation for our moral nature *squandered* through the development of extremely dishonest, alienated, mad biological theories to 'explain' our species' altruistic instincts—a tragic journey that has now culminated in the immensely dangerous Social Ecological Model, and its most recent incarnation, the Self-Domestication Hypothesis.

[482] In short, what has just been outlined illustrates how, even though mechanistic science had no choice other than to practice denial of any truths that brought the issue of the human condition into focus until we found the true explanation of the human condition, such denial was an extremely dangerous practice. Yes, what has been described illustrates the dangerous 'trap' that was described in chapter 2:4 where, as necessary as they have been, the longer denials are practiced, the more refined they become, so that in the end, after many decades of development, they inevitably become so sophisticated, so cleverly refined and so entrenched they effectively lock humanity onto a path to terminal alienation—to total derangement, death and extinction; the **'universal ruin to the world'** that Plato said **'more and more forgetting** [denial]**'** leads to.

## Chapter 6:3  <u>The nurturing origins of our moral soul is an obvious truth</u>

[483] Like the truths of the origin of our corrupted human-condition-afflicted state and of Integrative Meaning, the importance of nurturing in both the maturation of our own lives and in the maturation of our species—that nurturing created our moral instincts—*is* an obvious truth. While we have had to live in denial of it, <u>we all intuitively know that a mother's love is crucial to the creation of a well-adjusted human and that</u>

we are *all* born with an instinctive expectation of receiving unconditionally selfless love from our mother. And it's *also* obvious that such powerful instincts to nurture with love and be nurtured with love can't have come from nowhere. To be so strong in us they must have played a significant role in our species' development. So yes, if we weren't living in denial of the importance of nurturing in human life, it wouldn't be hard to work out that our unconditionally selfless moral instincts were borne out of the mother-infant relationship.

[484] While, as pointed out in par. 189, Charles Darwin was not a denial-free thinker who could confront the issue of the human condition, he was nevertheless a remarkably honest, truthful and thus effective thinker who could see that our **'social instinct seems'** to be **'developed'** from **'parental'** **'affections'**, writing in chapter 4 of his 1871 book, *The Descent of Man*, that **'The feeling of pleasure from society is probably an extension of the parental or filial** [family] **affections, since the social instinct seems to be developed by the young remaining for a long time with their parents; and this extension may be attributed in part to habit, but chiefly to natural selection.'** Further on in chapter 4, Darwin affirmed his belief that **'the foundation of morality'**, our **'moral sense'**, which he said **'affords the best and highest distinction between man and the lower animals'**, is these **'social instincts'** that **'I have so lately** [in the just referred to quote] **endeavoured to shew'** the origins of; adding that these **'social instincts'** were **'aid[ed]'** by **'intellectual powers** [that appreciate the importance of moral behavior] **and the effects of habit'**—earlier he had said that **'love, sympathy and self-command become strengthened by habit'**. So, while Darwin didn't go on and develop the idea into a full account of the origins and consequences of humans' moral nature, he did canvas the idea that the **'natural selection'** of an **'exten[ded]'**, **'long'** infancy allowed **'parental'** **'affections'** (which he also refers to as **'maternal instincts'**) to **'develop'** our **'moral sense'**—he recognized the nurturing, love-indoctrinating origins of humans' instinctive moral soul.

[485] (I should mention that while Darwin clearly recognized that humans have a nurtured-with-love, cooperative, selfless, loving **'moral'** **'instinct[ive]'** heritage—even saying in chapter 4 of *The Descent of Man* that **'the social instincts, which must have been acquired by man in a very rude state, and probably even by his early ape-like progenitors, still give the impulse to many of his best actions'**—elsewhere in his writing he contradicts himself by saying we have a divisive **'evil' 'devil[ish]'** ape ancestry, writing, for instance, that **'Our descent then, is the origin of our evil passions!!—The Devil**

**under form of Baboon is our grandfather** [ape ancestor]!' (1838; *The Complete Work of Charles Darwin Online*, ed. John van Wyhe, 2002, *Notebook M*, line ref. 123). So how are we to explain these contradictory positions? I begin the answer by pointing out that in *The Descent of Man* Darwin issued a disclaimer at the beginning of the two chapters in which he dealt with our moral nature: at the beginning of chapter 4, before discussing the **'great question'** of the origin of the **'naked law in the soul'** [Darwin was here quoting Immanuel Kant] of our **'moral sense or conscience'**, Darwin said **'my sole excuse for touching on it, is the impossibility of here passing it over'**; and at the very beginning of chapter 5, which is titled **'On the development of the intellectual and moral faculties during primeval and civilised times'**, Darwin conceded that **'The subjects to be discussed in this chapter are of the highest interest, but are treated by me in an imperfect and fragmentary manner.'** As I explained in par. 189, Darwin didn't feel secure enough in self to address the human condition, which is what is required if you are to truthfully and thus effectively look into human behavior, so despite making an attempt in *The Descent of Man* to look into the origins of human behavior, it was only a tentative step in that all-important exploration. Yes, by his own admission it was **'an imperfect and fragmentary'** attempt. Even in one of his letters, Darwin admitted that **'I have never systematically thought much on religion in relation to science** [as explained in par. 925, to understand religion from a scientific basis requires recognizing that religions are at base a way of coping with the insecurity caused by the human condition]**, or on morals in relation to society** [the question of how selfless, cooperative, moral behavior came about]**; and without steadily keeping my mind on such subjects for a *long* period, I am really incapable of writing anything worth sending to the *Index*** [a religious publication he had been asked to contribute to]' (Letter to F.E. Abbot, 16 Nov. 1871; *The Complete Work of Charles Darwin Online*, ed. John van Wyhe, 2002). Clearly Darwin oscillated from not being **'prepared personally to deal'** with the **'wider' 'issues' 'of man's mental evolution'**, as Harrison said in her quote included in par. 189, to attempting to tackle those issues but only **'in an imperfect and fragmentary manner'** because he wasn't going to [wasn't able to?] **'steadily keep…my mind on such subjects for a long period'**. Darwin oscillated from attempting to do some human-condition-confronting, honest, effective thinking, such as recognizing that **'natural selection'** of a **'long'** infancy allowed **'parental' 'affections'** to **'develop'** our **'moral sense'**, to complying with human-condition-avoiding, dishonest, mechanistic thinking by agreeing that our **'evil passions'** come from our **'Baboon'**/ape **'grandfather'**/

ancestors. This swinging between human-condition-confronting and human-condition-avoiding thinking was very evident when, in the *Origin of Species*, Darwin initially truthfully left it undecided as to whether individuals who manage to reproduce are better or 'fitter' than those who don't, but later caved into Spencer and Wallace's suggestion to incorporate the patently dishonest human-condition-avoiding concept of 'survival of the fittest' (see par. 195). Yet another example of this dishonest thinking occurred when, as mentioned in par. 203, Darwin flirted with the biologically unsound idea of group selection to explain our moral instincts. Yes, Darwin was sufficiently sound and thus capable of brave, truthful effective thinking to come up with the idea of natural selection to explain the origin of species, but not sound enough to confront the human condition and explain the origin of human behavior. I have written a more in-depth analysis of Darwin's thinking at <www.humancondition.com/freedom-expanded-darwin>.)

[486] Although the renowned anthropologist Richard Leakey wasn't thinking honestly when he wrote that the **'bond between mother and infant'** occurs so that her infant can have a period of **'prolonged learning'** about its **'environment'** (the dishonesty of this thinking will be explained shortly in par. 505), he *was* thinking truthfully when he then emphatically asserted that **'the basis of all primate social groups is the bond between mother and infant. That bond constitutes the social unit out of which all higher orders of society are constructed'** (Richard Leakey & Roger Lewin, *Origins*, 1977, p.61 of 264). In his acclaimed television series and accompanying book that was dedicated to explaining **'the ascent of man'**, the great science historian Jacob Bronowski also recognized that the **'real vision of the human being'**—of an unconditionally selfless, loving, sound, integratively behaved individual, which Christ so exemplified—is a direct product of the nurturing that takes place between a mother and her child, saying that **'But, far more deeply, it** [a sound mind] **depends on the long preparation of human childhood... The real vision of the human being is the child wonder, the Virgin and Child, the Holy Family'** (*The Ascent of Man*, 1973, pp.424-425 of 448). It's true, the **'family'**, especially the **'Virgin** [uncorrupted, soul-intact, innocent, psychologically sound and secure mother] **and Child'**, *is* **'Holy'**; **'that bond'** *does* lie at the heart of what makes us, and made us as a species, truly human—namely loving and cooperative.

[487] Yes, the crucial role played by the nurturing, loving **'bond between mother and infant'** in **'the ascent of man'** *is* a truth we are all intuitively aware

of. That awareness is apparent when, for example, Africa is described as **'the cradle of mankind'**—Africa *is* where humanity was *nurtured* into existence. Similarly, when the anthropologist Loren Eisely wrote that **'Man is born of love and exists by reason of a love more continuous than in any other form of life'** (*An Evolutionist Looks at Modern Man*, c.1959), he was recognizing the truth that humanity *was* **'born of love'**.

[488] In fact, the nurturing explanation for our extraordinary unconditionally selfless, all-loving, social, moral instinctive self or soul is so obvious that only three years after Darwin tentatively ascribed the origin of our **'social instinct'** to **'parental' 'affections'**, it was put forward as a developed theory by the aforementioned philosopher John Fiske in his 1874 book, *Outlines of Cosmic Philosophy: based on the Doctrine of Evolution*. Indeed, if we were to prioritize the information we humans need to truly understand our world and our place in it, the first item would be to explain the origins of the variety of life, which is what Darwin did with his idea of natural selection. And the second would have to be to explain the origins of the particular variety of life that is the cooperative organism we call humanity, which is what Fiske did with his nurturing explanation for our species' original instinctive unconditionally selfless, loving, moral, social sense. *However*, while both these fundamentally important insights *were* made available to us way back in the mid-1800s, the only one we have been taught at school and university was Darwin's idea of natural selection. It wasn't until 2004 when I chanced upon a comment about Fiske that I learnt of his remarkable contribution—which was many years after I had worked out that nurturing was the obvious explanation for our moral instincts. (As I have already pointed out, I first presented the nurturing, love-indoctrination explanation that is described in chapter 5 to the scientific community in 1983 in a submission to *Nature* journal. I have submitted it elsewhere many times since, but to no avail, with each submission either rejected or ignored—something I will talk further about in ch. 6:12.) The trail I followed in 2004 that led me to Fiske began with a reference in a 1992 paper by the linguist Robin Allott to **'human love evolv**[ing]**'** from the **'mother/infant bond'** ('Evolutionary Aspects of Love and Empathy', *Journal of Social and Evolutionary Systems*, Vol.15, No.4). Following the source provided for this concept directed me to the work of the scientist and evangelist Henry Drummond—via the historian Dorothy Ross, who had written that **'To Darwin's principle of natural selection by means of the struggle for survival, he** [Drummond] **added another principle that he considered**

far more important—"the Struggle for the Life of Others," or "altruistic Love," which developed in the course of evolution from the necessities of maternity' (*G. Stanley Hall: The Psychologist as Prophet*, 1972, p.262 of 482). Going then to the 1894 writings of Drummond, I found that in responding to the question as to why humans have such long infancies, Drummond had written that 'The question has been answered for us by Mr. John Fiske, and the world here owes to him one of the most beautiful contributions ever made to the Evolution of Man. We know what this delay means ethically—it was necessary for moral training that the human child should have the longest possible time by its Mother's side' (*The Ascent of Man*, 1894, ch. 'The Evolution of a Mother'). Progressing then to Fiske, in 1874 he had written that 'Throughout the animal kingdom the period of infancy is correlated with feelings of parental affection...The prolongation [of infancy] must...have been gradual, and the same increase of intelligence to which it was due must also have prolonged the correlative parental feelings, by associating them more and more with anticipations and memories. The concluding phases of this long change may be witnessed in the course of civilization. Our parental affections now endure through life...I believe we have now reached a...satisfactory explanation of...Sociality' (*Outlines of Cosmic Philosophy: based on the Doctrine of Evolution*, Vol. IV, Part II, ch. 'Genesis of Man, Morally'). The 'prolongation' of 'infancy' was not, however, 'due' to the 'increase of intelligence', rather, as will be explained in chapter 7:3, the prolonged infancy and its nurturing of selflessness *liberated* the fully conscious, intelligent mind, which only developed strongly *after* the love-indoctrination process was well established; this is evidenced by the emergence of the large brain that appears in the fossil record of our ancestors *after* the stage when we resembled the existing non-human great apes. Nevertheless, nurturing, 'parental affection' was put forward by Fiske as a developed theory to explain 'human lov[ing]', 'altruistic', 'ethical', 'moral', 'social' instincts way back in 1874.

[489] Since Drummond has provided us with an absolutely wonderful description of Fiske's nurturing explanation for our moral nature, the following is a condensation of that account: 'The...pinnacle of the temple of Nature...is...The Mammalia, THE MOTHERS...[It is] That care for others, from which the Mammalia take their name...All elementary animals are orphans... they waken to isolation, to apathy, to the attentions only of those who seek their doom. But as we draw nearer the apex of the animal kingdom, the spectacle of a protective Maternity looms into view...[the] love of offspring...That early world, therefore, for millions and millions of years was a bleak and loveless world. It was a world without children and a world without Mothers. It is good to realize

how heartless Nature was till these arrived...the ethical effect...of this early arrangement was nil...There was no time to love, no opportunity to love, and no object to love...Now, before Maternal Love can be evolved out of this first care...Nature must...cause fewer young to be produced at a birth...make them helpless, so that for a time they must dwell with her...And...she...dwell with them...In this [Mammal] child...infancy reaches its last perfection...On the physiological side, the name of this impelling power is lactation; on the ethical side, it is Love...Millions of millions of Mothers had lived in the world before this, but the higher affections were unborn. Tenderness, gentleness, unselfishness, love, care, self-sacrifice—these as yet were not, or were only in the bud...To create Motherhood and all that enshrines...required a human child...The only thing that remains now is...that they [human mother and child] shall both be kept in that school as long as it is possible to...give affection time to grow...No animal except Man was permitted to have his education [in love] thus prolonged...We know what this delay means ethically—it was necessary for moral training that the human child should have the longest possible time by its Mother's side...A sheep knows its lamb only while it is a lamb. The affection in these cases, fierce enough while it lasts, is soon forgotten, and the traces it left in the brain are obliterated before they have furrowed into habit [Note here recognition that the training in love wears off with age, which, as was explained in ch. 5:9, is why there was the selection for neotenous youth in the love-indoctrination process]...To her [the human mother] alone was given a curriculum prolonged enough to let her graduate in the school of the affections...unselfishness has scored; its child has proved itself fitter to survive than the child of Selfishness... A few score more of centuries, a few more millions of Mothers, and the germs of Patience, Carefulness, Tenderness, Sympathy, and Self-Sacrifice will have rooted themselves in Humanity...However short the earliest infancies, however feeble the sparks they fanned, however long heredity took to gather fuel enough for a steady flame, it is certain that once this fire began to warm the cold hearth of Nature and give humanity a heart, the most stupendous task of the past was accomplished...[And here Drummond quotes Fiske] "From of old we have heard the monition, 'Except ye be as babes ye cannot enter the kingdom of Heaven'; the latest science now shows us—though in a very different sense of the words—that unless we had been as babes, the ethical phenomena which give all its significance to the phrase 'Kingdom of Heaven' would have been non-existent for us. Without the circumstances of Infancy...we should never have comprehended the meaning of such phrases as 'self-sacrifice' or 'devotion'" ' (*The Ascent of Man*, 1894, ch. 'The Evolution of a Mother').

[490]The outstanding question, it follows, is, *why* did Fiske's fundamentally important explanation for the origins of our moral instincts that created **'humanity'** — **'one of the most beautiful contributions ever made to the Evolution of Man'** — virtually vanish from scientific discourse? *Why* weren't we taught the nurturing explanation for our altruistic moral nature at school, or when we studied biology at university. *Why* did I have to work the idea out myself? *Why* was this **'altruistic' 'principle'** that was **'considered far more important'** than the **'principle of natural selection'**, and which Fiske explained was able to be developed in our forebears by **'the necessities of maternity'**, allowed to *so* disappear from biological discourse that in the 140 years that have elapsed since Fiske presented his explanation a veritable mountain of books have been published presenting all manner of unaccountable, dishonest theories for the origins of our species' extraordinary moral nature? *Why*, when we had the truth, has there been such a colossal amount of tragically misguided effort that, as we will see, has now resulted in the dangerously dishonest, misleading Social Ecological/Self-Domestication explanations for our moral soul? And *why*, in turn, has *my* nurturing, love-indoctrination explanation for our moral soul — and, indeed, *all* my work — not just been rejected and ignored, but (as I will document in ch. 6:12) *so* ruthlessly attacked that I was made a pariah, and those helping disseminate these insights ostracized?

## Chapter 6:4 <u>The problem has been that the nurturing origin of our moral soul has been devastatingly, unbearably, excruciatingly condemning</u>

[491]As mentioned in chapter 5:7, the answer to all these *'whys'* is that the nurturing explanation for our moral soul has been devastatingly, unbearably, excruciatingly condemning of humans' present inability to nurture children with the real, unconditional love that their instincts expect. Indeed, in his aforementioned paper, Allott noted that when the nurturing explanation for our moral instincts was put forward by Fiske, and supported by a few others, it **'attracted a good deal of opprobrium [abuse]'**. But as I also pointed out in chapter 5:7, since the upset state of the human condition emerged some 2 million years ago, *no* child has been able to be given the amount of love its instincts expect, and unable, until now, to explain the human condition and thus provide the explanation for why this provision of love has been so compromised, the human race has

had no choice but to deny the role nurturing has played in the development of humanity and in the maturation of our own individual lives. The
great difficulty we have admitting the importance of nurturing in human
development is evident in the comments that were referred to in chapter 5:7,
that **'The biggest crime you can commit in our society is to be a failure as a
parent and people would rather admit to being an axe murderer than being
a bad father or mother'**; and that **'For a lot of women the only really important
anchor in their lives is motherhood. If they fail in a primary role they feel should
come naturally it is devastating for them.'** To these two comments I might
add these others that, read one after the other, capture the full horror of
the difficulties and consequences of parenting under the duress of our
alienated, soul-estranged, insensitive, loveless human condition. Firstly,
from the child's point of view: **'The greatest terror a child can have is that
he is not loved'** (John Steinbeck, *East of Eden*, 1952, p.268 of 640), **'They fuck you up,
your mum and dad. They may not mean to, but they do. They fill you with the
faults they had, and add some extra, just for you'** (Phillip Larkin, *This Be The Verse*,
1971), **'There is no great event I can pinpoint; just years of feeling unknown...by
my mother. She doesn't have the language of emotional connection'** ('The Ties that
Unwind', *The Weekend Australian Magazine*, 1 Mar. 2014) ; and from the parent's point
of view: **'Parents who fail to produce a well-adjusted child carry a terrible
burden of guilt'** ('The parent trap', *The Australian*, 12 Jan. 1999), **'Ultimately, you are
only as happy as your most unhappy child'** (Jamie Oliver, 'Please, Sir, I want some more',
*Good Weekend*, *The Sydney Morning Herald*, 20 Feb. 2010); and, finally, with regard to
the overall effect: **'The greatest chasm between two people is love withheld
by a parent'** (Nikki Gemmell, 'Body blows of love', *The Weekend Australian Magazine*,
12 May. 2012), and this revealing joke: **'be sure to have at least four kids. Why?
So you'll have at least two you can talk about'** ('A parent's nightmare', *The Weekend
Australian Magazine*, 6 Sep. 2014).

[492] These quotes are remarkable in their honesty; in fact, in my 40 years
of constant thinking and writing about the human condition, I have only
been able to assemble a small collection of such rare occasions when
the human-condition-avoiding, denial-practicing, dishonest world we
live in momentarily dropped its guard and let some truth through. The
following two passages are cases in point—they are by far the most honest
admissions I have come across of both the importance of nurturing in
human life and the now dire inability of mothers to adequately nurture
their children due to the corrupting effects of our species' heroic search
for self-understanding. Firstly, consider this quote from the author Olive

Schreiner that also featured in chapter 5:7: 'They say women have one great and noble work left them, and they do it ill...*We* bear the world, and *we* make it. The souls of little children are marvellously delicate and tender things, and keep for ever the shadow that first falls on them, and that is the mother's or at best a woman's. There was never a great man who had not a great mother—it is hardly an exaggeration. The first six years of our life make us; all that is added later is veneer...The mightiest and noblest of human work is given to us, and we do it ill.'

[493]Then there is this powerful extract from the anthropologist Ashley Montagu's extraordinarily honest 1970 paper, 'A Scientist Looks at Love': 'love is, without question, the most important experience in the life of a human being...One of the most frequently used words in our vocabulary...[yet] *love* is something about which most of us are still extremely vague...There is a widespread belief that a newborn baby is a selfish, disorganized wild creature who would grow into a violently intractable savage if it were not properly disciplined. [However,] The newborn baby is organized in an extraordinarily sensitive manner...He does not want discipline...he wants love. He behaves as if he expected to be loved, and when his expectation is thwarted, he reacts in a grievously disappointed manner. There is now good evidence which leads us to believe that not only does a baby want to be loved, but also that it wants to love; all its drives are orientated in the direction of receiving and giving love. If it doesn't receive love it is unable to give it—as a child or as an adult. From the moment of birth the baby needs the reciprocal exchange of love with its mother...It has, I believe, been universally acknowledged that the mother-infant relationship perhaps more than any other defines the very essence of love... survival alone is not enough—human beings need and should receive much more...We now know that babies which are physically well nurtured may nevertheless waste away and die unless they are also loved. We also know that the only remedy for those babies on the verge of dying is love...The infant can suffer no greater loss than deprivation of the mother's love. There is an old Eastern proverb which explains that since God could not be everywhere he created mothers...maternal rejection may be seen as the "causative factor in... every individual case of neurosis or behavior problem in children.".. .Endowed at birth with the need to develop as a loving, harmonic human being, the child learns to love by being loved...To love one's neighbor as oneself requires first that one must be able to love oneself, and the only way to learn that art is by having been adequately loved during the first six years of one's life. As Freud pointed out, this is the period during which the foundations of the personality

are either well and truly laid—or not. If one doesn't love oneself one cannot love others. To make loving order in the world we must first have had loving order made in ourselves…Nothing in the world can be more important or as significant…love is demonstrable, it is sacrificial, it is self-abnegative [self-denying]. It puts the other always first. It is not a cold or calculated altruism, but a deep complete involvement with another. Love is unconditional…Love is the principal developer of one's capacity for being human, the chief stimulus for the development of social competence, and the only thing on earth that can produce that sense of belongingness and relatedness to the world of humanity which is the best achievement of the healthy human being…Scientists are discovering…that to live as if to live and love were one is the only way of life for human beings, because, indeed, this is the way of life which the innate nature of man demands. We are discovering that the highest ideals of man spring from man's own nature… and that the highest of these innately based ideals is the one that must enliven and inform all his other ideals, namely, *love*…Contemporary scientists working in this field are giving a scientific foundation or validation to the Sermon on the Mount and to the Golden Rule: to do unto others as you would have them do unto you, to love your neighbor as yourself…In an age in which a great deal of unloving love masquerades as the genuine article, in which there is a massive lack of love behind the show of love, in which millions have literally been unloved to death, it is very necessary to understand what love really means. We have left the study of love to the last, but now that we can begin to understand its importance for humanity, we can see that this is the area in which the men of religion, the educators, the physicians, and the scientists can join hands in the common endeavor of putting man back upon the road of his evolutionary destiny from which he has gone so far astray—the road which leads to health and happiness for all humanity, peace and goodwill unto all the earth' (*The Phi Delta Kappan*, Vol.51, No.9).

[494] But while these quotes *are* incredibly honest, with understanding of the psychologically upset state of the human condition finally found (the explanation of which was presented in chapter 3), we can at last explain what is, in fact, fundamentally wrong with what Schreiner and Montagu have written, which is that rather than loving our infants being **'the mightiest and noblest of human work'**, and of there being **'nothing in the world…more important'** than being loved, the incursion of the human condition saw a **'mightie[r]'** and **'more important'** task assigned to humans, which was to persevere with humanity's corrupting, love-destroying

search for knowledge until we found the understanding that would finally liberate the human race from that condition and allow the practice of nurturing to once more regain its place as the **'mightiest and noblest of human work'**.

[495] So there has been a *very* good reason for why humans **'have literally been unloved to death'**, but until we could compassionately explain that reason we had no choice but to leave **'the study of love to the last'**. It is *only* now that we can explain the human condition, explain that humanity has had to be preoccupied with its soul-corrupting, love-destroying, anger-egocentricity-and-alienation-producing heroic search for knowledge, that we can explain *why* we have been so alienated as parents that we have been unable to give our offspring anything like the alienation-free, sound, secure, unconditional love needed to create **'The real vision of the human being'** of the sound child, the **'child wonder'**. Yes, it is *only* now that we can afford to admit that the playwright Samuel Beckett was only slightly exaggerating the brevity today of a truly loved, soulful, happy, innocent, secure, sane, human-condition-free life when he wrote, **'They give birth astride of a grave, the light gleams an instant, then it's night once more'** (*Waiting for Godot*, 1955), or that the psychiatrist R.D. Laing was right when he wrote that **'To adapt to this world the child abdicates its ecstasy'** (*The Politics of Experience* and *The Bird of Paradise*, 1967, p.118 of 156).

[496] Such is the enormous paradox of the human condition: we humans appeared to be horribly bad but are, in fact, heroically wonderful—but until that reconciling biological understanding was found we had no choice but to be prepared to create and live in a world that was devoid of truth—and love! As the song *The Impossible Dream* described this predicament, we had to be prepared **'to march into hell for a heavenly cause'**. So what is really needed to balance Schreiner's and Montagu's honest but unfairly condemning revelations about how inadequate parents have been in their ability to provide their children with the unconditional love their instincts expect, and, as a result, how hurt and alienated children have been, is a presentation emphasizing just how incredibly, amazingly, extraordinarily heroic *all* parents have been to have even had children while they were living under the horrific duress of the human condition. Yes, a balancing presentation was needed, which this book now supplies, about how the human race has had to live with 2 million years of unjust condemnation—about how every day humans

have had to get out of bed and face a world that, in effect, hated them, that considered them to be horrible mistakes, blights on this planet, defiling, bad, awful, even evil, sinful creatures, when, as explained in chapter 3, humans are nothing less than the heroes of the whole story of life on Earth!!

[497] It follows then that there has been a justifiable reason for each of the '*whys*' listed earlier—*except* for the last one of why I and those advocating my work have been so thoroughly persecuted for present-ing the nurturing explanation for our moral instincts. As emphasized, it was *only* when we could explain the human condition and thus finally understand our inability to nurture our offspring that it would be safe to admit the critically important understanding of our species' nurtured origins and, as Montagu said, put **'man back upon the road of his evolu-tionary destiny from which he has gone so far astray'** and restore **'health and happiness for all humanity'**, and since it is precisely that explanation of the human condition that I have presented, inclusive of that nurturing explanation, there is *NO* justification for the rejection, ostracism and persecution I have been subjected to. Quite the reverse, in fact—such a response represents the very height of irresponsibility and an abuse of science's mandate to support endeavors that seek to understand and ameliorate the plight of man. The seriousness of this ill-treatment will be revisited shortly (in ch. 6:12).

[498] From the perspective, however, of mechanistic science, this need to deny the importance of nurturing in our human origins until we could explain the human condition has meant that biologists had to find some way of supporting this denial. And, as we are now going to see, this need to deny the truth that our distant ancestors lived a nurtured-with-love, all-loving life, which led to the corruption of Darwin's idea of natural selection into a 'survival of the fittest' process, through to the development of E.O. Wilson's dishonest Multilevel Selection theory for eusociality to deny that we lived an all-loving existence, also led to the dishonest Social Intelligence Hypothesis to deny the nurturing-with-love *origin* of that all-loving life. (Obviously, the entire need for denial should have been eradicated 30 years ago when I first presented the nurturing explan-ation in accompaniment with the explanation of the human condition, but, again, that is an issue I will return to in ch. 6:12.)

## Chapter 6:5 <u>To deny the importance of nurturing, the Social Intelligence Hypothesis was invented</u>

[499] Since humans are primates, the obvious area of research that has the most potential to shed light on our origins is the field of primatology, but it is in this most enlightening of fields that some of the most dishonest thinking about the origins of our species' moral sense has been taking place. Despite John Fiske having presented the nurturing explanation for our moral nature way back in 1874, the great majority of primatologists have been so fearful of the truth of nurturing that they have persevered along the habituated path of denial, serving up completely dishonest interpretations of primate behavior. This denial is particularly palpable if we compare their dishonest studies with the work of the rare few honest primatologists who have dared to recognize the role that nurturing plays in primate society.

[500] For instance, the obviousness—if you're not practicing denial—of the nurturing, love-indoctrination process, and how extremely confronting a truth it is, is apparent in Dian Fossey's study of gorillas. As described in pars 446-448, Fossey was an extraordinarily strong-willed woman for whom the universal practice of denial in mechanistic science held no sway. Few, if any, however, have been able to cope with the honesty of her studies, and, as a result, she has been misrepresented as merely a fanatical gorilla conservationist—such as in the 1988 film of her life, *Gorillas in the Mist*. A read, however, of her wonderfully insightful treatise on gorilla behavior—the 1983 book *Gorillas in the Mist* upon which the film was unfaithfully based—shows just how courageous a scientist Fossey was. She watched the lives of troops of gorillas over many generations and gave a denial-free, honest account of what she saw, which was the whole love-indoctrination process at work. Fearlessly, she wrote that **'Like human mothers, gorilla mothers show a great variation in the treatment of their offspring...Flossie was very casual in the handling, grooming, and support of both of her infants, whereas Old Goat was an exemplary parent'**. Old Goat's offspring, the **'exemplary parent[ed]'** 'Tiger', **'was taking his place in Group 4's growing cohesiveness. By the age of five, Tiger was surrounded by playmates his own age, a loving mother, and a protective group leader. He was a contented and well-adjusted individual whose zest for living was almost contagious for the other animals of his group...[**However,**] The immigrant...menace...Beetsme...developed an unruly desire to dominate...I found myself strongly disliking Beetsme as I**

**watched his discord destroy...**[the group's] **cohesiveness'**. On reading this, one can appreciate why the whole nurturing, love-indoctrination process has been so determinedly denied—Old Goat was an **'exemplary parent'** who created a **'well-adjusted'** offspring with a wonderful **'zest for living'**, while the **'menac**[ing]**'**, **'unruly'**, **'discord'**-creating, **'cohesiveness'-'destroy**[ing]**'**, non-**'loving'**, and by inference unloved, Beetsme was **'dislik**[able]**'**; the implication for humans being that if you don't give your child love you're a bad person, and, as has been emphasized, humans **'would rather admit to being an axe murderer than being a bad father or mother'**.

[501] In his 1989 book, *Peacemaking Among Primates*, the primatologist Frans de Waal describes a meeting that was held between an unnamed psychiatrist and the aforementioned Harry F. Harlow, a psychologist who, in the 1950s, studied the extremely damaging effect isolation and touch deprivation had on rhesus monkey infants. Their discussion reveals just how unbearable and confronting both the concept and the importance of nurturing love has been for those studying primates, and just how fearless Fossey was, by comparison, in her honesty: **'For some scientists it was hard to accept that monkeys may have feelings. In** [the 1979 book] ***The Human Model...*** [authors Harry F.] **Harlow and** [Clara E.] **Mears describe the following strained meeting: "Harlow used the term 'love', at which the psychiatrist present countered with the word 'proximity'. Harlow then shifted to the word 'affection', with the psychiatrist again countering with 'proximity'. Harlow started to simmer, but relented when he realized that the closest the psychiatrist had probably ever come to love was proximity"'** (pp.13-14 of 294). Yes, despite our species' instinctive need to give and receive love, humans' present human-condition-afflicted, unloved and unloving lives—where the **'closest'** the immensely upset human race has **'probably ever come to love'** is **'proximity'**—has meant that our ability to even acknowledge the existence of unconditional selfless-ness/love, which is what nurturing essentially is, and the ramifications of not receiving it, has been nigh impossible. Since this *is* the end play time predicted in the Bible when **'the love of most will grow cold'** (Matt. 24:12), it is no wonder mechanistic science practices the *extreme* dishonesty that Allot described in his aforementioned paper, when he reported that **'Love has been described as a taboo subject, not serious, not appropriate for scientific study'**—this despite, as Montagu acknowledged, **'love'** being **'One of the most frequently used words in our vocabulary'**!

[502] It follows then that the bonobos—whose extraordinarily maternal, nurturing, loving treatment of their infants and the resulting remarkable

integrative, loving behavior they exhibit as adults was described at length in chapter 5—have been *extremely* exposing, confronting and condemning of the unloved and unloving human race. As I mentioned in par. 416, the biologist, psychologist and bonobo authority Sue Savage-Rumbaugh has bravely admitted that nurturing is the focus of bonobo society; like Fossey, she has let the truth out of the bag that the cooperative behavior of bonobos *is* a product of the infant-focused, nurturing of love, love-indoctrination process—writing, with the assistance of her co-author, the writer Roger Lewin, that **'Bonobo life is centered around the offspring. Unlike what happens among common chimps, all members of the bonobo social group help with infant care and share food with infants. If you are a bonobo infant, you can do no wrong…Bonobo females and their infants form the core of the group.'** But *also* like Fossey, Savage-Rumbaugh's honesty appears to have made her the target of the human-condition-avoiding, nurturing-denying, mechanistic scientific establishment, because in 2012 a campaign was launched to discredit and marginalize her ground-breaking work with bonobos. I have been told by a scientist whom I respect and who knows Savage-Rumbaugh personally that she has become somewhat erratic, but if that is the case I strongly suspect the genesis of such instability would be years of unfair and undermining criticism from the mechanistic establishment.

[503] So the question now is, how was denial of the obvious role nurturing plays in the lives of more developed/integrated/social mammals achieved, especially of primates, and most especially bonobos—which the studies of Fossey and Savage-Rumbaugh bear witness to, and which provide such powerful evidence for how we acquired our cooperative, unconditionally selfless, moral instincts? Clearly, such a denial wasn't going to be easy, but looking an obvious truth in the face and finding a way to deny it—such as finding a way to deny the extremely obvious truths of Integrative Meaning and of our corrupted human condition—is something we humans are masters at!

[504] So yes, how did mechanistic science manage to look the obvious truth of the importance of nurturing in the face and deny it? There have been two ways. The first was to portray maternalism as nothing more than a mother providing her dependent offspring with food and protection. As was explained in chapter 5:4, the truth is that mothers' maternal instincts to nourish and protect their offspring *did* provide the base from which the love-indoctrination process was able to develop, however, in love-indoctrination, maternalism became about much more than a

mother looking after her infant—it became a case of actively *loving* that infant. Again, it is not insignificant that we speak of 'motherly love', not 'motherly protection'. The problem, however, with this method of denying the nurturing, loving significance of maternalism is that in the case of the extremely exposing-of-the-truth, infant-focused, maternal bonobo society, their environment has historically been food-rich and competitor-and-predator-free, so it doesn't make sense to argue that their exceptionally maternal behavior has been driven by the need to either source food or provide protection.

[505]The second method used to deny the significance of nurturing in bonobo life, and, by inference, its significance in the lives of our ape ancestors, is the one Richard Leakey referred to in his and Roger Lewin's 1977 book, *Origins*—that the extended infancies in primates is due to the infant's need for **'prolonged learning'** about their **'environment'**. A more complete rendition of this alleged explanation for the need for the nurturing that we see in the society of more developed mammals, particularly in bonobos and humans, is that **'The more sophisticated species also exhibit longer infant and juvenile stages, which are probably related to the time required for their more advanced mental development and their integration into complex social systems'** (*Encyclopedia Britannica*; see <www.wtmsources.com/136>). As will be explained at some length in chapter 7:3 (in pars 660-669), this so-called 'Social Intelligence Hypothesis' (sometimes referred to as the 'Machiavellian Intelligence Hypothesis'), and its more sophisticated version, the 'Ecological Dominance-Social Competition (EDSC) Model', essentially maintain that the long mother-infant association is needed to ensure the infant learns the skills necessary to manage **'complex social'** situations, and that it was this need that also led to **'more advanced mental development'**, ultimately the fully conscious, intelligent mind in humans.

[506]As will be described in chapter 7:3 when the truthful explanation of the origins of consciousness is presented, even human-condition-avoiding mechanistic scientists have raised serious concerns over the viability of both the Social/Machiavellian Intelligence Hypothesis (S/MIH) and the EDSC Model, but from the human-condition-confronting, truthful view of biology some very obvious flaws with both models can be pointed out immediately—particularly in regard to their core argument that our fully conscious intelligent mind emerged as a result of having to learn to manage complex social situations, a process that supposedly required and explained the long infancy.

[507] Social problem-solving is, of course, an obvious benefit of being conscious, but *all* activities that animals have to manage would benefit enormously from conscious intelligence—from the ability to reason how cause and effect are related, to understand change, to make sense of experience, to be insightful—so it is completely illogical to suggest that it wasn't until the need arose to manage complex social situations that consciousness developed. No, any sensible analysis of how and when consciousness emerged must be based on the question, 'What has *prevented* its development in other animals?' A lack of social situations doesn't explain why the fully conscious mind hasn't appeared in non-human species because there was ample need for a conscious mind prior to the appearance of complex social situations.

[508] No, the only accountable explanation for the emergence of the fully conscious mind in humans and for what is blocking its emergence in other species is the nurturing, love-indoctrination explanation—which, to recap very briefly, states that the nurturing of selflessness *liberated* the fully conscious, intelligent mind from the block that exists in non-human species' minds against thinking selflessly and thus truthfully and thus effectively. (Again, this will all be fully explained in chapter 7:3.)

[509] Another obvious flaw with both the S/MIH and the EDSC Model is that if it wasn't for the psychologically upset state of the human condition there would be no need to learn, and become intelligent enough to master, the art of managing **'complex social systems'**. Through the process of love-indoctrination, we humans became *so* instinctively integrated that there was no disharmony/conflict/discord/**'complex**[ity]' to have to manage. Prior to the emergence of the human condition some 2 million years ago our species lived instinctively as one organism. What did the Greek poet Hesiod say? **'Like gods they lived, with calm untroubled mind, free from the toils and anguish of our kind...They with abundant goods 'midst quiet lands, all willing shared the gathering of their hands.'** As Plato wrote, this **'was a time...most blessed, celebrated by us in our state of innocence, before we had any experience of evils to come, when we were admitted to the sight of apparitions innocent and simple and calm and happy...and not yet enshrined in that living tomb which we carry about, now that we are imprisoned'**—the time when we lived a **'blessed and spontaneous life...**[where] **neither was there any violence, or devouring of one another, or war or quarrel among them...In those days...there were no forms of government or separate possession of women and children; for all men rose again from the earth, having no memory of the past**

[we lived in a pre-conscious state, obedient to our loving instincts]. **And…the earth gave them fruits in abundance, which grew on trees and shrubs unbidden, and were not planted by the hand of man. And they dwelt naked, and mostly in the open air, for the temperature of their seasons was mild; and they had no beds, but lay on soft couches of grass, which grew plentifully out of the earth'**. The pre-human-condition-afflicted, integrated, cooperative, 'shar[ing]', loving, social, **'calm and happy'**, **'blessed and spontaneous'** Specie Individual state, such as that which largely exists in bonobo society today, simply wasn't a situation that called for infants to be skilled in social management techniques. As Plato said, there was no **'quarrel[ling]'** and **'no** [need for] **forms of government'**.

[510] And the even more blatant flaw with the two arguments that have been used to dismiss the important role nurturing played in the lives of our forebears, and continues to play in the development/integration/ socialization of other mammals, especially other primates, and most especially bonobos, is that they suggest nurturing is nothing more than a mother providing her dependent infant with food and protection, and that the long mother-infant association is needed only to allow the time to impart social skills. But, as stated earlier, humans can't have developed such powerful instincts to nurture our offspring with love, or such powerful expectations of receiving unconditional love as children, if nurturing hadn't played a fundamental role in the history of our species' development—as I said, such instincts and expectations do not appear out of thin air. The response to this statement, however, from advocates of the S/MIH and the EDSC Model would be that mothers simply don't have powerful instincts to nurture their offspring with love, and children simply don't have expectations of being loved; rather mothers have powerful *instincts to teach* their offspring how to manage **'integration into complex social systems'**, and children have *instinctual expectations of being taught* such skills. In other words, it is not a case of instincts to love and be loved 'coming out of thin air', but a case of such instincts never existing in the first place. But that is absurd; indeed, it is offensively dishonest, because everyone does intuitively know that what Schreiner and Montagu wrote about infants' need for love is true. Yes, what an infant needs from its parents—and from its mother in particular—*is* unconditional love, *not* training in the management of complex social situations! Certainly, when the need for denial is critical any excuse will do, and the art of denial is to then stick to that excuse like glue, but that does *not* mean we are so

unaware we are practicing denial that we are unable to recognize and admit the truth when that denial is no longer needed—which, with the human condition now explained, it no longer is.

[511] We all do actually know that to achieve the **'loving order in the world'** that Montagu recognized, the **'cold'**, **'bleak and loveless'** (as Drummond described it), ruthlessly selfish, competitive, must-reproduce-your-genes 'animal condition' had to be overcome, and the only means by which that could be achieved was through the mother-infant, nurturing-of-love situation: the love-indoctrination process. As was emphasized in chapter 5:4, the problem is that love-indoctrination is an extremely difficult process to develop and maintain to the point where the fierce competition to reproduce your genes is contained and integration achieved. Only our ape ancestors managed to develop love-indoctrination to the point where competition amongst males especially was contained and unconditional love and integration developed, something the fossil record is now confirming (as described in chapter 5:5). Bonobos are well on their way, but all the other relatively developed/integrated/social mammals are still battling to develop love-indoctrination to the point where it has over-come selfish competition amongst males to reproduce their genes. But they *are* trying to do so; they *are* trying to indoctrinate their infants with love to the degree their circumstances allow. Reports from anyone who has worked with the relatively developed/integrated/social mammals and who is not under the control of the thought police—the truth-denying, mechanistic scientific constabulary—give accounts of the development of love through nurturing, such as this of the nurturing, loving behavior of elephants: **'After years of research and scientific observation it has been shown that elephant's social structure and familial bonds are similar, if not deeper, than the bonds developed among** [present immensely psychologically upset] **human beings. There are deeply stirring accounts, by such scientists as Joyce Poole, Cynthia Moss and Dr. Dame Daphne Sheldrick, of elephants weeping and expressing grief at the loss of their calves...and other herd members. There are recorded behaviors of near spiritual proportion...Calves frequently die of heartbreak from the loss of their mothers and abuse by human beings...There are also great displays of affection and mutual respect rarely viewed in the social structure of humans. Joyce Poole, internationally known expert on elephants, states, "I have never seen (wild) calves 'disciplined'. Protected, comforted, cooed over, reassured and rescued, yes, but punished, no. Elephants are raised in an incredibly positive and loving environment"'** ('The Heart of Africa', Sacred Wildlife.org;

see <www.wtmsources.com/102>). Note that Poole's comment that elephant calves are never **'disciplined'** echoes Savage-Rumbaugh's observation that bonobo infants **'can do no wrong'**. Revealingly, like elephant orphans, orphaned **'Gorillas and bonobos…just die. They see their mothers killed and they give up'** (Vanessa Woods, *Bonobo Handshake*, 2010, p.67 of 278) — because, as stated above, they suffer **'heartbreak'**; their emotional desire for, their instinctive expectation of receiving, and their attachment to, a loving true (Integrative-Meaning-compliant) world is so great they literally cannot survive without it. In par. 452, a quote from Savage-Rumbaugh was included that described the rapturous joy expressed by the bonobos Matata and her adopted son, Kanzi, at their *loving* reunion. And when the Friends of Bonobos charity designed their fundraising **'A Bonobo Mother's Love T-shirt'** to have a picture of a mother bonobo cradling her infant, they weren't 'anthropomorphizing' or inappropriately humanizing bonobos, as those mechanistic thought police would argue, they were unwittingly conveying the simple truth. Bonobo mothers aren't merely giving infants training in how to manage **'integration into complex social systems'**, that is an absurd suggestion — they are giving them **'love'**. But, again, when the need for denial is desperate, any excuse will do. The *truth* is that in the relatively developed/integrated/social species of mammals, nurturing *has* moved beyond the primitive, pre-love-indoctrination, 'must nourish and protect' maternal situation to the 'must love' maternal situation — they *are* attempting to develop love-indoctrination.

## Chapter 6:6 <u>Dismissing maternal love as training to manage complex social situations still left the extraordinarily cooperative lives of bonobos, and of our ape ancestors, to somehow be explained</u>

[512]While both the dishonest S/MIH and the EDSC Model have been relied upon to dismiss the mother-infant bond as nothing more than a mother nourishing and protecting her offspring, and training them in the art of managing complex social situations, a big problem remained: how to account for the remarkable cooperative behavior of bonobos, and the light they shed on our own unconditionally selfless moral instincts? So the question now is, what nurturing-of-love-denying 'explanation' did human-condition-avoiding mechanistic scientists come up with to 'solve' this problem?

[513] The answer is that mechanistic scientists initially tried to portray the competitive aggression and violence that can be found in all ape species (except bonobos) as evidence of what our ape ancestors were supposedly like. But when it was found that the peace-loving bonobos didn't fit this model, they attempted to ignore the anomaly they represented altogether — and even maintain that despite the evidence, bonobos were no different to the other, competitive, aggressive primate species. And *then*, when these scientists could no longer ignore the extraordinary integration that is so apparent in bonobo society, they conceded that bonobos *are* cooperative but found a way to explain how they became cooperative that did not invoke, or credit in any way, nurturing. (At this point, it should be stated that, just as our fear of the human condition and resulting denial of it has been so great that we, the upset human race, have hardly been aware that we are living in denial of it — recall in chapter 2:2 how Resignation needed to be described to re-connect us to the real horror of the human condition — our fear of the truth of the importance of nurturing in human development and resulting denial of it has also become *so* developed and entrenched in us that we are hardly aware that we are practicing it. It is almost instinctive in us now to avoid the significance of nurturing in human history, as though it's a rule we live by but with only a subliminal awareness that we are abiding by it.)

[514] In regard to the initial strategy stated above — of relating our aggressive behavior to that of apes — as was explained in chapter 2:9, ever since Darwin presented his idea of natural selection, humans have been misrepresenting it as a 'survival of the fittest' process, using that misinterpretation to support the reverse-of-the-truth lie that, just like other animals, we humans have competitive, selfish and aggressive instincts that our intellect has to heroically control. Chimpanzees appeared to support this lie — they were obviously human-like, and so were used as a model for our ancestors, with anthropologists such as Raymond Dart and Robert Ardrey pointing to chimpanzees' intense and violent male competition, rape, infanticide and inter-group warfare as indicative of the behavioral heritage of our ancestors. Dart argued for the **'predatory transition from ape to man'** ('The Predatory Transition from Ape to Man', *International Anthropological and Linguistic Review*, 1953, Vol.1, No.4), while Ardrey was even more emphatic, saying, **'Man had emerged from the anthropoid background for one reason only: because he was a killer'** (*African Genesis*, 1961, p.29 of 380). More recently, in 1999, a leading anthropologist and the author of *Demonic Males*, Richard

Wrangham, put forward the so-called 'Chimpanzee Violence Hypothesis', which claimed that **'selection has favored a hunt-and-kill propensity in chimpanzees and humans, and that coalitional killing has a long history in the evolution of both species'** ('Evolution of Coalitionary Killing', *Yearbook of Physical Anthropology*, 1999, Vol.42). Obviously these theories were immensely popular because the idea that our instincts are wildly aggressive made our intellect's supposed role as mediator seem all the more heroic. While it all amounted to a reverse-of-the-truth lie (because, as explained in chapter 2, our instincts are loving while our intellect is the offending, divisive influence), it was, nevertheless, a very human-condition-relieving thesis.

[515] (Incidentally, even though we do now have fossil evidence supporting the love-indoctrination explanation for our moral instincts, without the living evidence that the bonobos provide, it would be very difficult to prove the nurturing of love explanation for our unconditionally loving moral nature. Thank goodness for bonobos!)

[516] Yes, with their peaceful and gentle society, the bonobos exposed this initial strategy for the lie it was, but in doing so exposed themselves to the wrath of mechanistic science as an unbearably exposing and confronting reminder of our now immensely angry, egocentric and alienated, unloving and unloved lives. So, as stated, mechanistic science's strategy to deal with *this* problem was to simply ignore the anomaly that bonobos represented. Indeed, this strategy was so successful that the first in-depth study of bonobos, which only occurred in 1954, was **'ignored and forgotten by the scientific community'** (Frans de Waal & Frans Lanting, *Bonobo: The Forgotten Ape*, 1997, p.11 of 210) because it dared to describe them as **'an extraordinarily sensitive, gentle creature, far removed from the demoniacal primitive force of the adult chimpanzee'** (E. Tratz & H. Heck, 'Der africkanische Anthropoide "Bonobo": Eine neue Menschenaffengattung', *Säugetierkundliche Mitteilungen*, Vol.2). In fact, it was this ongoing denial that led de Waal and the photographer Frans Lanting to title their 1997 collaboration, *Bonobo: The Forgotten Ape*. This book, which acknowledged rather than ignored the extraordinary sensitivity of bonobos, was, not surprisingly, vilified: **'De Waal's bonobo research came under sustained attack'** ('The Future of Bonobos: An Animal Akin to Ourselves', Alicia Patterson Foundation, 2002; see <www.wtmsources.com/122>) from primatologists such as Craig Stanford who argued that bonobos are not, in fact, extraordinarily gentle and cooperative, but competitive and aggressive like chimpanzees. Stanford wrote that **'It is clear that much of the research on these two intensively studied apes** [in the case of bonobos, I would argue

'superficially' studied] **remains fraught with untested assumptions'** and that **'reported differences have been inflated'** ('The Social Behavior of Chimpanzees and Bonobos: Empirical Evidence and Shifting Assumptions', *Current Anthropology*, 1998, Vol.39, No.4). Responding to Stanford's criticisms, de Waal insightfully wrote that **'Two strategies have emerged to keep bonobos at a distance so as to preserve chimpanzee-based scenarios of human evolution, which traditionally emphasize warfare, hunting, tool use, and male dominance. The first strategy is to describe the bonobo as an interesting but specialized anomaly that can be safely ignored as a possible model of the last common ancestor (see Wrangham and Peterson 1996). The second strategy, adopted by Stanford, is to minimize the differences between the two *Pan* species: if bonobos behave, by and large, like chimpanzees, there is no reason to question the latter species' prominence as a model'** (ibid). (Another illustration of the use of this second strategy was provided in par. 208, when it was documented how E.O. Wilson attempted to suggest that bonobos and chimpanzees were indistinguishable because both **'do not share'** and both **'hunt in coordinated packs'**. However, as was demonstrated there, even the most basic research shows that Wilson had to fudge the evidence to support his lie.)

[517] The problem that emerged with these dismissive strategies was that modern technology has increasingly made the bonobos more accessible, and their extraordinarily integrative behavior, that is so different to chimpanzees, almost impossible to ignore. *Bonobos*, the French documentary that was referred to in par. 418, is a case in point. The fact is, with their extraordinarily loving behavior, bonobos have represented an ever-growing thorn in the side of a mechanistic scientific fraternity that desperately wanted to avoid their significance. And since bonobos couldn't be ignored or misrepresented forever, something had to be done to at least minimize their confronting presence. What happened was that while their **'extraordinarily sensitive, gentle'**, peaceful, cooperative behavior could not be credibly denied (and, in any case, there was, as will be explained shortly in ch. 6:9, a growing desire among ideal-behavior-emphasizing-but-human-condition-avoiding left-wing biologists to be able to emphasize cooperativeness and gentleness), it was hoped that at least a way could be found to explain why bonobos were cooperative in a manner that still avoided acknowledging the unbearable significance of their remarkable nurturing, maternal behavior. And the way that was found was through a theory known as the 'Social Ecological Model' that sought to explain social behavior in terms of ecological factors that influence social interactions.

[518] Before describing the Social Ecological Model it should be documented how mechanistic science has dismissed the fossilized evidence of our species' cooperative past by employing a strategy of evasion almost identical to that which has been applied to the bonobos.

## Chapter 6:7 <u>Fossil evidence of our species' cooperative past has also been dismissed, ignored or misrepresented by mechanistic science</u>

[519] Yes, just as the evidence provided by bonobos of our species' cooperative past has been denied by mechanistic science—first by claiming that our ancestors behaved competitively and aggressively 'like most ape species', and, when bonobos clearly didn't fit that model, by simply ignoring or misrepresenting their anomalous existence altogether—the fossilized evidence of our cooperative past has *also* been dismissed as irrelevant and, when that strategy became untenable, it too was simply ignored, misrepresented or accounted for in a dishonest, nurturing-avoiding way.

[520] For instance, it was stated in chapter 5:5 that the fossil record reveals that our ape ancestors had small canine teeth. Although it was explained there that the only accountable explanation for the reduction in canine size in our ape ancestors is that it was caused by female sexual selection against male mating aggression—that small canines are **'indicative of minimal social aggression'**—mechanistic scientists initially tried to avoid the implications these small canines raised by maintaining that they were only a recent development, and that the fossil record would inevitably produce evidence of an aggressive heritage. For example, in 1915 it was written, **'That we should discover such a race [a human race in which the canine teeth were pointed, projecting, and shaped as in anthropoid apes], sooner or later, has been an article of faith in the anthropologist's creed ever since Darwin's time'** (Sir Arthur Keith, *The Antiquity of Man*, p.459 of 519). However, as subsequent fossil discoveries pushed back, by millions of years, the age at which our ape ancestors still had small canines it became increasingly difficult to maintain this **'article of faith'** of an aggressive **'human race'**. As noted in chapter 5:5, there have also been attempts to account for our ancestors' canine reduction that did not cite a reduction in aggression, but they have also been rendered untenable by these ongoing discoveries, such as by what we now know of their diet.

[521] So, through the recent discoveries of *Ardipithecus*, *Orrorin*, and the 7-million-year-old *Sahelanthropus*, and the conclusive evidence they provide of our species' cooperative heritage, it appears that human-condition-avoiding, mechanistic scientists have been cornered into simply ignoring this evidence. For example, Richard Wrangham published the Chimpanzee Violence Hypothesis in 1999, and yet failed to mention the discovery, only 7 years earlier, of *Ardipithecus* — a find that had confirmed the existence of small canines at least 4.4 million years ago. Similarly, E.O. Wilson's attempt in his 2012 book, *The Social Conquest of Earth*, to portray war as a **'universal and eternal'** presence in human history (a claim that was repudiated in chapter 2:11) makes no mention of the fact that small canines characterize our ape ancestors, despite Wilson discussing both *Australopithecus* and *Ardipithecus* in his book. In 2009, the popular science magazine *Scientific American* (a subsidiary of the leading journal *Nature*) failed to make a single reference to a suite of papers — which had been some 15 years in the making — on *Ardipithecus ramidus* that had been published that year in a special edition of *Science*, **'the most extensive special issue of *Science* since Apollo 11'** (Tim D. White, Letters, *Scientific American*, 2010, Vol.302, No.1). As Tim White, team leader of the *Ardipithecus* researchers, asked with justifiable perplexity, **'How and why did the *Scientific American* editorial miss that story?'** (ibid). It is as if the fossil evidence of a cooperative past has become too strong to refute but the implications too daunting to acknowledge, rendering the majority of mechanistic scientists speechless. While a handful of scientists have broken the silence since the *Ardipithecus* discoveries were published, predictably most have done so to deny their cooperative implications on the only grounds left, which is by arguing that *Ardipithecus*, *Orrorin* and *Sahelanthropus* are not part of the human lineage, despite all the evidence indicating that they are.

[522] And even those who do recognize that these early hominids are part of the human lineage only do so by contriving implausible explanations to account for the characteristics that their fossils reveal, specifically those that were described in chapter 5:5 — small canines and bipedality. For example, C. Owen Lovejoy (the primary anthropologist on the *Ardipithecus* research team) and Tim White have put forward what is really an absurdly improbable explanation for the emergence of these traits in the form of the so-called 'male provisioning model'. This model suggests that our ancestors adopted a monogamous social structure, where females

selected for less aggressive males who would provide for them and their offspring in exchange for mating exclusivity, arguing that this selection for less aggressive males accounts for the reduced canines evident in our ancestors, and that their need to carry provisions led to the development of bipedalism. Even denial-compliant, nurturing-avoiding mechanistic scientists are able to point to the weaknesses of the model, with Wrangham, for example, writing that **'obstacles** [to the male provisioning model] **include skepticism that a male who left his mate to find food for her could guard her from rival males, the absence of any evidence for home bases, the matter of why females became bipedal, and evidence that australopith life histories resemble those of apes, not of humans as Lovejoy's scheme implied they should. In addition, no living nonhuman primates exhibit monogamy-within-social-communities'** (*Tree of Origin: What Primate Behavior Can Tell Us about Human Social Evolution*, ed. Frans de Waal, 2002, pp.134-135 of 320). I could add that in addition to the flaws Wrangham points out, the only monogamous primate in our species' direct lineage, the gibbon, has very large canines, while bonobos, the species of great ape that most closely resembles our ape ancestor, have the smallest canines and are the most bipedal, characteristics the male provisioning model is meant to explain—and yet they are polygamous.

[523] Yes, mechanistic science's strategy for dealing with the fossil record's evidence of a cooperative past has followed a pattern similar to that which was employed to deal with the bonobos—first denounce the evidence as irrelevant on the false basis that our past is aggressive, and then, when that position becomes untenable, simply ignore the evidence—or, failing that, come up with a fanciful, nurturing-avoiding explanation for it.

[524] A further point of significance, and one that is raised above, as well as in par. 411, is that the fossil record, particularly the 1992 discovery of *Ardipithecus*, clearly shows the physical and behavioral similarities between our pre-australopithecine ape ancestors and bonobos. But if bonobos themselves needed to be ignored, it follows that any evidence that links *our* ancestors to them will be treated the same way—as indeed it has. It should be no surprise that, as de Waal explains in his 2013 book, *The Bonobo and the Atheist*, **'a scientist on the Ardi** [*Ardipithecus*] **team...Owen Lovejoy, could think only of chimps as a comparison...**[and yet] **The bonobo's body proportions—its long legs and narrow shoulders—seem to perfectly fit the descriptions of Ardi, as do its relatively small canines. Why was the bonobo overlooked?'** (pp.60-61 of 289).

## Chapter 6:8 <u>The Social Ecological Model</u>

[525] To return now to mechanistic science's development of the theory known as the Social Ecological Model (SEM) as a means of avoiding the unbearable significance of bonobos' remarkable nurturing, maternal behavior.

[526] In essence, the SEM holds that the abundance of food available to the bonobos was what gave rise to their extraordinary harmony and cooperation. The model maintains that a plentiful supply of food meant that groups of bonobos no longer had to split up to feed, which gave females the opportunity to socialize more and, in time, form coalitions to protect themselves against aggressive male competition for mating opportunities, thus forcing males to adopt more cooperative strategies. To quote from a description of the SEM in a 2001 paper by the anthropologists Richard Wrangham and David Pilbeam: **'The absence of gorillas made high-quality foliage more available for proto-bonobos than for chimpanzees. As a result, proto-bonobos experienced a reduced intensity of scramble competition compared to chimpanzees. Reduced scramble competition allowed more stable parties, which then made several forms of aggression more dangerous and costly, and less beneficial, to the aggressors. This change in the economics of violence led through various social consequences to female-female alliances, concealed ovulation, and reduced individual vulnerability to gang attacks. All these favored a reduction in the propensity for male aggressiveness'** ('African apes as time machines', *All Apes Great and Small: Volume 1: African Apes*, eds Biruté Galdikas et al., 2002, p.12 of 316). So, under the dictates of the SEM, bonobos were able to become, to quote Wrangham again, **'a species biologically committed to the moral aspects of what, ironically, we like to call "humanity": respect for others, personal restraint, and turning aside from violence as a solution to conflicting** [social] **interests'** (*Demonic Males*, 1997, p.230 of 350). Wrangham is also reported to have referred to **'those loving bonobos'** ('Chimps and Chumps', *National Review*, 1999, Vol.51, No.18).

[527] In essence, this model claimed to explain bonobos' extraordinary **'respect for others'**, **'personal restraint'**, **'loving'**, **'moral'**, cooperative nature without a single mention of the importance of nurturing! Its appeal, however, can be judged by its adoption, in one form or another, by almost all the leading bonobo researchers (including Alison and Noel Badrian, Ben Blount, Christophe Boesch, Barbara Fruth, Brian Hare, Gottfried Hohmann, Takayoshi Kano, Suehisa Kuroda, Toshisada

Nishida, Amy Parish, David Pilbeam, Barbara Smuts, Randall Susman, Frans de Waal, Frances White and Richard Wrangham), and by the fact that it remains virtually uncontested since being put forward in the early 1980s. While not all these scientists necessarily agree with Wrangham that bonobos behave in a **'loving'** way—with some maintaining bonobos are capable only of reciprocal selflessness—all acknowledge that bonobos are extraordinarily cooperative and all use the SEM to explain that cooperation.

[528] But while the SEM serves the purpose of denying the significance of nurturing in bonobo society, the arguments it uses to do so—namely that stable parties allow females to form coalitions to counter male aggression, and that those coalitions are successful in dominating males—are, in fact, seriously flawed.

[529] Firstly, with regard to bonobo females being able to form coalitions to protect themselves against male aggression, the SEM argues that the opportunity to spend more time together (which, in the case of bonobos, they say was made possible by an abundance of food) allowed bonobos to form **'stable parties'** that in time allowed the formation of **'female-female alliances'** to protect themselves against **'male aggressiveness'** arising from competition for mating opportunities. However, while it is not uncommon amongst mammals for situations to occur where **'stable parties'** are able to and do form, it is rare for such **'stable parties'** to lead to **'female-female alliances'** that have the power to prevent **'male aggressiveness'**—which is surprising because such an alliance would certainly seem beneficial since male aggression does come at a cost to females, as these quotes indicate: **'Sexual harassment has significant negative consequences for females even when it doesn't end up in coerced sex; it can increase stress and interfere with normal activities, reducing health, fitness and longevity. Some forms of harassment may even culminate in injury or death'** (Linda Mealey, *Sex Differences: Developmental and Evolutionary Strategies*, 2000, p.130 of 480); and **'Females in many mammalian species experience both sexual aggression and infanticide by males'** (Barbara Smuts & Robert Smuts, 'Male Aggression and Sexual Coercion of Females in Nonhuman Primates and Other Mammals: Evidence and Theoretical Implications', *Advances in the Study of Behavior*, 1993, Vol.22). If **'stable parties'** were all that were required to form **'female-female alliances'** that could stop **'male aggressiveness'** then such alliances would be common amongst mammals, but they aren't. In short, the formation of such alliances *do* occur within primate societies (especially among bonobos), but if the SEM is correct it should *also* be

regularly occurring in the societies of non-primate mammals, but it isn't. To quote the anthropologists Barbara and Robert Smuts: **'the use of female coalitions to thwart aggressive males appears to be rare in other mammals compared with nonhuman primates'** (ibid).

[530] The only conclusion for why the strategy of forging **'female-female alliances'** to prevent **'male aggressiveness'** is rare outside primate species is that the strategy is not genetically successful—in other words, it is not in the interest of the female's chances of reproducing her genes, otherwise natural selection would have 'discovered' it and the practice would be common amongst all mammals. As such, there has to be some other factor aside from **'stable parties'** enabling females to form these types of coalitions, and the regularity with which these alliances appear within primates compared with non-primate mammals suggests it is something unique to primates. But while the SEM cannot account for this peculiarity, the love-indoctrination explanation certainly does. As was explained in chapter 5:4, primates, whose arms are semi-freed from walking and are thus able to hold a helpless infant, are uniquely placed to develop love-indoctrination should conditions be favorable—and, as we've established, the bonobos have the most favorable environmental conditions amongst the primates and, as a result, *have* been able to develop the most love-indoctrinated integration. Yes, it would appear that it is only amongst primate species, where more nurturing is able to be selected for and both males and females are able to be indoctrinated with unconditional selfless love, that the **'economics of violence'** changes enough for it to be possible to form **'female-female alliances'** to help rein in **'male aggressiveness'**.

[531] Secondly, in attempting to explain how male aggression could be quelled by **'female-female alliances'**, the SEM relies on the superficially persuasive but biologically flawed argument that biologists have desperately been resorting to more and more (especially left-wing biologists, who are wanting to promote cooperation and gentleness), which is that cooperation is more advantageous than competition and can therefore be selected for. This reliance is evident in Wrangham's argument that **'female alliances'** made male **'aggression more dangerous and costly, and less beneficial, to the aggressors'** and that **'turning aside from violence'** was **'a solution to conflicting** [social] **interests'**. While it may appear persuasive to suggest that males could stop competing for mating opportunities once female coalitions formed—because aggression became more **'dangerous**

**and costly, and less beneficial'** than cooperation, or simply because **'turning aside from violence'** is a **'solution to conflicting** [social] **interests'** — the biological reality is that, without love-indoctrination, genetic selfishness will see males *continue* to aggressively compete for any and all mating opportunities. As mentioned in par. 407, this fact of life was pointed out by C. Owen Lovejoy when, in discussing our ape ancestors, he wrote that **'Loss of the projecting canine raises other vexing questions because this tooth is so fundamental to reproductive success in higher primates. What could cause males to forfeit their ability to aggressively compete with other males?'** The same **'vexing'** question arises in regard to bonobos, something that was pointed out by the primatologist Gottfried Hohmann when he observed that **'The** [bonobo] **males, the physically superior animals, do not dominate the females, the inferior animals?...It is not only different from chimpanzees but it violates the rules of social ecology'** ('Swingers: Bonobos are celebrated as peace-loving, matriarchal, and sexually liberated. Are they?', *The New Yorker*, 30 Jul. 2007). Frans de Waal made a similar observation when he wrote about bonobos that **'dominance by the "weaker" sex constitutes a huge violation of every biologist's expectations'** (*Bonobo: The Forgotten Ape*, 1997, p.76 of 210).

[532] This biologically flawed argument that cooperation can be selected for by natural selection because it is more advantageous than competition is, in fact, the same biologically flawed argument that E.O. Wilson relied upon in his Multilevel Selection theory for eusociality. Recall in chapter 2:11 how Wilson argued that a group composed of selfless members who consider the welfare of the group above that of their own will be more cooperative and thus successful when competing against groups who have selfish, non-cooperative members, and, as such, competition between groups can lead to the selection of unconditionally selfless traits. Again, while it is superficially persuasive to suggest that a group with cooperative members will defeat a group with competitive members, the genetic reality is that whenever an unconditionally selfless, altruistic trait appears those that are selfish will take advantage of it, thus negating the establishment of a group of cooperators in the first place. Any selflessness that *might* arise through group selection will be constantly exploited by individual selfishness from *within* the group; as the biologist Jerry Coyne pointed out, **'group selection for altruism would be unlikely to override the tendency of each group to quickly lose its altruism through natural selection favoring cheaters** [selfish individuals]' ('Can Darwinism improve Binghamton?', *The New York Times*, 9 Sep. 2011).

[533] The flawed reasoning of this second aspect of the SEM can be seen even more clearly in this description of it by the anthropologist Brian Hare: **'there are some species that outcompeted others by becoming nicer...[because] it's very costly to be on top. Often in primate hierarchies, you don't stay on top very long. Everyone is gunning for you. You're getting in a lot of fights. If you don't have to do that, it's better for everybody'** ('Why Some Wild Animals Are Becoming Nicer', *Wired*, 7 Feb. 2012). While it is superficially persuasive to argue that **'it's better for everybody'** if **'you don't have to' 'fight'** to stay **'on top'**, the idea **'violates'** the fundamental fact about the gene-based natural selection process, which is that you do have to be constantly **'gunning'** to **'stay on top'** because if you are not others will be and you won't reproduce your genes. That's the reality of the 'animal condition': fierce competition exists between sexually reproducing individuals seeking to reproduce their genes.

[534] It should be mentioned that advocates of the SEM argue that the bonobo practice of **'concealed ovulation'** (where females do not give signs of their fertile period) and of making sex continually available have contributed to the reduction of male competition for mating opportunities, *however*, such practices can *only* develop once competition for mating opportunities has begun to subside because while ever competition to mate remains intense, if a female makes mating available to every male, and/or doesn't advertise she is ready for fertilization, she can't ensure she mates with the strongest, most virile male and, therefore, that her offspring will be successful competitors. Making sex continually available and concealing ovulation could *assist* love-indoctrination, but not initiate the development of love/unconditional selflessness. No, these practices do not *lead* to a reduction in competition, rather they are a *result* of male competition having largely *been* contained, which, as explained, 'selection against aggression' can't achieve.

[535] So although it is seductive to run the argument that cooperation can be selected for by natural selection because it is more advantageous than competition, it is so blatantly biologically incorrect that for biologists to employ it means they are deliberately lying—resorting to a desperate form of bluff—but such has been the extent of the need to avoid the nurturing explanation for bonobos' unconditionally selfless behavior, and, by extrapolation, for humans' **'moral'** nature. So it is not surprising that this now insatiable need to account for our moral instincts in a non-confronting way has led to the development of a somewhat more refined form of the SEM.

## Chapter 6:9  A brief history of left-wing dishonest mechanistic biology

[536]Before introducing this new, 'improved' version of the SEM, it should be mentioned that this seductive but patently untrue biological argument that cooperation is more advantageous than competition and can therefore be selected for has a long history of use by cooperation-not-competition-supporting, selflessness-not-selfishness-emphasizing, knowledge-opposing left-wing biologists (the pseudo idealistic philosophy of the left-wing was briefly explained in ch. 3:9 and will be more fully dealt with in chs 8:15 and 8:16). For example, in 1880, the zoologist Karl Kessler said that **'the progressive development of the animal kingdom... is favoured much more by mutual support than by mutual struggle'** (Address titled *On the law of mutual aid* to the St Petersburg Society of Naturalists, Jan. 1880). This so-called 'naive' misrepresentation of natural selection as being socialistic rather than individualistic was still occurring even up to the 1960s, with, for instance, the behaviorist Konrad Lorenz writing frequently of *behavior* having **'a species-preserving function'** (there are many mentions of this phrase in his 1963 book, *On Aggression*). As was explained in chapter 4, the development of species *is* part of the integrative process, but the *behavior* of a species is characterized by extremely selfish competition between its sexually reproducing members. In fact, it was the biologist George Williams' exasperation with this misrepresentation of natural selection as not being a selfish process that motivated him to write his famous 1966 book, *Adaptation and Natural Selection*—a publication that laid the foundations for the selfishness-justifying, right-wing theory of Sociobiology/ Evolutionary Psychology.

[537]When, however, Williams' theory was dishonestly used to misrepresent our cooperative, moral instincts as being nothing more than a product of kin-selection-based selfish reciprocity (assisting only those who share your genes in order to propagate your own), left-wing biologists *then* tried to maintain that we do have unconditionally selfless moral instincts by arguing that they are derived from by-products of natural selection—what the biologists Stephen Jay Gould and Richard Lewontin described in 1979 as **'a lot of [build**ing] **cranes'** acting in conjunction with **'natural selection'** ('Darwin Fundamentalism', *The New York Review of Books*, 12 Jun. 1997, in which Gould elaborated upon his and Lewontin's by-products or 'Spandrels' theory). But when Gould and Lewontin were unable to specify what the particular 'by-products'/

'spandrels'/'cranes' were that achieved this feat of creating our genuinely moral instincts (the by-product was nurturing but they couldn't confront and admit that truth), the left-wing was left with nowhere to go but to loop back to the now highly discredited 'cooperation is more advantageous than competition and can therefore be selected for', group-selection-type argument. In 1994, despite the situation where **'group selection has been regarded as an anathema by nearly all evolutionary biologists'** (Richard Lewontin, 'Survival of the Nicest?', *The New York Review of Books*, 22 Oct. 1998), the biologist David Sloan (D.S.) Wilson desperately tried to **'re-introduce group selection…as an antidote to the rampant individualism we see in the human behavioral sciences'** (David Sloan Wilson & Elliot Sober, 'Re-Introducing Group Selection to the Human Behavioral Sciences', *Behavioral and Brain Sciences*, 1994, Vol.17, No.4).

[538] In fact, it was D.S. Wilson's theory of Multilevel Selection that argued that natural selection operated at the group level as well as the individual level that (as described in chapter 2:11) E.O. Wilson commandeered to re-assert the right-wing emphasis on selfishness by claiming that even though multilevel selection supposedly confirmed we have unconditionally selfless instincts derived, he said, from warring between groups, it still allowed for the existence within us of selfish instincts derived from individual-level selection. But since the idea that our moral instincts were derived from warring between groups was unpalatable to the cooperation-not-competition-supporting, selflessness-not-selfishness-emphasizing left-wing, they came up with an alternative, non-warring group-level explanation, which was that our moral instincts arose from having to cooperate to defend ourselves against predators. This argument was first put forward by the anthropologist Robert Sussman in his 2005 book, *Man the Hunted: Primates, Predators, and Human Evolution*. In summarizing his theory, Sussman said, **'Our intelligence, cooperation and many other features we have as modern humans developed from our attempts to out-smart the predator'** (presentation at the American Association for the Advancement of Science's Annual Meeting, 19 Feb. 2006). But while the threat of predators would have encouraged cooperation, the biological reality is that it was not going to *make* us social; it was not going to overcome the fundamental problem of genetic selfishness, which is that wherever selflessness develops it is going to be subverted by selfish opportunists. The fact is, species have been living with the threat of predators since life first emerged and that threat has never been able to bring about full integration. What we see instead is the eventual development of dominance hierarchy as a means

to try to contain the rampant selfish competition and opportunism. The whole reason E.O. Wilson put forward the argument that warring *between* groups was the full integration/eusociality threshold breaker was because there had to be an *extreme* need for cooperation if selfish opportunism was going to be defeated; there had to be a situation of conflict where groups of cooperators would defeat groups of non-cooperators. Arguing that groups of cooperators survived the threat of predators better than groups of non-cooperators doesn't create anything like the same selection pressure as actual conflict between groups.

[539] Indeed, this **'defence against predators'** argument overlooks the basic reason Wilson put forward the supposedly plausible, but actually still flawed, between-group warfare argument, and, apparently sensing this deficiency, what left-wing biologists *then* did was try to bolster it by adding the old 'by-products/many cranes/matrix of influences' illusion that Gould and Lewontin had first used. In a desperate attempt to somehow find a cooperation-emphasizing biological account of human origins, the left-wing were now being forced to throw everything into the pot—group selection *and* a multitude of vague 'influences'! This is evident in the 2011 book *Origins of Altruism and Cooperation* where, for example, Robert Sussman and Robert Cloninger wrote in the Foreword that **'Research in a great diversity of scientific disciplines is revealing that there are many biological and behavioral mechanisms that humans and nonhuman primates use to reinforce pro-social or cooperative behavior. For example, there are specific neurobiological and hormonal mechanisms that support social behavior. There are also psychological, psychiatric, and cultural mechanisms'** (viii of 439). Yes, it was being alleged that a matrix of **'many biological and behavioral mechanisms'** created **'pro-social or cooperative behavior'**, but the question remains, *how exactly did it do it?* The *illusion* is that the origin of our moral instincts has been explained when it hasn't—but, again, in the desperation to counter the right-wing's selfishness-emphasizing doctrine such extreme illusion was deemed necessary!

[540] (As was mentioned in chapter 2 when the human-condition-avoiding, fundamentally dishonest right-wing biological theories for human origins were first presented, a much more complete description of all these right-wing and left-wing dishonest biological theories about human origins can be found in *Freedom Expanded* at <www.humancondition.com/freedom-expanded-the-denials-in-biology>.)

[541] Clearly, avoiding the nurturing explanation for our moral instincts meant that neither the right-wing nor the left-wing was ever going to provide an accountable explanation of our moral instincts. While the selfishness-emphasizing-but-still-human-condition-avoiding right-wing at least had the advantage of being able to truthfully emphasize that natural selection is a selfish process to support their position, the selflessness-emphasizing-but-human-condition-avoiding left-wing had to rely on the patently dishonest 'cooperation beats competition' group-selection-type argument. And since both camps were, in any case, avoiding the entire issue of the human condition itself and, therefore, denying Integrative Meaning, it wasn't possible for either to explain what was explained in chapter 4, which is that even though the gene-based natural selection process is dedicated to developing the order of matter it couldn't, outside of the love-indoctrination scenario, select for the self-eliminating, unconditionally selfless traits that would allow full integration. And unable to access this human-condition-confronting reconciling explanation of the paradoxical nature of the gene-based natural selection process, these two positions became more and more desperate, with the right-wing position, which stressed the fact that genes are selfish, producing the selfishness-justifying theories of Social Darwinism $\rightarrow$ Sociobiology/Evolutionary Psychology $\rightarrow$ and Multilevel Selection that were described in chapter 2; and the idealism-stressing left-wing position, which attempted to emphasize the greater truth that natural selection is an integrative process, embarking on the journey just outlined, where group selection was put forward as the explanation for the origin of our moral nature, then, when group selection was dismissed as unsound biology, a multitude of influences was proposed, and then, when that failed, group selection was once again brought back into play, and then, when *that* inevitably fell short, a completely desperate combination of both group selection and a matrix of influences was resorted to! Of course, without the reconciling explanation as to why the right-wing and the left-wing emerged in the first place, both strategies were bound to become sillier and sillier, and, in the end, completely mad—and, as is being described, that is what happened: the right-wing ended up developing the extremely mad and dangerous theory of Multilevel Selection, while the left-wing ended up developing the extremely mad and dangerous theory of the SEM.

[542] (I have written more about just how farcical science has become in *Freedom Expanded* at <www.humancondition.com/freedom-expanded-science-a-farce>. Based around the shocking revelations in the 2010 documentary, *Secrets of the Tribe* (directed by José Padilha), I describe the equally dishonest and thus never can be reconciled and thus now totally polarized right-wing and left-wing anthropological theories that are reviewed in the documentary and which seek to account for the behavior of the relatively innocent Yanomamö Indians of the Amazon. This ridiculous situation that so characterizes science now, where no one cares about the pursuit of truth-based understanding anymore, only about imposing their own twisted philosophy on the world, was perfectly captured at the end of the documentary when it played George and Ira Gershwin's well-known song *Let's Call the Whole Thing Off*, which features the lyrics, '**Things have come to a pretty pass…It looks as if we two will never be one…You like potato and I like potahto…Let's call the whole thing off**' (1937). Yes, science is so dishonest and thus inept now it may as well '**call the whole thing off**'; abandon its task of finding understanding of human behavior! I might also mention that right-wing anthropologists' portrayal of so-called 'primitive' 'races' as not being relatively innocent, but fierce and aggressive—as described in the program in relation to the Yanomami—will be exposed for the lie that it is later in pars 862-868.)

[543] Significantly, in relation to the two arguments employed by the SEM to explain bonobo behavior—that stable parties allowed bonobo females to form coalitions to counter male aggression for mating opportunities, and that those coalitions are successful in dominating males and eliminating aggression—in 2009 the leading architect of the SEM, Richard Wrangham, admitted that '**The circumstances in which females are able to form effective alliances among each other, and the frequency and effectiveness of this strategy, remain important** [unexplained] **problems for detailed examination in bonobos, chimpanzees, and other primates**' (*Sexual Coercion in Primates and Humans*, 2009, p.464 of 504). So the whole basis of the SEM, of the '**circumstances in which females are able to form effective alliances among each other**', and the '**effectiveness of this strategy**' in stopping male aggression, is being undermined by its leading architect and proponent! This is somewhat like E.O. Wilson conceding there were serious problems with his theory of Sociobiology when he moved on to develop his Multilevel Selection theory for eusociality! But where else could nurturing-avoiding biologists go in their efforts to explain bonobos? Nowhere—so, despite

its **'important problems'**, support of the SEM continued. What will now be described is how nurturing-avoiding, mechanistic biologists tried to make the SEM more accountable of bonobos', and our ape ancestors', extraordinarily integrative, moral behavior.

## Chapter 6:10 <u>The Self-Domestication Hypothesis</u>

[544] In terms of providing a nurturing-avoiding, human-condition-escaping explanation for bonobo behavior, the problem with the SEM is that it only offers a supposed explanation for bonobos' *lack* of aggression, and so still falls well short of being able to provide a supposed explanation for bonobos' extraordinary **'personal restraint'**, **'respect for others'**, **'loving'**, **'moral'**, cooperative, harmonious, gentle state. <u>And there is a *big* difference between not being aggressive and being loving</u>. Given this shortfall, nurturing-avoiding, mechanistic biologists clearly needed to come up with a more sophisticated version of the SEM, one that could supposedly account for the bonobos' extraordinary cooperative, gentle, peaceful, loving nature. This supposed solution was provided in 2012, with the presentation of the so-called <u>Self-Domestication Hypothesis</u> (SDH) by anthropologists Brian Hare, Victoria Wobber and Richard Wrangham (one of the originators of the SEM) in a paper titled 'The self-domestication hypothesis: evolution of bonobo psychology is due to selection against aggression' (*Animal Behaviour*, 2012, Vol.83, No.3).

[545] The first point to note is the use up front in the title of this paper of the idea that the **'evolution of bonobo psychology is due to selection against aggression'**, as if being able to **'select...against aggression'** is a normal, acceptable biological principle, a fait accompli, when it isn't. As emphasized, genes are selfish; outside of the love-indoctrination situation they don't allow for **'selection against aggression'** between sexually reproducing individuals. What is being put forward is the superficially persuasive but biologically flawed 'cooperation is more advantageous than competition and can therefore be selected for' argument that the SEM relies on, but to be putting it up front in their title is an outrageous bluff, a desperate deception, an all-out effort to create the illusion that **'selection against aggression'** is sound, acceptable biology.

[546] Another point that should be made before examining the soundness or otherwise of the SDH is that by concluding their 2012 paper with the following statement, its proponents suggest that it not only explains

bonobo cooperation but potentially human morality as well: **'The self-domestication hypothesis is therefore a potentially powerful tool for understanding the processes by which selection shapes both psychological and other seemingly unrelated traits, including those in humans.'** (Incidentally, the **'psychological'** **'traits'** they refer to are *behaviors*, such as tolerance and playfulness, *not a psychosis*—so like the Multilevel Selection theory for eusociality, the SDH does not address the psychology of the human condition, rather it is just another desperate attempt to deny it.) And indeed, by 2014 the idea of using the SDH to explain human behavior, which the authors only suggest in this 2012 paper, had taken hold to the extent that in October of that year a symposium called 'Domestication and Human Evolution' was hosted by the prestigious Center for Academic Research & Training in Anthropogeny (CARTA). A subsequent report on the symposium in *Science* magazine was subtitled **'"Self-domestication" turned humans into the cooperative species we are today'** (Ann Gibbons, 'How we tamed ourselves - and became modern', 2014, Vol.346, No.6208).

[547] Which brings us to the accountability of the SDH. Supposedly inspired by research into domestic dogs and the experiments of the Russian scientist Dmitri Belyaev in domesticating silver foxes for the fur industry, the SDH proposes that **'selection against aggression'** inadvertently involves selection for youthfulness or juvenileness, so that adults in subsequent generations end up retaining **'pro-social'** juvenile behavioral traits such as **'increased tolerance'**, **'increased adult play'**, a **'decrease in predatory motivation'**, and **'decreased xenophobia** [fear of outsiders]'. (Hare, Wobber and Wrangham describe these traits as **'correlated by-products'** of the original SEM-derived **'selection against aggression'** process.) This 'juvenilization' is known as the **'domestication syndrome'** because it is thought to account for changes between wild animals and their domestic descendants, such as changes between wolves and dogs.

[548] Essentially, when the proponents of the SDH say that **'In addition to showing less severe forms of aggression compared to chimpanzees, bonobos show differences…that appear analogous to the domestication syndrome'** (ibid), what they are claiming is that the increase in **'pro-social'** behavior that characterizes the **'domestication syndrome'** bridges the gap between the mere **'lack of aggression'** that the SEM could only hope to account for, and bonobos' extraordinary **'personal restraint'**, **'respect for others'**, **'loving'**, **'moral'**, cooperative, harmonious, gentle behavior. However, juvenile

'**pro-social**' behavior *does not* replace or override the selfish genetic need to aggressively compete for the fundamental biological needs of food, shelter, territory and mates. In fact, species that have been domesticated, like dogs and foxes, still aggressively compete for food, territory and mating opportunities, something that is almost entirely absent in bonobo behavior. So the SDH's claim to bridge the gap and explain bonobos' selfless, loving behavior is simply another giant bluff. The truth is that without the involvement of love-indoctrination to first establish unconditionally selfless love in the system, retarding stages of maturation alone can't create a state of unconditionally selfless love.

[549] Certainly, humans have domesticated dogs and even silver foxes by selecting for tamer and more social juvenile characteristics, the effect of which has been to retard the development of these animals so that they retain the tamer, more tolerant and more '**pro-social**' behavior of juveniles into adulthood—with the juvenile physical characteristics of floppy ears, more neotenous faces, etc, also carrying through into adulthood. However, while retarding development does bring the tamer, more tolerant, more '**pro-social**' characteristics of the juvenile stages into adulthood, it doesn't free the genes from their need to be selfish, and so doesn't eliminate selfish competition and aggression—*only* love-indoctrination can do that. Juvenileness is a form of more tolerant socialness but it isn't a selfless state. In fact, as stated, dogs and foxes who have been 'puppyfied' still aggressively compete for the resources of food, territory and mating opportunities, behavior that lies in stark contrast to bonobos' selfless and loving behavior. Being prepared to mingle and socialize, even being more playful—like domesticated dogs—is an improvement on the SEM's reduced aggression theory for bonobo behavior, but the truth is it still falls well short of being able to account for bonobos' unconditionally loving behavior.

[550] A brief summary of the love-indoctrination process that was described in chapter 5 may help clarify this failure of both the SEM and the SDH.

[551] The love-indoctrination process states that by selecting for longer infancies (which primates, with their arms semi-freed from having to walk on all fours, were able to do because they could hold a helpless infant), and for more maternal mothers, all within an ideal nursery environment of ample food and few predators, an infant's brain is able to be inscribed or

indoctrinated with unconditionally selfless love, thus allowing it to grow up to behave selflessly. An accidental, but fortuitous, side-effect of this indoctrination or training of a mind in selfless, truthful, effective thinking, however, was the emergence of consciousness, for once liberated, the conscious mind could then support the development of selflessness by consciously favoring (especially in mate selection) more selfless individuals, thus greatly speeding up the development of selflessness. Since the training in selfless love tended to wear off with age, selection for selflessness became, to a degree, a selection for youthfulness, resulting in more youthful, neotenous characteristics in adults. Both the SEM and the SDH, in effect, describe this process *but without* the key element of the involvement of the nurturing of unconditional selflessness; they omit the whole process of love-indoctrination—a glaring omission that created two problems. First, it necessitated the development of the flawed, dishonest **'selection against aggression'** argument to attempt to explain how selfless cooperation could emerge without love-indoctrination. And, second, since this **'selection against aggression'** could not, in fact, create unconditionally selfless love, only a supposed reduction in aggression, it could *only* ever lead to less aggressive, tamer, more tolerant, more social characteristics in adults.

[552]The point is, unconditionally selfless love is *not* produced by the situation espoused by either the SEM or the SDH, whereas love-indoctrination *does* produce love, which *can* then be actively selected for. So neither the SEM nor the SDH explains the extraordinary **'personal restraint'**, **'respect for others'**, **'loving'**, **'moral'**, unconditionally selfless, cooperative, harmonious, gentle behavior we see in bonobos and in our own moral instincts. The selection against aggression that the SEM describes and the delaying of the onset of adult competitiveness and aggression that the SDH proposes do *not* produce an unconditionally loving individual; to produce that you have to be selecting for individuals that have been nurtured with love, but that is a process neither the SEM or the SDH recognizes.

[553]To suggest that selecting for juvenileness can lead to less aggressive juvenile **'psychology'** being carried through to adulthood to the point of eliminating selfishness was simply a deception—another bluff—by the proponents of the SDH. In interviews Hare has conducted about the SDH we can see him trying to bridge this gap between the extraordinarily

cooperative, gentle, peaceful, unconditionally selfless, **'loving'** behavior that bonobos display, and what the SDH is able to supposedly explain, namely **'pro-social'** traits such as tameness and **'increased tolerance'**, when he describes bonobos as **'peaceful'** ('Why Bonobos Don't Kill Each Other', *The New York Times,* 5 Jul. 2010) and **'kind'** creatures (Brian Hare speech at Poptech 2010; see <www.wtmsources.com/139>) who **'absolutely are upset if there is any hint of aggression in the group'** ('Bonobos–Making Love Not War', *Catalyst*, ABC-TV, 20 Sep. 2007), and who find **'joy in working with others'** ('Dogged', *Smithsonian*, Oct. 2007)—as if those traits and emotions are what his hypothesis is able to explain the origins of.

[554] This attempt by Hare to bridge the gap between the extraordinarily cooperative, selfless behavior that bonobos display, and what the SDH is allegedly able to explain, is very similar to Wrangham's earlier claim that the SEM is able to account for bonobos' extraordinary (and these are his words) **'personal restraint'**, **'respect for others'**, **'loving'**, **'moral'** behavior. Furthermore, it was also described earlier how, in their 2012 paper, Hare, Wobber and Wrangham said their **'self-domestication hypothesis is...a potentially powerful tool for understanding the...psychological...traits... in humans'**—**'a potentially powerful tool for understanding'** the origin of our unconditionally selfless moral nature no less! Hare has also proposed that **'bonobos display...what might be thought of as our better angels'** ('"Hippie chimp" genome may shed light on our dark side', *Science* on NBCNews.com, 13 Jun. 2012; see <www.wtmsources.com/141>), which again is our unconditionally selfless moral nature!

[555] The truth is, domestication or juvenilization cannot *create* this type of behavior without there having been love-indoctrination, it can only stymie the growth of adult types of behavior, and so Hare and Wrangham are having to exaggerate its effect to account for bonobos' love and gentleness. Yes, it is only as *part* of the nurturing, love-indoctrination process that juvenilization can produce real **'loving'**, **'moral'**, **'peaceful'**, **'kind'**, **'joy'** in cooperation, **'personal restraint'**, and **'respect for others'**, and an abhorrence **'of aggression'**.

[556] All my publications have included a description of the love-indoctrination, mate selection process—with an account of humans' domestication of dogs appearing in my 1988 book, *Free: The End Of The Human Condition* (see <www.humancondition.com/free-love-indoctrination>), and a description of humans' domestication of both dogs and foxes appearing in the 2009 edition of my book *The Great Exodus* (see <www.

humancondition.com/exodus-mate-selection>). (Wrangham was sent *Free* in 1988, and over 2005-2006 all three SDH authors were sent another of my publications, *The Human Condition Documentary Proposal*, which also contains a description of the love-indoctrination, mate selection process (see <www.humancondition.com/doco-maternalism>)). The reason I referred to how **'domesticated dogs are derived from their common ancestral wild type by neoteny—retarding development at some juvenile stage'** (*Free: The End Of The Human Condition*, p.142 of 228) was because the domestication of dogs and foxes does dramatically illustrate some of the aspects involved in the love-indoctrination, mate selection process, particularly how powerfully effective conscious selection can be in producing a change (it **'Explains [the] speed of human development'** (ibid. p.142)), and how the development of stages of maturation is retarded by selecting for youthfulness (it **'is a marvellous illustration of the development of neoteny'** (ibid. p.141)). However, I explained that **'self-selection'** (as I originally termed the process that bonobos and our ape ancestors employed to assist in the development of unconditionally selfless behavior) differs to the selection we employed to domesticate dogs and foxes in that without love-indoctrination to create the unconditionally selfless love that could then be selected for, **'self-domestication'** (or, again, as I termed the process in all my books, **'self-selection'**) can *only* achieve tamer, more tolerant and more social characteristics in adults, *not* unconditionally selfless love. As I emphasized in *Free*, **'On their own genes could not develop selflessness but once there was love-indoctrination [they could]'** (p.47).

[557] An illustration of the difference between the effects of love-indoctrination and the effects of domestication put forward by the SDH, which is merely selecting against aggression, can be seen in the work of the famous 'dog whisperer' Cesar Millan. Millan is forever informing dog-owners that the mistake they are making in trying to control their dogs is that they are attempting to love them into behaving less aggressively when what they have to do to achieve control and reduce aggression is impose dominance. Millan is, in effect, recognizing that domesticated dogs haven't overcome the 'animal condition' of selfishly having to ensure their genes reproduce, which is why they are still highly com-petitive for food, shelter, territory and a mate—a competitiveness that can only be partially overcome through the imposition of a dominance hierarchy, where each individual accepts its position in a hierarchy that

is determined according to the competitive strengths of the various individuals involved.

[558] I might mention that in 2013 Hare published a bestselling book titled *The Genius of Dogs*, in which he again exaggerated the effects of domestication, this time in regard to dogs, writing that in contrast to wolf packs where the leaders are the dominant breeding pair, **'the leader of a feral dog pack is the dog who has the most friends'** (B. Hare & V. Woods, 2013, p.174 of 367). This statement is supposedly supported by a report titled 'Effect of affiliative and agonistic relationships on leadership behavior in free-ranging dogs', which found that **'formal dominance in free-ranging dogs may be a more consistent predictor of leadership than agonistic dominance'** (Roberto Bonanni et al., *Animal Behaviour*, 2010, Vol.79, No.5). Although this finding distinguishes between **'formal dominance'** (in which the subordinate animal signals its submission during a greeting ceremony), and **'agonistic dominance'** (in which the subordinate animal signals its submission following a fight), it still makes very clear that feral dogs are highly competitive and operate within a **'dominance'** hierarchy.

[559] In summary, there is a quantum difference between the claimed reduction in aggression that both the SEM and the SDH can supposedly produce and the very real love we see in bonobos. Neither the SEM or the SDH *begin* to offer an accountable explanation of that species' extremely loving behavior, whereas the nurturing, love-indoctrination explanation *fully* accounts for it. Although genes are a tool for developing order, they are limited in the sense that they can't normally develop unconditional selflessness, which means that genetics is a selfish, cold, loveless process; it is not going to produce bonobos' warm, gentle, cooperative, loving behavior—unless the love-indoctrination path is taken, for it alone has the power to superimpose love on an essentially selfish system. As Drummond said of nurturing love, it was only **'once this fire began to warm the cold hearth of Nature and give humanity a heart, the most stupendous task of the past was accomplished'**. In contrast, the SEM and the SDH are desperate and hopelessly flawed attempts to explain bonobo behavior and the origins of our moral nature without admitting the critical role of nurturing.

[560] (A more complete description of the SEM and the SDH and their limitations can be found in *Freedom Expanded* at <www.humancondition.com/freedom-expanded-social-ecological-model>.)

[561] Having now analyzed the mechanisms of both the SEM and the SDH, we now need to describe the immense danger they present to the human race.

## Chapter 6:11 <u>End play for the human race</u>

[562] As pointed out in chapter 2:4 and emphasized again at the beginning of this chapter, the great danger of the practice of denial is that, in the end, it becomes *so* entrenched and sophisticated that it locks humanity onto a path to terminal alienation, to total madness and extinction—to the **'universal ruin to the world'** that Plato said **'more and more forgetting** [denial]' leads to. The development of the denial of the truth that nurturing created humanity, firstly in the form of the SEM, and in its most recent and most sophisticated incarnation, the SDH, dramatically illustrates this dangerous potential.

[563] The bonobos offer the most powerful evidence of the nurturing origins of our unconditionally selfless moral soul, but the SEM and, to an even greater extent, the SDH attempt to not only deny that evidence but bury it under seductive yet totally false explanations for their gentle, loving, cooperative nature. Misappropriating aspects of the truth about bonobos, such as their extraordinarily loving cooperation, their neoteny, and even burgeoning intelligence (which the SDH does using a version of the Social/Machiavellian Intelligence Hypothesis, the dishonesty of which was briefly explained in ch. 6:5), and using that to evidence the SDH, is a very sophisticated way of giving credibility to the lie that nurturing had no role to play in the development of our moral soul. But to bury such evidence of the origins of our unconditionally selfless moral soul that created the cooperative, integrative state that is humanity, is to threaten the human race with permanent estrangement from the truth about our all-loving true self or soul, which is the truth we need if we are to properly understand and, by so doing, heal our psychologically alienated condition. Burial of the truth about our soul stood in the way of us ever gaining an honest, ameliorating understanding of ourselves— of our origins, our present condition and future potential.

[564] Further, to deny the importance of nurturing is to deny the importance of the main activity we need to practice if we are to produce humans who are sound and secure in self. Indeed, it is *only* through the nurturing of our offspring that the human race can hope to become healthy,

psychosis-and-neurosis-free humans once again—to, as Montagu said, put **'man back upon the road of his evolutionary destiny from which he has gone so far astray'** and restore **'health and happiness for all humanity'**.

[565] Yes, if we refuse to admit the critical role nurturing has played in the emergence of humanity and, as a direct result of that heritage, in the sound upbringing of humans today, then levels of alienation will only increase and terminal alienation *will* soon destroy the human race. Like E.O. Wilson's use of the Multilevel Selection theory to deny the true nature of the human condition, the SEM's and the SDH's denial of the importance of nurturing is such an alienating lie that it *will* lead the human race into a state of irretrievable madness.

[566] Indeed, as if the danger just described wasn't enough, the even greater threat posed by the SEM/SDH is that it is now being developed into *another* denial of the true nature of the human condition that is even more seductive and thus dangerous than Wilson's because it appears to take into account, albeit in a totally distorting manner, the most crucial evidence we have about human origins from our closest relatives, the bonobos! As the lead author of the SDH, Brian Hare, has said, **'They** [bonobos] **have done something in their evolution that even humans can't do. They don't have the dark side we do...If we only studied chimps, we'd get a skewed view of human evolution'**, and **'bonobos display...what might be thought of as our better angels'** ('"Hippie chimp" genome may shed light on our dark side', *Science* on NBCNews.com, 13 Jun. 2012; see <www.wtmsources.com/141>). So, according to Hare, chimpanzee-like instincts in humans account for our **'dark side'**, while the instincts allegedly accounted for by the SEM/SDH, as demonstrated by the bonobos, gave rise to our goodness, our unconditionally selfless moral instincts, our **'better angels'**. This argument presents a model for our 'good and evil'-afflicted human condition that is similar to that provided by Wilson's Multilevel Selection theory for eusociality, except that our 'good' instincts are supposedly derived from factors espoused by the SEM/SDH, rather than cooperation forged through warring with other groups, as Wilson suggested. As was emphasized when the Multilevel Selection 'explanation' for the human condition was presented and exposed, we humans *do* have unconditionally selfless, loving, 'good' instincts but they were derived from the nurturing love-indoctrination process that bonobos are developing; and we *do* practice divisive, 'bad' behavior, but, again, that is derived not from genetic competitiveness but from a psychosis that emerged when our conscious mind was liberated by the

love-indoctrination process, and, once liberated, rapidly developed to challenge our *completely* loving (not partially loving and partially selfish) 'good' instincts for the role of managing our lives. So, what Hare, Wobber and Wrangham are doing is precisely what Wilson was doing with his Multilevel Selection theory, which was to bring the human condition to our attention, but only to trivialize it—to subvert the truth about the all-important issue of our psychologically troubled condition!

[567] There are so many dangers associated with the SEM/SDH that it hardly bears thinking about. With E.O. Wilson's all-out atomic bomb attack on the truth about the human condition, and now Hare, Wobber and Wrangham's no-holds-barred asteroid attack on the best evidence we have for the origins of our moral soul, and also on the truth about the human condition, the honesty and resulting insight needed to save the human race is being buried at the bottom of the deepest, darkest ocean trench that has been filled with reinforced concrete for good measure. It is truth-hating behavior of the highest order. Basically, mechanistic science has become completely deranged—and since science is the facility charged with delivering us from the human condition, this state of affairs represents end play for the human race!

[568] Without the exposé being presented in this book all hope for humanity would be lost. But, as will now be described, this exposé is yet to be recognized by the scientific establishment—indeed, for 30 years now all it has received from that corner is obscenely irresponsible, stone-wall resistance.


## Chapter 6:12 *The* **great obscenity**

[569] Early in this book, in chapter 2:4, a warning was given about the extremely dangerous 'trap' that lay in wait for the human race in its quest to find understanding of the human condition. It was explained that unable to face the truth of our corrupted condition while we couldn't truthfully explain it, humans had no choice but to avoid any truths that brought the unbearable issue into focus and instead live in Plato's dark, **'living tomb' 'cave'** of alienated denial of it, with science, practiced as it is by humans, having to comply with this strategy. Science has necessarily been **'reductionist'** and **'mechanistic'**. It has avoided the overarching whole view of life that required having to confront the issue of the human condition and instead *reduced* its focus to only looking down at the details of the

*mechanisms* of the workings of our world, in the hope that understanding of those mechanisms would eventually make it possible for someone who wasn't afraid of the human condition to come along and synthesise the actual explanation of that condition from those hard-won insights—at which point there would no longer be any need for humanity to live in a dark and horrible state of alienating denial.

[570] As was pointed out about this strategy, the very dangerous trap inherent in this mechanistic, resigned-to-living-in-denial-of-the-human-condition, fundamentally dishonest approach is that it could become *so* entrenched that those practicing it could resist the human-condition-confronting, truthful explanation of the human condition when it *was* finally found and continue to persevere with the dishonest strategy to the point of taking humanity to terminal alienation and extinction. The potential trap is that the established *dishonest* scientific paradigm might not tolerate the arrival of the *truthful* scientific paradigm, even though facilitating its arrival has been science's great objective and fundamental responsibility—and the only means by which the human race can be liberated from its condition, and thus transformed. Yes, despite putting forward an *extremely* dishonest and dangerous 'explanation' of the human condition, E.O. Wilson did truthfully recognize this great objective in the human journey of solving the human condition when he wrote that **'There is no grail more elusive or precious in the life of the mind than the key to understanding the human condition'** (*The Social Conquest of Earth*, 2012, p.1 of 330).

[571] And most alarmingly, the evidence so far is that science *has* fallen into this human-race-destroying trap of becoming *so* habituated to living in denial of the human condition that it is refusing to acknowledge the fully accountable, truthful explanation of the human condition that has been synthesized from the insights it was tasked with finding—in particular, as explained in chapter 3, the insights it discovered into the difference in the way genes and nerves process information.

[572] For over 30 years now, I and the 50 Founding Members of the World Transformation Movement (the WTM, the organization I established in 1983 to research and promote understanding of the human condition because mechanistic science wouldn't approach this all-important subject) have been trying to interest the scientific community in the world-saving insights into the human condition that are being presented in this book, but,

save for a handful of supportive scientists and some positive responses from other eminent scientists and thinkers, all of our submissions have so far, as I will shortly document, been either ignored or rejected by the scientific establishment. And not just rejected, but allowed to be *brutally* persecuted.

[573] In 1995, instead of attracting interest, debate and support, I and the WTM were so ferociously attacked by the two biggest, left-wing (dogmatic, pseudo idealistic, 'let's pretend there's no human condition that has to be solved and the world should just be ideal', dishonest) media organizations in Australia, namely our national public broadcaster, the Australian Broadcasting Corporation (ABC), and Fairfax Media, that I was made a pariah in the Australian community and the WTM was completely marginalized. We endured this situation until, after 15 long years of emotionally exhausting, and, for such a small group, financially taxing, defamation actions taken by us (which, we've been told by legal experts, involved what was at the time the biggest defamation case in Australia's history), we finally managed to right the extremely serious wrong. This vindication came in 2010 when three judges in the New South Wales Court of Appeal unanimously found that the earlier 2008 lower court ruling—which found my work to be of **'such a poor standard that it has no support at all from the scientific community'**—did **'not adequately consider'** **'the nature and scale of its subject matter'**, in particular **'that the work was a grand narrative explanation from a holistic approach, involving teleological elements'**, and that other important submissions **'were not adequately considered by the primary judge'**, including that the work can make **'those who take the trouble to grapple with it uncomfortable'** because it **'involves reflections on subject-matter including the purpose of human existence which may, of its nature, cause an adverse reaction as it touches upon issues which some would regard as threatening to their ideals, values or even world views'**! (A documentation of this attack in the Australian media and the subsequent court cases can be read at <www.humancondition.com/persecution>.)

[574] This situation, where rather than *receiving* support from the scientific establishment for producing the all-important breakthrough synthesis of the human condition that science has been charged with finding, I have been not just ignored and rejected but viciously attacked, was actually fully predicted long, long ago, some 360 years before Christ, by Plato during Greece's Golden Age. Recall in par. 81 how the acclaimed philosopher A.N. Whitehead described the history of philosophy as merely **'a series**

**of footnotes to Plato'**? Well, that greatest of all philosophers was not only sound and secure enough in self to admit humans have metaphorically been hiding deep underground in a cave of denial (as was described in par. 83), he was *also* able to warn of the great danger posed by the cave prisoners' determined opposition to the arrival of the truth that frees them from their horrible existence in that dark 'cave'.

[575] While some of this description was referred to in par. 83, the inclusion here of more of what Plato wrote shows just how prophetic he was in anticipating the persecution that would meet the arrival of understanding of the human condition. Plato wrote: **'I want you to go on to picture the enlightenment or ignorance of our human conditions somewhat as follows. Imagine an underground chamber, like a cave with an entrance open to the daylight and running a long way underground. In this chamber are men who have been prisoners there'** (*The Republic*; tr. H.D.P. Lee, 1955, 514; or see <www.wtmsources.com/227>). In this cave allegory Plato went on to describe how the cave's exit is blocked by a fire, such that if one of the prisoners were **'to stand up and turn his head and look and walk towards the fire; all these actions would be painful...and he would** [have to] **turn back and take refuge'** in the cave of **'shadows'**, which are only an **'illusion'** of the real world outside the cave (515). The allegory makes clear that while **'the light of the fire in the cave prison [corresponds] to the power of the sun'** (517), and **'the sun...makes the things we see visible'** (509), such that without it we can only **'see dimly and appear to be almost blind'** (508), having to hide in the **'cave'** of **'illusion'** and endure **'almost blind'** alienation was infinitely preferable to facing the searing, **'painful' 'light'** of the **'fire'/'sun'** that would make **'visible'** the unbearably depressing issue of **'the imperfections of human life'** (517).

[576] Having described how living in a cave-like state of **'almost blind'** alienated denial has, tragically, been absolutely necessary, Plato wrote that it was ultimately only by being **'illuminated by truth and reality'** (508) that **'the enlightenment...of our human conditions'** could be achieved and the cave prisoners be **'released from their bonds and cured of their delusions'** (515). He then described the initial resistance the cave prisoners would have to being **'released from their bonds and cured of their delusions'** when someone achieved **'the enlightenment...of our human conditions'** through being **'illuminated by truth and reality'**. To quote from a summary of the cave allegory that appeared in the *Encarta Encyclopedia*'s entry for 'Plato': **'Breaking free, one of the individuals escapes from the cave into the light of day. With the aid of the sun** [necessarily living free of denial of the

glaring, burning, searing truth of Integrative Meaning and the issue it raises of **our human condition**—and assisted by the understandings that science has found of the differences in the way genes and nerves process information], **that person sees for the first time the real world and returns to the cave with the message that the only things they have seen heretofore are shadows and appearances and that the real world awaits them if they are willing to struggle free of their bonds. The shadowy environment of the cave symbolizes for Plato the physical world of appearances. Escape into the sun-filled setting outside the cave symbolizes the transition to the real world, the world of full and perfect being, the world of Forms, which is the proper object of knowledge'** (written by Prof. Robert M. Baird; see <www.wtmsources.com/101>). To return to *The Republic* and Plato's own words: **'if he** [the cave prisoner] **were made to look directly at the light of the fire** [again the fire represents the searing truth of Integrative Meaning and the issue of **our human condition** it raises], **it would hurt his eyes and he would turn back and take refuge in the things which he could see** [take refuge in all the denials and dishonest explanations and arguments that he has become attached and accustomed to], **which he would think really far clearer than the things being shown him. And if he were forcibly dragged up the steep and rocky ascent** [out of the cave of denial by the person who has broken free of the cave] **and not let go till he had been dragged out into the sunlight** [shown the truthful all-liberating—but at the same time all-exposing and confronting—explanation of **our human condition**], **the process would be a painful one, to which he would much object, and when he emerged into the light his eyes would be so overwhelmed by the brightness of it that he wouldn't be able to see a single one of the things he was now told were real** [as described in ch. 1:4, this inability to absorb discussion of the human condition because it has been such an unbearable, off-limits subject for so long is what we in the WTM regularly encounter and refer to as the 'deaf effect'—as one reader of my books admitted, **'When I first read this material all I saw were a lot of black marks on white paper'**]' (515-516). Plato continued: **'they would say that his visit to the upper world had ruined his sight** [they would treat the person who tries to deliver understanding of **our human condition** as if he was mad, which is how I have been treated for many years, as I talk more about below], **and** [they would say] **that the ascent** [out of the cave] **was not worth even attempting** [such assertions that it's not worth attempting have been regularly made against our work, with one of the architects of the mid-1990s public campaign of persecution against myself and the WTM saying, **'You know you are encroaching on the personal unspeakable inside people and you won't** succeed' (WTM records, 12 Feb. 1995), and the other architect of the attacks, the church minister who produced the defamatory articles and television program about my work, saying, **'You realise you are attempting the impossible, you will be fighting to have this material accepted right down to the last person on the planet'** (WTM records, 16 Feb. 1995)]' (517). Plato didn't stop there, going on to say, **'And if anyone tried to release them and lead them up, they would kill him if they could lay hands on him'** (ibid). (I might mention here that this idea that humans have been living as prisoners in a mind-controlled state of denial of their reality, and of attacking the person who tries to liberate them from this state, forms part of our collective subconscious awareness because it periodically crops up in our mythologies. The same essential myth is found in the ancient Native American legend of *The Story of Jumping Mouse*, and, in more contemporary times, in the film *The Truman Show*. The deep resonance of the myth is also evidenced by the various science fiction films that have been based on Plato's cave allegory, such as *Dark City*, *City of Ember* and, most notably, *The Matrix*.)

[577] So Plato predicted that the cave prisoners would not just resist adopting the human-race-liberating explanation of the human condition **'which is the proper object of knowledge'**, they would be murderously angry towards it—hence the brutal campaign of persecution against me and those supporting these understandings of the human condition. What the attackers did that was *so* destructive of myself and the WTM was to base their whole ferocious and malicious media fear campaign against me and our work at the WTM on the accusation that I am a deluded, megalomanic leader of a dangerous anti-social, mind-controlling cult. But how could I possibly be so unsound as to be a deluded, megalomanic leader of a dangerous anti-social organization when I have been sound enough to look into the human condition, as even the statements made by the architects of the attack acknowledged when they said that I was **'encroaching on the personal unspeakable [of the human condition]'** and **'attempting the impossible [of confronting people with the human condition]'**? Quite simply, as Christ pointed out when he was similarly accused of being **'possessed by…the prince of demons'** (Mark 3:22) for his honest, human condition confronting and penetrating words, **'How can Satan drive out Satan?'** (3:23). Christ was making the same point when he said that **'A good tree cannot bear bad fruit, and a bad tree cannot bear good fruit'** (Matt. 7:18). No, the real motivation for the whole vicious media campaign was

clearly prejudice against addressing the issue of the human condition. In fact, with regard to the involvement of mind control, as the psychologist Arthur Janov pointed out (in par. 221), the *human-condition-avoiding*, resigned, alienated, neurotic, cave-dwelling, mechanistic mind is the one that is being **'subject**[ed] **to indoctrination and brainwashing—because neurosis** [blocking out] *is* **brainwashing'**. Yes, by *confronting* instead of *avoiding* the human condition the information being presented in this book is the soundest, least neurotic, least brainwashed information that exists! This information is the very opposite of the mind*less* dogma that characterizes a mind-controlling sect—it is mind*ful* understanding; brain food *not* brain anesthetic. To accuse those supporting this information of being involved in mindlessness was an absolute reverse-of-the-truth lie. Again, the *real* problem was precisely that we were daring to address the historically forbidden subject of the human condition—that we were daring to think, and think very deeply, not *not* think. With regard to the tactic of using a reverse-of-the-truth lie to counter truth (examples of which are littered throughout this book), Hitler once said that the most convincing lie is the absolute lie, **'because in the big lie there is always a certain force of credibility'** (*Mein Kampf*, 1925; tr. James Murphey, 1939, p.185 of 525). Yes, the reverse-of-the-truth lie is the ultimate unscrupulous and malicious way to attack truth!

[578] Some idea of the prescience of Plato's anticipation of the *ferocity* of the attack that would be made on the person who **'escapes from the cave into the light of day'** and then **'dragged** [humanity] **out into the sunlight'** can be gained by the first ruling that the court made *against* my work in the defamation action we took to redress the horror media campaign. As mentioned above, it ruled that my work is of **'such a poor standard that it has no support at all from the scientific community'**! It was certainly an immense relief then when, in 2010, the three judges in the New South Wales Court of Appeal unanimously recognized the real problem was that the primary judge clearly found my work unbearably confronting, ruling that he did **'not adequately consider'** **'the nature and scale of its subject matter'**, in particular **'that the work was a grand narrative explanation from a holistic approach, involving teleological elements'**, and that other important submissions **'were not adequately considered by the primary judge'**, including that the work can make **'those who take the trouble to grapple with it uncomfortable'** because it **'involves reflections on subject-matter including the purpose of human existence which may, of its**

nature, cause an adverse reaction as it touches upon issues which some would regard as threatening to their ideals, values or even world views'! Yes, the problem was *not* that I am a deluded monster, but that I am daring to address the human condition!

[579] Those scientists who came out to Australia and gave evidence at the initial 2007 trial, specifically that my work is not at all of **'a poor standard'**, namely Professor Harry Prosen, Professor Scott Churchill, Professor Walter Hartwig and Dr William Casebeer, played an absolutely crucial role in humanity's great journey to achieve liberation from the human condition. As the philosopher John Stuart Mill wrote in his famous 1859 essay *On Liberty*, **'the dictum that truth always triumphs over persecution, is one of those pleasant falsehoods which men repeat after one another till they pass into commonplaces, but which all experience refutes. History teems with instances of truth put down by persecution. If not suppressed for ever, it may be thrown back for centuries'** (ch.2). And in the case of these biological explanations of human nature, such **'persecution'** has already **'thrown back'** its recognition for 30 years! The science historian Thomas Kuhn similarly warned that the only guarantee truth will survive prejudice is the determination of its advocates, writing that **'In science…ideas do not change simply because new facts win out over outmoded ones…Since the facts can't speak for themselves, it is their human advocates who win or lose the day'** (Shirley Strum, *Almost Human*, 1987, p.164 of 297—Strum's references are to Thomas Kuhn's *The Structure of Scientific Revolutions*, 2nd edn, 1970). Interestingly, Kuhn also recognized that in science **'revolutions are often initiated by an outsider—someone not locked into the current model, which hampers vision almost as much as blinders would'** (ibid). Even Charles Darwin was **'a lone genius, working from his country home without any official academic position'** (Geoffrey Miller, *The Mating Mind*, 2000, p.33 of 538). While there are certainly advantages to not being **'hamper[ed]'** by **'the current model'**, the inherent danger of not being part of the establishment is that the **'outsider'** is an easy, undefended target for those in the establishment who feel threatened by their new ideas. Thank goodness for those scientists who travelled to Australia to provide expert reports at our trial. (Again, these reports, and documentation detailing the 30 years of terrible persecution that we have been subjected to, can be read at <www.humancondition.com/persecution>.)

[580] As I mentioned earlier, the ferocious and malicious fear campaign conducted in Australia in the 1990s and early 2000s against me and our work at the WTM has only been part of the malicious campaign

of persecution that has been waged against my work. As I said, I and the WTM have been trying to interest the scientific community in the world-saving insights into the human condition for over 30 years now, but so far all our submissions have been irresponsibly and unjustifiably either ignored or rejected by the scientific establishment. So the question is, how could such a fully accountable and well argued and evidenced explanation have been dismissed?

[581]The main method of rejecting the innumerable presentations and submissions that have been made of the explanation (which will be listed shortly) has been to simply ignore them, and—beyond that—to make the most outrageously dishonest and irresponsible claim that my synthesis presents no new data, is an untestable hypothesis, and is therefore not even science! This was the main argument used to claim my work was of 'a poor standard' at the 2007 trial, and, as was mentioned in par. 458, it was also used to reject my 2005 submission about the nurturing origins of our moral conscience that I made to the International Primatological Society's (IPS) Congress in Uganda. Most recently, in 2014, the Executive Editor of *Scientific American* rejected a pitch for a feature story about this book and an excerpt for possible publication on their website. In each case, the rejections were vigorously protested; in the case of the IPS and *Scientific American*, complaints were sent to *every* member of their administrations. Having received virtually no response from the administrators of *Scientific American*, that complaint is now being taken to the highest possible level of appeal in the world. In all these complaints it is pointed out that the same ridiculous accusation was used against Darwin's Natural Selection synthesis. For instance, Bishop Wilberforce, the opponent of natural selection in the great debate about Darwin's theory at Oxford in 1860, said it was a **'theory which cannot be demonstrated to be actually impossible'** (Wilberforce's review of *Origin of Species* in *Quarterly Review*, 1860, p.249), while the geologist and bishop Adam Sedgwick said it was **'not a proposition evolved out of the facts'** ('Objections to Mr Darwin's Theory of the Origin of Species', *The Spectator*, 7 Apr. 1860) and that it was **'based upon assumptions which can neither be proved nor disproved'** (Letter from Sedgwick to Darwin, 24 Dec. 1859; *The Complete Work of Charles Darwin Online*, ed. John van Wyhe, 2002). The paleontologist Louis Agassiz similarly complained that **'absolutely no facts...can be referred to as proving evolution'** (William Penman Lyon, *Homo versus Darwin: A judicial examination of statements recently published*

*by Mr Darwin regarding 'The Descent of Man'*, 1872, p.140), while more recently, the philosopher Karl Popper commented, before later changing his mind, that **'Darwinism is not a testable scientific theory'** (*Unended Quest*, 1976, p.168). At our 2007 trial, however, Professor Scott Churchill, then Chair of the Psychology Department at the University of Dallas, pointed out—and has since summarized in regard to my work—that **'Griffith's ideas have been criticized for not presenting the field of science with "new data" and "testable hypotheses." But such a complaint is disingenuous since evolutionary processes are not subjectable to the same kind of "hypothesis testing" that one finds in the other sciences. An hypothesis is a "smaller, more compact thesis" that is "deduced" from a larger idea or thesis in such a way that one can test that larger idea piece by piece. Whereas, the kind of synthesis offered in Griffith's book is presented both conceptually and metaphorically with an aim to tie together existing data, while correcting and expanding upon the more limited existing interpretations of those data...Such a perspective comes to us not as a simple opinion of one man, but rather as an inductive conclusion drawn from sifting through volumes of data representing what scientists have discovered'** (Review of *FREEDOM* submitted to *New York Magazine*, 26 Sep. 2014). In his Introduction to this book, Professor Prosen similarly points out that **'not only has Jeremy's work been treated as heretical by mechanistic science because he dares to look at the real "psychological" nature of the human condition, it has also been resisted because of the two reasons referred to in the ruling by the...three judges of the New South Wales Court of Appeal. Firstly, rather than the usual, more mechanistic and less thinking dependent, deduction-derived theories, Jeremy, like Darwin did with his theory of natural selection, puts forward a wide-ranging, induction-derived synthesis, a "grand narrative explanation" for behavior—which, incidentally, very wrongly led to both Darwin's and Jeremy's work being criticized by some for not presenting "new data" and a "testable hypothesis"; even for "not being science at all"! Secondly, Jeremy's enormously knowledge-advancing (and "science" literally means "knowledge", derived as it is from the Latin word scientia which means "knowledge") thinking is based on "a holistic approach involving teleological elements". As Jeremy beautifully explains in chapter 4, the reason that the teleological, holistic purpose or meaning of existence of developing the order or integration of matter into ever larger and more stable wholes...has been denied by human-condition-avoiding mechanistic science, is because it implies humans should behave in an ordered, integrative, cooperative, selfless, loving way.'**

582 Yes, the reason a more **'thinking dependent' 'wide-ranging, induction-derived synthesis' 'drawn from sifting through volumes of data representing what scientists have discovered'** has not been something that human-condition-avoiding, mechanistic scientists have been comfortable with is because, as the cartoon in chapter 2:4 illustrates, it requires being able to confront such historically unbearable truths as **'the teleological, holistic purpose or meaning of existence of developing the order or integration of matter'**. Essentially, **'induction-derived'** knowledge requires fearless, free, imaginative thinking—as Arthur Koestler has pointed out: **'Max Planck, the father of quantum theory, wrote in his autobiography that the pioneer scientist must have "a vivid intuitive imagination for new ideas not generated by deduction, but by** *artistically* **creative imagination"'** (*The Act of Creation*, 1964, p.147 of 751). And with regard to the human-condition-confronting fearlessness of such a free **'imagination'**, Berdyaev said (in par. 237) that **'Knowledge requires great daring. It means victory over ancient, primeval terror. Fear makes the search for truth and the knowledge of it impossible. Knowledge implies fearlessness... Particularly bitter is moral knowledge, the knowledge of good and evil...Moral knowledge is the most bitter and the most fearless of all for in it sin and evil are revealed to us along with the meaning and value of life.'** So declaring **'induction-derived' 'new ideas'** and **'knowledge'** are not science is simply mechanistic science saying it doesn't want to participate in **'the search for truth'** and **'knowledge'**; basically, it doesn't want to practice science—because **'"science" literally means "knowledge"'**!

583 Another malicious device that has been used to try to dismiss my work is claiming it puts readers in a **'non-falsifiable situation'** where if you oppose this information you are said to be suffering from denial, leaving you no way to disprove or falsify the explanation being put forward—but the problem only really exists at the superficial level because the ideas being put forward can be tested as true or otherwise. These are not untestable hypotheses that must be accepted on blind faith. For instance, the existence of denial of the issue of the human condition can easily be established by scientific investigation. In fact, since humans are the subject of this particular study, each person can experience and thus know the truth or otherwise of what is being put forward. Once the explanations are presented and applied you will discover they are able to make such sense of human behavior that your own and everyone else's becomes transparent. This new-found transparency confirms that this understanding is the long-sought explanation of the human condition.

[584]Other equally disingenuous arguments that have been used to try to undermine this synthesis include accusing me of **'overly appealing to authority to try to persuade the reader'**, **'cherry picking evidence to make it fit a preferred theory'**, **'false interpretation of quotes'**, **'faulty generalization'**, **'repetition designed to indoctrinate'**, **'mind projection fallacy where you project your own false view of the world and declare others irrational if they don't hold that view'**, **'dogmatically asserting ideas to be true'**, being **'full of hyperbole about humans being selfish egomaniacs'**, and **'claiming that humans are fallen and heading for disaster when humans are only getting better, for example, look at how the quality of life has improved in China'**! I would argue, however, that the accountability of the arguments and the quality of the evidence I refer to in support of all my various theses are undeniable—but again, when the need for denial is desperate, any excuse will do.

[585]The main point is that this *denial*—this ignoring, rejecting and persecuting—has been completely and utterly unjustified; in fact, it has been *totally* irresponsible, *obscenely* irresponsible. *Certainly* all the world-saving insights that appear in this book, insights that I have been putting forward for 30 years, bring the historically unbearably confronting issue of the human condition into stark focus, *but all are presented within the framework of the compassionate explanation of the human condition, which means the need for denial of them has been removed.*

[586]As emphasized from the beginning of this book, the human race has always lived in hope, faith and trust that one day the redeeming explanation of our psychologically distressed and insecure human condition would be found, and the most fundamental reason for freedom of expression to be maintained in the world is to keep the door open to that possibility. The 2007 court case in Australia about my work was at base a fight to maintain this freedom of expression because what my detractors were doing was using lies and misrepresentation to try to deny me the freedom to express my ideas about the human condition and have them fairly evaluated. Of course, when a person resigned themselves to living in denial of the human condition that in itself was an act of oppression of their own freedom to think about the human condition. The psychological process of denial, which we humans have had to practice on a vast scale while we couldn't truthfully explain and thus safely confront the human condition, has entirely been about denying our minds the freedom to think freely. It has been about oppression. Yes, alienation, which the human race is entering a state of terminal sickness from, is all about

repressing the freedom to think. So the truth is when we have spoken about 'freedom of expression' we have only used it in a relative sense, because while we allow people to oppress the freedom within their own mind to think about the human condition, we seek to stop the oppression of free thought everywhere else in society. It is all very contradictory. But despite the existence of this conundrum, the *ultimate* freedom of expression humans have been fighting for is the freedom for all people to not have to resign, to not have to live in Plato's horrible, dark, **'living tomb'** **'cave'** of alienated denial of the human condition, which means we have been fighting to maintain the freedom to express analysis of the human condition—because *only* that freedom could allow the answers about the human condition to emerge and end the need for that cave-dwelling state of denial.

[587] Yes, the ultimate knowledge that science—that part of our society that was assigned the special task of searching for knowledge—had to find was self-knowledge, understanding of our corrupted human condition. Certainly science, being practiced by insecure, human-condition-afraid humans, had to go about that search in a denial-complying, mechanistic way, avoiding any confronting truth about the human condition, but that was always only the first stage of science's search for understanding of the human condition. It always had to remember that at some point the second stage had to take place, where someone secure enough in self could confront the human condition and synthesise the explanation of it from those hard-won insights into the mechanisms of the workings of our world that its practitioners had found. Science *had* to remain open to the possibility that the human-condition-confronting truthful explanation of the human condition would one day be found. *Yes, science holds the ultimate responsibility to consider, not ignore, or worse, falsely dismiss and even persecute, well-reasoned and evidenced scientific analysis of the human condition!* What has transpired *has been* obscenely irresponsible.

[588] What now needs to be explained is that there have been two aspects to humans' resistance to new ideas. Firstly, there is the just described fear we have of encountering ideas that confront us with the historically (but no longer) condemning truth of our corrupted condition. The second aspect of our resistance to new ideas is a consequence of this first aspect. As was explained in chapter 3, being insecure about our fundamental goodness and worth we became egocentric, our minds became centered

or focused on trying to prove we are good and not bad, forever searching for reinforcement and validation. The result of this insecure, egocentric existence meant that as we grew up and searched for ways to validate ourselves we each built a world around us that gave us this feeling of worth. We became head of a company, or the parents of a new generation, or the captain of our football team, or a success at making ourselves look pretty, etc, etc, and no matter how superficially meaningful these forms of validation actually were, they were all that kept at bay the underlying insecurity of our condition that arose from wondering whether we were bad, vile creatures for having departed so incredibly far from our species' original unconditionally selfless, all-loving, Godly, 'ideal' way of living and the suicidal depression such thinking could lead to. It follows that we each became extremely attached to the fragile little ego castles we had built and, as a result, we found it difficult accepting any change that might threaten its structure, with the most threatening change being new ideas that required us to fundamentally change our philosophical world view that we had built our 'castle' upon. Being thinking creatures, our primary need was to have some philosophy about life, some way of making sense of our existence, because from there we could build a larger meaningful structure, so to have to change those philosophical foundations could be *very* destabilizing and thus difficult to accept.

[589] The point is we humans have resisted new ideas both because of how much they confronted us with the issue of the human condition *and* because of how much we felt they threatened our way of justifying ourselves. Yes, humans have been almost as afraid of change as they have been of being confronted with the issue of the human condition, and since science deals with knowledge—understanding and insights that can affect the philosophical foundations of our 'ego castle'—advances in science could be very difficult for humans to accept and adapt to. It is no wonder Thomas Kuhn said that **'the old scientists who became established within the dominant paradigm have to die off first: they will virtually never accept the new paradigm. Only the younger generation of scientists, who don't have the emotional attachment to the old paradigm, will be willing to change their minds'** (a reference to the work of Kuhn by Marilyn Ferguson, *New Age* mag. Aug. 1982; see <www.wtmsources.com/174>). The physicist Max Planck put it more succinctly when he said that **'Science progresses funeral by funeral'** (Marilyn Ferguson's reference to a comment by Planck in his *Scientific Autobiography*, 1948; *New Age* mag. Aug. 1982; see <www.wtmsources.com/174>). It follows that since the arrival of the truthful explanation of the human

condition brings *the* most self-confrontation and the greatest change—in fact, a whole paradigm shift from living in denial to living honestly—it will be resisted the most, which is what Plato predicted.

[590] So the reality is that any meritorious new idea in science has typically gone through stages of resistance and even persecution before gaining universal acceptance—a process the philosopher Arthur Schopenhauer articulated when he **'said that the reception of any successful new scientific hypothesis goes through predictable phases before being accepted'**. First, **'it is ridiculed'** and **'violently opposed'**. Second, after support begins to accumulate, **'it is stated that it may be true but it's not particularly relevant'**. Third, **'after it has clearly influenced the field** [including members of the establishment quickly remodeling/plagiarizing the ideas as their own discoveries, something we in the WTM have, unfortunately, also experienced with the ideas in this book] **it is admitted to be true and relevant but the same critics assert that the idea is not original'**. Finally, **'it is accepted as being self-evident'** (compiled from two references to Schopenhauer's quote—*New Scientist*, 15 Nov. 1984 and *PlanetHood*, B. Ferencz & K. Keyes, 1988). Note how each stage of recognition is achieved in a way that protects the ego of the onlookers. So much for science being a rigorously objective enterprise where the 'scientific method' dictates that explanations for phenomena are accepted by proven test; in reality science is *extremely* subjective—scientists will only accept a proven new idea if they see it as reinforcing them somehow. In fact, they have totally ignored ideas that confront them with the issue of the human condition, they have lived **'a long way underground'** in Plato's **'cave'** of denial. I put forward insights that are fully accountable and human race-saving but because they are self-confronting they are rejected!

[591] The famous playwright George Bernard Shaw was another who warned of the true nature of progress when he wrote that **'All great truths begin as blasphemies'** (*Annajanska*, 1919). It should be explained that the reason great truths so often began as **'blasphemies'** is because the search for truth was very often not a case of finding it but of uncovering it from epochs of denial—an exposure that seemed highly *blasphemous* to all those who had been practicing the denial. We have and will see how finding the great truths presented in this book of the explanation of the human condition, of the meaning of existence, of the origin of our moral soul, and, in chapter 7, the explanation of the origins of our conscious mind, all required the defiance of deeply entrenched denials, which means all these great truths *are* going to **'begin as blasphemies'**.

[592]With regard to the greatest of all **'blasphemies'**, that of the **'truth'** about the human condition, when Alvin Toffler wrote in his famous 1970 book *Future Shock* of **'the shattering stress and disorientation that we induce in individuals by subjecting them to too much change in too short a time'** (p.4 of 505), he was actually anticipating the immense shock that the arrival of understanding of the human condition would inevitably bring. Yes, given the arrival of understanding of the human condition does represent a *massive* paradigm shift for humans, especially a shift from denial to honesty, it was always an inevitability that the initial **'violently opposed'**, treated-as-**'blasphemies'** stage would have to be negotiated—*but the fundamental truth remains that science holds the ultimate responsibility to consider, not ignore, or worse, falsely dismiss and even persecute, well-reasoned and evidenced scientific analysis of the human condition.* What has transpired certainly *has been* obscenely irresponsible.

[593]Basically, while most, indeed almost all, scientists who have become attached to the old denial-complying mechanistic paradigm will initially **'violently oppose'** and treat as **'blasphemies'** these world-saving new understandings about the world of humans (and have done so), there needs to be some who are sufficiently secure and sound to support the new paradigm. As John Stuart Mill said, **'History teems with instances of truth put down by persecution'**, which means that, as Kuhn said, **'In science…ideas do not change simply because new facts win out over outmoded ones…Since the facts can't speak for themselves, it is their human advocates who win or lose the day'**. And indeed, Professor Prosen and the other scientists who defended my work at the trial were just such critically important, secure and courageous **'advocates'**. Yes, as the three judges in the New South Wales Court of Appeal, led by Justice David Hodgson, recognized as being critically important considerations, my work can make **'those who take the trouble to grapple with it uncomfortable'** because it **'involves reflections on subject-matter including the purpose of human existence which may, of its nature, cause an adverse reaction as it touches upon issues which some would regard as threatening to their ideals, values or even world views'**! What an absolutely courageous and precious ruling for the human race! Blessed are the supportive scientists and Appeal Court judges for they are very great heroes in the human journey—especially Justice Hodgson, a Rhodes Scholar who **'from an early age…had been fascinated by what went on inside the** [human] **head'** and had written books on **'consciousness'**, and was said to be **'blessed with flawless logic'** and to **'fit the**

**description of Plato's "philosopher king"'** (Obituary, *The Sydney Morning Herald*, 4 Sep. 2012; see <www.wtmsources.com/184>).

[594]So, despite the **'threatening'** and **'uncomfortable'** effects of the world-saving insights in this book, there *are* people in the world who are capable of supporting their *arrival*. The great prophet Daniel understood that there would be people who could cope with the truth about the human condition when it arrived when he said that **'the people who know their God** [those who are sound and secure] **will firmly resist'** **'those who have violated the covenant** [become so unsound that they can't tolerate any truth]' (Bible, Dan. 11:32). (More is said about these quotes by Daniel in par. 1123.)

[595]The reason I put 'arrival' in italics above is because once this initial support emerges, others will hear about it and support will grow from there. In fact, as will be described in chapter 9, once this project of bringing understanding to the human condition gets through the very difficult preliminary, 'arrival' stage where these insights are **'violently opposed'** and treated by most as **'blasphemies'**, the whole human race will discover that *there is not only a way to cope with this massive paradigm shift, there is a completely satisfying and utterly exciting and extremely easy way to now live*. It is only at the 'arrival' stage that there is extreme resistance.

[596]So while support *has* begun for these world-saving understandings, much more is needed because, as described, advances in science of this magnitude have to overcome a *great* deal of resistance. History has taught us that since the beginning of scientific enquiry those who pioneered the demystification and clarification of the human situation have particularly been persecuted. It is only through the immense courage of the advocates for free thinking that knowledge has been able to replace mysticism, abstract description, delusion and denial.

[597]At the beginning of the formalization of the discipline of science, Socrates, Plato's great teacher, and the fearless pioneer in replacing superstition and dogma with logical explanation, faced such hostility he was forced to take his own life by drinking poison!

[598]The journalist Robert Howard described the backlash the great demystifications of our historical ways of fortifying ourselves against the insecurity of our condition have provoked when he wrote, **'Three major blows have dented humanity's self-esteem: Copernicus showing that the Earth was not the centre of the universe, Darwin showing descent from animals and Freud arguing that the rational, conscious mind is not master'** (*The Bulletin*, 11 Aug.

1992). Yes, each of these ideas was widely rejected and their proponents persecuted by society, with eminent scientists and/or religious leaders of the day leading the charge.

[599] Copernicus delayed publication of his theory revealing that the Earth is not the center of the universe until the last days of his life in 1543 because he feared persecution, and indeed, 57 years later Giordano Bruno was burnt at the stake for teaching Copernican theory. Furthermore, 10 years after Bruno's death when Galileo upheld the same belief, **'jealous philosophers joined forces with ignorant fanatics in denouncing Galileo to the Inquisition'** (*Reader's Digest: Great Lives, Great Deeds*, 1966, p.306 of 448). This church court said the beliefs were **'contrary to Scripture'** (ibid) and Galileo was forced to recant his support of Copernican theory and live under house arrest for the rest of his life.

[600] Darwin so feared that his theory showing descent from animals would be seen as offensive that he avoided publishing his ideas for eight years. As mentioned in par. 195, under pressure, he also allowed the evasive, but competition-excusing, 'survival of the fittest' phrase to be introduced into subsequent editions. And yet despite these efforts to comply with the principles of denial, Darwin's concept was **'greeted with violent and malicious criticism'** (*The Origin of Species*, 1968 Penguin edn, title page), so much so that Darwin said, **'I have got fairly sick of hostile reviews...I can pretty plainly see that, if my view is ever to be generally adopted, it will be by young men growing up and replacing the old workers'** (*Charles Darwin*, ed. Francis Darwin, 1902, p.244). In the famous debate at Oxford in 1860, Bishop Wilberforce said that Darwin's concept of natural selection was **'contrary to the revelations of God in the Scriptures'** (ibid. p.238) and mockingly asked, **'Is the gentleman [Darwin] related by his grandfather's or grandmother's side to an ape?'** (*Reader's Digest: Great Lives, Great Deeds*, 1966, p.335 of 448).

[601] Freud gave major impetus to the process of de-throning evasive intellectualism and re-emphasizing instinctualism by exposing and emphasizing the existence within humans of a subconscious, innate self that is not under the control of our rational mind. He was a pioneer in exposing the limited nature of mechanistic ways of thinking and opened the door to the much repressed holistic paradigm and yet for his efforts suffered extreme vilification. As Sir Laurens van der Post observed, **'One could perhaps better have measured the originality of Freud's achievement by reason of the numbers of the highly intelligent, well-informed men who instantly mobilised to attack him'** (*Jung and The Story of Our Time*, 1976, p.108 of 275).

[602] Each of these giant strides in the journey of demystification of our human situation were met with so much resistance they *were* lucky to survive, so it is clear that with the arrival of actual demystification and exposure of the human condition, human insecurity and nervousness is going to be extreme—which means that these cautionary words from John Stuart Mill have never been more applicable: **'We have now recognised the necessity to the mental well-being of mankind (on which all their other well-being depends) of freedom of opinion, and freedom of the expression of opinion'**, for **'the price paid for intellectual pacification, is the sacrifice of the entire moral courage of the human mind'** (*On Liberty*, 1859). Yes, for this ultimate enlightenment to be allowed the human race is going to have to adhere scrupulously to democratic principles which allow freedom of expression, a responsibility the New South Wales Court of Appeal judges respected and practiced, but which many others on many fronts have ignored.

[603] The starkest example of how deeply science has fallen into the trap of being so habituated to living in denial of the issue of the human condition that it refuses to recognize the human-race-saving, truthful explanation of the human condition can be seen in the establishment of all the so-called Brain Initiatives that have been launched recently, the prime examples of which are outlined below:

- in December 2012 it was announced that an American billionaire had pledged $200 million to Columbia University's **'accomplished scholars whose collective mission is both greater understanding of the human condition and the discovery of new cures for human suffering'** (*The Educated Observer*, Winter 2013); and,

- in January 2013 the European Commission **'announced that it would launch the flagship Human Brain Project with a 2013 budget of €54 million (US$69 million), and contribute to its projected billion-euro funding over the next ten years'** ('Neuroscience: Solving the brain', *Nature*, 2013, Vol.499, No.7458) with the goal of providing **'a new understanding of the human brain and its diseases'** (Press Release, The Human Brain Project, 28 Jan. 2013; see <www.wtmsources.com/111>) to **'offer solutions to tackling conditions such as depression'** ('Scientists to simulate human brain inside a supercomputer', *CNN International Edition*, 12 Oct. 2012; see <www.wtmsources.com/129>); and,

- in April 2013 the President of the United States, Barack Obama, announced a **'Brain Initiative'**, giving **'$100 million initial funding'** (*The Sydney Morning Herald*, 4 Apr. 2013) to mechanistic science, in **'a research effort**

**expected to eventually cost perhaps ten times that amount'** ('Neuroscience: Solving the brain', *Nature*, 2013, Vol.499, No.7458), to also find **'the underlying causes of... neurological and psychiatric conditions'** afflicting humans in order to **'develop effective ways of helping people suffering from these devastating conditions'** (US National Institutes of Health; see <www.wtmsources.com/114>); and,

- in April 2013 *BBC News Business* reported that **'Lord Rees, the Astronomer Royal and former president of the Royal Society, is backing plans for** [Cambridge University to open] **a Centre for the Study of Existential Risk** [meaning risk to our existence]. **"This is the first century in the world's history when the biggest threat is from humanity," says Lord Rees'** ('How are humans going to become extinct?'; see <www.wtmsources.com/120>). The article then referred to Oxford University's Future of Humanity Institute that was established in 2005, which is **'looking at big-picture questions for human civilization...**[and] **change...**[that] **might transform the human condition'** (Future of Humanity Institute, University of Oxford; see <www.wtmsources. com/118>), quoting its Director and advisor to the Centre for the Study of Existential Risk, Nick Bostrom: **'There is a bottleneck in human history. The human condition is going to change. It could be that we end in a catastrophe or that we are transformed by taking much greater control over our biology'**!

[604] We can see in these initiatives that the human race has finally been forced to at least attempt to address **'the human condition'**, because the underlying, core problem involved in *all* of the **'big-picture questions'** and the **'Existential Risk'** of **'catastrophe'** is our **'neurological and psychiatric'**-troubled **'human brain'**—*that's* what had to be addressed. The **'biggest threat is from humanity'** itself, our species' inability to **'discover' 'the underlying causes of'** our psychologically distressed lives and, by so doing, end this **'bottleneck in human history'** and achieve the way-overdue **'cures for human suffering'**, which will **'transform' 'our biology'**. It certainly is a case of self-discovery or self-destruction—understand the human condition or it's **'catastrophe'** for the human race! <u>BUT, the mechanistic scientific paradigm, including all of its universities, has proved completely incapable of adopting a denial-free approach and truthfully addressing the human condition, and yet now, when the world is absolutely on its knees and desperate for the neurosis-and-psychosis-addressing-and-healing real insight into the human condition, that same mechanistic paradigm is being given millions and millions and millions and millions of dollars to turn around</u>

and do what it has already shown it can't, but what I and the WTM have already done! And, I might add, we have done this, 40 years of work in all, without *any* outside financial support—from academic institutions or from public or private benefactors. We have funded the whole effort from our own self-sufficient initiatives, efforts, and contributions. Indeed, as has been mentioned, rather than receiving any encouragement or financial support from the world at large, we have been *attacked* by the establishment and had to generate the wherewithal ourselves to fight and defeat that enormously powerful institution-backed attack! For example, the principal media attack against us was carried out by the Australian Broadcasting Corporation, a body which receives well over a billion dollars annually in government funding. We have never been invited to participate in any major scientific or public forums, or been given a wage or grant by an academic institution to investigate any of the subjects that we have so effectively studied; quite the reverse—we've been brutally ostracized and forced to sustain and pay our own way to carry on our work and to fight the massive forces opposing it. Talk about David and Goliath! And it is not as though these huge financial initiatives that are now, in desperation, being made to specifically investigate the human condition can be oblivious to our existence and work because, for one thing, the WTM has the domain name '**humancondition.com**' and any planned attempt, such as these, to supposedly properly address the human condition would, one would assume, have to be aware of our existence. So the mechanistic paradigm is saying it is finally going to bring '**greater understanding of the human condition**', but has not changed tack at all and acknowledged those who have done just that! I should add that in 2014 a special scientist-orientated edition of this book, which was given its own title, *IS IT TO BE Terminal Alienation or Transformation For The Human Race?*, was sent to all the Directors and scientists involved in each of the above initiatives (except the European Commission's Human Brain Project which wasn't approached because this book is presently only available in English) to directly inform them of this breakthrough work on the human condition—however, so far (and I'm writing this passage in early 2015) we have had virtually no response from any of them, and certainly no appreciative response. The hypocrisy and obscenity of what is going on is *absolutely astronomical!*

[605] In regard to how utterly dishonest and completely self-defeating the human-condition-*avoiding*, reductionist, mechanistic paradigm's

attempt is to solve the mental problems that now threaten to overwhelm humanity, I should re-include the reference Professor Prosen made in this book's Introduction to an opinion piece by Benjamin Y. Fong of the University of Chicago that was published in *The New York Times* in 2013. Fong wrote that **'The real trouble with the Brain Initiative is…the instrumental approach…**[such **biological reduction** is] **intent on uncovering the organic "cause"…of mental problems…rather than looking into psychosocial factors…By humbly claiming ignorance about the "causes" of mental problems… neuroscientists unconsciously repress all that we know about the alienating, unequal, and dissatisfying world in which we live and the harmful effects it has on the psyche, thus unwittingly foreclosing'** the ability to **'alleviate mental disorder'** (see <www.wtmsources.com/151>). Fong is right; as he says here and elsewhere in his article, mechanistic science's **'synthetic'** focus on the **'organic'** rather than the **'psychological'** nature of mental problems can only end in denying humans **'the possibility of self-transformation'**. As Professor Prosen recognized in his Introduction, it is *only* the *psychosis-addressing, human-condition-confronting* position that I have taken that could hope to find, and now *has* found, the reconciling and human-race-transforming understanding of the human condition—and yet it is this approach that has been treated as heretical, an anathema and a threat by the mechanistic scientific establishment!!

[606] In a further horrific example of the obscene ill-treatment of these human-race-saving, breakthrough understandings in this book, the reader can imagine our shock when, in 2012, the WTM learnt that the three leading anthropologists responsible for putting forward the Self-Domestication Hypothesis (SDH), Wrangham, Hare and Wobber—all of whom, as has been mentioned, were informed many years ago of my love-indoctrination explanation for the origins of our and the bonobos' moral instincts—made no acknowledgment or even mention of my synthesis in their 2012 paper, despite acknowledging the work of many other researchers in a detailed section on **'evolutionary explanations for reduced aggressiveness in bonobos relative to that in chimpanzees'**. Worse, it would appear that since they were each made aware of my synthesis (in the case of Wrangham, on four separate occasions, the earliest being in 1988), what they have done is take virtually all the elements from my synthesis—such as the bonobos' ability to throw light on our origins, and specifically the origins of our morality; that their social groups are much more stable than those of chimpanzees; the role of females in taming male

aggression; the liberation of consciousness; the role of self-selection; the neotenizing, juvenilization process; the use of the domestication of dogs and foxes as an illustration of the neotenizing, juvenilization process; the significance of ideal ecological conditions; the use of sex as a device to reduce tension; the reduced dimorphism between the sexes; the reliance of males on their mothers for social standing; the lack of aggression between groups of bonobos; the lack of routine hunting by bonobos, etc, etc—and, leaving out anything to do with nurturing, presented it as **'A new hypothesis'** (Ed Yong, 'Tame Theory: Did Bonobos Domesticate Themselves? A new hypothesis holds that natural selection produced the chimpanzee's nicer cousin in much the same way that humans bred dogs from wolves', *Scientific American*, 25 Jan. 2012). While it is *extremely* irresponsible to ignore and reject world-saving insights into the human condition, it is so, so much worse to actually take those insights and wantonly subvert or misappropriate the truth they contain. If that is indeed the case, and we believe there is no other plausible interpretation, then that is the very greatest of crimes against humanity.

[607] The following then is the very brief summary of the 30 years' worth of submissions to the scientific establishment of these world-saving insights into the human condition, into the integrative meaning of existence, and into the origins of our moral nature and conscious mind. Also included are other instances of the 30 years of mistreatment these insights have received. (The full presentation can be accessed at <www. humancondition.com/full-history-of-rejection>. Incidentally, this link provides details of the presentations of the whole synthesis, including the nurturing explanation for our moral soul, that have been made to Wrangham, Hare and Wobber.) These 30 years' worth of submissions represent an *extraordinarily* long and determined (and yet, so far, futile) effort to have all these fully accountable insights properly considered by the scientific establishment and conveyed to the wider world. As has been emphasized, in the case of the origins of our moral instincts, there has actually been 140 years of dishonest biological thinking on this issue, mountains of books written and oceans of wasted effort since John Fiske first solved the problem with the nurturing explanation—and 30 years now since I presented the nurturing answer to the problem in accompaniment with the explanation of the human condition, which is 30 years of terrible

human suffering and acts of atrocity that have occurred on Earth from a
lack of self-understanding in humans that should never have happened!!!

- In 1983 I wrote to Sir David Attenborough and Professor Stephen Jay
Gould, presenting these insights, but received no real response from
either.

- As described in par. 24, in 1983 I personally submitted an 8,000 word
summary of this synthesis (<www.humancondition.com/nature>) to the then
editor of *Nature* journal, Sir John Maddox, and the then Features
Editor of *New Scientist* magazine, Colin Tudge—both of whom
rejected it.

- In 1988, 800 copies of my book *Free: The End Of The Human
Condition* (<www.humancondition.com/free>) with first-rate publicity
packages were sent to every relevant scientist and journal in the world
for review, but despite a few significant commendations there was no
real response.

- In 1989 Professor John Wren-Lewis personally presented *Free* to 10
science journals including *Nature*, *New Scientist* and *Endeavour*, none
of which responded.

- In 1989 a booklet titled *Reconciliation* that contained a summary of
the synthesis was circulated to 600 scientists, scientific journals and
other relevant parties—to little response.

- In 1991 over 1,000 copies of my book *Beyond The Human Condition*
(<www.humancondition.com/beyond>) with first-rate publicity packages
were circulated to scientists, journals, universities, and relevant institutions and media—to little response from the scientific establishment
overall.

- In 1992 *Beyond* was launched at the National Museum of Kenya and
over 70 copies were given and sent to eminent scientists and influential
people—however, no lasting support eventuated.

- In 1992 Professor Wren-Lewis published a paper in which he
plagiarized my work, claiming the insights were his own! Redress
was achieved and Wren-Lewis ceased his involvement (including a
directorship) with our project.

- In 1993, 76 first-rate publicity packages with copies of *Beyond* were sent to all the leading literary agents and publishers in the world, but all declined to represent or publish the book.

- In 1995 two highly defamatory publications—an Australian Broadcasting Corporation (ABC) television program, and a full page *The Sydney Morning Herald* newspaper article—were made about me, my work and its supporters (<www.humancondition.com/persecution>).

- In 2002, 70 copies of my book *A Species In Denial* (<www.humancondition. com/asid>) were sent to the leading literary agents in the world, and another 70 copies were sent to the major international publishers, but all declined to represent or publish the book.

- In 2003 a further 800 copies of *A Species In Denial* with first-rate publicity packages were circulated to scientists, journals, universities, and relevant institutions and media. Despite a foreword by Professor Charles Birch, a commendation from Professor John Morton, and becoming a bestseller in Australia and New Zealand where it sold more than 10,000 copies, the scientific community all but failed to respond.

- Between 2004 and 2006, 2,500 copies of a documentary proposal (<www.humancondition.com/doco>) on the human condition (in which I outlined all the main biological explanations that are presented here in *FREEDOM*), were sent to scientists, scientific publications and organizations. While the proposal received over 100 commendations from leading scientists and thinkers, and engaged Professor Harry Prosen's ongoing support of these explanations, it did not produce any substantial, long-term interest from the scientific community.

- In 2005 my proposal to present 'The Citadel Of The Darwinian Revolution—The Biology Of Our Human Condition—At Last Explained', at the Annual Meeting of the American Association for the Advancement of Science (AAAS) titled *Grand Challenges, Great Opportunities*, was rejected by the Program Committee.

- In 2005 an abstract of my paper 'Nurturing as the Prime Mover in Primate Development and Human Origins' (<www.humancondition.com/ ips-2006-congress>) was submitted for presentation at the International Primatological Society's (IPS) 2006 Congress in Uganda, but, as just mentioned in par. 581, was rejected.

- In 2006 journalist Jo Sandin was unable to include reference to my nurturing explanation for human and bonobo moral behavior in her 2007 book about the bonobos at the Milwaukee County Zoo, *Bonobos: Encounters in Empathy*.

- In 2006 my book *The Great Exodus: From the horror and darkness of the human condition* (<www.humancondition.com/exodus>) was published online, and sent to many relevant scientists—to no response from the scientific community.

- In 2008 the World Transformation Movement began presenting its online Introductory Videos—however, as yet there has been little response from the scientific community.

- In 2009 my book *Freedom Expanded* (<www.humancondition.com/ freedom-expanded>) was published online, but has not yet generated any real response from the scientific community.

- In 2011 my book *The Book of Real Answers to Everything!* (<www. humancondition.com/real-answers>) was published online, but has so far failed to attract significant support from the scientific community.

- In 2012 Wrangham, Hare and Wobber published the paper 'The self-domestication hypothesis: evolution of bonobo psychology is due to selection against aggression', making no acknowledgment or even mention of my love-indoctrination synthesis, despite each being previously informed of it (in the case of Wrangham on four separate occasions)—even appearing to subvert and misappropriate virtually all the elements of my synthesis, while leaving out anything to do with nurturing.

- In 2014, despite the explanation of the human condition that is presented in my books being the fulfillment of the core vision of Geelong Grammar School of cultivating the sensitivity needed to achieve that specific, all-important-if-there-is-to-be-a-future-for-the-human-race task, the school chose *not* to include an essay on my life's work that was commissioned by its publishers for possible inclusion in its Corio anniversary book *100 Exceptional Stories* which **'celebrates the lives of 100 exceptional past students'**—see <www.humancondition. com/100-exceptional-stories>.

- From July to September in 2014 a special edition of this book that was orientated to scientists (it was even given its own title that focused on the very serious plight of the world: *IS IT TO BE Terminal Alienation or Transformation For The Human Race?*) was sent to 930 leading science organizations, scientists and science commentators in the English-speaking world, including the scientists involved with the main Brain Initiatives. But despite each copy being accompanied by a personal appeal for support for the book's insights from Professor Harry Prosen, and undertaking two trips to the US and UK to discuss the book with interested scientists and commentators, the situation in early 2016 is that while there has been a few positive responses from individual scientists, our publishers are still waiting for appreciative responses from the scientific establishment.

[608] The reader should now have some idea of the scale of the persecution that my work has had to endure, and also some idea of the immense relief that the final court case ruling brought to those supporting these understandings — in particular to myself; to my partner of 30 years Annie Williams, whose loving support enabled me to cope with all the years of persecution (in particular it was Annie's care that enabled me to recover from 10 years of the debilitating illness of Chronic Fatigue Syndrome [CFS] that I developed in 1999 after endless years of fighting the persecution, which, incidentally, were 10 of what should have been the most productive years of my life); to my brother Simon, who started the base group of people supporting these insights into the human condition; to my very close friend Tim Macartney-Snape, whose high standing in the Australian community (Tim is a twice-honored Order of Australia recipient) had been jeopardized because of his courageous, unwavering support of my work; and to every one of the other 50 Founding Members of the WTM who have each also had to endure horrible persecution, ostracism and suffering. I might mention that our struggle against persecution has been so monumental that we have had to forgo having children to ensure we have sufficient resources of time, energy and funds to effectively resist the persecution and maintain our efforts to ensure these ideas lead to the transformation of the human race. After all that we have been put through, the fact that our little, but mighty, band of brothers and sisters supporting these human-race-liberating insights is still standing and able

to mount, with this book, a further assault on the citadel of denial/lying (mechanistic science) to try to crack it open and free the human race, *is* an absolute miracle. (With regard to my CFS, although the syndrome wasn't officially identified as an illness in Darwin's day, we know it is what he suffered from because of the accurate description he gave of his condition, such as **'I believe to a stranger's eyes, I should look quite a strong man, but I find I am not up to any exertion, and I am constantly tiring myself by very trifling things'** (Letter to Charles Lyell, 1841; *The Correspondence of Charles Darwin*, Vol.2, p.298 of 603). Indeed, there has even been some talk of renaming CFS the 'Charles Darwin Syndrome' (Roger Burns, 'Chronic Fatigue Syndrome Changing the Name', Sep. 1996; see <www.wtmsources.com/123>). Presumably Darwin suffered from CFS because he too was under enormous pressure for, as mentioned earlier, having his revolutionary ideas **'greeted with violent and malicious criticism'**.)

[609] Of course, since the subject of the human condition and its resolution is both a new and confronting issue for humans to have to think about and adjust to, the initial resistance stage that Schopenhauer said new ideas in science are typically subjected to was always going to be exceptional, and it certainly has been. However, we in the WTM hope that after having successfully fought the terrible public campaign of persecution against us in the law courts, which is the proper, civilized place to seek redress, that the initial **'violently opposed'**, treated-as-**'blasphemies'** stage will soon pass and these world-saving understandings will move on to the next stage in the journey to acceptance where they are evaluated by the scientific establishment for their accountability and thus truth—a stage that will naturally involve skepticism, but of a healthy rather than brutal nature. And since these explanations of human behavior are *so* accountable and thus truthful, this last stage should not take long. As Professor Prosen says in his Introduction, **'My sincere hope is that with our species' predicament now so dire, the scientific establishment will finally acknowledge and support these critically important insights.'**

[610] History certainly teems with examples of those who have been persecuted for telling the truth, for daring to defy the great denial enshrouding our wonderful world—Christ's crucifixion being the most famous—but we now have another to match that ultimate example, which is what has been done to me; what's that rhetorical question in the Bible: **'was there ever a prophet** [a denial-free thinker] **your fathers did not persecute'** (Acts 7:52).

HOWEVER, when all the truth about human existence is accompanied by the greater, dignifying, compassionate, redeeming and rehabilitating full truth of the explanation of the human condition, *that denial is no longer justified*. And, of course, that denial is *so* habituated—our fear of the truth has soaked into our bones—but, nevertheless, that door to the possibility of understanding the human condition *must be kept open*. There *has* to be sufficient soundness and strength of character left on this planet for the full truth to be supported when it emerges, which (as has been fully evidenced) it now has—because if there isn't, then the consequences of terminal alienation for the human race that Michelangelo's and Blake's paintings at the beginning of chapter 3 horrifically portray are unthinkable.

[611] Dishonest intellectualism, as opposed to truthful instinctualism, is incredibly dangerous when it becomes overly deluded and carried away with its imagined authority, when it is **'given…to utter proud words and blasphemies'** (Bible, Rev. 13:5). No wonder Christ angrily called the denial-practicing academics of his day, **'You blind fools!… You blind guides!… you hypocrites!…You snakes! You brood of vipers! How will you escape being condemned'** (Matt. 23:17-33), and said, **'Woe to you experts in the law, because you** [your dishonesty] **have taken away the key to knowledge'** (Luke 11:52). Yes, as Einstein recognized, **'Science** [knowledge] **without religion** [denial-free truthfulness] **is lame, religion without science is blind'** (*Out of My Later Years*, 1950, p.26 of 286). To find understanding of human behavior required both intellectualism *and* instinctualism. As my headmaster at Geelong Grammar School, Sir James Darling, put it, **'conscience is the executive part of consciousness'** (*The Education of a Civilized Man*, 1962, p.96 of 223); without input and guidance for the intellect from our truthful moral instinctive self or soul, the voice of which is our conscience, then total dishonesty and terminal alienation occurs. Again, as explained in par. 138, intellectual, human-condition-avoiding, dishonest mechanistic science had to find all the details of the workings of the world that make the explanation of the human condition possible, and then instinctual, human-condition-confronting, honest thinking had to use those insights to assemble the actual explanation of the human condition—which is what has occurred, but the problem is science is not acknowledging that explanation. Yes, professors promenade across their campuses full of conviction of their own importance, the students too are certain they are where it's 'at', but universities have become one big dishonest castle of lies; in fact, rather than centers of learning that seek truth as they are supposed to be, universities have become hideous,

human-race-destroying places that indulge and glorify dishonesty! Christ foresaw this danger of the delusion of denial-committed academia when he further cautioned, **'Beware of the teachers of the law. They like to walk round in flowing robes and love to be greeted in the market-places and have the most important seats in the synagogues and the places of honour at banquets'** (Luke 20:46).

[612] Since the great achievement of life on Earth has been the development of consciousness, it follows that for those charged with the responsibility of overseeing its maturity to enlightenment and sanity—namely the scientific establishment—to abuse that responsibility is the greatest of obscenities; worse, it is a spear through the very heart of all of life and meaning on Earth. It is the meanest, most bitter, most selfish, most bloody-minded, most hateful of the truth, most unnecessary behavior this planet has ever witnessed.

[613] There is one last comment I would like to make about this awful business of science having had to practice denial. Much of this book has had to be dedicated to dismantling and exposing all the dishonest human-condition-avoiding mechanistic accounts of human behavior, so it will be an immense relief when, in the future, post-human-condition world, it will not be necessary to have to wade through all that material and just be able to read the true, and, in truth, obvious, explanations of human behavior it contains. I think that after presenting Plato's truthful account of our condition in chapter 2:6—of how **'our ancestors'** lived in a pre-human-condition-afflicted, **'innocent'**, **'blessed'**, **'divine'**, **'upright'**, **'cleanly made'**, **'pure'**, **'noble'**, **'good'**, **'modest'**, **'honour**[able]**'**, **'spirit**[ed]**'**, **'simple and calm and happy'** state, and how horrifically that contrasts with our present corrupted, **'fallen'**, **'evil'**, **'ignoble'**, **'bad'**, **'crooked'**, **'terrible'**, **'unlawful'**, **'insolent'**, **'pride**[ful]**'**, **'lumbering'**, **'disorder**[ly]**'**, **'chaos'**-causing, increasingly **'forget**[ful]**'**, **'deaf'**, threatening **'universal ruin to the world'**, **'imprisoned in'** a **'living tomb'** lives—it was, from that point on in this book, very obvious how completely committed to evasion, denial and delusion science, and indeed the whole human race, was going to become—which means that all the patient and laborious exposés of all those contrived, implausible, fanciful, ridiculous, blatantly biologically incorrect 'explanations' for human behavior, really became so predictable that in the end I think it must become tiresome to read. As mentioned in pars 541-542, science certainly has become farcical. It *is* an immense shame that, while the human condition wasn't able to be truthfully explained, scientists

had to turn themselves inside out, perform incredible contortions, in order to invent a non-confronting world of supposed explanations of our behavior for humans to live in. That's what has been so clever about the resigned upset state—how it managed to invent all these lies. *So* much brain power has been spent on that activity; a whole universe of bullshit has been created. So imagine how relieving and peaceful it is going to be to be able to leave all that behind? In fact, one day, a version of this book will be available that has all the now redundant denials printed in grey highlight, so readers can just bypass them and go straight to the truthful insights it contains into our lives. Yes, one day all the lying will stop and go forever from planet Earth, and blessed will be that day—and now, with the truth revealed, it should not be far off! I might say that, given how confronting this new honest world inevitably is for the current, first generation to encounter, I can envisage that in the not-too-distant future, the truth of the explanations of human behavior contained in this book, all anyone from this current generation will need to do is read chapters 1 and 3 for the explanation and defense of our horrifically corrupted lives, and the final chapter that explains how every human can be completely transformed from that terrible condition. Not only can the dishonest 'explanations' be avoided, but, except for the explanation of the human condition, all the other truthful explanations can also be avoided! Chapter 9 explains how this strategy of supporting the truth, without overly confronting it, works.

[614]Finally, I would be remiss not to observe how amazingly truthful and accurate Plato was in describing in his cave allegory the human condition and the difficulties involved in its eventual resolution. Plato was certainly one of the soundest men in recorded history. Indeed, along with the other two very great denial-free thinking prophets in recorded history, namely Moses and Christ, Plato made the most important contributions to humanity's great journey to enlightenment; specifically, that through the Ten Commandments Moses gave humanity the most effective form of Imposed Discipline for containing the ever-increasing levels of upset in humans; that Christ gave humanity the soundest and thus most effective corruption-and-denial-countering religion; and that Plato gave philosophy—the actual business of studying **'the truths underlying all reality'**, in particular studying and finding the all-important understanding of the human condition—the best possible orientation and assistance. So,

you could say that 'the beauty and taste of roses, rice and potato saved the human race'!! And I should add 'a leg of lamb'—the prophet Abraham—to that world-saving feast because Abraham contributed the precious foundations of real, effective religion with his emphasis on monotheism, the need to revere the fundamental truth about our existence of there being one true 'God', which we can now acknowledge is Integrative Meaning. Indeed, in what has to be the most truth-filled and thus greatest poem ever written (extraordinarily insightful extracts from it are referred to throughout this book), *Intimations of Immortality*, the poet William Wordsworth probably best described the critical role denial-free thinking prophets have played in leading humanity by terming them the **'Eye[s] among the blind'**: **'Thou best Philosopher, who yet dost keep / Thy heritage, thou Eye among the blind / That, deaf and silent, read'st the eternal deep / Haunted for ever by the eternal mind / Mighty Prophet! Seer blest! / On whom those truths do rest / Which we are toiling all our lives to find / In darkness lost, the darkness of the grave'** (1807). The **'darkness lost, the darkness of the grave'** where no truth is accessible equates perfectly with Plato's **'living tomb' 'cave'** of alienating denial where virtually the entire human race has had to live incarcerated against the glaring truth of our condition. Yes, Wordsworth's reference to prophets being the **'Eye among the blind'** perfectly describes how denial-free, truthful thinking had to act in partnership with human-condition avoiding, **'blind'**, **'toiling'** mechanistic science for the truth about the human condition to be found. In the words of the *Encarta* summary of Plato's cave allegory, **'the proper object of knowledge** [science]**'** has been to achieve **'the transition to the real world'**—that is, end the alienated state of denial that humans have had to live in, get out of our species' horrible **'darkness of the grave'**, **'cave'** existence. It is only the **'enlightenment'** of our **'imperfect' 'human condition'** that enables us to be **'released from'** the **'bonds'** of our **'cave'-'like' 'prison'** of **'almost blind'** alienated denial and, as a result, be **'cured of'** our **'illusion'** and **'delusions'**. I should reiterate that Plato also emphasized how relieved the cave prisoner would be to be free of, and transformed from, his old dishonest existence by saying, **'when he thought of his first home and what passed for wisdom there, and of his fellow-prisoners, don't you think he would congratulate himself on his good fortune and be sorry for them?'** (*The Republic*, 516).

[615] (I should mention that I have written much more in *Freedom Expanded* (see <www.humancondition.com/freedom-expanded-Abraham-Moses-Plato-Christ>) about Plato's extraordinary contribution to the human journey

to enlightenment of his insightful thinking about the human condition—
and about the absolutely incredible contribution <u>Christ</u> made to the human
journey; indeed, it is a measure of the precious way of coping with the
human condition that Christ gave the human race that he is considered
**'the most famous man in the world'** (*Jesus Revealed*, National Geographic Channel, 2009)
and that most of the world dates its existence around his life, as either BC
or AD, 'Before Christ' or 'Anno Domini', which translates as 'in the year
of our Lord', referring to the year of Christ's birth. I also explain more
about the absolutely incredible contribution <u>Moses</u> made to the human
journey with his first five books of the Bible being fundamental teaching
in Christianity, Judaism and Islam, and about the absolutely incredible
contribution of monotheism that <u>Abraham</u> gave to the human journey.)

[616]The remaining three chapters in this book explain the emergence of
consciousness in humans, the story of humanity's journey from ignorance
to enlightenment, and, finally, how the fabulous transformation of the
human race takes place. In regard to the next chapter's explanation of how
consciousness emerged, unlike the mysteries of the human condition, of
our meaning, and the origin of our moral soul, our ability to understand
consciousness depended on recognizing so many historically denied
truths that mechanistic science can't be as easily criticized for ignoring its
discovery as it can be for ignoring the other key insights—which is why
the main damnation of science for its ill-treatment of these world-saving
insights appears here in chapter 6.

## Chapter 6:13 <u>Nurturing now becomes a priority</u>

[617]Before progressing to those final chapters it needs to be emphasized
that having found the compassionate biological understanding of the
upsetting battle that humans have had to wage against the ignorance
of our instinctive self as to the fundamental goodness of our conscious
self, our species' preoccupation with that battle can at last end and focus
return to the nurturing of our children. Further, now that we have found
that *very* good reason *why* parents have been unable to nurture their
children with anything like the unconditional love children instinctively
expect and need, it is no longer necessary to deny the importance of
nurturing both in the maturation of our species and in our own lives. Yes,
with the arrival of understanding the terrible guilt that parents have felt

for not being able to adequately nurture their offspring departs. Parents can now understand that their inability to adequately love their offspring is *not* their fault, but the product of the ever-accumulating levels of upset in the human race as a whole that resulted from our species' heroic 2-million-year search for knowledge, specifically for self-knowledge, understanding of why we conscious humans are good and not bad. *So yes, we can finally now admit the importance of nurturing and turn our attention back to the nurturing of our offspring—in fact, it now becomes a matter of great urgency that we do both these things.*

[618] As stated in par. 564, the nurturing of children is what is required to produce adults who are sound and secure in self. Ultimately, it is *only* through the nurturing of our offspring that the human race can hope to become healthy—namely psychosis-and-neurosis-free—once again; to, as Ashley Montagu said, put **'man back upon the road of his evolutionary destiny from which he has gone so far astray'** and restore **'health and happiness for all humanity'**. As will be explained in chapter 9, with understanding of the human condition found all humans can *immediately* be free of the agony of the human condition by taking up the Transformed Way of Living (which involves adopting a mental attitude where you leave that agony behind as dealt with), but to produce humans who are *completely* free of psychosis and neurosis and not needing to take up the Transformed Way of Living, they need to be nurtured with unconditional love in their infancy and childhood—a process that will take some generations to achieve while all the accumulated upset within our species subsides.

[619] The initial difficulty then for virtually all parents is that since they are so psychologically upset, especially so alienated from their natural, loving, instinctive true self or soul as a result of humanity's 2-million-year upsetting search for knowledge, it is simply not possible for them to give their children, or indeed anyone for that matter, unconditional love. HOWEVER, while virtually all parents will have to accept that they will not be able to give their offspring unconditional love, the opportunity now to actually understand the very good reason for *why* they are unable to do so will bring so much calming, reassuring relief that nurturing won't be nearly as fraught with guilt and difficulty as it has been. So again, while it will take a number of generations before the majority of parents will be able to generate real, natural, unconditional love, <u>the ability now to understand our species' upset state, and the realization now that each and</u>

every person no longer has to continue with the upsetting battle to prove their worth (which, as will be explained in chapter 9, is what makes the fabulously exciting new Transformed Way of Living possible), is going to bring *enormous* relief to all men and women, which will help them immensely with the whole activity of nurturing. Indeed, the overall relief and excitement alone that will now emerge in the human race from the realization that we have finally ended our 2-million-year struggle against ignorance will mean that children are going to find themselves in, if not an unconditionally loving world, then at least an anxiety-free, truthful, compassionate, happy one. And significantly, since the adult world can finally break the silence of their denial and tell children the truth about their horrifically upset condition, that honesty will also make an immense difference to the psychological wellbeing of children.

[620] Basically, finding understanding of the human condition makes possible a *whole new world* that is relieved of the horrible frustration and agony of that condition; the terrible siege state where humans couldn't understand themselves can at last be lifted, and, with its lifting, life can return to every aspect of human existence, including to the impossible situation parents have been in trying to nurture their children. Yet again, we see how important it is that science recognizes these denial-free understandings of the human condition to help the public at large become aware of them. The ignoring, and, when pushed, attack and repression, of these answers MUST stop, and they must instead be supported — because through that support this great transition for the human race will be able to begin in earnest. Yes, this end to humanity's great, heroic but horrifically upsetting battle to find knowledge, ultimately self-knowledge, understanding of the human condition, together with this change from having to live in denial of the importance of nurturing and of our extremely corrupted condition, changes the whole direction of the human race. Rather than plunging towards ever greater levels of upset, especially towards terminal levels of denial/alienation, suddenly humanity can be heading towards a world free of the human condition.

[621] (Much more will be said in chapters 8:16B, 8:16C and 8:16D about the devastating consequences of parents' inability to nurture their infants while under the duress of the human condition, and how parents are to cope with the truth about the importance of nurturing.)

# Chapter 7

# What is Consciousness, and Why, How and When Did Humans' Unique Conscious Mind Emerge?

rolffimages / 123RF Stock Photo

## Chapter 7:1 <u>Summary</u>

[622] Earlier in this book, four great outstanding questions were identified as having to be solved for peace to finally come to our species' troubled existence. They were to explain the human condition itself (the psychosis-addressing-and-solving explanation for which was presented in chapter 3); to explain the meaning of our lives (which was done in chapter 4); to explain how humans acquired our

extraordinary unconditionally selfless moral sense (the answer to which was presented in chapter 5); and, finally, to explain how humans became conscious when other animals haven't, the answer to which will now be presented in this chapter.

## Chapter 7:2 <u>What is consciousness?</u>

[623] To address the question of the origins of humans' conscious mind it is obviously first necessary to explain what consciousness actually is, but as with the truths of our corrupted human condition, of Integrative Meaning and of the nurturing origins of our moral sense, the truth about the nature of consciousness has been so unbearably condemning of our corrupted condition that we humans have had no choice but to deny it until the human condition was truthfully explained, which it now has been.

[624] The fact is, the subject of consciousness is *so* closely correlated with the unbearably depressing subject of the human condition that 'consciousness' has become synonymous with—indeed, code for—the problem of the human condition. The science writer Roger Lewin articulated the problem precisely when he described the great difficulty humans have had of trying to **'illuminate the phenomena of consciousness'** as **'a tough challenge...perhaps the toughest of all'** (*Complexity*, 1993, p.153 of 208), using the philosopher René Descartes' own disturbed reaction when he tried to **'contemplate consciousness'** to illustrate the nature and extent of the difficulty: **'So serious are the doubts into which I have been thrown...that I can neither put them out of my mind nor see any way of resolving them. It feels as if I have fallen unexpectedly into a deep whirlpool which tumbles me around so that I can neither stand on the bottom nor swim up to the top'** (p.154).

[625] My professor of biology when I was at Sydney University, Charles Birch, also acknowledged the difficulty mechanistic science has had looking at the subjective, psychological, terrifying-**'deep'**-**'whirlpool'**-of-self-**'doubts'**-causing issue of the human condition dimension to the study of consciousness when, in an interview conducted in 2007 (two years prior to his death at the age of 91), he said: **'there are two aspects of consciousness—[mechanistic] science deals with the objective facts, in other words with what happens in the cells of your brain when you have a conscious thought, but it leaves unanswered the question about the feeling I have of consciousness; and there is a tremendous gap between what I experience and**

**what science tells me, and this is the gap that somehow or other has eventually to be filled...Science has great difficulty in dealing with that [subjective side]. It is highly successful in dealing with the objective events in the brain, and an awful lot of people are working on consciousness right now, but they don't touch on the problem of the subjective side because it's too hard...you have to be a philosopher...**[and be able to deal with] **the metaphysical side of things...Biology is still a very mechanistic science...most scientists don't think at all, anyway** [about the deeper subjective issues—the deepest of all being the issue of our horrifically corrupted human condition]; **they think on their subject about as much as a bank clerk thinks about them!'** (*Religion Report*, ABC Radio National, 19 Dec. 2007). Yes, as the philosopher Benjamin Fong pointed out about all the 'brain initiatives' that are currently taking place, mechanistic science's **'synthetic'** focus on the **'organic'** rather than the **'psychological'** aspect of being conscious can only end in denying humans **'the possibility of self-transformation'** (see par. 605). These comments by Birch about the limitations of mechanistic science in understanding human behavior, reinforce his other exceptionally honest comments that were mentioned in par. 225, specifically that '[mechanistic] **science can't deal with subjectivity...what we were all taught in universities is pretty much a dead end'**; and, **'the traditional framework of thinking in science is not adequate for solving the really hard problems'**; and, **'Biology has not made any real advance since Darwin'**; and, **'Biology right now awaits its Einstein in the realm of consciousness studies.'** Yes, progress in biology has been stalled at the issue of human behavior, because understanding our behavior depended on confronting the psychological issue of the human condition that has arisen in humans' conscious brain.

[626] So while anyone who has searched the term 'consciousness' on the internet will have found it to be a subject cloaked in mystery and confusion, it's *not* because consciousness is an impenetrably complex subject, as we may think—it is *because it raises the unbearable issue of the human condition* and, as a result, has been *deliberately* left in an obscure, cryptic state. Resigned, human-condition-avoiding humans simply haven't wanted to know what consciousness actually is.

[627] There have been two particular reasons why the subject of consciousness raised the unbearable issue of the human condition and why, therefore, examination of it led to such a fearful, all-our-moorings-taken-from-under-us, **'deep whirlpool'** of terrible depression for upset humans— and why, it follows, the subject has been deemed a 'no-go zone'.

[628]The first particular reason is that even beginning to *vaguely* contemplate the nature of our human situation invariably led to the initial obvious question of 'What makes humans unique?', with the answer clearly being that we are conscious—but thinking about *that* marked the start of a very slippery, dangerous slope because it quickly led to the depressing accusation: 'Well, if we are such a clever species why do we treat each other and our planet so appallingly?' Yes, trying to think about consciousness meant trying to fathom what—when we humans are the only fully conscious, reasoning, intelligent, extraordinarily clever, 'can-get-a-man-on-the-Moon' animal—is so intelligent and clever about being so competitive, selfish and aggressive; in fact, *so* ruthlessly competitive, brutal and even murderous, that human life has become all but unbearable and we have nearly destroyed our own planet?! It is worth recalling here the powerful words of the polymath Blaise Pascal and the playwright William Shakespeare that were included in chapter 1:2. When it came to articulating the bewildering dichotomy of our species' condition, neither man minced his words, with Shakespeare writing, **'What a piece of work is a man! how noble in reason! how infinite in faculty!...in action how like an angel! in apprehension how like a god! the beauty of the world! the paragon of animals! And yet, to me, what is this quintessence of dust? man delights not me'** *(Hamlet,* 1603)!! Pascal followed suit, despairing, **'What a chimera then is man! What a novelty, what a monster, what a chaos, what a contradiction, what a prodigy! Judge of all things, imbecile worm of the earth, repository of truth, a sewer of uncertainty and error, the glory and the scum of the universe!'** *(Pensées,* 1669). Each was daring to ask the same unbearable question: how is it that humans can be the most brilliantly clever of creatures, the ones who are **'god'-'like'** in our **'infinite' 'faculty'** of **'reason'** and **'apprehension'**, a **'glor**[ious]**'**, **'angel'-'like' 'prodigy'** capable of being a **'judge of all things'** and a **'repository of truth'**, and yet also be *so* mean, *so* vicious, *so* capable of inflicting pain, cruelty, suffering and degradation? *Why* are humans so choked full of volcanic frustration, anger and hatred—the species that behaves *so appallingly* that we seem to be **'monster**[s]**'**, **'imbecile**[s]**'**, **'a sewer of uncertainty and error'** and **'chaos'**, the **'essence'** of **'dust'**, **'the scum of the universe'**?

[629]Yes, the unbearable prospect that our intelligent, rational, insightful, aware, understanding human mind has, it seems, unintelligently, irrationally, thoughtlessly, indifferently and stupidly almost destroyed the whole planet we live on and brought human existence to a state of

unbearably lonely, alienated, selfish, aggressive and egocentricity-crazed dysfunctionality has been an *extremely* confronting matter to think about. While our fully conscious, intelligent mind *is* surely the culminating achievement of the grand experiment in nature that we call life, it *also* appeared to be to blame for all the devastation and human suffering in the world! Rather than being wonderful, our conscious mind appeared to be the plague of the planet! *That* is how 'serious are the doubts' that thinking about consciousness produced within us! *And*, our inability—until now—to understand this dichotomy of being the most clever and brilliant but also the most apparently destructive, evil and stupid force to have ever appeared on Earth has meant that we humans have, understandably, been *extremely* insecure and defensive about, and thus hesitant to explore the inner workings of, our supposedly wonderful conscious mind.

[630] The second particular reason why the subject of consciousness has been so confronting is because thinking about the nature of consciousness quickly brought us into contact with the truth of the selflessness-dependent, loving, integrative meaning of existence, the unbearably depressing implications of which were described in chapter 4. The explanation of what consciousness actually is will reveal the problem because, as we will see, while there is a simple and obvious explanation for consciousness itself (like there is for the human condition, the meaning of our existence and the origin of our moral soul), that explanation has had very confronting implications.

[631] As briefly described in chapter 3:3, humans can be distinguished from other animals by the fact that we are fully conscious—that is, sufficiently able to understand and thus manage the relationship between cause and effect to wrest management of our lives from our instincts, and even to reflect upon our existence, specifically the issue of our immensely upset human condition that wresting management from our instincts brought about. This consciousness is a product of the nerve-based learning system's ability to remember, for it is memory that allows understanding of cause and effect to develop.

[632] To elaborate, nerves were originally developed as connections for the coordination of movement in multicellular animals. An incidental by-product of the development of nerves was that of memory. While the actual mechanism by which nerves are able to store impressions is not yet fully understood, we do know it involves chemical processes. What is important, however, is that nerves have the capacity for memory

because once you have the ability to remember past events, you are then in a position to compare them with current events and identify regularly occurring experiences. This knowledge of, or insight into, what has commonly occurred in the past enables the mind to predict what is likely to occur in the future and to adjust behavior accordingly. <u>Thus, the nerve-based learning system (unlike the gene-based learning system) can *associate* information, *reason* how experiences are related, learn to *understand* and become CONSCIOUS of the relationship of events that occur through time</u>.

[633] In the brain, nerve information recordings of experiences (memories) are examined for their relationship with each other. To understand how the brain makes these comparisons, think of the brain as a vast network of nerve pathways onto which incoming experiences are recorded or inscribed, each on a particular path within that network. Where different experiences share the same information, their pathways overlap. For example, long before we understood the force of gravity we had learnt that if we let go of a weighted object it would invariably fall to the ground—falling to the ground was a shared characteristic. The value of recording information as a pathway in a network is that it allows related aspects of experience to be physically related. In fact, the area in our brain where information is related is called the 'association cortex'. Where parts of an experience are the same they share the same pathway, and where they differ their pathways differ or diverge. All the nerve cells in the brain are interconnected, so with sufficient input of experiences onto a nerve network of sufficient size, similarities or consistencies in experience show up as well-used pathways, pathways that have become highways. (In terms of illustrating the brain's capacity to associate information, it has been found that in the convolutions of our brain's cerebral cortex there are about 20 billion nerve cells with 10 times that number of interconnecting dendrites which, if laid end to end, would stretch at least from Earth to the Moon and back.)

[634] An 'idea' represents the moment information is associated in the brain. Incoming information could reinforce a highway, slightly modify it or add an association (an idea) between two highways, dramatically simplifying that particular network of developing consistencies to create a new and simpler interpretation of that information. For example, the most important relationship between different types of fruit is their edibility. Elsewhere, the brain has recognized that the main relationship

connecting experiences associated with living organisms is that they appear to try to stay alive, at least for a period of time. Suddenly it 'sees' or deduces ('tumbles' to the idea or association or abstraction) a possible connection between eating and staying alive which, with further experience and thought, becomes reinforced as 'seemingly' correct. 'Eating' is now channeled onto the 'staying alive' highway. And it is at this point of information association that the issue of the human condition appears, for any subsequent thought would try to deduce the significance of 'staying alive' and, beyond that, the meaning of life and the relevance of selfishness and selflessness. Ultimately the brain would arrive at the unbearable truth of Integrative Meaning, which, as explained in chapter 4, upset humans have not wanted to face because it confronted us with our *lack* of integrativeness, with our divisiveness.

[635] The process of forgetting also plays a part in forging the relationship between experiences. Since duration of nerve memory is related to use, our strongest memories will be of those highways, those experiences that have the greatest relativity. Our experiences not only become related or associated in the brain, they also become concentrated because the brain gradually forgets or discards inconsistencies or irregularities between experiences. Forgetting serves to cleanse the network of less consistently occurring information, preventing it from becoming cluttered with meaningless (non-insightful) information.

[636] Once insights into the nature of change are put into effect, the self-modified behavior starts to provide feedback, refining the insights further. Predictions are compared with outcomes, leading all the way to the deduction of the meaning of all experience, which, as emphasized, is to order or integrate matter.

[637] Our language development took the same path as the development of understanding. Commonly occurring arrangements of matter and commonly occurring events were identified (became clear or stood out). Eventually all the main objects and events became identified and, as language emerged, named. For example, we named those regularly occurring arrangements of matter with wings 'birds' and what they did we termed 'flying'.

[638] Consciousness, therefore, is the ability to understand the relationship of events sufficiently well to effectively manage and manipulate those events; it represents the point at which the confusion of incoming information clears, starts to fit together or make sense and the mind becomes

effective, a master of change. Chimpanzees, for example, demonstrate consciousness when they effectively reason that by placing boxes one on top of the other they can create a stack that can then be climbed upon to reach a banana tied to the roof of their cage. It should be pointed out, however, that it is one thing to be able to stack boxes to reach bananas—to manage immediate events—but quite another to effectively manage events over the long term, to be secure managers of the world. In fact, as will be explained in chapter 8, bonobos (and, to a lesser degree, the other great apes) demonstrate the same degree of consciousness that humans exhibit during infancy, the stage between the ages of two and three when we develop sufficient consciousness to recognize that we are at the center of the changing array of experiences around us. It is when we become aware of the concept of 'I' or self and discover conscious free will, the power to manage events. This is followed by the childhood stage, during which we revel in this free will, 'play' or experiment with it, while adolescence is when we encounter both the sobering responsibility of free will and the agonising identity crisis brought about by the dilemma of the human condition—of whether or not we are meaningful beings. Adulthood, it follows, is the mature stage that humanity is now in a position to enter, for we finally have the understanding of the human condition that allows us to end our species' insecure state of adolescence forever.

[639] In summary, 'insight' was the term given to the nerve highways, the correlation our brain made of the consistencies or regularities it found between events through time. Once humans could deduce these insights—these laws governing events in time past—we were in a position to predict or anticipate the likely turn of events. We could learn to understand—be conscious or aware of—what happened through time. Our intellect could understand the design inherent in, and the process behind, changing information; it could *learn* the predictable regularities or common features in experience, and thus ultimately distil the integrative theme or meaning or purpose or direction of existence.

[640] So, yes, consciousness has been an immensely difficult subject for humans to investigate—not because of the practical issues involved in determining how our brain works, but because *resigned humans didn't want to know how it worked*. While we couldn't explain our divisive, psychologically upset state of the human condition we had to avoid admitting too clearly how the brain functions because admitting that information could be associated and simplified—admitting to

insight—was only a short step away from realizing the *ultimate* insight of Integrative Meaning, a truth that immediately confronted us with our own inconsistency with that meaning.

[641] As emphasized in chapter 4, to admit to Integrative Meaning meant having to face the fact that our competitive and aggressive behavior is seemingly totally at odds with the integrative direction of life, no less. The development and maintenance of the order of matter requires that the parts of developing wholes cooperate *not* compete. *Integrative* Meaning confronts us squarely with our *divisive* human condition. Better then to evade the existence of purpose in the first place by avoiding the possibility that information could be associated, refined and simplified. It is the same reason we sidestepped the term 'genetic refinement' for the process of the genetic refinement of the integration of matter on Earth, preferring instead the much vaguer term 'genetics'. We had to evade the possibility of the refinement of information in *all* its forms because, again, admitting that information could be simplified or refined was admitting to an ultimate refinement or law, confronting us with our inconsistency with that law, namely with the law of Integrative Meaning, against which we previously had no defense.

[642] In fact, not only have we avoided the idea of meaningfulness, we have also shied away from any deep, meaningful thinking that might *lead* to confrontation with Integrative Meaning—as Birch said, **'most scientists don't think at all, anyway; they think on their subject about as much as a bank clerk thinks about them'**! But while ensuring deeper insights remained elusive may have saved us from exposure, in the process we buried the truth, becoming, as a result, *extremely* superficial in our thinking. In short, we became masters of *not* thinking—deeply-alienated-from-the-truth beings. Demonstrating our masterful evasion of the nature of consciousness we used words like 'conscious', 'intelligent', 'understanding', 'reason' and 'insight' regularly without ever actually identifying what we are conscious of, intelligent about, understanding, reasoning or having an insight into, which is how events or experiences are related. The conventional obscure, evasive definition of intelligence is **'the ability to think abstractly'**. The other imprecise, obscure, evasive phrase used whenever we wanted to refer to the uniqueness of our intelligence without actually saying what our conscious, understanding, insightful intelligence is, was to say that 'We are the species that is able to reflect upon itself.' So to name the area of the brain that associates and simplifies information as the 'association

cortex' was, in fact, a slip of our evasive guard. Of course, when we weren't 'on our guard' against exposure few would deny that information can be associated, simplified and meaning found. In fact, most of us would say we do it every day of our lives—if we didn't, we wouldn't have a word for 'insight'. That is the amazing aspect about our denial of anything that brings the dilemma of the human condition into focus: it is not unusual for resigned humans to accept an idea up to a point, but as soon as it starts to lead to a confronting conclusion, pretend it doesn't exist—and do so without batting an eyelid.

[643] To illustrate how we avoided acknowledging the fundamental ability of the brain to associate and reduce information to essentials (and thus be forced to deduce the integrative meaning or theme or purpose in experience), consider the following case from my files of a *Newsweek* magazine cover story (7 Feb. 1987). While the title and subject of the nine-page article raised the crucial question of 'How the brain works', the author referred to the association capability of the brain in such an unnecessarily complicated way it was effectively buried: **'Productive thought requires not just the rules of logic but a wealth of experience and background information, plus the ability to generalise and interpret new experiences using that information.'** The **'ability to generalise and interpret'** is the ability to associate information, but the meaning is all but lost in the sentence.

[644] In case it is thought this convoluted description may have been due to poor expression rather than deliberate evasion on the part of the author, it should be pointed out that apart from a mention of **'chunking or grouping of similar memories together'** and one unavoidable mention of the **'association cortex'**, the article contains no other reference to the brain's fundamental ability to associate information. The entire nine-page piece, on how our brain works, hangs on this one inept sentence. If you are not intending to be evasive it is not difficult to clearly describe the mind's ability to associate information, as has just been done.

[645] More recently, *National Geographic* magazine's February 2014 edition carried a feature story on **'how the human mind really works'**, but, again, the article didn't go anywhere near the issue of **'how the human mind really works'**. Titled **'Secrets of the Brain – New technologies are shedding light on biology's greatest unsolved mystery: how the brain really works'**, the article referred to **'President Barack Obama's 'BRAIN Initiative'** and mentioned **'disorders such as schizophrenia, autism and…depression'**, however, there was not even an *attempt* in this article to address and explain the

all-significant feature of **'the human mind'** of consciousness; in fact, there was no mention of 'consciousness', and certainly not of the dilemma of the human condition associated with being a conscious species. The article was completely mechanistic, totally evasive of any even oblique reference to the crippling psychosis and neurosis involved in the **'disorders'** now plaguing our conscious **'human mind'**. The denial practiced in the 1987 *Newsweek* story is chicken feed compared to the level of denial in this 2014 story. Benjamin Fong was certainly right when, as mentioned in par. 605, he predicted that **'The real trouble with the Brain Initiative'** would be its **'synthetic'** focus on the **'organic'** rather than the **'psychological'** nature of mental problems.

[646] I should emphasize that our evasion and denial is often obviously false and yet we believed it, because we had to. For instance, in the case of Integrative Meaning, we are surrounded by examples of integration everywhere we look—every object, enduring or otherwise, is a hierarchy of ordered matter, testament to the development of order of matter—and yet we have denied it, just as mechanistic science couldn't even provide a definition for two of humanity's most commonly used and important words/concepts: 'love' and 'soul'. The hypocrisy inherent in denial is palpable yet understandable.

[647] The most honest and yet still somewhat intellectually evasive description I have found in the denial-complying mechanistic paradigm of what gives rise to consciousness is one that appears in Wikipedia: **'Memory...allows sensory information to be evaluated in the context of previous experience...and...allows information to be integrated over time so that it can generate a stable representation of the world'** (see <www.wtmsources.com/128>). Yes, consciousness *is* the ability to **'generate a stable representation of the world'**, the ability to generate a representation of the world that is accurate and can therefore be effectively used to base anticipations of future events upon.

[648] So it's not difficult to explain consciousness if you are not needing to avoid what it is. To look at someone with a track record of thinking truthfully, it was mentioned in par. 187 that the naturalist Eugène Marais wrote these amazingly insightful words about the human condition: **'The highest primate, man, is born an instinctive animal. All its behavior for a long period after its birth is dominated by the instinctive mentality...it has no memory, no conception of cause and effect, no consciousness...As the... individual memory slowly emerges, the instinctive soul becomes just as slowly**

submerged…For a time it is almost as though there were a struggle between the two.' Notice the ease with which Marais describes consciousness—'memory' makes possible 'conception of cause and effect', which gives rise to 'consciousness'. It really is as obvious and straight forward as that! Indeed, when the writer Aldous Huxley commented that 'Non-rational creatures do not look before or after' (par. 250), or when Plato referred to our original innocent pre-conscious ancestors as 'having no memory of the past' (in par. 170), they were simply stating the obvious truth that to be 'rational' is to have the ability to 'look before or after', to have 'memory of the past', in order to make sense of cause and effect. Which again raises the question: 'Why on earth was this not explained to us by science, why weren't we taught this at school?' But as has now been explained, the issue of our conscious intelligent mind prompted the unbearable self-realization, 'Well, if I'm so cleverly insightful why can't I manage my life in a way that is not so mean and indifferent to others; indeed, why, if I am such a brilliantly intelligent person, am I such a destructively selfish, angry, egocentric, competitive and aggressive monster?' And as was emphasized, the other problem was that explaining the nature of consciousness very quickly brought us into contact with the unbearably depressing truth of Integrative Meaning. An appropriate definition of 'consciousness' is 'the ability to make sense of experience', but applying such a definition immediately highlights the problem with the issue of consciousness, because, due to the depressing implications, we humans haven't wanted to 'make sense of experience' because we have not wanted to recognize the truth of Integrative Meaning. So, to ask people to look into the issue of consciousness was to expect them to confront the issue of their own less-than-ideal human condition, which is why 'consciousness' has become a code word for the issue of the human condition. We can, therefore, appreciate why, when Descartes tried to 'contemplate consciousness', it caused him such fearful depression that he said, 'It feels as if I have fallen unexpectedly into a deep whirlpool which tumbles me around so that I can neither stand on the bottom nor swim up to the top.'

[649]The great psychiatrist R.D. Laing acknowledged both the importance of the issue of consciousness (the human condition), and how truly difficult a 'realm' it has been for upset humans to study when he wrote (the underlining is my emphasis): 'The requirement of the present, the failure of the past, is the same: to provide a thoroughly self-conscious and self-critical human account of man…Our alienation goes to the roots. The realization of this

is the essential springboard for any serious reflection on any aspect of present inter-human life [pp.11-12 of 156] ...We respect the voyager, the explorer, the climber, the space man. It makes far more sense to me as a valid project—indeed, as a desperately urgently required project for our time—to explore the inner space and time of consciousness. Perhaps this is one of the few things that still make sense in our historical context. We are so out of touch with this realm [so in denial of the issue of the human condition] that many people can now argue seriously that it does not exist. It is very small wonder that it is perilous indeed to explore such a lost realm [p.105]' (*The Politics of Experience* and *The Bird of Paradise*, 1967).

[650] Clearly, the need to find a way to avoid confronting what consciousness entailed was immense, and the initial obvious excuse that mechanistic science devised to do so was to simply assert that the phenomenon of the human mind is just too extraordinary for science to be able to explain; that, as a mechanistic scientist might say, 'While science might expect to be able to explain the physical mechanisms operating in our brain it is beyond our present powers of explanation to account for how a collection of neurons could produce the subjective, uniquely personal experience of the human mind.' Viewed superficially, it *is* incredible that the mental state, where we have our own personal feelings and particular sense of self, can arise from a bunch of nerves, but really it is simply a consequence of there having been sufficient development of nerves' ability to understand cause and effect. In the same light, it *is* amazing that 92 elements operating under the law of Negative Entropy could give rise to the amazing variety and complexity of life that we see around us, or that the simple combustion engine could give rise to the amazing phenomenon that is a Ferrari sports car, but they have. We can try to argue that knowing how the combustion engine works doesn't explain the mystique that surrounds a Ferrari, but, in fact, it does. Consciousness *is* amazing but it is simply a result of memory which allows us to understand cause and effect. So while the claim that it was beyond our current powers to explain the phenomenon of consciousness no doubt helped upset humans avoid the issue of the human condition, the argument was in truth nothing more than intellectual bluff.

[651] Given his aversion to the subject, it is not surprising that the whole body-mind dichotomy excuse for not being able to explain consciousness appears to have begun with Descartes when, in 1641, he wrote that the mind was **'entirely and truly distinct from my body'** (*Meditations on First*

*Philosophy*, Meditation VI, Part 9). This idea of consciousness as a sort of **'ghost in the machine'** (Gilbert Ryle, *The Concept of Mind,* 1949, p.15 of 334) of the brain, which could not be explained in physical terms, was embraced by scientists such as Thomas Huxley, who in 1868 wrote, **'how it is that any thing so remarkable as a state of consciousness comes about as the result of irritating nervous tissue, is just as unaccountable as the appearance of the Djin when Aladdin rubbed his lamp'** (T.H. Huxley & W.J. Youmans, *The Elements of Physiology and Hygiene: A text-book for Educational Institutions*, 1868, p.178 of 420). By the 1990s this excuse that the state of consciousness cannot be accounted for had become so established it was actually referred to as the 'hard problem' of consciousness, as distinct from the 'easy problem' of explaining the brain's neurological circuitry and activities. The philosopher who coined these terms, David Chalmers, wrote that **'It is widely agreed that experience arises from a physical basis, but we have no good explanation of why and how it so arises. Why should physical processing give rise to a rich inner life at all?** ('Facing up to the Problem of Consciousness', *Journal of Consciousness Studies*, 1995, Vol.2, No.3); and, **'How does the water of the brain turn into the wine of consciousness?...How is it that all of this matter adds up to something as complex, as interesting and as unique as consciousness?'** ('David Chalmers: The Conscious Mind', *Thinking Allowed with Jeffrey Mishlove*, TV series and DVD collection, #H230). In contrasting the 'easy problem' with the supposed 'hard problem' of explaining our subjective mind, the cognitive scientist Steven Pinker wrote that **'The Hard Problem...is why it feels like something to have a conscious process going on in one's head—why there is first-person, subjective experience...The problem is hard because no one knows what a solution might look like or even whether it is a genuine scientific problem in the first place...To appreciate the hardness of the Hard Problem, consider how you could ever know whether you see colours the same way that I do. Sure, you and I both call grass green, but perhaps you see grass as having the color that I would describe, if I were in your shoes, as purple'** ('The Brain: The Mystery of Consciousness', *TIME*, 29 Jan. 2007). In truth, it is intellectual nonsense to argue that we can't reach a shared understanding of the nature of our world. The philosopher John Searle spoke honestly about consciousness when he wrote that **'Brain processes cause consciousness but the consciousness they cause is not some extra substance or entity. It is just a higher level feature of the whole system'** ('The Problem of Consciousness', *Social Research*, 1993, Vol.60, No.1).

[652] The cognitive scientist Daniel Dennett was approaching the truth about the real problem surrounding consciousness when he wrote that **'With consciousness, however, we are still in a terrible muddle. Consciousness**

**stands alone today as a topic that often leaves even the most sophisticated thinkers tongue-tied and confused. And, as with all the earlier mysteries, there are many who insist—and hope—that there will never be a demystification of consciousness'** (*Consciousness Explained*, 1991, p.22 of 528). Yes, the psychologically upset and insecure human race hasn't wanted to know what consciousness is, because it didn't want to confront the human condition! What was required for humans to tolerate the **'demystification of consciousness'** was for the human condition to be compassionately explained, at which point the contrived **'muddle'** could be resolved; as Pinker wrote, **'I admit that the theory** [that consciousness is beyond our powers of explanation] **could be demolished when an unborn genius—a Darwin or Einstein of consciousness—comes up with a flabbergasting new idea that suddenly makes it all clear to us. Whatever the solutions to the Easy and Hard problems turn out to be, few scientists doubt that they will locate consciousness in the activity of the brain'** ('The Brain: The Mystery of Consciousness', *TIME*, 29 Jan. 2007). When Birch observed that **'Biology right now awaits its Einstein in the realm of consciousness studies'**, he too was referring to the need for a biologist to find understanding of the human condition. But as I explained in par. 479, it wasn't **'genius'** in the sense of extreme cleverness, a high IQ, that was required to **'suddenly make... it all clear'** and explain how consciousness arose from the basic **'activity of the brain'**, namely memory, it was simply a sound, denial-free, honest approach to thinking about the human condition. Indeed, as will now be explained, focusing on the need for a high IQ to solve problems was just another expression of the insecurity surrounding the whole issue of consciousness—just another way to laud our intelligence and gloss over its corrupting effects.

[653] The response taken by adolescents when faced with the unbearable issue of the human condition perfectly illustrates the response virtually all humans have had to confronting the nature of consciousness—so, if the reader will bear with me, I will now briefly reiterate just how determinedly adolescents blocked out the issue of the human condition.

[654] When the concept of Resignation was explained in chapter 2:2 it was described how when upset humans were around 14 or 15 years of age they each tried to face down the issue of the imperfection in the world of humans around them and in their own behavior—the issue of the human condition—only to find it a suicidally depressing exercise. As Carl Jung said, **'When it** [our **'shadow'**, the negative aspects of ourselves] **appears...it is quite within the bounds of possibility for a man to recognize the relative evil**

**of his nature, but it is a rare and shattering experience for him to gaze into the face of absolute evil.'** To avoid subjecting themselves to that **'shattering experience'** ever again, adolescents were forced to resign themselves to never revisiting the issue of the human condition. In fact, every moment thereafter was spent carefully avoiding any encounter with the issue, and denying any truths that brought it into focus. What was not described in chapter 2:2, however, was the strategy a resigned person adopted *after* Resignation. While living in denial of their corrupted condition brought some relief, resigned humans still needed to find ways to feel good about themselves. Yes, not only did resigned humans have to find ways to avoid and deny the issue of the human condition, they also had to find ways to convince themselves they were the opposite of flawed and corrupted and, it would seem, **'the face of absolute evil'**. So, unable to refute the negative view of themselves with compassionate understanding, they became focused on emphasizing and developing whatever positive views of themselves they could find or create. In practice this meant they became preoccupied competing for power, fame, fortune and glory. Winning a football match, or behaving in some way that brought praise—basically carrying out any activity that would relieve the insecurity of their condition—became all that mattered. Once the full extent of the imperfection of their corrupted condition became apparent at the time of Resignation, the need for reinforcement to counter that uncertainty about their worth became extreme, desperately so for those who were more insecure. So after Resignation, the 'ego', which the dictionary defines as **'the conscious thinking self'** (*Concise Oxford Dictionary*, 5th edn, 1964), became extremely selfishly focused or centered on the need for reinforcement and relief from criticism. Thus, in addition to practicing complete denial of any confronting truths, and becoming very angry towards any criticism, extreme ego-centricity was one of the main characteristics to come out of Resignation. Basically after Resignation, the angry, egocentric and alienated responses to the dilemma of the human condition that had emerged in infancy and which had been gradually developing throughout childhood and early adolescence suddenly greatly amplified. Indeed, the extremely self-worth-embattled, heroic mantra of the resigned person became one of, as the adages go, **'Give me liberty** [from criticism] **or give me death'**, **'No retreat, no surrender'**, **'Death before dishonor'**; in essence, 'I don't believe the criticism is deserved, and in any case it's too unbearable to accept, so I will *never* tolerate any insinuation that I am a bad person'! (A more complete description of the process of Resignation can be found in *Freedom Expanded* at <www.humancondition.com/freedom-expanded-resignation>.)

[655]We can visualize the situation faced by resigning adolescents by imagining them in a room, with one end containing all manner of depressing truths about the world of humans around them and, most depressing of all, about their own lives, and the other end the few positive thoughts and views about the world and about themselves. Well, not surprisingly, resigned humans chose to live entirely at the end of the room where they could be surrounded by the few positives they could identify or develop—especially positive reinforcements for themselves gained through achieving power, fame, fortune and glory. And not only did they stay jammed right up against the wall at that positive end, as removed as humanly possible from any negative truths, they made sure they stood with their nose flat against the wall so they couldn't even *see* the other side of the room. That is how narrow and limited the existence of a resigned human has been. The upset, resigned mind has been fixated on a few positives about themselves while blocking out a whole 'room'—in fact, a whole universe of subjects and thoughts and awarenesses. In this light, we can appreciate the accuracy of Plato's analogy of humans **'enshrined in that living tomb which we carry about, now that we are imprisoned' 'a long way underground'** in a cave where all they could see were shadowy illusions of the real world.

[656]This 'jammed-down-one-end-of-the-room' analogy of how the upset, resigned human race has had to live in denial of the whole issue of the human condition also applies to the aspect of the human condition that involves our species' apparently destructive conscious intelligence. Unless we were exceptionally free of upset behavior, exceptionally nurtured with unconditional real love in our infancy and childhood and thus exceptionally secure in self and thus able to avoid Resignation, thinking truthfully about what consciousness is was only ever going to raise the question of why aren't we cognizant of change and insightful enough to behave in an ideal way? And so to avoid raising those depressing questions resigned humans had to avoid any truthful analysis of what consciousness actually is. And not only that, they needed to shift their focus *entirely* away from the issue of what consciousness entails onto any positives they could find about their intellect. The resigned mind had to focus on convincing itself that its intellect is a brilliant talent,

capable, for instance, of inventing a machine to send man to the Moon and back. To cope with the whole issue of the goodness or otherwise of their intellect resigned humans maximized all the positive aspects about their conscious intellect, which is their cleverness, and minimized all the negative aspects, which is their state of alienation—namely the extent of their soundness or lack thereof—with the best method of doing so to simply not think about and acknowledge what the nature of conscious intelligence really is.

[657] And so in keeping with that methodology, the resigned world started to regard, or more specifically, measure, people according to their level of mental dexterity, with the creation of IQ tests and societies whose membership was restricted to the most clever, such as Mensa. The resigned world only allowed people into university who had high IQs and could pass exams that tested for a person's intellectual brilliance, never for their soulful soundness. It tested children for their ability to remember endless streams of ridiculously superficial and irrelevant facts such as Queen Isabella the 5th married King Arnold the 12th in 1522 and together they fought The War Of The Itchy Armpits in 1591, or something like that; never asking the *real* questions of why there were kings and queens and poor people—selfishness, inequality and indifference to others—and why humans fought and killed each other in wars. And it created game shows that glorified those with the best memories for mundane, superficial facts or for successfully spelling words or completing mathematical sums. It was an extremely escapist, evasive *intellectual* world, not a sound, soulful *instinctual* world. The emphasis was entirely on intellectual brilliance, not on soulful soundness. It never measured people for how alienated they were, or for the speed at which their minds could block out confronting truth, or the speed at which they could override their instinctive moral sense and exploit others, or for how mentally insecure and thus egocentrically self-preoccupied and thus indifferent to others they were. It only stressed how smart you were, never how corrupted and destructive your intellect was. And it wasn't as though the resigned world didn't know who was soul-corrupted, upset and alienated and who was innocent; after all, to ignore, deny, repress and, in the extreme, persecute to the point of, in the case of Christ, *crucifying* honest, truthful innocent soundness because it found it too confronting, it had to first be able to recognize it. It would have been as easy—indeed, probably much easier—to design exams that tested a person's level of alienation or

soundness or soulfulness quotient, their SQ, than it was to design exams that tested their intelligence quotient, or IQ.

[658] The truth is, our intelligence has been an *extremely* insecure and defensive entity. It has been an instrument of dishonest denial, not of honest thoughtfulness. It has been preoccupied with escapist, superficial, **'phony'**, **'fake'**, sophisticated intellectualism, not with confronting, penetrating, truthful, thoughtful instinctualism. Such has been the human condition. Thank goodness then that it has at last been explained and the horrifically debilitating practice of denial can end.

## Chapter 7:3 Why, how and when did consciousness emerge in humans?

[659] As stated at the beginning of this chapter, our self-adjusting conscious mind is a 'fabulous' phenomenon, the culminating achievement of the grand experiment in nature that we call life. Unlike the gene-based natural selection system of processing information, where different arrangements of matter with their different properties are compared for their integrative potential, the conscious mind is able to make these comparisons with the information abstracted from its source. The gene-based natural selection process *is* a form of thinking, of comparing the integrative potential of different arrangements of matter which can be thought of as different 'ideas' — but that 'thinking' or 'learning' or 'refinement', that experimentation, that 'selection' of one 'idea' over another, has to be carried out in practice. The conscious mind, however, has the incredible capacity to be able to do this 'thinking', this comparing of 'ideas', without having to carry them out in practice. Yes, for all the problems it has given rise to, the power of the conscious mind to separate information from its source *is* fabulous, it *has to be* nature's greatest invention.

[660] The obvious question then is how did our wondrously powerful conscious mind emerge from the gene-based natural selection process? To solve problems it's first necessary to ask the right questions, and in the case of consciousness, the obvious initial question is, 'Why did humans develop consciousness when other animals haven't?' Since consciousness occurs at a certain point in the development of a mind's efficiency in associating information, and since conscious intelligence — the ability to reason how cause and effect are related, to understand change, to be insightful — would obviously be a very great asset for any animal to

acquire, one would assume that fully developed conscious intelligence would have been actively selected for as soon as animals were able to develop a reasonably elaborate central nervous system, and would have thus appeared in many species—and yet it hasn't. As was described in chapter 6:5, despite this being an obvious assumption, the conventional mechanistic explanation for the emergence of conscious intelligence in humans, and its absence in other animals, is that it occurred as a result of the need to manage complex social situations—for example, in *The Social Conquest of Earth*, E.O. Wilson wrote that in order **'to feel empathy for others, to measure the emotions of friends and enemy alike, to judge the intentions of all of them, and to plan a strategy for personal social interactions... the human brain became...highly intelligent'** (2012, p.17 of 330). Termed the <u>Social Intelligence Hypothesis</u>, and occasionally the <u>Machiavellian Intelligence Hypothesis</u>, this theory (S/MIH) was first formally put forward by the psychologist Nicholas Humphrey in 1976: **'In broad terms, the MIH was originally developed to explain the special intelligence attributed to monkeys and apes (Humphrey 1976) as adaptations for dealing with the distinctive complexities of their social lives, such as volatile social alliances. The term "Machiavellian" was used by Byrne & Whiten (1988) to capture the central concept of adaptive social manoeuvring within groups made up of companions subject to similar pressures to be socially smart, and the spiralling selection pressures this implies'** (Andrew Whiten & Carel van Schaik, 'The evolution of animal "cultures" and social intelligence', *Philosophical Transactions of The Royal Society B*, 2007, Vol.362, No.1480).

[661] The first point to make is one that was explained in chapter 5 and re-emphasized in chapter 6 (pars 505-509)—that if it wasn't for the psychologically upset state of the human condition there would be no need to learn, and become intelligent enough to master, the art of managing **'social' 'complexities'**. Through the process of love-indoctrination, humans became *so* instinctively integrated that there was no disharmony/conflict/discord/**'complex**[ity]**'** to have to manage. Before the emergence of the human condition some 2 million years ago our species lived instinctively as one organism—as the Greeks Hesiod and Plato said (respectively), **'Like gods they lived, with calm untroubled mind, free from the toils and anguish of our kind...They with abundant goods 'midst quiet lands, all willing shared the gathering of their hands'**, and this **'was a time...most blessed, celebrated by us in our state of innocence, before we had any experience of evils to come, when we were admitted to the sight of apparitions innocent and simple and calm and happy...and not yet enshrined in that living tomb which we carry**

**about, now that we are imprisoned'**; the time when we lived a **'blessed and spontaneous life...**[where] **neither was there any violence, or devouring of one another, or war or quarrel among them...In those days...there were no forms of government or separate possession of women and children; for all men rose again from the earth, having no memory of the past** [we lived in a pre-conscious state, obedient to our universally loving instincts]. **And...the earth gave them fruits in abundance, which grew on trees and shrubs unbidden, and were not planted by the hand of man. And they dwelt naked, and mostly in the open air, for the temperature of their seasons was mild; and they had no beds, but lay on soft couches of grass, which grew plentifully out of the earth'.** The pre-human-condition-afflicted, integrated, cooperative, social, **'calm'**, **'happy'**, no **'quarrel[ling]'**, **'no forms of government'** necessary, free of the divisive selfish and competitive **'evils to come'**, Specie Individual state that the fossil record of our ancestors now evidences, and which largely exists in bonobo society today, simply wasn't a situation that involved **'social'** **'complexities'**. No, the bonobos' harmonious society, and humanity, was borne out of a genuinely altruistic, unconditionally selfless, all-loving, empathetic-towards-others, non-devious, opposite-of-**'smart'**-cunning-**'manoeuvring'-'Machiavellian'** situation.

[662] However, coming back to the obvious question of 'Why did humans develop conscious intelligence when other animals haven't?', as I pointed out in par. 507, while the ability to solve social problems *is* an obvious benefit of having a conscious mind, *all* activities that animals have to undertake would benefit enormously from being able to understand cause and effect, so it is completely illogical to suggest that consciousness developed as a result of the need to manage extremely complex social situations. No, any sensible analysis of the question of the emergence of consciousness must be based on the question of 'What has *prevented* its development in other animals?' Consciousness is such a powerful asset for an animal to acquire that something must have blocked its selection in other species. The lack of social situations doesn't explain why the fully conscious mind hasn't appeared in non-human species because there was ample need for a conscious intelligent mind prior to the appearance of complex social situations.

[663] It should also be pointed out that while **'Most contemporary workers in animal cognition, particularly those working with primates, are enthusiastic about the social intelligence hypothesis'** (Kay Holekamp, 'Questioning the social intelligence hypothesis', *Trends in Cognitive Sciences*, 2007, Vol.11, No.2), the hypothesis has at

least two flaws that have actually been recognized by human-condition avoiding, mechanistic scientists. Firstly, research has shown that there *isn't* a correlation between more complex social groups and greater social learning, which you *would* expect if the reason for the long mother-infant association was to teach the skills needed to live in complex social groups. For instance, **'several phenomena have been identified...for which the social intelligence hypothesis cannot account...Evidence of this sort has also accumulated in the literature on primate social intelligence. For example, a comparative analysis found that innovation, tool use and the incidence of social learning in primates co-vary across species and that the frequency of occurrence of social learning is not correlated with group size among primate species'** (ibid). The **'comparative analysis'** in question says that in regard to primates, **'social group size and social learning frequency are not correlated'** (Simon Reader & Louis Lefebvre, 'Social Learning and Sociality', *Behavioral and Brain Sciences*, 2001, Vol.24, No.2).

[664] Secondly, evidence that having to manage more complex social situations has *not* led to greater intelligence is also being furnished through mechanistic science's studies of relative brain size (which is widely regarded as an indicator of greater intelligence), with research showing that highly social species such as meerkats and hyenas have not developed a larger brain in proportion to body size beyond that of less social species: **'no association exists between sociality and encephalization** [brain size in proportion to body size] **across Carnivora** [which include meerkats and hyenas] **and that support for sociality as a causal agent of encephalization increase disappears for this clade** [group]**'** (John Finarelli & John Flynn, 'Brain-size evolution and sociality in Carnivora', *Proceedings of the National Academy of Sciences*, 2009, Vol.106, No.23).

[665] But rather than accepting these fundamental flaws and abandoning the S/MIH, a more sophisticated version of it was actually put forward in 1989 by the biologist Richard Alexander. This refinement, which became known as the 'Ecological Dominance–Social Competition' (EDSC) model, holds that a species must first somehow overcome or dominate its environment *before* the S/MIH can apply, arguing that **'as our hominin ancestors became increasingly able to master the traditional "hostile forces of nature," selective pressures resulting from competition among conspecifics** [companions] **became increasingly important, particularly in regard to social competencies. Given the precondition of competition among kin–and reciprocity–based coalitions (shared with chimpanzees), an autocatalytic** [self-fuelling] **social arms race was initiated, which eventually resulted in the**

**unusual collection of traits characteristic of the human species, such as…an extraordinary collection of cognitive abilities'** (M.V. Flinn et al., 'Ecological dominance, social competition, and coalitionary arms races: Why humans evolved extraordinary intelligence', *Evolution and Human Behavior*, 2005, Vol.26). While it does not answer the obvious question of why *all* activities that animals have to manage wouldn't profit from a conscious intellect, this prerequisite of **'ecological dominance'** is meant to explain why other species in complex social situations haven't developed intelligence—apparently because they hadn't dominated their environment first.

[666] The concept of **'ecological dominance'** put forward by the EDSC model describes a supposed mastery of the environment achieved, it says, through developments such as tool use, projectile weapons and controlled use of fire that **'roughly coincided with the appearance of *Homo erectus*, 1.8 mya** [million years ago]' (ibid). While it is true that conscious intelligence was increasing rapidly during the reign of *Homo erectus*, as will be revealed in chapter 8:2, conscious intelligence first emerged in our ape ancestor, prior to the emergence of the australopithecines (who appeared some 4 million years ago) and so well before *Homo*; and further, the dramatic increase in intelligence evident in *H. erectus* was *not* a result of **'ecological dominance'** but, as described throughout this book, a product of *psychological* factors—as has been the continual increase in conscious intelligence throughout the genus *Homo*. (It is important to differentiate here that the **'ecological dominance'** described by the EDSC model is *not* the same ecologically dominant/beneficial situation our ape ancestors found themselves in when living in the 'ideal nursery conditions' that allowed love-indoctrination to develop, which, as was briefly described in chapter 5:8 and as is about to be described in more detail, *did* have the side effect of liberating conscious intelligence.)

[667] This flaw in the EDSC model's theory that conscious intelligence did not arise until the emergence of *H. erectus* because until that time we hadn't dominated our environment is also evidenced by the fossil record, which shows that brain size dramatically increased in humans *before* they became **'ecologically dominant'**. Yes, the model's argument that **'ecological dominance should arise prior to or along with increases in brain size'** (ibid) is a flaw that even mechanistic, human-condition-avoiding scientists recognize, as this study points out: **'a great deal of encephalization** [brain size relative to body size] **occurred before humans were dominant…The EQ** [encephalization quotient] **of the first instance of *Homo*, *Homo habilis*, had already doubled relative**

to our nearest relatives today, chimpanzees (*Pan troglodytes*). These hominins were still largely foragers, scavengers (not yet organized hunters), and prey for more powerful predators' (R.D. Horan et al., *A Paleoeconomic Theory of Encephalization*, selected paper presented at the annual meetings of the American Economic Association, San Francisco, Jan. 2009). It certainly doesn't make sense that developments such as tool use, projectile weapons and the controlled use of fire enabled us to become intelligent because such advancements would have *required* conscious intelligence to both invent and subsequently manage. Something must have enabled consciousness to emerge, which *then* allowed us to become clever enough to invent and manage these early technologies.

[668] There are several other theories for how and why humans developed consciousness, including models based around the need to solve problems resulting from climate change, sexual selection for consciousness as an indicator of fitness, and the discovery of cooking, but without being able to acknowledge the human condition, they—as with the S/MIH/EDSC models—are unable to explain in a fully accountable way why consciousness is *not* a powerful, fitness-assisting asset for an animal to have in all situations (not just social situations).

[669] I should note here that there has been an attempt, using what is called the Expensive Brain Hypothesis, to explain why consciousness has not been widely developed. This theory first equates brain size with intelligence and then claims the reason intelligence has not evolved is because **'The high proportion of energy necessarily allocated to brain tissue may therefore constrain the response of natural selection to the beneficial impact of increased brain size on an animal's survival and/or reproductive success'** (Karin Isler & Carel van Schaik, 'Metabolic costs of brain size evolution', *Biology Letters*, 2006, Vol.2, No.4). The fact is, natural selection has the ability to select for an asset if the benefits outweigh the costs, and since consciousness is such an *extremely* valuable asset, the costs in energy of developing a conscious mind would not, you would expect, be great enough to inhibit its development—after all, natural selection has selected for assets that require an *immense* amount of energy, such as the males of many species being able to expend colossal amounts of energy during the mating season; consider, for instance, the peacock, which grows an extravagant new train each season, or antelopes engaging in a period of endless and ferocious rutting each year. Another point is that bonobos, who are verging on having developed consciousness, don't have huge brains like we have. As will be described in pars 707 and 714, what has driven *our* brain and

conscious mind to become so developed is the difficulty of having to try to manage the extreme complexities of *our* human condition, as evidenced by the fact that the very large brain developed when the human condition developed, which was *after* the time we became conscious. (Note that in chapter 3:1 I describe how it wasn't until approximately 2 million years ago that we humans became *fully* conscious in the sense of being fully cognizant of the situation we humans have been in of having to live under the duress of the human condition.)

[670]The fact is, other animal species have been able to develop all manner of extraordinary mental abilities, many superior to our own, but never full consciousness. (The reason we can know that other animals like dolphins and elephants haven't developed full consciousness was explained in par. 425.) For instance, in the United States, the nutcracker bird buries around 30,000 nuts throughout the summer months, each in a different location, but come winter and the cover of snow it can recall the location of 90 percent of them. The goby fish can memorize the topography of the tidal flats at high tide so that when the tide retreats it knows the exact location of the next pool to flip to when the one it is in evaporates. And then there is the male common canary, which has a specific part of its brain that expands dramatically every spring so it can learn new mating songs, only to shrink again once the need for it ends at the conclusion of the mating season (just as peacocks shed their tattered tail feathers at the end of the season). So again the question is, if other animals have been able to develop such extraordinary mental abilities, what's stopping them from developing *full* consciousness? As will now be explained, the simple and very obvious explanation—if you are not living in denial of the truth of selflessness-dependent, Integrative Meaning—is that genetics is such a selfish process that it will normally block the development of a selflessness-recognizing, truthful, effective-thinking conscious mind.

[671]To now explain more fully this obvious—if you are not living in denial—explanation for what has blocked the development of consciousness in almost all species. The explanation begins by re-stating what was pointed out in chapter 4:4, which was that one of the limitations of the gene-based learning system is that it normally can't select for unconditionally selfless, altruistic, self-sacrificing behavior because altruistic traits tend to self-eliminate; they tend not to carry on and so normally can't become established in a species. The effect is that the gene-based learning system actively resists altruistic behavior. For instance, whenever

a female kangaroo comes into estrous, the males pursue her relentlessly. Despite both parties almost falling with fatigue, the chase continues. It is easy to see how this behavior developed: if a male relaxed his efforts he would lose his opportunity to reproduce. Self-interest is fostered by natural selection with the result that genetic selfishness has become an extremely strong force in animals. It is clear then that there would be no chance of a variety of kangaroo that considered others above itself developing. Unless, of course, they could develop love-indoctrination, but while a kangaroo can look after a joey in its pouch, the pouch is more an external womb, allowing little behavioral interaction between mother and infant. It is the selfless treatment—the active demonstration of love—that trains the infants in selflessness or love. Also, since grass, which they live on, is not very nutritious, kangaroos have to spend most of their time grazing, which leaves relatively little time for social interaction between mother and infant and thus limited training in love.

[672] Genetic refinement normally acts against any inclination towards selfless behavior because selflessness disadvantages the individual that practices it and advantages the recipients of the selfless treatment—such is the meaning of selflessness. Selflessness normally can't be reinforced by genetic refinement; indeed, it is emphatically resisted by it. It follows then that in terms of the development of consciousness, the gene-based learning or refinement system was, in effect, totally opposed to any altruistic, selfless thinking. In fact, genetic refinement developed *blocks* in the minds of animals to prevent the emergence of such thinking. And it is this block against truthful, selflessness-recognizing-thinking in the minds of almost all animals that prevents them from becoming conscious of the true relationship or meaning of experience.

[673] To explain more fully how these blocks against selflessness-recognizing-thinking developed, an example of how genes resist self-destructive behavior will be helpful. In what are termed 'visual cliff' experiments, newborn kittens are placed on a table and while they will venture towards the edge, they won't allow themselves to go beyond the edge and fall—a sheet of glass is actually placed over the table to prevent them from accidentally slipping off the edge, but the point is the glass is unnecessary because the kittens instinctively know not to travel beyond the table's edge. Presumably, this instinctive orientation against doing so evolved because any cat that did venture too close to a precipice invariably fell to its death, leaving only those that happened to have an

instinctive block against such self-destructive practices. Natural selection or genetic refinement develops blocks in the mind against behavior that doesn't tend to lead to the reproduction of the genes of the individuals who practice that behavior.

[674] And so just as surely as cats were eventually selected for their instinctive block against self-destruction, most animals have been selected with an instinctive block against selfless thinking because such thinking also tends not to lead to the reproduction of the genes of the individuals who think that way. The effect of this block was to prevent the developing intellect from thinking truthfully and thus effectively.

[675] As pointed out when Integrative Meaning was explained in chapter 4, selflessness or love is the theme of existence, the essence of integration, the meaning of life. While the upset, alienated human race has learnt to live in denial of this truth of the selfless, loving, integrative meaning of existence, it is, in fact, an extremely obvious truth and one that is deduced very quickly if you are able to think honestly about the world. We are, as mentioned, surrounded by integration—every object we look at is a hierarchy of ordered matter, witness to the development of order. It follows then that if you aren't able to recognize and thus appreciate the significance of selfless Integrative Meaning you are not in a position to begin to think straight and thus effectively; you can't begin to make sense of experience. All your thinking is coming off a false base and is therefore derailed from the outset from making sense of experience. As stated in par. 220, the philosopher Arthur Schopenhauer wrote that **'The discovery of truth is prevented most effectively...by prejudice, which...stands in the path of truth and is then like a contrary wind driving a ship away from land'**. You can't think effectively with lies in your head, especially with such important lies as denial of Integrative Meaning. Your mind is, in effect, stalled at a very superficial level of intelligence with little ability to understand the relationship of events occurring around you.

[676] To elaborate, any animal able to associate information to the degree necessary to realize the importance of behaving selflessly towards others would have been at a distinct disadvantage in terms of its chances of successfully reproducing its genes. It follows then that those animals that *don't* recognize the importance of selflessness are genetically advantaged, which means that eventually a mental block would have been 'naturally selected' to prevent the emergence of the ability to make sense of experience, to prevent the emergence of consciousness. At this point

in development, genetic refinement favored individuals that were not able to recognize the significance of selflessness, thus ensuring animals remained incognizant, unconscious of the true meaning of life.

[677] Having denied the truth of Integrative Meaning and the importance of selflessness, it is not easy for the alienated human race to appreciate that conscious thought depends on the ability to acknowledge the significance of selflessness/love/Integrative Meaning. However, our own human-condition-induced mental block or alienation is, in fact, the perfect illustration of and parallel for this block in the minds of animals. Unable to think truthfully about the selfless, loving integrative theme of existence, all our thinking has been coming off a false base and, as a result, we too have been unable to think effectively. Alienation has rendered the human race almost stupid, incapable of deep, penetrating, meaningful thought.

Wallace & Gromit by Nick Park and Bob Baker of Aardman Animations

[678] So when it comes to thinking truthfully, and thus soundly, humans are now almost as mentally incognizant as animals—a state of affairs that is parodied in the popular animated cartoon *Wallace & Gromit*. In the series, Wallace is a lonely, sad—alienated—human figure whose dog, Gromit, is very much on an intellectual par with him in his world. Both wear the same blank, stupefied expression as together they muddle their way through life's adventures. Yes, as R.D. Laing was quoted as saying in par. 123, **'Our alienation goes to the roots'**, there is now **'fifty feet of solid concrete'** between us and our condemning instinctive self or soul. But

to be alienated is to not know you are alienated because if you did you wouldn't be alienated, you wouldn't be blocking out the truth—all of which means it will be difficult to accept that humans are now 'almost as mentally incognizant as animals', so these further references to just how alienated we have become may help to reveal the extent of our alienation.

[679] Plato, that extraordinary denial-free thinking prophet, wrote that **'when the soul** [our instinctive orientation to Integrative Meaning] **uses the instrumentality of the body** [uses the body's intellect with its preoccupation with denial of Integrative Meaning] **for any inquiry...it is drawn away by the body into the realm of the variable, and loses its way and becomes confused and dizzy, as though it were fuddled** [drunk]**...But when it investigates by itself** [free of intellectual denial]**, it passes into the realm of the pure and everlasting and immortal and changeless, and being of a kindred nature, when it is once independent and free from interference, consorts with it always and strays no longer, but remains, in that realm of the absolute** [Integrative Meaning]**, constant and invariable'** (*Phaedo*, c.360 BC; tr. H. Tredennick, 1954, 79). He also wrote that the **'capacity** [of **a mind...to see clearly**] **is innate in each man's mind** [we are born with an instinctive orientation to Integrative Meaning]**, and that the faculty by which he learns is like an eye which cannot be turned from darkness** [the state of living in denial] **to light** [the denial-free truth] **unless the whole body is turned; in the same way the mind as a whole must be turned away from the world of change until it can bear to look straight at reality, and at the brightest of all realities which is what we call the Good** [Integrative Meaning or God]**'** (*The Republic*, c.360 BC; tr. H.D.P. Lee, 1955, 518). Yes, if the human race is to begin to think effectively it *has* to stop living in denial of Integrative Meaning, **'the Good'**—otherwise our species will stay in the situation where our collective mind, in perpetuity, **'loses its way and becomes confused and dizzy, as though it were fuddled** [drunk/stupid/alienated]**'**.

[680] While our **'capacity' 'to see clearly'** is, as Plato said, **'innate'**, denial and its alienating effects came about through our own corrupting search for knowledge and through our encounter with the already upset, human-condition-afflicted, corrupt world. And just as this search for knowledge and our encounter with the upset world began at birth and continued throughout our lives, the extent of our insecurity about our corrupted state and associated block-out or alienation *also* increased throughout our lives, until eventually we were walking around free, in effect, of criticism but totally inebriated in terms of our access to truth and meaning. So it follows that it was when humans were very young

that they were most able to think truthfully. The famous psychoanalyst Sigmund Freud recognized this when he said, **'What a distressing contrast there is between the radiant intelligence of the child and the feeble mentality of the average adult'** (*The Freud Reader*, ed. P. Gay, 1995, p.715). Christ also recognized the mental integrity of the young when he said, **'you have hidden these things from the wise and learned, and revealed them to little children'** (Bible, Matt. 11:25). Albert Einstein famously echoed these sentiments when he noted that **'every child is born a genius'**, while the philosopher Richard Buckminster Fuller acknowledged that **'There is no such thing as genius, some children are just less damaged than others'** (NASA Speech, 1966), and that **'All children are born geniuses. 9999 out of every 10,000 are swiftly, inadvertently de-geniused by grown-ups'** (Mario M. Montessori Jr, Paula Polk Lillard & Richard Buckminster Fuller, *Education for Human Development: Understanding Montessori*, 1987, Foreword). R.D. Laing also observed that **'Each child is a new beginning, a potential prophet [denial-free, truthful, effective thinker]'** (*The Politics of Experience* and *The Bird of Paradise*, 1967, p.26 of 156) and pointed out that **'Children are not yet fools, but [by our treatment of them] we shall turn them into imbeciles like ourselves, with high I.Q.'s if possible'** (ibid. p.49).

[681] Many exceptionally creative people have also made statements to the effect that genius is the ability to think like a child. For example, one of the most accomplished artists of all time, Pablo Picasso, famously said about his struggle to paint well that **'It's taken me a lifetime to learn to paint like a child.'** The poet Charles Baudelaire similarly wrote that **'genius is no more than childhood recaptured at will'** (*The Painter of Modern Life*, 1863), while the Chinese philosopher Mencius said, **'The great man is he who does not lose his child's heart, the original good heart with which every man is born'** (*The Works of Mencius*, Book 4 ch.12, c.371-289 BC). And just like the already mentioned quotes about our alienated condition by Samuel Beckett that **'They give birth astride of a grave, the light gleams an instant, then it's night once more'**, and R.D. Laing that **'To adapt to this world the child abdicates its ecstasy'**, the artist Francis Bacon said **'the shadow of dead meat is cast as soon as we are born'** (*The Australian*, 15 Jun. 2009, reprinted from *The New Republic*). Similarly, in the 1993 film *House of Cards*, one of the characters makes the following intuitive comment about how sensitive and vulnerable innocent children have been to the horror of the alienated world of adults: **'I used to watch Michael** [a character in the film] **about two hours after he was born and I thought that at that moment he knew all of the secrets of the universe and every second that was passing he was forgetting them** [he was having to live in denial of them]' (based on a screenplay by

Michael Lessac). The poet Percy Bysshe Shelley was another who decried the fate of adult humans when he wrote that **'Our boat** [our being] **is asleep on Serchio's stream, its sails are folded like thoughts in a dream'** (*The Boat on the Serchio*, 1821). The Bible also offers this account of our estranged, resigned adult condition: **'This people's heart has become calloused** [alienated]**; they hardly hear with their ears, and they have closed their eyes'** (Isa. 6:10 footnote). And in his incredibly honest poem *Intimations of Immortality from Recollections of Early Childhood*, William Wordsworth provided this description (which was referred to in par. 182) of how quickly humans become alienated from our all-sensitive and truthful, innocent instinctive self or soul: **'There was a time when meadow, grove, and streams / The earth, and every common sight / To me did seem / Apparelled in celestial light / The glory and the freshness of a dream / It is not now as it hath been of yore / Turn wheresoe'er I may / By night or day / The things which I have seen I now can see no more // The Rainbow comes and goes / And lovely is the Rose / The Moon doth with delight / Look round her when the heavens are bare / Waters on a starry night / Are beautiful and fair / The sunshine is a glorious birth / But yet I know, where'er I go / That there hath past away a glory from the earth // ...something that is gone / ...Whither is fled the visionary gleam? / Where is it now, the glory and the dream? // Our birth is but a sleep and a forgetting / The Soul that rises with us, our life's Star / Hath had elsewhere its setting / And cometh from afar / Not in entire forgetfulness / And not in utter nakedness / But trailing clouds of glory do we come / From God, who is our home / Heaven lies about us in our infancy! / Shades of the prison-house begin to close / Upon the growing Boy / ...And by the vision splendid / Is on his way attended / At length the Man perceives it die away / And fade into the light of common day / ...Forget the glories he hath known / And that imperial palace whence he came'** (1807). The prophet Job similarly recognized that under the duress of the human condition **'Man...springs up like a flower and withers away; like a fleeting shadow, he does not endure'** (Job. 14:1-2).

[682] So it is widely appreciated that while children have wonderful imaginations, that creativity is often lost by the time they reach adulthood—but with understanding of both the human condition and the different roles played by the left and right hemispheres of our brain we can understand why. In the human brain, one hemisphere (the right) specializes in general pattern recognition while the other specializes in specific sequence recognition. The right is lateral or creative or imaginative, while the left is vertical or logical or sequential. The right has even been described as intuitive, thoughtful and subjective and the

left as logical, analytical and objective. One stands back to 'spot' any overall emerging relationship while the other goes right in to take the heart of the matter to its conclusion. We need both because logic alone could lead us up a dead-end pathway of thought. For example, we can imagine that for a while our thinking mind could have assumed that the most obvious similarity between fruits was that they were brightly colored; however, with more experience the similarity that proved to have the greatest relevance in the emerging overall picture was their edibility. When one thought process leads to a dead-end our mind has to backtrack and find another way in: from the general to the particular and back to the general, in and out, back and forth, until our thinking finally breaks through to the correct understanding. The first form of thinking to wither from Resignation and alienation was imaginative thought because wandering around freely in our mind all too easily brought us into contact with unbearable human-condition-confronting truths such as Integrative Meaning. On the other hand, if we got onto a logical train of thought that at the outset did not raise criticism of us there was a much better chance it would stay safely non-judgmental. So children have always had wonderful imaginations because they had yet to learn to avoid free/open/ adventurous/lateral thinking; they had yet to resign themselves to living a separated-from-the-truth, in denial of the issue of the human condition, alienated existence.

[683] So adults have become immensely alienated, not wanting to think truthfully and thus *unable* to think effectively. Indeed, we have become *so* alienated that we will readily intellectually focus on a safely sectioned-off area of inquiry or activity—such as solving a math equation, or mastering a computer problem, or debating whether God has been destroyed by the big bang theory of the origins of the universe, or ordering our wardrobe, or polishing our car, or making a cake, or even sending man to the Moon—and yet we won't go beyond those safety limits and risk encountering *anything* to do with the issue of 'self', the depressing subject of the human condition. In fact, mechanistic science is just such a reduced, deeply afraid, don't-want-to-look-at-the-big-picture, lose-yourself-in-tiny-little-details, sharpen-your-pen-scratch-your-ear-fix-that-little-problem-over-there-don't-think-don't-think-don't-think, neurotic activity. The more neurotic a person, the smaller their world becomes. And as will be explained in chapter 8:16C, in the extreme case of upset, that of the autistic mind, it becomes completely detached.

[684]The genesis of this disconnection in the resigned, human-condition-avoiding, neurotic mind from any form of truthful, meaningful, integrated, effective thinking was perfectly described by the psychologist Arthur Janov in the extract of his that was included earlier in par. 221, but which is so relevant here I am re-including it again in part (underlinings are my emphasis): **'As the child becomes split by his Pain** [caused by his encounter with the human condition], **he will develop philosophies and attitudes commensurate with his denials. He will have a warped view of the world…Thus, intellect becomes the mental process of repression…The more reality a person is forced to hide in his youth, the more likely it will be that certain areas of thinking will be unreal. That is, it is more likely that thought process will be constricted so that generalised extrapolations cannot be made about the nature of life and the world. Conversely, to be free to articulate one's feelings while growing up will lead to becoming an articulate, free-thinking person, unhampered by fear, which paralyses thought…A young child can split from his feelings** [from his **Pain**, the human condition] **and learn every aspect of engineering. He can be a "smart" engineer or scientist…His intellect is something apart from his feelings…Neurotic intellect is an order superimposed on reality…He is truly a specialised man, living in his head because his body** [where his feelings/pain/hurt soul lives] **is out of touch and reach. He will deal with each piece of news he hears as an isolated event, unable to assemble what he sees and hears into an integrated view. Life for him is a series of discrete events, unconnected, without rhyme or true meaning'** *(The Primal Revolution*, 1972, pp.158-160 of 246). In short, the human-condition-afflicted brain is one that **'paralyses thought'**.

[685]It's an extraordinary—indeed, mad—situation, this one where humans are unable to think on any substantial, effective scale, and one that General Omar Bradley saw the ramifications of when, as mentioned in par. 226, he said that **'The world has achieved brilliance…without conscience. Ours is a world of nuclear giants and ethical infants.'**

[686]Yes, we can wrestle with and assemble this bit and that bit of our world but we can't look at and deal with the big subject of the human condition. So in terms of the *all-important* issue of what needs to be done about the state of the world, and in particular our species' plight, while we will apply all our vigor to protesting an environmental cause or the rights of an indigenous 'race' or the demand for peace, or any one of a number of other 'makes-you-feel-you-are-doing-good-but-actually-totally-superficial' so-called politically correct causes, we will *not* look at the nightmare of angst in ourselves, the *real* devastation and issue of

our own condition and, beyond that, the human condition that needs to be addressed if we are to bring about a caring, equitable and peaceful world—because the fact is, no matter how much we try to restrain and conceal our upset eventually our world *will* become an expression of ourselves and thus as devastated as we are. To fix the world we have to first fix ourselves.

[687] As will be made very clear in chapter 8:16O, the truth is that the main function of politically correct causes has been to allow upset humans to feel that they are doing good when they are actually *avoiding* what is required to make a positive difference—namely confronting the issue of the human condition. Human life has been preoccupied with maintaining the many delusions and false ways of making us feel good about ourselves and with all manner of escapisms from reality rather than with the meaningful thinking and progressive actions we claim it is.

[688] In short, the human condition is *the* all-important issue that had to be looked at to free ourselves from our condition and truly bring peace to our world and yet it is the one issue we have refused to look at. As Carl Jung recognized, **'Man everywhere is dangerously unaware of himself. We really know nothing about the nature of man, and unless we hurry to get to know ourselves we are in dangerous trouble'** (Laurens van der Post, *Jung and the Story of Our Time*, 1976, p.239 of 275). R.D. Laing also recognized this when he wrote that **'The requirement of the present, the failure of the past, is the same: to provide a thoroughly self-conscious and self-critical human account of man'** (*The Politics of Experience* and *The Bird of Paradise*, 1967, p.11 of 156). The human condition *is* the proverbial elephant in our living rooms that we pretend not to see, the all-important issue that we assiduously practice denying.

[689]The examples of our extremely escapist, extremely superficial, indeed, extremely separated or alienated from any deep, meaningful, truthful, effective thinking are endless, but what all this alienation that now exists in the minds of adult humans shows is that the human mind has been *twice* alienated from the truth in its history: once when we were like other animals, instinctively blocked from recognizing the truth of selflessness, and again in our present state, in which we are terrified of the issue of our selfish and divisive human condition and as a result are living **'a long way underground'** in Plato's dark **'cave'** of denial of the significance of the selfless, loving integrative meaning of existence. Yes, other animals are unable to think truthfully and thus effectively because they can't recognize the truth of selflessness, and under the duress of the human condition we too have been incapable of effective thought because we haven't been able to recognize the truth of the selfless meaning of existence.

[690]But while humans have gradually retreated from consciousness into virtual unconsciousness because of our insecurity about our non-ideal, soul-corrupted, 'fallen', selfish, competitive and aggressive human condition, we are, to our knowledge, the first species to become fully conscious. So, the next question is, how were our ape ancestors able to overcome the block that exists in the minds of the great majority of animals and become capable of making sense of experience, become conscious? (As mentioned in par. 511, all animals are trying to develop love-indoctrination and to what degree they have been able to develop it will dictate to what degree they have been able to develop at least a rudimentary level of consciousness, but no other existing species has developed full consciousness like humans have, and bonobos almost have.)

[691]Understanding how the nurturing love-indoctrination process was able to develop selfless, moral instincts in our ape ancestors (and to some degree in some other primates today) allows us to answer this crucial question, because the reason we were able to become fully conscious is that the nurturing of selfless instincts breached the block against thinking truthfully by superimposing a new, truthful, selflessness-recognizing mind over the older, effectively dishonest, selfless-thinking-blocked one. Since our ape ancestors could develop an awareness of cooperative, selfless, loving meaning, they—and, by extension, humans—were able to develop

truthful, sound, effective thinking and so acquired consciousness, the essential characteristic of <u>mental infancy</u>.

[692] To use a comparative example, chimpanzees are currently in a relatively early stage of mental infancy—they have the conscious mental powers of approximately a two-year-old human and demonstrate rudimentary consciousness, in that they are beginning to relate information or reason effectively and make sufficient sense of experience to recognize that they are at the center of the changing array of events they experience. Experiments have shown that they have an awareness of the concept of 'I' or self and, as mentioned earlier, are capable of reasoning how events are related sufficiently well to know that they can reach a banana tied to the roof of their cage by stacking and climbing upon boxes.

[693] In the case of bonobos, evidence suggests that this species is the most intelligent or conscious next to humans—as is apparent in these quotes that were referred to in pars 426-428: **'Everything seems to indicate that [Prince] Chim [a bonobo] was extremely intelligent. His surprising alertness and interest in things about him bore fruit in action, for he was constantly imitating the acts of his human companions and testing all objects. He rapidly profited by his experiences...Never have I seen man or beast take greater satisfaction in showing off than did little Chim. The contrast in intellectual qualities between him and his female companion [a chimpanzee] may briefly, if not entirely adequately, be described by the term "opposites"'** (Robert M. Yerkes, *Almost Human*, 1925, p.248 of 278). Sue Savage-Rumbaugh reinforced this view when she wrote that **'Individuals who have had first hand interactive experience with both *Pan troglodytes* [chimpanzees] and *Pan paniscus*** [bonobos or pygmy chimpanzees] **(Yerkes and Learned, 1925; Tratz and Heck, 1954) have been left with the distinct impression that pygmy chimpanzees are considerably more intelligent and more sociable than *Pan troglodytes*'** and that **'Each individual who has worked with both species in our lab is repeatedly surprised by their [bonobos'] communicative behavior and their comprehension of complex social contexts that are vastly different from anything seen among *Pan troglodytes*'** (*'Pan paniscus* and *Pan troglodytes*: Contrasts in Preverbal Communicative Competence', *The Pygmy Chimpanzee*, ed. Randall Susman, 1984, pp.396, 411-412 of 435). The following extract demonstrates how extraordinarily aware, cooperative, empathetic and intelligent bonobos are: **'Barbara Bell...a keeper/trainer for the Milwaukee County Zoo...works daily with the largest group of bonobos... in North America..."It's like being with 9 two and a half year olds all day," she [Bell] says. "<u>They're extremely intelligent</u>...They understand a couple of hundred**

words," she says. "They listen very attentively. <u>And they'll often eavesdrop</u>. If I'm discussing with the staff which bonobos (to) separate into smaller groups, if they like the plan, they'll line up in the order they just heard discussed. If they don't like the plan, they'll just line up the way they want." "<u>They also love to tease me a lot</u>," she says. "<u>Like during training, if I were to ask for their left foot, they'll give me their right, and laugh and laugh and laugh</u>. But what really blows me away is <u>their ability to understand a situation entirely</u>." For example, Kitty, the eldest female, is completely blind and hard of hearing. Sometimes she gets lost and confused. "They'll just pick her up and take her to where she needs to go," says Bell. "That's pretty amazing. <u>Adults demonstrate tremendous compassion for each other</u>"' (*Chicago Tribune*, 11 Jun. 1998). More recently, the bonobo researcher Vanessa Woods described bonobos as **'the most intelligent of all the great apes'** ('Bonobos – our better nature', blogs.discovery.com, 21 Jun. 2010; see <www.wtmsources.com/134>).

[694] Again, as was explained in par. 429, there are some scientists who suggest that chimpanzees are as intelligent as bonobos, but there is an undeniable freedom in the mind of a bonobo that is apparent in their capacity to be interested in the world around them, and in their empathy, compassion and even simple, childish humor, as evidenced by the quotes above. While it is true that chimpanzees show a mental dexterity, it is a narrow, opportunistic, self-interested mental focus (similar to a very primitive version of the alienated, deadened minds of humans today), not the broad, free, open, curious, aware, all-sensitive, loving, truly thoughtful, conscious mind that bonobos have. It does have to be remembered that mechanistic science doesn't even have an interpretation of the word 'love', so we can't expect it to be capable of showing interest in or acknowledging the kind of open, curious, aware, all-sensitive and loving conscious mind that bonobos have. Indeed, the extraordinarily cooperative, unconditionally loving and truly aware characteristics of bonobos would motivate, and indeed (as we saw in chapter 6) *have* motivated, human-condition-avoiding, alienated mechanistic scientists to find a way to demote them at every opportunity, just like they found a way to demote my work and that of Dian Fossey and Sue Savage-Rumbaugh.

[695] <u>So how did the process of nurturing overcome the instinctive block against selfless thinking/behavior?</u> It makes sense that at the outset the brain of our ape ancestors was relatively small with a limited amount of association cortex, the brain matter in which information is associated.

These brains had instinctive blocks preventing the mind from making deep meaningful/truthful/selflessness-recognizing perceptions. At this stage, however, these small, inhibited brains were being trained in selflessness, so although there was not a great deal of unfilled cortex available, what was available was being inscribed with a truthful, effective network of information-associating pathways. The mind was being taught the truth and given the opportunity to think clearly, in spite of the existing instinctive blocks or 'lies'. While at first this truthful 'wiring' would not have been very significant due to the small size of the brain, it had the potential for much greater development. Further, as was explained in par. 390, with this selfless training of the brain occurring over many generations, the selfless 'wiring' in the brain would have gradually become instinctive or innate. Genes would inevitably follow and reinforce any development process—in this they were not selective. The difficulty lay in getting the development of unconditional selflessness to occur, for once it was regularly occurring it would naturally become instinctive over time, which it did—our instinctive moral soul, the 'voice' of which is our 'conscience', was formed. We are born with a brain that has instinctive orientations that incline us to behave unconditionally selflessly, and to expect to be treated in the same way—as Ashley Montagu wrote, **'to live as if to live and love were one is the only way of life for human beings, because, indeed, this is the way of life which the innate nature of man demands'.**

[696] Thus, the mind was trained or programmed or 'brain-washed' or 'indoctrinated' with the ability to think in spite of the blocks working against such training; it had, at last, been stimulated by the truth. Of course, it must be remembered that in this early stage of development the emphasis was on training in *love*, not on the liberation of the conscious ability to *think*, which was incidental to Negative Entropy's push for our ape ancestors to become an integrated group of multicellular animals; but, once fully liberated, the conscious mind takes on a life of its own, it is free to develop—and that development follows an inevitable path.

[697] Yes, after assisting the love-indoctrination process by allowing for the conscious selection of less aggressive mates, the fully conscious mind progresses down its own particular path of development. As summarized earlier, and this will be elaborated upon in some detail in chapter 8, this journey kicks off in the infancy stage, during which the conscious mind is sufficiently aware of the relationship of events that occur through time

to recognize that the individual doing the thinking is at the center of the changing array of experiences around it. It is during infancy that the conscious individual becomes aware of the concept of 'I' or self, which is what bonobos and, to a lesser degree, the other great apes are capable of. Infancy is also when the individual discovers conscious free will, the power to manage events. Childhood is the stage when the conscious individual revels in this free will, 'playing' or experimenting with it, while adolescence is the stage during which the conscious individual encounters both the sobering responsibility of free will and the agonising identity crisis brought about by the conflict with the already established instinctive orientations and the dilemma that results, which, in the case of humans, we call the human condition—the question of whether or not we are meaningful beings. Adulthood is when the conscious mind ends the insecure stage of adolescence by finding understanding of that corrupted condition, which is the mature, transformed state that humanity can, if it chooses correctly between terminal alienation and transformation, now enter.

[698] Of course, as was pointed out at the end of chapter 6, this explanation of the origin of consciousness was always going to be an extremely difficult explanation for mechanistic science to appreciate because it depends on recognizing so many truths that have historically been denied. To arrive at this explanation of how consciousness emerged and then developed into the highly intelligent brain humans have today hinges on being able to recognize the truth of Integrative Meaning and its theme of unconditional selflessness—and from there why animals would have developed blocks in their minds preventing selfless, truthful, effective thinking and thus consciousness—and from there how the nurtured training of selflessness in humans would have liberated truthful thinking and thus consciousness—and from there how the emergence of consciousness would have led to a terrible battle with our instinctive self—and from there how the psychological upset that resulted from the battle would have demanded a more developed, intelligent, bigger brain in order to find understanding of why we had become divisively behaved. The journey has involved so many unbearably confronting truths that it is not surprising that the human-condition-avoiding mechanistic paradigm has found it difficult to acknowledge. It also reveals how if you are living in denial of truth you have no chance of making sense of our world and place in it—as evidenced by the mountain-high pile of books that have

been written about consciousness without ever managing to penetrate the subject. So, sadly, the President of the United States Barack Obama's 2013 **'Brain Initiative'**, which will see **'$100 million initial funding'** going to *human-condition-avoiding*, *denial-committed*, *mechanistic science* to find **'the underlying causes of...neurological and psychiatric conditions'** and **'develop effective ways of helping people suffering from these devastating conditions'**, namely the human condition (and as described in par. 603, there are very similar initiatives occurring in Europe), is so self-defeating it *is*, in effect, nothing more than an act of pure desperation.

[699] In summary, it was the process of nurturing love-indoctrination that not only gave our species its instinctive orientation to behaving cooperatively—our moral soul—it *also* liberated consciousness in our forebears. As pointed out in chapter 5, since nurturing is largely a female role and females controlled the selection of cooperative mates, it is true to say that the female gender created humanity. Throughout humanity's infancy and childhood, a period of time that, as will be explained in the next chapter, lasted from some 12 to 2 million years ago, nurturing played the most important role in the group; it was a matriarchal society in which males had to support this focus on nurturing and protect the group from external threats. However, as will also be described in the next chapter, humanity's matriarchal structure came to an end when our ignorant, ideal-behavior-demanding instinctive self began to criticize our conscious mind's search for knowledge and men, in their role as group protectors, had to take up the battle of resisting that naive, idealistic criticism from our instinctive self that was threatening to stop the maturation of humanity from ignorance to enlightenment. At this point, the patriarchal society that has existed for the last 2 million years came into being.

[700] What now needs to be presented is a description of all the psychological stages that our conscious mind progressed through in that heroic journey from ignorance to enlightenment.

[701] (A more complete description of the nature and origins of consciousness can be found in *Freedom Expanded* at <www.humancondition.com/freedom-expanded-consciousness>.)

# Chapter 8

# The Greatest, Most Heroic Story Ever Told: Humanity's Journey from Ignorance to Enlightenment

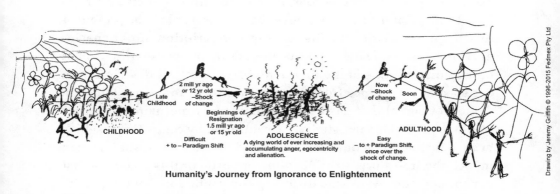

**Humanity's Journey from Ignorance to Enlightenment**

## Chapter 8:1 <u>Summary</u>

[702] With the desperately needed answers to the great outstanding mysteries facing the human race of the human condition, of the meaning of our existence, and of the origins of our unconditionally selfless moral instincts and conscious mind, now presented in this book, the full story of life on Earth can at last be told. In particular, we can finally explain where all species appear on the 'ladder' of integration of matter, and why, thus far, it is only the human race that has managed to progress to the 'rung' of integrating the members of our species into the Specie Individual. Indeed, it was that unique achievement that led to the emergence of our conscious mind that then had to search for knowledge, ultimately self-knowledge—the result of which was the breakdown of the integrated Specie Individual into the divisive, human-condition-afflicted state. But with the human condition now understood, humanity is finally in a position to return to the fully integrated Specie Individual situation once more, but *this time* in a knowing or understanding position—as the exceptionally denial-free, effective-thinking prophet Moses predicted, we can finally become *'like God* [the integrated state], *knowing'* (Bible, Gen. 3:5).

[703]What remains to be presented in this book is a description of the actual stages that our conscious mind progressed through from ignorance to enlightenment of our human condition, and with that understanding now found, to the transformation of the human race into the *'knowing'* or enlightened form of the integrated Specie Individual. Yes, we can now explain and describe the *entire psychological journey* our species has been on—and since that psychological journey is the *real* journey we fully conscious, human-condition-afflicted humans have been on, and since it has never before been able to be presented, this chapter provides the first ever true description of our species' development. While there have been libraries of books written about human origins, this is the first denial-free, truthful description of our species' emergence from our ape ancestors. And, since the main dishonest denial that has been maintained by mechanistic science was that our distant ancestors were controlled by competitive, selfish, brutish instincts that our conscious mind had to try to restrain, when, in fact, the true story is the absolute opposite of this account—that our ape ancestors were actually loving and gentle and that it was with the emergence of our conscious mind that we *became* competitive and selfish brutes—what is going to be presented is the completely new, never before told story of the human race!

[704]Yes, what follows is the most amazing and epic journey of any species to have existed on Earth—and it's our own story, the story of the incredibly, phenomenally brave and heroic human race!

## Chapter 8:2  <u>The stages of humanity's maturation from ignorance to enlightenment</u>

[705]We begin this most amazing story of humanity's journey from ignorance to enlightenment by meeting with our ancestors through the wonderfully fortuitous fossilized remains we have of them.

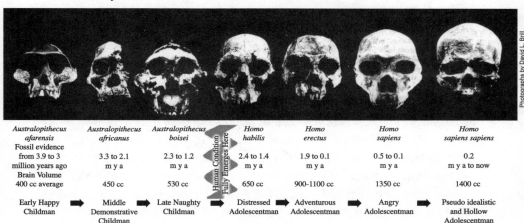

Photographs by David L. Brill

| *Australopithecus afarensis* | *Australopithecus africanus* | *Australopithecus boisei* | | *Homo habilis* | *Homo erectus* | *Homo sapiens* | *Homo sapiens sapiens* |
|---|---|---|---|---|---|---|---|
| Fossil evidence from 3.9 to 3 million years ago | 3.3 to 2.1 m y a | 2.3 to 1.2 m y a | Human Condition Fully Emerges Here | 2.4 to 1.4 m y a | 1.9 to 0.1 m y a | 0.5 to 0.1 m y a | 0.2 m y a to now |
| Brain Volume 400 cc average | 450 cc | 530 cc | | 650 cc | 900-1100 cc | 1350 cc | 1400 cc |
| Early Happy Childman ➡ | Middle Demonstrative Childman ➡ | Late Naughty Childman ➡ | | Distressed Adolescentman ➡ | Adventurous Adolescentman ➡ | Angry Adolescentman ➡ | Pseudo idealistic and Hollow Adolescentman |

[706]The above photo sequence of fossil hominin (human ancestor) skulls, dating back to the beginning of the australopithecines, appeared in the November 1985 edition of *National Geographic* magazine. Of course, as explained in chapter 2, mechanistic science hasn't recognized—indeed, has determinedly avoided—the psychological nature of our human condition, so the psychological descriptions for each of the stages of our development given above of 'Early Happy Childman', then 'Middle Demonstrative Childman', and so on, are my additions. Further, anthropologists no longer consider the robustly built *Australopithecus boisei* to be one of our direct ancestors but part of a now extinct branch line of development. However, as I will explain in pars 733-737, I disagree that there was any branching in the progression from the ape state to humans today, because for branching to occur there has to be deflecting influences—such as when Darwin's finches gradually became adapted to different food niches on the Galápagos Islands—and in our case there was only one major development going on and that was the *psychological* one. In a situation where there is only one all-dominant influence causing change there is no opportunity for divergence to develop, and from our species' infancy we have been under the all-dominant influence of what was occurring in our

brains, namely the development of consciousness and its psychological consequences. Any other influence was so secondary as to be ineffectual in causing our path to branch. In the case of *Homo habilis* fossils shown here as predating *A. boisei*, I strongly suspect that future fossil finds will push back the date of *A. boisei* and confirm the order depicted above, with *A. boisei* emerging from *Australopithecus africanus*, and giving rise to *H. habilis*. The reason for the overlap in the stages is explained in par. 735. Also, while anthropologists have since discovered more varieties of *Australopithecus* and *Homo* than those depicted here, these remain representative of the main varieties. There is also argument within the scientific community as to when to change the name as one variety develops into the next; for example, the varieties of *Homo* designated here as *Homo sapiens* and *Homo sapiens sapiens* may be referred to in other texts as the more 'archaic *Homo sapiens*' and as the more 'anatomically modern *Homo sapiens*' respectively. So the names used here are simply the traditional ones. Finally, the fossils that have been found to belong to our forebears who existed prior to the australopithecines are not included in this picture, but they too will be accounted for shortly.

[707] In examining this sequence it is apparent that a sudden increase in the size of the brain case, and by inference the brain's volume, occurred around 2 million years ago. A larger brain case was needed to house a larger 'association cortex'. As explained in pars 633-639, the ability to 'associate' information is what made it possible to reason how experiences are related, learn to understand and become conscious of, or aware of, or intelligent about, the relationship between events that occur through time. It follows that the development of a *larger* association cortex meant that a greatly increased need for understanding had emerged, which we are now able to explain would have resulted from the emergence of the dilemma of the human condition, where only self-understanding could bring an end to all the psychological upset that condition has produced and which has been so crippling of our species' development and that of our own lives. The inference we can take from this evidence is that the human condition became a full-blown problem some 2 million years ago with the emergence of *Homo*.

[708] The descriptions I have provided below the picture of the various fossil hominid skulls document the stages humanity progressed through as this liberated consciousness developed. The names ascribed to each stage indicate parallels with our own human life-stages, because the stages

that we, as conscious individuals, progress through are the same stages our human ancestors progressed through—'ontogeny recapitulates phylogeny': our individual consciousness necessarily charts the same course that our species' consciousness has taken as a whole. Eugène Marais recognized this when he wrote that **'The phyletic history of the primate soul can clearly be traced in the mental evolution of the human child'** (*The Soul of the Ape*, written between 1916-1936 and published posthumously in 1969, p.78 of 170). Wherever consciousness emerges it will first become self-aware (the 'Infancy' stage of the development of a conscious mind), then, inevitably, it will start to experiment with its power to effectively understand and thus manage change (the 'Childhood' stage), following which it will seek to understand the meaning behind all change ('Adolescence'), and from there it will obviously try to comply with that meaning ('Adulthood').

[709] In the case, however, of consciousness developing within us individually now and in our ancestors, that journey was significantly disrupted at adolescence by our search for the understanding of why we—individually and as a species—were not behaving in accordance with the integrative, cooperative meaning of existence; adolescence being the stage in the development of consciousness where the search for identity takes place, where the search for understanding of the meaning behind change, and the conscious organism's relationship to that meaning, occurs. In the case of our individual lives now, and in this stage in our species' journey, the particular identity we needed to find understanding of was why we were behaving divisively when the meaning of existence is to behave cooperatively and lovingly. Until we could answer that question, our development, both individually and collectively, was stalled (or what psychologists refer to as 'arrested') in an insecure adolescent state, unsure of our identity, particularly the reason why we haven't been ideally behaved—and preoccupied trying to validate our existence, prove that we are good and not bad, find *some* relief from the insecurity of the human condition. So, to mature from insecure adolescence to secure adulthood depended on finding understanding of our divisive human condition.

[710] To go over this important point again, without understanding of the human condition, humans haven't been able to properly enter adulthood. When stages of maturation aren't properly completed it doesn't mean subsequent growth stages don't take place, they do, but if a previous stage isn't properly fulfilled those subsequent stages are greatly compromised by the incomplete preceding stages. People do grow up, but in a state of

arrested development. Without the explanation of the human condition, humans have been insecure, not properly developed—in fact, completely preoccupied justifying themselves and finding ways to escape feeling unworthy. As the actress Mae West once famously said about men, **'If you want to understand men just remember that they are still little boys searching for approval.'** So it is only with understanding of the human condition now found that humans will be able to properly complete their adolescence and grow into secure adults, and the human race as a whole will be in a position to mature from insecure adolescence to secure adulthood. This is why descriptions such as 'The Angry Adulthood Stage of Humanity's Adolescence' appear in the following stages to describe our species' progression, for while we became adults in a physical sense, without the explanation of the human condition we were still psychologically stranded in adolescence.

[711] In summary, infancy is 'I am', childhood is 'I can', adolescence is 'but who am I?', while adulthood is 'I know who I am.'

[712] To apply these stages involved in the development of consciousness to the journey through which our species progressed, the term 'Infantman' pertains to the ape ancestor who first developed the nurturing training in selflessness that produced the fully cooperative state and, in doing so, liberated consciousness. As described in chapter 5:5, fossils from this period, which dates from 12 to 4 million years ago, are rare; however, scientists believe that our ape ancestors from this period include *Sahelanthropus tchadensis* (who lived some 7 million years ago and is thought to be the first representative of the human line after we diverged from humans' and chimpanzees' last common ancestor); *Orrorin tugenensis* (who lived some 6 million years ago); and the two varieties of *Ardipithecus*: *kadabba* (who lived some 5.6 million years ago), and *ramidus* (who lived some 4.4 million years ago). The various stages of 'Childman', who, it follows, developed from 'Infantman', were, as just summarized under the picture of the skulls above, the australopithecines who began to experiment with the power of conscious free will: 'Early Happy Childman' (*Australopithecus afarensis*), who developed into 'Middle Demonstrative Childman' (*Australopithecus africanus*), who then developed into 'Late Naughty Childman' (*Australopithecus boisei*). Again, at each stage greater experimentation in conscious self-management was taking place—from demonstrating the power of free will in mid-childhood, to beginning to challenge the instincts for the right to manage events in late childhood.

[713] As has now been explained, this challenging of our instincts led to criticism from those instincts, which in turn upset our conscious mind. In late childhood this emerging upset expressed itself in the physical flailing out at the 'unjust world' — the **'naughty nines'**, as parents and teachers describe this stage. What occurs to bring an end to childhood at about the age of 12, and what would have occurred at the end of the childhood stage in our ancestors around 2 million years ago, is the realization that physically protesting and flailing out doesn't achieve anything and that what we need to do instead is calm down, take stock of our situation, and try to understand what is causing us to be so upset; try to find the explanation for *why* we are being criticized. This significant change from being a frustrated, protesting, boisterous extrovert to a sobered, deeply thoughtful introvert signals the beginning of adolescence. Although unaware of the underlying reason, anthropologists have, nevertheless, recognized that a significant change did take place around 2 million years ago because it was at this juncture that they changed the name of the genus from *Australopithecus* to *Homo*; 'Childman', the australopithecines, became 'Adolescentman', *Homo*.

[714] As stated, adolescence is the stage when the search for identity takes place, and the identity that 'Adolescentman', *Homo*, particularly sought to understand was their lack of ideality — the reason why they were not ideally behaved. And once begun, this march of upset could *only* be brought to an end by finding sufficient knowledge to explain why the instincts' 'criticism' was undeserved. As such, there was an ever increasing need for mental cleverness to explain ourselves — specifically to find the liberating understanding of the human condition — hence the rapid increase in brain volume from 2.4 million years ago onwards.

[715] As with our species' progression through the various stages of its childhood, when our species entered its adolescence we necessarily went through a similar progression of stages, starting with the early sobered-by-the-emerging-problem-of-the-human-condition Adolescentman stage (early *Homo habilis*), followed by the distressed-by-the-human-condition Adolescentman stage (late *Homo habilis*), through to the adventurous Adolescentman stage (*Homo erectus*, who first left Africa), the embattled angry Adolescentman stage (*Homo sapiens*), and the pseudo idealistic and hollow Adolescentman stage that we currently occupy (*Homo sapiens sapiens*). But with the finding of understanding of why we have been divisively behaved now found, humans individually,

and humanity collectively, can finally mature from this insecure adolescence to a secure adulthood: 'Adolescentman' *can* become transformed 'Adultman'. (Note, 'man' is an abbreviation for 'human' or 'humanity', however, the use of 'man' also denotes a recognition that while humanity's infancy and childhood was matriarchal or female-role-led because that was when nurturing of infants was all-important, with the emergence of the egocentric, male-role-led need to defy the instincts, search for knowledge and prove humans are good and not bad, humanity's adolescence became patriarchal—a transition that will be more fully explained shortly (in pars 769-770). Since humanity's adulthood will be neither female or male led—because our species' maturation is complete—'Adultman' should more properly be described as 'Adulthuman'. By the same logic, since humanity's childhood was female role-led, the description for that stage should be 'Childwoman' rather than 'Childman', but to avoid unnecessarily complicating the matter we will leave all as 'man' for now.)

[716]We will now examine these various stages that a conscious mind has to progress through from infancy to childhood to adolescence and adulthood, looking at the individual living in the world today, and how these stages manifested within humanity's journey overall.

## Chapter 8:3 <u>INFANCY</u>

**The genus: our early ape ancestors, including *Ardipithecus* — 12 to 4 million years ago**
**The individual now: 0 to 3 years old**

Photograph of Matata by Manny Rubio

Drawing by Jeremy Griffith © 2006 Fedmex Pty Ltd

Bonobos Matata and her adopted son, Kanzi

[717] As explained in chapter 5, it was through the process of love-indoctrination, the exceptional nurturing of infants, that our ape ancestor was able to develop unconditionally selfless, fully altruistic, cooperative, moral, integrative, Specie Individual-forming behavior—a process the bonobos are currently perfecting, as illustrated by the above photo of the bonobo Matata hugging her adopted son, Kanzi. Nurturing is what made us human, a truth the image of the Madonna and child has been the archetypal representation of. And, as was explained in chapter 7:3, with the development of selflessness consciousness was able to emerge, and with consciousness came all the stages of maturation that consciousness itself had to progress through—namely infancy, childhood, adolescence, and adulthood. (It should be mentioned that while the main stages of maturation in non-human species are generally described as infancy, then adolescence when individuals are on the threshold of sexual maturity, and finally the sexual maturity of adulthood, in the case of fully conscious humans our stages of maturation involve a *psychological* journey, so our stages of maturation—from infancy to childhood, adolescence and finally adulthood—are fundamentally different to the infancy, adolescence and adulthood stages of other species.)

[718] As stated, the first stage of infancy occurs when humans become sufficiently conscious, sufficiently aware of cause and effect to realize that 'I exist'; it is when we realize that we are at the center of constantly changing experiences. This emerging consciousness and resulting self-awareness is apparent in bonobos who are the most intelligent of all non-human primates; indeed, some of that intelligence is apparent in Matata's expression. Yes, as was explained and evidenced in pars 426-429 and 693-694, bonobos are **'the most intelligent of all the great apes'**; and **'Everything seems to indicate that Chim** [a bonobo] **was extremely intelligent. His surprising alertness and interest in things about him bore fruit in action, for he was constantly imitating the acts of his human companions and testing all objects. He rapidly profited by his experiences...Never have I seen man or beast take greater satisfaction in showing off than did little Chim. The contrast in intellectual qualities between him and his female companion** [a chimpanzee] **may briefly, if not entirely adequately, be described by the term "opposites"'**; and, **"'It's like being with 9 two and a half year olds all day,"** she [zoo keeper Barbara Bell] **says. "They're extremely intelligent...They understand a couple of hundred words,"** she says. **"They listen very attentively. And they'll often eavesdrop. If I'm discussing with the staff which bonobos (to) separate into smaller groups,**

if they like the plan, they'll line up in the order they just heard discussed. If they don't like the plan, they'll just line up the way they want." "They also love to tease me a lot," she says. "Like during training, if I were to ask for their left foot, they'll give me their right, and laugh and laugh and laugh. But what really blows me away is their ability to understand a situation entirely."'

[719] As was explained in detail in chapter 5:5, recently published reports on the 'Infantman' fossils, *Sahelanthropus*, *Orrorin* and *Ardipithecus*, provide powerful validation of the love-indoctrination process, and also confirm that 'Infantman' closely resembled bonobos in many important respects. To briefly reiterate, there are, as has been emphasized, three main requirements for the nurturing-of-infants, love-indoctrination process to take place—upright walking, or bipedalism; ideal nursery conditions; and the presence and influence of more maternal mothers—and these fossil discoveries now show that at least as far back as 7 million years ago our ancestors were adapted to walking upright, as are bonobos, who walk upright more often than any other non-human primate. With regard to ideal nursery conditions, these fossils also confirm that 'Infantman' lived in forests and woodlands, not savannahs, and was semi-arboreal like the bonobos, who are the most arboreal of the African apes. Importantly, with regard to the need for more maternal mothers, it was also described in chapter 5:5 that the fossil record now strongly suggests that 'Infantman' lived in a cooperative society that was characterized by female sexual selection for less aggressive males, indicating a matriarchal society, which is consistent with a society where the raising of infants had become the central focus, as it is with bonobos. Based on these fossil finds, the anthro-pologist C. Owen Lovejoy reinforced both the reality and the duration of our cooperative past, writing that '**Even our species-defining cooperative mutualism can now be seen to extend well beyond the deepest Pliocene** [well beyond 5.3 million years ago]' ('Reexamining Human Origins in Light of *Ardipithecus ramidus*', *Science*, 2009, Vol.326, No.5949).

[720] A comparative look at the physique of bonobos and the fossil evidence provided by 'Lucy', the 3.5-million-year-old 'Childman', *Australopithecus afarensis*, who was discovered in the Rift Valley of Africa in 1974 by a team headed by the paleoanthropologist Donald Johanson, is also informative, as Lucy's bone structure shows an amazing similarity to that of a bonobo. While Lucy's skull indicates she had a slightly larger brain than a bonobo, the two are very similar in stature

and in the length of the lower limbs, and are fairly similar in overall body proportions. Lucy's pelvis shows that she walked fully upright. The pelvis of a bonobo, while not quite as adapted to bipedalism/upright walking as Lucy's, is significantly more adapted than that of a chimpanzee. Incidentally, 'Infantman' *Ar. ramidus*, who it has already been shown is at a stage of development comparable to that of the bonobo (recall Frans de Waal's observation that **'The bonobo's body proportions—its long legs and narrow shoulders—seem to perfectly fit the descriptions of Ardi, as do its relatively small canines'**), has a brain volume measuring between 300 and 350cc, which is the same size as a bonobo.

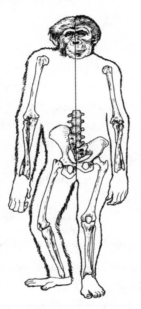

Left side: Bonobo skeleton. Right side: Early australopithecine, *Australopithecus afarensis*, 'Lucy', skeleton. Drawing by Adrienne Zihlman from 'Pygmy chimps, people, and the pundits', *New Scientist*, 15 Nov. 1984.

[721] In summary, the fossil record is now providing clear evidence that love-indoctrination had been established in 'Infantman' at least as far back as 7 million years ago, and this time frame will no doubt increase as more fossils emerge. The fossils also evidence that in bonobos, the most conscious and intelligent of the great apes, we have a living comparison of our 'Infantman' ancestor on the verge of 'Childman'.

## Chapter 8:4  CHILDHOOD
**The genus:** *Australopithecus* — **3.9 to 1.2 million years ago**
**The individual now: 4 to 11 years old**

Drawing by Jeremy Griffith © 1996 Fedmex Pty Ltd

[722] As stated, childhood is the stage when consciousness begins to experiment in self-adjustment and manage events to its own chosen ends. It comprises three phases—'Early Happy, Innocent Childhood', 'Middle Demonstrative Childhood', and 'Late Naughty Childhood'—with each stage involving a greater degree of experimentation in conscious self-management, from 'playing' with the power of free will in early childhood, to demonstrating the power of free will in mid-childhood, through to beginning to challenge the instincts for the right to manage events in late childhood.

## Chapter 8:5  Early Happy, Innocent Childman
**The species: the early australopithecines including** *Australopithecus afarensis* — **3.9 to 3 million years ago**
**The individual now: 4, 5 and 6 years old**

Drawing by Jeremy Griffith © 2006-2013 Fedmex Pty Ltd

[723] 'Early Happy, Innocent Childhood' encompasses the time when the intellect becomes sufficiently able to understand the relationship between cause and effect to start actively experimenting—'playing'—with the conscious power to self-manage and self-adjust. At this stage we are still, as it were, holding onto our mother's apron strings with one hand while carrying out short experiments in conscious self-management with the other; we are still depending on our instinctive nurtured orientations to love for the overall management of our life. In the case of humans today, it is the 'Look at me, Daddy, I can jump puddles' stage where reinforcing admiration from parents of the emerging conscious ability to manage events is so important.

[724] As mentioned, 'Lucy', the 3.5-million-year-old *Australopithecus afarensis* whose skeleton so closely resembles that of a bonobo, is an example of 'Childman', or more specifically, 'Early Happy, Innocent Childman'.

## Chapter 8:6 <u>Middle Demonstrative Childman</u>

**The species: *Australopithecus africanus* — 3.3 to 2.1 million years ago**
**The individual now: 7 and 8 years old**

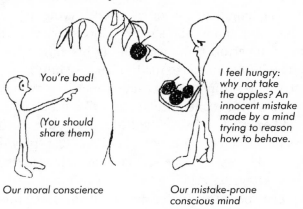

You're bad!

(You should share them)

I feel hungry: why not take the apples? An innocent mistake made by a mind trying to reason how to behave.

Our moral conscience

Our mistake-prone conscious mind

Drawing by Jeremy Griffith © 1991 Fedmex Pty Ltd

[725] It is during the 'Middle Demonstrative Childman' stage that the intellect starts demonstrating the power of free will and experiences its first encounter with the frustrations of a conflict with the instincts, which is the human condition.

[726] By mid-childhood the conscious mind is sufficiently able to make sufficient sense of experience to successfully manage and thus plan

activities for not just minutes ahead, but for hours and even days—a development that empowers the individual to be both outwardly marveling at, and demonstrative of, its intellectual power. It is at this stage of active self-management that the results of some experiments in self-adjustment begin to get the child into trouble. If we consider the behavior of children today who have reached this stage, we can appreciate the kind of trouble that entails. Imagine a young boy sitting at a table with a birthday cake on it. Being new to this business of reasoning, he thinks, innocently enough, 'Well, why shouldn't I take all the cake for myself', before duly doing so. While many mothers actually witness these grand mistakes of pure selfishness that young children make when they first attempt to manage their lives from a basis of understanding, they still have to be reasonably lucky to do so because, once done, the child generally doesn't make such a completely naive mistake again due to the criticism that experiment in self-management attracts from the child's instinctive moral conscience and from his conscious mind's awareness of the, in truth, very obvious integrative, selfless theme of existence—as well as from others present. But despite the nasty shock from all the criticism and his desire to not make such a mistake again, the boy, while unable to explain his actions, does feel that what he has done is not something bad, not something deserving of such criticism. In fact, by this stage in the child's mental development, he has become quite proud of the effort he's taken during his early happy innocent childhood stage to self-manage his life, successfully carrying out all kinds of tentative experiments in self-adjustment—drawing attention to his achievements with excited declarations like 'Look at me, Daddy, I can jump puddles', and so on. So the child is only just discovering that this business of self-adjusting is not all fun and that 'playing' with the power of free will leads to some serious issues. Indeed, the frustrated feeling of being unjustly criticized for some of his experiments gives rise to the precursors of the defensive, retaliatory reactions of anger, egocentricity and alienation; some angry, aggressive nastiness creeps into the child's behavior. (As was explained in chapter 3:5, this retaliatory behavior brings about the 'double and triple whammy' of condemnation that humans experienced when we searched for knowledge.) Furthermore, in this situation of feeling unfairly criticized, it follows that any *positive* feedback or reinforcement begins to become highly

sought-after, which is the beginning of egocentricity—the conscious
thinking self or ego starts to become preoccupied trying to defend its
worth, assert that it is good and not bad. At this point, the intellect also
begins experimenting in ways to deny or deflect the unwarranted criticism, which, in this initial, unskilled-in-the-art-of-denial stage, takes
the form of blatant lying: 'But, Mum, Billy told me to do it', or 'But,
Mum, the cake accidentally fell in my lap.' These apparent misrepresentations weren't *actually* lies, rather they were inadequate attempts at
explanation. Lacking the real excuse or explanation, it was at least
*an* excuse, a contrived defense for the child's mistake. The child was
evading the false implication that his behavior was bad, in the sense
that a 'lie' that said he wasn't bad was less of a 'lie' than a partial
truth that said he was. Basically, the child has started to feel the first
aggravations from the horror of the injustice of the human condition—
and we can expect that exactly the same kind of mistakes in thinking
and resulting frustrations with the ensuing criticism would have also
occurred in the lives of our 'Middle Demonstrative Childman' ancestors, namely *Australopithecus africanus*. Experiments in thinking,
such as 'There is some fruit; why shouldn't I take it all for myself?',
would have occurred, which would have resulted in criticism from
their moral instincts and conscious mind's awareness of the integrative,
selfless theme of existence—criticism that would have led to the beginnings of the psychologically upset behaviors of anger, egocentricity
and alienation in *Au. africanus*.

[727] While children today have to—just as our *Au. africanus* ancestors
would have had to—negotiate this middle demonstrative stage where at
times they behave in a way that is 'disobedient' of their moral instincts
and 'defiant' of the integrative theme of existence, their conscious mind
still doesn't know why it has been disobedient and defiant; it isn't able
to understand and explain that it has become a conscious being. Also,
now that they are capable of thought they can't stop thinking, which
means mistakes in self-management are going to continue to occur, as
will the criticism those mistakes attract, and, it follows, the upset with
that criticism. Yes, from demonstrating the power of free will, the child
has started to feel the first aggravations from the horror of the injustice
of the human condition. Of course, for our 'Middle Demonstrative
Childman', *Au. africanus* ancestors, love would still have very much been

the dominant influence in their lives overall, which means any distress from upset would have been quickly healed with love. Furthermore, the defensive expressions of anger, egocentricity and denial would have been restricted to feelings and actions rather than expressed in words; in fact, there would not have been a strong call for language until the adolescent state emerged some 2 million years ago when the battle of the human condition developed and, with it, alienation, because it was only when we became variously alienated in self and thus variously alienated from each other that a strong need to try to justify and explain ourselves to one another arose. The anthropologist Richard Leakey's study of brain cases in fossil skulls for the imprint of Broca's area, the word-organizing center of the brain, evidences this development: **'*Homo* had a greater need than the australopithecines for a rudimentary language'** (*Origins*, 1977, p.205 of 264). Prior to the emergence of alienation we were all instinctively aware of and in sync with each other; apart from contact, orientating and warning calls, and expressions of excitement and joy, there was no need to develop a sophisticated, complex language. As Plato described life during **'our state of innocence, before we had any experience of evils to come'**, we lived a **'blessed and spontaneous life…**[where] **neither was there any violence…or quarrel'** and **'there were no forms of government'**; it was a **'simple and calm and happy'** life (see ch. 2:6).

[728] While discussing the effects of alienation, it should be pointed out that infants would have stopped being quiet when mothers stopped being able to properly respond to their needs due to their alienated, soul-devastated condition. For instance, even the infants of relatively innocent, less alienated 'races' of humans today, such as the Yequana of Venezuela, the Australian Aborigine and the Bushmen of the Kalahari, rarely cry, as these remarkably similar quotes confirm: **'[Yequana] babes in arms almost never cried'** (Jean Liedloff, 'The Importance of the In-Arms Phase', *Mothering*, Winter 1989; see <www.wtmsources.com/150>); and **'!Kung [Bushmen]…infants hardly ever cry'** (Dr Harvey Karp, 'Cultures without Colic: Breastfeeding & Other Baby Lessons from the !Kung San'; see <www.wtmsources.com/117>). Yes, motherese language developed as a way for alienated humans to try to pacify their distressed innocent infants.

## Chapter 8:7 <u>Late Naughty Childman</u>

**The species: the robust australopithecines (*Au. robustus* and *Au. boisei*) — 2.3 to 1.2 million years ago**

**The individual now: 9, 10 and 11 years old**

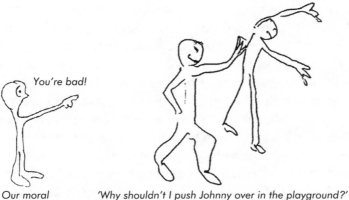

Our moral conscience

'Why shouldn't I push Johnny over in the playground?' –the frustrated, naughty, bullying final stage of childhood.

[729] The 'Late Naughty Childman' stage represents the time when the intellect naively lashes out at the increasing unjust criticism it is encountering as a result of its first tentative experiments in self-adjustment.

[730] Since school teachers become very aware of the changing behavior of the many children they have under their care, I asked a teacher to describe what she and her colleagues knew of the stages children and early adolescents go through. These are the main points from the response she collected: **'Six and seven-year-olds are considered to be very compliant, but by eight children are starting to test the waters and challenge the world a little.'** She continued, **'the eight-year-olds can be annoying and a little naughty'**, while **'nine and ten-year-olds can be hard to handle as they seem to hit a phase of recklessness'** and **'are considered naughty'**. She noted that **'Teachers love teaching 11 and 12-year-olds because it is during this stage that children become civilised'** but that **'Teachers consider students who are 14, 15 and 16 years old the most difficult to teach. The adolescents seem to be at complete odds with what is expected of them. Most teachers are terrified of these extremely uncooperative mid-teenage ages'** (WTM records, 1997). These insights support the explanations being given of the stages of maturation of consciousness

through childhood—and confirm the agonising stage of Resignation that virtually all adolescents now succumb to when they are about 15 years old, a process that was described in chapter 2:2 and in pars 654-655, and will be further explained within the context of humanity's development of consciousness shortly.

[731] Of course, it was inevitable that as the human condition developed children were going to become increasingly influenced by the psychologically upset, angry, egocentric and alienated world in which they were raised, and that it would become harder to differentiate what upset in them was a result of those external circumstances or from their own experiments in understanding. However, even in the original situation, where there was little or no upset in the world, we can expect that by the age of eight, children would justifiably be feeling resentful towards the 'criticism' emanating from their instinctive self of their tentative efforts to self-manage their life using understanding. And, unable to adequately cope with this 'criticism' with understanding of it, we can expect that each child would begin to retaliate against the criticism as the only form of defense available to them. The problem then, however, was that these early, relatively mild experiments in retaliation—of anger, selfishness and dishonest excuse-making in mid-childhood—had the alarming effect of greatly compounding the 'criticism' from the child's perfectly integratively orientated, moral instinctive self and from their awareness of the integrative meaning of existence; they induced the double and triple whammy mentioned earlier. From being mildly insecure we can expect the child to now feel guilty and that this drastic escalation in criticism and thus frustration would be a contributing factor to the turbulent, boisterous **'naughty nines'** that parents and teachers have labeled this stage. By the end of childhood, at the ages of 10 and 11, we can expect the resentment and frustration to be such that it would express itself in the form of taunting and bullying. The child would be belligerently lashing out at the unjust world: 'Why shouldn't I feel resentful and retaliate?', 'Why shouldn't I shove you around if I can, especially since I'm bigger and stronger?', 'Why can't I have my way?', and 'What's wrong with being selfish and aggressive anyway?'

[732] In the situation that exists today, where the external upset is almost overwhelming, we can expect that almost all of the child's upset will have resulted from their encounter with external upset. The increasingly

thoughtful child can see the whole horribly upset world and would be understandably totally bewildered and deeply troubled by it. Eight-year-olds will only be *beginning* to be consciously troubled by the horror of the state of the world they have been born into, but by nine they will be overtly troubled by it and requiring a lot of reassurance that 'Everything is going to be alright.' In fact, nine-year-olds can be so troubled by the imperfection of the world that they go through a process of trying not to accept that it is true. By 10, this despair about the state of the world reaches desperation levels with nightmares of distress for children. It is a very unhappy, lonely, anxious, needing-of-love time for them. So at 11 some enter a 'Peter Pan' stage where they decide they don't want to grow up; they decide they want to stay a child forever, surrounded by all the things they love, and not ever become part of the horror world they have discovered. It is no wonder **'Teachers love teaching 11 and 12-year-olds'** who have **'become civilised'**—they're essentially tame compared with the **'reckless'**, **'naughty'**, flailing-out-at-the-world **'nine and ten-year-olds'**.

[733]The fossil record evidences the description that has been given of these stages of 'Childman'. The early *Australopithecus afarensis*, who have been described as occupying the early happy, prime-of-innocence stage, and the subsequent *Australopithecus africanus*, who have been described as being in the middle demonstrative childhood stage, are both finely built compared with the much more robustly built *Australopithecus boisei* and associated *Australopithecus robustus*, both of whom have been described here as being in the 'Late Naughty Childman' stage. As mentioned earlier, in attempting to account for this physiological discrepancy, anthropologists have even placed the more robust late australopithecines on a separate, dead-end branch to *Homo* (with some recently going so far as to reclassify them as an entirely separate genus, *Paranthropus*), but that has to be impossible because for branching to occur there has to be deflecting influences—such as when Darwin's finches gradually became adapted to different food niches on the Galápagos Islands—and in our case there was only one major development going on and that was the *psychological* one. In a situation where there is only one all-dominant influence causing change there is no opportunity for divergence to develop, and from our species' infancy, we have been under the all-dominant influence of what was occurring in our heads, namely

the development of consciousness and its psychological consequences. Any other influence was so secondary as to be ineffectual in causing our path to branch. So, contrary to what human-condition-avoiding, psychosis-denying mechanistic anthropologists have been telling us, there has been no branching off from either the australopithecines or *Homo*. But if the robust australopithecines weren't a separate branch, the question remains: why was there such a big physical disparity between them and the much more gracile or fine featured preceding *Au. afarensis* and *Au. africanus*, and the variety of early humans they gave rise to, the also much more gracile *Homo habilis*? The answer has to lie in the psychological differences between the much quieter, love-immersed *Au. afarensis* and *Au. africanus*, the extroverted, boisterous, bullying *Au. robustus* and *A. boisei*, and the introverted, sobered, quiet early adolescent *H. habilis*, who will be described shortly.

[734]To quickly elaborate on these size disparities, 'Late Naughty Childman', *Au. robustus* and *Au. boisei*, had comparatively large frames and skulls that were especially heavily built with very pronounced cranial and facial bone structures. Anthropologists recognize that these skull modifications came about to support the much stronger facial muscles that were required to work the heavy jaw and huge grinding teeth that characterize these late australopithecines. We know from such evidence as the wear patterns on their teeth that the australopithecines were vegetarian, but why did the later australopithecines need larger grinding teeth? What dietary change occurred, and why? Being extroverted, increasingly naughty and roughly behaved, the late australopithecines were like older children today who would rather be out playing than eating, but such an extremely physically assertive and energetic lifestyle required fuel. Not being sufficiently conscious to attempt self-management, all other animal species exert only enough energy to secure their necessary food, space, shelter and a mate—they are conservative energy users—but, with their rough, energetic play, late childhood humans became the first non-conservative energy users on Earth. In order to 'eat and run', 'Late Naughty Childman' would have needed a readily available food source that they could quickly ingest and, being vegetarian, they would have needed a lot of it because vegetables don't convert into as much energy as meat for instance, which was not to appear on humanity's dining table until the much greater levels of upset that characterized our

species' adolescence developed. (While the australopithecines, in their naughtiness, would have been capable of being rough and possibly even cruel at times to other animals, they were not yet sufficiently upset with innocence to be killing innocent animals on a regular basis, which, as will be explained shortly in pars 778-780, is what so-called 'hunting' was really all about for upset humans and what finally led to meat-eating.) This increased energy requirement is evidenced by recent studies of dental wear patterns in robust australopithecines that indicate their **'diet included more $C_4$ [grasses and sedges] biomass than any other hominin…[and** its massive jaw, grinding teeth and necessary facial structure] **represents an adaptation for processing large quantities of low-quality vegetation'** (Thure Cerling et al., 'Diet of *Paranthropus boisei* in the early Pleistocene of East Africa', *Proceedings of the National Academy of Sciences*, 2011, Vol.108, No.23) that was **'nearly identical'** (ibid) to the diet of its predecessor *Au. afarensis*. Interestingly, anthropologists have traditionally argued that *Au. boisei*'s massive facial structure developed to allow **'hominins to colonize increasingly seasonal and open environments'** that offered **'a diet of nuts, seeds and hard fruit'** (ibid) — but, as it turns out, *Au. boisei* was not eating anything different to *Au. afarensis*, just greater quantities. As will shortly be described, in contrast to these late australopithecines, *H. habilis* was an entirely different individual. Introspective, deeply thoughtful and sobered, *H. habilis* were no longer interested in physically intimidating the world, and did not, therefore, need great quantities of energy and thus food, hence their reversion to a more gracile frame.

[735] Another feature of the fossil record of early humans is the evidence it provides of the overlap between the different varieties, but now that we can take into account what was happening psychologically, the overlap becomes understandable. Just as there are very early models of cars still around today, long after they have been superseded, so groups of early, less intelligent and thus less human-condition-afflicted-and-adapted varieties of humans endured long after they had been superseded by those more human-condition-afflicted-and-adapted. Of course, the best example of such overlapping in the anthropological record is the existence today of equivalents of our early infant ape ancestors, specifically the non-human primates, and particularly chimpanzees, gorillas and orangutans, as well as bonobos, who are in a later stage of infancy. Apes are not, as is widely believed, a branched development from the human

line of development—they are on exactly the same development path, but at a much, much earlier stage. The reason that more recent varieties in the *Homo* lineage such as *H. Habilis* do not currently exist, whereas the earlier equivalents of our ancestors such as bonobos, chimpanzees, gorillas and orangutans still do, is that once consciousness was liberated, subsequent psychological development became inevitable, which meant that any species with a liberated consciousness would inevitably develop into the subsequent stages that a conscious mind undergoes, and, as will become very apparent, this development and thus supersession became increasingly rapid. There was somewhat of a stalled stage in this rapid development in the step from childhood to adolescence, with the australopithecines persisting long after the emergence of *Homo* (some 1.2 million years after), but that is plausible when we consider that the essential psychological feature of late childhood is procrastination about having to accept an increasingly upset, human-condition-afflicted, corrupted, imperfect world. Once varieties of late Childman appeared who were sufficiently intelligent to realize that they had no choice but to set out in search of understanding, then those varieties would have done so and progressed into adolescence; the process of adapting to an increasingly upset world would have picked up pace from there in those varieties and they would have rapidly left behind those who were psychologically still in the procrastinating childhood stage. Basically, once our ancestors crossed the threshold into the fully emerged upset state of the human condition, the development of that condition progressed rapidly. As occurred with the great apes being considered a branched development from the human line of development, the evidence that the australopithecines were still in existence up to 1.2 million years after *H. habilis* appeared has been used to support the argument that the late australopithecines branched away from the *Homo* line, but now that we can take into account what was happening psychologically the overlap *does* become understandable.

[736] With regard to the existing great apes, I should emphasize that they are described here as *equivalents* of our infant ape ancestor because they may not be actual 'living remnants' of our particular infant ape ancestor. The main issue of our relatedness to the great apes lies in the fact that they are on the same journey that primates have been variously able to follow of becoming integrated through the process of love-indoctrination,

a process that when sufficiently developed leads to the emergence of consciousness and the inevitable stages that consciousness then progresses through. Indeed, the *Ardipithecus* fossils indicate that the last common ancestor we shared with chimpanzees had physical differences to today's chimpanzee. However, what is clear from the fossil record is that the ancestors of the existing great apes were extremely 'ape-like', and that existing apes are therefore equivalents of our infant ape ancestors, stalled in development. With regard to the more developed bonobos, we can date their emergence from the chimpanzee line around 1 million years ago, when they chanced upon the ideal nursery conditions that allowed love-indoctrination to develop the selfless, cooperative behavior that liberates consciousness, and so there is no suggestion that they are *literally* a living remnant, but as the comments in par. 411 by Zihlman and de Waal indicate, they are a very close equivalent to our infant ancestor. In addition, they demonstrate the profound changes that inevitably occur when consciousness is liberated, as is clear from their intelligence and their continually changing physiology. Further, they are rapidly becoming more and more discrete from their chimpanzee cousins who are still stalled in early infancy, unable to develop love-indoctrination beyond an elementary degree.

[737] It is even possible that some varieties of apes might have developed from entirely different primate stock to those our ape ancestors developed from. For example, we know from DNA analysis that the orangutans, who live in South East Asia, split from the African apes some 16 million years ago. 'Parallel development' or 'evolutionary convergence' can occur where animals have the same form but come from entirely unrelated stock, such as, in an extreme example, the thylacine (Tasmanian Tiger), which is a marsupial mammal, had an almost identical build to a wolf, which is a placental mammal. So developments can be the same or very similar and yet be unrelated genetically. So while we do know that bonobos and chimpanzees share 98 percent of our DNA, in terms of understanding our behavior, which is the main objective of our search for knowledge, it wouldn't really matter if bonobos weren't at all related to us genetically because they would still represent an equivalent of our early ape ancestor, and as such what insights we can glean from their behavior can be extremely relevant in terms of understanding our own.

## Chapter 8:8  ADOLESCENCE

**The genus: *Homo* — 2.4 million years ago to the present**
**The individual now: 12 to 50 plus years old**

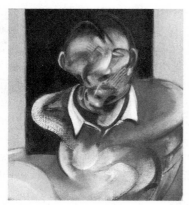

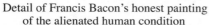

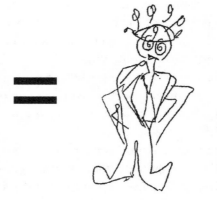

Detail of Francis Bacon's honest painting          Deluded, alienated 'Adolescentman'
of the alienated human condition

[738] Moving now into analysis of adolescence, it was emphasized in chapters 2:2 and 8:2 that adolescence encompasses the stage when we fully conscious humans search for our identity, for understanding of who we are — specifically for why we have not been ideally behaved. And since completing that upsetting search was to take some 2 million years, humanity's adolescence came to be characterized by ever-increasing levels of upset, which required the invention of more and more desperate means to try to contain that upset as well as find relief from it. Ultimately, during the final, agonising stages of this horror story, when upset reached such excruciating levels that we could no longer bear to confront and think about it, we came to a point where we had no choice but to resign ourselves to living in complete denial of the whole unbearably depressing issue of our corrupted condition. The result of this extremely dishonest block-out or denial was almost total separation from our all-sensitive and all-loving — but, at the same time, all-criticizing — true self or soul. So when this process of Resignation became an almost universal phenomenon amongst humans during the last 11,000 years (as will be explained, this was due to the development of agriculture, which allowed people to settle and multiply and form towns and cities where closer interaction amongst humans had the effect of rapidly escalating upset) it was indicative of

the fact that the human race was fast approaching the psychologically dead state of terminal alienation—a state that *has* now virtually been reached, and which all the apocalyptic-and–zombie-themed movies we are currently seeing in our cinemas are testament to. So, unfortunately, the true story of our species' journey through the various stages of adolescence that is now, for the first time, going to be told in this book is one that travels from soundness to insanity—keeping in mind that with the reconciling and rehabilitating understanding of the human condition now found, the human race can be transformed back to life, back to a state of universal sanity, happiness and togetherness, as will be explained in chapter 9.

## Chapter 8:9  Early Sobered Adolescentman

**The species: the first half of *Homo habilis'* reign — 2.4 to 1.4 million years ago**
**The individual now: 12 and 13 years old**

*You're bad!*

*Our moral conscience*

Drawing by Jeremy Griffith © 1991-2011 Fedmex Pty Ltd

[739]The Early Sobered Adolescent Stage (of Humanity's Adolescence) signals the end of childhood and represents the time when we fully encounter the sobering imperfections of life under the duress of the human condition.

[740]Throughout childhood we saw how the frustration with being criticized for searching for knowledge continued to increase until, in late childhood, the child's exasperation and resentment caused him to angrily lash out at the 'injustice of the world'. What happens at the end of childhood is that the child realizes that physical retaliation doesn't make any difference and that the only possible way to solve the frustration is to find the reconciling understanding of *why* the criticism he is experiencing

is not deserved. Of course, as just pointed out in par. 731, the more upset developed in the human race as a whole, the more children also became worried by the upset in the world around them. While resigned adults became accomplished at overlooking the hypocrisy of human life because of its human-condition-confronting implications, children in their naivety could still see it. They asked: '*Mum, why do you and Dad argue all the time?*' and '*Why are we always worried about having enough money?*' and '*Why are we going to a big, expensive party when the family down the road is so poor?*' and '*Why is everyone so lonely, unhappy and preoccupied?*' and '*Why are people so fake and artificial?*' and '*Why is the only thing people talk about when they meet each other is such superficial things as the weather or the football?*' and '*What is religion?*' and '*Why do people pray?*' and '*Who is God?*' and '*Why do people make awful jokes?*' and '*Why are there wars?*' (as the following cartoon by Bill Watterson poignantly depicts) and '*Why are there pictures about sex everywhere?*' and '*Why did those people fly those planes into those buildings?*' And the truth is, these are the *real* questions about human life, as this quote by the Nobel Prize-winning biologist George Wald acknowledges: **'The great questions are those an intelligent child asks and, getting no answers, stops asking'** (Introduction to *The Fitness of the Environment*, Lawrence J. Henderson, 1958, p.xvii). Children **'stop**[ped] **asking'** the real questions—stopped trying to point out the all-important and obvious-if-you-are-still-looking-at-the-world-truthfully, yet almost totally unacknowledged proverbial 'elephant in the living room' issue of the human condition—because they eventually realized that adults couldn't answer their questions; and, more to the point, they were made distinctly uncomfortable by them. The novelist George Eliot (the pen-name of Mary Ann Evans) wrote that **'Childhood is only the beautiful and happy time in contemplation and retrospect: to the child, it is full of deep sorrows, the meaning of which is unknown'** (1844; *George Eliot's Life, as Related in Her Letters and Journals*, 2010, p.126 of 518). The only reason **'the meaning of'** the **'deep sorrows'** of children was **'unknown'** to adults was because adults live in denial of the human condition. A *Newsweek* article that discussed childhood stress was bordering on this truth when it said, **'Parents are frequently wrong about the sources of stress in their children's lives, according to surveys by Georgia Witkin of Mount Sinai Medical School; they think children worry most about friendships and popularity, but they're actually fretting about the grown-ups** [and their world]**'** ('Stress', Jerry Adler, May

1999; see <www.wtmsources.com/177>). The author Antoine de Saint-Exupéry articulated the child's point of view beautifully in his celebrated 1945 book *The Little Prince*, when he had the Little Prince say, **'grown-ups are certainly very, very odd'** (p.41 of 91). Resigned and unresigned minds have lived in two completely different worlds; they have been like different species, each almost invisible to the other. The renowned writer Robert Louis Stevenson described the situation thus: **'And so it happens that although the paths of children cross with those of adults in one hundred places every day, they never go in the same direction; nor do they even rest on the same foundations'** (*Child's Play*, 1878; as quoted at the beginning of the 1997 film *The Colour of the Clouds*). So in the final stages of childhood it was not only the issue of the imperfections of their own behavior that so troubled children, but also the issue of the imperfections of the human-condition-afflicted world around them—a psychological collusion that sees children mature from frustrated, extroverted protestors into sobered, deeply thoughtful, introverted adolescents. It's a *very* significant transition in our psychological journey, from the relatively human-condition-free state to the very human-condition-aware state, that is actually recognized by the fact that we separate those stages into childhood and adolescence—a demarcation even our schooling system reflects by having children graduate from what is generally called primary school into secondary school at 12 to 13 years of age. As mentioned earlier, this critical juncture in our species' development was also recognized by anthropologists when they changed the name of the genus from *Australopithecus* (meaning 'southern ape') to *Homo* (meaning 'man'); 'Childman', the australopithecines, became 'Adolescentman', *Homo*. And so it was in the lives of the earliest *Homo*, *Homo habilis*, that this deeply introspective, thoughtful and sobered state of mind would have existed.

## Chapter 8:10 <u>Distressed Adolescentman</u>
**The species: the second half of *Homo habilis'* reign — 2.4 to 1.4 million years ago**
**The individual now: 14 to 21 years old**

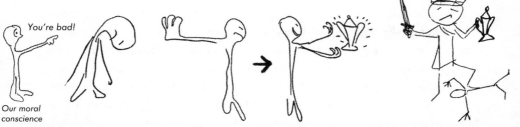

Approaching Resignation          The moment of Resignation          The selfish, power-fame-fortune-
                                                                    and-glory seeking resigned adult

[741] The Distressed Adolescent Stage (of Humanity's Adolescence) represents the time when adolescents struggled with the extreme distress brought about by their full engagement with the issue of the human condition.

[742] When the human species became excessively upset, as has occurred during the latter stages of the reign of *Homo sapiens sapiens*, the extreme distress experienced by an adolescent confronting the issue of the imperfection both of the world at large and within themselves became so great it led to a state of such unbearable depression that it forced the adolescent to resign to living in denial of the issue of the human condition and to never again thinking about anything that brought that issue into focus, which, as we have seen, was almost all thinking—an agonising process that resulted in the psychotic (psyche/soul repressed) and neurotic (neuron/mind repressed) state of extreme alienation. <u>This process of resigning to a life of living in denial of the human condition was described in chapter 2:2</u>, with this stage being the one that school teachers described as **'the most difficult to teach. The adolescents seem to be at complete odds with what is expected of them. Most teachers are terrified of these extremely uncooperative mid-teenage ages.'** <u>And in pars 654-655 the strategy a person adopted *after* Resignation to bring some relief to the insecurity of their situation was described</u>. It was explained there that not only did resigned humans find ways to avoid and deny the issue of the human condition, they also found ways to convince themselves that

they were the opposite of flawed and corrupted. It was described how, unable to refute the negative view of themselves with understanding, the resigned person could only counter the negatives by focusing on, emphasizing and developing whatever positive view of both the world and themselves they could find. In particular, they became preoccupied finding ways to feel good about themselves by competing for power, fame, fortune and glory. If their team could win a football match, or if they could do something that brought praise—basically any activity that would relieve the insecurity of their condition—that was all that mattered. And the more insecure they were, the more desperate was their need for reinforcement, so it follows that once the *full* extent of the imperfection of their corrupted condition became apparent at the time of Resignation, then their need for reinforcement to counter that uncertainty about their worth became extreme. After Resignation, the 'ego' or 'conscious thinking self' became *extremely* selfishly focused or centered on the need for reinforcement and relief from criticism. So, as well as practicing complete denial of any confronting truths, and becoming very angry towards any criticism, extreme ego-centricity was the main outcome of Resignation. Basically, after Resignation, the angry, egocentric and alienated responses to the dilemma of the human condition that emerged in infancy and had been gradually developing throughout childhood and early adolescence suddenly became greatly amplified. So much so that the self-worth-embattled, heroic mantra of the resigned person became one of **'Give me liberty** [from criticism] **or give me death'**, **'No retreat, no surrender'**, **'Death before dishonor'**—in essence, 'I don't believe the insinuation that I am a bad person is deserved, and in any case it's too unbearable to accept, so I will *never* tolerate any suggestion of it'!

[743] In the case of humanity's journey, Resignation would not have become the key feature it is in virtually all adult lives today during the time of *Homo habilis*, or the *Homo erectus* representatives of the next adventurous early adulthood stage of humanity's adolescence, or even the subsequent *H. sapiens* representatives of the angry adulthood stage of humanity's adolescence. Resignation would have only appeared during the 200,000 year reign of *H. sapiens sapiens*, and only become almost universal in the final 11,000 years of that reign. This is because it is the upset from the lack of nurturing in infancy and early childhood that makes self-confrontation during the thoughtful early adolescent

stage overwhelmingly depressing, and this lack of nurturing—basically, the alienated, detached-from-their-true-loving-soulful-self state of parents—was not an outstanding feature of human life until the *final* stages of humanity's adolescence. The upset from a developing mind's own efforts to self-adjust, while distressing and even depressing, was not sufficient to cause the mind to have to pay the very high price of blocking out our instinctive self or soul's happy, all-loving and all-sensitive world because of the condemning expectation that gave rise to that we should still be behaving in that all-loving, cooperative, selfless way. Upset and its effects had to be very great for Resignation to occur, and while upset and its effects *are* very great in modern humans and as a result Resignation *is* almost universal, the fact that even amongst modern humans there have been adults who didn't resign, such as the prophets Abraham, Moses, Plato and Christ, evidences that the extreme upset that leads to Resignation is not yet an intrinsic part of the human make-up. Further, for adult members of the relatively innocent representative 'races' of *H. sapiens sapiens* living today, such as the Bushmen of the Kalahari and the Australian Aborigines, to be as happy and full of the zest and enthusiasm for life and as generous, selfless and free in spirit as numbers of them are, or at least were when they were still living as hunter-foragers, means that many of them must not have resigned. The world of soundness and happiness is not far below the surface in humans today. So while there must be a degree of genetic adaption to a resigned, soul-destroyed existence in humans now, a sound, sensitive and happy life is retrievable for all humans. What is presented in chapter 9 will evidence that this is true. Resignation, with all the soul-dead insensitive and must-prove-yourself mean and unsatisfying life that went with it, is fundamentally a *mental, psychological* condition, not an immutable *genetic* condition, so you *can* choose to leave that insecure, embattled, insensitive, mean and unhappy life behind. In chapter 9 it will be explained that while it will take a number of generations for the psychological rehabilitation of the human race to be completed, all humans *can* immediately leave the soul-repressed, insensitive, denial-committed-and-thus-extremely-alienated, selfish and egocentric power-fame-fortune-and-glory-seeking resigned life, and become part of the secure, happy, human-condition-and-Resignation-free new world that understanding of the human condition now makes possible. While any human living today will be to a significant degree genetically adapted to living with upset, and will therefore be, to a significant degree, soul-destroyed and alienated, it was the process of Resignation that largely killed off access to our soul and made humans virtually totally mad—sufferers of the **'dead' 'fifty feet of solid concrete'** state of **'alienation'** from our soul that R.D. Laing spoke of (see par. 123). But, again, even though there has to be, after living in a state of upset for some 2 million years, a significant degree of alienation in humans now, the wonderful reality is that once a person is free from the alienating denial that accompanies Resignation, *and free from anxiety about the state of the world*, they *will* be immensely happy and well-adjusted—as is described in chapter 9.

[744] Yes, the existence of extreme sensitivity and soundness in modern humans like Christ, and the happiness and soundness of relatively innocent 'races' living today like the Bushmen evidences that once humans are freed from the escapist preoccupations that follow Resignation, they *will* be able to access a great deal of our soul's sound sensitivity and happiness. The aforementioned English explorer and philosopher Bruce Chatwin acknowledged the unresigned soundness and sensitivity of Christ and also of the relatively innocent 'races' when he wrote these extraordinarily honest words: **'There is no contradiction between the Theory of Evolution and belief in God** [Integrative Meaning] **and His Son** [the uncorrupted expression of our original instinctive orientation to Integrative Meaning] **on earth. If Christ were the perfect instinctual specimen—and we have every reason to believe He was—He must be the Son of God. By the same token, the First Man was also Christ'** (*What Am I Doing Here*, 1989, p.65 of 367). Yes, since the common dictionary definition of a 'prophet' is **'someone who speaks for God'** and Christ was an exceptionally uncorrupted expression of our original instinctive self or soul's orientation to Integrative Meaning, then Christ spoke for God; he was a prophet. In terms of being unresigned, Christ was certainly amongst **'the firstborn from among the dead** [resigned]**'** (Bible, Col. 1:18).

[745] I should mention that as a member of the Jewish 'race', Christ would have benefited from a degree of genetic toughness in that 'race' because it would have allowed his exceptionally well nurtured, **'Lamb of God'** (Bible, John 1:29) innocence to survive contact with the upset world where someone less genetically toughened may have not. As my headmaster at Geelong Grammar School, Sir James Darling (who in his full-page obituary in *The Australian* newspaper was described as **'a prophet in the**

**true biblical sense'** (3 Nov. 1995; see <www.wtmsources.com/165>)), acknowledged in one of his famous speeches about sensitive, innocent soundness not being enough for someone to be able to defy the alienated, dishonest world of denial and find the explanation of the human condition: **'he must be sensitive *and* tough'** (*The Education of a Civilized Man*, ed. Michael Persse, 1962, p.34 of 223)—which is why a member of the Bushmen 'race', a genetically relatively innocent 'race', could not have found the explanation of the human condition. (Note, later in par. 1032 I cite more of Sir James' speech, in which he spoke about the need to **'be sensitive *and* tough'**—and much more can be read about his incredibly visionary education program at Geelong Grammar School, where he deliberately set out to cultivate the innocent soundness needed to confront and solve the human condition, at <www.humancondition.com/darling>.) Sir Laurens van der Post was another who recognized that the Bushmen, although relatively innocent, did not have sufficient instinctive toughness to withstand the upset world when he described how **'mere contact with twentieth-century life seemed lethal to the Bushman. He was essentially so innocent and natural a person that he had only to come near us for a sort of radioactive fall-out from our unnatural world to produce a fatal leukaemia in his spirit'** (*The Heart of the Hunter*, 1961, p.111 of 233). A further difference between Christ and the more innocent so-called 'primitive' 'races' such as the Bushmen is the greater level of self-restraint that accompanies toughness, which Christ, being a member of the more upset adapted Jewish 'race', would have also possessed. How restraint has accompanied the rise in upset during the human journey is one of the main themes of this chapter.

[746] (Incidentally, understanding that Christ represents our **'instinctual'** self allows us to decipher the so-called 'Trinity' of main influences or forces on Earth that many religions recognize, with 'God the Father' being Integrative Meaning, while 'God the Son' and 'God the Holy Ghost or Spirit' are, respectively, the two great tools for developing integration or order, namely the gene-based and nerve-based learning systems that our instincts and conscious intellect represent. Yes, as a representation of **'the perfect instinctual specimen'**, Christ *did* represent 'God the Son', and our intellect, particularly a Godly, Integrative-Meaning-*acknowledging*, *inspired* and *guided* intellect, *is* 'God the Holy Ghost or Spirit'.)

[747] Ancient Greece must have also been home to quite a number of unresigned, denial-free, truthful, effective thinking so-called 'prophets'

for that empire to have been so extraordinarily innovative, establishing as it did in that golden era so many of the foundation ideas for the Western world, across politics, philosophy, science, psychology, astronomy, architecture and art. Certainly, the early Athenians Socrates and Plato were unresigned, denial-free thinking prophets; indeed, very early Athenian society must especially have been populated by relatively innocent people because they were sufficiently ego-free to both seek out uncorrupted, innocent shepherds to run Athens and, it follows, tolerate their authority. In fact, the prophet Muhammad observed **'that every prophet was a shepherd in his youth'** (Edward Rice, *Eastern Definitions*, 1978, p.260 of 433). It is the unnatural world of city living that is especially distressing to, and thus corrupting of, our original instinctive self or soul. As that exceptional denial-free thinking prophet of our time, Sir Laurens van der Post, also noted, during the turbulent period of Plato's time, Pericles, a close friend of Plato's stepfather, **'urged the Athenians therefore to go back to their ancient rule of choosing men who lived on and off the land and were reluctant to spend their lives in towns, and prepared to serve them purely out of sense of public duty and not like their present rulers who did so uniquely for personal power and advancement'** (Foreword to *Progress Without Loss of Soul*, Theodor Abt, 1983, p.xii of 389).

[748] And as to the relatively alienation-free, natural, loving, nurturing ability of existing relatively innocent 'races', in Australian Aboriginal society **'All observers agree upon the extraordinary tenderness which parents display towards their children, and indeed, to all children whether of their own family and race or not'** (Ashley Montagu, *Coming into being among the Australian Aborigines*, 1974, p.345 of 426). In the case of the Bushmen of the Kalahari, **'Their love of children, both their own and that of other people, is one of the most noticeable things about the Bushman'** ('Tribes of the Kalahari Desert'; see <www.wtmsources. com/103>). The **'Bushmen...mother carries her child with her at all times up to four years of age'** (Virginia Abernethy, *Population Pressure and Cultural Adjustment*, 2005, p.34 of 189). **'Children are breast-fed for up to 3½ years, and among the Bushmen lactation suppresses ovulation'** (G.N. Bailey, *Hunter-gatherer economy in prehistory: a European perspective*, 1983, p.114 of 247). And, as mentioned earlier, **'!Kung [Bushmen]...infants hardly ever cry.'** As for the Yequana Indians, a relatively innocent indigenous tribe of Venezuela, the author Jean Liedloff, who spent two and a half years living with them, wrote of their **'in-arms phase'**, that **'consists, simply, of the infant having 24-hour contact with an adult

**or older child'**, and that **'the notion of punishing a child had apparently never occurred to these people'** ('The Importance of the In-Arms Phase', *Mothering*, Winter 1989; see <www.wtmsources.com/150>). And just as with Bushmen infants, **'[Yequana] babes in arms almost never cried'**. On a similar note, a report on the studies of natural-living, more innocent and nurturing societies undertaken by the dentist and nutritionist Dr Weston Price offered these insights: **'For the next ten years** [during the 1930s]**, he** [Dr Price] **travelled to various isolated parts of the earth, where the inhabitants had no contact with "civilisation" in order to study their health and physical development...Price took photograph after photograph of beautiful smiles, and noted that "healthy primitives" were invariably cheerful and optimistic. Such people were characterized by "splendid physical development". The women gave birth with ease. Their babies rarely cried and their children were energetic and hearty. Many others have reported a virtual absence of degenerative disease, particularly cancer, in isolated, so-called "primitive" groups'** (Sally Fallon Morell, 'Nasty, Brutish and Short?', The Weston A. Price Foundation; see <www.wtmsources.com/110>).

[749] Yes, the human race is not so instinctively adapted to upset now that humans are no longer capable of being innocent enough to avoid the psychologically deadening state of Resignation; with sufficient nurturing and shelter from upset behavior an individual can avoid that path and state. The reality is that there does have to be a great deal of upset in humans for that upset to become so unbearable that they have no choice but to pay the extremely high price of blocking out all access to their all-sensitive, enthralled-with-all-of-life, all-loving, soulful true self. And the behavior of a resigned person *is* essentially a form of death-like dissociation or autism; indeed, it matches perfectly the description a former president of the British Psychoanalytical Society, the psychiatrist and pediatrician D.W. Winnicott, gave for behavior associated with autism: **'Autism is a highly sophisticated defence organization. What we see is invulnerability...The child carries round *the (lost) memory of unthinkable anxiety*, and the illness is a complex mental structure insuring against recurrence of the conditions of the unthinkable anxiety'** (*Thinking About Children*, 1996 posthumous publication of his writings, pp.220-221 of 343). Later in this chapter we will see that when upset became even more extreme how an even more dishonest, alienating, autism-equivalent psychological strategy than Resignation was invented to cope with the human condition. This was to take up born-again, pseudo idealism—an adaption that progressed from religion

through to the extremely dishonest forms of pseudo idealism, from socialism, to the New Age Movement, to feminism, to environmentalism, to the politically correct movement and, ultimately, to totally truthless postmodern deconstructionism.

[750] Since it requires a great deal of upset for Resignation to become unavoidable, we can, as mentioned earlier, expect that it has only become an *almost* universal phenomenon amongst adult humans from about 11,000 years ago when the advent of agriculture and the domestication of animals brought humans together in close proximity, the effect of which, as will be talked about in pars 848 and 905, was to rapidly spread and compound upset behavior. Moses' Genesis account of Noah's Ark is actually a metaphorical description of this time when Resignation 'flooded' the world and our soul and all its truths went under, 'drowned'—when our soul was pushed into our subconscious, out of conscious awareness, and our extremely superficial and artificial, living-only-on-the-meniscus-of-existence, highly competitive, 'I-only-care-about-proving-my-worth', egocentric way of living became all-dominant. The only creatures to escape the horror of Resignation, to survive this 'drowning' of our soul, were the animals and the very few well-nurtured-with-unconditional-love, sound and secure unresigned prophets, as symbolized in this case by Noah and his zoo. As Moses says in Genesis, **'Noah was a righteous man, blameless among the people of his time, and he walked with God** [he did not have to deny Integrative Meaning]... **God saw how corrupt the earth had become, for all the people on earth had corrupted their ways. So God said to Noah, "...make yourself an ark...I am going to bring floodwaters on the earth...Everything on earth will perish** [the soul and all the denial-free truths will perish when people resign to a life of denial]. **But I will establish my covenant with you** [but from here on prophets will have to preserve the truth of Integrative Meaning and all the other great truths that relate to it], **and you will enter the ark...Go into the ark** [don't resign], **you and your whole family, because I have found you righteous in this generation'''** (6:9, 12, 14, 17, 18; 7:1). In the following depiction of Noah's Ark we see dead people littered everywhere, and, as was pointed out above, Resignation *is* a form a death. The **'dove'** that Noah **'sent out' 'to see if the water had receded'** (8:8), kept coming back because the world was still flooded—Resignation was still universal—because it is *only now* with the human condition finally understood that the need to resign can end;

the time, that Moses prophesized would eventually be possible, when the **'dove'** will never need to **'return'** (8:12) to the Ark.

*The Dove Sent Forth From The Ark*, Gustave Doré, 1866

[751] So Moses knew all about Resignation, including how it became all but universal, and, as with his accounts of Adam and Eve, and Cain and Abel (the deeper significance of which will be explained in par. 906), he used a story to pass down that knowledge through generations of humans who were resigned to living in denial of the true nature of the human condition. Moses—who wrote not just Genesis but the first five books of the Bible—was certainly an exceptional unresigned, denial-free-thinking prophet. To think people have actually searched for remnants of an ark, and tested ice cores from glaciers and sea beds in the Black Sea for proof of a great flood in the past, as if the story depicted an actual event rather than the metaphor/allegory it really is! Understandably, however—in a development that will be elaborated upon in ch. 8:16I—the more upset and thus insecure humans became, the less they could afford to confront

the truths contained in religious scriptures and the more they needed to interpret their contents in literal and fundamentalist ways — 'God is actually a person sitting in the clouds somewhere', 'Christ was actually physically resurrected from death', 'Christ's mother was actually a virgin', 'Abraham actually considered murdering his son', 'Judgment day actually heralds an afterlife in which some unlucky souls will be judged as evil and burnt in a fiery pit', etc, etc. But with the upset state of the human condition now defended, all religious metaphorical descriptions, parables and symbols — in fact, all mythology — can be safely explained and demystified, as is shown throughout this book and in the compilation of all my work, the book *Freedom Expanded*. Indeed, with understanding of the human condition found the Bible's *entire* contents can be fully demystified now — an exercise that will likely amaze the reader because they will discover that, unlike all the libraries of denial-complying books in the world, the Bible is a repository of extremely rare denial-free truth. In fact, as will be described in par. 927, virtually the whole story of the human condition, bar its scientific explanation, is perfectly described in the Bible, albeit in the abstract, metaphysical and often metaphorical terms that denial-free, truthful-thinking prophets were limited to in those early pre-science times when the Bible was written. The reformist theologian Martin Luther recognized the comparative integrity of the Bible when he said, **'Homer, Virgil, and other noble, fine, and profitable writers, have left us books of great antiquity; but they are nought to the Bible'** (*Table Talk*, 1566; tr. William Hazlitt, 1857, p.1). No wonder it is the world's bestselling and most widely distributed book, with 6 billion copies having been printed so far. And I might mention that just as people practice regularly re-reading the Bible because its denial-free, out-of-Plato's-cave, soul-filled honesty is so aligning for soul-repressed, alienated, lost humans, so people will discover that the more they re-read this book the more it will re-connect them to their soul and clear their mind of alienation — but in an *infinitely more effective* way, because the denial-free truth in this book is both reconciling of our upset state *and* based on first-principle science, which means it is infinitely more explanatory, clarifying and understandable, and thus infinitely more soul-aligning and mind-clearing.

752 We now need to look at what happened during the second phase of the Distressed Adolescent Stage, in the years immediately *following* Resignation, as well as what happened in those same years, from 15 to 21 years of age, to those humans who *didn't have to resign* — a group

that included virtually all our forebears who lived prior to the advent of *H. sapiens sapiens*.

[753] Firstly, in looking at the situation of those who *did* resign to living a life of denial of the issue of the human condition and of any truths that brought the issue into focus, after resigning at about the age of 15 it normally took the individual another six years of procrastination to make sufficient mental adjustments to be able to embrace the new, post-resigned, extremely dishonest resigned way of living.

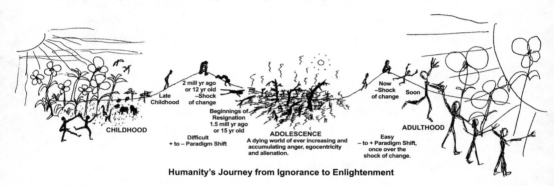

**Humanity's Journey from Ignorance to Enlightenment**

[754] To best describe the situation presented by this sobered and then depressed period leading up to Resignation—and the six-year period of procrastination over having to take up a dishonest, soul-dead resigned life that followed—imagine we're all sitting on a ridge between two valleys. Behind us lies the valley of humanity's enchanted childhood, the 'Garden of Eden' in which everyone lived happily and extremely sensitively in a non-upset, cooperative, all-loving state. Ahead of us, however, is a hell of smoldering wasteland of devastation and destruction, a wilderness of terrible upset and alienation. It was, of course, a wasteland we didn't *want* to enter, but retreat was not an option. To leave all that happiness, laughter and togetherness behind was heartbreaking, but we had no choice but to turn our back on it; we couldn't throw away our conscious mind, we couldn't stop thinking, and while we practiced thinking upset was an inescapable by-product that could only be brought to an end by finding understanding of our corrupted state—understanding that lay on the other side of that terrible wilderness of devastation, aloneness and alienation. While the lyrics of Joe Darion's classic song *The Impossible Dream* have already been referred to in chapter 1:3, they provide such a wonderful description of how awesomely courageous humans have been in undertak-

ing our species' corrupting search for knowledge they are worthy of inclusion again: **'To dream the impossible dream** [of one day, in the far future (which has now arrived), finding the redeeming understanding of the human condition], **to fight the unbeatable foe** [of our ignorant, ideal-behavior-demanding instincts], **to bear the unbearable sorrow, to run where the brave dare not go. To right the unrightable wrong** [of being unjustly criticized], **to love pure and chaste from afar, to try when your arms are too weary, to reach the unreachable star. This is my quest, to follow that star, no matter how hopeless, no matter how far. To fight for the right without question or pause, to be willing to march into hell for a heavenly cause. And I know if I will only be true, to this glorious quest, that my heart will lie peaceful and calm, when I'm laid to my rest. And the world will be better for this, that one man scorned and covered with scars, still strove with his last ounce of courage, to reach the unreachable star.'**

[755] In short, there was no retreat for the resigned adolescent; like all the fully conscious humans who had gone before them, they each *had to* find the courage to continue humanity's heroic search for knowledge. Procrastination got them nowhere, and so (and this behavior is more typical of recent times) after a few years' spent consuming lots of drugs and alcohol and partying long into the night to help them accept their fate, the adolescent had to 'get on with it' and take up the challenge of adulthood in a world where understanding of the human condition was yet to be found. In fact, it normally wasn't until they reached 21 that young resigned adults finally managed to orientate themselves to their extremely compromised resigned life—an orientation that involved two main adjustments: first, they had to block out the negative reality that living so falsely and thus so dead in soul and intellect meant they would have to endure living a completely corrupted, **'phony', 'fake'** life; and second, they had to train their mind to block out all memory of the innocent childhood state and focus on whatever meager positives they could eke out in the journey ahead. I describe these positives as 'meager' because the degree of happiness they provided did not compare to the happiness the human race enjoyed while living in the magic state of our soul's true, all-loving and all-sensitive world.

[756] The first, in truth, tiny positive was the prospect of the <u>adventure</u> involved in trying to avoid, for as long as possible, the inevitable disaster of complete self-corruption. We may have been about to 'go under'— become totally corrupted—but at least we could hope to make a good fight of it. In fact, as will be described shortly, by the age of 21 young

resigned adult men in particular could have so blocked out the truth of another ideal, soulful, integrative true world, and so adopted belief in a selfishness-justifying, competitive, survival-of-the-fittest meaning to life, that they deluded themselves that winning power, fame, fortune and glory would *genuinely* bring them validation, prove that they actually were good and not bad.

[757] The second tiny positive in the resigned existence was <u>romance</u>, the hope of 'falling in love', which (as will be described in par. 786) can be now understood as the hope of escaping reality through the dream of ideality that could be inspired by the neotenous image of innocence in women.

[758] Although these two positives were tiny, resigned adolescents gradually built them up in their mind to the extent that they became all-consuming. They had no choice but to mentally posture themselves and their resigned environment in such a way to be able to propel themselves off that ridge and take up humanity's journey to find liberating understanding of our species' upset, corrupted condition.

[759] We now need to look at the journey into adulthood, and beyond, of those individuals who *didn't* resign—a group, as stated earlier, that included virtually all those forebears who lived prior to the emergence of *H. sapiens sapiens*, and even most members of *H. sapiens sapiens* since Resignation only became almost universal during the last 11,000 years of *H. sapiens sapiens'* reign.

[760] Although not becoming so upset that they had to resign to a life of living in denial of the issue of the human condition, the lives of unresigned individuals still followed a parallel path to that of resigned individuals, who were living with the delusion that by winning power, fame, fortune and glory they could genuinely validate themselves, prove that they were actually good and not bad. The reality of the *resigned* path, however, was that those on it were inevitably bound to discover that power, fame, fortune and glory couldn't bring them any real validation, but merely resulted in them becoming *more* upset, and thus *more* insecure about their meaning and worth, and thus *more* dissatisfied. In contrast, while unresigned individuals were not living with the delusion that they could prove that they were champions and heroes of a competitive, survival-of-the-fittest, 'red-in-tooth-and-claw' world, they were, nevertheless, living with the naive illusion, the optimistic hope, that all the wrongness in the world could be righted; that they could make the world a better,

more ideal place. The reality then for the unresigned person was that the situation all around them, and even in themselves, only got worse as the upsetting search for knowledge continued—which means *everywhere* humans were becoming more upset and thus cynical about the goodness of other humans, mean-spirited and destructive. So while their disappointments and frustrations were coming off vastly different bases, both the resigned and the unresigned faced overwhelmingly difficult paths.

[761] In the situation, however, where most of the population were resigned, as has been the case during the last 11,000 years, the overwhelming problem for the unresigned person was that the resigned state of complete dishonesty and aggressive competitiveness was a total mystery to them—because those who *were* resigned to living a life of extreme dishonesty and deluded competitiveness could never admit they were doing so. It was the 'silence' of the resigned state that was the most destructive of innocence, be that in children or unresigned adults—in the words of Simon and Garfunkel's 1964 song *The Sound of Silence*, '**Fool, said I, you do not know—silence like a cancer grows**'. The reason Roald Dahl's children's books have been so immensely popular—his sales exceed 100 million, making him one of the bestselling fiction authors of all time— is that even though his stories of child-eating giants, etc, etc, seem ghastly to resigned, dishonest, upset-denying adults, to children their admission of the extreme imperfection of the world of adults is phenomenally relieving. In effect, resigned people assumed everyone else was *also* resigned and, therefore, that it was self-evident as to why they behaved so dishonestly, aggressively and competitively, but their behavior was, in fact, a complete mystery to the unresigned. And since the idealistic innocence of the unresigned mind was so trusting and thus codependent to the resigned state, they were brutalized to the point where, almost invariably, their innocence was destroyed by the extreme dishonesty and defensiveness of the resigned state. The psychiatrist Wilhelm Reich wrote honestly about this effect of upset on innocence when he described how '**The living** [those relatively free of upset]**...is naively kindly...It assumes that the fellow human also follows the laws of the living and is kindly, helpful and giving. As long as there is the emotional plague** [the flood of upset in the world], **this natural basic attitude, that of the healthy child or the primitive...**[or the unresigned adult, is subject to] **the greatest danger...For the plague individual also ascribes to his fellow beings the characteristics of his own thinking and acting. The kindly individual believes that all people are kindly and act accordingly. The**

**plague individual believes that all people lie, swindle, steal and crave power. Clearly, then, the living is at a disadvantage and in danger'** (*Listen, Little Man!*, 1948, p.8 of 109). Recall Sir Laurens van der Post's description of how the relatively innocent Bushmen struggled to cope with upset: **'mere contact with twentieth-century life seemed lethal to the Bushman. He was essentially so innocent and natural a person that he had only to come near us for a sort of radioactive fall-out from our unnatural world to produce a fatal leukaemia in his spirit.'** In truth, this description also applies to the catastrophic effect today's terminally alienated world is having on many children, the result of which is the epidemic proportion of anxiety-related disorders like ADHD and autism that we are now seeing, a development I will describe in greater detail later (in chs 8:16B and 8:16C).

[762] The dilemma for the unresigned person has been that no matter how much idealism, no matter how much selfless behavior they threw at a problem, the bottom line truth was that *only* understanding of the human condition could stop the resigned from behaving the way they were behaving—simply because the resigned had no other way of coping in the meantime. We will see later in chapters 8:15 and 8:16 how various mechanisms, like religion, were developed to try to contain the dishonesty and devastation of the resigned way of living, but ultimately such measures were limited in their effectiveness. (I might mention here that I am able to describe the agonies of the unresigned life because that has been my personal situation—as evidenced by my ability to think freely about, and write at length about, the human condition, for if I was resigned to living in denial of the human condition I obviously wouldn't be able to confront the subject.)

[763] So while, in the final 15-to-21-year-old stage of adolescence, the resigned person procrastinated over having to take up such a dishonest, soul-less life, the unresigned person (or, in humanity's case, virtually all members of *H. habilis*, *H. erectus* and *H. sapiens*, and most members of *H. sapiens sapiens* prior to 11,000 years ago) had to adjust to the prospect of having their idealism disappointed, resisted and thwarted at every turn. For the unresigned, facing that valley of devastation was just as difficult in its own way as it was for the resigned, and, *like* the resigned, they had no choice but to accept that fate. And just as the resigned used those years between 15 and 21 to condition themselves to taking up the challenge of **'march[ing] into hell for a heavenly cause'**, so too did the unresigned. The 'adventure' for the unresigned was to see how much they could resist the

corruption in the world, and if not change it then at least contain it. And the unresigned also used romance to inspire their horrifically difficult—and as Resignation became all but universal—increasingly lonely undertaking. And so, after an initial period of mental adjustment, both the resigned and the unresigned entered their 20s with a determination to make a difference, even if both realities were bound to become overwhelmed by the horror of life under the duress of the human condition.

[764]The courage of all humans who have lived during humanity's heroic 2 million years in adolescence, during which time they had to face the inevitability of total self-corruption and frustrated despair by the end of their lives, has been so immense it is, and possibly will be for all time, out of reach of true appreciation. And thank goodness all that heroic effort has finally produced the understanding of the human condition that ends that 2-million-year journey of horror. In particular, we can see that since it was our silence, our denial, our alienation, that was the main cause of the destruction of innocence in both new generations of humans and in more innocent 'races', it follows that the primary way to end the destruction of the innocence in new generations and in the more innocent 'races' (bring to an end the alcoholism and other degradations in the lives of Australian Aborigines, the Bushmen of the Kalahari, Amazonian Indians, etc) is to end all that denial about the corrupted state of humans. It is the *full* truth about the human condition—which can now at last be explained and thus safely admitted—that not only sets those who are corrupted free from the human condition, it also saves those who are *becoming* corrupted from the **'fatal leukaemia'** of their **'spirit'** that Sir Laurens van der Post wrote about. The **'catcher in the rye'** that Holden Caulfield wanted to be in order to save innocent children from the human condition (see par. 118) is the *same* **'catcher'** that is needed to save more innocent races. As the saying goes, **'honesty is therapy'**, and it is the explanation of the human condition and the honesty it makes possible that cleanses *all* human situations *everywhere*—as Cat Stevens wrote and sang, **'breaking down the walls of silence** [the denial], [is what brings about the] **lifting** [of the] **shadows from your** [everyone's] **mind'** *(Changes IV,* 1971). Yes, this explanation is *the* ultimate form of therapy for each human and for the situation that exists *between* humans of all genders, generations, 'races', countries and cultures. On this note, it should be pointed out that while manufacturing concern for indigenous 'races' was one of the favorite ways supporters of the pseudo idealistic left-wing used to make themselves 'feel good' and thus relieve

themselves of the agony of their unbearably corrupted condition, the whole act was based on dishonesty, on deluding yourself you were good and, by inference, uncorrupted. In fact, it was just such dishonesty about the corrupted state of humans that was *so* destructive of innocence! As such, supporters of the left-wing approach were doing the very *opposite* of trying to help and support indigenous people! Later in chapter 8:16Q when the extreme danger of pseudo idealism is made more apparent, we will see Nietzsche's reference to **'many sickly people'** who **'have a raging hate for…honesty'**, and Sir Laurens van der Post's reference to **'liberal socialist elements…[that] are not honest with themselves…They feel good by being highly moral about other people's lives, and this is immoral'**. Also included in that section is Christ's warning about those who conceal their extreme upset behind pseudo idealistic causes in order to delude themselves that they are sound and ideally behaved people who are leading others to a sound and ideal world: **'Beware of false prophets. They come to you in sheep's clothing, but inwardly they are ferocious wolves'**. But to return to my point, the fact that humans have had to resort to such extreme delusion and dishonesty as this to cope with their corrupted condition is really just a further reflection of how incredibly courageous the whole human race has been in **'march[ing] into hell for…[its] heavenly cause'**!

## Chapter 8:11A Adventurous Adolescentman

**The species: *Homo erectus* — 1.9 to 0.1 million years ago**
**The individual now: 21 to 30 years old**

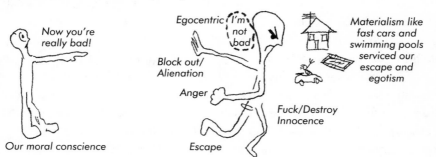

[765] The Adventurous Early Adulthood Stage (of Humanity's Adolescence) is the time we took up the battle to overthrow our idealistic instinctive self or soul's ignorance as to the fact of our conscious self's fundamental goodness.

[766] In the journey of the resigned individual, by 21 years of age, after about six years of blocking out the negatives and focusing only on the tiny positives available to them, they each finally adjusted to life in resignation. In fact, by 21 both resigned and unresigned young adults were able to arm themselves sufficiently well with a positive attitude to commit themselves to the battle that humanity as a whole was involved in of gradually, step by step, generation by generation, working towards one day accumulating sufficient knowledge to be able to explain and liberate the human race from the human condition. Indeed, by 21, resigned men in particular had become *so* focused on the positive of the adventure of attempting to make a good fight of the battle to validate themselves through winning power, fame, fortune and glory that they were cavalier and swashbuckling and raring to go. In the case of the unresigned, they were also boldly and defiantly believing they could, if not change, then at least contain all the corrupt behavior that was occurring. Naive about just how quickly overwhelming the battle was going to become, both resigned and unresigned males had plenty of strength and resilience—plenty of 'rock-n-roll'. For their part, both resigned and unresigned 21-year-old women had also become firmly focused on the few positives they had of the reinforcements they could receive from men for their physical, youthful beauty and of the satisfaction of being able to support men and nurture another generation of brave humans to carry on humanity's heroic struggle. (The different roles of men and women will be described more fully shortly.) Hence the significance of the long-held tradition in Western societies to hold a 'coming of age' party for offspring when they reached this milestone, at which they were typically given a 'key' symbolizing that they were at last ready to leave home and 'face the world', and so with a big kiss from Mum and a slap on the back from Dad the young adult set off 'to see what life held for them'. Interestingly, the fact that young adults were considered sufficiently adapted to life under the duress of the human condition to be considered independent at 21 rather than at the round figure of 20 is an indication of how precisely all these stages with ages occurred, and also how uniform and powerful the effects of the human condition have been.

[767] So it was essentially during our 20s in the lives of humans today where Resignation is almost universal, or in the case of humanity during its adventurous 20s-equivalent adolescence where everyone was un-resigned, that all the techniques needed to cope with living with the horror of the human condition began to be refined. In our teens we agonizingly

adjusted to having accepted a life of living with upset, but it was in our 20s that we took up the challenge of living out that life. Our forebear who lived in humanity's adventurous 20s-equivalent early adulthood stage was 'Adventurous Adolescentman', *Homo erectus*, who existed between 1.9 million and 0.1 million years ago. Consistent with the description that has been given for this stage, fossil evidence has revealed that it was *H. erectus* who first adventured out from our ancestral home in Africa around 1.9 million years ago and initiated the spread of humans throughout the world.

[768] While the many methods for coping with the human condition that were perfected during the over 1-million-year reign of 'Adventurous Adolescentman' have been part of human life for so long now we tend to think of them as having always been part of our species' make-up, as part of our 'human nature', the fact of the matter is they all had to be invented. Yes, in truth, our behavior underwent an *immense* transition— especially amongst those who became resigned and changed from living cooperatively, selflessly and lovingly, to living extremely competitively, selfishly and aggressively.

### Chapter 8:11B  Men and women's relationship after the emergence of the human condition

[769] As has been mentioned, the dominant aspect of the transition that took place when humanity matured from childhood to adolescence was that humanity went from being a matriarchal society to a patriarchal one. Since it was this shift that had the most dramatic impact on the lives of men and women during this adventurous adolescence stage, I will now describe how it came about and its repercussions.

[770] Throughout humanity's infancy and childhood our society was matriarchal or female-role led because the suckling of young and the nurturing of infants it led to provided the basis for the whole love-indoctrination process, and because females had to become sufficiently assertive to rein in the aggressive competition amongst males for mating opportunities. The honest-thinking philosopher Jean-Jacques Rousseau acknowledged women's nurturing role when he wrote: **'The first education is the most important, and this first education belongs incontestably to women; if the Author of nature had wanted it to belong to men, He would have given them milk with which to nurse the children'** (*On Education*, 1762, p.37 of 501). And when

Olive Schreiner wrote that **'They say women have one great and noble work left them...*We* bear the world and *we* make it. The souls of little children are marvellously delicate and tender things, and keep for ever the shadow that first falls on them, and that is the mother's or at best a woman's'** (*The Story of an African Farm*, 1883, p.193 of 300), she too was acknowledging that it has been women's role to nurture infants. Thus, with women preoccupied nurturing the next generation it made sense that when our instinctive self threatened to stop our conscious mind's search for knowledge it was men who took on the role of resisting that threat. It also made sense for men to take up this role because historically men had always been the group protector, from marauding leopards, and so on, and since this resistance by our ignorant instinctive self of our mind's need to search for knowledge was a group threat, namely a threat to humanity, it especially made sense that men went out to meet it. Indeed, since the threat of ignorance that emerged some 2 million years ago was *such* a serious threat to our species—if it prevailed we would never find understanding and never fulfill our responsibility to master intelligence—winning the battle against it became our species' priority, which is why our society became patriarchal or male-role led. Of course, this role differentiation where men took up the task of fighting against the ignorance of our instinctive self also made sense because nurturing depends on love, whereas fighting is an aggressive, non-loving behavior; far better then to leave women out of the upsetting battle to preserve as much upset-free, loving innocence as possible to nurture the next generation. The Bible acknowledges that human society changed from being a matriarchal, soul-centric, nurturing society to a patriarchal, egocentric, embattled society after the upsetting search for knowledge began when, after describing how Adam and Eve took the **'fruit' 'from the tree of...knowledge'** (Gen. 3:3, 2:17), Moses wrote, **'To the woman he [God] said...Your desire will be for your husband, and he will rule over you'** (3:16). Wives and children in virtually all cultures have not adopted their husband or father's surname because of some cultural coin toss, but because they were living in a patriarchal world.

[771] So, in keeping with their already established roles as the group protectors, it was men who took up the loathsomely upsetting job of championing the ego over the ignorance of our original instinctive self, leaving women to their loving, nurturing role. The problem this role differentiation gave rise to, however, was that in not being responsible for or participating in the terrible battle to overthrow ignorance women

were naive or unaware of the ramifications of fighting the battle, and, as a result, were unsympathetic to both the battle and the frustrated anger and egocentricity it produced in men—a situation that placed men in the awful predicament of being misunderstood and unjustly condemned by women. Women, not responsible for the fight against ignorance, and so not partaking in the battle itself, did not and could not be expected to understand what happened in the battle or the effect it had on men. Women could understand the *search* for knowledge, but not what the battle involved, as this comment reveals: **'Shirley MacLaine can't find a man to love. The 48-year-old actress...**[said she] **longs for a "close and warm relationship" but hasn't met a suitable partner. "Most men I meet seem to be too involved in trying to be successful or making a lot of money," she said. "I feel sorry for all of them. Men have been so brainwashed into thinking they have to be so outrageously successful—to be winners—that life is very difficult for them. And it's terribly destructive, as far as I am concerned, when you are trying to get a serious relationship going"'** (Sydney's *Daily Mirror*, 14 Dec. 1982). The journalist, businesswoman and 2013 'Australian of the Year', Ita Buttrose, acknowledged women's comparative lack of ego when she said, **'That's the difference between men and women. Men have egos and we don't'** (*Enough Rope*, ABC-TV, episode 47, 21 Jun. 2004). Yes, contrary to what feminists would have us believe, men and women are different—indeed, as the author Camille Paglia said herself to her feminist sisters, **'Wake up, men and women *are* different'** (*The Australian*, 4-5 Jul. 1992).

[772] Life has certainly been **'very difficult'** for men; they have had the absolutely horrible job of having to be strong enough to, in effect, kill soul—to search for knowledge and determinedly defy our beautiful, co-operatively orientated, original instinctive self. And since that defiance resulted in becoming angry, egocentric and alienated, which are all divisive, un-Godly traits, men were, in effect, in violation of God, the integrative ideals of life! So from an initial state of upset, men had then to contend with a sense of guilt, which very greatly compounded their insecurity and frustrations and made them even more angry, egocentric and alienated. (As explained in chapter 3:5, this avalanche of criticism is the 'double and triple whammy' of condemnation that humans experienced when we searched for knowledge.) How tough were men going to have to be to continue to do their job without receiving any respect or appreciation for why they were having to do what they were having to do! No wonder they have become so incredibly upset.

[773] But what could men do in the face of such a diabolical situation?

They couldn't explain themselves to women because they couldn't explain the human condition, and not able to explain the human condition, they weren't able to defend their immensely upset, corrupted state. Indeed, instead of being able to explain that they were, in fact, <u>the heroes of the whole story of life on Earth</u> because they had to succeed (and have now succeeded) in championing the cause of nature's greatest invention, namely the conscious mind's battle to establish itself on Earth, <u>men have had to endure being *completely* misunderstood and misrepresented as the villains of the piece</u>. Yes, there was *very* significant meaning in the sparse, tightly written prose that the great American novelist Ernest Hemingway used to describe the stoic lives of men; just consider the titles alone of some of his books: *Death in the Afternoon, For Whom the Bell Tolls, Winner Take Nothing, To Have and Have Not, The Old Man and the Sea* and *Islands in the Stream*. The following comment about the feminist movement describes the immensely frustrating situation endured by men:
**'One of the reasons that men have been so quiet for the past two decades, as the feminist movement has blossomed, is that we do not have the vocabulary or the concept to defend ourselves as men. We do not know how to define the virtues of being male, but virtues there are'** (Asa Baber, *Playboy*, Jul. 1983). The truth about men was certainly not as it appeared: they weren't selfish, world-destroying monsters—quite the reverse. So when the leading feminists Germaine Greer and Gloria Steinem said, respectively, **'As far as I'm concerned, men are the product of a damaged gene. They pretend to be normal but what they're doing sitting there with benign smiles on their faces is they're manufacturing sperm'** (*The Sydney Morning Herald*, 14 Nov. 1991), and **'A woman needs a man like a fish needs a bicycle'** (*TIME*, 18 Sep. 2000), they were making two of the most embarrassingly wrong statements in history, and yet, in her secret self, every woman has been making them for 2 million years! (Incidentally, Greer's assertion that all men are doing is **'manufacturing sperm'** reflects human-condition-avoiding, psychosis-denying mechanistic science's dishonest explanation for men's competitive nature, which is that they are competing to reproduce their genes.)

[774] Yes, most unfortunately, women have not appreciated our species' battle—they have not been able to empathize with what has been going on, nor respect the corrupting effect it had on men. They tended to be soul-sympathetic, not ego-sympathetic. For example, while the embattled egos of men needed to build <u>towering buildings</u> symbolizing their will and determination to defy and defeat the unjustly condemning world that

surrounded them, **'If civilization had been left in female hands, we would still be living in grass huts'** (Camille Paglia, *Sexual Personae*, 1990, p.38 of 718). This famous statement from Paglia can be understood both literally and metaphorically, because the fundamental situation was that if the soul (which women represented) had its way the intellect would never have been allowed to search for knowledge. Our instinctive self or soul's ignorance of our conscious mind's need to search for knowledge *had to be* defied if knowledge, ultimately self-knowledge, understanding of the human condition, was to be found. To give in to soul was to go nowhere, to remain in **'grass huts'**.

[775]The gulf between men and women—acknowledged in the title of John Gray's bestselling 1992 book, **'Men are from Mars, Women are from Venus'**—is palpably clear in this conversation I once overheard: 'She: **You men are wholly monstrous, foreign bodies, in fact cancers on this planet.** He: **Yes, well, haven't you heard that women are so meaningless as to not even exist'**—adding, while walking away: **'The trouble with women is they come with a brain because their brain is missing a cog, they have no idea what's going on.'** This bitter exchange provides a true measure of the extent of 'the war between the sexes'. The fact is, women have not understood men at all, as this quote admits: **'Men are a knot, I'll never untie, around a box I long to peer into'** (Kate Llewellyn, poem 'Men', *The Sydney Morning Herald*, 20 Apr. 1996). In his painting *The Creation of Adam* (shown at the beginning of ch. 4) Michelangelo recognized the fundamental schism between men and women when he depicted a woman siding with God (the personification of integrative idealism) while looking aghast at man.

[776]The following cartoon by Michael Leunig, titled *Men and Women, War and Peace*, recognizes the immense gulf that has existed between the situation men and women faced under the duress of the human condition and their subsequent view of the world—with Leunig indicating the lack of empathy women have had for the battle and its corrupting effect on men by the woman's tears of disappointment and sadness. Women have not been, as it were, 'mainframed', as intuitively understanding of the battle as men—in the same way men have never been as mainframed to the role of nurturing as women intuitively are. While humans have not been able to understand and thus explain and talk about the upsetting battle against ignorance in the search for understanding of the human condition (especially unable to understand, explain and talk about the different roles each sex was playing in that battle), being charged with fighting that upsetting battle meant men at least retained an awareness

of the battle the human race has been involved in, whereas women were in a blind position when it came to understanding what it all meant. Sir Laurens van der Post described this limitation of women in his 1976 book, *Jung and the Story of Our Time*, when he related a dream Carl Jung had about a blind woman named Salome. Sir Laurens wrote that **'Salome was young, beautiful *and* blind'**, explaining the symbolism of Salome's blindness with the following words, **'Salome was blind because the anima** [the soulful, more feminine side of humans] **is incapable of seeing'** (p.169 of 275). And in his classic 1902 novel, *Heart of Darkness* (which is a metaphorical journey into the heart of the dark horror of what the human condition *really* is), Joseph Conrad recognized that it made sense to leave women **'out'** of men's battle against ignorance when he had a man say, **'Oh, she is out of it—completely. They—women I mean—are out of it—should be out of it. We must help them to stay in that beautiful world of their own, lest ours gets worse'** (p.84 of 121). Conrad also recognized how this omission left women out of touch with reality by having another man say, **'It's queer how out of touch with truth women are. They live in a** [idealistic] **world of their own...It is too beautiful altogether, and if they were to set it up it would go to pieces before the first sunset'** (p.39). Yes, as Paglia similarly inferred, to give in to soul would be to **'go'** nowhere! Regarding the caption on Leunig's cartoon, paradoxically *real* **'peace'** could only come by men winning their **'war'** against the ignorance of our instinctive self or soul.

*Men and Women, War and Peace* by Michael Leunig, *The Age*, 2 Sep. 1989

[777] It was because women lacked empathy with the battle taking place on Earth that they have had to work through men; as it states in the Bible, women were man's **'helper'** (Gen. 2:20). Soul helped the intellect; it did not lead it. Not being mainframed to the battle going on in the world has made it very difficult for women to create profound works of literature, music and art, as evidenced by almost all the quotes employed in this book's analysis of the human condition coming from males. Christ did not choose 12 men as disciples because of the cultural conditioning of his day, as some have claimed. Christ was never influenced by arbitrary tastes and attitudes. He was only influenced by the truth, and the truth is women have not been in the best position to mediate in a battle they were not directly engaged in, which is why in religions such as Christianity women have been excluded from the priesthood. The poet Charles Baudelaire's infamous comment that **'I have always been astonished that women are allowed to enter churches. What conversation can they have with God?'** (*My Heart Laid Bare*, 1864) was not actually meant to imply that women have no place in the Christian church, only this truth that women don't wrestle with the human condition like men have had to. Humanity's 2-million-year adolescent search for its identity, search for understanding of itself, particularly for the reason for its divisive nature, *has* been a patriarchal journey, just as humanity's infancy and childhood *was* a matriarchal stage. The following passage from the philosopher Friedrich Nietzsche acknowledges what has just been explained about women—their greater ability to nurture; that they have had to work through men; and that they haven't been aware of, or mainframed to, the deeper battle that has been waging on Earth: **'Woman understands children better than a man...The man's happiness is: I will. The woman's happiness is: He will. "Behold, now the world has become perfect!"— thus thinks every woman when she obeys with all her love. And woman has to obey and find a depth for her surface. Woman's nature is surface, a changeable, stormy film upon shallow waters. But a man's nature is deep, its torrent roars in subterranean caves: woman senses its power but does not comprehend it'** (*Thus Spoke Zarathustra: A Book for Everyone and No One*, 1892; tr. R.J. Hollingdale, 1961, p.92 of 342).

[778] What now needs to be described is that without the ability to explain the all-important role that men were having to play, and thus defend themselves against women's lack of appreciation of that role—a situation that led to such unappreciative criticisms as those from Greer and Steinem—what men in their anger, frustration and desperation did, was turn on women and attack their condemning innocence by violating

it through sex, as in the 'fucking' or destroying of innocence. But before describing that extremely tragic development I first need to explain that women were not the *original* victims of men's upset. *That* unfortunate distinction went to animals because their innocence—in fact, the innocence of *all* of nature—also criticized men. Nature was a friend of our original instinctive self or soul because we grew up with it—humanity spent all its infancy and childhood alongside nature in the 'Garden of Eden' that was Africa—and so by association the natural world, especially the innocent world of animals, *also* criticized men. Yes, the hunting and killing of animals was the first great expression of men's upset anger and egocentricity. One of the first adaptions to living in the embattled angry, egocentric and alienated state, ultimately epitomized by Resignation, was that we changed from being a relatively peaceful vegetarian species to a ruthless hunter of animals. It has always been claimed that the hunting in the 'hunter-forager' lifestyle that characterized virtually the entire 2-million-year period of humanity's adolescence was primarily driven by the need for protein-rich food—a mechanistic denial-complying belief that has so far protected upset humans from the condemning truth of the extreme aggression involved in hunting. But, in fact, research shows that 80 percent of the food consumed by existing hunter-foragers, such as the Bushmen of the Kalahari, is supplied by the women's foraging (*Kalahari Hunter-Gatherers*, eds Richard B. Lee & Irven DeVore, 1976, p.115 of 408). So if providing food was not the reason, why did men hunt? This obvious question has led to the development of the 'show-off' hypothesis by some human-condition-avoiding, psychosis-denying, mechanistic scientists, such as the anthropologist Kristen Hawkes, which suggests that men developed hunting in order to display their worth as potential mates and so increase their reproductive chances. However, the honest, denial-free answer is that hunting was men's earliest ego outlet: men attacked animals because their innocence, albeit unwittingly, unfairly criticized men's lack of innocence; it condemned their upset aggressive lives. Also, by attacking, killing and dominating animals, men were demonstrating their power, which was a perverse way of demonstrating their worth as humans. If men could not rebut the accusation that they were bad, they could at least find some relief from that guilt by demonstrating their superiority over their accusers. The exhibition of power was a substitute for explanation. This 'sport' of attacking animals, which were once our species' closest friends, was one of the earliest expressions of our upset. One of the definitions given for 'sport' is **'the pastime of hunting, shooting, or fishing with reference to the pleasure achieved: "we had good sport today"'** (*Encylopedic World Dictionary*, 1971). The **'pleasure'** of hunting was of the perverse, sadistic, sick kind, of attacking animals for their innocence and its implied criticism of us.

[779] Yes, the *real* reason men needed to go hunting was to get even with innocence for its unjust criticism of men's *lack* of innocence. I remember seeing a cartoon that depicted two cement truck drivers gleefully dumping their load of concrete over a tiny road-side daisy. Such behavior is merely an adult version of children burning ants and tormenting pets. The extent of the satisfaction corrupted humans could derive from retaliating against the unjust condemnation that innocence represented was revealed in par. 276 when a 'legendary' hunter's dispatch of **'966 Rhinos'** over a two year period was cited, but in this comment by W.D.M. Bell, an African big-game hunter of the early 1900s, the pleasure is explicit: **'There is nothing more satisfactory than the complete flop of a running elephant shot in the brain'** (P. Jay Fetner, *African Safari*, 1987, p.113 of 678). And in the following passage another sport hunter makes perfectly clear the cathartic satisfaction gained through being able to 'get even' with unjustly condemning innocence: **'Next thing I knew, a large male chimpanzee had hoisted himself up out of the underbrush and was hanging out sideways from the tree trunk, which he was clutching with his left hand and left foot. Looking down my barrel at ten yards was man's closest relative, an ape, which, when mature, has the intelligence of a three-year-old child. Wouldn't I feel like a murderer if I shot him? I had some misgivings as my globular front sight rested on the ape's chest and my finger on the trigger. But then, gradually, insidiously, my thinking took a different turn. I thought of the gorge-lifting sentimentality—most of it commercially inspired—that has come to surround chimpanzees. I thought of the long list of ridiculous anthropomorphic books about the "personalities" of these apes. I thought of that chimp who fingerpainted on TV and sold his "works" for so much money he wound up having to pay income tax. I thought of one ape who was recommended for a knighthood, the ape who was left his master's yacht, the ape who was elected to parliament in some banana republic; and various other apes who were made astronauts and honorary colonels. Gathering like storm clouds in my mind, these thoughts roused me to such a pitch of indignation that there appeared to be only one honorable course of action. I blasted that ape with downright enthusiasm and have felt clean inside ever since'** (ibid. pp.117-118).

[780] Anthropological evidence supports the notion that hunting is an aspect of fully conscious, upset 'Adolescentman', because it was during

the time of *H. erectus* that the first signs of hunting appeared in the fossil record. All the anthropological evidence indicates 'Childman' was a vegetarian, but with big game hunting came <u>meat-eating</u>, an adaption that would have revolted our original instinctive self or soul since it involved *killing* and *eating* our soul's friends. Even today, the act of killing animals, or just *seeing* animals get slaughtered, produces feelings of deep revulsion within us. But we weren't to be put off and in time, as our increasingly upset and driven (to find ego relief) lifestyle developed, we became somewhat physically dependent on the high energy value of meat. (With respect to whether the lifestyle of 'Adolescentman' should be described as hunter-forager or forager-hunter, since the priority was the heroic search for knowledge and hunting provided men with some retaliatory relief from the criticism of the innocent world they had to live in, it should be described as hunter-forager.)

Photo on the front page of *The Australian* (27 Oct. 2011) of a shooter posing with a scimitar-horned oryx, which he shot to death at a game ranch in the Northern Territory of Australia. The species is officially listed as extinct in the wild. Note the pride and happiness on the face of the shooter and his partner.

[781] Returning to the overall situation that the human race was faced with, in particular the issue of how the different roles that men and women had taken up played out. As upset increased, so too did humanity's insecurity about being corrupted—as did, it follows, the need to combat that insecurity with whatever form of relieving reinforcement we humans could find. And since it was men who had to especially take on the responsibility of championing the conscious thinking self or ego over the ignorance of our original instinctive self, it was men who particularly

came to need reinforcement of their worth through the acquisition of power, fame, fortune and glory. In the soundtrack to the 1986 African musical *Ipi Tombi*, the female narrator says, **'The women had to do all the work because the men were so busy being big, strong and brave'** (Narration: Sesiya Hamba, *Drinking Song*, lyrics by Thandi Lephelile). This quote acknowledges just how preoccupied men eventually became in trying to prove their worth, in defeating the implication that they weren't worthy. In the end men became *so* insecure/ego-embattled that it really did become a case of **'Give me liberty or give me death'**, **'No retreat, no surrender'**, **'Death before dishonor'**, **'Death or glory'** — they wouldn't do anything that didn't bring them glory and adulation, which meant someone else (namely women) had to do all the menial work if it was going to get done. The following two photographs from Richard B. Lee and Irven DeVore's *Kalahari Hunter-Gatherers* (1976), of the relatively innocent Bushmen members of the present *Homo sapiens sapiens* variety of humans, perfectly illustrate the situation. In the first image, women are shown gathering the aforementioned 80 percent of the food, in addition to nurturing the children — basically doing all the practical work — while the other image, titled *Telling the Hunt*, shows the men sitting around together with their backs contemptuously shunning innocent nature's condemning presence as they boast about their heroic conquests over innocent animals. Clearly, the basic adaptions humans made to the human condition are well established in these relatively innocent Bushmen members of the present *H. sapiens sapiens* variety of humans; these could as easily be photos of women shopping and businessmen discussing a company takeover.

*Telling the Hunt*

Women with infants digging roots

Photographs by Marjorie Shostak; and Laurence Marshall

[782] Of course, to properly context hunting, it has to be appreciated that the destruction of innocence has been going on at *all* levels. After all, in resigning, which became almost universal in its occurrence some 11,000 years ago, humans also destroyed the innocent soul in themselves by repressing it. Yes, our ignorant soul and any innocent representation of it, has been unbearably condemning, especially of men, which is why men, having already turned on and attacked their innocent animal friends, *then* turned on and attacked the relative innocence of their partners in life, women. Unable to explain their behavior to women, men were left in an untenable situation: they couldn't just stand there and accept women's unjust criticism of their behavior—they had to do *something* to defend themselves—but because women reproduced the species, men couldn't kill women the way they destroyed animals, and so instead men violated women's innocence or 'honor' through rape. Men perverted sex, as in 'fucking' or destroying, making it discrete from the act of procreation. What was being fucked, violated, destroyed, ruined, degraded or sullied was women's innocence. The feminist Andrea Dworkin recognized this underlying truth when she wrote that **'All sex is abuse'** (*Intercourse*, 1987; reported in *The Sydney Morning Herald*, 25 Jul. 1987; see <www.wtmsources.com/172>). If we immerse ourselves in what was really happening with hunting and progress from there we can see something of the truth and horror of what has just been said about sex being a way of attacking the innocence of women. While the hunting (actually murdering) of our instinctive self or soul's friends, the animals, was shockingly offensive to our all-loving and all-sensitive soul, humans eventually became so upset that the feeling of condemnation from the innocence of animals became greater than our love for them, at which point we started murdering them to relieve ourselves of that condemnation. So when we shoot animals (and this reaction would have also occurred when our ancestors ritually sacrificed the lives of animals) the shock to our soul of what we are doing temporarily relieves the unjust condemnation we feel, which is why some people became addicted to hunting. In fact, in order for professional hunters of wildlife to shoot accurately they first have to learn to overcome the momentary mental 'blackout' that is brought about by the shock of what they are about to do. All hunters—indeed, anyone about to kill an animal—are aware, if they are honest, of the momentary 'blackout' their mind experiences when they are about to kill an animal. Well, sex as humans have been practicing it has similarly been extremely offensive to our instinctive self or soul, and

has caused the same **'emotion-induced'** shock to our soul and thus temporary **'blackout'** in our mind, as this study found: **'Research suggests that when shown erotic or gory images, the brain fails to process images seen immediately afterward. This phenomenon is known as "emotion-induced blindness."**...[or] **short-vision blackout'** ('New study: Sexy images can cause temporary blindness', sourced from *New Scientist*, 20 Aug. 2005; see <www.wtmsources.com/176>). The **'emotion'** referenced here is our soul communicating extreme distress to our mind. Humans don't remember sexual episodes very well and the reason we don't is because sex, as currently practiced, is a violation of our soul and we don't want to remember such violation. Incidentally, understanding this psychology also allows us to explain the ritual of <u>human sacrifice</u>, a practice that has been found to have occurred in nearly all cultures. While hunting and later sacrificing (murdering) our soul's friends, the animals, was shockingly offensive to our soul, sacrificing (actually murdering) a fellow human was *astronomically* offensive to our soul. However, the upset in humans eventually became so great that only such astronomically shocking acts as murdering our fellow humans could exceed our astronomical levels of upset, and by exceeding the upset temporarily quell it. The feeling of shock and revulsion overrode the feeling of upset and, through that, temporarily eliminated the latter. As will be described later in par. 1289, the terrible bloodletting that took place during the Second World War represented such an immense valving off of upset that it brought forth a period of freedom from upset, a freedom that gave rise to the freshness of the 1960s post-war generation, the irrepressible 'Baby Boomers'. The 'valving off of upset' can be better understood as the souls of those involved being so revolted and shocked by all the bloodletting that the upset in those involved was, for a time, nullified. The extraordinary extent of the innocence of our soul, and the extraordinary extent of the upset in humans now, especially in men, are two immensely confronting truths that we can now at last safely understand and admit. The main point being made here, however, is that sex became a way of attacking the innocence of women, the result of which was that women's innocence was oppressed and, to a degree, they tragically came to share men's upset.

[783] The consequences of this horrific development—which will now be looked at—have obviously been immense. Prior to the perversion of sex as a way of attacking the condemning innocence and naivety of women, women weren't viewed as sex objects and so nudity had none of the problems of attracting lust and so there was no need to conceal

our nakedness with <u>clothes</u>. To quote Moses in the Bible, when Adam and Eve took the fruit from the tree of knowledge (set out in search of understanding), **'the eyes of both of them were opened, and they realized that they were naked; so they sewed fig leaves together and made coverings for themselves'** (Gen. 3:7). Plato similarly described our original **'state of innocence, before we had any experience of evils to come'** as the time when we were **'simple and calm and happy'**, and lived a **'blessed and spontaneous life...[where] neither was there any violence, or devouring of one another'** and no **'possession of women [and we]...dwelt naked, and mostly in the open air'** (see ch. 2:6). Clothing was not *originally* designed to protect the body from cold as children have been evasively taught at school, but to restrain lust, to the extent that once we became extremely upset even the mere sight of a woman's ankle or face became dangerously exciting to men, which is why in some cultures that cater for extreme upset women are completely shrouded and persecuted if any part of their body is revealed in public. It was a reverse-of-the-truth lie to say, as it is frequently argued, that this concealment of women was introduced out of 'respect for women'. The truth is, it was enforced because women were being *disrespected*.

[784] In the case of the convention of <u>marriage</u>, this institution was invented as one way of containing the spread of upset. By confining sex to one life-long, <u>monogamous</u> relationship, the souls of the couple could gradually make contact and coexist in spite of the sexual destruction involved in their relationship. As stated in the Bible, in marriage **'a man will leave his father and mother and be united to his wife, and the two will become one flesh. So they are no longer two, but one'** (Mark 10:7, 8). Brief relationships, on the other hand, kept souls repressed and spread soul repression. However, the more upset, corrupted, insecure and alienated humans became, the more they needed sexual distraction and reinforcement through sexual conquest (in the case of men) and sex-object attention (a development in women that will be explained shortly), and thus the more difficult it became for both sexes to remain content in a single, monogamous relationship. The saying **'the first cut** [the first falling out of love] **is the deepest'** is an acknowledgment of the deep and total commitment humans make to their first love. It reveals that the original, relatively innocent relationship between a man and a woman was monogamous. If you want a truly beautiful description of the depth of commitment men especially (because they are more innocent in this regard than women) can make to their first love, watch Jeff Nichols' 2012 film *Mud*, or read the prophet Kahlil Gibran's 1957 story *The Broken Wings*.

Since sex killed innocence, ideally (although impractical for the majority of the human race who had to ensure the continuation of the species) if we wanted to free our soul from the soul-destroying hurt sex caused it we needed to be <u>celibate</u>; as Christ explained it, some priests **'renounce marriage [for] the kingdom of heaven'** (Matt. 19:12).

[785] BUT, while sex *was* an attack on innocence, an act of aggression, it was also one of the greatest distractions and releases of frustration and, on a nobler level, it became an inspirational <u>act of love</u>, an act of faith in and affection for men. A sublime partnership between men and women did develop, for when all the world disowned men for their unavoidable divisiveness, women, in effect, stayed with them, bringing them the only warmth, comfort and support they would know. As Moses says in Genesis, **'The Lord God said, "It is not good for the man to be alone. I will make a helper suitable for him"…Then the Lord God made a woman…and he brought her to the man'** (2:18, 22). Sir Laurens van der Post offered this sensitive attempt by a man to explain to a woman the greater significance of sex; it is a conversation that takes place on the eve of a Second World War battle: **'Touched by her concern for her honour, in his imagination he would have liked to tell her that he could kneel down before her as a sign of how he respected her and beg her forgiveness for what men had taken so blindly and wilfully from women all the thousand and one years now vanishing so swiftly behind them. But all he hastened to say was: "I would have to be a poet and not a soldier to tell you all that I think and feel about you. I can only say that you are all I imagined a good woman to be. You make me feel inadequate and very humble. Please know that I understand you have turned to me not for yourself, not for me, but on behalf of life. When all reason and the world together seem to proclaim the end of life as we have known it, I know you are asking me to renew with you our pact of faith with life in the only way possible to us"'** (*The Seed and the Sower*, 1963, p.238 of 246). Friedrich Nietzsche gave an honest description of the roles that developed for men and women in humanity's heroic journey to overthrow ignorance when he famously wrote that **'Man should be trained for war and woman for the recreation of the warrior: all else is folly'** (*Thus Spoke Zarathustra: A Book for Everyone and No One*, 1892; tr. R.J. Hollingdale, 1961, p.91 of 342). The 1960s sex symbol Brigitte Bardot was of a similar view when she said, **'A women must be a refuge for the warrior. Her job is to make life agreeable'** (*The Australian*, 11 Oct. 1999); although shortly after making this statement Bardot encapsulated the paradox of life for women when she declared that all men are **'beasts'** (ibid)—the more usual rendition from women of this misunderstanding

of men being that **'all men are bastards'**.

[786] Yes, innocence was two-sided: it condemned and upset men, who, therefore, had to attack it, but it was *also* an inspirational reminder of the soulful, true world that they were fighting to reinstate by finding the understanding that would stop the upsetting criticism of them and the human race as a whole. So women's innocence could both condemn and inspire men, which, as will now be explained, is why the image of innocence—that Bardot's beauty was such a magnificent representation of—was so inspirational. The image of innocence in women could inspire the dream of the human race's return to living in a cooperative, loving, upset-free, ideal state, a state free of the human condition; it could lead to 'romance'. Men could dream that the *image* of innocence in women meant women were actually innocent and that through their partnership with women they could share in that innocent state; and for their part, women could use the fact that men were inspired by their image of innocence to delude themselves that they actually *were* innocent. Men and women could 'fall in love', let go of reality and dream of an ideal, cooperative, loving world. Cole Porter's 1928 song *Let's Fall In Love* contains lyrics that reveal how falling in love is about allowing yourself to dream of the ideal state, of 'paradise': **'Let's fall in love / Why shouldn't we fall in love? / Our hearts are made of it / Let's take a chance / Why be afraid of it / Let's close our eyes and make our own paradise.'**

[787] The effect of the 'attraction' of innocence—which has been the preserve of youth because the young were innocent, they hadn't yet been exposed to all the upset in the world—for both dreaming through and for sexual destruction was that through the course of the 2-million-year journey through our species' adolescence our physical features became increasingly youthful looking or neotenous, as the increasingly child-like features of the skulls of the varieties of our *Homo* ancestors pictured in ch. 8:2 clearly evidence. The dramatic increase in neoteny from *Homo habilis* to *Homo erectus* reflects the dramatic increase in upset that took place once humanity determinedly set out on its search for understanding at the age-equivalent of 21, and the dramatic increase in neoteny from *H. erectus* to *H. sapiens sapiens* reflects the dramatic increase in upset that occurred when humanity entered the rapidly dis-integrating stage of the last quarter of the exponential growth of upset's development. Women were especially selected for their more innocent looking, neotenous, youthful, childlike features of a domed forehead, large eyes, snub nose and hairless body.

Just how adapted women have now become to being <u>sex objects</u> can be seen in women's magazines, which are almost entirely dedicated to instructing women how to be 'attractive', which really means just better able to imitate the *image* of innocence. Women are now habituated and codependent to the reinforcement that men, for over 2 million years, have given their object self rather than their real self—for instance, they love to adorn themselves with beautiful objects, use make-up on their faces to increase their neotenous appearance, and wear high-heel shoes to give themselves the leggy, youthful, ultra-innocent look of pubescents.

| | | |
|---|---|---|
| Innocent looking, long-legged pubescent body proportions. Drawing by Gladys Perint Palmer, *Vogue Australia*, May 1994 | Innocent, child-like, youthful, neotenous features. German supermodel Claudia Schiffer, *The Daily Telegraph-Mirror*, 1 Nov. 1994 | Stages of women's adaption to, and acceptance of, having to be a sex object—from 2 million years ago to the present, and from ages 12-18. |

"I'm finally getting attention for what I have upstairs."

As these popular cartoons illustrate, women believe it is their real self, not their object self, that men are in love with. The sooner the horror and suffering of the human condition ends the better!!

[788] What now needs to be explained is how the <u>beauty of women</u> came to be *so* powerfully attractive and inspirational for men. It was explained in chapters 5 and 6 that throughout humanity's infancy and childhood we self-selected integrative traits by consciously seeking out love-indoctrinated mates—members of the group who had experienced a long infancy and were closer to their memory of infancy (that is, younger). Since the older we became the more our infancy training in love wore off, we began to recognize that the younger an individual, the more integrative he or she was likely to be. We began to idolize, foster and select for youthfulness—represented as it was by neotenous (infant-like) features such as the aforementioned large eyes, dome forehead, snub nose and a hairless body—because of its association with cooperativeness or integrativeness. But what happened around 2 million years ago when the upset state of the human condition emerged is that instead of selecting for neotenous features because they signaled a *cooperative* individual, such features began to be selected for because they signaled an *innocent* individual who was 'attractive' for sexual destruction. What this means is that while the motivation behind the selection changed *significantly*, the neotenous features signifying soundness and innocence have been selected for *throughout* humanity's infancy, childhood and adolescence. Also of significance is the fact that while all other forms of innocence were being attacked and destroyed throughout humanity's adolescence, this *image* of innocence—'the beauty of women'—was the *only* form that was actually being cultivated during that time. So the image of innocence in women has both a very long and, since 2 million years ago, unique history of development, and that is why it is *so* powerfully attractive.

[789] <u>Women's representation of innocence</u>, their representation of our now lost pure world, <u>has been the only form of that purity that has been continuously cultivated since we were apes</u>, so it is little wonder men 'fell in love' with women. The following quotes reveal just how inspiring women's image of innocence became for men: **'we lose our soul, of which woman is the immemorial image'** (Laurens van der Post, *The Heart of the Hunter*, 1961, p.134 of 233); **'I believe hers to have been the kind of beauty in which the future of a whole continent sings, exhorting its children to renounce what is out of accord with the grand design of life'** (ibid. p.86); **'Woman stands before him [man] as the lure and symbol of the world'** (Pierre Teilhard de Chardin, *Let Me Explain*, 1966; tr. René Hague et al., 1970, p.67 of 189); **'Women are all we [men] know of paradise on earth'** (Albert Camus, *The Fall*, 1956, p.73 of 108); **'You give me a reason to live'** (Joe Cocker, *You*

*Can Leave Your Hat On*, 1986); **'I, I who have nothing / I, I who have no one** [because the whole world has hated men's upset behavior] / **Adore you and want you so'** (Jerry Leiber & Mike Stoller, *I Who Have Nothing*, 1963); **'Sex is life'** (graffiti on a granite boulder at Meekatharra in Western Australia). In the 1996 film *Beautiful Girls*, one of the male characters is criticized for plastering pictures of supermodels all over the walls of his room, to which he responds: **'Look, the supermodels are beautiful girls. A beautiful girl can make you dizzy, like you've been drinking bourbon and coke all morning, she can make you feel high for the single greatest commodity known to man—promise. Promise of a better day, promise of a greater hope, promise of a new tomorrow. This particular awe can be found in the gait of a beautiful girl, in her smile and in her soul; in the way she makes every rotten thing about life seem like it's going to be okay. The supermodels, that's all they are, bottled promise, scenes from a brand new day, hope dancing in stiletto heels.'** The ever-insightful Nietzsche also recognized the role women played in inspiring the world with their *illusion* of innocence when he wrote, **'her great art is the lie, her supreme concern is appearance and beauty. Let us confess it, we men: it is precisely *this* art and *this* instinct in woman which we love and honour'** (*Beyond Good and Evil*, 1886; tr. R.J. Hollingdale, 1972, p.145 of 237).

[790] So while it certainly *is* of little wonder that men fell in love with women, the great 'mystery of women' was that it was only the physical *image* or object of innocence that men were falling in love with. The illusion was that women were psychologically as well as physically innocent. This passage from the writings of Leo Tolstoy captures the whole amazing paradox that this situation gave rise to (the underlining is my emphasis): **'so** [this is how] **I lived till I was thirty...I weltered in a mire of debauchery** [destroying soulful purity] **and at the same time was on the lookout for a girl pure enough to be worthy of me** [marrying]. **I rejected many just because they were not pure enough to suit me, but at last I found one whom I considered worthy... One evening...I was sitting beside her admiring her curls and her shapely figure in a tight-fitting jersey, I suddenly decided that it was she! It seemed to me that evening that she understood all that I felt and thought, and that what I felt and thought was very lofty. In reality it was only that the jersey and the curls were particularly becoming to her and that after a day spent near her I wanted to be still closer. It is amazing how complete is the delusion that beauty is goodness. A handsome woman talks nonsense, you listen and hear not nonsense but cleverness. She says and does horrid things, and you see only charm...I returned home in rapture, decided that she was the acme of moral perfection, and that therefore she was worthy to be my wife, and I proposed to her next day. What a muddle it is!'**

(*The Kreutzer Sonata*, 1889, ch.V; tr. L. & A. Maude). As intimated in what Tolstoy has written, for their part, women were able to fall in love with the dream of their own 'perfection' that men projected—of their being truly innocent. Men have been in love with the image of innocence and women have been in love with the idea of their innocence. As has been pointed out, men and women *fell* in love, we abandoned the reality in favor of the dream; it really was the one time in our life when we could romance, when we could be transported to how it once was and how it could be again—to heaven. And like with Cole Porter's song *Let's Fall in Love*, it was a dream the composer and lyricist Irving Berlin spoke explicitly of in his classic 1935 song *Cheek to Cheek*: **'Heaven, I'm in heaven / And my heart beats so that I can hardly speak / And I seem to find the happiness I seek / When we're out together dancing cheek to cheek'**. The lyrics of the song *Somewhere*, written by Stephen Sondheim for the 1956 blockbuster musical (and later film) *West Side Story*, also provide a wonderful description of the dream of the heavenly state of true togetherness that humans allow themselves to be transported to when they fall in love: **'Somewhere / We'll find a new way of living / We'll find a way of forgiving / Somewhere // There's a place for us / A time and place for us / Hold my hand and we're halfway there / Hold my hand and I'll take you there / Somehow / Some day / Somewhere!'** Thank goodness the **'day'** of reconciliation of the lives of men and women has finally arrived and all the perverse destruction of women's souls can end, and instead of dreaming of a loving, ideal world, the *real* loving, ideal world for men and women can emerge.

[791] While different cultures have different perceptions of female beauty, essentially men are 'attracted' by innocent looks, which are youthful neotenous features. The popular saying '<u>blondes</u> have more fun' illustrates the tendency in Caucasian cultures to regard blond women as more attractive because many young Caucasians have blond hair, a sign of youth/innocence. The writer Raymond Chandler acknowledged this appeal when he wrote, **'It was a blonde. A blonde to make a bishop kick a hole in a stained glass window'** (*Farewell, My Lovely*, 1940, p.93 of 292). <u>Long, healthy hair</u> is also associated with youth, which is why men find long hair on women attractive. In general, any feature unique to women will be attractive and signal a sex object to men, hence the desirability of <u>breasts, shapely hips and a narrow waist</u>. The different cultural definitions of beauty can also be explained in terms of what has historically signified innocence. For instance, in times when few could afford to eat or live well, fat women

were considered beautiful because their appearance generally indicated that they had been well cared for, better nurtured and were thus more innocent. Today, however, the attraction of a long, ultra-thin female shape can be explained by the increase in alienation amongst humans. The duration of innocence now is very brief and, as has been mentioned, what is now deemed attractive is that pubescent age when young girls first start to develop physically and have the slender, long-legged frame of young animals, such as foals. So for women to be perceived as attractive they have to endeavor to look like an innocent pubescent teenager, which explains the long-held obsession with extremely uncomfortable high-heels that elongate women's legs. This ultra-thin body shape is, of course, completely unnatural for most adult, child-bearing women and to achieve it necessitates a starvation diet, a situation that has led to debilitating disorders such as anorexia and bulimia. Moreover, **'Salons say clients** [girls] **as young as 14 or less are requesting Brazilians** [removal of all their pubic hair]**'** (Nikki Gemmell, 'Going bush', *The Weekend Australian Magazine*, 15 Feb. 2014), in order to make themselves look even more nubile, innocent! Such is the level of perversion/sickness that has developed in the human race!

[792] Of course, human-condition-avoiding, psychosis-denying reductionist, mechanistic scientists have been totally committed to avoiding any recognition that sex as men practice it is about attacking the psychologically innocent state of women for its unjust criticism of men. They maintain that the attractiveness of younger women has nothing to do with them being less psychologically corrupted (or at least the appearance of that state), claiming instead that it is due to a genetic reproductive strategy—that men want to **'mate with women who look like…Barbie—young with small waist, large breasts, long blond hair, and blue eyes…**[because] **they are healthier and more fertile than other women'** (Alan Miller & Satoshi Kanazawa, 'Ten Politically Incorrect Truths About Human Nature', *Psychology Today*, 1 Jul. 2007). This is the old Social Darwinist/Evolutionary Psychology strategy of blaming our competitive, selfish and aggressive human condition *not* on an upset psychosis in humans, which is its real source, but on non-existent savage, aggressive and competitive animal instincts within us! No beautiful, unconditionally loving, altruistic, selfless soul in humans, just brutal, aggressive, make-sure-you-reproduce-your-genes animal instincts! No psychological upset in the human race now! What rubbish. No wonder 'fuck' is such a good swear word—everyone actually knows sex as humans practice it is about attacking the psychological innocence of women,

that it's about 'fucking', but no one is admitting it, and (as I will explain more fully in par. 870) such extreme dishonesty provides the motivation for swearing; that swearing is a way of tearing through the overburden of dishonesty that we are all being forced to endure, in this case tearing through the dishonesty that 'sex' isn't about destroying innocence. *Of course* 'sex' is about attacking, 'fucking', innocence; as Dworkin said, it's **'abuse'**, it's psychotic behavior—but mechanistic scientists weren't going to admit that. No, as emphasized in par. 217, mechanistic scientists' human-condition-avoiding attitude was: 'What psychosis? What inner insecurity? What sense of guilt? What original 'Golden Age' of innocence? What 'fallen' condition? What alienation? What great elephant in the living room of our lives that we can't acknowledge? To hell with your psychological garbage!' Yes, they just blame all the psychological upset that humans very obviously suffer from on savage animal instincts within us—'Humans kill each other in wars not because of serious philosophical differences but to reproduce their genes in the same way bulls fight and kill other bulls to ensure they win the mating opportunities'! But such 'explanations' are completely absurd because our human condition *obviously* involves our fully conscious mind; humans suffer from a psychological *human condition*, not a non-psychological, genetic-opportunism-based *animal condition*. Again, Arthur Koestler summarized mechanistic, reductionist science's deliberate blindness to the issue of the **'mental disorder'** of our **'unique'** human condition when he wrote that **'symptoms of the mental disorder which appears to be endemic in our species...are specifically and uniquely human, and not found in any other species. Thus it seems only logical that our search for explanations** [of human behavior] **should also concentrate primarily on those attributes of *homo sapiens* which are exclusively human and not shared by the rest of the animal kingdom. But however obvious this conclusion may seem, it runs counter to the prevailing reductionist trend. "Reductionism" is the philosophical belief that all human activities can be "reduced" to – i.e., explained by – the behavioural responses of lower animals – Pavlov's dogs, Skinner's rats and pigeons, Lorenz's greylag geese, Morris's hairless apes...That is why the scientific establishment has so pitifully failed to define the predicament of man'** *(Janus: A Summing Up*, 1978, p.19 of 354). Yes, turn on the television and there will be endless programs presenting completely bullshit—**'pitiful'**—biological explanations of human behavior, such as this claim that men prefer younger women because they are better able to reproduce their genes. Just this week on television (3 Jun. 2013, at time of writing) the first program of a *National Geographic* series titled *Ape Man* began with these words: **'In this series I'll introduce you to your inner ape…to expose our most basic instincts and explore their origins in the ape world. So get ready to learn the laws of the jungle because, like me, you are ape man…I've learnt that there's actually an ape living inside us humans and that inner ape is hungry for power. For our primate cousins being the alpha male gives them the first pick of the three things that matter most: food, territory and of course females. And we humans are just the same: world leaders, celebrities, even your boss—alphas rule the world!'** The truth is 100 percent the opposite of this—our instinctive heritage is of being cooperative, selfless and loving, and *then* we became *psychologically* upset and as a result of that we became competitive and aggressive. Yes, it was fucking mechanistic biological crap like this that was leading humanity straight to terminal alienation and extinction!

[793] The very obvious truth is that sex as men practice it has, at base, been all about attacking the innocence of women, and unfortunately this destruction of women's souls and the cultivation of their image of beauty has been occurring for some 2 million years. As such, men and women have become *highly* adapted to their roles; while men's magazines are full of competitive battleground sport and business, women's magazines are, as has been pointed out, dedicated to enhancing beauty, to becoming more 'attractive', and to maintaining that attractive beauty for as long as possible. Just how aware women are of the importance of their sex object self, and how much women's sense of self-worth and self-esteem is now dependent on their sex appeal, is apparent in the 'twinning phenomena', in which you see women walking around together who are virtually identical in their level of sex appeal or lack thereof. Clearly women feel too exposed and confronted if they are with another woman who is even slightly more 'attractive', so they end up befriending their sex appeal 'twin'. It is extremely common to see, and *very* revealing of what men have done to women! Sex has played a *huge* role in the lives of men and women.

[794] Indeed, lust and the hope of falling in love have assumed such importance that many people, including the psychoanalyst Sigmund Freud, have been deceived into believing that sex actually rules our lives. Two of the best examples of this conviction are found in interpretations of Moses' story of the Garden of Eden, in which Eve is blamed for tempting Adam to take the apple from the tree of knowledge, and in interpretations of Plato's two-horsed chariot allegory that was described in chapter 2:6,

in which the lustful behavior of the **'bad' 'dark'** horse that is described as *eros* is seen as a natural response to the object of desire. The truth is that women, and even boys in the case of the homosexual situation that Plato was referring to in his *eros* description in *Phaedrus*, were the *victims* of upset in men, not the cause, but lust became such a strong force that people have been misled into believing it seduced humans into behaving in an upset way. As was explained in chapter 2:6, Moses and Plato were simply using sexual desire as the most obvious indication of our corrupted, upset state, as evidenced by the fact that Moses referred to our original state as being like an innocent **'Garden of Eden'** where **'Adam and his wife were both naked, and they felt no shame'**, while Plato similarly referred to our instincts as **'innocent'** and **'pure'** and that in our pre-corrupted state there was no **'possession of women'**, no **'devouring of one another'**, and **'they dwelt naked'**. As just pointed out in par. 792, and as was emphasized in chapter 2:5, what needs to be appreciated is just how determinedly humans have sought to avoid the human condition by blaming our divisive behavior on savage, we-have-to-reproduce-our-genes instincts—which was so clearly evidenced in all the dishonest biological 'explanations' for human behavior that were described in chapter 2. Blaming our upset behavior on supposed selfish, survival-of-the-fittest, we-have-to-reproduce-our-genes has been the main lie humans have employed to cope with the horror of our corrupted condition.

[795] The problem for women, of course, is that while their image of innocence was being cultivated throughout those 2 million years, their soul, their actual innocence, was being destroyed—women are *only* the image of innocence. But without the understanding necessary to explain themselves, men had no choice other than to repress the relative naivety of women, which in turn tied women's corruption inextricably to men's. It has been an extremely difficult situation for women. They have had to try to 'sexually comfort' men but also preserve as much true innocence in themselves as possible to nurture the next generation. Their situation, like men's, worsened at an ever-increasing rate, in that the more women 'comforted' men, the less innocence they retained and the greater comforting the following generation needed. Had humanity's battle continued in this exponential pattern for a few centuries more, all women would have eventually become like <u>Marilyn Monroe</u>, complete sacrifices to men. At this point men would have destroyed themselves and the human species, for there would be no soundness left in women

to love/nurture future generations. Olive Schreiner emphasized this point in her already referred to amazingly truthful 1883 book, *The Story of an African Farm*. When talking of men persuading women to have sex, Olive Schreiner's female character stated that men may say, **"Go on; but when you [men] have made women what you wish, and her children inherit her culture, you will defeat yourself. Man will gradually become extinct..."** Fools!' (p.194 of 300). Incidentally, we again see here how calling men 'fools' was an expression of women's lack of understanding of men, and why they were being attacked sexually by men in the first place.

[796] Yes, tragically, across every generation, individual women have had a very brief life in innocence before being soul-destroyed through sex, following which they have had to try to nurture a new generation, all the time trying to conceal the destruction that was all around and within them. Mothers have tried to hide their alienation from their children, but the fact is if a mother knows about reality/upset her children will know about it and will psychologically adapt to it. Alienation is invisible to those alienated, but to the innocent—and children are born innocent—it is clearly visible. For example, Christ's mother Mary must have been innocent because Christ was. Since women become upset through sex, Mary must have had virtually no exposure to sex. The symbol for women's innocence/purity is virginity, hence the description of Christ's mother as the 'Virgin Mary' (see Matt. 1:23 & Luke 1:26-34). The renowned writer D.H. Lawrence recognized the essential innocence of the 'Virgin' Mary when, in reference to her, he wrote, **'Oh, oh, all the women in the world are dead, oh there's just one'** (Lawrence Durrell & Earl Ingersoll, *Lawrence Durrell: Conversations*, 1998, p.178 of 261).

[797] Having to inspire love when they were no longer loving or innocent, and attempt to nurture a new generation—all the while dominated by men who couldn't explain why they were dominating, what they were actually doing or why they were so upset and angry—was, in truth, more than extremely difficult, it was an altogether impossible task, and yet women have done it for 2 million years. Indeed, it is because of women's phenomenally courageous support that men, when civilized, treat them with such chivalry and deference. Men have had an impossible fight on their hands, but at least they had the advantage of intuitively understanding that battle. To be a victim of a victim, as women have been, is an insufferable situation, because while a primary victim knows what the primary source offence is, a victim of a victim does not. This is why, when men became overly upset they became mean, even brutal, but when women

became overly upset they became nasty, even venomous. Not knowing what it is they are flailing at, women's fury is unsourced, untargeted and unbounded. The proverb **'hell knows no fury like a woman scorned'** recognizes this potential extremity in the nature of women. Nietzsche wrote, **'Let man fear woman when she hates: for man is at the bottom of his soul only wicked, but woman is base'** (*Thus Spoke Zarathustra: A Book for Everyone and No One*, 1892; tr. R.J. Hollingdale, 1961, p.92 of 342). Women have historically had to carry so much unsourced frustration and hurt that their psychological situation is very fragile; so fragile, in fact, that the hormonal upheaval accompanying menstruation is enough to destabilize this delicate balance, hence pre-menstrual tension (PMT).

[798] So while men's situation has been horrible, so has women's; and, just as men have yearned for freedom from their oppressor, ignorance, so women have yearned for freedom from their oppressors, men, as this comment by Olive Schreiner describes: **'if I might but be one of those born in the future; then, perhaps, to be born a woman will not be to be born branded...It is for love's sake yet more than for any other that we [women] look for that new time...Then when that time comes...when love is no more bought or sold, when it is not a means of making bread, when each woman's life is filled with earnest, independent labour, then love will come to her, a strange sudden sweetness breaking in upon her earnest work; not sought for, but found'** (*The Story of an African Farm*, pp.188, 195). Thank goodness with the battle to defeat the ignorance of our instinctive moral soul now won, the horror of both men's and women's existence can end, and this dreamed-of **'new time'** where society will be neither matriarchal or patriarchal but gender-neutral and at peace can begin.

[799] We can see that with men defying and repressing their own souls, women became representative of soul in their partnership with men. Further, because of men's unexplained oppression of women and the world of the soul, and women's own inability to understand this oppression, women put increasing trust and reliance in their soul and instincts, rather than in their ability to understand. As a consequence, women are more intuitive or dependent on their soul's guidance than men—hence the well-known term, **'women's intuition'**. Other common sayings are that **'women feel while men think'** and **'women are more right-brained than men'** (our left and right brain hemispheres were explained in par. 682). It has also often been said that women talk more than men—indeed, a study has found that **'women speak some 13,000 more words a day than the**

**average man'** ('Women really do talk more than men', *Daily Mail Online*, 20 Feb. 2013; see <www.wtmsources.com/168>)—and we can now explain why. Because men have been preoccupied with championing the ego while women were left doing the support duties, women have had less opportunity to develop their own egos—which, while not as developed and embattled as men's, obviously exists because ego is just humans' conscious mind trying to understand the world and its place in it. While they were preoccupied with support duties, the one way women could try to understand the world, and justify and measure themselves was through talking to each other; they couldn't act out their experiments in self-adjustment but they could think their ideas through out aloud, evaluate and try to establish their worth through words.

[800] The following honest description of the different roles of men and women from Sir Laurens van der Post illustrates these very distinct attitudes: **'The sword was, he would suggest, one of the earliest images accessible to us of the light in man; his inborn weapon for conquering ignorance and darkness without. This, for him, was the meaning of the angel mounted with a flaming sword over the entrance to the Garden of an enchanted childhood to which there could be no return. He hoped he had said enough to give us some idea of what the image of the sword meant to him? But it was infinitely more than he could possibly say about the doll. The doll needed a woman not a man to speak for it, not because the image of the sword was superior to the image of the doll. It was, he believed, as old and went as deep into life. But it was singularly in women's keeping, entrusted to their own especial care, and unfortunately between a woman's and man's awareness there seemed to have been always a tremendous gulf. Hitherto woman's awareness of her especial values had not been encouraged by the world. Life had been lived predominantly on the male values. To revert to his basic image it had been dominated by the awareness of the sword. The other, the doll, had had to submit and to protect its own special values by blind instinct and intuition'** (*The Seed and the Sower*, 1963, p.193 of 246).

[801] Note that while men and women are different, <u>sexist notions</u> of men being 'evil' or of women being irrelevant have no credibility. While the main device for avoiding prejudice was to deny that there was any difference between men and women, another was to maintain that any difference between men and women was simply a product of cultural conditioning—of girls being given dolls and boys swords as infants, for example. But as Sir Laurens acknowledges, our differences are the product of a very real distinction—**'a tremendous gulf'**—between the sexes.

[802] So while at a more noble level sex has become an expression of love, at its fundamental level it is an attack on the innocence of people; it is rape. The more upset and corrupted the human, the more sexually destructive and thus sexually perverted they are inclined to be, and the more innocent (or innocent-looking) the human, the more attracting of that destruction they have been. Understanding this makes it possible to explain homosexuality in men. As the victims of sex, women have historically been more exposed to and thus become, through natural selection over hundreds of thousands of years, more adapted to sex than men. In most cases, if a male was not interested in sex then sex did not occur, whereas women have been exposed to sexual advances regardless of their interest or lack thereof. Teiresias, the prophet mentioned in Homer's Greek legend, *The Odyssey*, recognized that women are more sexually aware than men. When asked **'whether the male or the female has most pleasure in intercourse'**, he replied that **'Of ten parts a man enjoys one only; but a woman's sense enjoys all ten in full'** (Hesiod, *The Melampodia*, c. eighth century BC; *Hesiod, The Homeric Hymns and Homerica*, tr. H.G. Evelyn-White, 1914, p.269 of 657). Evidence that women have been more exposed to sexual attention than men is the fact that they tend to have more neotenous facial features and have far less body hair. Since neotenous features and loss of body hair are consequences of sexual selection for innocent looking features, to have more neotenous facial features and less body hair means women must have been exposed to more sexual attention than men. (Incidentally, since women are now highly adapted to sex it means a virgin is not truly a virgin, she is not truly an innocent girl and thus completely 'attractive', because all women are now instinctively aware of 'sex'.) Yes, while women have had to hide their sexual awareness in order to present an attractive image of innocence, the fact is they *are* more sexually aware than men. In the *Happy Days* television series (set in the 1950s and first broadcast in 1974), girls are much more attracted to the sexually aware Fonzie character than to the naive, relatively innocent Richie Cunningham character. While only fictional characters, the viewing audience would not have responded with such empathy over the years if the characters did not resonate with truth. In the book, *Big Bad Wolves: Masculinity in the American Film*, by the academic Joan Mellen, the caption accompanying a picture of the 1920s sex symbol actor Rudolph Valentino reads: **'Rudolph Valentino in the film *Son of the Sheik*. Rape is the central visual metaphor'** (1977, p.54 of 365).

[803] Yes, as explained, 'the mystery of women' is that after 2 million

years of having been sexually used by men, women now only represent the physical image or object of innocence. It is this *image* of innocence that men have been falling in love with; the *illusion* that women are psychologically as well as physically innocent. A well-known African fable tells of a woman who agrees to marry a man on the condition he never looks inside a precious basket that she keeps. She warns him that if he does she will vanish. He agrees, but some time after they are married and his wife has gone to the river for water he cannot resist and peers in the basket. On his wife's return she finds the basket open and him laughing. When she accuses him of looking in the basket he says, 'You silly woman, there was nothing in the basket', at which point she vanishes into thin air. The basket is symbolic of the mystery of women—they are only the image of innocence; it is an 'empty basket' that men are looking at, and once men see through the illusion, women's attractiveness diminishes. It follows then that the more corrupted a man is, the less naive he is, and thus the more he is aware that women are not innocent. Therefore, if a man is extremely hurt and corrupted in his infancy and childhood, when he becomes sexually mature he will not be naive enough to believe that women are still innocent and will not, therefore, find women sexually attractive. The last bastion of 'attractive' innocence for such men is younger men, because they are not as exposed to sexual destruction as women have historically been. To explain the <u>effeminate mannerisms</u> particular to male homosexuality, if you have had your soul, which is your core strength, destroyed in childhood, then taking on the extremely difficult male role of having to fight against the ignorance of the soulful, idealistic world would be an untenable position that would make the female position of not having to fight a much more preferential option. You would rather adopt the female role of being an object of adoration and service than the male role of having to take on the loathsome job of championing ego over soul. The transsexual professional tennis player of the late 1970s, Renée Richards—who went so far as to have a sex change operation to become a woman—alluded to the difficulty of life as a male, and by inference the appeal of being a woman, when she once said, **'women don't realize the horror of the strife-torn world that men live in'**. Having to live with the condemnation that you are an evil monster, when you know you are not but cannot explain why you are not, has been a living hell for men. To be a man and have to oppress your all-magic soul without being able to explain why has been the most wretched of tasks. The following quote serves to illustrate how pressured men's lives have become: **'If women are so oppressed, how come they live much longer than men?'** (Don Peterson, review of *The Myth of Male Power* by Dr Warren Farrell, *The Courier Mail*, Jun. 1994). Homosexuality amongst women results from women's understandable disenchantment with men. Homosexuality is simply another level of perversion to heterosexuality. They are both psychologically corrupted states of sexuality that developed under the horror of the duress of the human condition.

[804] With regard to whether homosexuals are 'born' or 'made', even without the ability to explain the human condition and thus defend the corrupted state of humans (that is, explain that humans' various states of corruption are not 'bad' or evil but, in fact, immensely heroic states), a decade-long research project completed at the Institute for Sex Research in Bloomington, Indiana, found that **'a quarter of the gays interviewed believe** [are prepared to acknowledge?] **homosexuality is an emotional disorder'** (*TIME*, 17 Jul. 1978). In his 1992 book, *Health & Survival in the 21st Century*, Ross Horne referred to studies that show **'That the highest incidence of homosexuality coincides with the general level of stressful influences in a community and that the lowest incidence coincides with the degree of happiness and health in remote and unstressed populations indicates that, like many conditions of physical disease, it is just as unnatural as the mental breakdowns, depression and neuroses so common in civilization. Studies of primitive natives reveal that while in some populations homosexuality is non-existent or rare, in other populations it is fairly common; but the same pattern still holds—among the placid, happy, untroubled people homosexuality did not occur, while among fighting tribes and headhunters it did'** (p.206). After 25 years of clinical experience helping homosexual men and women, Dr Robert Kronemeyer of New York concluded that **'Homosexuality is a symptom of neurosis and of a grievous personality disorder. It is an outgrowth of deeply rooted emotional deprivations and disturbances that had their origins in infancy'** (*Overcoming Homosexuality*, 1980, p.7). While **'Researchers now openly admit that after searching for more than 20 years, they are still unable to find the "gay gene"'** (Dr Nathaniel Lehrman, 'Homosexuality: Some Neglected Considerations', *Journal of American Physicians and Surgeons*, 2005, Vol. 10, No. 3), the need to avoid blame has resulted in the irresistibility of the 'born gay' argument—as John D'Emilio, a professor of history and gay activist concedes, **'The idea that people are born gay—or lesbian or bisexual—is appealing for lots of reasons...If we're born gay, then it's not our fault...What's most amazing to me about the "born gay" phenomenon is that the scientific evidence for it is thin as a reed, yet it doesn't matter. It's an idea**

with such social utility that one doesn't need much evidence in order to make it attractive and credible' ('LGBT liberation: Built a broad movement', *International Socialist Review*, 2009, Issue No. 65). Yes, as Camille Paglia, herself a lesbian, famously stated, 'Our sexual bodies were designed for reproduction...No one is "born gay." The idea is ridiculous...homosexuality, in my view, is an *adaptation*, not an inborn trait' (*Vamps & Tramps*, 1994, pp. 71, 72, 76 of 560).

[805] While women are instinctively more sexually aware than men this does not exclude the fact that a woman can be more innocent and less sexually aware than a man. Girls who are nurtured and sheltered in their upbringing can be very innocent. However, because there has been no honesty about the existence of the different levels of upset and alienation amongst humans, they can be deceived by men who are much more upset and, therefore, much more sexually advanced down 'the rungs of the perversion ladder' (where one is holding hands, two is kissing, three is touching her breast, etc, etc, etc, to the extent that some people became so horribly psychologically sick and perverted that they derived sexual excitement from watching 'snuff movies' of people being killed—yes, sexual depravity is an accurate measure of alienation). Men who are more upset can be very attentive to women because sex for them is a distraction and a way of gaining reinforcement, and innocent women can be deceived—seduced—by this attention into a relationship. In her 1981 book, *African Saga*, the African photographer, and remarkably beautiful woman, Mirella Ricciardi gives an extraordinarily honest account of a relationship between a more innocent woman and a less innocent man. She wrote: 'We went to live in Rome, where I quickly began to taste the bitter-sweet agony of life with Lorenzo. I was young, unaware of the world, and ignorant of people and their behaviour. I married Lorenzo as easily as I had switched lovers. It was probably the most foolish, irresponsible, exciting thing I have ever done. Years later, I came to the conclusion that most of the men I had met fell into three categories—those prompted by their heads [presumably, upset men], those by their heart [presumably, less upset, more soulful, relatively innocent men] and those by their sex [presumably, extremely upset men]. Some—not many— were a combination of all three. Lorenzo belonged to the last category—these I have found are the most attractive. They are sexy, amusing, fun-loving, careless, irresponsible and lazy—they dress well and have a lot of style. Most people like them. They are excellent lovers and lousy husbands. Women usually find them irresistible or are terrified by them. Men either envy or despise them. No one can remain indifferent to them..."Lorenzo's mother died when he was seven,"

**Cesarino [Lorenzo's father] told me one day. "You will have to be more of a mother than a wife to him; do you realise this?" Then he laughed. "The only pleasure he ever gave me was nine months before he was born." But when his father died sixteen years later, Lorenzo's grief was immeasurable and I began then to understand the meaning of these words'** (p.136 of 300). Because women have lived through men, and because their means of healing the world is through nurturing, relatively innocent women involved in relationships with more upset men have often tried to change their partner, make him sounder and stronger through nurturing love, inspiration and motivation. BUT, since it is only understanding of hurt that can heal hurt, these efforts often only serve to further confront and criticize the man. Ricciardi's dedication in her book **'to Lorenzo, my magnificent obsession'** is an acknowledgment of her frustrated efforts to change Lorenzo. She wrote: **'When I married Lorenzo I had created an image of a giant in whose shadow I would live. I clung stoically to my belief in our union and waited patiently for ten years for him to cast his shadow, but he never did'** (ibid. p.138). Similarly, in the 1950 Broadway musical, *Guys and Dolls*, the Salvation Army innocent, Sarah, desperately protests about her upset gambler boyfriend Sky to her more upset-world aware nightclub singer friend, Adelaide, saying, **'Can't men like Sky ever change?'**, to which Adelaide says, **'They just can't change'**; and later Sky himself says, **'Change, change. Why is it the minute you dolls get a guy that you like, you take him right in for alterations?'** In a world of lies, the basis for relationships has often been unhealthy. Women have been seduced by men in so many ways and their innocence has been the casualty—another reason why women have historically become more sexually aware than men. Indeed, growing up in the countryside in Australia in the 1960s I saw many sheltered, relatively innocent country girls go off to Europe for a few years in their early 20s—it was considered the thing to do in those days, as it still is—only to return with 'knowing eyes', a different more sophisticated way of looking at the world. At the time, I could never understand the point of sheltering and nurturing young women if they were simply going to go off to Europe and, as I saw it then, 'cash in their innocence'. It was inevitable, of course, that with innocent women throughout the ages exposing themselves to upset by being attracted to upset men to try to rehabilitate them that there would not be much innocence left in women. Thank goodness that with reconciling understanding of the human condition now available, everyone can help everyone in an effective and constructive—not destructive—way.

[806] Given then that sex is an attack on innocence as well as an act of love the recent generations of humans who have been treating sex cheaply have been contributing significantly to the death of soul in the world, and thus contributing significantly to the level of alienation in the world. Queen Victoria was right to espouse the 'Victorian morality' of her era, a moral code that strongly encouraged people to treat sex with care and restraint.

[807] In light of this whole reality, during the 2 million years that women have endured the wretched situation of being unable to understand men's oppression of them, many must have found it impossible to accept and, as a result, there must have been a great deal of natural selection and thus genetic adaption to the role that women have had to play in the human journey to enlightenment. Olive Schreiner described women's resignation to their role in the following passage, again in her extraordinarily honest book, *The Story of an African Farm*. It is a dialogue between her young female character and that character's male friend, Waldo: **"'I know it is foolish. Wisdom never kicks at the iron walls it can't bring down,' she said. "But we are cursed, Waldo, born cursed from the time our mothers bring us into the world till the shrouds are put on us. Do not look at me as though I were talking nonsense. Everything has two sides—the outside that is ridiculous, and the inside which is solemn." "I am not laughing," said the boy sedately enough; "but what curses you?" He thought she would not reply to him, she waited so long. "It is not what is done to us, but what is made of us," she said at last, "that wrongs us. No man can be really injured but by what modifies himself. We all enter the world as little plastic beings, with so much natural force, perhaps, but for the rest—blank; and the world tells us what we are to be, and shapes us by the ends it sets before us. To you it says—*Work!* and to us it says—*Seem!* To you it says—As you approximate to man's highest ideal of God, as your arm is strong and your knowledge great, and the power to labour is with you, so you shall gain all that the human heart desires. To us it says—Strength shall not help you, nor knowledge, nor labour. You shall gain what men gain, but by other means. And so the world makes men and women. Look at this little chin of mine, Waldo, with the dimple in it. It is but a small part of my person; but though I had a knowledge of all things under the sun, and the wisdom to use it, and the deep loving heart of an angel, it would not stead me through life like this little chin. I can win money with it, I can win love; I can win power with it, I can win fame. What would knowledge help me? The less a woman has in her head the lighter she is for climbing. I once heard an old man say, that he never saw an intellect help a woman so much as a pretty**

ankle; and it was the truth. They begin to shape us to the cursed end," she said, with her lips drawn in to look as though they smiled, "when we are tiny things in shoes and socks. We sit with our little feet drawn up under us in the window, and look out at the boys in their happy play. We want to go. Then a loving hand is laid upon us: 'Little one, you cannot go,' they say; 'your face will burn, and your nice white dress be spoiled.' We feel it must be for our good, it is so lovingly said; but we cannot understand; and we kneel still with one little cheek wistfully pressed against the pane. Afterwards we go and thread blue beads, and make a string for our neck; and we go and stand before the glass. We see the complexion we were not to spoil, and the white frock, and we look into our own great eyes. Then the curse begins to act on us. It finishes its work when we are grown women, who no more look out wistfully at a more healthy life; we are contented. We fit our sphere as a Chinese woman's foot fits her shoe, exactly, as though God had made both—and yet He knows nothing of either. In some of us the shaping to our end has been quite completed. The parts we are not to use have been quite atrophied, and have even dropped off; but in others, and we are not less to be pitied, they have been weakened and left. We wear bandages, but our limbs have not grown to them; we know that we are compressed, and chafe against them. But what does it help? A little bitterness, a little longing when we are young, a little futile searching for work, a little passionate striving for room for the exercise of our powers,—and then we go with the drove. A woman must march with her regiment. In the end she must be trodden down or go with it; and if she is wise she goes'" (pp.188-189). Incidentally, note again here women's naivety; Schreiner believes men's sexualization of women is un-Godly, basically that men are evil, which is again why women were 'fucked' in the first place.

[808] With regard to women resisting **'march[ing] with her regiment'**, in the sense of accepting their role of inspiring men, it should be mentioned that it's not commonly acknowledged that suits were invented for men so they could hide their big guts, while dresses were invented for women so they could accentuate their waists and breasts and conceal their big bottoms and thighs, but, while men still wear coats, everywhere in Western society now women have forsaken skirts for trousers, *and even tights*, as if their role of inspiring men with their beauty no longer matters. This is, in truth, yet another illustration of women's lack of awareness of the nature of the struggle that the human race has been involved in—and of the irresponsibility of feminism, which encouraged women not to **'march with her regiment'**. Women's role has understandably become unbearable for them but the battle to find understanding still had to be won.

[809]Yes, while the feminist movement has improved 'a woman's lot' superficially, there has, in fact, been no real change to the situation that Schreiner so honestly described, as these quotes confirm: **'Nirvana hasn't happened. Although men are speaking about understanding** [the need for women's liberation from men's oppression] **on the surface, they're not doing anything about it'** (Carmel Dwyer, *The Sydney Morning Herald*, 22 Sep. 1993); and, **'What happened was that the so-called Battle of the Sexes became a contest in which only one side turned up. Men listened, in many cases sympathetically but, by the millions, were turned off'** (Don Peterson, review of *The Myth of Male Power* by Dr Warren Farrell, *The Courier Mail*, Jun. 1994). Until men could explain to women why they have had to be so egocentric, competitive and aggressive—explain the human condition—there could be no fundamental change to the situation where men found themselves with no choice other than to oppress women.

'Goddess' statue, Çatalhöyük, Anatolia (modern Turkey), 8,500-5,500 BC

[810]It should be pointed out here that women resisted men's oppression, and resisted it for a long time. Yes, having said that humanity has been patriarchal for some 2 million years, it has to be explained why women were still so powerful, even seemingly treated as goddesses, in central Europe during the Upper Paleolithic (50,000 to 10,000 BC) and Neolithic (10,000 to c.4,000 BC) periods, as evidenced by the many so-called

'Goddess' or 'Venus' figurines, such as the one pictured opposite, that have been found in this region from this period. In this typical example, we can see from the *extremely* regal stature of the very well-nourished figure seated on her throne of cheetahs just how powerful and in control of their societies such women must have been. In chapter 5:11, 'The importance of strong-willed females in developing integration', it was explained just how powerfully self-assured, secure-in-self, strong-willed and assertive females had to be during the development of love-indoctrination in order to rein in male aggression from competing for mating opportunities. Yes, what greatly contributed to the defeat of the historical patriarchal animal-condition situation where males relentlessly competed for dominance was the selection for strong-willed, authoritative females who would no longer tolerate male aggression; indeed, females became downright contemptuous of divisive behavior in males. It simply would not be tolerated. Quotes were included to illustrate this situation, such as: **'An impressively stern** [bonobo] **female enters and snaps a young sapling. Once she picks herself up she does something** [that would be] **entirely surprising for a female chimp, she displays, and the males give her sway. For this is the confident stride of the group's leader, its alpha female, whom Kano has named Harloo'**; and, **'She [Peggy] was the highest-ranking female in the [baboon] troop, and her presence often turned the tide in favor of the animal she sponsored. While every adult male outranked her by sheer size and physical strength, she exerted considerable social pressure on each member of the troop...another reason for the contentment in this particular family was Peggy's personality. She was a strong, calm, social animal, self-assured yet not pushy, forceful yet not tyrannical.'** As has been explained, the problem that emerged when our ancestors became fully conscious and men had to take up the task of defeating the ignorance of our instinctive self and champion our conscious thinking self or ego, was that women tended to be unsympathetic towards the angry, aggressive and egocentric upset that it unavoidably produced in men. In fact, this situation was made so much more difficult for men because women were *so* immensely strong-willed and authoritative as a result of this very powerful matriarchal heritage. It makes sense, then, that while humanity had become fundamentally patriarchal, the old matriarchy hung on. In fact, even today there are still many women who are so strong-willed and contemptuous of non-ideal behavior that men find it almost impossible to cope with them, with some even going to great lengths to try to break their strong-willed, but ignorant, resolve.

If you watch WTM Founding Member Stacy Rodger's affirmation at <www.humancondition.com/affirmations> you will hear about just such a tragic situation—where Stacy was psychologically traumatized because, she said, **'every time I lived out my strength and said what I really thought, especially about the insincerity of men's treatment of women—basically of men being self-centred, totally, seemingly unaware and insensitive towards others and even seemingly unaware and insensitive towards the whole world—it would always end up with men side-lining me and freezing me out'** in an effort to **'shut me out and shut me down'**. Unable to explain why they are so upset, embattled and egocentric, men have truly struggled to do their job of championing the ego over our ignorant instinctive self or soul in the midst of women who have been as strong-willed and yet as naive as Greer and Steinem have been. In hindsight (because understanding of the human condition finally brings to an end the horrible 'battle between the sexes'), it can be seen that it was an act of great generosity (actually not generosity but weakness, which I will talk more about in a moment) on the part of men to give women the vote! So this situation where women were seemingly in power in prehistoric times was a case of 'delayed ownership'—of situations where the new owner, patriarchy, wasn't able to take over because the old owner, matriarchy, refused to relinquish power.

[811] A 1999 BBC documentary titled *Ancient Voices: Tracking the First Americans* presents convincing evidence that some Australian Aborigines managed to reach South America long ago, and that the Fuegian Indians of Tierra del Fuego are their descendants. To prove that such immensely long sea journeys are possible the program referred to the fact that recently **'five African fishermen were caught in a storm and a few weeks later were washed up on the shores of South America'** with three still alive ('First Americans were Australian', BBC News Online Network, 26 Aug. 1999; see <www.wtmsources.com/161>). In terms of what is being explained here about women resisting male dominance, some of the commentary in this documentary is particularly enlightening (the underlinings are my emphasis): 'Narrator: **The tribal wisdom taught to the Fuegian initiates was secret. It was only revealed to the men, the women were kept in the dark. Any speculation about it was, and still is, strictly taboo.** Christina, one of the last of the Fuegian tribe, then says: **They said it was a very secret ritual, that's why we never talked about it. Only the men were supposed to know about it.** Narrator: **What was so secret that had to be kept from the women? Some of the chiefs confiding to ethnographers explained**

that <u>there was a time in the very distant past when women ruled society. The</u> <u>women must never know lest the men lose their grip on power</u>...Extraordinarily, <u>similar legends have been recorded amongst aborigine tribes in Australia</u>. It seems the traditions of the first peoples of Australia have been preserved here at the utmost end of the Earth by a small band of their descendants. But after surviving 50,000 years, the memory of those traditions is now at risk of being lost forever.'

[812]This narration provides powerful evidence of the aforementioned 'delayed ownership' analogy, of situations where the 'old order' is slow to relinquish its position, for it demonstrates that even though humanity had become patriarchal, male-role led, in so many ways—such as by women being treated as sex objects (fucked for being ignorant of the goodness of men) and having become highly adapted to this situation, and men leaving women to do the gathering and other practical work while they conducted the innocence-dominating task of killing (hunting) animals—such was women's strength of character that for a long time **'women ruled society'**. It took a great deal of effort, determination and time for men to finally gain a **'grip on power'**.

[813]The question this raises is, if women have been so unmainframed, and so duped by soul, <u>how do we explain such effective right-wing-</u> <u>supporting, ego-sympathetic females</u> like the political leaders Margaret Thatcher, Madeleine Albright and Golda Meir, and authors like Ayn Rand, and political commentators like Ann Coulter in the USA and Janet Albrechtsen in Australia? To explain this phenomenon I need to first point out that feminism hasn't been entirely unfounded. The more men fought to defeat ignorance and protect the group (humanity), the more embattled, upset and corrupted they became and thus the more they appeared to worsen the situation. The harder men tried to do their job of protecting humanity, the more they appeared to endanger humanity! As a result, they have become almost completely ineffective or inoperable, paralyzed by this paradox; made cowards by the extent of their self-corruption and its effects. At this point, women have had to step in and usurp some of the day-to-day running of affairs as well as attempt to nurture a new generation of soundness. Women, not oppressed by the overwhelming responsibility and extreme frustration that men felt, could remain effective. Further, when men crumpled, women *had* to take over, otherwise the family, group or community involved would fall apart. A return to matriarchy, such as we have

recently seen in some parts of society, is a sign that men in general have become almost completely exhausted. However, it is important to understand that *total* matriarchy has not emerged because men could not afford to stand aside completely while the fundamental battle still existed. They needed to stay in control and remain vigilant against the threat of ignorance. So while some elements in the recent feminist movement have seized the opportunity to avenge men's oppression, the movement was, to a degree, borne out of necessity. The tragedy is that like all pseudo idealistic, politically correct movements, feminism is based on a lie: in this case, that there is no real difference in the roles of men and women.

[814] To address the question then of how are we to explain the existence of such right-wing, ego-sympathetic women as Baroness Thatcher and the others mentioned, what needs to be considered is the important sentence that was included in the previous paragraph: 'Women, not oppressed by the overwhelming responsibility and extreme frustration that men felt, could remain effective.' Men have been overly corrupted for at least half a million years and, as such, have lived with extreme frustration, even self-loathing, of the immense destruction they have inflicted upon the planet. After such a long time, it can be expected that women now have a strong instinct for an opportunity to participate up-front in the battle and even—in situations where men became totally destructive, disdainful of themselves, paralyzed by their predicament and inoperable—take control from men. We can now expect women to anticipate the opportunity for greater power within personal relationships, and in larger economic and political spheres, and to some degree be adapted to and thus appreciative of what is required to effectively take up the male role of championing the ego. After all, if men had not been available to take on the battle to champion ego over ignorance, women would have had to take it on fully and become as aware as men now are of what happens when you fight that battle. Camille Paglia once wrote that **'It is woman's destiny to rule men'** (*Vamps & Tramps*, 1994, p.80 of 560). This comment is an expression of the expectation that now exists in women that men eventually crumple, at which point women have the opportunity to take over. The truth is, many sensitive 'New Age' guys (so-called SNAGs) and 'metrosexuals' are actually crumpled men. Women's nagging of their menfolk is also a case of women chiseling away at, attempting to break, men's ability to keep fighting and defying the ignorance of the world of the soul—

a situation made much of in the BBC sitcom *Keeping Up Appearances*, where the character of poor old Richard is constantly hen-pecked by his wife Hyacinth.

[815]But while women have learnt that men often lose the ability to keep persevering with their job of defying soul, and eventually 'crumple' giving women the opportunity to gain power, for the most part women have learnt to, as Schreiner said, **'march with…[their] regiment'**—resign themselves to a life of being a sex object and accepting the subjugation to men that entails. This is because women haven't been 'mainframed' to the battle of the human condition, they couldn't 'read the play' of what was really going on; as Stacy Rodger said in her affirmation, **'every time I lived out my strength and said what I really thought…it would always end up with men side-lining me'**. This limited existence left women *very* frustrated—so it is no wonder they ended up throwing up their arms in despair and resorting to the saying, **'girls just want to have fun'**! As I said earlier, the sooner the reconciling understanding of the human condition is found, as it now is, and all this horror can end, the better.

[816]So, yes, while dogmatically imposing an ideal situation where men and women treat each other as though there is no difference between them—as the politically correct culture has attempted to do—could disguise and contain upset to a degree, it could *not* remove or resolve it. *Only* understanding the world of men, and why they have been so divisively behaved, could subside the anger, alienation and egocentricity that caused them to victimize virtually everyone and everything they encountered. The boot that was screwing men into the dirt had to be lifted for the horrible war between men and women to end. And an all-out war it has been, lived to the full extent of what was possible under the limitation that men and women were forced to coexist if they were to reproduce and nurture a new generation. Yes, as men have long lamented about women, **'You can't live with them and you can't live without them!'**

[817]And, as has been emphasized, in finding the liberating under-standing of why men have been divisively behaved, what it reveals is that men are the heroes of the story of life on Earth. From being thought of as the villains one day, they become the absolute heroes the next—a turn of events that is long overdue. Everywhere men have become wretchedly oppressed by the politically correct dogma that denies them any real meaning in the world. It has reached the point where there are

now books being published like *Are Men Obsolete?*, which has on its cover the label 'TERMINATION PENDING' stamped across the face of a man. Like the aforementioned comments from the feminists Greer and Steinem, an article about this book contains this comment from one of its authors, the Pulitzer Prize-winning commentator Maureen Dowd: '**Norman Mailer used to be terrified that women were going to take over the world as a punishment for being bad to them over the centuries…All women needed, he said, were about a hundred semen slaves that they could milk every day…Dream on, Norman! All women need is a few cells in the freezer next to the cherry-flavoured vodka and we're all set**' ('Are men obsolete in the modern world?', *The Telegraph*, 2 Jun. 2014). Indeed, men have become so intimidated by pseudo idealism that many have come to believe they *are* useless. The science writer Bob Beale wrote that while the concession pained him, he was prepared to concede that, except for '**the baby business [reproduction]**', '**males are largely useless**' (*Men: From Stone Age to Clone Age*, 2001, p.vii of 369). Showing less remorse, however, was the anthropologist Melvin Konner, who completely turned on men, writing in his 2015 book, *Women After All: Sex, Evolution, and the End of Male Supremacy*, that '**maleness**' was a '**syndrome**', '**a birth defect**', '**a disorder**', and the result of '**androgen poisoning**' (p.8 of 400), and that '**Humans in a future world, could perhaps stay all female, designating one of them to become male only when collectively wanted or needed**' (p.66). No wonder there has been a proliferation of men's movements that aim to counter men's horrific situation of being totally misunderstood!!

818 It follows that boys growing up in this current, men-are-worthless world are having their self-esteem destroyed even before they have a chance to enter manhood. A 2014 article titled 'The Sexodus' about '**a large-scale exit from mainstream society by males**' reports that '**among men of about 15 to 30 years old, ever-increasing numbers are checking out of society altogether, giving up on women, sex and relationships and retreating into pornography, sexual fetishes, chemical addictions, video games…all of which insulate them from a hostile debilitating social environment**'. '**Rupert, a young German video game enthusiast…says…"My generation of boys are f\*\*cked… Marriage is dead. Divorce means you're screwed for life. Women have given up on monogamy, which makes them uninteresting to us for any serious relationship or raising a family"…In schools today across Britain and America, boys are relentlessly pathologised…Boyishness and boisterousness have come to be seen as "problematic", with girls' behaviour a gold standard against which these defective boys are measured. When they are found wanting, the solution is often drugs…**

Millions will be prescribed a powerful mood stabiliser, such as Ritalin, for the crime of being born male. The side effects of these drugs can be hideous and include sudden death. Meanwhile, boys are falling behind girls academically... Never before in history have relations between the sexes been so fraught with anxiety, animosity and misunderstanding...One professional researcher...puts it spicily: "For the past, at least, 25 years, I've been told to do more and more to keep a woman. But nobody's told me what they're doing to keep me...the message from the chicks is: 'It's not just preferable that you should fuck off, but imperative. You must pay for everything and make everything work; but you yourself and your preferences and needs can fuck off and die'."...The media now allows radical feminists to frame all debates...Women can basically say anything about men, no matter how denigrating, to a mix of cheers and jeers... modern feminists...[are] parading around in t-shirts that read: "I BATHE IN MEN'S TEARS."...Men created most of what is good about the world. The excesses of masculinity are also, to be sure, responsible for much of what is bad. But if we are to avoid...a world in which men are actively discriminated against, we must arrest the decline in social attitudes towards them before so many victims are claimed that all hope of reconciliation between the sexes is lost' (Milo Yiannopoulos, Breitbart.com, 4 Dec. 2014; see <www.wtmsources.com/146>).

[819] Yes, a very real concern for the future was that there would be a dearth of men psychologically strong enough to fight ignorance, and indeed, if the human condition had not been solved the human race would be facing very dark times. As will be talked more about in ch. 8:16, that is how dangerous the politically correct, postmodernist culture has become.

[820] The reason for the just mentioned phenomena of school boys not performing as well in their studies as girls is not only because they have had their immense relevance in the world denied, but also because they are not as duped by the historical denial of the issue of the human condition as women. Women tend to believe the world we are living in is the real world, whereas men, being mainframed, are intuitively aware that it is a fraudulent existence, and as a result don't take our current artificial, fabricated world too seriously. In the television series, *The Simpsons*, the young boy, Bart, has little respect for school whereas his sister Lisa applies herself completely and excels. Nietzsche was alluding to the trusting naivety of women when, as mentioned earlier, he said, '"Behold, now the world has become perfect!"—thus thinks every woman when she obeys with all her love.'

[821] While on Nietzsche, his much debated and misunderstood concept 'the will to power' (the title of his last work before his psychological breakdown—which, since he was such an honest thinker, would have happened when he eventually tried to confront more truth than his level of soundness of self could cope with) can now be interpreted as man's will to achieve power over humans' idealistic instinctive self or soul's ignorance of the true goodness of corrupted humans. In his now much acclaimed (and rightly so) book *Thus Spoke Zarathustra: A Book for Everyone and No One*, Nietzsche recognized that, unlike other animals, humans have had to fight a psychological demon, the human condition: **'Man, however, is the most courageous animal: with his courage he has overcome every animal. With a triumphant shout he has even overcome every pain; human pain, however, is the deepest pain'** (p.177 of 342). Yes, humans, especially men, had to have the courage to triumph over their deepest pain, the pain of not being able to know whether they were fundamentally evil beings or not. *Humans have had to learn to love themselves*—and now they truly can.

[822] Incidentally Nietzsche's subtitle to *Thus Spoke Zarathustra*, **'A Book for Everyone and No One'**, was an open acknowledgment that to speak the truth to people who are resigned and living in denial of the human condition was to court total rejection—what I call the 'deaf effect'. **'No One'** would hear, understand and accept his words. What Nietzsche knew, however, was that only the truth could liberate humans and that, in time, people would hear and understand his words, and that eventually his book *would be* for **'Everyone'**. Denial-free thinkers or prophets such as Nietzsche have historically not been appreciated in their own lifetime, but their honesty has always led the way to a better world for humans. In fact, their honesty is the purest form of love the human race has ever known. The ultimate example of this is Christ, whose words were also not understood in his lifetime, or indeed in the ensuing years; as he said, **'Why is my language not clear to you? Because you are unable to hear what I say… The reason you do not hear is that you do not belong to God** [you live in denial of such fundamental truths as Integrative Meaning]' (Bible, John 8:43-47)—and yet he gave humanity a home to live safely in while the corrupting search for knowledge had still to be completed. I will come to the role that religions such as Christianity have played shortly (in ch. 8:15).

[823] In summary, for 2 million years women have stood by their men, just as for some 10 million years prior to that, men supported their women.

With understanding of the human condition now found, men and women can at last stand side by side—the 'war of the sexes' can finally be resolved. Yes, the human journey can have the happy ending we always hoped it would.

[824] (A note to the reader: I have taken the description of what happened in the relationship between men and women much further than the adventurous Adolescentman stage because it made practical sense to follow this very important relationship through to its resolution. Given how many subjects have to be looked at, the presentation in this chapter is bound to be somewhat disjointed; the intention can only be to convey a general appreciation of the sequence of events that have occurred in humanity's incredible journey from ignorance to enlightenment. More is said later in pars 897-901 about the horrifically difficult situation women have had to endure under the duress of the human condition, while more can be read about the relationship between men and women at <www.humancondition.com/freedom-expanded-men-and-women>.)

### Chapter 8:11C Other adjustments to life under the duress of the human condition that developed during the reign of Adventurous Adolescentman

[825] The overall point being made here is that since all forms of innocence unfairly criticized humans during our species' insecure adolescence, all forms of innocence have been attacked by upset humans during this stage. As stated earlier, animals also fell victim to the human condition; and not only animals, but nature in the broader sense, because it too was a friend and 'ally' of our instinctive soul and, therefore, an 'enemy' of our apparently 'bad' conscious mind. Attacking nature, be it chopping down trees or setting fire to vegetation, brought a retaliatory sense of satisfaction to the upset within us. Even the wearing of dark glasses, ostensibly as sunshades, was often an effort to block or alienate ourselves from the natural world that was alienating us.

[826] Indeed, if we take a moment to extrapolate this situation, we can see that eventually our upset was likely to increase so much, and our associated resentment of any criticism of it would become so great, that disputes with other humans were inevitably going to break out, at which point people would eventually start grievously attacking and even murdering each other, which would lead to, and, of course, has

led to, outright, organized <u>warfare</u>—but, as will be explained during humanity's 40-year-old equivalent stage, such extremely destructive behavior didn't emerge until the latter period of our 2 million years in adolescence.

[827] Of course, the more upset we became, the more we needed ways to escape and relieve the trauma of our condition. And so to compensate for the extremely unhappy state of becoming corrupted, we began to seek out the material rewards of luxury and comfort, with this <u>materialism</u> becoming one of the main driving forces or motivations in life when upset became extreme. The accumulation of wealth and what it could offer us—the land, the staff, big houses, hordes of gold, glittering dresses, sparkling diamonds and shiny, pretentious cars—gave us the fanfare and glory we knew was due us, but which the world in its ignorance would not give us. From being bold, challenging and confrontationist, the heroic 21-year-old eventually became embattled, cynical and exhausted, greatly in need of escapism and relief, and thus an increasingly superficial, material and artificial person. We personally abandoned any idealistic hope of winning the battle to overthrow ignorance as to the fact of our true goodness and became realists, concerned only with finding material relief and bestowing glory upon ourselves.

[828] As mentioned in par. 727, while innocent 'Childmen' were instinctively coordinated and connected, once upset, especially alienation, developed, <u>language</u> became a necessity. With alienation differing from one person to another, the need emerged to try to explain ourselves, to explain why we were behaving differently, in such a seemingly non-ideal manner. In fact, talking became the key vehicle for justifying ourselves, both in our minds and to others. But since we couldn't speak directly about the human condition, or about other people's particular states of alienation without overly confronting and condemning them, <u>stories</u> became a way of passing on knowledge, or what we call <u>wisdom</u>, about the subtleties of life under the duress of the human condition. Much later, with the development of <u>the written word</u> about 6,000 years ago, the fundamental quest for self-justification became greatly assisted because the wisdom acquired during each generation could be more accurately recorded, which meant that quite suddenly the accumulation of knowledge gained real impetus. *But*, it follows that throughout the upsetting journey through humanity's adolescence our increasing need to *somehow* explain and justify ourselves with words, both oral and written, also led to the

development and dissemination of all kinds of increasingly sophisticated excuses and lies for our behavior. The industry of <u>denial</u> became one of the main features of our lives; indeed, the extreme denials that have taken place in science about our species' innocent, upset-free, psychologically secure and happy past bear stark witness to just how sophisticated the art of denial has become.

[829] At this point in our journey, other forms of self-expression, such as <u>art</u> and <u>music</u>, became particularly useful because, unlike language and stories, their often deep and important message wasn't as clear and, therefore, as potentially confronting—as the writer Victor Hugo said, **'Music expresses that which cannot be said and on which it is impossible to remain silent'** (*William Shakespeare*, 1864). Each person could derive as much meaning from the art or the music or even the <u>dance</u> and other cultural rituals as they could personally cope with. Of course, once humans became extremely alienated and had overly repressed their all-sensitive, beautiful world of their original instinctive self or soul because it was so condemning and confronting, then art, music and dance and other forms of cultural expression could also serve to reconnect them back to the soul's true world. For example, it is often said of great art that it **'can make the invisible visible'**; it can cut a window into our alienated, effectively dead state and bring back into view some of the beauty that our soul has access to. After years of developing his skills, Vincent van Gogh was able to bring out so much beauty that resigned humans looking at his paintings find themselves seeing light and color as it really exists for possibly the first time in their life: **'And after Van Gogh? Artists changed their ways of seeing…not for the myths, or the high prices, but for the way he opened their eyes'** (*Bulletin* mag. 30 Nov. 1993). On the whole, <u>culture</u> essentially encompassed the various ways people passed on, from one generation to the next, the knowledge they had learnt about living under the duress of the human condition.

[830] Although the oldest known cave paintings date back just 35,000 years, archaeologists working in Zambia announced in 2000 that they had found pigments and paint grinding equipment believed to be between 350,000 and 400,000 years old. At the time of the discovery it was reported that the find showed that **'Stone Age man's first forays into art were taking place at the same time as the development of more efficient hunting equipment, including tools that combined both wooden handles and stone implements…**[and that it was evidence of] **the development of new technology,**

**art and rituals'** (*BBC World News*, 2 May 2000; see <www.wtmsources.com/162>). The British archaeologist Lawrence Barham, a member of the team in Zambia, described the find as the **'earliest evidence of an <u>aesthetic sense</u>'** and that **'It also implies the use of language'** (ibid). As just explained, language would have emerged with alienation because people would have then needed some way to account for their unnatural behavior, and since we can expect alienation to have begun soon after the emergence of *Homo* we can assume that at least a rudimentary language would have been practiced by *H. habilis* who emerged approximately 2.4 million years ago. With regard to other expressions of aesthetic sensitivity—a sensitivity we have had since we became instinctively immersed in love and in tune with all of nature during our love-indoctrinated past, but have not been able to reflect upon or express until our conscious mind and associated state of alienation reached a certain level of development—the oldest musical instruments found so far, phalange (bone) whistles, show that Neanderthals, the early variety of *H. sapiens sapiens*, were making music around 80,000 to 100,000 years ago, while a Neanderthal burial site at the Shanidar Cave in Iraq, estimated to be around 50,000 years old, contains traces of pollen grains, indicating that bouquets of flowers were buried with the corpses. The creative and aesthetic sense of our ancestors of nearly half a million years ago, as indicated by the pigments and paint grinding equipment, suggests that the creative and spiritual sensitivities demonstrated by the Neanderthals were in existence long before their time.

[831]The extreme sensitivity that is particularly apparent in the rock paintings of the Bushmen of southern Africa and Australian Aborigines, and in the cave paintings of early humans in Europe, is especially revealing of how much innocence the human race has lost in relatively recent times. The Chauvet Cave in southern France, for example, contains a wealth of cave drawings that date from around 30,000 years ago (some of which are reproduced opposite) that have inspired such descriptions as **'miraculous'**, **'overwhelming in density, humbling in sophistication, and awe-inspiring in sheer beauty'** ('The Goddess Bites'; see <www.wtmsources.com/131>). The drawings are three dimensional, even animated; in short, the animals appear so real it is as if they are alive! You can almost feel what it is like to be those animals, the whole struggle of their lives is revealed.

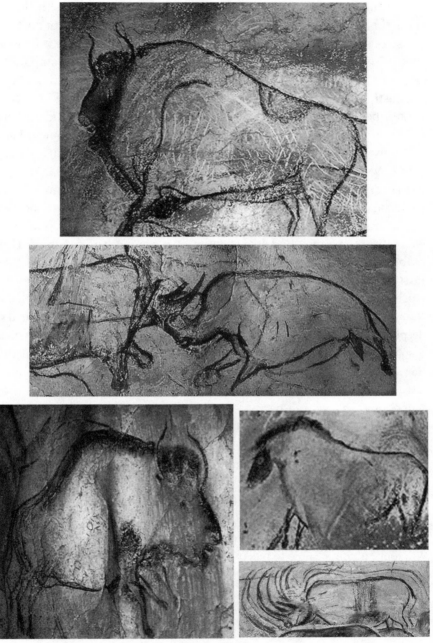

Extraordinarily empathetic renditions of animals in
the Chauvet Cave in southern France, c.30,000 years old

[832]I might mention that I learnt long ago that to draw the little pictures that are included throughout this book, I had to disconnect my conscious mind and just let my instinctive sensitivity express itself, and that if I didn't do that I simply couldn't draw at all. For example, the drawing of the three 'Childmen' happily embracing that I used in ch. 8:4 to illustrate humanity's childhood stage was done so quickly I shocked myself because I could hardly believe that such an empathetic drawing could be produced from an almost instant scribble. At that moment I saw just how much sensitivity we humans once had, and how much alienation now exists within us 2-million-years'-embattled humans. Yes, the extraordinary empathy and accuracy of the paintings of animals in the rock and cave paintings shown above are similarly incredibly indicative of the amount of sensitivity we humans once had and have since lost; we *truly* are an embattled species now, so worn out, so brutalized. How extremely sensitive must early humans have been! Sir Laurens van der Post wasn't exaggerating when he wrote about the relatively innocent Bushmen that **'He and his needs were committed to the nature of Africa and the swing of its wide seasons as a fish to the sea. He and they all participated so deeply of one another's being that the experience could almost be called mystical. For instance, he seemed to *know* what it actually felt like to be an elephant, a lion, an antelope, a steenbuck, a lizard, a striped mouse, mantis, baobab tree'** (*The Lost World of the Kalahari*, 1958, p.21 of 253). Plato made a similar observation when, as mentioned in par. 174, he described our innocent ancestors as **'having…the power of holding intercourse with brute creation** [being able to relate to other animals]'.

[833]When all the upset in humans heals the world is going to open up for us humans. Our long repressed all-loving and all-sensitive original instinctive self or soul is going to come back to the surface. We are going to be able to feel everything around us. We are going to have so much kindness and love and empathy for each other and our fellow creatures because we will, once again, be able to feel everything they are experiencing, including just how embattled the lives of animals are; they suffer *enormously* from the 'animal condition', from the unrelenting need to compete for food, shelter, space and a mate. While, through the nurturing, love-indoctrination process, *our* ape ancestors were able to break free from the tyranny of genes having to ensure their own reproduction, other animals remain stuck in

a continuous cycle of competition. Unlike humans (and bonobos, who are in the midst of developing love-indoctrination), other animals can't develop full unconditionally selfless cooperative instincts. And so above all else, it is this empathy with, this feeling for, the relatively short, brutish, forever-having-to-fight-for-your-chance-to-reproduce lives of animals that those who made these drawings have so sensitively expressed. To use Sir Laurens' words, they **'seemed to *know* what it actually felt like to be'** a bison, horse or rhinoceros. You can sense the whole internal struggle of the animals' lives in these drawings. Their huge chests heave with their brutal and tough battle to survive and reproduce—they are struggling so much to endure their lot it is as if they have asthma! Yes, now that humans can get over the terrible agony of our 'human condition', we will again be able to empathize with the terrible agony of the 'animal condition'. It's not very nice to have to belt the living daylights out of others to ensure your genes reproduce, let alone other members of your own species—in fact, your cousins, uncles and even your own father! No, it is not at all easy being a non-human animal, and that is an extreme understatement, just as it has not been at all easy being an upset human, which is, of course, another extreme understatement!

[834] While the Paleolithic artists clearly weren't as alienated as humans are today, they were still much, much more alienated than humans *originally* were. I think this is revealed by the fact that these cave artists almost completely avoided depicting humans. For instance, in the entire Chauvet Cave complex there is only one representation of a human, and even that is limited to a drawing of only the lower half of a woman's torso. On the few occasions when these cave artists tried to depict humans they almost invariably ended up drawing stick figures. The human face, in particular, which you would think would be the most interesting and relevant of subjects for these artists to depict, seems to have been totally beyond their ability. It seems clear that the facial expressions of humans were by then so alienated, so devoid of the innocence that they must have once exhibited, that our instinctive self or soul couldn't relate to it; it couldn't, and perhaps didn't want to, draw us. What did R.D. Laing say about our present alienated state: **'between *us* and It [our true selves or soul] there is a veil which is more like fifty feet of solid concrete'**. The artist Francis Bacon revealed just how corrupted and alienated upset humans *really* are in his honest painting of

the psychologically-contorted-smudged-human-condition-afflicted-face that was included after par. 124. Indeed, the weird, kidney-shaped blob for the human face that the Aboriginal artist drew in the rock painting shown below, from Ubirr in the Kakadu National Park in the Northern Territory of Australia, is very Bacon-like! Revealingly, when I was looking at this painting at Ubirr, which is thought to be some 2,000 years old, I asked a guide, who was accompanying a tour group, whether she thought the reason the paintings of wildlife were so accurate while the paintings of the humans were so pathetic was because we are now too alienated for our soul to be able to empathize with us, the guide, and everyone else, reacted with a real shudder and audible choking noise. What I had said was just too close to the truth.

The author at Ubirr, Kakadu, 2010

Author's photo of fish, spear thrower and tortoise at Ubirr

A 27,000 year old Bushmen rock painting of an eland hunt, Kamberg, Drakensberg Mountains, South Africa

A 17,300 year old cave painting of a wounded bison, a dead man with a broken spear beside him, a bird and what appears to be a rhinoceros in Lascaux, France. While there are many extraordinary empathetic paintings of animals in the Lascaux caverns there is only this one crude image of a man.

[835] It is truly an insight into how sensitive and loving humans once were that our instinctive self or soul can't relate to the way we are now. Consider the tenderness in the expression on the face of the Madonna in the drawing of the Madonna and child that was included at the beginning of the infancy stage. My soul drew that—I, my embattled conscious self, had nothing to do with it. Truly, as William Wordsworth wrote, **'trailing clouds of glory do we come, From God** [the integrated, loving, all-sensitive state], **who is our home'** (*Intimations of Immortality from Recollections of Early Childhood*, 1807). And people say humans have brutish, aggressive instincts! No, it's the world we humans *currently* inhabit that is mad. It is just so traumatized with psychological upset that it hasn't been able to deal with the fact that it is deeply, deeply dishonest; horrifically alienated. What did the great Spanish artist Pablo Picasso famously say about his ability to paint: **'It's taken me a lifetime to learn to paint like a child.'** And again, what did R.D. Laing say, **'between** *us* **and It** [our true selves or soul] **there is a veil which is more like fifty feet of solid concrete'**. Turn on the television and find any wildlife documentary and I bet it will show pictures of crocodiles on the Mara River tearing wildebeest apart, or white sharks devouring seals, or snakes striking at the camera lens, or some equally 'brutal' interaction. All the beauty in nature has been reduced to representations of butchery and horror because we humans have become so upset that all we can cope with are pictures of animals 'being' as aggressive as we are—everything else in nature is far too confronting. I have been to natural Africa and seen its spectacle, and the sheer magic of it surpasses all imaginings; it is just achingly beautiful, the most sacred realm on Earth—'sublime amnesia' are the only words I can think of to describe it and they don't even make sense. In 1992, Annie and I were fortunate enough to join a small reconnaissance party that was being sent in on foot into the northern end of the Tsavo East National Park in Kenya, an area that had been shut off from the public for many years due to the prevalence of dangerously-armed poachers from Somalia. I remember sitting hidden downwind amongst the trees on the banks of the Tiva sand river there and seeing dust rise above the tree line in the shimmering midday heat and then watching as a vast herd of black Cape buffalo, led by an old crooked horn cow, quietly materialized from the bush, cautiously coming down to drink at pools in the river bed. I really felt like a spy in heaven. It was all just unbelievable. Earth at its primal, spiritual, authentic, soulful, magical very best. I think God was there

beside us sitting on his heels like a little Bushman smiling at all that he had created. That visit to the Tiva river remains the highlight of my life. With our sophisticated communication technology, why oh why don't we have documentaries sensitively immersing us in all of that. It is *so* sad. We haven't been able to cope with any truth. Our world has shrunk to the size of a pea. All the beauty and magic that is out there escapes us, we don't see it; worse, *we don't want to see it*. No wonder our soul can't relate to us and just draws stick figures with weird blobs for faces.

© 1992 Fedmex Pty Ltd

Annie Williams and I in Samburu National Park in Kenya in 1992. Those giraffes behind us are just walking around as free as a daisy. In natural Africa, animals like giraffes and elephants and rhinoceroses (well, terrifyingly, there are actually almost no rhinos left!) aren't in cages; there are no fences over there. Animals—and the place is teeming with them, all sorts of weird shapes and sizes—just walk all around the place. It's amazing. They can go wherever they want. They can stop here for a while and then go over the hill if they want to. They just mooch about everywhere; walk around a bush and there is another one, this time with great spiral horns coming out of the top of its head, big eyes looking at you as if to say, 'So, who are you, what's your problem?' '*My* problem! Have you had a look at what's coming out of the top of your head!?' It takes some getting used to I can tell you. I don't know who made them all, and was he just having fun making them all in such weird and different shapes—and, more to the point, who let them all out!

[836]Yes, humans now are immensely alienated, extremely psychologically separated from our true self or soul. Similar to what happens when I draw, in my writing I have also learnt to, as I describe it, 'think like a stone', or 'think like a child' — say the simplest, most elementary thought — because I learnt that such a thought will be the most truthful and accurate and accountable and explanatory. Absolutely every time I encounter a problem I have to solve in my thinking about the human condition I go into a routine where I say to myself, 'Just go into yourself and think like a stone, just let the truth come out that's within and you will have the answer.' Basically, I learnt to trust in and take guidance from my truthful instinctive self or soul. I learnt to think honestly, free of alienated, intellectual bullshit, and all the answers, all the insights that I have found, and there are many hundreds of them, a breakthrough insight in almost every paragraph, were found this way. I have so perfected the art of thinking truthfully and thus effectively that you can put any problem or question in front of me to do with human behavior and I can get to the bottom of it, answer and solve it. It has been astonishing to me to watch my mind work, the freedom it has and where it is capable of going in its thinking. It wears me out keeping up with it. This wearing out problem is especially so because there is so much suffering in the world that simply has to be brought a stop to. Yes, I know that every sentence I write is truth-laden, in complete contrast to the billions of sentences being churned out every second everywhere else on Earth. It is the innocent instinctive child in us that knows the truth. Christ, as usual, put it perfectly when he said, **'you have hidden these things from the wise and learned, and revealed them to little children'** (Bible, Matt. 11:25). A comment that was mentioned in par. 240 by George Seaver reiterates what I have just said about natural thinking: **'The ultimate thought, the thought which holds the clue to the riddle of life's meaning and mystery, must be the simplest thought conceivable, the most natural, the most elemental, and therefore also the most profound.'** Yes, as Plato was recorded as saying in par. 679, when we use our intellect with its preoccupation with denial **'for any inquiry...it is drawn away by the body into the realm of the variable, and loses its way and becomes confused and dizzy, as though it were fuddled** [drunk]**...But when it investigates by itself** [free of intellectual denial/bullfuckingshit]**, it passes into the realm of the pure and everlasting.'**

[837] The human condition has certainly been a cruel incursion. It was bad enough to have acquired a fully conscious brain, the marvelous computer we have on our heads, and not be given the program for it and instead be left to wander this planet searching for that program/understanding in a terrifying darkness of confusion and bewilderment, most especially about our worthiness or otherwise as a species, but to *then* be disconnected from access to the integrative, Godly, cooperatively orientated, all-loving and all-sensitive, ideal world of our original instinctive self or soul—having to block it out because it unjustly condemned us—means we have been enduring an *extraordinarily* lonely, sad, soul-destroyed, alienated existence! We have had to put on a brave, positive face to carry on with our horribly upsetting job of searching for knowledge, ultimately for self-knowledge, but that is the true description of our existence during that search. It follows then that it became a matter of great urgency for the increasingly upset human race to find ways to cope with such a torturous and lonely existence.

[838] While upset was rapidly increasing through humanity's adventurous adolescence stage and beyond, we eventually became so horrifically alienated from our all-loving and all-sensitive true self or soul that a way simply had to be found to reconnect with it. We had to find a way back to some purity and sanity, but in a safely non-confrontational manner, and one of the ways we managed to do so was by creating one of the earliest forms of <u>religion</u>, which was <u>ANIMISM AND NATURE WORSHIP</u>—religion being the strategy of putting our faith in, deferring to, and looking for comfort, reassurance and guidance from something other than our overly upset and overly soul-estranged conscious thinking egoic self. Unlike our upset soul-destroyed self, the natural world remained in an innocent state, and since nature was also associated with our original instinctive self because our species grew up with nature, it could also reconnect us to the innocent, true world of our soul. So, despite our upset state's often violent repudiation of nature's condemning innocence, nature *could* still link us back to repressed 'spiritual', soul-infused sensitivities, feelings and awarenesses within us that we had lost access to. Again, we see the two-sided aspect of innocence: it could condemn us and hurt us terribly, but it could also heal and inspire us.

[839] Another way that eventually developed to counter the loneliness of our situation, and this was also one of the earliest forms of religion, was <u>ANCESTOR WORSHIP</u>. Having managed to survive our mind's loneliness and our soul's estrangement, our ancestors were a source of great reassurance and comfort. In our uncertainty and distress, we could look to them for the hope that we too might survive the horror of life under the duress of the human condition. We could look to them for 'spiritual' guidance, for inspiration for our troubled minds. If we tried to imagine how they coped and what they would have done in situations that we now faced, we could be inspired to reach potentials within ourselves that our troubled minds might not otherwise have allowed us access to. By revering them and enshrining their memories, our ancestors could remain a presence in our lives to look after and guide us.

[840] Of course in addition to the practical need for inspired guidance from our ancestors, we also wanted to perpetuate our love for them and theirs for us. Humans now are so toughened—so soul-destroyed—by the levels of anger, egocentricity and alienation in human life today that it is hard for us to imagine how loving and empathetic humans once were, but the truth is the emotional need to put flowers on the graves of those 'near to us' would have been *so* much stronger and purer in earlier times when humans were stronger and purer in soul. The emotion of love would have been so powerful that everyone would have been 'near to us' and remained 'near to us' after they died. A truly loving universe is such a different universe to the one we inhabit now. Basically people didn't 'die' in earlier times—they died physically of course, but their entire spirit *lived on* with us. Love and feeling and emotion and togetherness was everywhere and in everyone. It is only the extremely alienated disconnection from our all-sensitive and all-loving soul in humans of more recent times that has left us needing to believe in a physical '<u>afterlife</u>'—needing to construct pyramids as a vehicle to supposedly carry us on to another life after we died, or needing to hold onto a belief in reincarnation, etc, etc. But it is love, and our love of love, that is what is truly universal and eternal. So it wasn't so much ancestor worship that humans once practiced but ancestor love. For instance, Stonehenge in Britain (and other such

similar sacred sites) would have originally been a place to let all the love of, and from, our ancestors not only look after us but surround and embrace us, and we them in return. Likewise, the big old oak trees that the druids revered were similarly sacred beings that were 'near to us' when we were still open to the loving life that they were so full of. Love (which, as explained in chapter 4, is the unconditionally selfless theme of the integration of matter) was everywhere we walked, but as humans became more upset and alienated we began to need to create places like Stonehenge and sacred groves of oak trees to remind ourselves of that loving, connected state. Later, however, as upset and alienation became even more extreme, the deeper sensitivities of our soul became more and more inaccessible, so that nowadays there is no real spirituality left in life, only festivals of fake, imitated spirituality and sensitivity. This state of terminal alienation is one of the main subjects of the latter part of this book. (I should explain that having said that unconditionally selfless love is everywhere and that nature, such as oak trees, are full of loving life, we, of course, have always been aware that there is conflict in nature, that animals especially fight with and kill each other, but we considered that this occurs because living things sometimes, in effect, lose sight of love. In fact, we were aware that many animals struggle to be loving and can only manage it for periods, and so we forgave them for that—'there's my unfortunate friend Mr Crocodile, dressed in armor, anticipating, even provoking battle, and with a massive extended mouth full of ferocious teeth, lying there in the swamp ready to tear to pieces any creature that comes close'. Love *is* everywhere even though some creatures struggle to, in effect, appreciate it. [This truth that we once knew was explained in chapter 4 when the integrative limitation of the gene-based natural selection process that produced the competition and aggression we see in nature was described.] In more innocent times, we were magnanimous towards the sometimes divisive behavior that occurs in nature, such as in our animal friends, because we could feel and see the greater truth that love is universal; that it is the one fabulously wonderful, great force in the world. Again, this was before the upset state of the human condition became so developed that our shame killed off this awareness, at which point we invented all manner of false truths or 'gods', such as gods for war, and for sexual love, and for imperfections that we no longer had the generosity of spirit to cope with—such as gods for lack of rain, and for violent weather. And once these <u>multiple gods</u> were invented it took a long time, and some exceptionally sensitive, soulful thinkers, such as Abraham, to return us to the truth that there is only one God—love or unconditional selflessness.)

[841] Alongside the development of these spiritual supplements, humans were also becoming more <u>technologically</u> advanced. For instance, tools including sharpened stones, choppers, hand axes and scrapers, cudgels, spears, harpoons and bone needles appear in the archaeological record from 3.3 million years onwards, while there is evidence that *H. erectus* (who lived some 1.9 to 0.1 million years ago) made refined tear-drop shaped flint axe heads and that even the earliest of this variety of humans were using <u>fire</u> (as indicated by the remnants of hearths at Koobi Fora in Kenya). However, it is only within the final 14,000 of the 2 million years of humanity's adolescence that the *most* dramatic improvements occurred. It was during this period that the bow and arrow, fish basket traps and crude boats first appeared, while the practice of <u>agriculture and the domestication of animals</u>, which both began around 11,000 years ago, brought with it the production of earthenware pottery, looms, hoes, ploughs and reaping-hooks. (This acceleration of technology can be seen in the relatively swift succession between the great ages that define our modern history, with the Stone Age being replaced around 5,000 years ago by the Bronze Age, which in turn was replaced around 3,000 years ago by the Iron Age.)

[842] It needs to be emphasized that throughout these epochs of time the whole development of upset was being driven by increasing levels of intelligence, and vice versa; this is because the more intelligent we were, the more we searched for understanding, and the more we engaged in that corrupting, soul-destroying search, the more upset we became—and with each new level of upset, a new psychological and accompanying physical existence and state emerged, including increased alienation. The following two graphs chart the psychological journey that humanity has been on, with the top graph charting the development over time of mental cleverness, as indicated by brain volume, and the bottom graph charting the development of cooperativeness or integration.

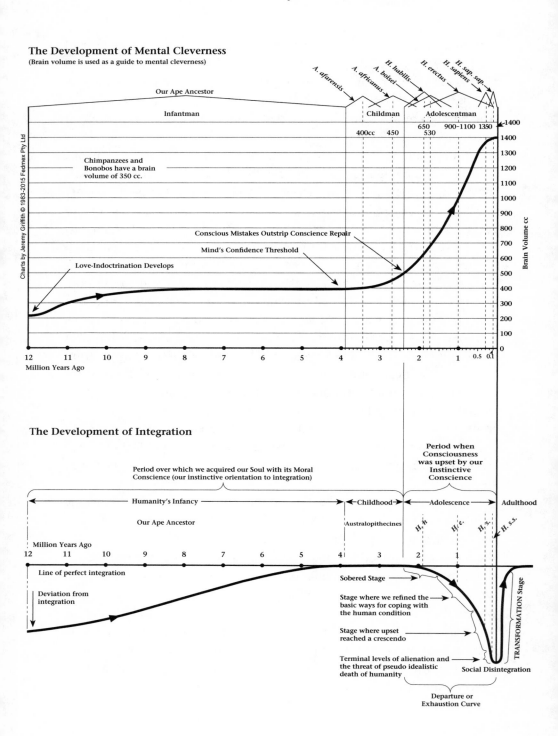

The Development of Mental Cleverness
(Brain volume is used as a guide to mental cleverness)

The Development of Integration

[843] We can see that while the brain size of 'Childman' (the australo-pithecines) was not much bigger than 'Infantman' (represented today by species such as chimpanzees and bonobos), a sudden increase in brain size occurred with the emergence of the first 'Adolescentman', *H. habilis*, when the need to think and understand began in earnest as a result of the emergence of the dilemma of the human condition. This dramatic growth continued through 'Adventurous Adolescentman' (*H. erectus*) and 'Angry Adolescentman' (*H. sapiens*) before finally plateauing with 'Pseudo Idealistic and Hollow Adolescentman' (*H. sapiens sapiens*). Anthropologists have long wondered why this growth stopped in the last 200,000 years or so. The reason is that in 'Pseudo Idealistic and Hollow Adolescentman' a balance was struck between the need for cleverness and the need for soundness; between knowledge-finding yet corrupting mental cleverness and conscience-obedient yet non-knowledge-finding lack of mental cleverness, with the average IQ today representing that relatively safe conscience-subordinate, not overly upset, not too selfish, egocentric and dysfunctional, compromise. Conscious mental cleverness is what caused us to challenge our instincts, the result of which was we became corrupted in soul, so it follows that the more mentally clever we humans became, the more corrupted in soul we also became, and that eventually we became too clever and too corrupted. Cleverness and alienation have been related. That has been the elementary truth about the human condition.

[844] The bottom graph, which indicates the development of coopera-tiveness or integration, shows that by 5 million years ago nurturing had enabled our ancestors to live in an utterly cooperative state. However, with conscious self-management, and with it the upsetting battle of the human condition, becoming fully developed some 2 million years ago, we see that the graph marks a rapid increase in upset from that time to the present, where we now face the prospect of terminal levels of alienation and social *dis*-integration.

## Chapter 8:12 <u>Angry Adolescentman</u>

**The species: *Homo sapiens* — 0.5 million (500,000) to 0.1 million (100,000) years ago**

**The individual now: 30 to 40 years old**

Extreme frustration and anger

[845] It was during the Angry Adulthood Stage (of Humanity's Adolescence) that we encountered the reality, frustration and anger of trying but failing to defeat the ignorance of our idealistic instinctive self or soul and had to impose upon ourselves <u>SELF DISCIPLINE</u> in order to contain or civilize our now overly upset state.

[846] Throughout our 20s we individually now, or, in the case of humanity, *Homo erectus*, settled into the long, corrupting journey to find understanding, ultimately understanding of why we became corrupted in the first place. But, tragically, the more we searched for knowledge, the more upset we became, and the more upset the human race as a whole became, and the more new generations had to contend with that ever-accumulating upset. It was an extremely upset-compounding situation. As the just included graph charting humanity's increase in upset over the last 2.4 million years shows, for the first three-quarters of the journey (to the end of 'Adventurous Adolescentman'/*H. erectus'* reign), the rate of increase in upset, while beginning to accelerate, was not yet extreme. However, in the last quarter of that time period (during the reign of 'Angry Adolescentman'/*H. sapiens*), the graph descended markedly, and then, in the final 200,000 years (during the reign of 'Pseudo Idealistic and Hollow Adolescentman'/*H. sapiens sapiens*),

it entered into free fall, with upset beginning to compound at an *extremely* rapid rate—a rate that can only be brought to an end by the rise of the human-condition-understood-and-ameliorated, transformed 'Adultman', or what could be termed 'Triumphantman', or even Integrative-Meaning-compliant 'Godman'.

[847] A contributing factor to the speeding up of this progression in upset was the hardship and confinement of life throughout the four great ice ages that occurred during the Pleistocene epoch, the period from 1.8 million years ago to 10,000 years ago. These ice ages greatly contributed to the increase in upset because, in forcing close habitation between people of varying degrees of upset, they dramatically accentuated the difficulties encountered by humans coexisting under the strain of the human condition, to the point where life today, towards the bottom of the graph, has become so difficult that even coupling has proved untenable for many, with marriage breakdown a common occurrence.

[848] The closer humans lived during humanity's adolescence and/or the more difficult the living conditions, the greater the occurrence and spread and thus increase in upset. Innocence doesn't last long in New York's Times Square or Sydney's Kings Cross where drug pushers, prostitutes, muggers and beggars work the streets. And as those from cold climates will attest, winters *are* particularly confining and testing, so each great ice age did, in effect, represent one *very* long, trying winter. It is not surprising then that out of the hardship of each of the great ice ages came the next more upset/soul-exhausted/embattled/alienated stage of humans. From the rigors of the first great ice age, called the Günz Ice Age, came the flowering of *H. erectus*. *H. sapiens* emerged after the second ice age, the Mindel Ice Age, while Neanderthal man, a precursor of *H. sapiens sapiens*, appeared after the third ice age, the Riss Ice Age. The Cro-Magnons, described as the first behaviorally modern humans, emerged after the Würm Ice Age, the fourth ice age. Each ice age also contributed significantly to the culling of the human race in terms of humans' ability to adapt to life under the duress of the human condition. This is because as upset increased throughout humanity's adolescence many individuals must have, in effect, quit the great battle humanity was waging against the ignorance of our instinctive self or soul through their inability to withstand the degree of compromise to their soul that was increasingly being demanded of them, leaving only the most courageous, determined and enduring—but, it follows, also the most soul-destroyed.

Sir Laurens van der Post once described how a member of the relatively innocent Bushmen 'race' (*relatively* because they are still members of the extremely upset stage of humanity, *H. sapiens sapiens*) found it impossible to cope with having his innocent, natural spirit compromised: **'You know I once saw a little Bushman imprisoned in one of our gaols because he killed a giant bustard which according to the police, was a crime, since the bird was royal game and protected. He was dying because he couldn't bear being shut up and having his freedom of movement stopped. When asked why he was ill he could only say that he missed seeing the sun set over the Kalahari. Physically the doctor couldn't find anything wrong with him but he died none the less!'** (*The Lost World of the Kalahari*, 1958, p.236 of 253). Sir Laurens was more specific when (as mentioned in par. 745) he stated that **'mere contact with twentieth-century life seemed lethal to the Bushman. He was essentially so innocent and natural a person that he had only to come near us for a sort of radioactive fall-out from our unnatural world to produce a fatal leukaemia in his spirit.'** Given how **'radioactive'** our present **'unnatural world'** is for the innocent souls of children, it's little wonder there is an epidemic of distressed, non-coping, dissociating symptoms like ADHD and autism breaking out amongst children now. Yes, the human race has suffered 2 million years of soul-destroying, toughening, upsetting, anger-egocentricity-and-alienation-producing adaption to the horror of life under the duress of the human condition! But just how toughened the human race has become is now hidden under layer upon layer of self-restraint, or, as stated in par. 276, what we call 'civility'. This restraining, civilizing process will be elaborated upon shortly, but the point being made here is that beneath our facade of restraint and manufactured positivity, which was so necessary to cope with the horror of the human condition, lies a highly genetically toughened and resilient individual.

[849] But to return to humanity's journey through its adolescence, by the age of 30 in the case of the individual now, or by some half a million years ago in the case of humanity, the exponential increase in upset meant that the levels of upset, namely anger, egocentricity and alienation, had exceeded the inflection point on the graph charting our development of (dis)integration/rise in upset, and had entered the stage where upset increased both dramatically and *rapidly*, which it has continued to do ever since. So while all the adjustments that were made during the Adventurous 20s had served us well—both as individuals today and, in the case of humanity, as members of *H. erectus*—there had emerged an

urgent need to take more specific measures to manage the new extreme levels of upset.

[850] If we consider what happened to the 21-year-old more closely we can see why management of upset had become such a serious matter, for despite their bravery and sheer optimism, it wasn't long before the reality of, in the case of the resigned, trying to win the battle of proving you were good and not bad—or, in the case of the unresigned, trying to reform upset behavior—started to sour. Gradually he or she came to experience and appreciate just how truly difficult it was to self-manage and contain upset without the ameliorating understanding of that upset.

[851] The problem for those who were resigned was the harder they fought to validate themselves, the more criticism they attracted from their idealistic soul, and thus the more upset they became. Also, throughout their 20s, they were increasingly encountering the upsetting difficulty of trying to survive and compete alongside *other* embattled humans who were *also* trying to prove their worth. The resulting compounding of upset meant that by the time they reached 30 they were becoming very frustrated and angry, and by the time they reached their mid-30s they were becoming seriously upset, embattled individuals. While resigned 20-year-olds were naive about the difficulties of living under the duress of the human condition, resigned 30-year-olds had become realists about such an existence. Rod Stewart's song *I Was Only Joking* contains lyrics that vividly describe the reality check of reaching 30: **'Me and the boys thought we had it sussed. Valentinos all of us...running free, Waging war with society...But nothing ever changed...What kind of fool was I. I could never win... Illusions of that grand first prize, are slowly wearing thin...I guess it had to end'** (lyrics by Gary Grainger & Rod Stewart, 1977).

[852] While our inability—until now—to defend the corrupted state of the human condition has meant that it hasn't been possible to admit it, the following Japanese proverb does, at least, acknowledge the stages of the development of upset in a resigned person: **'At 10 man is an animal, at 20 a lunatic, at 30 a failure, at 40 a fraud and at 50 a criminal.'** But with understanding of the human condition now found we can finally decipher these stages. Ten-year-olds were **'animals'** in the sense that they had yet to learn any methods of restraint for the upset that they were beginning to experience from the frustrations and agony of the human condition. Twenty-year-olds—and young men in particular—were **'lunatics'** in the

sense that they were swashbuckling cavaliers who deludedly believed they could take on and overthrow the ignorant world of the soul and prove they were good and not bad by winning power, fame, fortune and glory. Thirty-year-olds (and, again, men in particular) were **'failures'** in the sense that, although they were still determinedly trying to defy the inevitable, they were being forced to accept that the corrupting life of seeking power, fame, fortune and glory was not going to be a genuinely reinforcing, meaningful and satisfying way of living. As will be described shortly when the resigned 40-year-old stage is explained, when this stage was reached, men in particular were **'frauds'** in the sense that they had become so corrupted and disenchanted with their efforts to 'conquer the world' that they suffered a 'mid-life crisis'—a crisis of confidence that resulted in their decision to take up support of some form of 'idealism' to make themselves feel better about their corrupted condition. Having had enough of the critically important, yet horribly corrupting, battle to champion the ego over soul, they effectively 'changed sides' to become 'born-again' supporters of the soul's 'idealistic' world. This 'born-again' conversion made them **'frauds'** because in taking up support of some form of idealism they were deluding themselves that they were at last on the side of good when, in truth, they were working *against* good, in that the good, right and responsible path depended on defying and defeating—not supporting—the ignorant 'idealistic' world of the soul. They were being *pseudo* idealistic, not *genuinely* idealistic. And, finally, as will be described when the 50-year-old stage of a resigned person's life is explained, at this age men, in particular, were **'criminals'** in that they had become so disillusioned with the extreme dishonesty of the born-again state that many returned to the battle of championing the ego over ignorance, but were, by this stage of their journey, *so* deeply upset that they were extremely angry and cynical about life—basically, they knew they were beaten on every front and *had* become bitter and vengeful **'criminals'**.

[853]To return, however, to the 30-year-old stage or, in the case of humanity, the life of *H. sapiens*, we can see that this period was characterized by extreme frustration and anger. Thirty-year-olds/*H. sapiens* had entered the rapidly deteriorating stage in the development of upset where they were brought into contact with the destructive and depressing horror of their **'failure'**, of being either, in the case of the resigned, excessively upset, or, in the case of the unresigned, having had their innocence so destroyed

that they too had *also* become overly upset. At this stage in the human situation and journey, upset was becoming overwhelming *everywhere*.

[854]It follows then that it was at this point that the radical measures alluded to earlier had to be implemented to contain the growing upset in the world, with the first solution being, as mentioned, to practice self discipline.

[855]Fully aware that upset was not desirable we, as individuals, had been trying to, with varying success, practice self-restraint of our upset ever since it first appeared in our childhood. But during our 30s, when upset started to become seriously destructive of the fabric of our society, self discipline became an *essential* part of our behavior, something that everyone *had to* make sure they practiced. And so we learnt to manufacture a calm, controlled, even compassionate and considerate exterior, and to conceal the real extent of our, by now, inner savage fury from being so unjustly condemned by the Godly ideals of life. We *civilized* our upset; we brought it under control. Since this self discipline, and its civilizing effect, has been the primary way of managing our extremely upset state and has been practiced since the emergence of the human condition some 2 million years ago, it has become, to a large degree, an automatic, instinctive element of human behavior, so much so that we now hardly notice we are practicing it—the consequence of which is that we are barely aware of just how upset we all really are underneath our restrained exteriors. Our civility has disguised the volcanic upset that exists within us, both as individuals and as a species, from living for so long with the injustice of being condemned as evil, bad and worthless when we intuitively knew we weren't but couldn't explain why we weren't. As mentioned in par. 271, the writer Morris West offered a rare honest insight into the extent of the upset that exists in *all* humans today when, in his memoir, he confessed that **'The disease of evil** [now able to be understood as upset] **is pandemic; it spares no individual, no society, because all are predisposed to it...I know that, given the circumstances and the provocation, I could commit any crime in the calendar.'**

[856]And not only did civility conceal the extent of our anger and egocentricity, it also masked the degree to which we had departed from our original, upset-free, happy, innocent soulful selves. In order not to be overcome by the true negativity of life under the duress of the human condition we have had to, as it's said, 'put on a brave face', 'keep our chin up', 'stay positive', and 'keep up appearances'. And so we manufactured

smiles and politely greeted acquaintances with 'Good morning' and asked 'How are you?' and talked about totally non-confronting subjects, such as the weather and sporting results. But while such civility and positivity made living together possible, it was an extreme form of pretense—of being what we were not. But, in turn, although this falseness in adults *was* highly corrosive of any young unresigned innocent looking on who found it unbearably **'phony'** and **'fake'**, it was far less destructive than allowing our real upset to express itself and has, therefore, been a very necessary and effective tool. But, again, after millennia of use, our civilized facades now hide the extent to which we have blocked out the truth of our upset, corrupted, alienated condition—so when, for instance, we donned scary masks, items that have been used in the ceremonies of almost all cultures, we were, ironically, not disguising who we were but actually exorcising our *real* upset self; we were being honest about ourselves; we were admitting that '*This* is what I am *really* like, this is who I've *become*.'

Bhairav Mask, Nepal        Maori Koruru (gable mask), New Zealand,
c.1880, Peabody Essex Museum

[857] Yes, R.D. Laing certainly spoke the truth about just how corrupted our species has become when he said, **'The condition of alienation...is the condition of the normal man...between *us* and It [our true selves or soul] there is a veil which is more like fifty feet of solid concrete.'**

[858] At this point, it needs to be emphasized that adopting self discipline in our/humanity's 30s did *not* mean we/the human race had stopped our/its corrupting search for knowledge—we had just decided to try not to allow any expression or manifestation of the effects of that search, of our corruption, to show. However, when the 40-year-old stage is explained in more detail shortly we will see that when upset developed even further some individuals were, as just mentioned, forced to abandon, and even side against, the corrupting search for knowledge in a far more drastic attempt to stem their ever-increasing upset by becoming 'born again' to pseudo idealistically supporting some form of idealism—this being the **'fraud'** stage that featured in the aforementioned Japanese proverb.

[859] Again, the overall essential feature of the human journey since we first became conscious is that of the accumulation of knowledge at the expense of our innocent soul—the more we searched for knowledge, the more upset and soul-destroyed we became. Certainly we learnt to restrain our upset, civilize it, but underneath that disguise we were becoming more and more angry, egocentric and alienated. This means that while children, adolescents and young adults—and their early human equivalents—could at times behave very angrily and aggressively (such as children deliberately torturing insects by burning them, or when their disagreements turn into rowdy, physical altercations), older adults who had learnt to civilize and thus hide their upset were actually far more upset, angry, egocentric and alienated than those younger than themselves. Throughout history, however, older and thus more civilized people and 'races' (ethnic groups) have misused this *appearance* of being 'better behaved' to denigrate other more innocent people and 'races' by referring to them as 'savages'—or 'barbarians', as the Romans did when they spoke this way about the northern Scandinavian and Germanic tribes (such as the Goths, Vandals, Franks and Lombards). The adventurous soulful soundness, enthusiasm, vitality and energy of these supposedly backward, savage, primitive, 20-to-30-year-old equivalent northern barbarians was apparent in their ability to conquer the more technologically advanced but more soul-exhausted southern 'races' who lived around the Mediterranean—including the Romans—during the fourth to sixth centuries AD. Their victories, however, do not mean that all conquerors were more innocent than those they conquered; when humans become extremely upset—when they reach the 50-plus-year-old equivalent **'criminal'** stage—they could go on extremely angry rampages, murdering

everyone in sight. It is this variety of extreme upset that has characterized most of the warfare that has occurred in the last thousand years of human history. There is a very big difference between a high-spirited Viking-like adventurous mindset and the massively angry and massively egocentric Genghis Khan/Napoleon/Hitler/Mussolini-like vengeful, conquering and murdering psychopathic mindset. The true story of the ever-increasing levels of upset anger, egocentricity and alienation in humans has not been told, but with the upset state of the human condition now explained and defended, it can, and is, at last being revealed—as you will especially see when the last 200-year stage of humanity's journey from ignorance to enlightenment is presented (in ch. 8:16).

------

[860] It needs to be explained here how adventurous 20-to-30-year-old equivalent 'races' could be said to exist during the last 2,000 years when that period is described later as being part of the born-again Pseudo Idealistic 40-to-50-year-old **'fraud'** stage and the horrifically angry, punch-drunk, bitter and vengeful Hollow Adolescentman 50-plus-year-old **'criminal'** stage of humanity's maturation. The answer is that what is being referred to here by 'adventurous 20-to-30-year-old equivalents' indicates a further level of refinement within the already established stages of maturation. To elaborate, while the first T-model Ford car (a development that replaced the horse and buggy form of transport) had all the basic elements of a car in place, that didn't mean those elements could not become much more refined over time, as in the variety of cars we see today—and even *highly* refined as in the form of a Ferrari. In the same way, the relatively innocent hunter-forager Bushmen people who live in the Kalahari desert today, for example, have all the basic adjustments in place for managing extreme upset. They are, for instance, civilized, instinctively restrained from living out all their upsets; they don't generally attack when they feel frustrated and angry. They have a form of marriage to artificially contain sexual adventurousness. They clothe their genitals to dampen lust. The women love to wear adornments such as jewelry; they are adapted to being sex objects. The men love hunting animals; they find relief from attacking innocence. They employ fatigue-inducing dance to access their repressed soul. In short, they are members of the Pseudo Idealistic and Hollow Adolescentman stage that all humans living

today occupy. *But*, while they have these basic adjustments for managing extreme upset firmly in place, they are still a *relatively* innocent 'race' compared with other more human-condition-embattled-and-adapted 'races' living today—they could be described as a 15-year-old equivalent variety of Pseudo Idealistic and Hollow Adolescentman. In the same way, the northern 'barbarians' were members of the Pseudo Idealistic and Hollow Adolescentman stage but were still *relatively* innocent compared with other more human-condition-embattled-and-adapted people—they were 20-to-30-year-old equivalent varieties of Pseudo Idealistic and Hollow Adolescentman. We are talking about levels of refinement occurring *within* the main stages of refinement. It should be emphasized that while we couldn't explain and defend the upset state of the human condition we couldn't afford to differentiate individuals, 'races', genders, generations, countries, civilizations and cultures according to how upset they were because it would have left the more upset condemned as bad, unworthy and inferior. It would have led to unfair, destructive and dangerous racist, ageist and sexist prejudice and discrimination against the more upset—and so a dishonest attitude of not allowing any differentiation was maintained.

[861] The truth is, the only significant difference between humans—the acknowledgment of which makes it possible to truthfully explain and understand much of human behavior—is the difference in upset anger, egocentricity and alienation between individuals, 'races', genders, generations, countries, civilizations and cultures, but until the human condition was explained and upset was defended as a good, heroic state we couldn't admit and talk about that all-important, clarifying difference.

[862] Some appreciation of just how condemning innocent 'races' have been of those more upset can be gained through looking at the determination with which mechanistic science has sought to deny that so-called 'primitive' peoples such as the Bushmen, Australian Aborigines and the Yanomamö of South America are in any way more 'innocent' than the more alienated 'races'. For example, as discussed in pars 205-208, to support the idea of a history of **'universal and eternal'** warfare, E.O. Wilson portrays Bushmen and other 'primitive' societies as **'violent'** and **'aggressive'**, even saying **'Rousseau claimed** [that humanity] **was originally a race of noble savages in a peaceful state of nature, who were later corrupted...**[but what] **Rousseau invented** [was] **a stunningly inaccurate**

**form of anthropology'** (*Consilience*, 1998, p.37 of 374). In contrast to that view is of course all our mythology that attests to an innocent past, including Plato's reference to our distant ancestors having lived a **'blessed' 'life'** where **'neither was there any violence, or devouring of one another, or war or quarrel among them'**, as well as the honest research of Elizabeth Marshall Thomas, who wrote that in her and her mother's (Lorna Marshall's) accounts of the Bushmen in the 1950s **'we both emphasized the absence of violence and competition. Indeed, we were struck by it…The relatively few outbreaks of violence seemed isolated and were discussed over and over, since they caused such distress'** (*The Harmless People*, 1989. p.286 of 303). But, as explained above, the acknowledgment of the existence of relative innocence in one 'race' would unfairly condemn the more upset, less innocent 'races' as 'bad', unworthy and inferior, so such reports of the relative innocence of the Bushmen could not be tolerated. Marshall Thomas recounts how mechanistic science rapidly mobilized to neutralize the threat her reports represented: **'In the ten to twenty years after we started our work, many academics developed an enormous interest in the Bushmen. Many of them went to Botswana to visit groups of Kung Bushmen, and for a time in Botswana, the anthropologist/Bushman ratio seemed almost one to one. Yet although the investigators were numerous, the range of some of their investigations seemed narrowed to an emphasis on questions of violence and aggression'** (ibid. p.284). Yes, to escape the agony of the human condition, some excuse *had to be* found—*some* evidence of aggression had to be identified and then a case determinedly built around it, regardless of how transparently false the case really is! As has been repeatedly pointed out, when the need for denial is critical any excuse will do, and the art of denial is to then stick like glue to that excuse because doing so saves you from suicidal depression.

[863] One of the more prominent anthropologists whose **'investigations [into the Bushmen] seemed narrowed to an emphasis on questions of violence and aggression'** was Melvin Konner, who stressed **'findings that seemed to confirm what might be called the darker side of !Kung life'** (Melvin Konner & Marjorie Shostak, 'Ethnographic Romanticism and the Idea of Human Nature', *The Past and Future of !Kung Ethnography*, eds Megan Biesele with Robert Gordon & Richard Lee, 1986, p.73 of 423). When Konner wrote a clearly biased review of Marshall Thomas's 2007 book *The Old Way: A Story of the First People*, Marshall Thomas felt compelled to respond, writing that **'the moment I saw Konner as the reviewer,**

I knew we were back where we started. I measured the length of his review—
141 inches or 11¾ feet in all—and saw he was averaging four attacks per foot of
column. And in the barrage, I'd say only one criticism had substance. Even then
he distorted what I'd said' ('Response to Dim Beginnings', *The New York Review of Books*,
29 Mar. 2007). Another who focused on violence was anthropologist Richard
Lee, author of *The Dobe Ju/'hoansi*, who argued for a Bushman 'past that
was decidedly not "noble" and that was out of kilter with the harmless image [put
forward by Marshall Thomas]' (2013, p.125 of 294). Despite his own argument,
however, Lee recognized that 'the Ju/'hoansi [Bushmen] managed to live in
relative harmony with a few overt disruptions. How the Ju/'hoansi and people
like them could live as peacefully as they did has puzzled and mystified observers
for decades' (ibid. p.121). Yet another who 'attack[ed]' Marshall Thomas was
the primatologist Richard Wrangham (who, as described in par. 514, put
forward the Chimpanzee Violence Hypothesis), who accused Marshall
Thomas as having 'conjured' the idea of 'Peaceful primitives' (*Demonic Males*,
1997, p.76 of 350).

[864] And just as Marshall Thomas was attacked for her honest description of the peaceful nature of Bushmen society, so too was Sir Laurens
van der Post for his recognition of their relative innocence. A prime
example of this persecution occurred after Sir Laurens' death, when
the journalist J.D.F. Jones wrote a book that set out to denigrate Sir
Laurens as a charlatan, with one of the focal points of his attack being
Sir Laurens' depiction of the Bushmen; for instance, he accused Sir
Laurens of having 'a romantic and no doubt inaccurate portrait of this dying
social group' (*Storyteller*, 2001, p.230 of 505). Jones' deep allegiance to the world
of denial is also apparent in this emotionally charged comment he made
on the topic: 'the academic experts on the Kalahari [Bushmen] are absolutely
berserk with rage about the things he [Sir Laurens] said, because, if you read
*The Lost World of the Kalahari*, you must not believe that this is the truth about
the Bushmen; it's not' (*Late Night Live*, ABC Radio, 25 Feb. 2002). Unable to defend
our immensely corrupted human condition, to have it exposed naturally
made humans 'berserk with rage'. And just as Marshall Thomas pointed
out the 'narrow', superficial, mechanistic, confrontation-avoiding studies
anthropologists were making of the Bushmen, so too did Sir Laurens,
who wrote that 'It seemed a strange paradox that everywhere men and women
were busy digging up old ruins and buried cities in order to discover more about
ancient man, when all the time the ignored Bushman was living with this early

spirit still intact. I found men willing enough to come with me to measure his
head, or his behind, or his sexual organs, or his teeth. But when I pleaded with
the head of a university in my own country to send a qualified young man to live
with the Bushman for two or three years, to learn about him and his ancient way
he exclaimed, surprised: "But what would be the use of that?"' (*The Lost World of
the Kalahari*, 1958, p.67 of 253).

[865] Again, while there *is* violence in primitive peoples, the true explanation for the aggression apparent in their societies is that while they are
undoubtedly more innocent than the majority of humans in the world
today, they are, as mentioned above, still members of the extremely
upset stage of humanity, *H. sapiens sapiens*, and are, therefore, *nowhere
near* as innocent as humans were some 2 million years ago when the
battle of the human condition first emerged. Moreover, while basic
levels of restraint are instinctive in primitive hunter-forager people such
as the Bushmen, as will be explained in ch. 8:16E, they do not possess
the more sophisticated levels of self-discipline that more upset 'races'
adopted following the advent of agriculture and herding some 11,000
years ago, and which has subsequently become, to a degree, instinctive.
As a result, to draw upon data on homicide rates, as E.O. Wilson was
shown to do in par. 206, where he equated Bushmen homicide rates to
those present in the more upset-populated cities of Detroit and Houston,
and use that comparison to argue that more primitive peoples are not
more innocent than more alienated 'races', is to totally ignore the effect
increasing levels of restraint have on upset behavior. As any mother
will attest, a nine-year-old child is more innocent than an adult and yet,
as was described in ch. 8:7, during the 'naughty nines' phase they will
lash out at the world in a way that a more restrained or 'civilized' adult
would not.

[866] The effect of restraining violence was well demonstrated by the
successful 'Iroquois Confederacy' of the North American Indians. As
will be described shortly in par. 916, by the time Europeans arrived in
North America, a grand league of American Indian tribes had been
established to prevent, through adherence to certain restraining rules
that were enforceable through punishment, the endless rounds of payback warfare that had been occurring between and within the tribes.
The absurdity of evaluating a peoples' level of innocence through their
display of violence is apparent if we were to imagine anthropologists

measuring homicide levels the month before and after the Confederacy was established. While homicide rates would have dropped dramatically, the only difference or reason for that change would be that the levels of restraint had increased dramatically, not the degree of innocence, which would have, of course, remained unchanged. The fear of punishment was simply preventing each member from expressing or living out their upset.

[867] The whole story of the human journey during the last 2 million years that has been described in this chapter is really the story of the emergence of ever-increasing levels of upset, and the development of ever more sophisticated ways to restrain and contain each new level of upset. The recognition in all our mythologies and in the work of our most profound thinkers of a wonderful, all-loving, innocent past for the human race isn't some **'romantic'**, fanciful dream of some impossible, unrealistic, idyllic, utopian existence, nor is it, as was mentioned in pars 184-185, nostalgia for the security and maternal warmth of infancy, as mechanistic science has tried to dismiss it as— no, it is a *completely real* time in our species' distant past that recently discovered fossil evidence is now confirming, and that bonobos provide ample living evidence of. The *true* story of human life over the last 2 million years *is* that of the loss of innocence—our 'fall from grace', our departure from the 'Garden of Eden', the corruption of our soul, our ever-increasing levels of anger, egocentricity and alienation! Everyone does, in truth, know that under the duress of the human condition we each, and our species as a whole, started life in an innocent state and ended up in a variously psychologically upset, embattled, soul-corrupted state. Innocence is associated with youth, not old age. As I pointed out in par. 185, it is ridiculous to claim that 'advanced' 'races' are more innocent than 'primitive' 'races'.

[868] So yes, tribal warfare and outbreaks of individual violence have been occurring for a long time, but that doesn't mean that the relative innocence of hunter-forager tribes still living, like the Bushmen of South Africa, the Australian Aborigines and the Yanomamö of South America, don't reveal a great deal about how extremely upset the great majority of the human race has become, and how much civility, and other pseudo idealistic means of restraint, humans now rely upon to mask and contain that extreme upset. But again, the psychological agony of our

human condition has been so great that while we couldn't truthfully explain our condition all we had to protect ourselves from the vicious, unbearable self-confrontation of such relative innocence were equally vicious, retaliatory, denial-based lies. Yes, it is *only* now that we can explain and defend upset that we can admit and talk about different states of upset and by so doing finally make sense of human behavior. Again, much more will be explained about the differences in upset between 'races' of humans in ch. 8:16E.

---

[869] So civility has been a marvelous tool for helping to restrain upset and allow individuals of various states of self-corruption to co-exist; however, it obviously necessitated repressing or bottling-up our frustrations and angers, which produced another problem of how then to relieve that pent-up state. And so, in our inability to be honest about our internal upset, we had to learn to valve off and relieve ourselves in ways that weren't destructive of those around us, such as the mask ceremonies already mentioned, or the fatigue-inducing, soul-accessing dances of the Bushmen. The origin of our developed sense of humor, for example, has never been able to be properly explained, but once it is understood how false resigned humans became, the source of so much of our sophisticated sense of humor becomes very clear. For the most part, adults maintain a carefully constructed facade of denial, but every now and then a mistake is made, we 'slip-up', and the truth of our real situation is revealed, providing the basis for humor. Occasionally situations occurred where the extreme denial, self-deception, delusion, artificiality, alienation became apparent and transparent, and in those moments the truth of that immense falseness was exposed for what it really was—so farcical it *was* funny; in fact, a 'joke'. When someone tripped or fell over, for instance, or had their 'bluff' called, it *was* humorous because suddenly their carefully constructed, civilized image of togetherness disintegrated. We take our developed sense of humor for granted now as being a natural part of our make-up, but there was a time, prior to becoming alienated and false, when the only comic situations were those that children—and bonobos—find amusing, such as when a harmless trick is played on someone. The power of consciousness makes it possible to recognize and even create such silly and funny situations that delight

our emerging minds, but it is only when the absurdly dishonest situation of the resigned, alienated human-condition-afflicted state emerged that *extremely* ridiculous, fraudulent situations appeared in human life to laugh about and make fun of.

[870] Swearing has been another way of tearing down and breaking free from the extreme dishonesty of our condition. Indeed, it is a stark measure of just how dishonest humans have become that we don't even have an everyday word for all the evasions and dishonest denials and delusions we practice every minute of the day, *except* for the swear word 'bullshit', or 'BS' or 'bull' or 'crap'! To understand why 'fuck' is such a powerful swear word we only have to recall the truth of what sex really is. As explained, while sex at its noblest level was something that marvelously complemented the human journey and as such has truly been an act of love, it has, nevertheless, at base been about attacking innocence (which women represent) for innocence's unjust condemnation of humans' (especially men's) lack of innocence. 'Fuck' means destroy or ruin, and what is being destroyed or ruined or sullied or degraded or violated is innocence or purity. Sex has been such a preoccupation of humans and yet everyone lives in denial of the truth that it is, at base, an attack on innocence. This makes sex one of the biggest lies and thus jokes of all, which is why using the word 'fuck' is such a powerful attack on the world of lies, and thus such a powerful swear word.

[871] Returning to the main stages of maturation again: as emphasized, since civilizing our upset didn't stop its development—it could only ever conceal and help contain it—it was inevitable that, as the corrupting search for knowledge continued, levels of upset were only going to escalate until eventually, by our late 30s, we/*H. sapiens* were embroiled in a rage of hate and anger. Even though we were for the most part still containing and concealing our upset, the compounding effect of upset meant that underneath that civility we became immensely embattled, saturated with upset, and thus absolutely despairing about our situation. On reaching this state of extreme anger and destructiveness we began to hate even ourselves. Life had become both personally and socially unbearable, an untenable position that produced a crisis, the well-known 'mid-life crisis' of the early 40-year-old individual now, or, in the case of humanity, the emergence some 200,000 years ago of *H. sapiens sapiens*.

## Chapter 8:13  Pseudo Idealistic Adolescentman

**The sub-species:** *Homo sapiens sapiens* — 0.2 million (200,000) years ago to the present day

**The individual now: 40 to 50 years old**

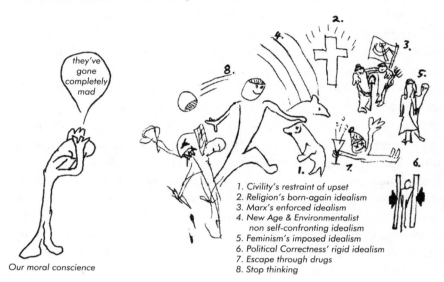

1. Civility's restraint of upset
2. Religion's born-again idealism
3. Marx's enforced idealism
4. New Age & Environmentalist non self-confronting idealism
5. Feminism's imposed idealism
6. Political Correctness' rigid idealism
7. Escape through drugs
8. Stop thinking

*Our moral conscience*

Drawing by Jeremy Griffith © 1991-2011 Fedmex Pty Ltd

[872] The Born-Again, Pseudo Idealistic Late Adulthood Stage (of Humanity's Adolescence) encompasses the time of the 'mid-life crisis' and, for those who had resigned, the adoption of Pseudo or False Idealism—the 'born-again in support of cooperative idealism' lifestyle.

[873] In this 40-year-old stage, upset compounded at such a rapid rate that the graph charting its intensity did go into free fall, rendering social disintegration an imminent risk. But with upset intensifying throughout humanity's adolescence, we always knew that if we didn't find the relieving understanding of the human condition in time then eventually the human race *would* enter a final end game stage where the levels of upset anger, egocentricity and alienation *would* threaten to destroy humanity. And this fear did play out: the 2-million-year race our species has been involved in between self-destruction and self-discovery *did* finally enter this crisis stage some 200,000 years ago and the variety of humans involved was *Homo sapiens sapiens*—us, the anatomically modern humans who emerged from *Homo sapiens* around that time. In the case of the individual growing up during humanity's human-condition-afflicted, insecure

adolescence, this was the time when, at about 40 years of age, we entered our so-called 'mid-life crisis'.

[874] At this point in our personal journey, our upset had become so great that, on one hand, we were hating the condemnation from the cooperative idealistic world of our soul with such fervor that we were beginning to become murderously behaved—while on the other, we were despising ourselves for being so upset and destructive of the world. Although we had, through the measures taken through our 30s, developed a very great deal of instinctive capacity to restrain and conceal—civilize—our upset, it was becoming so great that it all too readily broke out, revealing the extremely angry person that lay beneath. So despite all our efforts to 'conquer the world' if we were resigned, or 'fix the world' if we were unresigned, all we had to show for ourselves at this stage was an *immensely* embattled, overly upset, despairing individual. In the words of *Man of La Mancha*, we had finally **'march**[ed] **into hell...**[and become] **scorned and covered with scars'**—that was the price humans had to pay for pursuing the **'heavenly cause'** of trying to prove that our species is, in fact, fundamentally good and not bad. But what could we do to resolve such an untenable position? The 'mid-life crisis' long associated with turning 40 was certainly upon us.

[875] To answer this question of what we could do now that we had entered this overly upset 'mid-life crisis' stage, we need to look closely at the situation both of the unresigned and the resigned individual. As has been explained, the upset in humans that resulted from searching for knowledge didn't become so great that humans needed to resign until the reign of *H. sapiens sapiens*, which began 200,000 years ago, and even during this time it only became an almost universal phenomenon during the last 11,000 years when the advent of agriculture had the effect of greatly compounding upset. So during the vast majority of *H. sapiens sapiens'* reign humans weren't resigned—but they were now so upset they were bordering on having to resign when they were adolescents and fully engaging with the issue of the human condition. Every human born during the reign of *H. sapiens sapiens* who lived a long life grew up progressing through the stages of childhood innocence, early adolescent distress about the human condition, the adventurous 20s, the angry 30s, to becoming an extremely overly upset 40-to-50-year-old. And with all humans now maturing to this 40-to-50-year-old *immensely* embattled, overly upset, despairing 'mid-life crisis' state, the overall condition of

the human race during this time has been of that *immensely* embattled, overly upset, despairing 'mid-life crisis' condition. During humanity's childhood australopithecines stage, all members of the australopithecines only progressed psychologically to that childhood stage, so childhood was the mental state of adults then. Today the state of adult humans is the extremely upset 'mid-life crisis' 40-year-old-equivalent state, and, as we will see later, even the 'hollow' 50-year-old-equivalent state. To grow up to be an extremely upset individual and to encounter a world of extremely upset mid-life-crisis-afflicted, and even hollowness-afflicted adults, was the destiny of members of *H. sapiens sapiens*. Basically, there was so much upset in the human race by now that every human growing up was *almost having to*, or, in the case of virtually all those in the last 11,000 years, *having to*, become resigned to living in denial of the human condition when they were adolescents.

[876]To describe the situation of those who almost had to resign but didn't, they were encountering so much upset in the world and in themselves they were finding their ideal-world-and-behavior-expecting instinctive self or soul very difficult to live with, and, as a result, they too were encountering the need to block out of their mind that condemning soul part of themselves. Such resignation was mightily resisted because, firstly, it would mean becoming a soul-repressed and thus soul-dead person; secondly, it would mean becoming a person who uses all manner of denial to hide from the truth that they are an extremely upset, angry and psychologically distressed individual, and, as a result of all that denial, a 'fake', 'phony', deluded, dishonest, superficial, artificial, escapist person; and thirdly, to try to maintain their insecure self-esteem, it would mean becoming an extremely selfish, competitive and egocentric power-fame-fortune-and-glory-seeking person. While Resignation was therefore a state that was being mightily resisted, these individuals were nevertheless *almost* having to give in and resign to adopting that extremely compromised life. Significantly, however, since they *didn't* resign, they *didn't* lose access to the inspired, enthusiastic, alive, sensitive and vital world of our soul. So the essential characteristic of the nearly-resigned-but-still-unresigned, pre-11,000-years-ago *H. sapiens sapiens*, 40-year-old-equivalent-stage person—or a rare person during that 11,000 years who managed to avoid Resignation—is that throughout their life they were living in a denial-free state of honest awareness of the world's and their own extreme state of upset, *and* living in a state that still maintained access to the all-loving

and all-sensitive world of our soul. We saw how much soulful sensitivity
humans still retained in this pre-resigned state in the exquisitely empathetic drawings of animals that were done by presumably unresigned
humans in the Chauvet Cave some 30,000 years ago, and we can also
see something of their awareness of their own and their fellow humans'
extremely upset, corrupted condition in their apparent lack of desire to
draw themselves.

[877] So the life of someone living in the extremely upset but not yet
so upset that they had to resign state, was one that was characterized by
great sensitivity, but also great distress about the corrupted condition of
humans. The great prophets Moses and Christ were very rare examples of
unresigned individuals living during the last 11,000 years, so in their lives
we can see something of the unresigned state where there is still great
sensitivity as well as extremely despairing awareness of the corrupted
state of humanity. Being unresigned and thus still able to access all the
sensitivities of our soul, the full horror of the *extremely* upset world
of humans around them, and also in them, would have been very clear
to them; so much so they realized there was an absolutely paramount
need to do something about it—as Christ said about all those who were
resigned around him, **'I stood in the midst of the world and...found all men
drunken** [behaving in a mad way]**...And my soul grieves over the sons of men,
because they are blind in their heart, and see not** [they are alienated from the
truth of their horrifically corrupted condition]' (*Gospel of Thomas*, saying 28).
(This is the same soul-sensitive insight the great prophet Plato had
when he described resigned, **'cave'**-dwelling humans as being **'fuddled
[drunk]'** and unable **'to see clearly'**—see par. 679.) In the case of Moses,
this absolutely desperate need to do something about all the suffering
his sensitivity allowed him to see in the world led him to establish the
ground rules—his Ten Commandments—for living in an overly upset
world. And in the case of Christ, it led him to realize that he needed to
create a religion around his soundness for people to defer to and live
through. Earlier Bruce Chatwin was quoted as saying Christ was **'the
perfect instinctual specimen'**, meaning he was free of upset, as a member
of the completely innocent early Childman variety of humans would
be, but he wasn't. While he was exceptionally innocent and sound, he
nevertheless was a member of extremely upset and despairing, 'mid-life
crisis' *Homo sapiens sapiens*. Prophets were only *relatively* innocent. For
example, all the frustration and despair in modern humans was apparent

when, as young men, Moses **'killed the Egyptian'** he saw **'beating a Hebrew'** (Exod. 2:11-12), and Christ angrily **'overturned the tables of the money-changers'** in the temple (Matt. 21:12 & Mark 11:15). Moses' and Christ's relatively corrupted condition was also apparent when they found they had to fast for **'forty days and forty nights'** (Deut. 9:9 & Matt. 4:2) to break up the alienation in their minds and gain the deep access to their all-sensitive souls that they needed in order to think completely truthfully and thus effectively.

[878] We now need to look at what has been happening in the minds and lives of the desperately overly upset and distressed *H. sapiens sapiens*, the 40-year-old-equivalent resigned person—and since this resigned state is the almost universal condition of humans during the latter years of the reign of *H. sapiens sapiens*, it is this state that will be described from here on.

[879] The situation for these desperately overly upset and distressed humans was that they were unable to resist Resignation when they were adolescents. And once resigned, their lives were then characterized by the three features of life in resignation that were described above—of becoming soul-dead; and **'fake'**, **'phony'**, deluded, dishonest, superficial and artificial; and extremely selfish, competitive and egocentric seekers of power, fame, fortune and glory. Living in this extremely corrupted state meant that by the time they arrived at their immensely distressed 40-year-old 'mid-life crisis' stage, their crisis was severe. In fact, *so* severe it has parallels with the situation they encountered in their early adolescence when they were approaching Resignation and becoming overwhelmed with the extent of the corruption both in the world and within themselves. To elaborate, the most masterful writing on the mid-life crisis has to be that penned by the great Italian poet Dante in his epic poem, the *Divine Comedy*, which famously begins: **'Midway through our life's journey, I found myself in dark woods, the right road lost. To tell about those woods is hard—so tangled and rough and savage that thinking of it now, I feel the old fear stirring: death is hardly more bitter. And yet, to treat the good I found there as well I'll tell what I saw'** (c. 1308-1320; tr. Robert Pinsky, 1995). Yes, by the time the resigned adult approaches 40, the power, fame, fortune and glory trip has lost all its luster, leaving them **'in dark woods, the right road lost'**. And **'To tell about those woods is hard'** because the transparency of the **'fake'**, **'phony'** world they have been living in reconnects them to the issue of the human condition that they blocked out at Resignation. To see that their world and the world around them has become so corrupted forces them to question

why it is like that; honesty shatters the denial they have been using to block out the issue of the human condition. So they're now confronting the human condition *again*, which is an extremely **'hard'** position to be in because there *is* so much **'tangled and rough and savage'** truth to face. And so they *are* taken back to **'the old fear'** they faced as an adolescent during Resignation, an experience so terrible **'death is hardly more bitter'**—which is how Kierkegaard described confrontation with the human condition: **'that despair is the sickness unto death, this tormenting contradiction** [of our 'good and evil'-afflicted lives]' (see par. 119). So **'what'** Dante **'saw'** when he plunged down into the pit of terrible self-confrontation was an **'Inferno'** of horror, which was followed by terrible self-condemnation, or **'Purgatory'** as he titled the first two sections of the *Divine Comedy*. The incredible courage Dante demonstrated in confronting the human condition when he was obviously a resigned person and thus not sound enough to do so safely was eventually rewarded when he fought his way through all the doubts, until, with a great lift of spirit, he reached the ultimate realization of what life for humans would be like when they finally completed their heroic painful journey through ignorance and were reconciled with the ideals or God—namely **'Paradise'**, as he titled the third and final section of his poem; the time when the souls of all humans would be rehabilitated and everyone would be at one with everyone and everything. So Dante experienced mid-life crisis to the absolute full, a test of endurance few sufferers of the mid-life crisis would inflict upon themselves. No, as will now be described, rather than continuing to try to confront the human condition, the resigned mind in general found a very different way to again avoid the whole unbearable subject.

[880] In an article titled 'Turning 40 and Frantic, Mid life crisis', the journalist Ali Gripper acknowledged this parallel between the mid-life crisis state of mind and the crisis experienced at Resignation, writing that **'Mid life is undoubtedly a recycling of adolescent issues. It is as if the psyche goes back and picks up the threads of what we were dealing with as teenagers'** (*The Sydney Morning Herald*, 29 Mar. 1996). Yes, as with the situation that occurred in their early adolescence when they were faced with extreme states of despair and depression about their circumstances, at 40 resigned individuals also encountered variously extreme states of desperation about their situation. And, most significantly, like what happened when they were adolescents struggling with extreme despair and depression, the resigned 40-year-old's mind also searched frantically for a way to solve the problem of their

now untenable situation. And just as their unbearably upset and psych-ologically desperate adolescent selves came up with a desperate solution to completely put aside the reality of their circumstances by resigning themself to living in denial of cooperative ideality and thus the depressing issue of the human condition, so too did the, by now, extremely corrupted, resigned 40-year-old—BUT *this* time the evasion was achieved through focusing on the positive, guilt-relieving effect or feeling that came from being civilized. The angry 30-year-old had learnt to restrain/civilize their upset, but what the desperate resigned 40-year-old realized in their frantic search to find a solution to their problem was that being civilized or 'well-behaved' or 'good' produced a guilt-relieving positive feeling and that this was the one positive in their life that they could derive some reinforcement from.

[881] When humans are psychologically cornered they typically 'scan the horizon' for any positive, no matter how small, and make a huge deal of it, and this situation was no different. Frantically scanning for any positive that could be employed to escape condemnation and depression, it was the *side effect* of feeling good when we behaved in a civilized way that the resigned 40-year-old latched onto to develop. Indeed, in the case of the extremely upset, resigned 40-year-old, *so* desperate were they for relief from the horror and guilt of their situation that their mind decided to focus *so* completely on the positive that they were good when they behaved in a cooperative, civilized, ideal, loving way that they deluded themselves that they weren't actually corrupted, that they weren't massively upset human-condition-afflicted people. They convinced themselves that the mask or facade of civility was not actually a mask or facade at all, but a true representation of their real self: 'I am behaving in a cooperative, loving way, therefore, I *am* an upset-free, guilt-free, human-condition-eliminated, thoroughly good, cooperative, loving, sound human.' It was an *extraordinarily* false/dishonest/**'phony'**/ **'fake'**/deluded interpretation, but the depression from feeling guilty/bad/ worthless about being *so* upset was *so* great that their mind was well and truly capable of making, accepting and living with such a grand delusion. The situation was similar to the resigned person being so overwhelmed by the depression caused by their predicament that they were capable of making the extremely false interpretation that instead of having integra-tive, Godly, unconditionally selfless, moral instincts, we actually live in a non-integrative world of random, directionless change, and have

selfish, competitive, survival-of-the-fittest animal instincts that make us competitive and aggressive, and as such there is no psychological dilemma of the human condition to have to contend with, let alone have to explain or ameliorate.

[882] This resigned 40-year-old, 'do good in order to delude yourself that you actually *are* good, that you actually *are* free of corruption and thus the dilemma of the human condition' strategy was an extremely deluded way of coping with the problem of the now massively corrupted human-condition-afflicted state, but it has, nevertheless, been so seductive that it developed into an industry so huge and so influential that the dishonesty involved now threatens to destroy humanity.

[883] To elaborate, while being civilized—that is, using self discipline to restrain and contain your upset so it didn't show—*did* help contain destructive behavior and provide its practitioners with immense relief from doing so, what happened during the resigned 40-year-old stage was that this relieving, 'feel good', 'warm inner glow', 'blissed out' positive of having restrained your upset and behaved in a 'good'/ideal/cooperative way became the *entire* focus of existence. In the end, as we will see, when humans became extremely upset—saturated with the problem of the corrupted state of the human condition—their whole mental pre-occupation became one of searching for situations and opportunities where, through doing 'good', they could derive 'the rush' of relief from the condemning issue and truth of their corrupted state. So again, while the 30-year-old used civilizing self discipline to restrain and conceal their upset, they—unlike the resigned 40-year-old—*weren't* using it to delude themselves that they were an ideally behaved, upset-free, guilt-free, human-condition-eliminated, sound person.

[884] The immense danger of this preoccupation with relief-hunting through 'doing good' was that it could become so consuming, so addictive and thus so selfishly indulged that it could stop the all-important search for knowledge, because if there was too much preoccupation with 'doing good' it could result in insufficient tolerance of the corruption that un-avoidably resulted from pursuing humanity's heroic search for knowledge. If there was too much emphasis on cooperative idealism humanity would be denied the freedom necessary to find the liberating understanding of the human condition, and if it didn't find that liberating self-knowledge humanity would be condemned to the eventual emergence of terminal levels of upset—in particular, the unbearable levels of the psychosis and

neurosis of alienation that result from having to adopt excessive amounts of psychological denial and delusion. In short, the *dogma* of doing good could oppress and even stop the all-important search for knowledge by denying the *freedom* to be, to a degree, corrupted. As we will see, this extremely dangerous situation did arise; humanity *did* face a death by dogma, a fate only the finding of the liberating understanding of the human condition—that science has made possible and which is being presented here—can save humanity from.

[885]The danger of excessive oppression of freedom has been particularly great because of the massively seductive nature of relief-hunting. If we return to the Adam Stork analogy for describing the human condition, at any time Adam could surrender to his criticizing instinctive self and 'fly back on course', obey his instinctive orientation, and by so doing stop and thus relieve the criticism emanating from his instinctive self, but that meant abandoning the all-important search for knowledge. And, of course, in the case of humans, when we 'flew off course' and became angry, egocentric and alienated the sense of guilt we accrued from defying our cooperatively orientated, all-loving, Godly, moral instincts was immense, so 'flying back on course' was an *extremely* guilt-relieving, and thus an *extremely* tempting, option.

[886]There was, however, a further, very significant dimension to the problem of 'flying back on course'—being, as we revealingly say, 'born again' to supporting instead of resisting the cooperative ideals our instinctive self dogmatically demanded—which was that since our instinctive self was orientated to behaving cooperatively, when we abandoned the search for knowledge by taking up support of cooperative idealism we were not only abandoning the battle to champion our ego or conscious thinking self over our idealistic instinctive self or soul, *we were also* siding against those fighting the battle. We weren't just 'taking a rest' to recuperate, we had actually switched camps/allegiances to side with the 'enemy'. It was completely subversive, mad behavior—an act of cowardice and treachery—because in switching sides the individual was basically saying, 'I don't care about humanity anymore. I only care about making myself feel good and relieving my own guilt.' They were being totally selfish, the complete opposite of the selfless and ideal and cooperative person they were deluding themselves to be. And since the lie they were maintaining was so great, they had to work very hard at convincing both themselves and others of it, which meant they were

typically a strident, extremely intolerant, belligerent even fanatical advocate of their position.

[887] Yes, in choosing to be 'born again' you had to work *very* hard at maintaining the conviction that what you were doing was right because, although we haven't been able to explain, confront or talk about it, the truth is all humans who have lived during humanity's adolescence have intuitively been aware of the battle of having to overthrow the ignorant idealism of our soul; when we shook our fist at the heavens we were saying, 'One day, one day we are going to prove that we humans are good and not bad.' We *knew* that to give up the battle against our idealistic soul, and not just give it up but side with the 'enemy' and against those trying to win the battle, was a crime against all those still fighting for understanding, and against humanity as a whole. So despite how tempting it was, siding against humanity and those fighting the battle was, in reality, such a repulsive course of action that it took a great deal of despair and fear of depression about being overly corrupt to actually do so. The delusion, dishonesty and betrayal involved *was* extreme, but for ever-increasing numbers of people the need for relief from feeling loathful/guilty/bad about themselves became so great it could not be resisted.

[888] Again, the great danger was that, since upset was the price of searching for knowledge, if everyone became addicted to selfishly indulging their need for relief through 'doing good to feel good' there would be no tolerance of non-ideal upset behavior, and humanity's all-important search for knowledge, ultimately liberating self-knowledge, would be shut down, condemning humanity to terminal levels of alienation and thus extinction. So while Resignation to living in denial of the issue of the human condition and taking up a competitive, egocentric, selfish power-fame-fortune-and-glory-seeking lifestyle was, in itself, extremely desperate and mad behavior, it did at least involve participating in humanity's great battle to overthrow ignorance.

[889] Certainly, the upset behavior that resulted from participating in humanity's heroic search for knowledge was increasingly causing immense human suffering and environmental devastation, but if we didn't continue the search for knowledge then there was simply no hope. To put it in political terms, although the harsh, brutal reality associated with what became known as 'right-wing' politics *was* bringing about immense human inequality, hardship and suffering, and *was* destroying the planet, it was the search-for-knowledge-oppressing so-called 'left-wing' politics

that posed the real threat to the survival of the human race—because *only* the search for knowledge could lead to the finding of understanding of the human condition and the liberation of humans from that totally unbearable, crippling, soul-sickening, black-dog-depressing, real-person-extinguishing, deadening, human-life-denying condition! While, as the journalist Geoffrey Wheatcroft recognized, **'the great twin political problems of the age are the brutality of the right, and the dishonesty of the left'** (*The Australian Financial Review*, 29 Jan. 1999), and, as the scientist-philosopher Carl von Weizsäcker also recognized, **'The sin of modern capitalism is cynicism (about human nature), and the sin of socialism is lying'** (included in a speech by Prof. Charles Birch that was reproduced in the Geelong Grammar School mag. *The Corian*, Sep-Oct. 1980), it was NOT the **'cynicism'** and **'the brutality of the right'**, BUT THE **'lying'**, **'dishonesty of the left'** that stood like a colossal ogre over the human race, threatening to destroy it!

[890] To return to the Adam Stork analogy once more, Adam knew from the outset that he had to continue with the upsetting search for knowledge, that he could never afford to stop until the liberating understanding of his corrupted condition was found. That was his fundamental reality, *and it has remained our fundamental reality*. Again, the Statue of Liberty is as good a symbol as any of the fundamental responsibility humans have had to maintain *freedom*, which we can now understand means freedom from the cooperative ideals in order to continue the upsetting search for the knowledge that would allow humanity's *ultimate* liberation. Yes, paradoxically, *real* idealism, the real path to an ideal world, depended on continuing the corrupting search for knowledge until we found the human-race-liberating understanding of the human condition. The strategy of hunting for guilt-relieving, feel-good causes was, in fact, pseudo or false idealism, because it meant abandoning, and, worse still, oppressing, and—even worse still—actively opposing that all-important search for knowledge.

[891] In summary, while humans *have* had to counter the effects of the extreme upset that now plagues human life, and thus our world, with a degree of idealistic, concern-for-others-and-concern-for-the-world behavior—and, in truth, becoming civilized did involve a degree of abandonment of the upsetting battle in favor of being idealistic and showing concern for others and the world—what has happened is that the feel good aspect of behaving 'ideally' has evolved into what is now an *extremely* dangerous industry.

[892] Yes, given it has been *so* hard to explain and argue why not being ideally behaved is good, and *so* easy to argue that being ideally behaved can't be anything but good, 'the left' has had a field day mocking 'the right' as selfish, immoral and evil. As such, this quote by Nietzsche, which was referred to in par. 302, stands out as a brave and rare pronouncement on the need to hold our nerve and continue our species' great heroic battle to champion the ego over the ignorance of our instincts: **'There have always been many sickly people among those who invent fables and long for God** [ideality]**: they have a raging hate for the enlightened man and for that youngest of virtues which is called honesty...Purer and more honest of speech is the healthy body, perfect and square-built: and it speaks of the meaning of the earth** [to face truth and one day find understanding of the human condition]**...You are not yet free, you still *search* for freedom. Your search has fatigued you...But, by my love and hope I entreat you: do not reject the hero in your soul! Keep holy your highest hope!...War** [against the oppression of dogma] **and courage have done more great things than charity. Not your pity but your bravery has saved the unfortunate up to now...What warrior wants to be spared? I do not spare you, I love you from the very heart, my brothers in war!'** The author and journalist George Orwell was another who bravely recognized the very real danger of humanity losing its nerve when he famously predicted that **'If you want a picture of the future, imagine a boot stamping on a human face** [freedom]**—for ever'** (*Nineteen Eighty-Four*, 1949, p.267 of 328). And, in fact, as will soon be documented, this end play, death-by-dogma fate for the human race has all but descended upon humanity and can only be prevented by the eleventh-hour arrival of this understanding of the human condition.

[893] As pointed out, all humans have been intuitively aware that when they took up the born-again, pseudo idealistic way of coping with the human condition they were siding against humanity, siding with the enemy, and that doing so was a loathsome act of cowardice and treachery. So while the desperately upset, mid-life crisis of the resigned 40-year-old stage made taking up the born-again, pseudo idealistic way of coping a tempting option that was worth trying, the revulsion of living so treacherously caused many to change sides yet again and return to the upsetting battle of searching for knowledge. However, in returning to the battle they could only expect to become even more upset, angry, egocentric and alienated, which introduces the final stage that living under the duress of the human condition has resulted in: the extremely tragic 'Hollow Adolescentman' stage.

## Chapter 8:14  <u>Hollow Adolescentman</u>

**The final years of the sub-species: *Homo sapiens sapiens'* 0.2 million (200,000) year reign (Pseudo Idealistic and Hollowman stages are both characteristic of *Homo sapiens sapiens'* reign)**
**The individual now: 50 plus years old**

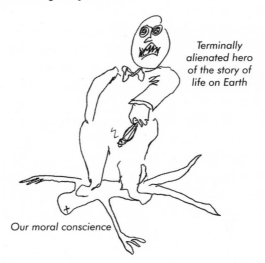

Terminally
alienated hero
of the story of
life on Earth

Our moral conscience

Drawing by Jeremy Griffith © 2011 Fedmex Pty Ltd

[894]The 'Hollow Final Adulthood Stage' (of Humanity's Adolescence) represents the time when many resigned post-40-year-olds become disillusioned with the treacherous, weak and cowardly, completely selfish, extremely dishonest and deluded 'do good to feel good' born-again existence and return to the upsetting battle to champion the ego over the condemning instincts. But as a result of this return to participating in the upsetting battle, they become even *more* upset, embattled and frustrated than they were when they were driven to adopt the pseudo idealistic way of living. In the context of the aforementioned extraordinarily honest Japanese proverb that described the stages of maturation under the duress of the human condition—**'At 10 man is an animal, at 20 a lunatic, at 30 a failure, at 40 a fraud and at 50 a criminal'**—this is the 50-year-old **'criminal'** stage where men in particular become so soul destroyed, so horrifically angry, punch-drunk, ego-unsatisfied, bitter and vengeful that they brutally and completely repress the condemning voice of their ideal-behavior-demanding, cooperatively orientated soul, leaving themselves adrift in an empty, hollow, soul-less wilderness.

[895] T.S. Eliot perfectly described this 'grumpy old man', vengeful, burnt-out, empty, sad existence that men typically inhabit when they reach 50 and beyond in his 1925 poem *The Hollow Men*: **'We are the hollow men / We are the stuffed men / Leaning together / Headpiece filled with straw. Alas! / Our dried voices, when / We whisper together / Are quiet and meaningless / As wind in dry grass / Or rats' feet over broken glass / In our dry cellar // Shape without form, shade without colour / Paralysed force, gesture without motion //...This is the dead land / This is cactus land / Here the stone images / Are raised, here they receive / The supplication of a dead man's hand / Under the twinkle of a fading star // Is it like this / In death's other kingdom / Waking alone / At the hour when we are / Trembling with tenderness / Lips that would kiss / Form prayers to broken stone // The eyes are not here / There are no eyes here / In this valley of dying stars / In this hollow valley / This broken jaw of our lost kingdoms // In this last of meeting places / We grope together / And avoid speech / Gathered on this beach of the tumid river //...Between the desire / And the spasm / Between the potency / And the existence / Between the essence / And the descent / Falls the Shadow /...This is the way the world ends / Not with a bang but a whimper.'**

[896] Before commenting on Eliot's extraordinarily honest poem, it needs to be explained that no amount of power, fame, fortune or glory could truly satisfy men's egos, because, being artificial forms of reinforcement, such 'success' was never going to genuinely make men feel they were good and not bad—it was never going to truly relieve them of the agony of their insecure, human-condition-afflicted existence—*only* understanding of their fundamental goodness could and now does achieve that. It can be appreciated then that the problem for older men was that there came a cross-over point in their lives when, no matter how much power, fame, fortune or glory they had achieved, they still hadn't satisfied their egos, but life had moved on—another generation of men trying desperately to validate themselves through the artificial, material forms of success of power, fame, fortune and glory had taken over—and they suddenly found themselves ignored. They were silently screaming for attention but no one was listening; no one was interested anymore. I was once told by the son of a retired politician how his father had bought numerous copies of a newspaper that mentioned him and strategically placed them around the house, open at the relevant page. Throughout their life men fought harder and harder to be a 'success' and relieve themselves of the insecurity of the human condition, and this

ever-increasing preoccupation meant they lost more and more access to the beautiful all-loving and all-sensitive true world of their soul. So in the end men became just monstrously ego-unsatisfied, ego-embattled, ego-infuriated volcanoes that couldn't afford to explode for fear of all the destruction it would cause. They were soul-less, empty, **'hollow'**, **'dead'**, **'broken'**, **'lost'**, **'stuffed'**, **'dry'**, **'cactus'** men with **'paralysed'** egos. Each was **'alone'** having to **'avoid speech'** for fear of either betraying their own immense need to be glorified or offending another older man's immense need to be glorified. This is how it always **'end**[ed]**'** for men under the duress of the human condition, **'Not with a bang but a whimper.'** Truly, what phenomenal heroes, what **'stars'** men have been—they **'march**[ed] **into hell for a heavenly cause'**, sacrificed themselves, contributed to the search for understanding that might one day, in the far future, but not in their lifetime, liberate humanity from the horrific insecurity of the human condition. Thankfully their corrupting search for dignifying understanding of humans has been completed, it now only has to be recognized and then absorbed.

[897] Of course, ageing during humanity's adolescence has been, in its own way, similarly horrific for women because it meant the inevitable loss of the image of innocence that women depended on for reinforcement, the loss of their sex-object 'attractiveness', and with it, the loss of their meaning in the world—a source of meaningfulness that all women's magazines that focus entirely on how to be 'attractive' are testament to. When women are young their beauty is generally so empowering it is as if they own the world—as ABBA's 1976 song *Dancing Queen*, which *Rolling Stone* magazine has rated as one of the greatest songs of all time, says, **'you are the Dancing Queen, young and sweet, only seventeen…having the time of your life…you turn them on, leave them burning, and then you're gone'**, and as Chuck Berry wrote and sang in 1958, **'They're really rockin' in Boston, in Pittsburgh, P.A., deep in the heart of Texas, and 'round the Frisco Bay, all over St. Louis, and down in New Orleans, all the cats wanna dance with Sweet Little Sixteen. Sweet Little Sixteen, she's just got to have about half a million framed autographs'** (*Sweet Little Sixteen*). But when women become older and their beauty/'attractiveness'/innocence fades they discover that they have become invisible; when they walk down the street they are no longer noticed. This quote from the French beauty therapist Diane

Delaheve describes how devastating it can be for women to lose their sex appeal: **'Her eyes, the mirror of her soul, speak nothing but despair. Her face may have kept its beauty, but it has become a picture of affliction. For some women, the prospect of age is sheer tragedy, worse than death, which might be seen as an escape'** (*The Sydney Morning Herald*, 4 Sep. 1988).

[898] An added dimension to the situation faced by older women is that in not being as responsible for the main battle of having to champion the ego over ignorance as men are, women find that their role of living in support of the battle is limited. It has been observed that a woman's life progressed from **'bimbo, breeder, babysitter to burden'**. Men, on the other hand, are directly participating in the battle of championing the ego and aren't faced, as such, with the prospect of one day feeling they are a **'burden'** to the extent that women are. In his 1993 book, *The Fisher King & The Handless Maiden*, the Jungian analyst Robert A. Johnson relates the myth of the Handless Maiden, which tells of a miller who makes a deal with the devil in return for the ability to complete more work with less effort. The devil demands the miller's daughter as payment: **'The miller is desolate but unwilling to give up his much expanded mill, so he gives his daughter to the devil. The devil chops off her hands and carries them away'** (p.59 of 103). Waited on by her newly prosperous family, the handless maiden is content for a time, until her growing sense of desperation sends her out to the forest alone. Johnson explains that the cry of women, like that of the handless maiden, is, **'What can I *do*? I feel so useless or second-rate and inferior in this world that puts its women on the rubbish heap when they are through with courtship and childbearing!'** (p.56). Indeed, **'one woman set up a blog called The Plankton. She says women over 45 are made to feel like the lowest life form – "flimflam, a nuisance, an embarrassment of landfill"'** (Angela Mollard, 'The invisible years', Sydney's *Sunday Telegraph Magazine*, 14 Aug. 2011).

[899] Yes, being cast **'on the rubbish heap' 'of landfill'** because you are not mainframed, part of the main battle that men have been waging against the ignorance of our instinctive self, and because you no longer have the incredible attractiveness of youthful innocence, has been a truly horrible situation for women. I once heard the actor Robert Duvall say on radio **'what man would be with an older woman if he didn't have to be'**, and it really is true, as high divorce rates and the existence of prostitutes, mistresses,

'first wives clubs', 'trophy wives', polygamy and, in earlier times, harems, all bear witness to. While almost every song written is dedicated to the attractiveness of youthful innocence in women, older women become invisible—they go from a situation where, in their youth, they own the world, as the songs *Dancing Queen* and *Sweet Little Sixteen* so powerfully evidence, to a situation where they are treated as being virtually irrelevant. Thank goodness then with the battle to overthrow ignorance and find understanding of the human condition finally won, this, in truth, absolutely horrific existence can come to an end and women, the young, the aged—*everyone*—can now be fully involved in the human journey! Again, Olive Schreiner provided an excellent description of the dream women have had of this **'new time'** that has now finally arrived: **'if I might but be one of those born in the future; then, perhaps, to be born a woman will not be to be born branded…It is for love's sake yet more than for any other that we [women] look for that new time…Then when that time comes…when love is no more bought or sold, when it is not a means of making bread, when each woman's life is filled with earnest, independent labour, then love will come to her, a strange sudden sweetness breaking in upon her earnest work; not sought for, but found'** (*The Story of an African Farm*, 1883, pp.188, 195 of 300). It is truly amazing what women have endured, as Sir Laurens van der Post has written of their wonderful, *immensely* heroic patience and tolerance: **'She is content, confident and unresentful because she is also the love that endureth and beareth all things even beyond faith and hope. She knows that, in the end, the child will grow and all shall be well'** (*Jung and the Story of Our Time*, 1976, p.158 of 275).

[900] So, men become 'hollow', and women become 'invisible'; when observing older couples walking together in the park you can see how united they are by their respective afflictions. Indeed, the following drawing by cartoonist Ralph Steadman depicts the full horror of the human condition, for both men and women; in fact, the main dragon in this cartoon provided the inspiration for my drawing that appears at the beginning of this 50-year-old-plus stage, for his eyes show the hollowness that T.S. Eliot spoke of: **'This is the dead land / This is cactus land.'** The desperately tragic, sickly state of older women is also explicit in the two crow-like creatures leaning against the bar in the background.

[901] Humans *truly* have been the great heroes of the story of life on Earth!

Ralph Steadman's *The Lizard Lounge*, 1971

## Chapter 8:15  The last 11,000 years and the rise of Imposed Discipline, Religion and other forms of Pseudo Idealism

[902] As explained in par. 750, Resignation to a life of finding escapist relief from the agony of the human condition through winning power, fame, fortune and glory became an almost universal phenomenon amongst adult humans around 11,000 years ago following the advent of agriculture and the domestication of animals. What *now* needs to be explained is how the domestication of plants and animals accelerated the development of upset and the adoption of the resigned position—and what transpired when this extremely competitive and selfish, and now universal, resigned way of living became unbearably destructive.

[903] Clearly, when the level of upset in the human race as a whole reached this final 'Hollowman' stage, humanity entered end play or end game in the race between self-discovery and self-destruction. The levels of upset within humans had become stupendous—but what could we do? We were stranded between two increasingly flawed options: adopting, or

in some cases returning to, the extremely irresponsible treacherous and fraudulent and thus dangerous born-again, pseudo idealistic, 'do good to feel good' way of living; or persevering with the ever more brutal, destructive and corrupting power-fame-fortune-and-glory-seeking, egocentric, knowledge-finding, resigned competitive existence. In fact, our lives, both individually and collectively as societies, for all of the last 11,000 years have been marked by the oscillation between these two extremely flawed strategies of coping with the human condition, strategies that were eventually refined into what we now know as the 'left-wing' and 'right-wing' in <u>politics</u>.

[904] <u>Of these two strategies, we'll look first at the impact that upset reaching this crescendo had on the increasingly brutal and destructive power-fame-fortune-and-glory-seeking, egocentric, knowledge-finding, resigned competitive strategy—the approach that became known as 'right-wing'</u>.

[905] The overall significance of the agricultural revolution that took place some 11,000 years ago was that it led to such a rapid increase in upset that, on the graph charting the intensification of our upset, it meant humanity was fast approaching rock-bottom—we were hurtling towards the cynical, bitter and vengeful, burnt-out, out-of-control, all-restraints-thrown-out, rampaging, warring level of upset. This was because the domestication of plants and animals allowed people to live a more sedentary, less nomadic existence, in closer proximity and in greater numbers, the effect of which was to greatly increase the spread and growth of upset in humans. As explained in par. 848 when describing the effect the ice ages had on our species, living closely under the strain of the human condition dramatically accentuated the difficulties encountered by humans who were living with upset. So it follows that the closer humans lived during humanity's adolescence and/or the more difficult the living conditions, the greater the spread of and increase in upset, and that isolation from such encounters with the battle of the human condition served to minimize the spread of upset or soul-exhaustion. In short, if we were each left alone with our personal level of exhaustion, we would not be criticized by fresher souls or corrupted by those more battle-hardened. The philosopher Jean-Paul Sartre succinctly summed up just how difficult it is for upset, alienated people of all degrees to coexist when he wrote, **'Hell is other people'** (*Closed Doors*, 1944). We only need to look at the extreme situation to see

the principle in action—as mentioned, innocence isn't going to survive long in New York's Times Square where criminals and beggars work the streets. Of course, while living in closer proximity in more organized societies did greatly increase the spread and accumulation of upset, it also assisted the spread and accumulation of knowledge. In terms then of the race between self-discovery and self-destruction, there was at this time a speeding up in the development of *both* aspects.

[906]This explanation of how the advent of agriculture and the domestication of animals led to a rapid increase in upset allows us to understand why, in Moses' Genesis story of Cain and Abel, Cain became more upset than his brother, Abel: **'Abel kept flocks** [he lived the nomadic life of a shepherd, staying close to nature and innocence], **and Cain worked the soil** [he cultivated crops and domesticated animals and as a result was able to become settled and through greater interaction with other humans became increasingly upset]…**Cain was** [became] **very angry, and his face was downcast** [he became depressed about his upset state and so]…**Cain attacked his** [relatively innocent and thus unwittingly exposing, confronting and condemning] **brother Abel and killed him'** (4:2, 5, 8). As with his stories of Adam and Eve and Noah's Ark, Moses, in his denial-free soundness, was able to recognize, admit and then, through simple narrative, summarize the essential features of humanity's entire journey from innocence to upset. And, as was mentioned earlier and will be referred to again shortly, being able to confront the truth about the extent and nature of humans' corrupted condition, Moses was in a position to clearly see that something needed to be done about it, which led him to formulate a set of rules, the Ten Commandments, for humanity to live by to contain their out-of-control upset. What a phenomenally great denial-free, effective-thinking prophet Moses was! Early in the Old Testament of the Bible it says, **'no prophet has risen in Israel like Moses, whom the Lord knew face to face'** (Deut. 34:10). Moses was certainly someone who was able to **'delight in the fear of the Lord'** (Bible, Isa. 11:3); he was someone who, unlike most people, was able to confront the truth of Integrative Meaning that God is the personification of. Moses himself described how **'The Lord spoke to you** [the Israelite nation] **face to face out of the fire** [as explained in par. 332, fire is a metaphor for the searing truth of Integrative Meaning] **on the mountain.** [This was only possible because] **At that time I stood between the Lord and you to declare to you the word of the Lord, because you were afraid of the fire'** (Deut. 5:4-5). Moses had an exceptional ability to mediate between the truth and all the denial that humans

practice. (Jacob was another unresigned prophet able to confront the truth of Integrative Meaning and the resulting dilemma of the human condition and survive, saying, **'I have seen God face to face and yet I am still alive'** (Gen. 32:30).) In my book *Freedom Expanded* (see <www.humancondition. com/freedom-expanded-Abraham-Moses-Plato-Christ>), I present an in-depth description and explanation of not only Moses' critical contribution to the human journey, but the critical contributions made by whom I consider to be the three other outstanding prophets in human history: Abraham, Plato and Christ.

[907] So, from around 11,000 years ago, those involved in the agricultural revolution (represented by Cain who **'worked the soil'**) began to live sedentary lives and as a result became extremely upset (**'very angry'** with **'face' 'downcast'**), which caused them to **'attack'**, **'kill'** and replace the more innocent, confronting and unwittingly condemning people who were still living more naturally as either hunter-foragers or nomadic herders (represented by Abel who **'kept flocks'**). Anthropology now evidences this progression in upset, with the more innocent Neanderthals, Denisovans, Flores man and Red Deer Cave people being replaced around this time by more upset modern humans. We can tell that these archaic people were more innocent because the skulls of the modern humans who re-placed them were far more neotenous, and, as explained in pars 787 and 802, neoteny is by now associated with increased upset. Archaeologists maintain that **'organised conflict between groups** [of people is]**...typically asso-ciated with agriculture** [because] **people settled into land and used it for growing crops, so they became very defensive about it'** (Rainer Grün, quoted in *The Australian*, 21 Jan. 2016), but this defense-of-resources explanation is just another of the excuses denial-complying scientists have employed to avoid the real, psychological interpretation of human development, which Moses was able to truthfully recognize. In his 1955 book *The Inheritors*, the Nobel Prize-winning author Sir William Golding gave an honest depiction of a meeting between modern man and a band of Neanderthals; in reviewing the book, Penelope Lively wrote that **'their** [modern man's] **objective was extermination'**, and **'the terrible ending, when Lok** [the Neanderthal] **is alone, the last of his kind, and dies of grief, is the death of innocence'** ('O unlucky man', *The Guardian*, 11 Jan. 2003).

[908] By some 4,000 years ago, the <u>development of villages</u>, the move-ment by people into <u>specialized occupations</u>, the beginnings of <u>trade and industry</u>, and the close personal interaction that each development

inevitably brought, resulted in humans becoming so upset that some could no longer contain their upset and had to live that upset out; as the story of Cain and Abel describes, they had to allow some expression of their upset if they were to find any relief from the pressure of being so hurt. Men especially began to feel the periodic urge to go on a rampage of <u>raping and pillaging</u>. Tragic examples abound of what eventually developed from this ever-escalating conflict between peoples. There is, for example, the thirteenth century Mongol conqueror Genghis Khan; he was certainly someone who lived out his upset to the full, every day satisfying his anger with bloodletting, his egocentricity through the domination of others, and his mind or spirit by blocking out any feelings of guilt or remorse coming from the moral instincts within himself. As Genghis Khan is reputed to have said, **'Happiness lies in conquering one's enemies, in driving them in front of oneself, in taking their property, in savouring their despair, in outraging their wives and daughters.'**

[909] To understand how warlords like Genghis Khan came into being we need to examine more closely the development of conflict between groups of people, and the extremely compounding effect it had on the development of upset. It was inevitable that the need to go on the rampage and express, indeed purge, unbearable levels of upset prompted endless rounds of <u>payback warfare</u>, where warriors from one tribe or village would raid another tribe in retribution for earlier attacks on their material goods and maidens, which in turn would provoke counter raids, and so on and so on, in wave after wave of ever-increasing ferocity and brutality. But if we consider the nature of these early raids and counter-raids—which, incidentally, were still taking place until very recently in isolated places like New Guinea—what was happening was that despite the raids everyone was basically still trusting in the fundamental goodness of humans to get along; the pillaging, plundering and maiden-snatching was like a sport, a little valving off of upset on the side of the main agenda of everyone behaving cooperatively and getting on with their lives and with each other. Essentially, our species' original instinctive self or soul's capacity for unconditionally selfless love and cooperation was still in ascendancy— the levels of upset anger and aggression hadn't yet completely broken free from those natural restraints. By this stage in the human journey, people were, of course, certainly extremely upset and mostly resigned to living in denial of their soul, but they were still being influenced by their soul's inclination to behave lovingly and cooperatively, which meant

that upset was still able to be kept in check; *civilized*. However, as upset inevitably increased, it is obvious that this situation was going to flip—at a certain point the amount of upset anger and aggression was going to become *greater* than our soul's capacity to contain it, at which point an 'arms race' was going to break out in which the focus became locked upon who could be the strongest and the most aggressive the fastest. Once you could no longer trust in the basic goodness of others—in the certainty that they would treat you properly—then you'd better forget trying to be good yourself and just go all out to make sure you were not going to lose in the race that had now emerged to see who could be the toughest! And that is what happened: at a certain point in the escalating conflict between groups of people, some groups realized earlier than others that, as the saying goes, 'It was on for young and old'—that if they didn't get their act together and toughen up and train themselves for battle and develop the best defenses and weaponry and means for waging war, others would at their expense!

[910] When the escaped horses of the Spanish Conquistadors spread up into the Great Plains of North America in the 1500s, the Sioux Indians, who were initially just a small tribe living at the edge of the Great Plains, showed great presence of mind to grasp the opportunity that the mobility of horses offered to become winners in the 'arms race' that had developed in their region: **'They** [the Sioux] **developed a culture of pride and superiority—and arrogance, if you will—that overcame all opposition by tribes that claimed these hunting grounds initially. And so they simply rolled over them, one tribe after another, until they were the most powerful, most numerous tribe on the northern plains'** (historian Robert M. Utley, *Crazy Horse: The Last Warrior*, A&E Television Networks, 1993). In Central America, the Aztecs similarly had the presence of mind to develop and refine what was originally a small tribe in northern Mexico into a winning warrior culture that between the fourteenth and sixteenth centuries ruthlessly conquered and dominated all the peoples in southern Mexico.

[911] Of course, such presence of mind depended on having become sufficiently upset to no longer be naive enough to believe that 'right' was stronger than 'might', and from there be willing to carry out any measures necessary to ensure you were going to be among the surviving 'mighty'. And as to which group was going to be the most upset first and thus the first to abandon trusting in 'right', that depended on who became the first to domesticate plants and animals and establish large sedentary

communities because, as has been explained, it was this closer living environment that greatly accelerated the increase in upset. However, while a degree of upset was necessary for this 'arms race' to get underway, *once* underway it very rapidly produced more upset anger and aggression in humans—so we can expect that while those at the beginning of the arms race would have been bold, energetic and adventurous 20-to-30-year-old equivalent varieties of 'Hollow Adolescentman', it wouldn't have taken long for that degree of innocence to be bled dry and evolve into the immensely upset 50-year-old equivalent varieties of 'Hollow Adolescentman', warlords like Genghis Khan.

[912] This development can clearly be seen in what took place on the biggest battlefield of all, that of the Eurasian landmass. Here history tells us there were two groups of people who were the first to set out to be the mightiest: the Indo-European language speakers and the Semitic language speakers. Living in the grasslands of the Western Eurasian Steppes, north of the Caucasus Mountains, the Indo-European speakers were apparently the first in northern and north-western Eurasia to develop agriculture and the practice of herding animals, especially horses, which allowed them to live in large settled groups and become sufficiently upset to realize they should develop a warrior culture to ensure they would be frontrunners in the coming arms race. These Indo-Europeans have been referred to as 'Caucasians' because of their origin north of the Caucasus', however, **'They called themselves Aryas** [hence 'Aryans'], **which meant "noble of birth and race". These** [were] **proud, tall, fair-skinned people, with…great herds of lowing cattle and high-spirited chestnut horses, their flocks of sheeps and goats'** (*Barbarian Tides: Time-Life History of the World 1500 to 600 BC*, 1989, p.127 of 176). While I don't know of any evidence for this, it seems quite possible that they deliberately selected for their fair skin as a way of distinguishing themselves. They may have even selected for height and strength by killing any offspring deemed weak, as the Spartans of Greece reputedly did later to produce their warrior elite—as this account indicates: if **'some of the elders of the** [Spartan] **tribe…**[whose] **business it was carefully to view the infant…found it puny and ill-shaped,** [they] **ordered it to be taken…**[to] **a sort of chasm'** and thrown to its death (*Plutarch's Lives Volume 1*, c. 100 AD; tr. John Dryden, 1683, p.67 of 784). (There are records of the Vikings ruthlessly carrying out similar selection for toughness.) In the case of the Semitic speakers, we know little of their origins, only that they came from somewhere in Arabia or North Africa.

[913]The following then is a description from the book *Barbarian Tides* of the horrific arms race and resulting rapid increase in upset and soul-destroyed toughness that developed in Eurasia between these opposing forces: **'On the great Eurasian landmass...**[from] **the beginning of the third millennium BC...two major groups of peoples were on the move...**[the] **Indo-European speakers...from the Eurasian Steppes, and the...Semitic speakers...from somewhere in...Arabia...**[or] **North Africa. Both possessed the means of mobility...horses for the Indo-Europeans, asses and camels for the Semitic peoples...and from about 1500 to 600 BC, few civilized parts of the world failed to feel the tremors that resulted** [from their movements. In the case of]**...the migrations of the peoples from the Steppes...**[the] **Indo-Europeans who would come to be known as the Mycenaeans moved into Greece and created a dazzling Aegean civilization...To the north and west, Indo-European peoples – Celts and Germans, Balts and Slavs among them – were to penetrate almost every inhabitable area of the European continent and to cross the waters to Britain and other offshore lands. Some who settled on the Italian peninsula, most notably a people called the Latins, would eventually eclipse the Etruscans – whose civilization was one of the crowning glories of this age – by building an even greater civilization for a later era: Rome. Other Indo-Europeans...called the Hittites, came from the Steppes to Anatolia** [present day Turkey, where they]**...would found a mighty empire. Others went from the Steppes to the Iranian plateau, and from here some groups trekked east over the Hindu Kush mountains to north India. These people, the Aryans, would bestow their social institutions on the Indian subcontinent and spawn a unique spiritual culture, Hinduism. But for all those sweeping large-scale shifts of people...nowhere were the effects...more visible – or more violent – than in the relatively small region that encompassed the Middle Eastern birthplace of civilization, Mesopotamia, and the adjacent lands along the Mediterranean's eastern shore, later known as Syria and Palestine. In this area resided tribes and nations that had been fighting territorial battles for centuries. Here, too, were cities large and rich enough to lure ravening armies from afar...And here, finally, was where Indo-Europeans pushing down from the north collided not only with settled populations but also with Semitic peoples thrusting in from the south and west. Most of the established inhabitants of the region were Semitic, among them the Assyrians...**[and later] **Chaldeans, Aramaeans, Phoenicians and Hebrews... The result was a furnace roaring hotly'** (*Barbarian Tides: Time-Life History of the World 1500 to 600 BC*, 1989, pp.9-11 of 176).

[914]This account describes the **'roaring hot'** arms race that then ensued: **'By about 1700 BC, the ambitious...Hittites had...settled in Anatolia. There, over**

the course of the next three centuries, they gradually imposed their rule as a
warrior elite...Their culture was notable for its energy and adaptability...they
were pioneers in the craft of diplomacy...Within the velvet glove, however, was a
heavy fist, strengthened by fiercely valiant Hittite warriors and by a mechanical
innovation: the Hittite battle chariot...By 1353 BC, the Hittite empire was rivalled
in size and power only by Egypt...[In] 1285 BC...Ramses II, Egypt's aggressive
young pharaoh, marched the four divisions of his 20,000-man army...[north
against the Hittites, but] the Hittite army...savagely attacked the...Egyptians
who shortly headed for home...[But then] the empire was helpless before the
onslaught of a new wave of invaders, seafarers who appeared in the eastern
Mediterranean in the late 13th century BC – the Sea Peoples...an alliance...
including Phrygians – Indo-Europeans from the west coast of the Black Sea...the
attackers obliterated the Hittite world...slaughtering much of the population and
driving the rest into exile. And so, in 1200 BC, the Hittite empire vanished, creating
a power vacuum...that...was filled by...the Assyrians [who] were the descendants
of the Semites who had settled along the middle Tigris [river]...fierce warriors...
[who created] an empire that in terms of military might would stand second
to none...it featured deliberate terror and atrocity as instruments of foreign
policy...[where, for example, they] carried 14,000 defeated enemy soldiers off
to Assyria as slaves – after first securing their docility by blinding them...Their
immense armies, sometimes numbering as many as 200,000 troops...could march
50 kilometres a day along roads, some paved with stone, that had been constructed
throughout the empire...From the tribes they had subjugated, the victorious
Assyrians exacted an enormous annual tribute consisting of 12,000 horses and
2,000 cattle...During this time, too, the fires of conflict...burnt particularly hot
as aggressive peoples such as the Philistines, the Aramaeans, the Hebrews and
the Canaanites pushed and shoved one another for position...Once in battle, the
Assyrians asked for and gave no quarter; they rejoiced in butchery...By the start
of the ninth century BC...the Assyrians...waged a campaign that would eclipse...
[their earlier] dread deeds...[They] took their [enemies'] warriors prisoner and
impaled them on stakes before their cities...stacked the corpses like firewood
outside the gate, then flayed the nobles...and spread their skins out on the piles...
[and had] captives...burnt in a fire...by deliberate design [they] practised and
proclaimed mutilations, flayings, impalements and other atrocities for the purpose
of spreading terror and thereby encouraging submission' (ibid. pp.11-24).

[915] These few passages give some indication of just how absolutely
gruesome the development of the great arms race of 'might over right' was!
In terms of the upset-increasing, soul-destroying toughening that occurred,

there was progression from the Hittites' more soulful **'energy and adaptability'** in **'1700 BC'** to the immensely upset, blood-thirsty Genghis Khan-like Assyrians who **'rejoiced in butchery'** by the **'ninth century BC'** — and there is still some 3,000 years of soul-destroying and toughening bloodshed that has since occurred in the Middle East! Later in chapter 8:16E I talk about how quickly the relative innocence of humans was bled dry after this tipping point in the development of the arms race, which began around 3,000 BC. It will be described there how only in the most remote parts of Eurasia, namely on the islands and fjords of the north-western fringe of Europe, did any of the original relatively innocent 20-to-30-year-old equivalent adventurous Indo-European speakers from the Steppes remain.

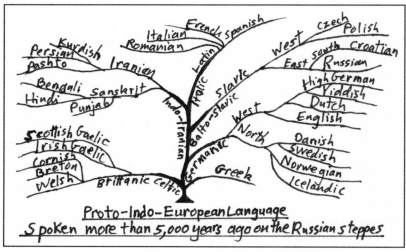

Diagram of the branches of the Indo-European language,
copied from the 2011 BBC documentary *Fry's Planet Word*

[916]So, to return to the main discussion, from the perspective of the right-wing power, fame, fortune and glory-seeking strategy, it was clear that self discipline could no longer be relied upon to contain or civilize the now completely out of control upset anger and aggression. A new means to restrain upset simply *had* to be invented; a further form of restraint for those participating in humanity's upsetting heroic search for knowledge *had* to be developed — and the solution that emerged was <u>IMPOSED DISCIPLINE</u>, the implementation of an agreed upon set of rules and laws that enforced social (integrative) behavior through threat of punishment. Once developed, this new form of restraint proved

significantly effective. For example, as briefly mentioned earlier, by the time Europeans arrived in north-eastern North America a grand league of American Indian tribes, known as the Iroquois Confederacy, had been established by two unresigned, denial-free, effective thinkers who had emerged from within their ranks. Recognized and described by their people as **'prophets'**, these two American Indians—known as 'The Great Peacemaker' and Hiawatha—with all their soulful sensitive feeling and denial-free clarity of thought, were able to realize that the warfare between the tribes could only be prevented by everyone agreeing to certain restraining rules that were enforced through punishment. The resulting discipline proved highly effective, as this quote illustrates: **'The Iroquois Confederacy was established before European contact, complete with a constitution known as the…"Great Law of Peace"…The two prophets, Ayonwentah [Hiawatha]…and Dekanawidah, The Great Peacemaker, brought a message of peace to squabbling tribes…Once they ceased infighting, they rapidly became one of the strongest forces in seventeenth and eighteenth century north eastern North America'** ('The Iroquois Confederacy and the Founding Fathers'; see <www.wtmsources.com/113>).

[917] Exactly the same scenario had played out some 3,000 years prior, when, in approximately 1,500 BC, Moses, the very great denial-free, effective thinking prophet from the Hebrew tribe—one of the groups of people who had **'pushed and shoved…for position'** in the **'roaring hot' 'furnace'** of horrific upset in the Middle East—brought order to the Israelite Nation (and eventually to much of the world) through the <u>Ten Commandments</u> he had etched on stone tablets. Yes, the moral code contained in those commandments became the basis of the constitutions, laws and rules that continue to govern much of modern society and proved vital in helping rein in the upset unleashed by the extremely upset 'Hollow Adolescentman'.

[918] <u>We now need to consider how the corrupting effects of the last 11,000 years affected the other strategy for coping with the human condition, which involved seeking out an idealistic cause to support to make yourself feel better about your, by now, immensely upset state—the approach that became known as 'left-wing'.</u>

[919] Yes, just as the resigned competitive practitioners developed a strategy for coping with their problem of excessive upset, which was Imposed Discipline, so born-again, pseudo idealistic practitioners developed a

strategy for coping with the problems associated with *their* approach to coping with the human condition, which was that in abandoning the all-important search for knowledge they were, in effect, siding *against* humanity's great battle and being excessively dishonest, and that solution was the adoption of <u>RELIGION</u> as it is practiced in most cultures today.

[920] To truly appreciate just how significantly humanity has benefited from religion we first need to look deeper into the very serious problems associated with the born-again, pseudo idealistic strategy.

[921] In addition to the already emphasized seriousness of siding against humanity's great battle to champion the ego over the ignorance of our instinctive self, the *other* very sinister effect of the born-again, pseudo idealistic, 'do good to feel good' strategy was that in *behaving* in a supposedly good, ideal way you were effectively asserting that you *actually were* a good, selfless, cooperative, gentle, loving, guilt-and-human-condition-free, ideal person. This was unlike the resigned competitive strategy where, firstly, you were still participating in humanity's heroic battle to find knowledge, and secondly, while you were using denial and lies to defend your upset state (asserting that there was no integrative purpose to existence, only random change, and that you were aggressive and selfish because humans had selfish instincts, and, therefore, there was no psychological dilemma of the human condition to have to explain), you were *not* deluding yourself you were a cooperative, loving, selfless, gentle, thoroughly good person free of upset. The born-again, pseudo idealist both abandoned the battle *and* took lying and delusion to a whole new level.

[922] It is worth mentioning that maintaining such extreme delusion was greatly assisted by the fact that we had *already* been practicing total denial of the issue of our corrupted, human-condition-afflicted state since resigning in our early adolescent years. So, in adopting a born-again, pseudo idealist strategy, all that its advocates were doing was adding another layer of delusion to the one that was already well entrenched. As mentioned in par. 749, while Resignation was a form of autism, of block-out/dissociation/denial, pseudo idealism was an even more extreme form of it; it was an even more extreme way of making yourself *'invulnerabl[e]'* by establishing **'a complex mental structure insuring against recurrence of the conditions of the unthinkable anxiety'**.

[923] Yes, the born-again, pseudo idealistic strategy was both treacherous and extremely dishonest—traits that totally undermined humanity's

search for knowledge—because in campaigning against the battle to find knowledge you were leading humanity towards an extreme state of denial/ alienation/separation from the truth/knowledge, when, in fact, humanity had to *continue* the battle to try to get closer to and ultimately reach the ultimate truth/knowledge/understanding of the human condition. Indeed, as we will see, when the born-again, pseudo idealistic state became fully developed in the form of 'postmodernism' even the existence of truth itself was denied! The fundamental objective of the human journey was to find the truth about ourselves, so adopting extreme denial of the truth, especially extreme denial of the truth about the human condition, was to undermine millennia of human progress and lead humanity *away* from its objective; it was to misguide humanity onto the path to oblivion, to total darkness in terms of enlightening ourselves about ourselves—which is why, as will be further explained shortly in ch. 8:16, pseudo idealism came to be described as **'the abomination that causes desolation'**.

[924] We can see then that since dishonesty was so dangerous, and pseudo idealism was the most dishonest strategy ever developed for coping with the human condition, <u>there was a very great need to find a form of pseudo idealism that somehow minimized or countered this extreme dishonesty</u>—a conundrum that eventually led to the development of the contemporary forms of <u>Religion</u> (the earlier ones being nature and ancestor worship).

[925] Religion is a form of pseudo idealism because it involved being 'born again' to the cooperative ideal state; instead of living through yourself, with all the associated overly upset angers, egocentricities and denials, religion required that you defer to someone exceptionally free of upset—namely one of the unresigned, denial-free-thinking, integrative-ideals-or-God-acknowledging, soulful, sound, innocent prophets around whom the great faiths were founded. Because you had become overly upset you decided to end your participation in humanity's heroic yet upsetting battle to find knowledge and instead put your <u>faith</u> in and lived through supporting the soundness and truth of a prophet's life and words. Rather than adhering to what your now overly upset self wanted to do and say, you adhered to the unresigned soundness and truth of the prophet's life and words—which was an *enormous* personal relief; as it says about Christianity in the Bible, **'if anyone is in Christ, he is a new creation; the old has gone, the new has come!'** (2 Cor. 5:17). You were reborn from your corrupted condition. But, in terms of humanity's heroic search for knowledge and

the need for honesty in that search, the immense benefit of deferring to religion is that while it allowed you to be born-again to a form of idealism and thus contain your upset and feel good about yourself, you were minimizing the dishonesty normally involved in the born-again, pseudo idealistic strategy, because you were acknowledging the soundness of the prophet and, by inference, your own lack of soundness. By recognizing, indeed worshipping, the integrity of the prophet, and his representation of another true, denial-free, integrative, soulful, sound state, you were indirectly being honest about or admitting your own immensely corrupted existence—your separation from the integrative, true, sound, soulful state. On this note, the concept of 'repentance' in religion recognizes the fundamental need in religion to be honest—self-reproachful—about your corrupted condition. It is an acknowledgment of your need to stop lying and start being truthful by supporting the unresigned, sound, honest words and life of the prophet around whom the religion is founded. As it says in the New Testament of the Bible, **'to fulfil what was said through the prophet Isaiah:…"[that] the people living in darkness…those living in the land of the shadow of death** [living in an extremely exhausted, upset, resigned, blocked-out-from-the-truth, alienated condition] **a light has dawned** [an unresigned, sound, truthful, non-alienated, soulful person has emerged]." **From that time on Jesus** [that sound person] **began to preach, "Repent"** [be honest about your corrupted condition and come and live through support of my truthful, sound world]' (Matt. 4:14-17).

[926]Religions also countered an aspect of dishonesty involved in the other strategy for coping—the resigned, competitive, right wing way of living—because most religions acknowledge the existence of a God who, as has been explained, is the personification of Integrative Meaning. Also, by acknowledging the soulful soundness of the prophet, you were recognizing the existence of a cooperative, unconditionally selfless, all-loving, moral soul in humans. The circular glow of light or 'halo' often drawn around the heads of religious prophets was used to indicate their soulful purity, innocence, soundness and holiness; indeed, the word 'holy', so often used to describe these great prophets, has the same origins as the Saxon word 'whole', which means 'well, entire, intact', and is thus a recognition of the prophets' wholeness or soundness or lack of separation or alienation from our species' sound, innocent, all-loving and all-sensitive original natural state. The religious concept of being able to be 'born again' from your corrupted condition back to a sound,

soulful state similarly recognizes humans' present alienation from an original, natural, happy, loving, innocent state.

[927] Another very important benefit of religion was that, on an individual level, it also helped assuage the guilt felt by pseudo idealists who were struggling with the fact that they were siding against humanity in its great battle. This is because in supporting your religion <u>you were also indirectly supporting humanity's heroic search for knowledge, because the truthful words of the prophet that are recorded in your religion's scriptures, which you were showing reverence for and deferring to, have represented the very font of knowledge; they have been the most denial-free expression of knowledge the human race has known</u>. Indeed, religious scripture has been the custodian of almost all the important truths about the nature of our world and of the human condition, albeit with all those truths safely represented in sufficiently abstract terms and parables to avoid directly confronting upset humans with the actual truth that they contain. In the case of the Bible, not only does it recognize the truth of Integrative Meaning in the concept of God, and the existence of our unconditionally selfless moral soul by the acknowledgment of the uncorrupted, soulful soundness of prophets, it also recognizes the truth that sex is an attack on innocence by attributing Christ's innocence to his mother being 'virgin'-like. Other insights contained in the Bible include how the concept of 'judgment day' recognized that one day reconciling understanding of the human condition would be found and all the dishonest denial and delusion humans are practicing would be exposed—and how Moses recognized the whole consciousness-induced, psychologically corrupted state of upset in humans in his parable of Adam and Eve, the soul-destroying process of Resignation in his parable of Noah's Ark, and how, after the advent of agriculture, the horrible 'arms race' developed in his parable of Cain and Abel, etc, etc. Christianity was *remarkably* aligning to the truth and thus supportive of our search for knowledge—which Carl Jung credited when he wrote that '[in Christianity] **the voice of God** [truth] **can still be heard'** (W.B. Clift, *Jung and Christianity*, 1982, p.114), and that **'The Christian symbol is a living thing that carries in itself the seeds of further development'** (Carl Jung, *The Undiscovered Self*, 1957, p.61 of 128). (Incidentally, given how precious all the denial-free truth in the Bible has been for the human race—as mentioned in par. 751, Martin Luther rated the significance of other **'fine'** books, written by the likes of Homer and Virgil, as being **'nought to the Bible'**—the Hebrews/Israelites/Jews who produced it deserve our deepest gratitude. It is amazing that

while extremely dishonest, denial-complying intellectualism holds sway everywhere in rational thought and debate today, with the proponents of such thinking celebritized, the ancient Hebrews collected only the words of their prophets. Humanity doesn't have any records of the great authors or poets or playwrights or composers or artists or singers or astronomers or academics or legal minds or politicians from the 4,000-year-history of the Israelites; instead what we have is the collection of the words of the few unresigned, honest, denial-free-thinking prophets who appeared amongst them during those millennia. That fabulously precious and incredible collection is the Bible's Old Testament.)

[928] In summary, religions offered humans a way to abandon living out their upsets and be born-again to a more soulful state of ideality, but in the least dangerously dishonest and most human-journey-sympathetic way. They enabled humans to indirectly continue to participate in humanity's heroic yet upsetting search for knowledge, *and* they provided a way for humans to significantly avoid being dishonest, because by deferring to a prophet you were able to indirectly admit the truth of your own corrupted state and the existence of another integrative, true, sound, soulful state. Religions provided a way for humans to be (to a degree) honest about their corrupted, false condition without having to openly admit and, therefore, nakedly confront it. In doing so, they helped minimize the truth-destroying levels of delusion and denial involved in the born-again, pseudo idealistic lifestyle. As such, religion has been a pseudo idealistic, 'do good in order to feel you are good', 'give up your overly upset life and be born again to a cooperative ideal life' way of living that allowed you to live in safe denial of your corrupted condition, but at the same time be honest about it—albeit indirectly.

[929] In the case of Christianity, the Bible actually referred to being **'born again'** (John 3:3), to having **'crossed over from death to life'** (John 5:24)—to, as mentioned, having become **'a new creation'** where **'the old has gone, the new has come!'** As Christ authoritatively said, **'I and the Father are one** [I am not a soul-devastated, resigned, alienated person having to live in denial of Integrative Meaning]**'** (John 10:30); **'I am the way and the truth and the life. No-one comes to the Father** [into alignment with the integrated state] **except through me'** (John 14:6); **'if I do judge, my decisions are right, because I am not alone. I stand with the Father** [I think truthfully, effectively and confidently in a holistic, inductive way because I don't live in a resigned, insecure, afraid, uncertain state of denial of such fundamental truths as Integrative Meaning—hence

Christ **'taught as one who had authority and not as their** [mechanistic, deductive] **teachers of the law'** (Matt. 8:29), and was able to **'get such learning without having studied'** (John 7:15), indeed, able to **'explain everything'** (John 4:25)]**'** (John 8:16); **'By myself I can do nothing; I judge only as I hear, and my judgment is just, for I seek not to please myself but him who sent me** [I am not a resigned, insecure, egocentric person, rather someone who follows what his Integrative-Meaning-acknowledging, truthful mind says is right]**'** (John 5:30); **'You are of this world; I am not of this world** [I am not resigned and living in denial of truth]**'** (John 8:23); **'I came into the world, to testify to the truth** [to counter all the dishonest denial]**'** (John 18:37); **'I have spoken openly to the world…I said nothing in secret** [I haven't been intimidated by all the dishonest denial that owns the world]**'** (John 18:20); **'The world…hates me because I testify that what it does is evil** [I tell the truth about the human condition]**'** (John 7:7); **'You are ready to kill me, because you have no room for my word** [unresigned truth]**'** (John 8:37); **'I am the light of the world. Whoever follows me will never walk in darkness** [in complete denial of the truth], **but will have the light of life'** (John 8:12); **'I am the resurrection and the life** [through me, your ideal, soulful true self can live again]**'** (John 11:25); and, **'I have overcome the world** [I have stood firmly by the truth and defied the world of denial]**'** (John 16:33). Christ wasn't a pseudo idealistic false prophet pretending to be a soft, sensitive and loving person, as pseudo idealists like to portray Christ as being in order to identify with him, rather he was a sound, strong person whose central talent was to fiercely defy all the dishonest denial in the world—a point the prophet Kahlil Gibran was making when he wrote, **'Humanity looks upon Jesus the Nazarene as a poor-born who suffered misery and humiliation with all of the weak. And He is pitied, for Humanity believes He was crucified painfully…And all that Humanity offers to Him is crying and wailing and lamentation. For centuries Humanity has been worshipping weakness in the person of the saviour. The Nazarene was not weak! He was strong and is strong! But the people refuse to heed the true meaning of strength. Jesus never lived a life of fear, nor did He die suffering or complaining…He lived as a leader; He was crucified as a crusader; He died with a heroism that frightened His killers and tormentors. Jesus was not a bird with broken wings; He was a raging tempest who broke all crooked wings. He feared not His persecutors nor His enemies. He suffered not before His killers. Free and brave and daring He was. He defied all despots and oppressors. He saw the contagious pustules and amputated them…He muted evil and He crushed Falsehood and He choked Treachery'** ('The Crucified', *The Treasured Writings of Kahlil Gibran*, 1951, pp.231-232 of 902).

[930] As was explained in par. 877, Christ, like Moses, was sound and secure enough in himself to fully confront and see the extent of the upset in the human race and therefore the extreme need for something to be done about it, which, in Christ's case, led him to realize that he had to create a religion around his soundness; he had to suggest to people that through supporting and living through his soundness they could be **'resurrect[ed]'** **'from death to life'**. It was a vision and act of extreme clarity of thought and extraordinary strength of character, especially since there was no science in his day, no first-principle-based insights into the workings of our world, that would have allowed him to find *reconciling* explanations for all the truths about the human condition that he could see, which meant that all he could do was offer his soundness as a place for upset humans to align themselves with and by so doing derive some relief from the human condition. It was also a vision of extraordinary strength of character because there were so many deluded false prophet charlatans misleading people and discrediting what he had to proclaim about himself for the sake of ensuring a future for the human race. Crucifixion was the price he had to pay for standing up so straight in a forest of bent and twisted timber. What a phenomenal, almost beyond comprehension prophet Christ was. Imagine the magnitude of what he was undertaking and then imagine how alone he must have felt not being able to share the weight of his task with those consumed with life in the everyday, messed up, self-preoccupied world around him—as he lamented, **'Foxes have holes and birds of the air have nests, but the Son of Man** [a sound expression of the integrated state that humans once lived in and could again align themselves with if he completed his mission] **has no place to lay his head'** (Matt. 8:20). It is no wonder that, as was mentioned in par. 615, most of the world dates its existence around his life, and he is regarded as **'the most famous man in the world'**. This essay captures the marvel of Christ: **'Here is a man who was born in an obscure village, the child of a peasant woman. He grew up in another village. He worked in a carpenter shop until He was thirty. Then for three years He was an itinerant preacher. He never owned a home. He never wrote a book. He never held an office. He never had a family. He never went to college. He never set foot inside a big city. He never travelled two hundred miles from the place He was born. He never did one of the things that usually accompany greatness. He had no credentials but Himself. While still a young man, the tide of public opinion turned against Him. His friends ran away. One of them denied Him; another betrayed Him. He was turned over to His enemies. He went through the mockery**

of a trial. He was nailed to a cross between two thieves. While He was dying His executioners gambled for the only piece of property He had on earth—His coat. When He was dead, He was placed in a borrowed grave through the pity of a friend. Nineteen long centuries have come and gone, and today He is the central figure of the human race and the leader of the column of progress. I am far within my mark when I say that all the armies that ever marched, and all the navies that ever sailed, and all the parliaments that ever sat, and all the kings that ever reigned, put together, have not affected the life of Man upon this earth as powerfully as has the One Solitary Life!' ('One Solitary Life', an essay adapted from a sermon by Dr James Allan Francis titled 'Arise Sir Knight', *The Real Jesus and Other Sermons*, 1926, pp.123-124). (More is said about Christianity in chapter 8:16I, and, as mentioned earlier, more can be read about the life of Christ at <www.humancondition.com/freedom-expanded-Abraham-Moses-Plato-Christ>.)

[931] So the criticism that could be leveled at someone extremely upset, like Genghis Khan or Adolf Hitler, is that they didn't take up religion—or, if they did claim to be religious, they weren't being *genuinely* religious. Indeed, for the exceptionally upset, the aspect of religion that made it so superior to the strategy of Imposed Discipline, which the Ten Commandments represented, is precisely that it allowed you to delude yourself that you were being **'born again'**, **'resurrect**[ed]**'** from your corrupted state. Rather than having good behavior forced upon you through fear of punishment, as was the case with Imposed Discipline, religion allowed you to feel that not only were you actively participating in goodness, you had actually *become* a good, selfless, loving, ideal person—that you were 'righteous'—which provided immense relief from the guilt of being overly upset. Possibly the best sales pitch for born-again religious life was that given by the apostle St Paul when he wrote, **'Now if the ministry that brought death, which was engraved in letters on stone** [Moses' Ten Commandments that were enforced by the threat of punishment]**, came with glory** [because they brought society back from the brink of destruction]**... fading though it was** [there was no sustaining positive in having discipline imposed on you]**, will not the ministry of the Spirit be even more glorious? If the ministry that condemns men is glorious, how much more glorious is the ministry that brings righteousness! For what was glorious has no glory now in comparison with the surpassing glory. And if what was fading away came with glory, how much greater is the glory of that which lasts!'** (Bible, 2 Cor. 3:7-11).

[932] Thus, in coping with the now raging levels of upset in humans, the first **'glorious'** improvement on destructively living out that ferocious

upset was Imposed Discipline, which was enforced through fear and punishment. But since discipline provided little in the way of joy or inspiration for the mind or **'spirit'** it was hard to maintain, it didn't **'last'**, it was **'fading'**, especially in comparison to the immensely guilt-relieving, **'righteous'**, 'do good in order to make yourself feel good' way of living offered by the next **'surpassing glory'**, religion.

[933] We can clearly see then that the 'do good in order to delude yourself that you actually are a kind, loving, selfless, good person and not horrifically corrupted' aspect of religion was very important, which means that only being *indirectly* honest about being extremely corrupted when you became religious was key to the effectiveness of religion—because if you had to be *directly* honest about being horribly corrupted you couldn't possibly delude yourself you were actually a kind, loving, self-less, good, not-horribly-corrupted person. This ability to not just *feel* good (because you were now behaving in a good way and not in the incredibly destructive way you had been), but to use this fact to delude yourself you *were* actually a loving, kind, selfless, good, upset-free, ideal, guilt-free, human-condition-solved, **'righteous'** person depended on this aspect of only *indirectly* acknowledging your corrupted state. So although in taking up religion you were being indirectly honest about being corrupted, you were still relying on being able to delude yourself that you weren't corrupted. In short, religions allowed people to admit to being horribly corrupted without having to suffer the confronting consequences of making such an admission. So just as humans could not directly acknowledge their corrupted state (because without the explanation of the human condition they couldn't defend and thus cope with that truth), religions similarly depended on not directly acknowledging/recognizing the soundness of the prophet around whom the religion was founded, even though acknowledging/recognizing his soundness was an intrinsic part of the honesty that made religions so special and effective. Religions depended on not recognizing—at least consciously, explicitly recognizing—that prophets were simply a sound variety of ordinary people, because that truth would directly confront their followers with the unbearable truth of their own lack of soundness. Instead, at least at the surface level of their conscious awareness, religious adherents viewed their prophets as being supernatural, divine, heavenly, from-another-world beings, because that way they could avoid any comparison with themselves. In fact, as will be described later in chapter 8:16I, the more upset and insecure the religious

person, the more fundamentalist/literal/superficial they had to be in their interpretations of religious scripture and the prophet himself—because too much honesty was impossibly confronting.

[934] A good example of how religions depended on not directly acknowledging the soundness of the prophet it was founded around, because it would be unbearably confronting of the truth of your own lack of soundness, is where Christ was said to have performed supernatural feats or 'miracles', such as his supposed ability to miraculously heal people. But with the human condition now understood, we can actually safely explain and demystify such 'miracles'. While the uncorrupted soundness of Christ and the denial-free, unresigned words he spoke could be extremely confronting, it could also center people and make them whole and well—it could release them from their psychosis, from their crippled state of living in disconnected denial of their soul. And, as physicians are increasingly recognizing, many, if not most, physical ailments *are* psychosomatic or soul-distressed in origin—recall the reference in par. 180 to Hesiod's description of humans before the psychologically upset state of the human condition emerged as being **'Strangers to ill'**; and, in par. 218, Buddhist scripture's anticipation that the arrival of understanding of the human condition **'removes all ill'**, that **'Human beings are then without any blemishes'**; and the Bible's anticipation that when understanding of the human condition arrives and **'the book of life'** is **'opened'** that **'There will be no more death or mourning or crying or pain'**. Yes, now that we can safely admit that humans' original instinctive state was one of living in a psychologically secure, happy and loving state, we can understand exactly what Christ meant when he said, **'Come to me, all you who are weary and burdened, and I will give you rest…for I am gentle and humble in heart, and you will find rest for your souls. For my yoke is easy and my burden is light'** (Bible, Matt. 11:28-30). Having obviously had a childhood that was *extraordinarily* sheltered from the upset in the world—so that he wasn't **'weary and burdened'** by the human condition and his **'yoke'** was **'easy'** and his **'burden'** **'light'**—Christ was still able to fully access all the love and security of our original instinctive soul's world, which meant he and his words were *immensely* realigning and reassuring for soul-damaged, human-condition-afflicted, soul-repressed-and-soul-disconnected resigned humans. Further, Christ's unevasiveness allowed him to see through people's denials and think truthfully and effectively about their situation—like the woman who had spoken to Christ at a well

and later said to her townspeople, **'Come see a man who told me everything I ever did'** (John 4:29). Christ could see into the human condition, and his ability to do so meant he could see where and why people were 'lost' or alienated. He could 'understand' them, not in terms of a first principle-based explanation of their condition—because in Christ's time there was no first principle scientific knowledge with which to explain the human condition—but in terms of being able to see into their situation, and this 'understanding' or appreciation represented true or pure love or compassion. To quote a reference to the work of psychoanalyst Carl Jung: **'Jung's statement that the schizophrenic ceases to be schizophrenic when he meets someone by whom he feels understood. When this happens most of the bizarrerie which is taken as the "signs" of the "disease" simply evaporates'** (R.D. Laing, *The Divided Self*, 1960, p.165 of 218). In par. 926 it was mentioned that the word 'holy' that has been used to describe the great prophets is derived from the word 'whole'. Well dictionaries inform us that the word 'healing' is *also* derived from the word 'whole'. Yes, the wholeness or soundness or lack of alienation of prophets is why prophets were considered holy and it is also why they were able to heal. So labeling Christ's ability to realign people with the true world as a 'miracle' protected humans from having to admit to Christ's soundness or innocence and, by inference, their lack of soundness or innocence.

[935]Instances of other supposed 'miracles' abound—consider, for example, the 'feeding the multitude with loaves and fishes' and 'walking on water' supernatural feats that supposedly took place when Christ spoke to a large gathering of people. Being unresigned, the penetrating truthfulness of what Christ had to say would have astonished the resigned minds of those listening. Christ was able to venture out into what the resigned mind knew as a terrifyingly dangerous minefield and, as it were, do somersaults and run and skip around out there—grapple with the human condition with impunity. As such, it would have been such a mesmerizing experience listening to Christ that even though the resigned mind would have soon afterwards begun to block out all the truths that were being brought to the surface as it realized their confronting implications, the listeners would have been so astonished and enthralled that their hunger after such a long talk would have been satisfied with the distribution of what little pooled food there was available. Later, after the event, unable to acknowledge the astonishing truth about what had really taken place, the resigned mind would have had to have found an evasive way

of recognizing the extraordinary nature of the occasion—which, in this instance, was achieved by saying, 'I remember a miracle where a few fish and some loaves of bread fed a mass of people.' Similarly, so overwhelmed would the audience have been by having so much truth emanate from someone, that, years later, some would evasively recall the impact of what happened by saying that, when Christ finally departed at dusk and walked out through the shallows of the lake to his disciples who were waiting in a boat to transport him back across the lake, he had 'walked on water'! The comedian Spike Milligan spoke the truth about Christ's miracles when he said: **'They made him do miracles..."Loaves and fishes, loaves and fishes, just like that!" This isn't indicative of the man. What he said and preached was enough. Why did he have to raise the dead? Did that make him holier? These are post-Jesus Christ PR stunts, raising the dead, walking on water. I find it an insult to the dignity of the man. I've written to *The Catholic Herald* about this. The outraged letters I've got! I said the Turin Shroud was a load of shit, I've said it for 15 years. Jesus Christ didn't need to do tricks'** (*Bulletin* mag. 26 Dec. 1989).

[936] Yes, under the duress of the human condition, the effectiveness of religion depended on not directly recognizing the soundness of the prophet around whom your specific religion was founded, because that would make the truth of your *lack* of soundness unbearably confronting, and obviously make it impossible to delude yourself that you had become a kind, loving, selfless, **'righteous'**, good, not-horribly-corrupted person—even though the recognition you were giving to the prophet was in itself an indirect acknowledgment of your corrupted condition. So religion involved maintaining a *very* delicate balance of delusion and honesty, for while it offered a way of only being implicitly honest about the corrupted state, its very existence depended on its adherents making at least a subconsciously relieving, honest acknowledgment of their own corrupted state, and on observers making at least, on a similarly subconscious level, a relieving, truthful recognition of that corruption; it was this honesty that made religions so special and effective. On the surface of conscious awareness, however, each individual adherent also depended on being able to maintain their facade and delusions about being an upset-free, **'righteous'** person, which meant there was still a great deal of dishonesty involved in religion.

[937] The cartoon series *The Simpsons* provides a wonderful illustration of the subtleties involved in religion. In the series, Ned Flanders is the

born-again religious character who is typically portrayed as having a
self-satisfied, 'I-occupy-the-moral-high-ground' attitude over the still-
human-condition-embroiled Homer Simpson. Ned's posturing drives
Homer crazy with frustration because Homer intuitively knows Ned is
deluding himself in thinking his Christianity gives him the moral high
ground—that he is the more together, sound person and is on the right
track—but Homer can't explain why Ned is so extremely deluded and
totally dishonest in his view of self. Homer can't explain and thus reveal
the truth that real idealism and the truly on track, moral high ground lay
with continuing the upsetting battle to find knowledge, and that Ned had
become so upset, so unsound, that he had to abandon that all-important
battle and leave it to others to continue to fight, including Homer. Worse,
in abandoning the battle, Ned has effectively sided against those still
trying to win the battle, adding substantially to the opposition they had
to overcome. But even Ned is intuitively aware that he is practicing delusion and so has to work hard at maintaining it. As explained in par. 886,
maintaining a delusion meant constantly persuading yourself, and others,
that you are right. Stridency and fanaticism characterized the behavior
of those maintaining a delusion, especially when, in becoming religious
for instance, you were practically admitting that you were being deluded
about being a sound, together, on track person yourself by having had to
defer to a sound prophet.

[938] In summary, the benefit of Imposed Discipline for the resigned,
competitive way of living over the born-again, pseudo idealistic way of
living was that it did not undermine a person's participation in humanity's
great battle—it simply provided a means to manage the upset associated
with that battle. However, since the religious born-again strategy both
minimized the irresponsibility of abandoning the battle, and (despite the
degree of delusion it still allowed) minimized the extreme denial involved
in becoming born-again, it provided a marvelous way of coping with
the by now extremely destructive and unbearable levels of upset and
associated guilt that affected nearly the entire human race, and which
Imposed Discipline could no longer contain. In fact, because of its degree
of honesty and indirect support of the search for knowledge, religion has
been by far the most special, the most wonderful form of pseudo idealism
to ever be developed. Indeed, it was religion that saved humanity from
destruction through the most difficult final stages of its journey. (It should
be pointed out that now that we can explain the incredibly important role

religions have played in the human journey, we can see how obscenely arrogant and wrong religion/truth-haters have been, when, for instance, people like Oxford University's Professor of Public Understanding of Science, Richard Dawkins, have said that **"'Faith is one of the world's great evils, comparable to the smallpox virus, but harder to eradicate. The whole subject of God is a bore"...those who teach religion to small children are guilty of "child abuse"'** ('The Final Blow to God', *The Spectator*, 20 Feb. 1999); and when E.O. Wilson, that quintessential exponent of dishonest, human-condition-avoiding mechanistic science, said that **'What's dragging us down is religious faith... I would say that for the sake of human progress, the best thing we could possibly do would be to diminish, to the point of eliminating, religious faith'** ('Don't let Earth's tapestry unravel', *New Scientist*, 24 Jan. 2015).)

[939] However, while religion *did* save the human race, at the very end of our species' journey through ignorance other forms of pseudo idealism evolved that have very nearly destroyed humanity. This final development will now be described and explained.

## Chapter 8:16A  The last 200 years during which pseudo idealism has taken humanity to the brink of terminal alienation

[940] We now need to consider the situation faced by humanity during the last 200 years of its now plummeting path to self-destruction through excessive upset, because it was during this final stage that the great benefit of religion, which was its honesty, became too confronting, forcing the development and adoption of other forms of pseudo idealism, including less honest forms of religion, that are *so* dishonest they are taking humanity to the brink of terminal alienation and extinction.

[941] What happened around 200 years ago to dramatically increase the levels of upset in society were further serious advancements on the factors that caused the sudden increase in upset 11,000 years ago. Firstly, due in part to improvements in medicine and sanitation, the world's population exploded, leaving people in many parts of the world living literally on top of each other: villages became towns, which became cities and even mega-city metropolises. Cities represent the most extreme congestion of people, and their development, along with their nature-eliminated, un-natural environment, which is so destructive of our innocent instinctive soul, greatly compounded the spread and increase of upset. Of course,

once humans became alienated, <u>cities</u> provided a refuge from the criticizing innocence of the natural world—they were not created as functional
centers where people could more efficiently work together, as often
claimed, but as <u>hide-outs for alienation</u>—however, for the souls of the
ensuing generations who had to grow up in such soul-*less* environments
cities were devastatingly alien places; as the historian Manning Clark
said, **'The bush** [wilderness] **is our source of innocence; the town is where the
devil prowls around'** *(The Sydney Morning Herald,* 18 Feb. 1985); and as the poet Percy
Bysshe Shelley wrote, **'Away, away, from men and towns, to the wild wood and
the downs—to the silent wilderness, where the soul need not repress its music'**
*(To Jane: The Invitation,* c.1820). No wonder Christ looked forward to the time
when understanding of the human condition would be found and alienating
cities could gradually be dismantled (as they now can be), saying, **'Do
you see all these great buildings?…Not one stone here will be left on another;
everyone will be thrown down'** (Bible, Matt. 24:2; Mark 13:2; Luke 19:44, 21:6).

[942] (Since it is such a bold, impacting statement to say cities are going
to be dismantled, I should explain it more fully. With understanding of
the human condition now found the great rehabilitation of the human
race from its alienated state can begin, and one of the important means to
achieve that rehabilitation will be to reconnect humans with the world of
their soul, which is the natural world. Nature isn't worth preserving simply
because it might contain useful drugs, or because it is needed to replenish
the atmosphere, or because of any of the other human-condition-avoiding,
psychosis-denying, mechanistic justifications. No, as important as those
evasive reasons are, we have to preserve nature because it is our original,
natural, instinctive self or soul's home and without it our soul, our being,
is destitute, a 'lost in the cosmos' itinerant wretch. As Sir Laurens van
der Post said in par. 472, **'We need primitive nature, the First Man in ourselves, it seems, as the lungs need air and the body food and water'**. Also, as
the historian Theodore Roszak wrote, **'"separation anxiety disorder"** [is
defined] **as "excessive anxiety concerning separation from home and from those
to whom the individual is attached." But no separation is more pervasive in this
Age of Anxiety than our disconnection from the natural world'** ('The Nature of
Sanity', *Psychology Today,* 1 Jan. 1996). Yes, as Yusuf Islam, or Cat Stevens as he
was formerly known, wrote and sang in 1970, **'Well I think it's fine, building
jumbo planes…roll on roads over fresh green grass…cracked the sky, scrapers
fill the air…I know we've come a long way, we're changing day to day, but tell
me, where do the children play** [how is our soul supposed to live in such an

alienated world]?' (*Where Do the Children Play*). In the 1960s many people tried to undertake the healing process that requires reconnecting with nature. They abandoned the alienated and alienating, soul-destroying cities and tried to return to the natural world. Of course, without the explanation of the human condition that would make it possible to stop having to live a denial-based, escapist, hide-from-condemning-nature-in-cities existence, this great **'alternative'**, **'hippie'**, **'flower power'**, **'let the sunshine in'**, **'back to nature'** revolution was a false start back to innocence and wholeness. Nevertheless, the initiatives that were being taken accurately anticipated the direction the rehabilitation of the human race needed to, and now can, take. Yes, we have to leave the alienated and alienating cities and dust off the *Whole Earth Catalog* and other 1960s masterpieces of new world planning and begin the *real* 'Age of Aquarius'. And I should mention that because we have understanding of the human condition there won't be the need for drugs to try to break open our alienated minds to access the authentic world of our soul and all the truth and beauty it has access to, or the need to indulge in 'free sex' to allow our upset self to feel artificially liberated from the human condition—both of which were activities that featured in the 1960s. We have found all the truth now that *genuinely* releases us from the binds of the human condition. As will be described in the final chapter of this book, what now emerges on Earth is not another messy, deluded false start to humanity's freedom from the human condition, but the real, genuinely loving one.)

[943]The second immense influence on the spread and growth of upset and thus alienation in the world has been the development of communication technology of such sophistication that, in terms of one upset human's access to another upset human, the world has basically shrunk down to one giant city—worse, one immense, dysfunctional household. Initially, there was the delivery of letters through a sophisticated postal system, and then mass printing of documents and newspapers, and then typewriters, and then the telegraph, and then the telephone, and then television, computers and faxes, and then emails and mobile phones, and then the world wide web and social media like Facebook and Twitter, and now 'smart' phones that enable around the clock access to this world of alienated superficiality. From birth, humans today are immersed in an ocean of upset behavior—especially dishonest, deluded, artificial, superficial, **'phony'**, **'fake'** behavior—leaving them utterly overwhelmed by anxiety and stress.

## Chapter 8:16B  The emergence of terminal levels of alienation in the 'developed' world

[944]The overall effect of such extremely congested living and the in-crease in the efficiency of communication technology is that it has led to a crescendo of upset—especially alienation from our natural, nurturing, all-loving and all-sensitive true self or soul. These quotes—which describe what is happening at the forefront of human progress, namely in the most materially so-called 'developed', mostly Western world, and specifically in the USA—provide frightening evidence of this epidemic of alienation: **'today's children are probably the least loved generation of all'** (Robert de Grauw, Letter to the Editor, *TIME*, 3 Apr. 2006); **'96 percent of American families are now dysfunctional'** (US counselor John Bradshaw quoted in *The Australian*, 8 May 1993); **'The 1990 US Census stated there will be more stepfamilies than original families by the year 2000, and that 66 percent of those stepfamilies break up when children are involved'** (The Stepfamily Foundation; see <www.wtmsources.com/104>); **'one in two US children will live in a single-parent family at some point in childhood'** (US Census Bureau of Household and Family Statistics, 2000); **'63 percent of the 18.5 million US children under 5 years of age were in some type of regular child care arrangement'** (US Census Bureau, 2005); **'the electronic age has ushered in electronic parenting. Kids spend far more time sitting passively before a device such as a computer or television than they do playing or speaking with their families'** (*The Commercial Appeal*, 26 Aug. 2001). As one journalist wrote, **'It is easier to get lost in a computer game than to deal with a dysfunctional family'** (David Boddy, 'The Facebook generation is in the grip of National Attention Deficit Disorder', *The Telegraph*, 4 Mar. 2013). Australia is one of the most sheltered and isolated and thus innocent countries left in the world, and yet our society is also unraveling with psychological suffering, as indicated by a 2011 study whose findings were reported on the front page of *The Sydney Morning Herald*: **'The well-being of Australia's children and young people has declined alarmingly in the past decade—and plunging marriage rates are partly to blame, a major study has found. Growing rates of child abuse and neglect, of children being placed in foster care, and of teenage mental health problems, includ-ing a rise in hospital admissions for self-harm, are rooted in the rise of one-parent families and de facto couples, violent and unstable relationships, and divorce, the report says'** ('Decline in marriage blamed for neglect', 6 Sep. 2011). The British parenting guru, Penelope Leach, recognized how hurtful to children parental sepa-ration is when she said, **'We have now reached a point where fewer than half of kids leaving high school will still have their parents living together. If we knew**

**for sure that up to half the nation's children were going to suffer something** [so] **damaging, we'd be moving heaven and earth to do something about it. This is a big issue that we just are not taking seriously—because we hate it, it's horrible'** ('10 Questions', *TIME*, 8 Jun. 2015). Yes, unable to explain the human condition, we haven't been able to confront and deal with it, or its implications. <u>The reality is that humans are now so alienated, so divorced from their true self, they can barely live with themselves, let alone anyone else</u>. The real reason for the breakdown in relationships is alienation, which in turn is a direct result of the human race's heroic but ever-increasing state of upset, a state that could only be ameliorated through the finding of compassionate understanding of the human condition.

[945] The result of all this rapidly accumulating **'dysfunction'**—the lack of a soulful, sound, alienation-free, nurturing and loving functional environment—is ever-increasing levels of psychological pain from dis-orientation and depression, and, to cope with that, even more block-out/denial/alienation, a process that is ending in the extreme disconnecting, dissociative behaviors of Attention Deficit Hyperactivity Disorder (ADHD) and autism, as these statistics confirm: **'someone born since 1945** [is] **likely to be up to 10 times more depressed than their grandparents'** (psychologist Michael Yapko, *Hand-Me-Down Blues* description on Amazon, 1999; see <www.wtmsources.com/173>); **'Depression…is the leading cause of disability in the US and abroad'** (Andrew Solomon, *The Noonday Demon*, 2001, p.25 of 560); there has been a **'41% increase in ADHD in US children during the past decade'** (*TIME*, 15 Apr. 2013); **'This** [today's US society] **is an ADD epidemic in the making'** (Katrin Bennhold, 'Generation FB', *The New York Times*, 23 Jun. 2011); and, **'Our country** [the UK] **is in the grip of a national attention deficit syndrome'** (David Boddy, 'The Facebook generation is in the grip of National Attention Deficit Disorder', *The Telegraph*, 4 Mar. 2013).

[946] In a 2013 article titled 'Once More on Attention Deficit Disorder', the creator of primal therapy, the aforementioned psychologist Arthur Janov, wrote that **'there are more and more cases of ADD among children** [in the US], **one in five to be exact…So what is going on?…The brain** [of the ADD child] **is busy, busy, dealing with the pain'** from **'childhood'**, **'where there was no** [real, unconditional] **love'**, only **'conditional love'** where, for example, **'parents wanted smart kids'** to fulfill their own insecure ego's desire for reinforcement; with the result that **'when there is stimulation from the out-side…it meets with a very active brain which says "Whoa there. Stop the input. I have too much going on inside to listen to what you ask for".'** Janov concluded, **'You know, ADD is also called the hyperactivity disorder (ADD HD: Hyperactive**

**Disorder). Of course, the kid is agitated out of his mind, driven by agony inside. We want her to focus on 18th century art and she is drowning in misery'** (4 Apr. 2013; see <www.wtmsources.com/132>). If we truly want some insight into what it *actually* is that is causing children and young adolescents to be **'agitated out of his [their] mind', 'drowning in misery'**, consider again the unresigned, denial-free honesty of these lyrics from the 2010 *Grievances* album of the young American heavy metal band With Life In Mind that were included in par. 229: **'It scares me to death to think of what I have become…I feel so lost in this world', 'Our innocence is lost', 'I scream to the sky but my words get lost along the way. I can't express all the hate that's led me here and all the filth that swallows us whole. I don't want to be part of all this insanity. Famine and death. Pestilence and war.** [Famine, death, pestilence and war are traditional interpretations of the 'Four Horsemen of the Apocalypse' described in Revelation 6 in the Bible. Christ referred to similar **'Signs of the End of the Age'** (Matt. 24:6-8 and Luke 21:10-11) and all of these descriptions are accurate because such extreme disintegration *is* the state of terminal alienation that occurs in the final stages of humanity's heroic search for knowledge.] **A world shrouded in darkness…Fear is driven into our minds everywhere we look', 'Trying so hard for a life with such little purpose…Lost in oblivion', 'Everything you've been told has been a lie…We've all been asleep since the beginning of time. Why are we so scared to use our minds?', 'Keep pretending; soon enough things will crumble to the ground…If they could only see the truth they would coil in disgust', 'How do we save ourselves from this misery…So desperate for the answers…We're straining on the last bit of hope we have left. No one hears our cries. And no one sees us screaming', 'This is the end.'**

[947] So it's no wonder children have a massive need for distraction from the overwhelming internal pain resulting from growing up in a fearful world without any real love or **'hope'** or **'answers'**, which in the more extreme cases has resulted in ADHD. (In the case of the most extreme state of dissociation from the **'screaming' 'agony inside'** of autism, I will look closely at that next in ch. 8:16C.)

[948] Regarding Janov's comment that **'We want her to focus on 18th century art and she is drowning in misery'**, in 2013 an art teacher at one of Sydney's leading private schools told me that **'while only two years ago students were able to sit through a half hour art documentary, I now know I lose them after only eight minutes; today's students' attention span is that brief!'** (WTM records, 26 Apr. 2013). This comment mirrors an observation made by the political scientist David Runciman in a 2010 BBC documentary series about the world wide

web: **'What I notice about students from the first day I see them when they arrive at university is that they ask nervously "What do we have to read?" And when they are told the first thing they have to read is a book they all now groan, which they didn't use to do five or ten years ago, and you say, "Why are you groaning?", and they say "It's a book, how long is it?"'** (*Virtual Revolution*, episode 'Homo Interneticus'). The same documentary also includes the following statement from Nick Carr, the author of *Is Google Making Us Stupid?*: **'I think science shows us that our brain wants to be distracted and what the web does by bombarding us with stimuli and information it really plays to that aspect of our brain, it keeps our brain hopping and jumping and unable to concentrate.'**

[949] Yes, what the internet is *really* providing is not access to more knowledge and understanding as many would have us believe, but distraction from pain. As Janov pointed out, the young person of today is increasingly **'agitated out of his mind, driven by agony inside…drowning in misery'**, with **'too much going on inside to listen'** and **'focus'**. As Carr said, their **'brain wants to be distracted'**. In the case of social media such as Facebook and Twitter, it allows people to be preoccupied/distracted (from the human condition) all day long with inane, frivolous, narcissistic, superficial self-promotion and gossip. The result of this extreme distraction from the **'agony inside'** is that **'The youth of today are living their lives one mile wide and one inch deep'** (Kelsey Munro, 'Youth skim surface of life with constant use of social media', *The Sydney Morning Herald*, 20 Apr. 2013). Yes, **'the net delivers this shallow, scattered mindset with a vengeance'** ('The effects of the internet: Fast forward', *The Economist*, 24 Jun. 2010). As one member of the 'Y generation' self-analyzed, **'Alone and adrift in what [Professor] de Zengotita calls our "psychic saunas" of superficial sensory stimulation, members of my generation lock and load our custom iTunes playlists, craft our Facebook profiles to self-satisfied perfection, and, armed with our gleefully ironic irreverence, bravely venture forth into life within glossy, opaque bubbles that reflect ourselves back to ourselves and safely protect us from jarring intrusions from the greater world beyond'** (Tom Huston, 'The Dumbest Generation? Grappling with Gen Y's peculiar blend of narcissism and idealism', *EnlightenNext*, Dec. 2008-Feb 2009). A 2012 article in the *Smithsonian* magazine summed up this shallow and superficial approach to thinking today, stating that **'Suddenly thanks to Google Books, JSTOR and the like, all the great thinkers…are one or two clicks away…And yet—here is the paradox—the wisdom of the ages is…buried like lost treasure beneath a fathomless ocean of online ignorance and trivia that makes what is worthy and timeless more inaccessible than ever'**; the **'digital revolution…[is] decapitating our culture, trading in**

the ideas of some 3000 years of civilization for BuzzFeed' (Ron Rosenbaum, 'Lewis Lapham's Antidote to the Age of BuzzFeed', Nov. 2012). A reviewer of my writing, who was deeply concerned about the profundity of my insights into the human condition not being heard in this superficial, trashy digital age, summed up the danger of this phenomenon when he wrote that 'We have a lot of competing noise for our attention these days, and it would be criminal to let that overwhelm our true potential, by masking [the] useful information [in Griffith's books] with hideous noise' ('Fitzy', *Humanitus Interruptus – Great Minds of Today*, 21 Oct. 2011; see <www.wtmsources.com/106>).

[950] The labels that have been given to recent generations are actually very telling of the rapidly escalating levels of alienation we are now seeing in society—and which the internet, while providing superficial relief for, is propagating and spreading at an exponential rate. While previous generations have been given the positive monikers the 'Builders' and the 'Baby Boomers', the generations who have followed are referred to as the 'X generation', the 'Y generation', and now the 'Z generation', which, according to Wikipedia, comprises 'people born between the mid-1990s and late 2000s'. The Canadian writer Douglas Coupland defined a Generation X'er as one who 'lives an X sort of life—cerebral, alienated, seriously concerned with cool' (*The Sydney Morning Herald*, 22 Aug. 1994)—qualities that are all associated with having had to adjust to an extremely soul-devastated existence. The adolescent psychologist Michael Carr-Gregg said that 'Generation Y is being ravaged by depression, anxiety and stress disorders' (Miranda Devine, 'Face it, we are all narcissists now', *The Sydney Morning Herald*, 3 Sep. 2009). So what exactly did we mean when we applied such valueless, end play lettering to these X, Y, Z generations? The answer is that they each represent an advanced state of terminal alienation. And the question left hanging is, what comes after Z?

[951] In 2013 *TIME* magazine featured a story on the Y-generation titled 'The Me Me Me Generation' (20 May 2013). Being one of the many frighteningly dangerous pseudo idealistic, left-wing magazines now—the left-wing being a phenomenon I will talk about shortly in chs 8:16H to 8:16Q—*TIME* tried to put an almost completely dishonest, positive spin on the present *extremely* narcissistic, *chronically* alienated '80 million strong…biggest age group in American history' Y-generation. In small text, the introduction to the article said, truthfully enough, that the Y-generation's 'self-centeredness could bring about the end of civilization as we know it', but then promoted them, in very large text, with the reverse-of-the-truth lie, as 'THE NEW GREATEST GENERATION: WHY MILLENNIALS WILL SAVE US ALL'! At least,

hidden away in this nine-page cover story about this **'narcissism epidemic'**, was the admission that narcissism is a **'personality disorder'** that occurs **'when people try to boost self-esteem'**, revealing that the *real* problem is lack of self-esteem, insecurity of self—alienation. What this article is *really* reporting on then, if it was to be honest, is how the present, **'80 million strong...biggest age group in American history'** Y-generation exists in a state of terminal alienation, and the dire implications such an epidemic poses for the future of our species. (One of the main psychological reasons for the loss of self-esteem that is causing this **'narcissism epidemic'** will be looked at shortly in ch. 8:16D.)

[952] A 2008 documentary on the destructive effects of all the consumer advertising directed at children, titled *Consuming Kids: The Commercialization of Childhood*, features a montage of powerful news clips from American television that provides a graphic snapshot of the symptoms of terminal levels of alienation being reached in society. In the first clip the newsreader reported that **'Forty times as many young people are now being diagnosed with bipolar disorder than 13 years ago** [bipolar disorder is the less confronting term now being used for what was once more truthfully termed 'manic depression', the state of extreme depression where the mind can only break free of it occasionally, resulting in oscillating bouts of euphoric relief then more depression].' The second clip reported that **'Almost 4.5 million children in this country have been diagnosed with ADHD.'** The third clip said that **'Doctors are writing a growing number of prescriptions for anti-depressants for children, as many as eight million a year.'** The fourth clip reported that **'One in three children born in the year 2000 will develop diabetes.'** The fifth clip said that **'For the first time in decades the rate of hypertension in children is rising** [with a medical journal cover in background saying] **2 million + American children may have high blood pressure.'** And, finally, the sixth clip reported that **'This generation of children is the heaviest in American history. An estimated 16% of all children and teenagers are overweight—four times as many since the 1960s. Life expectancy of children today will be shorter than that of their parents—the first such decline in modern times'** (Media Education Foundation).

[953] Yes, with upset becoming completely unbearable for humans, the distress it entails must be straining our bodies' ability to cope to the very limit, which is why diseases like cancer now riddle society. Indeed, as was mentioned in par. 934, descriptions of the healthiness of humans before the upset state of the human condition emerged reveal that most of our present sicknesses must be psychosomatic (emotional) in

origin: reference was made to Hesiod's description that **'When gods alike and mortals rose to birth / A golden race the immortals formed on earth...Like gods they lived, with calm untroubled mind...Nor e'er decrepit age misshaped their frame...Strangers to ill, their lives in feasts flowed by'** (*Works and Days*, eighth century BC); and there is Buddhist scripture's anticipation of what humans will be like once the ameliorating understanding of the human condition is found and absorbed, the time when humans **'will with a perfect voice preach the true Dharma, which is auspicious and removes all ill'**, and **'Human beings are then without any blemishes, moral offences are unknown among them, and they are full of zest and joy. Their bodies are very large and their skin has a fine hue. Their strength is quite extraordinary'** (Maitreyavyakarana; *Buddhist Scriptures*, tr. Edward Conze, 1959, pp.238-242); as well as the time anticipated in the Bible when **'Another book [will be]...opened which is the book of life** [the human-condition-explaining and thus humanity-liberating book]**...[and] a new heaven and a new earth** [will appear where]**...There will be no more death or mourning or crying or pain'** (Rev. 20:12, 21:1, 4). Sally Fallon's report that was referred to earlier when discussing the findings of Dr Weston Price (in par. 748), that **'Many others have reported a virtual absence of degenerative disease, particularly cancer, in isolated, so-called "primitive" groups'**, corroborates this conclusion. All the millions of health clinics in the world are not going to get humans back to full health; *only* the ameliorating understanding of the human condition being presented in this book can achieve that.

[954] Alongside ever-increasing anxiety-driven disorders like obsessive compulsive behavior, ADHD and autism, and unhappiness-driven excessive food consumption and physical sickness, the emotional lives of adolescents are also becoming shallower and cheaper, as this report indicates: **'the sexualisation of Western culture** [has meant] **that sex has been robbed of its emotional depth...For young men and women, it's increasingly a physical activity, with no real pleasure and no meaning at all...one hears of lipstick parties, where teenage girls wearing different coloured lipstick line up to give oral sex to boys with the aim of giving them a candy-striped penis'** (Clive Hamilton, co-author of the report 'Youth and Pornography in Australia', *The Australian*, 24 Jun. 2006). It has to be remembered that, as was explained in par. 782, sex is an attack on innocence, so treating it cheaply is another factor contributing to the escalating levels of hurt/upset/alienation in society. Also playing a huge role in the-death-of-soul-through-sex in the world is the frighteningly easy access to pornography that the internet allows, as well as the sexually predatory behavior it can facilitate, as this extremely alarming — worse, terrifying — article documents:

'Few things are certain in adolescence, but there's one thing upon which teenage girls agree; pubic hair is out. "Everyone shaves. Everything," says Sydney 16-year-old Anne. "If you've left it you are classified as disgusting. You'd be embarrassed for the rest of your life. Boys would pay you out, call you hairy. People start shaving in year seven [12 years old]." They know, or think they know, a few other things, too. That oral sex doesn't count as sex. That sending nude pictures via text or Facebook is the new flirting. That boys their age watch porn regularly, and demand from their girlfriends the sexual menu they see online – hairless, surgically-enhanced bodies, 'girl-on-girl action', and much, much more. They are learning from the 21st century's version of sex education class, the internet; a more enlightening and forthcoming source than nervous parents and teachers. But these lessons are a dangerous mix of misinformation and distorted images of sexuality, which is contributing to behaviour that can leave young women with deep psychological and physical scars.' The article goes on: 'girls are becoming women earlier than they used to. In the past 20 years, the age of a girl's first period has dropped from 13 years to 12 years and seven months, and as many as one in six eight-year-olds have periods. Children with 'precocious pubescence' can start menstruating at five or six. Reasons range from better nutrition and obesity to the break-down of the family unit. "When dads aren't around, they're more likely to move into puberty earlier," says parenting expert Michael Grose. "If it starts earlier, I imagine this would mean they are beginning to be sexually active earlier." In the past 60 years, the age at which girls lose their virginity has dropped from 19 (when many women were married in the 1950's) to 16, but many start much earlier...an Australian study...found the age of girls' first sexual experience ranged from 11 to 17 years, with a median age of 14...there has also been a marked increase in unwanted sex, an experience that can have a long-term effect on how a woman feels about herself and her sexuality. "The main reasons are being too drunk or high, and pressure from a partner,"...A Sydney study found that almost half of all adults, like Mike, first watched pornography between the ages of 11 and 13...[and that in pornography there is] a trend towards sex that is rough, aggressive, and idealises acts women don't enjoy in real life – gag-inducing fellatio, heterosexual anal sex, physical and verbal aggression...[porn] actors were required to be rough with the girl, and take charge. "He had moved from lovey dovey sex, towards material where the pornographists want to get more energy – 'f--k her to destroy her'". For many boys, porn is their sex education. They copy what they see, and expect their girlfriends to be like the women in the film' ('Lost innocence: Why girls are having rough sex at 12', *The Australian Women's Weekly*, 2 Jun. 2015). (The soul-destroying effects of sex in the lives of women especially will be looked at next in ch. 8:16C.)

[955] Yes, overall, hugely improved communication technology has spread upset everywhere, infected the world with horror, as this reference to violent computer games by the writer and broadcaster James Delingpole recognizes: **'What troubles me about Grand Theft Auto V — which has an 18 rating that will be ignored by thousands of younger teenagers — is not just the message it sends out to youngsters (drugs are cool; crime pays; violence is fun), but what it says about the coarsening, the decadence and the hopelessness of our modern culture. It's the electronic equivalent of those gladiatorial contests the Romans used to stage in the dying days of their empire, involving ever more exotic beasts and ever more elaborate sets. It may be entertaining, particularly to young men with a penchant for such nihilistic spectacle, but the sensibility to which it appeals is warped, jaded and riddled with the deepest, blackest despair. The fact that this is the most popular computer game on the market should make us all shudder'** ('Torture and murder with the addictive glamour of Hollywood', *Daily Mail*, 18 Sep. 2013).

[956] In March 2014 an article by the British psychiatrist Anthony Daniels, writing under the pen name Theodore Dalrymple, was published in *The Australian* newspaper. Aptly titled '<u>Next generation has apocalypse written all over it</u>', the piece provides further stark evidence of the emergence of terminal levels of alienation in the Western world: **'The end is nigh! French civilisation is on the verge of collapse! No one who attended the recent Mondial du Tatouage (the World Tattoo Fair) in Paris could be in any doubt about it. It was held, appropriately enough, in the old abattoir at the end of the Avenue Jean Jaurès...surrounded by a wasteland of concrete monstrosities, French modern architecture now being some of the worst in the world...Perhaps the rejection of beauty as a goal by French architects — their work seeming rather to exclaim "F..k off, humanity!" — accounts in part for the adoption as a style by so many of the young French of deliberate ugliness and self-mutilation. In a world of brutal ugliness over which you have no control you might as well admit defeat and join in. Thousands queued to enter the fair...Many of them had already permanently disfigured themselves and, since their tattoos would have quite likely cost hundreds or thousands of euros, poverty was not the explanation for their degradation. Inside the abattoir, there were about 250 stands from across the Western world...It was only to be expected that ugly loud tuneless rock music of the kind that makes thought impossible and speech difficult was poured into the atmosphere like poison gas. Customers lay down on couches to have dragons, skulls, vampires, insects, rats or Elvis Presley permanently inscribed on their legs, backs, arms, chests...One man held a baby in his arms as a tattooist inscribed a**

snake on his back. You are never too young to be indoctrinated with nihilism… Two…men…tattooed so heavily that only a few years ago they would have been considered degenerate freaks, patted small children on the head as their delighted parents proudly took photographs of them…[A] tattooed slogan…on the front of a young woman's upper thigh and knee, read "We are like roses that have never bothered to bloom when we should have bloomed and it is as if the sun has become disgusted with waiting."…Some of the names of the tattooists' enterprises were revealing: Evil from the Needle, Clod the Ripper, Perfect Chaos, Black Heart… Modern Butcher…Ten years ago there were 400 professional tattooists in France; now there are 4000. In England, where so much of the worst of modern culture originated, more than a third of young adults are tattooed. Civilisations collapse from within'** (17 Mar. 2014). Yes, this situation where the **'Next generation has apocalypse written all over it'** and **'civilisation is on the verge of collapse'** *is* because **'the sun has become disgusted with waiting'**, the liberating light of understanding of the human condition *has* been delayed for far too long. Science really must let these understandings through.

[957] On the subject of tattoos, this further masterpiece by Michael Leunig captures the whole horror of what new generations of humans have encountered coming into the now immensely upset, and embattled, and soul-devastated world of humans! Implicit in it is how, in their alienation-free, innocent state, infants can see right into the hearts of adults; they can see how alienated they *really* are!! God Almighty, the human condition *has* to end!!!

Baby wants to get some rest
But mum has tattoos on her chest:
A tiger's head, a dragon's tail
Baby's gone all sad and pale
A lightning bolt, a monster's eye
Baby starts to sob and cry
A spider's web, a Union Jack
And these good words: "To HELL AND BACK"

Cartoon by Michael Leunig, *The Sydney Morning Herald*, 24-25 Mar. 2012

[958] As a way of summarizing the terminal state of alienation and stress that now saturates humans, I want to mention two of the most powerful healing tools I know of, apart from massage and diet—and if you've suffered from Chronic Fatigue (as I have) you will have probably tried every method of healing known to man! Interestingly, these devices were developed by Soviet scientists and used to preserve the health of astronauts in the Soviet space program. The first is SCENAR, a device that delivers small electrical impulses that have the effect of an extremely powerful form of acupuncture, helping the body realign its electrical meridians that stress has left in unhealthy disarray. (A gentler form of this therapy is provided by the Frequency Specific Microcurrent device from the United States.) The second—and this is the tool that reveals just how alienated humans have become—is the Buteyko breathing technique, which involves learning to reduce your breathing rate and volume, the effect of which is to produce a profound healing of all manner of sicknesses. As has been explained, humans once lived in a pre-human-condition-afflicted, innocent, happy, loving, peaceful state, but that peaceful state is now so compromised with stress and anxiety that people today chronically overbreathe, which causes numerous detrimental physiological effects. In fact, overbreathing is an accurate measure of how alienated we have become from our original 'innocent **and simple and calm and happy'** and **'quiet'**, **'free from'** **'anguish'**, **'violence'** and **'quarrel'** existence, when we had a **'calm untroubled mind'**, as Plato and Hesiod described our pre-human-condition-afflicted lives (see chs 2:6 and 2:7). And just how much and how fast that level of alienation has increased in the last 50 years or so is particularly apparent in an article about hyperventilation that reported that **'modern people breathe about 2-3 times more air'** than humans did **'only 70-80 years ago'** (Dr Artour Rakhimov, 'Hyperventilation Causes'; see <www.wtmsources.com/145>). We are indeed at the deadly end of an exponential curve of increasing levels of alienation! (For those interested, a video of all the methods I have used to help maintain my health can be viewed at <www.humancondition.com/health>.)

[959] I might point out here the utter hypocrisy of the situation where humanity goes into a collective meltdown of concern when some people are killed in a plane crash or earthquake (which is, of course, understandable) but completely ignores the die-off of the *entire* human race from the massive increase in the levels of alienation that is now occurring from one generation to the next. As Kierkegaard noted, **'The**

**biggest danger, that of losing oneself, can pass off in the world as quietly as if it were nothing; every other loss, an arm, a leg, five dollars, a wife, etc., is bound to be noticed'** (*The Sickness Unto Death*, tr. A. Hannay, 1989, p.62 of 179). But, such has been our fear of making any reference to the *real* psychological nature of the human condition! How deep underground in Plato's cave of denial is the human race living!

[960] (A more complete description of how alienated humans, especially humans in the materially 'developed' Western world, have become can be found in *Freedom Expanded* at <www.humancondition.com/freedom-expanded-alienation>.)

## Chapter 8:16C <u>The dire consequences of terminal levels of alienation destroying our ability to nurture our children</u>

[961] A dire consequence of the arrival of terminal levels of alienation—especially at the forefront of human progress, namely in the materially 'developed' world, fuelled as it has been by advances in communication technology—is the near complete inability now of mothers to nurture their offspring.

[962] One of our WTM Founding Members, who for 15 years has been a nurse and midwife in major hospitals in Australia and England, has told me how, in just the last few years, there has been a shocking increase in the number of mothers who are unable to relate naturally to and connect empathetically with their newborn babies. She said that while it has long been the case that there were some mothers who couldn't access their natural, nurturing, loving instincts, the problem is now so common it is a relief to actually provide postnatal care for a mother who *can* relate naturally to her infant. A 2015 newspaper article carried this honest account of a mother who struggled to relate to her baby: **'When my baby was first born...I didn't experience the rush of love at first sight that so many new mums talk about. In fact, for the first three weeks of his life I was overwhelmed by him and what his presence meant. I looked at him and thought, "Who are you?" He was like a little impostor in our house. I didn't know him and I couldn't tell what he wanted...I felt hugely guilty...I wish I'd known what I was feeling wasn't so unusual. Forty per cent of first-time mums recall their predominant emotional reaction when holding their babies for the first time was one of indifference, according to a study in *The British Journal of Psychiatry***

[published 35 years ago (Apr. 1980), the situation now is apparently so much worse]…"It can take time to fall in love with your baby," confirms neonatologist Dr Howard Chilton…"There's a whole spectrum of how quickly women bond with their babies'" (Katherine Chatfield, 'Perfect Stranger', Sydney's *Sunday Telegraph's* mag. *Sunday Style*, 22 Feb. 2015). The result of this predominance of mothers who are unable to empathetically connect with (basically, love) their infants is the epidemic we are now seeing of highly distressed children—in the extreme, children who display the dissociative disorders of ADHD and autism.

[963]The reason why women have become so alienated from their natural, nurturing, all-loving true self or soul was provided earlier when, in chapter 8:11B, I described the different roles men and women have had to carry out in humanity's upsetting battle to find knowledge, ultimately self-knowledge, understanding of the human condition. Very briefly, in keeping with their already established roles as the group protectors, it was men who took up the loathsomely upsetting job of championing the ego over the ignorance of our original instinctive self, leaving women to maintain their loving, nurturing role. The problem, however, that this role differentiation gave rise to was that in not being responsible for or directly participating in the terrible innocence-destroying battle to overthrow ignorance, women were naive or unaware of the ramifications of fighting the battle, and, as a result, were unsympathetic towards both the battle and the frustrated anger and egocentricity it produced in men—a situation that placed men in the awful predicament of being both misunderstood and unjustly condemned by women. Unable to explain their behavior to women, men were left in an untenable situation: they couldn't just accept women's unjust criticism of their behavior, so what men did was turn on and attack the relative innocence of women. Men perverted sex, as in 'fucking' or destroying women's innocence, the result being that women came to share men's lack of innocence, their upset. Basically, through having to support men in their critically important task of championing the ego by sexually 'comforting' them, by inspiring them with their image of innocence and by giving them lots of attention, women were left with little innocent, soulful soundness and little time for the immensely emotionally and energy consuming task that is the nurturing of infants. The necessary consciousness-centered, 'ego-centric', male-led battle unavoidably intruded upon and compromised women's ability to nurture their offspring. Yes, having to nurture children with love/unconditional

selflessness has been an altogether impossible task for women while the battle of the human condition raged around and within them. And, obviously, as the upsetting battle intensified, the all-loving and all-generous souls of both men and women were eventually going to become completely exhausted, leading to the development of terminal levels of alienation—the result for women being that they would find themselves no longer able to emotionally relate to and connect with their infants, no longer able to give them the unconditional love they have come to expect from our species' love-indoctrinated, nurtured origins, which is the situation that has now developed, with disastrous consequences.

[964] Of course, the now *extremely* embattled, egocentric state of men is also having a devastating effect on children. Children need an *environment* of love and reinforcement and men's preoccupation with their egocentric battle, together with the now extremely strained relationship between men and women, have all but destroyed the possibility of creating that nurturing, loving, reinforcing environment for children. The effect of men's embattled, egocentric state on children will be described next in ch. 8:16D—what will be looked at here, however, is the effect of terminal levels of alienation on the primary nurturer of infants, the mother.

[965] Before doing so, it needs to be reiterated that this inability of parents to nurture their offspring is *not* their fault but the product of the ever-accumulating levels of upset in the human race as a whole, which in turn is due to our species' heroic *2-million-year* search for knowledge, specifically self-knowledge, understanding of the human condition.

[966] As just stated, the devastating result of the arrival of terminal levels of alienation amongst women and, with it, the loss of their ability to nurture their offspring, is an epidemic of highly distressed children—in the extreme, children who display the dissociative disorders of ADHD and autism. The epidemic of ADHD in society was just described in ch. 8:16B, but similar reports are occurring for the rising incidence of autism; for instance, **'Autism is a serious behavioral disorder among young children that now occurs at epidemic rates. According to a recent article, autism is diagnosed in America about 1 of every 120 girls and 1 in every 70 boys. In some areas, it is far more prevalent. It is now occurring around the world, especially in the developed nations of the world. According to California records, autism has increased 1,000% in the past 20 years! Autism used to account for 3% of the caseload of the California Department of Economic Security. As of 2012, it accounts for at least 35% of the caseload and costs taxpayers about $2 million**

**per child!'** (Dr Lawrence Wilson, 'The Autism Epidemic and Natural Solutions', Dec. 2012; see <www.wtmsources.com/130>), and this reference to **'the huge increase in autism that we see these days...**[with] **California's Silicon Valley** [being] **home to more autistic children than anywhere** [else] **in America'** (*Uncommon Genius*, ABC-TV, 24 May 2001). Also, **'in China it's estimated 7.5 million children are autistic, but many lack proper diagnosis'**, the inference being there are probably many more cases going undiagnosed within that massive population ('Children of the Stars', *Dateline*, SBS, 2 Apr. 2013).

[967] Since autism represents the extreme state of the distress that nearly *all* children are now suffering from, an analysis of what causes it will most clearly and graphically reveal the primary source of that now almost-universal distress within children.

[968] As with Janov's lack-of-real-love explanation for ADHD, autism is most often a disastrous consequence of a mother's inability to emotionally relate to and empathetically connect with her newborn baby. While autism can be caused by physical trauma to the brain, or a degenerative condition, most cases are the product of extreme instances of infants not receiving the unconditional love, in particular from their mother, that their instincts expect, which forces them to psychologically dissociate from their reality in order to cope with the violation and hurt to their instinctive self or soul. Able to safely admit now that our species' original instinctive orientation was to living in a harmonious, happy, loving, non-alienated, non-egocentric, non-angry state, and that children enter the world fully expecting to encounter such a state, we can understand the incredible shock it must be for them to be met instead with the extreme upset that now characterizes all of humanity. Understandably, many children can't psychologically bridge the gap and have to adopt all manner of coping mechanisms to block-out that anguish, including forms of compulsive and obsessive distraction, which is why we are seeing these epidemic rates of ADHD and autism. And the reason autism is, as mentioned in the report cited above, more prevalent in boys than girls is because, as was explained in chapter 8:11B, boys are both more aware of the problem of the human condition and responsible for solving it, so they are more attuned to, and thus more terrified and distressed by, the extent of the problem of the human condition in the world.

[969] In his 1996 posthumously published book *Thinking About Children*, the former president of the British Psychoanalytical Society, the psychiatrist and pediatrician D.W. Winnicott (who has been described as

'one of the most influential contributors to psychoanalysis since Freud...[and someone who] brought unprecedented skill and intuition to the psychoanalysis of children' (Adam Phillips, *Winnicott*, 1989, back cover)) provided the following honest description of the cause of autism. He wrote that in 'a proportion of cases where autism is eventually diagnosed, there has been injury or some degenerative process affecting the child's brain...[however,] in the majority of cases...the illness is a disturbance of emotional development...autism is not a disease. It might be asked, what did I call these cases before the word autism turned up. The answer is..."infant or childhood schizophrenia"** [p.200 of 343] [note, 'schizophrenia' literally means 'broken soul', derived as it is from *schiz* meaning **'split'** or **'broken'**, and *phrenos* meaning **'soul or heart'**]...**There are certain difficulties that arise when primitive things are being experienced by the baby that depend not only on inherited personal tendencies but also on what happens to be provided by the mother. Here failure spells disaster of a particular kind for the baby. At the beginning the baby needs the mother's full attention, and usually gets precisely this; and in this period the basis for mental health is laid down** [p.212] **...the essential feature** [in a baby's development] **is the mother's capacity to adapt to the infant's needs through her healthy ability to identify with the baby. With such a capacity she can, for instance, hold her baby, and without it she cannot hold her baby except in a way that disturbs the baby's personal living process...It seems necessary to add to this the concept of the mother's unconscious (repressed) hate of the child** [p.222] **...it is the quality of early care that counts. It is this aspect of the environmental provision that rates highest in a general review of the disorders of the development of the child, of which autism is one** [p.212] **...Autism is a highly sophisticated defence organization. What we see is** *invulnerability* [p.220] **...The child carries round** *the (lost) memory of unthinkable anxiety*, **and the illness is a complex mental structure insuring against recurrence of the conditions of the unthinkable anxiety** [that results from the mother's failure to provide her **full attention**] [p.221]'.

[970] The following is a typical case history of an autistic child from one of the many documented by Winnicott: '**When I first saw Ronald at the age of 8, he had very exceptional skill in drawing...Apart from drawing he was, however, a typical autistic child...The mother herself was an artist, and she found being a mother exasperating from one point of view in that although she was fond of her children and her marriage was a happy one, she could never completely lose herself in her studio in the way that she must do in order to achieve results as an artist. This was what this boy had to compete with when he was born. He competed successfully but at some cost...At two months the mother remembers smacking**

the baby in exasperation although not conscious of hating him. From the start he was slow in development...His slowness made him fail to awaken the mother's interest in him, which in any case was a difficult task because of her unwillingness to be diverted from her main concern which is painting' (ibid. pp.201-202).

[971] In his 1967 book, the revealingly titled *The Empty Fortress*, the psychologist Bruno Bettelheim similarly maintained that autism 'did not have an organic basis, but resulted when mothers withheld appropriate affection from their children and failed to make a good connection with them. The most extreme expression of this concept suggested that mothers literally did not want their children to exist' (Wikipedia; see <www.wtmsources.com/126>). As a result of his findings, Bettelheim was a prominent proponent of the 'refrigerator mother' theory for autism—named thus for the cold-heartedness of what we can, as emphasized, now understand is essentially *all* humans' unavoidable-after-2-million-years-of-struggle, human-condition-afflicted, immensely alienated, soul-disconnected, psychotic and neurotic state.

[972] Dialogue from the 1993 film *House of Cards* features this honest description of the dissociative behavior involved in autism: 'People say about the following categories that these kids have a problem or are disabled, or psychologically dumb, etc, but really they are children, through hurt or some kind of trauma, that have held onto soul, and not wanted to partake in reality— retarded, autistic, insane, schizophrenic, epileptic, brain-damaged, possessed by devils, crocked babies.' Yes, in their innocent vulnerability, the main way children have coped with the wounds to their instinctive self or soul has been to make themselves 'invulnerabl[e]' to the pain by splitting themselves off, dissociating themselves, from the world of that anxiety and trauma. And to maintain the denial and thus dissociation from the extreme pain of their circumstances required constant application of the denial/block, which is why autistic people tend to be compulsive and obsessive in their behavior; they escape into repetitive activities and tend to develop a one-track mind. In his 1997 book *Next of Kin*, the psychologist Roger Fouts described autism as 'a developmental disorder characterized by lack of speech and eye contact, obsessive and repetitive body movements, and an inability to acknowledge the existence or feelings of other people. The autistic child lives in a kind of glass bowl, inhabiting a separate reality from those around him' (p.184 of 420).

[973] Yet again, it has to be emphasized that since the emergence of the necessary battle with the human condition some 2 million years ago, *all* mothers have to some degree been 'refrigerator mothers'—indeed, the

truth is that *all* humans who are resigned are alienated or split off from their soul; they are all schizophrenic, they all have a **'split soul'** or **'broken heart'**. They are all, to a degree, autistic. What we are seeing now is just so much more of this **'split'**, **'broken heart**[edness]**'**. Winnicott even acknowledged this, that autism is the extreme degree of **'a universal phenomenon'**: **'It has always seemed to me that the smaller degrees of disturbance of the mind that I am trying to describe are common and that even smaller degrees of the disturbance are very common indeed. Some degree of this same disturbance is in fact universal. In other words, what I am trying to convey is that there is no such disease as autism, but that this is a clinical term that describes the less common extremes of a universal phenomenon'** (*Thinking About Children*, p.206).

[974] In pars 492-493, extracts were included from the writings of Olive Schreiner and Ashley Montagu that truthfully describe the importance of nurturing love in the lives of humans. Schreiner was quoted as saying, **'They say women have one great and noble work left them, and they do it ill... *We* bear the world, and *we* make it. The souls of little children are marvellously delicate and tender things, and keep for ever the shadow that first falls on them, and that is the mother's or at best a woman's. There was never a great man who had not a great mother — it is hardly an exaggeration. The first six years of our life make us; all that is added later is veneer...The mightiest and noblest of human work is given to us, and we do it ill.'** In the case of Montagu, he also spoke of the dire consequences of mothers' inability to give their infants real love: **'The newborn baby is organized in an extraordinarily sensitive manner... he wants love...If it doesn't receive love it is unable to give it — as a child or as an adult. From the moment of birth the baby needs the reciprocal exchange of love with its mother...The infant can suffer no greater loss than deprivation of the mother's love...maternal rejection may be seen as the "causative factor in...every individual case of neurosis or behavior problem in children."...In an age in which a great deal of unloving love masquerades as the genuine article, in which there is a massive lack of love behind the show of love, in which millions have literally been unloved to death, it is very necessary to understand what love really means...[and] its importance for humanity'.**

[975] With regard to this situation where **'a great deal of unloving love masquerades as the genuine article, in which there is a massive lack of love behind the show of love'**, given how upset and thus insecure about their sense of worth, in truth, hateful of themselves all resigned humans have become, how could mothers be expected to be able to give real, unconditional love to their children? You can't be loving when you are virtually devoid of

love—all you can do is try to imitate love—but infants, who are not yet alienated, can see through that pretense and somehow have to try to adapt to it. And the more upset and alienated the human race has become, the less real love there has been for the next generation, so that, in the end, there is virtually no real, unconditional love anywhere anymore—so, yes, almost everyone is now, unavoidably, to some degree, autistic! Indicative of just how needing of love infants are, recent **'Studies** [that] **show babies become anxious if ignored for even two minutes by** [their] **mother'** (*Daily Mail*, 25 Aug. 2010).

[976] Yes, children need and expect *so* much love and among the now rare, more innocent, less human-condition-embattled 'races' they have generally received it. Earlier (in par. 748), references were made to traditional Australian Aboriginal society where **'All observers agree upon the extraordinary tenderness which parents display towards their children, and indeed, to all children whether of their own family and race or not'**; to the Bushmen of the Kalahari, where **'Their love of children, both their own and that of other people, is one of the most noticeable things about the Bushman'**, the **'Bushmen...mother carries her child with her at all times up to four years of age'**, **'Children are breast-fed for up to 3½ years, and among the Bushmen lactation suppresses ovulation'**, **'!Kung** [Bushmen]**...infants hardly ever cry'**; and to the Yequana of Venezuela, who ensure their infants have **'24-hour contact with an adult or older child'**, and that **'the notion of punishing a child had apparently never occurred to these people'**, and that **'[Yequana] babes in arms almost never cried.'**

[977] In her 1975 book *The Continuum Concept* Jean Liedloff—whose insights into the nurturing behavior of the Yequana tribe feature above—emphasized the need to give infants the caring treatment **'which is appropriate to the ancient continuum of our species inasmuch as it is suited to the tendencies and expectations with which we have evolved'** (p.35 of 168) in order for them to have **'a natural state of self-assuredness, well-being and joy'** ('Understanding The Continuum Concept', The Liedloff Continuum Network; see <www.wtm-sources.com/105>). Liedloff also truthfully recognized that levels of alienation of people in the 'developed world' now is such that it has overwhelmed their natural instincts for nurturing, writing that **'We have had exquisitely precise instincts, expert in every detail of child care, since long before we became anything resembling *Homo sapiens*. But we have conspired to baffle this longstanding knowledge so utterly that we now employ researchers full time to puzzle out how we should behave towards children, one another and ourselves'**

(*The Continuum Concept*, p.34). Mothers now desperately read book after book about 'how to be a mother', how to imitate real, loving nurturing, when the real barrier to being able to do so naturally is the extent of their psychotic and neurotic alienation from their natural instinctive true self or soul, and beyond that the problem of the human condition, issues that none of these superficial, faddish books go near.

[978] Journalist Betty McCollister was another exception to this evasive phenomenon, someone who wasn't at all **'baffle[d]'** about **'how we should behave towards children'**. In an article titled 'The Social Necessity of Nurturance' (*The Humanist*, Jan. 2001), she wrote honestly about how infants need **'unconditional love'** from their mothers, and acknowledged the devastating consequences in society now of mothers not being able to provide it. While the bold attempt she made to explain why nurturing became so important during our species' development was not at all correct, McCollister did accurately and truthfully recognize that **'Our evolution has resulted in a species whose infants can't thrive without continual, loving attention. Here, then, is the clue to raising fewer unhappy, alienated, violent youth for jail fodder... Every human infant must have unconditional love; without it, an infant's health and growth will be stunted...Anthropologists, neurologists, child psychiatrists, and all other researchers into child development unequivocally agree and have sought for decades to alert society. For example: ...Ashley Montagu (anthropologist): "The prolonged period of infant dependency produces interactive behavior of a kind which in the first two years or so of the child's life determines the primary pattern of his subsequent social development." Alfred Adler (psychiatrist): "It may be readily accepted that contact with the mother is of the highest importance for the development of human social feeling..." Selma Fraiberg (child psychologist): A baby without solid nurturing "is in deadly peril, robbed of his humanity."... George Wald (biologist): "We are no longer taking good care of our young..." Ian Suttie (psychoanalyst): "...The infant mind...is dominated from the beginning by the need to retain the mother—a need which, if thwarted, must produce the utmost extreme of terror and rage."...James Prescott (neuropsychologist): Monkey juveniles "deprived of their mothers were at times apathetic, at times hyperactive and given to outbursts of violence** [the equivalent of children with ADHD]**... showed behavioral disturbances accompanied by brain damage..." Richard M. Restak (neurologist): "Scientists at several pediatric research centers across the country are now convinced that failure of some children to grow normally is related to disturbed patterns of parenting." Sheila Kippley (La Leche League): "It is obvious that nature intended mother and baby to be one..."**

[979] **In the face of such overwhelming, unanimous testimony, can we doubt that we are failing our children? The dismal truth is that, on the whole, babies received more and better care 25,000 years ago, 250,000 years ago, even 2.5 million years ago, than many do today…To correct this, we must first recognize that, while both parents play vital roles in an infant's development, the mother—like it or not—is the primary caregiver. Biologically, that's how the system works. And such an immeasurably important task cannot be sustainably carried out in her "spare time.".…Humanity was geared for females to cherish offspring in the womb, bond with them at birth, and lavish love on them at the breast. It isn't sexist to esteem motherhood. It is sexist to trivialize it…Grasping the connection between negligent infant care and adolescent violence…we are obliged to act…Alienated, with low self-esteem, pessimistic about the future, in schools that don't educate, the children who should be our hope for the future instead drink, smoke, take drugs, get pregnant, commit suicide, and commit crimes which land them in our awful jails.'**

[980] But for all her honesty, McCollister hasn't delved to the bottom of the problem and asked the question screaming to be addressed: 'But *why* have humans stopped loving their infants?' The upset state of the human condition was not being addressed, and without the compassionate explanation to that question of questions, trying to confront the truth of the importance of nurturing was unbearable for parents. So while over the years numerous movements have emerged that truthfully recognize the devastating consequences of not nurturing our children, and call for greater emphasis on nurturing, such as the '[Jean] **Liedloff Continuum Network**', the '**Touch the Future**' organization, and the '**Natural Child Project**', while we couldn't explain our inability to nurture/love our infants the ramifications of not doing so could clearly not be faced and admitted. Imagine being a mother of an autistic boy and trying to face the truth that your alienation caused his autism! Without the explanation for our upset condition you would find it impossible! No wonder any professionals in the field of child psychiatry who supported the belief that autism occurs as a result of inadequate parenting, such as Bettelheim, have been maligned and censured. While it wasn't possible to explain the human condition, to explain the very good reason *why* parents have been unable to adequately love their offspring, it simply wasn't possible to acknowledge the importance of nurturing in both the maturation of our species and in the maturation of our own individual lives.

[981] So while (as McCollister noted) some **'researchers into child development…have sought for decades to alert society'** to the fact that **'Every human**

**infant must have unconditional love; without it, an infant's health and growth will be stunted'**, the reality is that while humanity hasn't been able to explain our immensely soul-devastated, upset state and resulting inability to nurture our infants, denial has been the *only* way of coping. All the nurturing-avoiding mechanistic 'explanations' for the cooperative behavior of bonobos and our human ancestors that were described in chapter 6 evidence that—with the obscenity being that all that evasion was and is not justified given I presented the explanation for the human condition and with it the all-important relieving reason for why we have lost the ability to nurture more than 30 years ago. And worse, accompanying all the no-longer-justified, human-condition-evading, mechanistic 'explanations' for our moral sense that avoided admitting the importance of nurturing has been the raft of dishonest evasive excuses for the devastating *consequences* of not having been nurtured, the most prominent of which have been to apportion blame to a genetic predisposition, chemicals in our industrial world, childhood immunisation programs, or some contracted disease. Genes, in particular, have been blamed for every kind of ailment— for depression, drug addiction, violence, obesity, delinquency, learning and sleep disorders, suicide, lack of commitment/divorce, homosexuality, sex addiction, pedophilia, and almost every other human malaise and abnormality, including ADHD, and those more extreme dissociative states of autism and schizophrenia. Blaming biological, environmental and chemical factors has been infinitely more bearable than blaming our psychological alienation from our true, natural selves and our resulting inability to adequately nurture our children; as Winnicott cautioned, **'expect resistance to the idea of an aetiology [cause] that points to the innate processes of the emotional development of the individual in the given environment. In other words, there will be those who prefer to find a physical, genetic, biochemical, or endocrine cause, both for autism and for schizophrenia'** (ibid. p.219). Andrew Solomon, the author of the 2001 book about severe depression (including his own), *The Noonday Demon*, illustrated this preference when he admitted that **'Being told you are sick is infinitely more cheering than being told you are worthless'** ('Casting out the Demons', *TIME*, 16 Jul. 2001).

[982] Additionally, in his book *They F\*\*\* You Up: How to Survive Family Life*, which is **'devoted to making accessible the scientific evidence that early parental care is crucial in forming who we are'** (2002, p.3 of 370), the child psychologist Oliver James acknowledged that **'Our first six years play a critical role in shaping who we are as adults'** (p.7), and that **'One of our greatest**

**problems is our reluctance to accept a relatively truthful account of ourselves and our childhoods, as the polemicist and psychoanalyst Alice Miller pointed out'** (p.9), adding that **'Believing in genes** [as the cause of psychoses] **removes any possibility of "blame" falling on parents'** (p.13).

[983] So the whole so-called 'nature vs. nurture' debate has really been about wanting to, indeed *needing* to, attribute the formation of our character to the influence of 'nature', our genetic make-up, rather than to the confronting truth of the role nurturing (or the lack thereof) played; it has been a sophisticated deflection from the fact that our psychoses and neuroses and their many physical manifestations are *not* about our genes, but the hurt to our all-loving instinctive self or soul during infancy and childhood. Again, as John Marsden said, **'The biggest crime you can commit in our society is to be a failure as a parent and people would rather admit to being an axe murderer than being a bad father or mother.'** Yes, far better then are the 'blame-all-our-ailments-on-genetics-or-disease-or-immunisation' excuses that are flooding the scientific literature today.

[984] Just one of the unhealthy repercussions of not being truthful about the real cause of childhood madness is that treatment of it can be, and has been, dangerously misdirected. As stated in par. 72, the word **'psychiatry'** literally means **'soul-healing'**, but since we have never before been able to explain *why* we as a species have become so soul-destroyed and how that upset has impacted upon the health of our children's souls, we have, in our denial, relied instead on non-confronting, biomedical 'solutions'—as this article points out: **'Over the past decade, a condition known as...attention deficit hyperactivity disorder has been spreading among our children like an epidemic. During that time the prescription of drugs used to control this disorder, central nervous system stimulants chemically similar to the amphetamines that can be bought on the streets, has leaped by more than 1000%...Because it is so convenient and guilt-relieving to be able to attribute a child's difficult behaviour to a neurochemical problem rather than a parenting or broader social one, there is a risk that this problem will become dangerously over-medicalised'** (Dr Michael Gliksman, 'Social issues at the root of childhood disorder', *The Australian*, 8 Dec. 1997).

[985] Again, the problem stems from the immense vulnerability of children to all the human-condition-driven upset in the world today, a vulnerability we can gain an, albeit slight, appreciation of if we immerse ourselves in their situation. Since the upsetting battle of the human condition only emerged some 2 million years ago, and only became extreme towards the end of those 2 million years, the great majority of human history—from

the australopithecines through to the advent of *Homo sapiens sapiens* —
was spent living cooperatively, which means that infants now enter the
world, firstly, expecting it to be one of gentleness and love, and, secondly,
with almost no instinctive expectation of encountering a massively upset,
embattled world. And it is this *extreme contrast* between our species'
instinctive memory of a harmonious, happy, secure, sane, all-loving and
all-sensitive world, and our species' more recent massively embattled
angry, egocentric and alienated world, that makes the shock that infants experience entering the world now so psychologically damaging.
Resigned humans have been living in denial of both the truth that our
ancestors lived in a state of total love and that as a species we currently
live in a state of near complete corruption of that ideal instinctive world
of our soul, and that disconnect from our reality means we haven't been
attuned to how devastating it is for infants to encounter our world. But
the truth is, this contrast between what a child's innocent, love-saturated
instincts *expect* and what the child *encounters* in our human-condition-
afflicted, soul-butchered world is so great it is akin to a sunflower finding
itself having to grow in a dark cesspit. No wonder as adults we turned
out as gnarled thornbushes, ready to stunt the next generation.

[986] Indeed, the reality is that the levels of alienation in society today are
such that almost all humans are cardboard cut-outs of what they would
be like free of the human condition. And while resigned adults aren't
aware of their immensely alienated (virtually dead) state because they are
living in denial of it, new generations of children arriving into the adult
world who have yet to adopt this strategy of denial are acutely aware of
the difference between our species' original, ideal, innocent instinctive
state and our current immensely upset alienated existence and somehow
have to try to cope with the distress it causes them. Chapter 2:2 described
the agonising process of Resignation that adolescents have had to go
through adopting a life of denial of the issue of the human condition and
any truths that brought it into focus, but until a young person adopted
this blocked-out, resigned state they remained exposed and vulnerable
to the full horror of the dilemma of the human condition — horror that
is palpable in the lyrics of the heavy metal band With Life In Mind that
were included in pars 229 and 946.

[987] So, having not yet adopted this denial children have always struggled mightily with the imperfection of the upset world that surrounds
them, but with the gulf between humans' original innocent state and

our current *immensely* alienated state now so great, new generations *are* finding it impossible to bridge. The truth is, ADHD, and the more extreme states of autism and schizophrenia, are varieties of childhood madness, but as the bravely honest psychiatrist R.D. Laing famously said, **'Insanity—a perfectly rational adjustment to an insane world'** (Larry Chang, *Wisdom for the Soul: Five Millennia of Prescriptions for Spiritual Healing*, 2006, p.412).

[988] And what has made it especially difficult for new generations to cope with our corrupted adult world is that adults have been unable to admit to being corrupted in soul; in fact, as has been pointed out, adults haven't even been aware that they are corrupted—if they *were* aware they were corrupted and alienated they wouldn't be alienated, they wouldn't have blocked out and thus protected themselves from the truth of their upset condition. But with new generations able to clearly see the extent of the corruption and alienation in the world around them, this lack of any honesty by adults, their complete <u>silence</u> on the subject of our immensely corrupted human condition—in effect, their denial that there is anything wrong with themselves or their adult world—<u>left children dangerously prone to blaming themselves for the dysfunctionality of their environment</u>. And not only have resigned adults been silent about their corrupted condition, in the absence of the real reason for their divisive behavior, they have put forward all manner of dishonest reasons for their behavior, lies that have greatly offended the honest minds of children. In fact, in line with the big lie put forward by the resigned mechanistic adult world that our ancestors were aggressive, brutish and savage, children have even been treated as, and even told they are, 'just hateful little savages that have to be tamed'. Indeed, the purity and innocence of children has so confronted and exposed parents that they are typically referred to as 'kids'—a dismissive, derogatory, retaliatory 'put down', a way of holding their confronting innocence at bay. <u>And so, in encounters between the innocent and the alienated, where the alienated act as if there is nothing wrong with them or their world, in the innocents' instinctive state of total trust and generosity they are left believing there must be something wrong with them, that in some way or another they must be at fault</u>. In their immense naivety about the upset, alienated world, together with their soul's great love, trust and generosity, innocents question their own view, not the view being presented by the alienated. The innocent do not know people lie because lying simply did not exist in our species' original innocent instinctive world. In short, their trusting nature made

them *codependent* to the alienated, susceptible to believing the alienated were right rather than accepting and fighting to uphold their own accurate view of the situation. Psychologists coined the word 'codependent' to describe someone who is **'reliant on another to the extent that independent action is no longer possible'** (*Macquarie Dictionary*, 3rd edn, 1998), and, in the case of children, they are so trusting of, and thus **'reliant on'**, the adult world they are incapable of thinking **'independent**[ly]**'** enough to trust that they are not at fault or have somehow misread the play. Children come from *such* an innocent, wholesome, trusting, loving, generous, integrative instinctive world that they all too readily blame themselves in situations where they are faced with a denial. Then, when they decide they must be at fault, their sense of self-worth and meaning is completely undermined, and to cope with that **'unthinkable anxiety'**, as Winnicott accurately described it, they have no choice but to psychologically split themselves off from the perceived reality, adopt a state of **'invulnerability'**. More will be said about the devastating effects of children's codependency to adults next (in ch. 8:16D).

[989] So, able now to appreciate the extreme vulnerability of children to all the extraordinary upset in the adult world, and adults' silent denial of it, we can understand why children have died a million psychological deaths in the face of that upset. We can understand why Beckett wrote, **'They give birth astride of a grave, the light gleams an instant, then it's night once more'**, and why Laing wrote that **'To adapt to this world the child abdicates its ecstasy.'** *Thank goodness* we have found the understanding of the human condition that allows us to admit the importance of nurturing and make it a priority now for the human race.

[990] In summary, the human race is naturally extremely embattled after 2 million years of its heroic search for knowledge, which means that virtually no mother now can hope to love her infants as much as their instincts expect to be loved, and virtually no father now can hope to restrain their extreme egocentricity to the degree necessary to avoid oppressing their children somewhat. (Again, the devastating effect men's egos has on children will be explained in ch. 8:16D.) Thankfully, however, we can now explain why nurturing has been so compromised. And since, through that explanation, we can understand that virtually all humans now are so upset that virtually no one can hope to love their children as much as their children's instincts expect, parents do have to be realistic about their limitations—if too great an expectation is placed on parents

to be more loving than they are now genuinely capable of being no one will be prepared to run the risk of having children, which is obviously not the answer. However, with understanding of the human condition and the subsequent ability to admit how delicate our soul is, parenting can, and must, take on a whole new meaning and responsibility.

[991] Yes, to repeat what was said in chapter 6:13, having found the compassionate biological understanding of the upsetting battle that humans have had to wage against the ignorance of our instinctive self as to the fundamental goodness of our conscious self, our species' preoccupation with that battle can at last end and focus return to the nurturing of our children. Further, now that we have found that *very* good reason as to *why* parents have been unable to nurture their children with anything like the unconditional love children instinctively expect and need, it is no longer necessary to deny the importance of nurturing both in the maturation of our species and in our own lives. Yes, with the arrival of understanding the terrible guilt parents have felt for not being able to adequately nurture their offspring departs. Parents can now understand that their inability to adequately love their offspring is *not* their fault, but the product of the ever-accumulating levels of upset in the human race as a whole that resulted from our species' heroic 2-million-year search for knowledge, specifically for self-knowledge, understanding of why we conscious humans are good and not bad. *So yes, we can finally now both admit the importance of nurturing and turn our attention back to the nurturing of our offspring—in fact it now becomes a matter of great urgency that we do both these things.*

[992] The nurturing of children is what is required to produce adults who are sound and secure in self. Ultimately, it is *only* through the nurturing of our offspring that the human race can hope to become healthy—namely psychosis-and-neurosis-free once again; to, as Ashley Montagu said, put **'man back upon the road of his evolutionary destiny from which he has gone so far astray'** and restore **'health and happiness for all humanity'**. As will be explained in chapter 9, with understanding of the human condition found all humans can *immediately* be free of the agony of the human condition by taking up the Transformed Way of Living (which involves adopting a mental attitude where you leave that agony behind as dealt with), but to produce humans who are *completely* free of psychosis and neurosis and not needing to take up the Transformed Way of Living, they need to be nurtured with unconditional love in their infancy and childhood—a

process that will take some generations to achieve while all the accumulated upset within our species subsides.

[993] Of course, the difficulty for virtually all parents in the immediate future is that since they are so psychologically upset, especially so alienated from their natural, loving, instinctive true self or soul as a result of humanity's upsetting search for knowledge, it is simply not possible for them to give their offspring unconditional love. HOWEVER, while virtually all parents will have to accept this reality, their ability now to understand the very good reason for why they are unable to give their offspring unconditional love will bring so much calming, reassuring relief from that guilt that nurturing won't be nearly as difficult as it has been. Again, while it *will* take a number of generations before most parents can generate real, natural, unconditional love, being able to understand their upset state, and, significantly, not having to continue with the upsetting battle to prove their worth (which, as will be explained in chapter 9, is what makes the fabulously exciting new Transformed Way of Living possible) is going to bring *enormous* relief to parents and help them *immensely* with the whole activity of nurturing. The overall relief and excitement alone that will now emerge in the human race from the realization that we have finally ended our 2-million-year struggle against ignorance will mean that children are going to find themselves in, if not an unconditionally loving world, then at least an anxiety-free, truthful, compassionate, happy one. And since the adult world can finally break the silence of their denial and tell children the truth about their horrifically upset condition, that honesty will also make an immense difference to the psychological wellbeing of children. Basically, finding understanding of the human condition makes possible a whole new world that is relieved of the horrible frustration and agony of that condition; the terrible siege state where humans couldn't understand themselves can at last be lifted, and, with its lifting, life can return to every aspect of human existence, including to the impossible situation parents have been in trying to nurture their children.

[994] In the case of parents of extremely nurturing-deprived-and-thus-extremely-dissociated-from-the-world, autistic children, even with the ability now to understand the reason why they haven't been able to nurture their children, having lived in denial of the importance of nurturing, the revelation of its importance, and of the devastating effects of not receiving it, will come as a great shock. It is, in fact, an example of how the arrival of understanding of the human condition and all the truth that it

makes possible also brings with it the long-feared exposure day (or truth day or honesty day or revelation day or transparency day or 'judgment day'), which is an issue that will be addressed at length—including an explanation of how we are able to cope with that exposure—in chapter 9:7. It will be explained there that the way we cope with so much truth being suddenly revealed is by adopting the Transformed Way of Living, part of the strategy of which is to not overly confront all the truth that is suddenly exposed. What is important now is to support the truth, not seek to overly confront it.

———————————

[995] Before finishing this section, it is worth examining the comment from Winnicott, that the autistic boy, Ronald, **'had very exceptional skill in drawing'**. The established link between autism and savant abilities was described in the documentary *Uncommon Genius*, when it was stated that **'for more than half of all savants, the syndrome owes its origins to a familiar condition—autism'**. In describing savants, the narrator said, **'One of the great unsolved mysteries of the human mind is savant syndrome. This man cannot re-member how to clean his teeth, yet he recalls every zip code, every highway, every city in the United States. As a boy, this man was considered mentally handicapped but he can name the day of the week for every date on a forty-thousand-year calendar. And this man, blind, cerebral-palsied and barely able to talk, played Tchaikovsky's First Piano Concerto flawlessly after hearing it once on television. These genius-like abilities are the product of damaged minds...Most of us were introduced to savants through Dustin Hoffman's portrayal of Raymond in the Oscar-winning movie, *Rain Man*. Raymond had savant syndrome...Savants are people who produce awesome mental abilities from severely disabled minds'** (ABC-TV, 24 May 2001). The program mentioned that **'researchers discovered people who suffered dementia and then suddenly gained prodigious skills they'd never experienced before'**, and reported that **'savants may abandon their skills as they become more sophisticated socially...The really prodigious savants have a sort of memory super-highway that allows them to access and transfer enormous amounts of information. They develop that memory partly because they don't get side-tracked by thinking too much.'** It then concluded with the intuitively insightful question, **'Could savant genius lie dormant deep inside everyone's brain? Are savant skills merely obscured by layers of normal everyday reasoning?'** With understanding of the human condition it is not difficult to understand how those who suffer from autism and certain mental impairments can have savant abilities. If such people can

completely disconnect from or avoid **'get**[ting] **side-tracked by thinking too much'** about the **'normal everyday'** pain associated with the agony of the human condition, then their non-**'socially'-'sophisticated** mind can be freed to access some of the all-sensitive, all-loving, all-inspired, immensely imaginative and thus powerfully creative and able world of our soul. A window can be opened up to the fabulous potential that humans lost access to when they became so horribly preoccupied with worry about the hurtful, corrupted world around them and their own corrupted state.

[996] Earlier (in par. 832), I wrote about how I learnt that to draw the little pictures that are included throughout this book I had to disconnect my conscious mind and just let my instinctive sensitivity stream through, and that if I didn't do that I simply couldn't draw at all. For example, the drawing of the three 'Childmen' happily embracing that I used to illustrate humanity's childhood stage was done so quickly I shocked myself because I could hardly believe that such an empathetic drawing could be produced from an almost instant scribble. I wrote that at that moment I saw just how much sensitivity our species once had, and how much alienation now exists within us 2-million-years'-embattled humans, referring to Sir Laurens van der Post's observation of the relatively innocent Bushmen 'race', that **'He and his needs were committed to the nature of Africa and the swing of its wide seasons as a fish to the sea. He and they all participated so deeply of one another's being that the experience could almost be called mystical. For instance, he seemed to *know* what it actually felt like to be an elephant, a lion, an antelope, a…baobab tree'**. Yes, when R.D. Laing wrote that **'To adapt to this** [human-condition-afflicted, horrifically upset] **world the child abdicates its ecstasy'**, and **'between *us* and It [our true selves or soul] there is a veil which is more like fifty feet of solid concrete'**, and **'The outer divorced from any illumination from the inner is in a state of darkness. We are in an age of darkness. The state of outer darkness is a state of…alienation or estrangement from the inner light'**, and **'We are dead, but think we are alive. We are asleep, but think we are awake'**, he was emphasizing just how much soulful sensitivity we have lost access to as a result of our preoccupation with the agony of the human condition. As I said when describing how beautiful and amazing natural Africa is and how awful all our documentaries about natural Africa now are: 'Our world has shrunk to the size of a pea'! Yes, when humans become free of the agony of the human condition they will discover they have powers they have only ever dreamed of: they will see everything as if they have been blind up until now; they will feel everything; they will

be able to remember everything—their whole world will open right up! As William Blake famously prophesized in his poem, *The Marriage of Heaven and Hell*, which obviously refers to the time when the good and evil state of the human condition is reconciled: **'When the doors of perception are cleansed, man will see things as they truly are. For man has closed himself up, till he sees all things through narrow chinks of his cavern'** (1790).

[997] What occurs during a so-called <u>Near-Death Experience</u> actually illustrates what is occurring in some minds that have completely dissociated from reality. For instance, mountain climbers who survive falls that they were convinced would be fatal (they were saved, perhaps, by landing in a snow drift) often report that during those near-death moments they experienced a state of extraordinary euphoria in which the world suddenly appeared utterly beautiful and radiant and that they were flooded with a feeling of ecstatic enthrallment. With understanding of the horror of the human condition we can appreciate how, in such cases, the mind *would* give up worrying, and that all the facades—in particular the denial they adopted at Resignation—*would* become meaningless. If death is seemingly imminent, there is no longer any reason to worry or to pretend, at which point the struggle and agony and pretense of having to live under the duress of the human condition ceases and the true world of our all-sensitive soul suddenly surfaces. So yes, **'savant genius lie[s] dormant deep inside everyone's brain'**. The Superman mythology is an expression of humans' suppressed awareness of their alienated state—that resigned humans are Clark Kents with hidden Superman potential (news reporter Clark Kent being Superman's civilian identity).

[998] While it might be thought that the dissociation that occurred at Resignation, when adults adopted a life of denial of their reality, should have given them some access to the soul's truthful, sensitive world, resigned adults were not normally completely blocking out their reality—rather, they were constantly on the lookout to deflect any criticism arising as a result of that reality. Generally, the resigned did not completely dissociate themselves from the world in the same way that autistic people did. However, there were some people who although not sufficiently hurt in infancy and childhood to become autistic, were sufficiently hurt to need to live an extremely dissociative, alienated existence after they resigned; and such people *could* develop some access to the soul's truthful world. Sometimes when people became extremely upset/corrupted, their alienation, their mental blocks, their defenses, became disorganized and

through this 'shattered defense' the soul occasionally emerged; 'mediums' or 'psychics' or 'channelers' are examples of such individuals. Of course, such shattered-defense access of the soul's true world was not the natural, secure, balanced access that unresigned people have. For those people, whom we have historically referred to as prophets, the soul's world has always been an ultra natural place, not something apparently mystical or supernatural.

[999] R.D. Laing was describing the 'shattered defense' means of accessing the soul when he wrote that **'the cracked mind of the schizophrenic may *let in* light which does not enter the intact mind of many sane people whose minds are closed'** (*The Divided Self*, 1960, p.27 of 218). Interestingly, Laing immediately continued to say that the existentialist Karl Jaspers was of the opinion that the biblical prophet Ezekiel **'was a schizophrenic'**. While some biblical prophets may have accessed the soul's true world through a 'shattered defense', those who had full and natural access to the soul and were prophets in the true sense were exceptionally sound rather than exceptionally exhausted, alienated, separated from their true self, or schizophrenic. It should also be mentioned that when people prayed or chanted mantras or counted rosary beads they were trying to shut down their alienation-preoccupied mind in order to let through some of the truthful world of the soul. They were, in effect, trying to shatter their own defense. Fatigue, meditation, fasting, hallucinatory drugs, despair and faster-than-thought physical activities, such as scree-running, are other ways of achieving this breakthrough to the world of the soul.

[1000] (A more complete description of how nurturing has been seriously compromised by terminal levels of alienation can be found in *Freedom Expanded* at <www.humancondition.com/freedom-expanded-nurturing-now-a-priority> and <www.humancondition.com/freedom-expanded-autism>.)

## Chapter 8:16D The dysfunctional, extremely narcissistic 'Power Addicted' state

[1001] Another dire consequence of terminal levels of alienation emerging in the world, especially in the materially so-called 'developed' world, is the development of extreme egocentricity, especially in men. As will now be described, the effect of this egocentricity on children has also been catastrophic.

[1002] All around us are expressions of the punch-drunk egocentric state, this end play situation of the development of communication technology-fuelled upset in the world of having to get a win, a victory, success at any cost. We see parents on the sideline of sports fields, watching their children play and projecting their embattled need for a win onto those children, yelling, 'win, win, win'! These lyrics of Alanis Morissette's song *Perfect* describe how conditional love from parents passes the upset, embattled, soul-insensitive state of the human condition on to the next generation: **'Sometimes is never quite enough. If you're flawless, then you'll win my love. Don't forget to win first place, don't forget to keep that smile on your face. Be a good boy, try a little harder. You've got to measure up, and make me prouder...I'll live through you, I'll make you what I never was. If you're the best then maybe so am I...I'm doing this for your own damn good... What's the problem, why are you crying?...We'll love you just the way you are if you're perfect'** (1995). The effect this pressure has on their children is one of psychological devastation because, as I will now explain, it produces a new generation who are either psychologically crippled by that conditional reinforcement or turned into must-win 'power addicts', both of which are extremely dysfunctional states.

[1003] An article by the journalist Miranda Devine, truthfully titled 'Face it, we are all narcissists now', described how poor parenting is producing extremely narcissistic power addicts: **"Parents are becoming increasingly self absorbed [believing] 'the single most important thing in the world is for me to work like a dog and get the house, the car and the holiday house' and don't realise all their kids want is to be loved and to have one-on-one time with their parents." He** [the adolescent psychologist Michael Carr-Gregg] **says** [and part of this quote was referred to in par. 950] **an "epidemic of poor parenting" is to blame for a drastic rise in psychological problems in young people. "Generation Y is being ravaged by depression, anxiety disorders and stress disorders"** (*The Sydney Morning Herald*, 3 Sep. 2009). In the article Devine discussed the development of the narcissistic personality disorder in which adults develop **'a grandiose sense of self importance; preoccupation with fantasies of unlimited success, power, brilliance...a need for excessive admiration; a sense of entitlement; exploitive personal relationships; a lack of empathy; envy and arrogant, haughty behaviour...Everyone's a potential narcissist these days.'** The article mentioned **'a long-term study of...American college students** [that found]**...the incidence of narcissistic personality traits increased on a scale rivalling obesity'.** The reason, the article reported, for this recent explosion in narcissism is, to repeat Carr-Gregg, **'poor parenting'**. What Devine has written here about **'the incidence of narcissistic personality traits' 'rivalling obesity'** in American students is more than backed up by the aforementioned *TIME* magazine story, **'The Me Me Me Generation'**, which detailed the **'narcissism epidemic'** in the present **'80 million strong...biggest age group in American history'** Y-generation (see par. 951). This article even wondered if the Y-generation's **'self-centeredness could bring about the end of civilization as we know it'**, a very real concern that will now be explained.

[1004] As just described, having lived in denial of the truth that our species once lived in a cooperative, unconditionally-selfless, loving, harmonious state, there has been almost no recognition of the importance of nurturing children with unconditional love. The truth, however, is that children come into the world instinctively expecting to receive unconditional love and when they don't get it—receiving *conditional* love instead—they are so innocent, so trusting in a true, all-loving, upset-free world that the only conclusion they can come to for not being unconditionally loved is that for some reason *they* don't deserve it and are therefore a worthless, unlovable, bad person. Children are *so* naive, they are born *so* instinctively unaware of the existence of the upset state of the human condition, that they can't believe it is the adults' fault that they are not being given unconditional love; they are *so trusting* in the world around them, *so* codependent to it, they can only presume it's their fault. Again, 'codependent' describes someone who is **'reliant on another to the extent that independent action is no longer possible'**; in this case, children are so trusting of, and thus **'reliant on'**, the adult world they are incapable of thinking **'independent[ly]'** enough to know that they are not at fault, and trust in that knowledge. This conclusion by children that they are, in effect, bad either hurts them so deeply that they block all the pain of it out of their mind and become a psychologically detached, immensely alienated, crippled person—in the extreme, autistic—or they try desperately to prove they aren't bad or worthless, spending the rest of their lives obsessed with the need for reinforcement.

[1005] As already mentioned, the psychologist Arthur Janov developed the technique of 'primal therapy' in which adults seeking to lessen the hold their psychoses have on their lives are helped to work their way back to memories of the original (primal) hurt to their soul that occurred in their infancy and childhood as a result of growing up in the overly upset, insecure, have-to-somehow-establish-your-worth, massively egocentric world of today. In the following extracts (which come from—and serve as a condensation of—his famous 1970 book, *The Primal Scream*), Janov describes how children blame themselves for their parents' inability to love them and the dire consequences such recrimination has on their emotional development. He wrote (the underlinings are my emphasis): **'Anger is often sown by parents who see their children as a denial of their own lives. Marrying early and having to sacrifice themselves for years to demanding infants and young children are not readily accepted by those parents who never really had a chance to be free and happy** [p.327 of 446] **…neurotic parents are antifeeling, and how much of themselves they have had to cancel out in order to survive is a good index of how much they will attempt to cancel out in their children** [p.77] **…there is unspeakable tragedy in the world…each of us being in a mad scramble away from our personal horror. That is why neurotic parents cannot see the horror of what they are doing to their children, why they cannot comprehend that they are slowly killing a human being** [p.389] **… A young child cannot understand that it is his parents who are troubled…He does not know that it is not his job to make them stop fighting, to be happy, free or whatever…If he is ridiculed almost from birth, he must come to believe that something is wrong with him** [p.60] **…Neurosis begins as a means of appeasing neurotic parents by denying or covering certain feelings in hopes that "they" will finally love him** [p.65] **…a child shuts himself off in his earliest months and years because he usually has no other choice** [p.59] **…When patients** [in primal therapy] **finally get down to the early catastrophic feeling** [the 'primal scream'] **of knowing they were unloved, hated, or never to be understood—that epiphanic feeling of ultimate aloneness—they understand perfectly why they shut off** [p.97]**'.**

[1006] To reiterate, if a child is not given unconditional love, and is instead intimidated, frozen out and made to prove themselves all the time by arrogant, tyrannical, authoritarian, tough, extremely egocentric, overly self-worth-embattled, 'My way or the highway'-type parents, then that child will either be psychologically crippled by the situation, or made into a narcissistic <u>power addict</u> who has to win at all times and at all costs—where only success can keep at bay the terrifying conclusion that they are somehow unworthy. To gain power, the upper hand, these <u>powerpaths</u> or <u>psychopaths</u>, as they are otherwise known, will say anything, do anything, no matter how immoral—conduct that has been variously described as **'ruthless'**, **'manipulative'**, **'belligerent'**, **'bullying'**, 'totally self-centred', 'egocentric', 'amoral', 'cold', 'cruel', 'obsessed with wielding power over others' [in other words, controlling and dominating], 'deceitful', 'self-important', 'lacking any ability to empathise with others', with a mindset that is 'quick to blame others for their mistakes' (Robert Matthews, 'Are you living with a socialised psychopath?', *Focus*, May 1994 & *The Sunday Telegraph Review*, Apr. 1997; see <www.wtmsources.com/171>). As Morris West put it, 'brutalise a child and you create a casualty or a criminal' (*A View from the Ridge: The Testimony of a Pilgrim*, 1996, p.78 of 143).

[1007] Regarding the power addict's inability 'to empathise with others', the truth is, they are *so* preoccupied—in fact, utterly *consumed*—by their own pain that instead of empathizing with and being considerate of others as our instinctive self or soul is designed to be, they *impose* their condition on the world around them; they intimidate and frighten everyone with their angry, often silent embattled state; indeed, a common description of how they operate and their impact on those around them is that they have everyone 'walking on eggshells'. The psychiatrist Frank Lake articulated this problem of adults who, as a result of being under attack as infants and as children, treat everyone and every situation they encounter as if they were *still* under attack, writing that such an adult 'complains as if it remembered the bad times it had been through. It reacts to the world around it as if it were still in the bad place, still having to "feel its keenest woe". It reacts defensively as if the attack were still going on' ('Supplement to Newsletter No.39', *Clinical Theology Newsletter No.39*, Dec. 1981). Basically, the power addict's life is stranded or arrested in the childhood state of needing reinforcement.

[1008] You can quickly extrapolate just how absolutely desperate the power addict's mind is for reinforcement—how they aren't able to truly care about anyone or anything other than getting that reinforcement, how every moment of every day their mind can't afford to be anything other than singularly focused on how they can get that power or glory or attention or any other form of validation—if you imagine a child fighting with all their might to avoid the devastating conclusion that they are an unworthy, bad person. When all the evidence from the way that child is being treated seems to unequivocally indicate to that innocent, trusting, naive, ideal-world-expecting child that they are unworthy and bad, the child naturally fights back with all their being to resist that soul-destroying conclusion. If, however, that evidence is overwhelming, the child has no choice but to give in to that criticism and conclude that they are indeed worthless and bad. And then, to cope with that terrifying, absolutely unbearable conclusion, all the child can do is determine to never again allow their mind to connect with that conclusion that they are an unlovable, bad person, and since that core issue of their worthiness is where their true self is preoccupied, avoiding that issue amounts to separating or dissociating or splitting off from their true state and thus true self. This is the split, false, psychologically crippled state that was the focus of the previous sub-chapter (ch. 8:16C). *If, however*, the evidence that they are an unworthy, bad person is immense but not quite overwhelming the child will be stuck in a lifelong battle to avoid surrendering to the psychologically crippling conclusion that they are an unworthy, bad person. And that is where power addicts live—stranded in that state of terrible fear that they are unworthy, fighting with all their might to 'stay on their feet' and defy that implication. As mentioned, the actress Mae West famously articulated this situation of arrested development, within males in particular, when she said, **'If you want to understand men just remember that they are still little boys searching for approval.'**

[1009] Another of the many extremely dysfunctional features of the embattled power addict mind is its extraordinary ability to accumulate grievances. The narcissistic power addict is so unable to accept any criticism, so extremely insecure, so desperately preoccupied with **'searching for approval'** and avoiding any implication that they are unworthy, that all day, every day, they seek out only reinforcement, only 'wins', such as power, fame, fortune and glory. There is no room to accept any criticism and since most situations in life contain a spectrum of positives and negatives, of reinforcements and criticisms, of ups and downs, such minds are simply unable to fairly assess *any* situation because they cannot tolerate any of the negative, criticizing aspects of a situation. In fact, the power-addicted mind will find a way to avoid *all* criticism, no matter how much there is or how legitimate, and instead search for and find some fault or flaw in the situation, imagined or real, that brings reinforcement to them. Their totally defensive mind locks onto these 'grievances' as being all-significant at the exclusion of all other aspects of the scenario. Over time, any interaction with a power addict, no matter how fair or generous, will leave them harboring a mountain of grievances and totally incapable of recognizing the unfairness and irrationality of their view. The desperation of the power-addicted mind to avoid any threat to their power base is extremely difficult for the more balanced mind to comprehend. For instance, I have watched documentaries showing the former US President Franklin D. Roosevelt being overly cordial to the Russian dictator Joseph Stalin in the futile hope that Stalin would be fair-minded during their negotiations over the future of Europe at the conclusion of the Second World War. In public, Stalin always presented a benign, open and fair-minded countenance when underneath he was, in fact, a furious monster who, out of paranoia, oversaw the murder of millions and millions of his own people—including even his closest friends out of fear that they might one day challenge his power. Power addicts learnt to disguise their extraordinarily defensive mindset, but their behavior could often be absolutely, unbelievably selfish. Indeed, psychologists have learnt that it is all but impossible to reform power addicts because they are *so* desperate for success, *so* desperate not to fail, that they use what they learn in therapy to better disguise their weaknesses—for example, they learn to pretend they don't lack empathy and consideration of others to hide the fact that they are totally self-centered, ruthless and manipulative. As the article by Robert Matthews referred to above states, **'no amount of counselling seems to change them...Many people unwittingly try to change the behaviour of the psychopaths in their lives...Yet, as many psychiatrists have learned, it's almost invariably temporary. The psychopath's promise to mend his ways is usually just another pack of lies. Within a few months, the same traits will be back as strongly as before. The sad truth is that, if there is a psychopath in your office, your home or your bed, it will be you that has to change.'**

[1010] It has to be remembered, of course, that after 2 million years of the development of upset in the world, *all* adults are unavoidably stranded *somewhere* on the spectrum of insecurity of self due to the lack of reinforcement/love in their upbringing. The power addict state and, beyond that, the psychologically crippled state are simply the extreme states of the human condition. And again, it has to be emphasized that it is at last safe to admit this because we have now found the redeeming, dignifying understanding of how all the upset in the human species started, and, as will be explained in chapter 9, how it can be transformed and eventually ameliorated. All the upset in the world is now defended; the basis for our insecurity about being upset has gone.

[1011] I have drawn the following picture of the narcissism-producing and psychologically crippling situation to help children understand the world we live in. Since adults own the world there has been very little focus on the wellbeing of children. They can't fight for themselves. They can't write books. In fact, in their bewilderment children can't even tell adults what they are going through. If children were able to express themselves and be properly heard—for instance, if they were able to vote, if they had some power—the care and concern and respect for their wellbeing would have been much greater than it has been. As just emphasized in ch. 8:16C, unable to explain why humans have become so embattled, parents in the 'developed world' haven't been able to cope with any criticism of their inability to adequately nurture their children, but that incapacity to cope with any criticism of their inability to properly parent their children overlooks, indeed disregards, the fact that children in turn also can't cope with not being loved unconditionally. Parents have had it all their way. The powerlessness of children has made them the victims of the adults' powerfulness. The truth is, we live in an anti-child world today where 'kids' are irritants—and yet children are the next generation and to have no regard for them is to have no regard for the future.

Embattled, 'you-have-to-prove-yourself' parent.

Child trying to resist that it is unworthy of love.

*The child is turned into either a 'I-won't-back-down', 'must-win', power addict, or a psychologically crippled, broken person as an adult.*

[1012]The drawing shows a tyrannical father and the damaging effect his behavior is having on his child—the truth of which my professor of biology at Sydney University, Charles Birch, recognized when, after discussing with me the destructive effect power addict fathers have on their children, he made the extremely succinct (if somewhat blunt) comment: **'Haven't you heard Jeremy, the best thing that can happen in a man's life is that his father dies when he is born'** (WTM records, 12 Nov. 1998). Birch was homosexual and I think this comment is very revealing of the origin of the particular psychological upset that made him homosexual. (The psychological reason for homosexuality was explained in pars 802-804.) Some fathers claim it is necessary to toughen their children, kill off their inspired, truth-filled, happy, excited, nothing-is-too-difficult, loving original instinctive self or soul. A newspaper cartoon depicted this appalling point of view with the following exchange between a father and son: **'Son you're a liar, you're a bully, you're greedy, you're manipulative, you're self serving. You'll go far'** (Alan Moir, *The Sydney Morning Herald*, 11 Sep. 2009). Alienation is a tragically lost, dark, effectively dead state that the whole human race has been working desperately hard to escape from, *not* stay in! The *real* issue in the situation where parents want to toughen up their children is the extent of the upset anger, egocentricity and alienation in the parents, because it is *their* extreme upset that causes them to project onto their children the need to be a success at all costs. It is *their* need for glory, not the need to turn their children into 'survivors', that is the real motivation. So, no—parents saying they 'Want to toughen up their kids so they can face the real world', as though they are doing their children a favor, is just a convenient excuse to get away with **'poor parenting'** because all parents do intuitively know how critically important it is that children receive unconditional love if they are to grow into truly creative, functional and successful adults. As has been well and truly explained in this book, how to nurture children is one of the most ancient instincts in humans.

[1013]Earlier (in pars 493 and 974), reference was made to Ashley Montagu's 1970 paper, 'A Scientists Looks at Love', in which Montagu bravely admitted the importance nurturing plays in a child's well-being and future functionality, and the horrific consequences caused by a lack of unconditional love. His account is so powerful I am re-including some of what he wrote here, as well as some parts of the paper that weren't included earlier that refer to the **'psychopathic'** consequences of not receiving unconditional love: **'love is, without question, the most important experience in** the life of a human being…The newborn baby…wants love. He behaves as if he expected to be loved, and when his expectation is thwarted, he reacts in a grievously disappointed manner…If it doesn't receive love it is unable to give it—as a child or as an adult…Criminal, delinquent, neurotic, psychopathic, asocial, and similar forms of unfortunate behavior can, in the majority of cases, be traced to a childhood history of inadequate love…children who have not been adequately loved grow up to be persons who find it extremely difficult to understand the meaning of love…they tend to be thoughtless and inconsiderate. They have little emotional depth…they often seek ways of achieving power…Occasionally, when they attain prodigious political power as in the case of such unloved creatures as Hitler, Mussolini, and Stalin, they commit horrendous atrocities and subject humanity to irreparable damage. It is quite evident that the tragedy these men brought to the world was principally due to their incapacity to love…If one doesn't love oneself one cannot love others. To make loving order in the world we must first have had loving order made in ourselves…love is demonstrable, it is sacrificial, it is self-abnegative [self-denying]. It puts the other always first. It is not a cold or calculated altruism, but a deep complete involvement with another. Love is unconditional…Love is the principal developer of one's capacity for being human, the chief stimulus for the development of social competence'.

[1014] As the epidemic of the psychologically crippled state of ADHD and autism shows, what happens as upset anger, egocentricity and alienation increases is that eventually a threshold is reached where there is *so* much anger, egocentricity and alienation in the population that the great majority of children can no longer cope with the extent of the upset they encounter and a dysfunctional generation appears where there is a predominance of individuals who are either narcissistic power addicts or psychologically crippled. And it is this threshold, this terminal level of alienation, that Devine and the *TIME* magazine story have described as having now been reached at the forefront of human progress, which is mostly in the 'developed' Western world.

[1015] So yes, under the duress of the human condition there have basically been three varieties of resigned, alienated humans: those who received unconditional love; those who received conditional love (basically, non-love) and are still trying to resist the implication that they are unworthy and bad and as a result have become power addicts; and those who received either no love or so much conditional love that they were unable to resist the implication that they are unworthy and bad and as a result have had to dissociate from that terrifying, unbearable conclusion and, as a result of that, are psychologically crippled. The first category are those that our society has recognized as being 'functional' while the second and third categories are those that have been described as 'dysfunctional'. Another honest description we have had for these three states was, respectively, 'relatively secure', 'very insecure' and 'extremely insecure'. However, with the world reaching end play states of upset, the proportion of the population that is dysfunctionally alienated is becoming near total.

[1016] As a result, finding a secure, well-adjusted, functional 'soft handed' manager in a <u>company</u> now is almost unheard of, for they are nearly all power addicts. And once power addicts gain control of a company it really is the beginning of the end for that company—it will become dysfunctional because, as emphasized, power addicts are tyrannical and find anything that resembles criticism unbearable, such as criticism of their own ideas or having their mistakes pointed out. They are simply not effective free thinkers, which good business leaders need to be. And nor do they tolerate any perceived threat to their power base—as the adage goes, **'A's employ A's and B's employ C's'**, with A's being those who have been nurtured and are, as a result, relatively secure and therefore not threatened by talented people and eager to employ other A's. Insecure B's, on the other hand, will only employ C's because once in power they don't want anybody more functional than themselves usurping them. As a result, the company becomes dysfunctional.

[1017] And as with individuals in households and companies, the same process can be applied in the context of entire <u>countries</u>, just on a larger scale. What happens in this case is that when the proportion of power addicts in a country increases to a certain critical point 'right' suddenly no longer has the clout to win out over 'might', at which point dysfunctional tyrants are able to take control with terrifying consequences for both their country's population and, as history has shown, that of neighboring countries. Yes, just as children and wives of tyrannical men in family situations are unable to fight back and are crushed, tyrannical dictators can crush entire populations; the rise of Mussolini's totally ruthless, fascist 'Blackshirts' in Italy prior to the Second World War, and Hitler's ascension to power and invasion of eastern Europe are obvious examples. Tyrants such as these, and Napoleon and Stalin, have only been able to seize control when the number of power addicts in a population surpassed a critical point where the remaining soundness was no longer able to withstand their extreme intensity and defy them. As will be described shortly, the reason England has so often in the history of Europe been able to defeat tyrants there and elsewhere can be put down to its island isolation, which meant it was still sufficiently sheltered from much of the upset in the world, still sound enough, still free enough of power addicts, to stand against those tyrants who had taken over control of countries on the continent where, by inference, the tipping point of too many power addicts in the population had been surpassed.

[1018]Of course, it follows that the same process of a threshold of dysfunctionality eventually being reached within individuals, families, genders, generations, companies, 'races' and countries also applies to entire <u>civilizations</u>. When history books talk of great civilizations, such as the Greek and Roman Empires, and many others, becoming decadent, the main feature of that decadence was the dysfunctionality that resulted from having a majority of overly egocentric, insecure power addicts; there was simply not enough sound, functional, soulful generosity, fairness, honesty, sensitivity, humility and equanimity left in the system for it to remain functional. Once everyone wanted glory at all costs and there was no interest in truth or fairness—no interest in others, no moral strength—the system simply collapsed. And it is this threshold of power-addict-produced, decadent dysfunctionality that has now been reached in modern Western civilization, having, as our history books describe, already been reached in older civilizations around the world. So we can now understand exactly why and how the **'self-centred'**, **'narcissism'** of **'The Me Me Me Generation'** of the present **'80 million strong…biggest age group in American history'** Y-generation that was described earlier **'could bring about the end of civilization as we know it'**. The dire situation for our species has been reached where there are now no countries left that don't have a predominance of power addicts. Yes, recall the title of Miranda Devine's article: **'Face it, we are all narcissists now.'** God forbid, but it *has* come to this, a world of power addicts and cripples; a planet of chronically insecure humans; a complete madhouse. And since communication technology has shrunk the world in the sense of making it one civilization, the whole world has, in essence, reached the end play state of terminal alienation. This breakthrough understanding of the human condition and the transformation of the human race it makes possible has truly arrived without a moment to spare!

[1019]Yes, while, as mentioned earlier, psychologists have learnt that it is all but impossible to reform power addicts—that **'no amount of counselling**

**seems to change them'**—the new Transformed Way of Living that is made possible by the finding of understanding of the human condition enables all humans, *even power addicts*, to let go of their old competitive, power-fame-fortune-and-glory-seeking existence and transform their lives to a state free of that preoccupation. Having found understanding of the upsetting battle humans have had to wage against the ignorance of our instinctive self as to the fundamental goodness of our conscious self, preoccupation with that battle *can* at last end. Amongst the Founding Membership of the World Transformation Movement there are power addicts who have freed themselves of their affliction by taking up the Transformed Way of Living. The affirmations of the Transformed State given by WTM Founding Members John Biggs, James West and Nick Shaw, for example, that can be viewed and/or read at <www.humancondition.com/affirmations>, provide confirmation of this wonderful potential.

[1020] Again, the all-important fundamental truth that understanding of the human condition reveals is that upset is a *heroic* state, *not* a bad, evil, unworthy one. While all humans are variously upset as a result of their different encounters with humanity's heroic battle to overthrow ignorance, all humans are equally good and worthwhile. Understanding the cause of the upset state of the human condition eliminates the possibility of the prejudicial views of some people, genders, generations, 'races', countries or civilizations being good and therefore superior and others being bad, evil, unGodly and therefore inferior and unworthy. It is a simple and obvious truth that the longer and/or more intensely a person or a gender or a generation or a 'race' or a country or a civilization has been engaged in humanity's great battle to overthrow ignorance, the more upset and ego-embattled they unavoidably will have become. As mentioned, more will be said about the differences in alienation between 'races' and countries next, in chapter 8:16E.

[1021] So the power addict state is really only one of the extreme, end play results of the human condition, of 2 million years of generation after generation of humans trying to prove that they are good and not bad. Of course, any lack of reinforcement—and it has to be remembered that humans' original instinctive expectations are of receiving *complete* reinforcement during their infancy and childhood—can cripple or ego-centrically embattle a person; for example, children who are exposed to extreme ill-treatment, such as sexual abuse, are obviously likely to become horrifically psychologically crippled. But the biggest crippling influence on all children everywhere has been mothers' alienation from their natural nurturing instincts, and oppression from egocentricity, especially the extreme egocentricity of fathers. Yes, the upset state of the human condition in both mothers and fathers has had devastating psychological effects on children—especially mothers during their infancy, and fathers during their childhood. But *again*, this is all only another product of the end play state of terminal levels of alienation in the world.

[1022] (A more complete description of how egocentric the human race has become can be found in *Freedom Expanded* at <www.humancondition. com/freedom-expanded-poweraddict> and <www.humancondition.com/freedom-expanded-egocentricity>.)

### Chapter 8:16E  The differences in alienation between 'races' (ethnic groups) of humans

[1023] Having now described the emergence of terminal levels of alienation in the more materially 'developed' parts of the world (remembering that alienation from nurturing instincts in mothers and extreme egocentricity in fathers is having a devastating impact on the souls of all children everywhere), what now needs to be explained is the role greatly improved communication technology has played in the emergence of 'materialism envy' and the resulting unbridled greed, dysfunction and destitution in the materially poorer, so-called 'developing' world. But before presenting that explanation (which will be given next in ch. 8:16F), it is first necessary to explain what is really meant by the terms 'developed' and 'developing' worlds because, as will now be made clear, the explanation of what we really mean by these terms depends on recognizing the truth of the existence of different degrees of alienation between 'races' of people. (As explained earlier, since 'race' is a very imprecise term, what is meant by it in this book is a group of people whose members have mostly been together a long time and are thus relatively closely related genetically—basically people who have a shared history—so a more accurate term would be 'ethnic group'; however, *race* is what we use in our everyday language, so to indicate the interpretation of it here, I've placed inverted commas around it.)

[1024] The different states of material success in the world were once described in terms of the 'rich and the poor parts of the world', or the 'haves and have nots', then as the 'first, second and third worlds', then the 'developed and undeveloped worlds', before now being referred to as the 'developed' and 'developing' worlds. In the context of avoiding the unbearable issue of the human condition, the benefit of the terms 'developed' and 'developing' worlds is that they implied that any country could achieve material success—all that was needed was the time to develop it. The truth that will now be explained is that it wasn't possible to develop a functional society with ordered and well-run services where the majority of the population lived with a high standard of material comfort if the people in that society were either too innocent or too upset—once again, 'innocence' being lack of exposure to and familiarity with the angry, egocentric and alienated upset state of the human condition.

[1025] It has to be stressed that while we couldn't explain and thus defend the upset state of the human condition we couldn't afford to differentiate individuals, 'races', genders, generations, countries, civilizations and cultures according to how innocent or upset they were because it would have left the more upset condemned as bad, unworthy and inferior. It would have led to unfair, destructive and dangerous racist, ageist and sexist prejudice and discrimination against the more upset—and so a dishonest attitude of not allowing differentiation was silently enforced. But as was stated in par. 861, the truth is the *only* significant difference between humans—the acknowledgment of which makes it possible to truthfully explain and understand much of human behavior—is the difference in upset anger, egocentricity and alienation between individuals, 'races', genders, generations, countries, civilizations and cultures, but until the human condition was explained and upset defended as a good, heroic state we couldn't admit and talk about that all-important, clarifying difference. So yes, there are immense differences in upset between 'races' of humans which can—now that it can be understood that all humans are equally good though variously upset—at last be safely admitted. And the result of the differences in upset that exists between 'races' of people is that there are, in fact, more innocent 'races' who are relatively naive about the difficulties of living with the human condition, and other 'races' who are more instinctively adapted to upset and can thus cope better with it, and others still who are so instinctively adapted to upset that they are too aware of the reality of life under the duress of the human condition and thus overly cynical about being ideally behaved and thus overly selfish and opportunistic and thus socially uncooperative. The result of all this

variation in upset is that some 'races' have been effective in living with
the human condition while others have been either too innocent and naive,
or too upset, soul-exhausted and cynical.

<sup>1026</sup> Although we have had to avoid it, it is an obvious truth that humans
became increasingly adapted to life under the duress of the human condition, with some 'races' becoming more adept at that adaptation than others.
Just as individual humans vary in their degree of alienation from our
species' original instinctive, all-loving, selfless and trusting soulful true
self, so 'races' of humans naturally vary in their degree of alienation. The
longer an individual or a 'race' of people were subjected to life under the
duress of the human condition, the more they naturally became adapted
to that corrupt existence. While a relatively innocent person or relatively
innocent 'race' still behaved relatively ideally themselves and expected
others to do the same, other individuals and 'races' became so adapted to
the upset/corrupt world that they no longer behaved ideally themselves
and no longer expected others to behave ideally either. The longer humans
were exposed to the human-condition-afflicted state the more cynical
they became about human existence—a 'cynic' being **'one who doubts
or denies the goodness of human motives'** (*Macquarie Dictionary*, 3rd edn, 1998). As
mentioned in par. 761, when describing the Distressed Adolescentman
stage, the psychiatrist Wilhelm Reich wrote honestly about the effects
of the different levels of upset in the human race when he described how
**'The living** [those relatively free of exposure to upset]**...is naively kindly...It
assumes that the fellow human also follows the laws of the living and is kindly,
helpful and giving. As long as there is the emotional plague** [the flood of upset in
the world]**, this natural basic attitude, that of the healthy child or the primitive...**
[is subject to] **the greatest danger...For the plague individual also ascribes to
his fellow beings the characteristics of his own thinking and acting. The kindly
individual believes that all people are kindly and act accordingly. The plague
individual believes that all people lie, swindle, steal and crave power. Clearly,
then, the living is at a disadvantage and in danger.'**

<sup>1027</sup> The consequences for a society of its people becoming overly cynical
was that it meant that there would be too little soulful, selfless idealism
and too much upset-adapted cynicism-derived selfishness for the society
to function effectively. In the situation where it wasn't possible to explain
and thus defend the upset state of the human condition, the closest people
could come to admitting and talking about this fact that people became
adapted to the human condition was to describe individuals or families

or 'races' or countries or civilizations as having become **'dysfunctional'** and **'decadent'**, and—especially in the case of civilizations—as having **'passed their prime'** or **'peaked'** in terms of their creative powers.

[1028] Conversely, some 'races', like some individual humans, have, in fact, been too innocent to function effectively in the extremely upset-adapted, human-condition-afflicted, corrupt world. As mentioned in par. 848, Sir Laurens van der Post described how a member of the relatively innocent Bushmen people found it impossible to cope with having his innocent, natural spirit compromised: **'You know I once saw a little Bushman imprisoned in one of our gaols because he killed a giant bustard which according to the police, was a crime, since the bird was royal game and protected. He was dying because he couldn't bear being shut up and having his freedom of movement stopped. When asked why he was ill he could only say that he missed seeing the sun set over the Kalahari. Physically the doctor couldn't find anything wrong with him but he died none the less!'** And Sir Laurens was even more specific when he stated that **'mere contact with twentieth-century life seemed lethal to the Bushman. He was essentially so innocent and natural a person that he had only to come near us for a sort of radioactive fall-out from our unnatural world to produce a fatal leukaemia in his spirit.'** The honey-colored Bushmen are probably the most instinctively/genetically innocent group of people living today. They are more innocent, less soul-corrupted, less human-condition-adapted, less adapted to upset, less toughened, than dark-skinned Bantu Africans, but in turn Bantus are not as toughened and thus as operational and successful in the human-condition-afflicted corrupted world as Caucasians from Europe. For example, I once saw a documentary in which a Bantu African said something to the effect that **'My people can't compete with white people, you go to sleep at night only to wake up in the morning to find white people own everything.'** In turn, European Caucasians aren't as cynical, toughened and opportunistic—selfish—as people from even more ancient civilizations, like the Chinese from the ancient Yellow and Yangtze River valley civilizations, the Indians and Pakistanis from the ancient Indus and Ganges River valley civilizations, and the Arabs and Jews from the ancient Tigress, Euphrates and Nile River valley civilizations.

[1029] In another documentary I once saw, this time about a huge barge that travelled up and down the Congo River from Kinshasa to Kisangani, I could see the whole innocence-destroying, toughening process going on before my eyes. While most of the Africans on the barge were happy to pass the day innocently laughing and singing together, there were a few on-board who were buying produce from each port along the way and then selling it to their fellow passengers for a profit. In my mind it wasn't hard to extrapolate the situation and see that after only a few generations of this occurring that those who were innocently playing and enjoying life weren't going to survive as well as those who were less soulful and focused on 'making a living', which is code for 'achieving material success'—winning power, fame, fortune and glory; finding relief for their insecure human-condition-afflicted existence. Further, since their preoccupation with seeking that material relief will inevitably mean that their offspring will not receive the degree of nurturing and reinforcement they expect and require, that next generation will be even more insecure and thus even hungrier for the relief of power, fame, fortune and glory. And so it will go on, generation after generation of ever-increasing soul-less selfishness and psychological upset until there is either terminal alienation or reconciling understanding of our species' insecure condition.

[1030] The situation in Fiji provides a good case-study of what invariably took place when ethnic groups of varying degrees of upset cohabitated. In the late 1800s British colonists brought Indians to Fiji as indentured labor to farm sugar cane, and by the mid-1960s half the Fijian population was Indian. As a result, a serious conflict arose between the Indian and native Fijians, which we can now understand. The Indian Fijians, coming from an older and thus naturally more cynical, human-condition-toughened, human-condition-realistic and thus opportunistic civilization, have been so industrious and materially successful that they now monopolize the small business sector in Fiji to the extent that the native Fijians feel their country has been taken over by the Indian Fijians; for *their* part, however, the Indian Fijians also feel discriminated against. Indian Fijian sugar growers in particular feel this inequity, for while they produce 90 percent of the country's sugar, they are only allowed to lease land from the native Fijians (who own 90 percent of the land). Furthermore, since gaining independence in 1970 the native Fijians have ensured their domination of the political process—a state of affairs that was reinforced in 1990 when the Fijian constitution restricted the Indians to a maximum of 27 seats in the country's 71-seat Parliament. When this provision was amended in 1997 the Indians came to dominate the political scene, success-fully electing an Indian Prime Minister in 1999. This situation, however, was overthrown in 2000 when the native Fijians led a coup—and they have remained in power ever since. As mentioned, the Indian Fijians come from a very ancient civilization in India, one where innocence has long given way to more upset-adapted humans. In comparison, the native Fijians are still relatively innocent, yet to become embattled, hardened and upset-adapted. They aren't manically driven to win power and glory like more embattled, upset-adapted 'races', preferring to spend their day tranquilly occupied by such soulful activities as playing music, drinking the sedating kava and eating taro roots from their gardens — and trusting in soulful selflessness to care for each other. It is a situation where a 20-year-old, or thereabouts, equivalent 'race' is having to co-exist and compete with a toughened, cynical, more-upset-and-thus-more-insecure-about-their-goodness-and-thus-more-egocentrically-driven-to-try-to-prove-they-are-good-and-not-bad, competitive, selfish, opportunistic 50-year-old, or thereabouts, equivalent 'race'. (Note, it was explained in the Angry Adolescentman stage (in par. 860) how 15, 20 and 30-year-old equivalent stages could be said to exist during the last 2,000 years when that period is being described as the born-again Pseudo Idealistic 40-to-50-year-old **'fraud'** stage and the horrifically angry, punch-drunk, bitter and vengeful Hollow Adolescentman 50-plus-year-old **'criminal'** stage of humanity's maturation.)

[1031] While holidaying with Annie in Fiji in 1997 a local gave us this description of the structure of Fijian society: **'The Chinese** [who he said were **'the Jews of the East'**] **own all the big tourist resorts where the big money in Fiji is made, the Indians run all the shops and smaller businesses and produce all the sugar cane, and the Caucasians run the country in that they occupy so many of the important administrative positions, providing the good structure and order required for the whole society to function.' 'Fiji'**, he added, **'is one of the few countries in the world where the indigenous people still control the country even though they are the least materially productive and successful.'** When I asked other residents, including an Indian Fijian and a native Fijian, if they thought this was an accurate description they agreed it was but said that they would never say so publicly for fear of being labeled a racist. The human-condition-understanding-reconciled interpretation of this description is that as soon as you have an unavoidable and necessary battle such as the one that the human race has been involved in, it is inevitable that all those involved are going to become variously adapted to that battle depending on how long they have been exposed to it — with the result that the Chinese and Indians are the cynical 50-year-old equivalent 'races', the European Caucasians are the toughened, but not too toughened, too insensitive or too selfish, more operational 30-and-40-year-old equivalent 'race', while the native Fijians are the 20-year-old equivalent, overly innocent 'race'. Such differences are simply and obviously what manifest when you have an upsetting battle such as the one the human race has been involved in, where some people will have been engaged in the battle longer and/or more intensely than others. If we are going to have the truthful, meaningful, productive, effective discussion about human behavior—which is both possible *and necessary* now that the upset state of the human condition has been explained and defended—then the inevitable differences in upset (in particular differences in alienation from our species' all-trusting, sensitive, loving, selfless and sharing original instinctive self or soul) *have to be* acknowledged. The human race either stays living in Plato's terrible cave of alienated and alienating darkness until it self-destroys, or we get the full truth up and leave that awful cave existence for a world drenched in the sunshine of understanding, psychological freedom and togetherness.

[1032] As I mentioned in par. 745, Sir James Darling, my headmaster at Geelong Grammar School, recognized what is really an obvious truth, which is that for a person to be as functional as possible under the duress of the human condition they needed *both* human-condition-adapted toughness and sensitive, selfless, innocent, soulful soundness. While he was specifically talking about the qualities that education should strive to cultivate in an individual, what he said also applies to what a society of people needed if they were to be functional in the human-condition-afflicted world. In an address to The Royal Australasian College of Surgeons in 1960 Sir James said (the underlinings are my emphasis): **'The quality which, above all other, needs to be cultivated** [in education] **is sensitivity** [soul]…[Education's] **objective is a development of the whole man, sensitive all round the circumference…the future…lies not with the predatory** [selfish] **and the immune** [alienated] **but with the sensitive** [innocent]…**There is a threefold choice for the free man…He may** [be overly upset and selfish and] **grasp for himself what he can get and trample the needs and feelings of others beneath his feet: or he may try to withdraw from the world to a monastery** [find himself too innocent to cope with the upset world, like the Bushman who died in jail]…**or he may "take up arms against a sea of troubles, and by opposing** [all the dishonest, alienated denial and selfish, corrupt behaviors] **end them** [ultimately by solving the human condition]"…[and so] **There remains the sensitive, on one proviso: he must be sensitive *and* tough…Only by a growth** of sensitivity can man progress from the alpha of original chaos [upset] **to the omega of God's purpose for him** [to become sound and secure]...**Sensitivity is not enough. Without toughness it may be only a thin skin**...[only from] **an inner core of strength are** [you] **enabled to fight back** [defy all the dishonest denial and upset wrongness in the world]...**Can such men be? Of course they can: and they are the** [real] **leaders whom others will follow. In the world of books there are, for me, Antoine de Saint-Exupéry, or Laurens van der Post'** (*The Education of a Civilized Man*, ed. Michael Persse, 1962, pp.28-36 of 223). Yes, to be most operational under the duress of the human condition required a balance of innocent, soulful **'sensitivity'** and human-condition-adapted **'toughness'**, which is what the 30-and-40-year-old equivalent state represents.

[1033]The simple and obvious fact is that some 'races' are so relatively innocent and naive about life under the duress of the human condition that they lack the toughened self discipline and insecure egocentric drive to succeed of the more upset-adapted 'races' and, as such, can't legitimately compete with those people, so when they see an opportunity to obtain money and/or power and the material luxury both can bring they can't resist taking it, whether it's rightfully due or not. When Annie and I travelled through central Africa in 1992 everywhere we went, at every level of society, there was dysfunction, graft and corruption—even when we landed in Kenya we couldn't leave Nairobi airport until we paid certain 'fees' to various airport officials. At the top of such societies you invariably find completely despotic regimes—for instance, we were told that the reason the roads beyond the center of Nairobi weren't sealed and were in a terrible state was because all the money for such infrastructure had been syphoned off by the country's leaders to buy villas on the French Riviera and other luxuries. Indeed, an article titled **'Corruption tearing at the heart of Kenya'** revealed that **'only 1 per cent of government spending in Kenya can be properly accounted for, according to a report by the country's Auditor-General'** (*The Australian*, 1 Aug. 2015). And the simple and obvious truth is that at the opposite end of the spectrum of alienation there are 'races' where everyone is so upset-adapted, so soul-destroyed, so innocence-obliterated and manically egocentric and cynically selfish that graft and despotism is similarly endemic in their societies. Consider the events of early 2011, during the so-called Arab Spring, when the extreme despotism of almost every, if not every, Arab country right across North Africa and the Middle East provoked democracy-demanding uprisings throughout the region. During 2012 there were also reports in Australian newspapers every few days of the corruption that is endemic in China, where officials throughout the country are finding themselves unable to resist stealing whatever money they can lay their hands on. For example, under the heading **'Top to bottom, a culture of payola in China with 29,000 corruption convictions in a year'**, it was reported that **'China's corruption rate is astonishing even to its cynical citizens'** (*The Australian*, 13 Apr. 2012); while under the heading **'Lack of mining bribes "shocks China"'**, it was reported that **'Chinese mining officials are surprised when they learn that they do not need to bribe Australian officials to secure mining projects'** (*The Australian*, 28 Feb. 2012). The Chinese renowned love of gambling also indicates an extremely insecure need for a win, for material success. And in India it was reported that **'rampant corruption'** is so engrained there that the Indian government's **'Chief Minister's spokesman Wahid Parra readily concedes that…it's a virus in the society. It's in our blood'** (ABC News, 11 Jan. 2016; see <www.wtmsources.com/185>). In India there is little or no filter-down of wealth to the huge majority of poor people in the country—no viable middle class. The city streets are festooned with a tangle of illegal electricity wires, and everywhere there is chaotic traffic, pollution and poverty—there is little or no order or functionality. And since the only thing that keeps human society going is soul, innocence, soundness, moral strength—our species' original instinctive inclination to be selfless, loving and cooperative—once that goes all that remains is ruin. Yes, Tracy Chapman was right when she wrote and sang that **'All that you have is your soul'** (1989)—so thank goodness that through this ameliorating understanding we all now have the means to retrieve it, or, if not immediately retrieve it during our lifetime, to reconnect to it through the Transformed Way of Living. As was outlined in par. 743 and will be described fully in chapter 9, *all* humans can immediately leave the soul-repressed, insensitive, denial-committed-and-thus-extremely-alienated, selfish and egocentric power-fame-fortune-and-glory-seeking resigned life, and become part of the secure, happy, human-condition-and-Resignation-free new world that understanding of the human condition now makes possible.

[1034]It follows then that it is only at the middle of the spectrum of al-ienation, amongst 30-and-40-year-old, or thereabouts, equivalent 'races' where there is enough upset-adapted self-discipline and toughness, but not so much that there is excessive soul-less cynicism and thus selfishness, that you get maximum functionality and operable behavior in life under the duress of the human condition. The Anglo-Saxons are currently the stand-out example of such functionality, coming as they do from the more isolated and sheltered-from-upset north-western edge of Europe—they are actually more 30-year-old equivalents than 40-year-old equivalents. Although Anglo-Saxons come from a small, resource-deficient island country, they have been so operable and thus successful and thus influential that they have led the so-called 'globalization of the world' to the point where **'A quarter of the world's population speak English...English is increasingly becoming entrenched as the language of choice for business, science and popular culture. Three-quarters of the world's mail, for example, is currently written in English'** (*TIME*, 7 Jul. 1997). An article in the *National Review* titled 'Empire of Freedom' referred to the American internet entrepreneur James Bennett's use of the term **'Anglosphere'** to describe a coalition of English-speaking countries—the US, the UK, Australia, Canada, New Zealand and Ireland—that are **'characterised by a high degree of individualism and dynamism, and by a talent for assimilation'**. Bennett said it is **'no accident that it was in the Anglosphere that the industrial revolution and parliamentary democracy first emerged...Nor is it an accident that when French intellectuals and Malaysian prime ministers wish to denounce free markets, the phrase they use is "Anglo-Saxon capitalism"'** (Ramesh Ponnuru, 24 Mar. 2003). The Anglo-Saxons exhibit the same soulful enthusiasm, vitality and energy of the Vikings from Scandinavia who were referred to in par. 859. Indeed, like the Vikings, who sacked and settled in many parts of the British Isles, the Angles and Saxon tribes who colonized eastern England (and from whom the name Anglo-Saxon comes) were of southern Scandinavian stock. Also the Normans, who conquered England in 1066, were Scandinavians who had settled on the east coast of France—the name 'Norman' is actually derived from 'northman', meaning peoples from Northern Europe (Scandinavia). Of course, the basic Celtic stock of the British Isles are, like the Scandinavians, also an exceptionally isolated and thus relatively innocent 'race', especially coming, as a significant proportion of them did, from the early Celts of central Europe who were an *extremely* energetic, enthusiastic, relatively innocent/soulful 20-to-30-year-old equivalent adventurous 'race', but whose innocence had been (as was described earlier in pars 909-915) lost elsewhere in the less isolated parts of Europe where they had spread. This description of these ancient Celts from central Europe provides some indication of their 20-to-30-year-old equivalent personality: **'The Celts were high-spirited warriors – head-hunters no less – who loved drinking and brawling** [p.97 of 176] **...The flamboyant Celts originated in central Europe...**[having] **moved into central Europe from somewhere east**

**during the third millennium BC. They spoke a language that was probably the ancestral tongue of Indo-European...As the early Celts left no written words, it was the Roman and Greek observers who caught and preserved their fierce mien.** [Recording that] **"Nearly all the Gauls** [Celts from France] **are of a lofty structure...fair and ruddy complexion; terrible from the sternness of their eyes, very quarrelsome, and of great pride and insolence. A whole troop of foreigners would not be able to withstand a single Gaul if he called to his assistance his wife, who is usually very strong, and with blue eyes."..."The whole race...is war mad and both high spirited and quick for battle, although otherwise simple and not uncouth"** [p.110]' (*A Soaring Spirit: Time-Life History of the World 600-400 BC*, 1988).

---

[1035] Incidentally, I have often thought how inspirational and exciting it would be to meet some of those energetic, courageous, vital, still-full-of-our-soul's-enthusiasm-and-love-of-life, 20-to-30-year-old equivalent people, like the ancient Celts, or in more recent times, the Vikings who, from their Scandinavian home, bravely, fearlessly and adventurously wandered over most of the northern hemisphere, all through Europe, to Central Russia and North Africa and even to the then-unknown-to-Europeans land of the Americas. It is amazing to think about how afraid and timid humans can be when they become exhausted by the battle of the human condition, such as, in the extreme, a situation I read about where a person was too afraid of the world to venture beyond their home to post a letter at the nearby post office, when those Vikings ventured out in their long boats right across the then-unchartered face of our planet. Even though such exceptionally soul-infused-yet-toughened-but-not-yet-overly-toughened 20-to-30-year-old equivalent people have likely gone from Earth, I always thought it would be wonderful if we at least had a first-hand, detailed account of the lives of some of them. We have the life story of fearlessly courageous men like Lord Nelson, Winston Churchill and the polar explorers Amundsen, Nansen and Shackleton, but given the many generations and inevitable marriages with other 'races' that have occurred since the time when the original 20-to-30-year-old adventurous stock lived on the fringes of Europe, it seems unlikely that these men could be pure representatives of that high-spirited adventurous 20-to-30-year-old stock. However, I have discovered that there *are* contemporary, even first-hand, detailed accounts of the lives of those I believe have to be members of this amazingly vital original stock and stage in the human journey. As just mentioned, the Normans from Normandy in France came from Scandinavia, and in the eleventh century, not long after the Vikings were adventurously (not, as was emphasized in par. 859, 'angrily', like Napoleon and Hitler) marauding across Europe and beyond, there was a Norman named Tancred of Hauteville in Normandy who had 12 sons—and miraculously the absolutely amazing story of the lives of many of those sons remains in existence. Historians know from these first-hand accounts that one, Roger, became King of Sicily, where through **'tolerant' 'collaboration and assimilation'** he was able to **'build one of the most powerful kingdoms of the Medieval world'** (Prof. Robert Bartlett in his documentary series *The Normans*, BBC, 2010, Part 3). Another, the most courageous and adventurous of all the brothers, Robert Guiscard, conquered Italy and, according to the historian John Julius Norwich, **'proved to be a leader of the very, very first rank and I think probably is the greatest and most successful military adventurer between Julius Caesar and Napoleon—and if he hadn't died of typhoid when... leading an expedition against the Byzantine Empire, he might easily have taken Constantinople and the whole history of Europe and Christendom would have been changed'** (From the documentary series *The Normans: a dynasty that shaped the world*, narrated by John Morgan, 2004, Part 2). Comparing **'the two greatest Normans of their time'**, Norwich wrote in his 1992 book *The Normans in Sicily* that William the Conqueror, who boldly conquered England in 1066, **'used to screw up his own courage** [by] **reminding himself of the Guiscard's'** (p.249 of 793). Part 3 of *The Normans* documentary that was quoted from above also records that Robert Guiscard's offspring, along with the **'eldest son of William the Conqueror'**, **'were at the heart of'** the first Crusade to reclaim Jerusalem from its Muslim conquerors. In that Crusade, Robert's eldest son, Bohemond, led the re-taking of Antioch, **'one of the great holy cities of the Christian world'**, after which he became **'an independent Christian prince, the head of the Principality of Antioch'** that **'flourished under Norman rule for 200 years'**. Robert's grandson, Tancred, **'another fierce warrior'**, **'led the'** ultimately successful **'assault on'** Jerusalem and afterwards **'became Prince of Galilee'**. (You can learn more about these and other amazing offspring of Tancred of Hauteville in the book and two documentaries referred to in the quote sources above.)

[1036] So that's a window into how amazingly fearless, bold and adventurous the human race was during its 20-to-30-year-old stage. While the following may seem egocentric on my part, none of the insights in this book could have been found if I was insecure/egocentric so it is not prompted by egocentricity but my constant desire to know the truth. Obviously I've needed a lot of soundness and courage to find the insights into the human condition that are presented in this book, and while unconditional love from my mother when I was a child and having a father who was not at all oppressively egocentric when I was growing up would, in large measure, account for that soundness and courage, my genetic heritage *must* have also been influential; both nurture and nature form our character. And when I look into my paternal Celtic heritage there does appear to be evidence of remnants of that 20-to-30-year-old bold and adventurous strength there. 'Griffith' is a Welsh name, and while I don't know when my father's family originally ventured from Wales, I do know my great-great-great grandfather was an Anglican who lived in Cootehill, County Cavan in Ireland, and that his son adventured from there to the Caribbean (where the homestead on his sugarcane plantation in St Croix is now a museum), and that the generation that followed after him came to Australia where in the country town of Albury my great grandfather built the biggest stock and station agency outside of the capital cities, and that two generations after that, my father (who was a bold Z Special Unit commando sent to thwart the Japanese advance at Manus Island, north of New Guinea in the Second World War) set out without any financial inheritance and developed a sheep station near Mumbil in central west New South Wales and managed to send his four sons to the then soul-cultivating and prestigious (and expensive) Geelong Grammar School in Victoria. Looking at the origins of the name itself, Griffith, or Gruffydd as it was originally called, was, according to the **'tax roll' 'of 1292-3'**, the fourteenth most common name **'of Welsh people living in north-western Wales, in an area that had experienced relatively little influx of English people at that point** [so it was presumably a last stronghold of relatively pure Celtic stock]' (Heather Rose Jones, 'Constructing 13th Century Welsh Names', 1996; see <www.wtmsources.com/153>). Indeed, the most famous of all Welshmen, and he came from the remote mountainous region of **'north-western Wales'**, was Llywelyn ap Gruffydd, the only native King of Wales to be recognized by the English, or as the title was known at the time of his reign (from 1258 to 1282), the Prince of Wales. 'Ap' Gruffydd means 'son' of Gruffydd, with Gruffydd being the first name of his father since in those days there were no surnames, so while Gruffydd was not a family name and did not indicate a Gruffydd dynasty, he nevertheless had Gruffydd in his name, as did quite a number of members of his family tree. While I am probably no more related to Llywelyn ap Gruffydd than any other Griffith, he does serve as a good example of the relative innocence, soundness, enthusiasm and strength of the remnants of the original Celts of Europe from which my father's line emerged. To explain his fame I will quote from the 2012 BBC documentary series *The Story of Wales*: Llywelyn was **'a bold man, taking command of much of Wales and capturing land from English lords'**, but **'in 1277, King Edward I gather[ed] the biggest army seen in Britain since the Norman invasion'** to defeat Llywelyn who, in **'a cry of defiance'**, wrote to King Edward saying, **'He will never abandon his people who have been protected by his ancestors since the days of Brutus [the mythical ancestor of the Celts].'** Inevitably, Llywelyn was defeated, after which Edward **'bestowed the title Prince of Wales on his own heir, a tradition that continues to this day'** (presented by Huw Edwards, from Part 2). Llywelyn and his brother Dafydd were certainly very brave and defiant men—indeed, their whole family must have been extremely courageous and defiant considering the lengths Edward went to try to rid himself of their line by decreeing that all the **'child heirs of Llywelyn's dynasty are imprisoned'**. The program said that, **'like** [the legend of] **King Arthur'**, there is an ancient Welsh prophecy that says that one day a Welsh leader, like Llywelyn, will **'return to save his country'**, and while I'm not likely to be closely related to Llywelyn ap Gruffydd, my Celtic blood must have played a significant role in enabling me to defy the great overlord of all the dishonest denial of the resigned world and find the truth-based understanding of the human condition that, according to Professor Prosen, **'saves the world'**. In the classic 1941 film about a Welsh mining town, *How Green Was My Valley*, when the main character and priest, whose name, coincidentally, is Mr Gruffydd, says, **'I thought when I was a young man that I would conquer the world with truth. I thought I would lead an army greater than Alexander ever dreamed of, not to conquer nations, but to liberate mankind. With truth. With the golden sound of the Word'**, he is referring to the Bible's truthful **'Word[s]'**. However, at a deeper level, this is an extraordinary statement that really only makes sense when interpreted as an expression of the Welsh vision that one day one of them would *actually* liberate mankind with the *ultimate* golden truth, which is the science-based explanation of the human condition.

[1037]With regard to the prophesy of a Welshman one day defying all the denial and finding the truth about the human condition, a documentary about the legend of Merlin, the great Arthurian prophet/druid/priest/shepherd/guide, said that **'The newly prosperous age** [Britain in the 1700s]

needed an ancient past that could rival the empires of Greece and Rome. It looked
to the ruins of the British and Celtic world and in particular to the tales and traditions of the Druids and Bards…Mountainous [north-western] Wales becomes
a significant landscape in the British imagination. It was known as a place where
poets mattered, carrying the memory of the nation. One image had a huge impact,
the suicide of a lone bard who defied the invading armies of Edward I when he
defeated the Welsh and slaughtered their poets' (*Merlin: The Legend*, BBC, 2008).
Yes, a powerful tale in Welsh mythology holds that when King Edward I
completed his conquest of Wales and ordered all the Welsh bards—all
the truth-saying poet inheritors of the ancient druid-guided culture of the
Celts in Wales—to be put to death, there was one bard who defied the King
and, to quote Thomas Gray's 1757 poem, *The Bard*, with a **'prophet's fire'**
swore **'revenge'** on the King for having just killed **'soft Llewellyn'** and other
Celtic leaders of Wales, and prophesized the **'triumph' 'tomorrow'** of the
**'repair'** of the **'golden flood'** of **'the orb of day** [the sun]' which then **'warms
the nations** [of the world] **with redoubled ray** [with exposing but redeeming
truth]', before jumping off a cliff to his death. John Martin's absolutely
fabulously descriptive painting *The Bard*, opposite, depicts this particular
*defiant* stand of a prophet against tyranny, ultimately the tyranny of all the
dishonest denial imprisoning the human race. As William Blake wrote,
**'Hear the voice of the Bard, who present, past, and future, sees; whose ears have
heard the Holy Word…Calling the lapsed soul…light renew!…Night is worn, and
the morn rises from the slumberous mass…the break of day'** (*Songs of Experience*,
1789). The reference to **'soft Llewellyn'** relates to his reputation as being **'a
tender-hearted prince'** that **'though in battle he killed with fury, though he burnt
like an outrageous fire, yet was a mild prince'** (John Mitford, *The Poems of Thomas
Gray*, 1814, p.60 of 271). Like Roger Tancred who was extraordinarily **'tolerant'**,
these were soul-full but bold, adventurous 20-to-30-year-old equivalent
people, not soul-less, angry and egocentric 50-year-old equivalent people. (Incidentally, the tale of the Welsh bard who defied King Edward's
oppression by plunging to his death bears similarities with Australia's
greatest folk hero, the Celtic warrior Ned Kelly, who defied the overlords
here with his life—and also the swagman in Australia's unofficial national
anthem, 'Banjo' Paterson's famous poem *Waltzing Matilda*, who chose
death over compliance with the establishment. So Australians have held
a vision that is an extension of the Welsh prophesy—that it will supply
the defiant soundness needed to liberate humanity—and later in ch. 9:11
I provide more evidence of that wondrous Australian vision.)

John Martin's *The Bard*, 1817

[1038] So what people are really doing then through their efforts to either try to or actually remove tyrants/despots like Robert Mugabe, Saddam Hussein, Muammar Gaddafi, Hosni Mubarak, etc, etc, from power in the hope that those countries will become functional democracies is trying to make 20-year-old equivalents and 50-year-old equivalents behave like 30-and-40-year-old equivalents. But, as was explained, societies of 20-year-old equivalents and 50-year-old equivalents *are* going to revert to selfishness, at which point a selfish power struggle will occur where, in the end, the most ruthless will take over once again, a struggle we are currently witnessing in Syria. Having solved nothing at a fundamental level, those societies will invariably remain dysfunctional, resulting in further floods of refugees from those countries to others populated by more functional 30-and-40-year-old equivalent 'races'. Efforts to avoid this cycle, or at least contain it somewhat, in countries where there is too much cynical selfishness only led to the creation of authoritarian, dictatorial, freedom-and-democracy-denying, free-thinking-restricted, human-mind-oppressive regimes—which were therefore still fundamentally tyrannical and despotic—like those that have been established in China, and (to a degree) by Lee Kuan Yew in Singapore. As will be explained shortly in ch. 8:16I, the other form of tyranny that developed to contain excessive cynical selfishness was strict obedience to fundamentalist interpretations of religious teachings—as has occurred in many parts of the Arab world.

[1039] The fact is, a society that did not contain a significant proportion of functional 30-and-40-year-old equivalents was not going to be a functional one. Historically, **'the chief charge** [against Jews] **has been that Jews are somehow uniquely engaged—at other people's expense—in money making'** (*Ideas that shaped our World*, ed. Robert Stewart, 1997, p.55 of 223), and with compassionate understanding of the human condition we can understand that there was some truth in this view. The cynical 50-year-old equivalent Jews have managed to remain operational and, as a result, extremely materially successful by living amongst relatively selfless, functional 30-and-40-year-old equivalent 'races'—which is the real reason they have been persecuted in the predominately 30-and-40-year-old equivalent countries in which they settled. (The dishonest reason for their persecution being that they were the murderers of Christ). In this light it can be seen why the Pygmies and the Bushmen have resented the Bantu for being more operational and materially successful than they are, who in turn have resented the European Caucasians for being more operational and materially successful than they are—who in turn have resented and thus persecuted the Jews for being more operational and materially successful than they are. Everywhere the inevitable differences in upset between individuals, 'races', genders, generations, countries, civilizations and cultures have caused immense problems, so the greater truth is that it is a very great tribute to the character and courage of the human race as a whole that it has managed to maintain some semblance of functionality and cooperation under that almost impossible situation.

[1040] Indeed, indicative of the difficulty of the situation is the fact that during January and February 2011 the British Prime Minister David Cameron, German Chancellor Angela Merkel, the then French President Nicolas Sarkozy, former Spanish Prime Minister Jose Maria Aznar and former Australian Prime Minister John Howard all declared that **'multicultural policies'** have been a **'failure'** because **'immigrants'** had **'not successfully integrated'** (*Daily Mail*, 11 Feb. 2011). Yes, just as an individual person's lifestyle was inevitably going to largely be a response to that person's particular level of upset, so too a 'race's' culture was inevitably going to largely be a response to that 'race's' particular level of upset, which means different 'races' with their different cultures inevitably found it difficult co-existing. You don't very often see 30-year-olds forming close friendships with 50-year-olds, or even 20-year-olds with 30-year-olds. Most people relate much better to their own age group. In fact, the stages that occurred with different ages under the duress of the human condition changed so rapidly and were so dramatically different that 18-year-olds typically found it difficult relating even to 21-year-olds. Outside of family situations, everyone tends to fraternize with their own age group. The same situation of incompatibility obviously applied between 'races' of people. Different levels of upset had different needs. For example, as mentioned in par. 783, once humans became extremely upset even the glimpse of a woman's face or ankle became dangerously sexually exciting to men, which is why in some cultures women are completely shrouded and persecuted if any part of their body is revealed in public. Imagine then how difficult and provocative it has been for individuals from such extremely upset cultures to see young women from less upset cultures running around at liberty in bikinis and mini-skirts.

[1041] Again, as was stressed in par. 1025, such admissions of the relative innocence, or lack thereof, of different individuals, 'races', genders, generations, countries, civilizations or cultures could be very dangerous because they could lead to prejudiced views that some individuals, 'races', genders, generations, countries, civilizations or cultures are either good or bad, superior or inferior, when the truth is that while humans do vary in their degree of upset as a result of the necessary and heroic battle humanity had to wage to find self-knowledge, *all humans are equally good.* Upset is not a bad, evil state, but a good, heroic one. Trying to manage differences in upset between individuals, 'races', genders, generations, countries, civilizations and cultures has been extremely difficult, but once the prejudiced views arose of some individuals, 'races', genders, generations, countries, civilizations or cultures being either good or bad, superior or inferior, more worthwhile or less worthwhile, terrible atrocities and injustices very often followed. For instance, in the last century alone we have seen the Holocaust in which approximately 6 million European Jews were exterminated by the Nazis during the Second World War; the attempted 'ethnic cleansing' by the Bantu Hutu of an estimated 800,000 of the more upset-adapted Nilotic Tutsi in 100 days of bloodshed in Rwanda in 1994; and Idi Amin throwing out of Uganda, in 1972, all the Indians and Pakistanis, some 40,000-80,000 people, who owned and operated most of the businesses there because he claimed **'they** [were] **sabotaging the economy of the country'** (*Jet* mag. 14 Sep. 1972). No wonder then that when ascribing **'genes for negative traits'** to **'ethnic groups'** there has been concerned debate over **'whether some avenues of research are best left un-trodden because what they reveal is bound to be socially and culturally incendiary, whatever the outcome. Or is it intellectually dishonest, even cowardly, not to investigate all aspects of the human condition?'** (Andy Coghlan, 'Bun fight over warrior gene', *New Scientist* 'Short Sharp Science' blog, 10 Aug. 2006; see <www.wtmsources.com/147>). So again, until we could explain the upset state of the human condition it could be very dangerous acknowledging differences in where individuals, 'races', genders, generations, countries, civilizations or cultures were in their progression from innocence to upset in the human journey from igno-rance to enlightenment, but now that we have explained that all humans are equally good, it is both psychologically safe and necessary—if we are to truly understand ourselves—to acknowledge the differences.

[1042] By way of example of just how limiting it has been not being able to acknowledge the differences in upset between individuals, 'races', genders, generations, countries, civilizations or cultures, Plato quite sensibly wanted to have the least ego-embattled/most innocent—the **'philosopher kings'** or **'philosopher rulers'** or **'philosopher princes'** or **'philosopher guardians'** as he variously described them—lead society. He wrote, **'isn't it obvious whether it's better for a blind man** [an alienated person] **or a clear-sighted one** [an innocent, ego-unembattled, denial-free, honest person] **to keep an eye on anything'** (*The Republic*, c.360 BC; tr. H.D.P. Lee, 1955, 484), arguing that **'If you get, in public affairs, men who are so morally impoverished that they have nothing they can contribute themselves, but who hope to snatch some**

**compensation for their own inadequacy from a political career, there can never be good government. They start fighting for power...**[whereas those who pursue a life] **of true philosophy** [honest, unresigned, egocentricity-free thought] **which looks down on political power...**[should be] **the only men to get power...men who do not love it** [who don't egocentrically hunger for power, fame, fortune and glory]**...rulers** [who] **come to their duties with least enthusiasm'** (521, 520). But as completely **'obvious'** and right-minded as Plato's idea was of having the more innocent run society, such honest differentiation according to who was innocent, and who was not, wasn't possible because, once again, while the human condition wasn't able to be explained, any differentiation between individuals according to degrees of alienation or soundness left those no longer innocent unjustly condemned as bad and unworthy. In the absence of such honesty about who was innocent and who was not, functional societies did, however, try to avoid the overly insecure and egocentric from holding power by having regular democratic elections where such people could be voted out of office. (I should clarify that, as just explained in par. 1033, a degree of human-condition-adapted, toughened self-discipline has been needed to effectively manage the difficulties of life under the duress of the human condition, so 'races' could be too innocent to be effective **'philosopher guardians'**—which is why the phrase 'more innocent' rather than 'most innocent' was used above.)

[1043] But now that it *is* both possible and necessary to talk about who was innocent and who was upset, we can see and admit that colonization under the rule of 30-and-40-year-old equivalents did make significant sense. The 'races' who were most functional under the duress of the human condition tried to help the less functional become more materially developed—that is, advanced in the many arts of living with the human condition needed to progress humanity towards greater knowledge, ultimately understanding of the human condition. As Sir James Darling has written about the British Empire, **'the function of Empire is to educate rather than to oppress'**, and the British have **'an unbeaten record in the history of civilization'** (*The Education of a Civilized Man*, ed. Michael Persse, 1962, pp.134, 136 of 223). Sir Laurens van der Post similarly wrote that **'Great Britain'** created **'the largest, the greatest and, I still believe, the best-organised, and most civilized empire in the history of the world'** (*The Admiral's Baby*, 1966, p.108 of 340). All the history books written by truth-denying-in-order-to-artificially-make-everyone-equal postmodernists that condemned colonialism as the worst evil are going to have to be re-written truthfully.

[1044] In fact, I should mention here some of the human-condition-avoiding, denial-based reasons that have been put forward to explain the dysfunctionality of African countries like Kenya, including placing the blame on colonialism. Firstly, it is claimed that such countries are on the same journey as European nations, which went through their own dysfunctional stage before organizing themselves into upset-restraining, so-called 'civilized' democracies—that the Bantu African 'races' aren't any more innocent than European 'races' and will, in time, as mentioned earlier, be able to develop functional, materially successful democracies. In keeping with this theory, it was mentioned earlier that instead of using terms like 'first, second and third world countries', the current politically correct nomenclature for the different states of functionality of countries is to refer to them as being either 'developed' or 'developing'. But if time, rather than degrees of innocence, or lack thereof, was the issue then 'developed races' from ancient civilizations should not be dysfunctional—and yet they are. The Greeks gave us Plato, Socrates, Aristotle and the foundations of 'Western civilization' and yet, in 2011, through their present innocence-destroyed, burnt out, peaked, decadent, selfish greed and resulting dysfunction, brought the world's economy to its knees through their refusal to adopt responsible financial practices—a recalcitrance that continues to this day. The other Mediterranean Caucasian countries of Italy and Spain have similarly proved to be dysfunctional, peaked societies of people, having also had to be bailed out by the International Monetary Fund (IMF). The Celtic Irish, who also got themselves into financial difficulty, aren't burnt out so much as too innocent and naive—indicative of their comparative functionality is that in contrast to their European neighbors, Ireland recognized the need to implement and maintain austerity measures prior to and following the global financial crisis and, as a result, became the first Eurozone nation to exit the international bailout. As for blaming the dysfunctionality of African societies on colonialism, there is no doubt colonialism had negative, exploitive repercussions—some truly terrible, like the slave trade—and it did seriously disrupt the old tribal system that operated throughout most of Africa, but while tribalism, an authoritarian, dictatorial system in which the most powerful ruled, brought some peace and order (as it effectively does in all non-human societies, such as in wolf packs or in any herd animal species), it was still dysfunctional in that it oppressed the individual freedoms/liberties that humans' search for knowledge/creativity depends on. Of course, you *can* manage humans by tying them all down, as was done in tribal situations and Marxist and Confucianist regimes, but they will no longer be humans—they will no longer be conscious beings fulfilling their fundamental responsibility of exercising their minds and learning to understand existence. Overall, colonialism gave individuals many freedoms they hadn't had that the individual then had to manage—but the challenge for humans has always been to manage their consciousness-derived freedom *effectively*. As has been explained, the lack of effectiveness of that management across a social structure was due to 'races' being either too innocent, sensitive and naive about life under the duress of the human condition, or too toughened, soul/innocence-destroyed, cynical, opportunistic, greedy and selfish.

[1045] On the whole we can see that with understanding of the human condition it at last becomes possible to explain in psychological terms what was actually happening when history books spoke of civilizations having 'peaked' and become 'decadent'. As with each individual during his or her life, under the duress of the human condition all 'races' eventually became overly corrupted, corruption of our original instinctive self or soul being the price of humanity's heroic search for knowledge. In this journey from innocence to exhaustion of soul the most creative period was the toughened and disciplined, but not yet overly corrupted, 30-year-old equivalent stage. As each 'race' and its associated civilization passed through this stage it made its particularly creative contribution to the human journey. This was when civilizations were at their 'peak'; inevitably, however, they entered a more corrupted 'decadent' stage. To look at just the Mediterranean, Middle East, Indian and Eastern civilizations, they all made extensive contributions to the human journey during their energetic and creative 30-year-old equivalent stage. The Egyptians and peoples from the fertile crescent of the Tigris and Euphrates delta in the Middle East instigated the civilization of the 'known world'; for example, they invented the wheel, mathematics and writing, and divided time into minutes and seconds. As already mentioned, the Greeks and Romans laid the foundations for 'Western civilization' during this most creative stage of their journey through ever-increasing levels of upset. The great religions of the world, Hinduism, Buddhism, Judaism, Christianity and Islam, were developed in India and the Middle East when there was still enough soundness left in their populations to produce some exceptionally sound, truthful, unresigned, denial-free thinkers or prophets. And during their most creative stage, the Chinese contributed to the human journey such influential inventions as paper, moveable type, the compass and gunpowder. The truth is *ALL* civilizations right back through history made important contributions to the advancement of knowledge—but just where the leading edge in the advancement of knowledge was occurring at any one time depended on what stage in the human journey from innocence to exhaustion and decadence those various civilizations were at.

[1046] I might add here a balancing comment regarding the earlier statement about the Jewish 'race' having benefited from living amongst more innocent, soulful, selfless 30-and-40-year-old equivalent 'races'. At the end of the description of Adventurous Adolescentman, when analyzing the graph of *The Development of Mental Cleverness*, it was explained that the reason brain volume plateaued towards the end of the last 2 million years of its growth is because a balance was struck between the need for cleverness and the need for soundness—between knowledge-finding yet corrupting mental cleverness and conscience-obedient yet non-knowledge-finding lack of mental cleverness, with the average IQ today representing that relatively safe conscience-subordinate compromise. It is true that the ability to find answers didn't necessarily accompany increased intelligence because, as just described, increased intelligence tended to lead to an increase in upset and thus alienation, and alienation made thinking truthfully and thus effectively very difficult; however, a high degree of intelligence was still required to find knowledge, most especially in complex subject areas like higher mathematics and physics. Thus, if the human race couldn't develop exceptionally high levels of intelligence then many crucial understandings about the nature and workings of our world would not have been able to be found. The Jews are renowned for being exceptionally intelligent, which is consistent with them being an exceptionally upset 'race', and it is from within their ranks that some of the greatest minds and insights have emerged: Albert Einstein, with his breakthrough insights into the physical nature of our universe, is the most obvious example. It is true that Einstein must have had an exceptional degree of soundness to have been as an effective thinker as he was, but he also must have been exceptionally intelligent to so successfully grapple with the extremely complex subjects he was dealing with. I haven't ever tried to collate all the *intellectual* contributions to the human journey that the Jews have made (in addition to having, as was mentioned in par. 927, contributed the best collection by far of denial-free *instinctual* knowledge in the form of the Bible's Old Testament), but it would be very significant, as evidenced by this extraordinary statistic: **'just 0.2 per cent of the world population, Jews accounted for 29 per cent of Nobel Prizes in the late 20th century, and 32 per cent so far in 21st century'** (*The Spectator*, 6 May 2014). By, in effect, allowing exceptional cleverness/intelligence to develop by countering its corrupting effects with the presence of people who were not so intellectually clever and thus not so upset and thus not so upset-adapted was, in the bigger picture, a fortuitous outcome for the human race. But again, talking about 'races' having different levels of intelligence was very dangerous while we couldn't explain the human condition, explain that all humans are equally good—which is why James Watson, one of the Nobel Prize-winning co-discovers of the DNA double helix, was condemned and ostracized for **'daring to suggest that race and intelligence are linked'** (*The Independent*, 2 Jan. 2015; see <www.wtmsources.com/149>).

[1047] Of course, while it could be very dangerous admitting differences in upset between humans while we couldn't explain the human condition, complete denial of there being any difference was *also* very dangerous, because, as has been evidenced throughout this book, denial and the alienated, dark, death-like 'cave' existence that resulted from it, blocked access to the truthful understanding of our human condition; ultimately denial leads to the terminally alienated death of the human race. For there to be progress in the all-important human journey from ignorance to enlightenment there had to be some honesty. That is what democracy has really been about: trying to maintain a balance between unbearably confronting honesty and humanity-destroying dishonesty. We will see shortly (in parts 8:16H to 8:16Q of this chapter) how the completely dishonest denial and delusion that is currently being practiced by left-wing pseudo idealists is threatening humanity with extinction. In fact, Sir Laurens van der Post's honesty about the different degrees of upset between 'races' of humans provides a stark illustration of the negative and positive aspects of such honesty. As was described in par. 864, he was maliciously persecuted for daring to acknowledge the relative innocence of the Bushman people of the Kalahari—his honesty was so unbearable it made **'the academic experts...absolutely berserk with rage'**. But Sir Laurens' honesty about our species' lost state of innocence also saved the human race because, as will be explained in par. 1282, it was the crucial help I needed to continue my journey to truthfully confront and ultimately solve the human condition. Sir Laurens was actually acknowledging both the negative and positive aspects of his honesty about the differences in upset between 'races' when, in regard to the differences in upset between the Bushmen, Bantu and European Caucasians, he wrote that **'the indigenous people of Africa, although no longer primitive in the sense that the Stone-Age aboriginal Bushman of the Kalahari were, they were, for all their relative sophistication, far closer to their natural selves than we** [white people] **were. As a result they had become more and more something of a mirror reflecting a forgotten and fast-receding part of ourselves'** (*Jung and the Story of Our Time*, 1976, p.37 of 275). Yes, 'races' who are more **'natural'**/uncorrupted/innocent have been **'a mirror reflecting'** the unbearable truth of a corrupted **'fast-receding part of ourselves'**, *and* the truth that we are living in a dangerous state of dishonest denial or **'forgotten[ess]'**.

[1048] With regard to presenting a perfectly balanced account of all the different states of upset and their interactions and effects, such as that of the story of the Jewish 'race', it has to be recognized that the human journey has been *such* a complex story that a perfectly balanced view is beyond the powers of effective interpretation in this very early stage of viewing the history of the human race in a denial-free way. And such a detailed interpretation can actually wait because <u>what is so important now</u>—and this is all explained in the final chapter (9) of this book—<u>is that the human race can leave behind its whole upset history as compassionately dealt with</u>—our history *is finally*, as the saying goes, just that—it's 'all just history' now. <u>As will be explained in chapter 9, what brings all the horror of life under the duress of the human condition to an end is the Transformed Way of Living</u>.

[1049] My job, as the synthesizer of understanding of the human condition, is to get at least the main descriptions of all the hard truths up and dealt with so that humanity can move well out into the clear, free of the past—actually, my brief is to present the new world on a perfectly laid out platter, with the old world all bundled up and tied off in neat parcels for permanent storage. Yes, as will be emphasized in the final chapter, <u>we get the truth up and then we move on to a wonderful new world free of the agony of the human condition</u>. We leave the old effectively dead, dishonest, human-condition-afflicted world behind. It will be explained that this new world is brought about by the indescribable magnificence of the Transformed Lifeforce State that the songwriter and singer Bono (of the band U2) sang about: **'I've conquered my past** [found the dignifying, human race liberating understanding of the human condition] / **The future is here at last / I stand at the entrance to a new world I can see / The ruins to the**

**right of me / Will soon have lost sight of me'**; and similarly, as Beethoven's *Ninth Symphony* anticipated, **'Joy!'**, **'Joyful, as a hero to victory!'**, **'Join in our jubilation!'**, **'We enter, drunk with fire, into your** [human-condition-understood] **sanctuary…Your magic reunites…All men become brothers…All good, all bad…Be embraced, millions! This kiss** [of understanding] **for the whole world!'** (1824). (Incidentally, since the music of the *Ninth Symphony* gloriously elevates the magnificence of these words from Friedrich Schiller's 1785 poem *Ode to Joy*, it is no wonder **'it's generally considered to be the supreme artistic achievement of western civilisation'** (Paul Gambaccini, 'Your Desert Island Discs', BBC Radio 4, 11 Jun. 2011), and was adopted by the European Union as its anthem.) <u>Yes, it no longer matters who is more cynical, more human-condition-adapted, because the Transformed State leaves all that behind. The greater truth is we humans have *all* become well and truly corrupted/upset/messed-up/damaged/soul-devastated anyway, but none of that matters now because we are leaving that upset world, *we are out of there*; we have won the match and we can now all head for the showers and get ready for humanity's great victory party—and soon even the different scars we all carry from the match will be healed and gone, soon the human race will be rehabilitated. Even the upset that is now instinctive, in our genes, will not be an issue because when our capacity to love is finally liberated, as it can now be, it will effectively make all our upset—both psychological and instinctive—disappear without a trace. It *is* all history now; **'the ruins to the right of** [us]**, will soon have lost sight of** [us]**'**; **'all good, all bad'**, **'this kiss** [of understanding is] **for the whole world!'**</u>

[1050] (More can be read about how variously selfish and cynical humans have become at <www.humancondition.com/freedom-expanded-selfishness>.)


## Chapter 8:16F <u>The emergence of 'materialism envy' and with it unbridled greed, extreme dysfunction and destitution in the 'developing' world</u>

[1051] <u>Having now explained that the real difference between 'developed' and 'developing' countries is between the functional 30-to-40-year-old equivalents and the dysfunctional 20-year-old and 50-year-old equivalents, the issue of the 'materialism envy' and ensuing corruption and dysfunction that has been occurring in the less functional, 'developing' countries can and has to be looked at more closely.</u>

$^{1052}$The important factor that needs to be considered in the occurrence of 'materialism envy' in the less operational 'developing' parts of the world is the role that has been played by <u>immensely improved communication technology</u>. During the so-called colonization period of the world where the more functional, materially 'developed' countries established colonies in the more dysfunctional, materially 'undeveloped' countries, the populations of those nations had little awareness of just how materially successful people in the more functional, 'developed' world generally were, but with the advent of sophisticated communication technology this changed dramatically. With the introduction of the likes of television suddenly these colonized people could see what they were missing out on materially and when they found themselves unable to compete with those more operational and functional, not-too-innocent-but-also-not-too-cynical materially developed colonizers it naturally made them insanely envious. People in the materially impoverished parts of the world suddenly had access to programs like *Baywatch* (TV series 1989-2001), in which gorgeously groomed Californians in stunningly enticing swimsuits played in a fabulously ordered and materially wealthy world, and naturally they were overcome with distress about what they now knew they were missing out on. The psychiatrist Clancy McKenzie gave this stark description of, what might be termed, 'the Baywatch effect' or 'materialism envy' in action: **'While visiting Machu Picchu in Peru in 1979 I noted very poor persons, living in the mountains, who had only the clothes they wore and perhaps a lama or two, but had beautiful, warm smiles and seemed content and happy. Days later I was in Bogota in Colombia. It was a very hot day and we asked the driver to stop at an outdoor tavern to buy cold beer. The people were very impoverished, but there was a TV playing and they were able to view the "outside world" where everyone seemed to have more, and luxury was abundant. I offered to go in with the driver and he urged me to wait in the car. I soon learned why. The absolute hatred was so intense that it was palpable. These people did not have less than those in Machu Picchu but they saw others who had more, and their needs were intensified'** (Letter to Prof. Harry Prosen, 27 Mar. 2006). The result of this awareness of what materialism offers is that there is now rampant greed in all those populations who haven't had the material luxury that is so prevalent in the more functional, 'developed', mostly Western world. They see themselves as being on catch-up—they want the big house, a car, a fridge, a suit, a gown, a fancy watch, a beach holiday, Nikes, Coca-Cola and Big Macs, etc. The effect of this myopic drive for material wealth is that the

natural inclination of our original instinctive self or soul to be selfless
and loving has been thrown by the wayside with cynicism now reigning
across the less functional and materially fortunate 'developing' countries;
again, a 'cynic' is **'one who doubts or denies the goodness of human motives'.**
The following is another description of the consequences of 'materialism
envy' in Kenya, but it very much applies to what is happening throughout Africa, South America, China, India, south-east Asia and in parts of
the Middle East. In her 2000 book, *African Visions*, the aforementioned
Kenyan photographer Mirella Ricciardi wrote of her country's **'chaos
and confusion, the crime and lawlessness, the greed, the bribery and corruption
at all levels, the senseless wildlife killings, the environmental destruction, the
mindless power games, the hopeless destitution and hunger, the lack of concern
and forward thinking...The outrageous greed and disregard for anything and
anyone other than themselves and their immediate families in the top ranks of
governments, have created a financial drain on the country that filters down
and affects whole populations in varying degrees, threatening the fragile infrastructure of the land. All those, without exception, who have somehow attained a
position of command and authority, however humble, have but one idea in mind,
to emulate the example set by the head of state – to milk their positions for all
they are worth, for as long as it lasts. Because finances are siphoned off to Swiss
bank accounts, salaries are so low they need topping up by whatever means.
As a consequence bribery and corruption at every level is the only option. The
police force and judiciary, the ministers, the administrators, the teachers, the
doctors and the heads of government divisions, extract whatever they can, by
fair means or foul, under the guise of their uniforms and titles, while the country
collapses and the masses starve'** (pp.260-261 of 287). Yes, the end result of the, in
truth, very obvious difference in functionality between 'races' of humans
living under the duress of the human condition is that the less functional
and thus less materially successful 'races' naturally became extremely
jealous and envious of the material success of the more functional and
operational 'races', resulting in **'outrageous greed'**, **'bribery and corruption
at every level'**, **'while the country collapses and the masses starve'.**

[1053] As McKenzie's example shows, the material inequality in the world
that sophisticated communication technology has made visible to everyone has made those who are missing out not only extremely envious, but
resentful and thus angry, to the point of feeling **'absolute hatred'** towards
the more materially successful. Their self-esteem has suffered so much
that angry retaliations, like the flying of those planes into The Pentagon

and the World Trade Center on September 11, 2001, have occurred; Osama bin Laden actually said the 9/11 attacks **'were revenge for Western humiliation of Muslims'** (*TIME*, 7 May 2012). When the Muslim sympathizer Tessa Kum wrote, **'I am coming to hate you, white person. You have all the control, all the power, all the privilege'** (Sydney's *Daily Telegraph*, 18 Dec. 2014), she was expressing what psychiatrist Tanveer Ahmed was referring to when he wrote about **'aspects of Islamism, which is…resentful [of] humiliation, unable to accept the reality of the weak place of Islamic civilisation and determined to act destructively'** (ibid).

[1054] Yes, as a result of the emergence of overwhelming levels of upset in the last 200 years humanity has endured two massive world wars and countless other insurgencies, as well as the inglorious honor of inventing weapons of such ferocity that they could wipe out entire cities—and these weapons are now in the hands of *extremely* psychologically distressed rogue countries like North Korea, whose attitude toward the rest of the world is one of 'We don't care any longer, we just want relief for our egos'; 'It's either death or glory and if it's going to be death for us then it's going to be death for everyone because we're not going to back down and give up on our need for glory, which is, in effect, what you're asking us to do.' It should be pointed out that for the human race *as a whole* to have the technological capability it now has while we are, as has now been revealed, so deeply insecure and alienated is the equivalent of allowing a psychotic lunatic to play with an atomic bomb; as General Omar Bradley put it in his aforementioned quote, **'The world has achieved brilliance…without conscience. Ours is a world of nuclear giants and ethical infants.'**

[1055] So, the *real* debate about both the horrific material inequality and the terrorism and frightening instability in the world requires analysis of the differences in upset-adaption or alienation-from-soul between individuals, 'races', genders, generations, countries, civilizations and cultures, but until the human condition could be explained and the upset state of the human condition compassionately understood and thus defended that debate could not take place. The problem of selfishness in the world was not being addressed honestly and thus properly anywhere.

[1056] As will be explained shortly in ch. 8:16I, the distress from 'materialism envy' has also led to the adoption of extremely strident fundamentalist misrepresentations of religious teachings to try to manage the unbearable situation.

## Chapter 8:16G <u>So it is end play wherever we look</u>

[1057] It can be seen then that the world's situation is a dire one in which the more operational 'developed', materially affluent parts of the world are fast approaching a state of terminal alienation with anxiety disorders such as ADHD and autism now rampant and the **'biggest age group in American history'** in the grip of an extremely insecure, power addict-driven **'narcissism epidemic'**, while the less operational 'developing', materially poorer parts of the world are, as a result of 'materialism envy', fast approaching **'collapse'** and **'hopeless destitution'**.

[1058] So it is end play for the human race wherever we look, and while we humans have had no choice other than to stay positive in our thinking, pretend that everything is okay, underneath that facade everyone knows the situation is becoming *extremely* serious. Throughout history there have been 'The end of the world is nigh' doomsayers, but this is different. At the time of writing this passage, which is in early 2013, there is a plethora of end-of-the-world, post-apocalyptic programs appearing on television, such as **'Life After People'**, **'Evacuate Earth'**, **'Doomsday Preppers'**, **'Survivors Guide to the Mayan Apocalypse** [the prediction in the Mayan calendar that the world would end in 2012]', **'Apocalypse 2012 Revelation'**, **'Armageddon Outfitters'**, **'End Day'**, **'Omens of the Apocalypse: The End is Near'**, **'The Walking Dead'** and **'Zombie Apocalypse'**. Similarly, I happened to pick up this week's guide (11-17 April 2013) from my local cinema—and I haven't been near a cinema for years—and of the 12 films showing, these are the names of 7 of them: **'Oblivion'**, **'Scary Movie 5'**, **'Warm Bodies** [about zombies]', **'Escape from Planet Earth'**, **'Trance'**, **'The Host** [about the human race being infected by psychic parasites]' and **'A Good Day to Die Hard'**! And I've just heard of another movie that's about to be released, titled **'This Is The End'**!!

[1059] Since writing the above paragraph, the actor Brad Pitt's film **'World War Z'** has also been released. Described as **'the latest in a trend of apocalyptic thrillers...from a Hollywood convinced that the end is nigh'** (*The Australian*, 10 Jun. 2013), it's about **'a viral zombie disease that threatens world domination'** (*The Sydney Morning Herald*, 10 Jun. 2013). Yes, the **'zombie disease'** of alienation *is* threatening to take over the world, and the end of the world *is* nigh. Australia has just released its first **'post-apocalyptic zombie spectacular'** titled **'Wyrmwood'** (*The Sydney Morning Herald*, 15 Jun. 2013), while an article in *The Sydney Morning Herald* last month (3 Jun. 2013) titled 'Vacant stares become all the rage' contained the following photograph shown on the left with the caption, **'An enthusiast takes part in the Zombie Walk festival in Prague on the weekend. Zombie walks, in which horror fans dress as zombies and parade through city streets, have grown in popularity in recent years.'** The detail from a painting by Jean-Michel Basquiat shown below on the right appeared in *The Australian* newspaper the month before last (17 May 2013) accompanying a story about how it had just been sold at auction for a whopping $US48.8 million—a price no doubt indicative not only of its artistic merit but how much it resonates as an image of our time. Yes, Laing's **'fifty feet of solid concrete'** between us and **'our true selves'** or soul is now more like a thousand feet of **'solid concrete'**.

Depictions of the terminally alienated, zombie state
that humanity has arrived at (details in the preceding paragraph)

[1060] And the examples keep accumulating. An article in this week's *TIME* magazine (22 Jul. 2013) about **'*Pacific Rim*, the new Armageddon adventure'** observes that **'If our world isn't already post-mortem, as in Tom Cruise's *Oblivion* and Will Smith's *After Earth*, then zombies are over-running the planet (*World War Z*) or we are in search of a Superman to stop an alien attack (*Man of Steel*).'** Our obsession with superheroes at the moment—such as Superman, The Lone Ranger, The Phantom, Batman, Wonder Woman, Spider-Man, Captain America, Captain Marvel, Luke Skywalker of *Star Wars*, the wizard Harry Potter, The X-Men, Iron Man, and even Frodo Baggins of *The Lord of the Rings*—is an expression of an intuitive awareness humans have of what is now desperately needed to solve the human condition and **'stop' 'Armageddon'**. Yes, to the resigned, alienated mind, confronting the unbearably depressing issue of the human condition, which is what was needed to solve it, did seem an impossible, superhuman undertaking.

[1061] *No wonder* there is panic in officialdom with the desperate spate of initiatives that were described in par. 603 now occurring to try to find the only thing that could save the world, namely the reconciling understanding of the human condition—despite that reconciling understanding having been found and put forward in my books for the last 30 years! In fact, since what is happening is *so* obscene, I am going to re-include that list of initiatives:

- in December 2012 it was announced that an American billionaire had pledged $200 million to Columbia University's **'accomplished scholars whose collective mission is both greater understanding of the human condition and the discovery of new cures for human suffering'**; and,

- on 28 January 2013, the European Commission **'announced that it would launch the flagship Human Brain Project with a 2013 budget of €54 million (US$69 million), and contribute to its projected billion-euro funding over the next ten years'** with the goal of providing **'a new understanding of the human brain and its diseases'** to **'offer solutions to tackling conditions such as depression'**; and,

- on 2 April 2013, the President of the United States, Barack Obama, announced a **'Brain Initiative'**, giving **'$100 million initial funding'** to mechanistic science, in **'a research effort expected to eventually cost perhaps ten times that amount'**, to also find **'the underlying causes of... neurological and psychiatric conditions'** afflicting humans in order to **'develop effective ways of helping people suffering from these devastating conditions'**; and,

- on 24 April 2013, *BBC News Business* reported that **'Lord Rees, the Astronomer Royal and former president of the Royal Society, is backing plans for** [Cambridge University to open] **a Centre for the Study of Existential Risk** [meaning risk to our existence]. **"This is the first century in the world's history when the biggest threat is from humanity," says Lord Rees.'** The article then referred to Oxford University's Future of Humanity Institute that was established in 2005, which is **'looking at big-picture questions for human civilization...**[and] **change...**[that] **might transform the human condition'**, quoting its Director and advisor to the Centre for the Study of Existential Risk, Nick Bostrom: **'There is a bottleneck in human history. The human condition is going to change. It could be that we end in a catastrophe or that we are transformed by taking much greater control over our biology'**!

[1062] But even in the unlikely event that such desperate initiatives (unlikely and desperate as they are because the mechanistic paradigm they are part of is totally dedicated to *avoiding*, not confronting, the human condition) *did* somehow manage to find **'the underlying causes'** of our **'human condition'**, the very, very best they could hope to achieve would, again, be *exactly* what is already being presented in this book, and even then it would have to include the explanation of how the human race is supposed to cope with the terrifying exposure of our corrupted condition that such an understanding inevitably brings—the answer to which is presented in the concluding chapter (9) of this book. That the mechanistic paradigm is nourishing its own human-condition-avoiding, dishonest institutions while persecuting and blocking our human-condition-confronting, truthful World Transformation Movement is just another manifestation of the terminally alienated, deathly, sick state the world is in—because instead of attacking and blocking the work that we are doing, the scientific and wider community should be forming a steering committee composed of strong minded people, like Australia's former Prime Minister John Howard and the American political commentators Bill O'Reilly and David Brooks, to bring as much support as is humanly possible to our work, and thus to a future for humanity. Ideally, that exceptional denial-free-thinking prophet of the Irish rock band U2, Bono, would obviously also be on such a committee. And since HRH The Prince of Wales, the heir to the British throne, has throughout his life genuinely, not deludedly and falsely, been working on opening the door to a truthful human-condition-understood new world for humanity—something he has also been persecuted for—he is another who should be on such a committee.

### Chapter 8:16H **The progression of ever-more dishonest and dangerous forms of pseudo idealism to cope with the unbearable levels of upset**

[1063] In terms of the progression of the human journey that is being charted in this chapter, the repercussion of upset reaching this crescendo on our strategies for coping with the human condition was that <u>people could no longer cope with the honesty of religion</u>—it became too confronting, guilt-inducing and unbearably depressing. The great benefit of religion—of the honesty imbued in the prophet or prophets (in the case of Hinduism) that the religion was founded around—actually became its liability, because by retaining a presence of a prophet's soundness and truth, religions reminded humans of their own corrupted state and their alienation from truth, which in turn accentuated their sense of guilt; as the author Mary McCarthy once wrote, **'Only people who are very good can afford to become religious; with all the others it makes them worse'** (*Memories of a Catholic Girlhood*, 1957; Lloyd Reinhardt's rendition, *The Sydney Morning Herald*, 18 Jan. 1995; see <www.wtmsources.com/175>). It was at this point when the honesty of religion became too confronting that much less confronting and less guilt-emphasizing forms of pseudo idealism had to be found, with the *extremely* dangerous and negative effect being the loss of the precious honesty originally imbued in religion.

[1064] When the truthful lives and thoughts of religions' founding prophets became unbearably confronting and condemning, <u>GUILT-FREE EXPRESSIONS OF IDEALISM TO SUPPORT AND DERIVE 'FEEL GOOD' RELIEF FROM</u> became highly sought-after. These expressions took <u>two forms</u>: <u>Less Guilt Emphasizing Expressions of Religion</u>, and <u>Non-Religious Pseudo Idealistic Causes</u>.

## Chapter 8:161 <u>Less Guilt Emphasizing Expressions of Religion</u>

[1065] As humans became more upset religions were adapted and developed to be less guilt emphasizing. Within Christianity, for example, rather than following a denomination that focused on the study and acknowledgment of the integrity of the words and life of Christ, a person could select one that emphasized worship, adoration and ceremony, such as Catholicism. The emphasis in religion could be shifted away from self-confronting honesty, introspection and interpretation of the deeper truths contained in the religion's scriptures to simply worshipping the prophet as a more indirect way of glorifying and supporting the truth he stood for, and to simple, literal, non-interpretative, dogmatic propagation of the religion's scripture. And, overall, to counter the increasing waywardness of the greater levels of upset, to even more strict obedience to the cooperative ideals and principles espoused by the religion. The popularity of literal, <u>fundamentalist expressions of religion</u>—especially simplistic, emotional, euphoric, 'evangelical', 'charismatic' forms of Christianity and funda-mentalist representations of the Islamic faith—have increased so much in recent times that they are now the fastest growing forms of religion in the world. An article about the evangelical, charismatic, Christianity-based Hillsong Church that began in Australia in 1983 reports that it now **'oper-ates in 14 countries from Europe to South America'**, and describes **'Stadiums**

**packed with screaming fans'**, and an **'11-piece band that is the world's biggest Christian rock group, listened to by 50 million people around the world'**. In the article a former member describes how non-self-confronting, basically guilt-and-honesty-stripped-yet-emotionally-satisfying it is: **'Hillsong is a very easy religion to be a part of. They don't ask anything of you personally. They don't talk sin, forgiveness, they literally say we don't ask people to change. It's incredibly exciting and incredibly rewarding. There's a very big community element to spirituality…You're being told that you're part of changing the world, you're part of making the world a better place'** ('Hillsong Church to release Let Hope Rise in 2015', News.com.au, 1 May 2015; see <www.wtmsources.com/178>). In the case of fundamentalist representations of the Islamic faith, stories about its spread appear in newspapers daily, especially about the frightening expansion of ultra-fundamentalist and militant Islamic groups like Al-Qaeda and ISIS.

[1066] Basically, the more upset and thus insecure and guilt-feeling we humans became the more fundamentalist or literalist we needed to be. Mindlessness saved us from hurtful mindfulness. The Gnostics were an early Christian sect who encouraged self-confronting introspection and analysis and interpretation of the Gospels. Gnosticism stressed self-knowledge and the need, as they said, to **'find the Christ in yourself'**. (The word 'knowledge', which means 'to know', has the same roots as the word 'gnosis'.) The Gnostics also practiced living a frugal, ascetic, 'pure' life. But since the majority of people found such dedicated introspection and interpretation of the Gospels far too confronting and a life of frugal purity unbearably devoid of much needed self-distraction and escape from their upset reality, Gnosticism eventually died out. In fact, around 1230 AD a branch of Gnostics in France called the Cathars were viciously persecuted and eventually destroyed by emissaries from the Roman Catholic Church. As mentioned, the Catholic branch of Christianity especially caters for the more upset by emphasizing worship and a more literal interpretation of the Gospels. Rather than austerity and self-confrontation, Catholicism instead emphasizes self-distracting pomp, ceremony and ritual. As a result, they viewed the Cathars as heretics and burnt many of them at the stake. Catholicism also emphasizes, to the point of venerating, Christ's mother, the Virgin Mary, because, for the overly corrupt, it was obviously much easier to identify with the Virgin Mary and her world of gentle nurturing than relate to the strong, secure, confronting truthfulness of Christ. Other branches of Christianity, such as the Anglicans, aren't so upset-adapted and thus escapist, self-distraction-focused and fundamentalist in their orientation.

[1067]The more upset and thus insecure humans became the more literal and non-interpretative of scripture they needed to be; and the more they needed to emphasize deferment of self to the prophet through worship as opposed to emphasizing confronting self-analysis; and the more they needed to express their faith through the adornment of their churches and through elaborate ceremony, as opposed to expressing the commitment to the prophet's embodiment of the cooperative ideals of life through living humbly without extravagant adornments and ceremony. Sir Laurens van der Post observed how humans' religious images reflected their degree of alienation when he wrote: **'It seemed a self-evident truth that somehow the sheer geographical distance between a man and his "religious" images reflected the extent of his own inner nearness or separation from his sense of his own greatest meaning. If so this made the conventional Christian location of God in a remote blue Heaven just as alarming as, conversely, the descent of his Son to earth was reassuring'** (*Jung and the Story of Our Time*, 1976, p.31 of 275). While the increase in the artificiality of life that accompanied increased upset was extremely **'alarming'**—an issue I will return to shortly—when people became excessively upset, trying to be any less artificial (by adhering to a less artificial form of religion) was too confronting, guilt-inducing and depressing. It is only now that we have found the dignifying understanding of the human condition that the upset human race can afford to demystify 'God' as the personification of Integrative Meaning; and explain that Christ was, in effect, the **'Son'** of God, the uncorrupted expression of God; and that the 'virgin mother' of Christ was a metaphor for the exceptionally innocent mother needed to nurture an exceptionally innocent child who would be capable of becoming a denial-free thinking prophet; and that the 'resurrection' of Christ after his martyrdom was emblematic of the opportunity he gave upset humans to be 'born again' or raised up or resurrected from their corrupted, effectively dead state by living through support of him; and, as explained in par. 746, that the trinity of God the Father, Son and Holy Ghost (or Spirit) is a perfect pre-scientific representation of the three fundamental aspects of existence on Earth: God the Father is Integrative Meaning; God the Son is the first great tool for integrating matter, namely the gene-based learning system that gave rise to our integratively orientated instinctive self that Christ, for example, was an uncorrupted expression of; and

God the Holy Ghost (or Spirit) is the second great tool for integrating matter, namely the nerve-based learning system that gave rise to our conscious self or intellect—particularly a Godly, Integrative-Meaning-acknowledging, inspired and guided intellect.

[1068] Of course, for the religious faithful—and especially those who have held a more literal and non-interpretative, fundamentalist view of scripture—having God demystified, the prophets explained and humanized, and all their religions' articles of faith, such as, in the case of Christianity, the Resurrection, the Virgin Mary, Christ's miracles, the Biblical stories of Adam and Eve, Noah's Ark, Cain and Abel, etc, etc, explained scientifically (having religion and science reconciled), as this book does, will be a lot to have to adjust to. This sudden demystification of religious scripture forms part of the ultimate **'future shock'** that Alvin Toffler anticipated in his 1970 book by that name. The physicist Paul Davies summarized the problem when he said, **'A lot of people are hostile to science because it demystifies nature. They prefer the mystery. They would rather live in ignorance of the way the world works and our place in it…many religious people still cling to an image of a God-of-the-gaps, a cosmic magician'** ('Physics and the Mind of God: The Templeton Prize Address', 3 May 1995). Fortunately there is an absolutely wonderful way to cope with this sudden and extreme exposure, which is the subject of the final chapter (9) of this book.

[1069] With regard to Sir Laurens' observation that **'the sheer geographical distance between a man and his "religious" images reflected the extent of his own inner nearness or separation from his sense of his own greatest meaning'**, since God is representative of the cooperative ideals and 'his' presence, in effect, stands in judgment of humans' non-ideal state, religions, such as Buddhism, that didn't emphasize God, sin and guilt, and instead focused on extinguishing the mental trauma of upset through austere practices like meditation and recital of mantras, became increasingly popular in the latter stages of the last 200 years when upset became extreme. As one convert said of Buddhism, it's **'non-judgmental, there's no notion of sin, there's no notion of good and evil, you don't embrace negativity'** (*Light at Edge of the World: Science of the Mind of Buddhism*, National Geographic Channel, 2006). (While the thrust of this comment is true, to say there is *no* notion of good and evil in Buddhism is not entirely correct because Buddhism does recognize a sense of 'karmic' heaven and hell.) The problem, however, with focusing on ways, such as meditation, to extinguish the mental trauma of the upset, human-condition-afflicted state was that they undermined the essential responsibility of being a conscious being, which is to *think* and *understand*, ultimately to find understanding of the human condition—as Nikolai Berdyaev wrote: **'Man sought to escape from that terror** [of the truth of **man's...exile from paradise**] **by extinguishing consciousness and returning to the realm of the unconscious. But this is not the way to regain lost paradise'** (*The Destiny of Man*, 1931; tr. N. Duddington, 1960, p.41 of 310). The point being made, however, is that as upset increased, more and more escapist strategies for coping with the human condition simply *had* to be adopted, despite how irresponsible and destructive that trend was of humanity's heroic struggle to find knowledge.

[1070] Indeed, for some people, complete denial of God became the only acceptable option. Atheism, disbelief in God, and secularism, the rejection of all forms of religious faith and worship, gained popularity. In fact, in recent years the resentment and anger towards God for **'condemning people'** (*What the Bleep do We Know!?*, 2004) and for being a **'stupid'**, **'utterly evil, capricious and monstrous' 'maniac' 'who creates a world which is so full of injustice and pain'** (comedian Stephen Fry, *The Meaning of Life*, RTÉ TV, 1 Feb. 2015), and towards religion for being **'the church of perpetual misery'** (from a 2005 animated TV cartoon), has grown so much that secularism is on the rise everywhere—as commentator Phillip Adams wrote, **'More than ever He, She or It is a redundant notion. So it's time to dump your shares in religion...We're free to live our lives without risk of damnation...The only meaning our existence has...is the meaning(s) we choose to give it'** ('The Big Whimper', *The Weekend Australian Magazine*, 3 Dec. 2011). This now extreme resentment and anger is palpable in Richard Dawkins' statement referred to in par. 938, that **'"Faith is one of the world's great evils, comparable to the smallpox virus, but harder to eradicate. The whole subject of God is a bore"...those who teach religion to small children are guilty of "child abuse".'**

[1071] Yes, the extreme, **'alarming'** danger of guilt stripping religion—of making all forms of religious faith less confronting by avoiding emphasizing God, sin and guilt, and by trying to **'regain lost paradise'** through stopping thinking, and even by atheism's complete denial of God/ Integrative Meaning—is that the great benefit of religion of the presence of its honesty is lost. Sir Laurens pointed out how dangerously **'starved and empty'** of its truthfulness, guilt-stripped Christianity has become when he wrote that **'Yet the churches continue to exhort man without any knowledge of what is the soul of modern man and how starved and empty it has become...**

**They behave as if a repetition of the message of the Cross and a reiteration of the miracles and parables of Christ is enough. Yet, if they took Christ's message seriously, they would not ignore the empiric material and testimony of the nature of the soul and its experience of God that [Carl] Jung has presented to the world. He did his utmost to make us understand the reality of man's psyche and its relationship to God. But they ignore the call'** (*Jung and the Story of Our Time*, 1976, p.232 of 275).

[1072] This brings us back to the issue of 'materialism envy' and the resulting unbridled greed, dysfunction and destitution in the less functional, materially poor, 20-year-old and 50-year-old equivalent populated 'developing' countries.

[1073] While the traditional forms of religion have been losing favor in the more functional, affluent 'developed' parts of the world, religion, especially the simplistic, emotional, 'evangelical', 'charismatic' forms and the extremely strident, fundamentalist representations of religious teachings, has continued to grow in popularity and influence wherever the increasing levels of upset were also accompanied by impoverishment and/or lack of education. It requires a degree of material comfort and education to be able to engage the subtleties of life under the duress of the human condition, in particular the issue of how to manage the guilt associated with being an upset, non-ideal person. When mere survival is a struggle, thinking about your psychological state is a luxury. Also, without some education you lack the base of knowledge with which to think intellectually and sophisticatedly about the dilemma of being an upset human. As will be described shortly, while entirely new forms of pseudo idealism that are much more sophisticated than religion in avoiding guilt have developed in the more functional, materially wealthier, 'developed' parts of the world, for the impoverished and/or uneducated in the less functional, materially poor 20-year-old and 50-year-old equivalent populated 'developing' countries, religious faith has been the main way of countering upset. In fact, with ever greater numbers of people in the world and an ever greater frustration and resentment of the widening gap between the materially fortunate and the materially impoverished, the attraction and influence of religious faith, especially the simple charismatic and fundamentalist representations of religious teachings, has only increased amongst the materially less fortunate. Over time, however, with greater levels of upset developing everywhere—coupled with the growing anger (and even, as the Muslim sympathizer Tessa Kum said, **'hate'**, or as Clancy McKenzie said, **'absolute hatred'**) arising from 'materialism envy', and the loss of self-esteem amongst the less fortunate as the result of the burgeoning gap between the materially 'rich and poor'—extremely strident, even militarist and terrorist expressions of fundamentalism began to be adopted by those less fortunate. This can be seen in the rise of the Boko Haram Islamic militants in Nigeria, and Al-Qaeda and the Islamic State/ Daesh Islamic militants in the Middle East. <u>Since religion has essentially been about indirectly maintaining the search for knowledge, this strident form of fundamentalism amounts to a dangerous form of *anti*-religion, of degraded-and-debased-truth-and-honesty-obliterated misrepresentations of religious teachings</u>. It is threatening the essential need the human race has for there to be sufficient freedom of thought to continue the search for knowledge, ultimately self-knowledge—the liberating understanding of the human condition. In the race between self-destruction and self-discovery, the rapid and extreme degradation of religion has fast been leading humanity to self-destruction.

[1074] As will be described in chapter 9, ultimately the only thing that could end this race to terminal alienation and destruction was the finding of understanding of the human condition because that alone makes possible the great change of direction from living selfishly to living selflessly. Of special significance are pars 1254-1255, which point out that "it is only when the less fortunate see those who have been more materially fortunate making the great **'change of heart'** from living selfishly to living selflessly, [that] they will finally be released from feeling the injustice and humiliation of their situation. When selfless love finally comes to Earth, the hearts of humans everywhere will be flooded with enormous relief and incredible joy."

[1075] What will now be looked at is how the adoption of more guilt-free forms of pseudo idealism in the more functional, materially wealthy, 'developed' parts of the world has *also* been seriously threatening humanity's all-important journey to enlightenment. As we will see, extreme forms of pseudo idealism, such as politically correct postmodernism, are as stripped of honesty and thus as alarmingly dangerous as strident forms of religious fundamentalism.

## Chapter 8:16J  Non-Religious Pseudo Idealistic Causes

[1076] The other form of more guilt-free expressions of idealism to support and live through have been non-religious and in some cases atheistic, God/Integrative Meaning-denying Pseudo Idealistic Causes. These non-religious pseudo idealistic causes include communism or socialism, environmentalism, feminism, multiculturalism, aboriginalism, politically correct postmodernism, etc, etc—basically any pseudo idealistic cause you could find that allowed you to avoid having to think about and deal with the real issue behind all the destruction and imperfection in the world, namely your own and everyone else's corrupted condition. *TIME* magazine's former editor Richard Stengel perfectly captured this escapist trend when he wrote that **'The environment became the last best cause, the ultimate guilt-free issue'** (*TIME*, 31 Dec. 1990).

[1077] Since humanity was trending towards ever more degraded, guilt-and-honesty-stripped fundamentalist misrepresentations of religious teachings and, as we will now see, other more guilt-free but dishonest forms of non-religious pseudo idealism to cope with the exponentially increasing levels of upset in the world, there was clearly going to come a time—unless understanding of the human condition was found—when excessive dishonesty would herald the end of the all-important search for liberating understanding of the human condition and lead to terminal levels of alienation in humans. The danger was of becoming overly dishonest in denying the upset, corrupted state of the human condition—insisting on idealistic behavior at the expense of reality—and thus oppressing the need for some freedom from the ideals to search for knowledge. To become free of the human condition we had to be allowed to be, to a degree, free from the oppression of idealism if we were to search for knowledge, ultimately self-knowledge, understanding of the human condition. Again, as the journalist Richard Neville summarized, **'We are locked in a race between self destruction and self discovery'**—either we found understanding of ourselves or we faced self-destruction. If we gave up the search for knowledge there was no hope, and, as will now be described, the virtual abandonment of that search by taking up ever more dishonest/truthless forms of pseudo idealism *has* brought humanity to the brink of terminal alienation. By denying any confronting truth—that is, by taking the practice of guilt-stripping to the extreme—and simply dogmatically demanding we be ideal, pseudo idealism was only adding to alienation, burying humanity deeper in Plato's cave of denial of any truth, making it harder and harder to reach liberating understanding of ourselves.

[1078] In fact, as should be increasingly apparent, to find understanding of the human condition a veritable mountain of accumulated dishonesty, especially from pseudo idealists, had to be defied and corrected. AND, if that wasn't enough, as described in par. 573, the advocates of extreme pseudo idealistic dishonesty in my home country, in the form of our national broadcaster, the ABC, and Fairfax Media, then all but destroyed our efforts at the World Transformation Movement (WTM) to bring these humanity-saving insights to the world! The whole journey of getting all the truth up from its almost totally buried state and then bringing the humanity-liberating understanding of the human condition that that truth made possible to the world has been *so* difficult that, in hindsight, it seems all but a miracle that, as I said in par. 608, we are still on our feet making this further great effort with this book to have these world-saving insights recognized. The *Persecution of the WTM* section on our website (<www.humancondition.com/persecution>) documents the horrific treatment the WTM has had to endure and our final wonderful vindication in the law courts. However, until substantial support builds for these all-precious, liberating understandings, their survival hangs in the balance.

[1079] To drill down now into the progression of the increasingly guilt-stripped and thus ever more dishonest and thus ever more dangerous forms of non-religious pseudo idealism that humanity developed and adopted when <u>Religion</u> became too confronting, starting with <u>Socialism and Communism</u>.

## Chapter 8:16K  <u>Socialism and Communism</u>

[1080] These collectivist movements <u>denied the notion of a perfecting God and avoided the depressing recognition of a prophet's world of soundness, and instead dogmatically demanded an idealistic social or communal world</u>—and, in doing so, denied and oppressed the whole reality of the individualistic, freedom-dependent, knowledge-finding, creative, egocentric, corrupting, unavoidably-variously-upset, competitive, combative, materialism-compensation-needing, self-distraction-necessary, human-condition-afflicted world. As was pointed out in par. 299, Karl Marx, the political theorist whose mid-nineteenth century ideas gave rise to socialism and communism, argued that **'The philosophers have**

**only interpreted the world in various ways; the point is** [not to understand the world but] **to change it.'** By **'change it'** Marx meant just make it cooperative or social or communal, but he was wrong—the whole **'point'** and responsibility of being a conscious being *is* to understand our world and place in it, ultimately to find understanding of the human condition. The attraction—and inherent lie—of socialism/communism was that you could support and live the ideals without acknowledging the reality of the human condition and its struggle.

### Chapter 8:16L  The New Age Movement

[1081] The limitation of socialism and communism was that while there was no confronting prophet involved, there was an obvious focus on the condemning cooperative, communal, social *ideals*—and so in time, as levels of upset and thus insecurity increased, the need again arose for the invention of an even more guilt-free form of idealism through which to live, hence the development of the New Age Movement (the forerunners of which were the Age of Aquarius and Peace Movements). In this movement all the realities and negatives of our non-cooperative/ non-communal/non-social, corrupted condition were transcended in favor of taking up a completely escapist, think-positive, human-potential-stressing, self-affirming, motivational, feel-good approach. So, in truth, the New Age Movement was never going to be able to transport humanity to an Aquarian new age of peaceful freedom from upset, it was only ever going to lead to an even greater state of deluded, dishonest alienation than that espoused by socialism/communism. Talking about how he became **'a personal growth junkie'**, the comedian Anthony Ackroyd summed up the extremely deluded artificiality of the New Age Movement when he said: **'What are millions of us around the globe searching for in books, tapes, seminars, workshops and speaking events? Information to enhance our lifestyles and enrich our experience on this planet? Certainly…But I smell something else in the ether. Something more desperate and deluded. A worrying snake-oil factor that is spinning out of control. It's the promise of salvation. Salvation from the basic rules of human life. This is the neurotic aspect of the human potential movement. This hunger for a get-out-of-the-human-condition-free card'** (*Good Weekend, The Sydney Morning Herald*, 13 Sep. 1997). Yes, to **'get out of the human condition'** we had to confront and solve it, not deny and escape it; our **'desperate and deluded'** attempts to escape it only made it worse. As the philosopher Thomas Nagel recognized, **'The capacity for transcendence brings with it a liability to alienation, and the wish to escape this condition…can lead to even greater absurdity'** (*The View From Nowhere*, 1986, p.214 of 256).

## Chapter 8:16M  The Feminist Movement

[1082] The limitation of the New Age Movement, in terms of being an effective means of escaping the horror of the human condition, was that while it didn't stress the cooperative ideals like socialism and communism did, in seeking to transcend humans' upset state its very purpose served as a constant reminder and thus reprimand of our variously upset, embattled, troubled, estranged, alienated condition—a problem the *next* level of delusion sought to dispense with by simply denying its existence. Yes, the Feminist Movement maintained that there was no real difference between people—and *especially* not between the sexes. It particularly denied the legitimacy of the exceptionally egocentric, combative male dimension to life that had taken on the heroic frontline role in fighting the ignorance of our instinctive self. Women's lack of awareness of the all-important role that men have been playing is evident in this already mentioned comment from Germaine Greer, an icon of modern feminism: **'As far as I'm concerned, men are the product of a damaged gene. They pretend to be normal but what they're doing sitting there with benign smiles on their faces is they're manufacturing sperm.'** Beyond their necessary role in repro-duction men have been viewed as meaningless; as the leading feminist, Gloria Steinem, famously said, **'A woman needs a man like a fish needs a bicycle'**—but the truth that is now able to be revealed is that men were carrying out the most important and difficult task being undertaken on the planet! Based on extreme dogma, the feminist movement could not and has not brokered any real reconciliation between men and women, rather, as this aforementioned quote points out, **'What happened was that the so-called Battle of the Sexes became a contest in which only one side turned up. Men listened, in many cases sympathetically but, by the millions, were turned off.'** No, *only* by winning the battle to champion the ego—that is, explain the human condition and establish that our egocentric conscious thinking self is good and not bad—could the polarities of life of so-called 'good' and 'evil', that women and men are in truth an expression of, be reconciled.

## Chapter 8:16N  The Environmental or Green Movement

[1083] Again, as far as offering an effective way to transcend the realities of the human condition, feminism's flaw was that while it superficially dispensed with the problem of humans' divisive reality, humans remained the focus of attention and that was confronting. The solution that emerged to counter this limitation was the Environmental or Green Movement, which removed all need to confront and think about the human state because all focus was diverted from self onto the environment—as the aforementioned quote acknowledged, **'The environment became the last best cause, the ultimate guilt-free issue.'** Of course, the obvious truth is that by not addressing the cause of the destruction of the natural world, namely the issue of our human condition's massively upset, angry, egocentric and alienated state, there can be no real let up in the pace of our world's devastation, as these quotes emphasize: **'The trees aren't the problem. The problem is us'** (*Simply Living* mag. Sep. 1989), and **'We need to do something about the environmental damage in our heads'** (*TIME*, 24 May 1993). The bumper sticker **'SAVE THE HUMANS'** that parodies the green movement's **'Save the Environment'** slogan makes the point about how evasive of the real issue the environmental movement is—as does the following statement made by Ray Evans, an Australian business leader and political activist: **'Environmentalism has largely superseded Christianity as the religion of the upper classes in Europe and to a lesser extent in the United States. It is a form of religious belief which fosters a sense of moral superiority in the believer, but which places no importance on telling the truth** [about the real issue of our corrupted condition]' (*Nine Facts About Climate Change*, 2006; see <www.wtmsources.com/169>).

## Chapter 8:16O  The Politically Correct Movement

[1084] So for all its guilt-relieving benefits, the environmental movement still contained a condemning moral component: if we were not responsible with the environment, 'good', we were behaving immorally, 'bad'. Moreover, nature, in its purity, exists in stark contrast to humans' corrupted condition—hence our ruthlessness towards it in the first place.

[1085] At this stage in the march of upset yet another form of extreme pseudo idealism had to be manufactured where confrontation with the by now extremely confronting and depressing truth of the dilemma of the human condition could be totally sidestepped. What was required was a completely guilt-stripped dogma that was devoid of any need to confront and wrestle with the issue of soundness and Godliness; with whether you are a cooperative, social, integrative person; with the issue of your troubled self; with the morality issue of how men and women treat each other; and with the issue of whether or not you are being good to the environment. Upset had become so great that the need was to simply be ideal without question. This demand for a totally non-confronting form of relief from feeling 'bad' resulted in the establishment of the Politically Correct Movement, which has no other focus or requirement beyond simply choosing, from the two simplistic, fundamental, 'political' options in life—of being either 'good' or 'bad'—to be 'good'.

[1086] The politically correct culture was a pure form of freedom-denying dogma that fabricated, demanded and imposed ideality or 'correctness', specifically that of an undifferentiated world, which was in complete denial of the reality of the underlying issue of the existence of and reasons for humans' variously embattled and upset states, and beyond that of the deeper question raised by those 'non-ideal' states of the issue of the human condition. For instance, it argued that the children's nursery rhyme *Baa Baa Black Sheep* is racist and should instead be recited as **'Baa Baa Rainbow Sheep'** (*The Telegraph*, 24 Jan. 2008).

[1087] Within the politically correct culture the need for relief from guilt was all-pervasive; the mind was constantly on the hunt for opportunities and 'good causes' through which to be 'idealistic' and achieve that rush of psychological relief of feeling that at last you are a 'good' rather than a 'bad' person. Wherever there was a victim of humanity's battle, there was an opportunity to champion their cause and access that all-important relief. It was, as we will shortly see, a development that Christ described in much harsher terms when he said, **'Wherever there is a carcass** [the extremely upset]**, there the vultures** [the **false prophets**, the merchants of delusion and escapism] **will gather.'** But with the levels of upset in the world becoming so extreme, such relief-hunting became a huge industry, to the extent that we became, as the sociologist Frank Furedi recognized, **'a society that celebrates victimhood rather than heroism'** (*Culture of Fear*, 1997, p.13 of 205). Yes, as one critic complained, **'If you are a black vegetarian Muslim asylum-seeking one-legged lesbian lorry driver, I want the same rights as you'** ('Thought police muscle up in Britain', *The Australian*, 21 Apr. 2009). Indeed, the deluded arrogance of the left-wing politically correct culture has been absolutely extraordinary; virtually every left-wing magazine or newspaper or broadcaster will go out of its way now to lead with a story and picture or footage about one of the just referred-to categories in order to 'educate' us all in 'correct' thinking!!

[1088] Again, while there *was* an ever increasing need for more dishonest, guilt-free forms of idealism through which to live, for humanity to arrive at this desperately insecure state where people were concerned *only* with finding relief from their own guilt through supporting the cause of those who were suffering or less fortunate was an extremely dangerous development because it meant the human race had, in effect, abandoned the ongoing battle to find the all-important liberating understanding of ourselves. This is not to say that in a critical battle, such as the one humanity has been involved in, showing care and compassion towards those who were suffering from the effects of the battle was not important. It was very important, because although we have all been involved in the upsetting battle, selflessness is still, as has been repeatedly emphasized, what binds wholes together; it is the glue within humanity's army. *However*, while caring for those struggling to keep up was important, it was obviously more essential to support those on the frontline who were still carrying on the battle to ensure the war was ultimately won. In this light it can be seen how very dangerous and irresponsible the politically correct movement's focus and insistence only on caring for the victims of the battle was.

[1089] In fact, while showing care and compassion to those suffering from the battle was ('was', because the battle to find understanding of the human condition is now over) certainly the responsibility of any healthy society, doing so merely to delude yourself and others that you are an upset-free, ideal person seriously discredited the whole notion and practice of consideration and kindness itself. Indeed, using idealism to delude yourself that you were good gave idealism such a bad name that no relatively sound, secure person wanted to be part of the left-wing political culture where such relief-hunting had become endemic; so much so that, in the end, there was no longer any authentic, trustable, credible, healthy, meaningful idealistic movement in society to counteract any excessively selfish and destructive right-wing behavior—there were only moderate factions within the right-wing that the sane and rationally-behaved could join and support.

[1090] Yes, the whole democratic process that our society depended on for there to be effective progress and functionality was being destroyed by

mad desperados—by a group of people who were misusing democracy
to further their own selfish agenda of making themselves feel good,
rather than for what democracy was designed to be: a tool to decide
the best way to manage any particular course of action. How could you
possibly have an effective discussion about where the right balance lay
between selfishness and idealism if participants in that discussion were
only interested in whether their direct participation would make them
feel good and/or whether the course of action chosen would ultimately
make them feel good? The answer is you couldn't. It was a derailed,
ineffective, dysfunctional, highly imperfect, pointless—in fact, defunct—
debate. It was like being in mid-ocean on a life-boat, desperately trying
to find your way to the safety of land, when someone on-board decides
to hijack and destroy the mission by capsizing the boat because they had
become obsessed with wanting to cool off in the water. As has already
been carefully explained, it was totally irresponsible, selfish—in fact,
mad—behavior. The human race was trying to save itself from destruction
by finding knowledge, ultimately understanding of the human condition,
but the extreme practitioners of pseudo idealism were only interested in
making themselves feel good. Contrary to what their banners said, pseudo
idealists no longer cared about the future of the world. Their conduct
was completely selfish and not at all the selfless, idealistic behavior they
made it out to be and deluded themselves it was. Thank goodness the
arrival of understanding of the human condition exposes this madness for
what it really is, because it was only while it was not possible to explain
humanity's great heroic battle that it was possible to get away with such
mad and obscene behavior—'Can't you see our idealism is making the
world a better place.' What rubbish—such behavior is nothing more than
a selfish attempt to gain relief from and escape the agony of the human
condition! People who are not overly upset, not overly soul-corrupted,
are *naturally* concerned for other individuals, genders, 'races', cultures,
the environment, etc, while those who are overly upset practitioners of
pseudo idealism are merely *pretending* to be concerned for anyone and
anything other than themselves in order to relieve themselves of their guilt.
Instead of being the champions of selflessness they were, in fact, the champions of selfishness, the most selfish of people. Upset basically means
self-preoccupation—you are preoccupied with expressing *your* anger,
with satisfying *your* egocentricity and with maintaining *your* alienation.
Later (in par. 1126), it will be described how Christ got to the underlying

truth about relief-hunting pseudo idealists when he described them as having had their **'love' 'grow cold'** (Bible, Matt. 24:12).

[1091] Humans have had to do what they had to do, side with idealism to delude themselves that they were 'good'/ideal, relieve themselves of the truth about their corrupted, extremely selfish condition, but in terms of humanity's all-important journey to enlightenment, idealism was being extremely, horribly misused—and a particularly dysfunctional aspect of the misuse was the arrogant *extent* of pseudo idealists' delusion that they were actually ideal, that they held the moral high ground. In the situation that has existed, where the reality of the upsetting battle to find knowledge couldn't be explained, idealism—albeit the bastardized form of 'victim-hunting-to-make-yourself-feel-good, politically correct', *pseudo* idealism—certainly did have a field day mocking realism as evil. In the vacuum where the reason for humans' upset, corrupted state was not able to be explained, the **'intellectuals'/'liberal elites'/'chattering classes'/'left-wing trendy café society'/'chardonnay socialists'/'radical chic'/'Hollywood Left'/ 'CBS-*New York Times*, BBC-*Guardian*, ABC-*The Sydney Morning Herald*, *TIME* and *National Geographic* 'left-wing rags', etc, etc axis'/'high-minded do-gooders'/'rainbow extremists'/'strident bleeding hearts'/'feel good, warm inner glow, blissed out compassion junkies', 'so delighted by displays of your own sensitivity, so certain you hold the moral high ground'**, as the relief-hunting, pseudo idealistic left-wing have been variously referred to, conceitedly promenaded about with a holier-than-thou attitude, while the right-wing advocates of freedom (from the oppression of idealism in order to participate in the corrupting search for knowledge) were arrogantly and disdainfully vilified as morally bankrupt and contemptible. For example, the right-wing so-called 'Tea Party' that recently emerged in American politics was derided by the left-wing Democrats for being devoid of any sound arguments for their cause—they were accused of being nothing more than promoters of **'fear, xenophobia, cryptofascism, creationism, inequality and ignorance'** (cartoon by Turner in *The Irish Times* that was re-printed in *The Australian*, 3 Nov. 2010). It is no wonder politics has become so polarized—to the point where the two sides, rather than providing humanity with a healthy equilibrium, have existed in totally opposed philosophical continents, and may as well have lived on separate planets. The deluded arrogance of the extremely dangerous dishonesty of the left became an insufferable, unbearable, overwhelming, terrifying, sickening force. Listing the freedom-stifling **'keywords in PC's history'** as **'Identity, gender, gender-neutral,**

**diverse, inclusive, patriarchy, workplace harassment, multiculturalism, dead
white males, sexism, racism, organic, "privileged", hate speech, speech codes,
prayer in schools, affirmative action, respecting our differences, microaggressions,
trigger warnings'**, the ever-insightful Pulitzer Prize-winning journalist
Daniel Henninger was the first commentator to my knowledge to point
out that the extraordinary support for the hyper politically <u>in</u>correct US
Republican candidate for the 2016 Presidential elections, Donald Trump,
is due to people's **'revolt'** against **'years of political correctness they felt they'd
been forced to choke down in silence'** ('Donald Trump, Ben Carson and the revolt of the
politically incorrect', *The Wall Street Journal*, 8 Jan. 2016).

[1092] Again, the problem at base was the inability to explain the human
condition. While it was easy to argue the case for the idealism of the
left-wing, it was almost impossible to argue the case for the realism of
the right-wing. How could you justify any selfishness or inequality; how
could you defend behavior that appeared in every way to be inhumane;
how could you argue that not being ideally behaved was good? The answer
is that until we could explain the paradox of the human condition we
couldn't—well not sufficiently. Writers like Ayn Rand did well to mount
some sort of a case for right-wing free enterprise, but countering such
efforts were the dogmatic **'Capitalists are Pigs'** placards used in protests at
G8 summits, and left-wing advocates like Michael Moore, the filmmaker
and activist who, at the conclusion of his 2009 documentary, *Capitalism:
A Love Story*, smugly announced that **'Capitalism is evil.'** But while Moore
and **'Marxism designated capitalism as responsible for human misery...**[and, in
decrying its effects on the environment, the] **Bolivian President Evo Morales
declared in 2009 "Either capitalism dies or Mother Earth dies"'** (Pascal Bruckner,
'The Ideology of Catastrophe', *The Wall Street Journal*, 10 Apr. 2012), the truth is that it
was *not* communism that kept the human race going towards its goal of
ending the human condition and all the devastation that resulted from
it, but capitalism. Without the relief, reward, distraction and sustenance
of materialism/material goods that the exchange of money or capital
facilitated, humans would not have been able to cope with and carry on
their upsetting, idealism-defying, heroic search for knowledge, ultimately
self-knowledge.

[1093] Yes, with understanding of the human condition we can finally
now explain that socialism/communism, the pseudo idealistic dogmatic
insistence that everyone be social and communal—live for society
and the communal good rather than seek a degree of material relief

and reward for yourself—ignored the reality of the upsetting battle that humans have had to wage to find understanding of themselves, of their less-than-ideal human condition. As was emphasized in chapter 3:9, with understanding of the human condition found what is revealed is that it was, in fact, the left-wing that was morally bankrupt, *not* the right! The truth was not as it appeared. In the Adam Stork analogy, upset, angry, egocentric and alienated Adam is the hero of the story not the villain; it was his condemning, ridiculing, upsetting idealistic opponent, which the extreme pseudo idealist came to represent, who was actually the villain and not the hero he deluded himself to be and insisted to others he was. Yes, while the right-wing tended to be selfish and corrupting of soul, it was still participating in humanity's search for knowledge.

[1094] It should be mentioned that the overly upset were not only to be found on the side of left-wing pseudo idealism. Some people who were overly upset resisted taking up support of left-wing pseudo idealism because they didn't want to be perceived as, or consider themselves to be, weak and cowardly for having abandoned the battle, and/or because they were intuitively aware that too many had already done so, resulting in the now extremely high levels of delusion and dishonesty in the world that they didn't want to contribute any further to. It should also be mentioned that there have been people who supported the political right's promotion of the freedom needed to search for knowledge while adhering to fundamentalist misrepresentations of religious teachings that worked *against* that search. This dichotomy can be explained by the fact that while you could support right-wing policies without being particularly confronted by your own corrupted condition, supporting religion has been an unavoidably self-confronting practice, so while it has been possible for the more exhausted to support right-wing politics, it has not been as easy to be a supporter of religion unless it was in a literal, fundamentalist way, where, as was explained in ch. 8:16I, the emphasis could be shifted away from self-confronting honesty, introspection and interpretation of the deeper truths contained in the religion's scriptures to simply worshipping the prophet as a more indirect way of glorifying and supporting the truth he stood for. The so-called 'Bible Belt' in southern USA, for instance, is notoriously right-wing but also notoriously fundamentalist, such as in its advocation of creationism and opposition to the teaching of evolution. They support individual freedom but not analysis of the human condition; they despise the dishonesty and weakness of the left-wing for holding that, for example, there is no such thing as truth, and so want to support the truthfulness contained in religion, but not so much that it becomes personally confronting.

$^{1095}$ At this point we need to continue our analysis of the progression that the development of even more guilt/truth stripped forms of idealism to live through took, and look at what emerged after the development of the politically correct movement.

## Chapter 8:16P  <u>The Postmodern Deconstructionist Movement</u>

$^{1096}$ <u>While showing interest only in those who were suffering was extremely dishonest and dangerous</u>, from the perspective of the do-good, feel-good, politically correct supporter, gaining relief from guilt/ depression/the agony of the human condition was all that mattered — when the truth is killing you, you have no qualms about escaping it — and so <u>to help ensure no subversive questioning could creep in and undermine this strategy, a philosophical justification for truthlessness was bolted on to the politically correct culture</u>. This was <u>Postmodern Deconstructionism</u>, **'a bewilderingly complex school of continental philosophy, or pseudo-philosophy'** of **'intellectual assumptions — [that] truth is a matter of opinion, there is no real world outside of language and hence no facts independent of our descriptions of them'** (Luke Slattery, *The Australian*, 23 Jul. 2005). While language is artificial it nevertheless models a real world, so to say that just because language is artificial there can be no universal truths is ridiculous, but again, when the need to escape the truth becomes desperate, any excuse will do; just baffle the world, and yourself, with intellectual baloney, such as this from Jacques Derrida, one of the high priests of deconstructionism, who gave this highly intellectual (instinct/soul/truth-less) description of why truth supposedly doesn't exist: **'Every sign, linguistic or nonlinguistic, spoken or written, as a small or large unity, can be cited, put between quotation marks; thereby it can break with every given context, and engender infinitely new contexts in an absolutely nonsaturable fashion. This does not suppose that the mark is valid outside its context, but on the contrary that there are only contexts without any center of absolute anchoring'** (*Margins of Philosophy*, 1982, p.320 of 330).

$^{1097}$ In his 2001 book, *The Liar's Tale: A History of Falsehood*, Jeremy Campbell described **'postmodern theory'** as having elevated **'lying to the**

**status of an art and neutralised untruth'**. It **'neutralised untruth'**, made what's not true non-existent, because by denying the existence of the variously embattled and upset states of individuals, 'races', genders, generations, countries, civilizations and cultures, it made any discussion of such differences—any pursuit of insight into the human condition—impossible. Recall how the politically correct culture deemed the children's nursery rhyme *Baa Baa Black Sheep* racist, insisting it be recited as **'Baa Baa Rainbow Sheep'**? Well, the 1994 postmodernist teaching guide *From Picture Book To Literary Theory* similarly argued that school children shouldn't be read stories about witches on broomsticks because they were sexist for **'narrowly defining women's roles'**, or the *Three Little Pigs* fairy tale on the basis of its elitist promotion of **'the virtues of property ownership and the safety of the private domain which are key elements of capitalist ideology'**!

[1098] With levels of upset in people becoming extreme and the associated levels of guilt and self-loathing becoming unbearable for ever increasing numbers of people, hunting for opportunities to be 'idealistic' and feel that rush of relief from doing 'good' and feeling you were a good person at last, became all-consuming, to the point of ridiculously dangerous irresponsibility. The American columnist Ann Coulter described how **'Whenever the nation** [the USA] **is under attack, from within or without, liberals side with the enemy. This is their essence'** (*TIME*, 27 Sep. 2004), while a former US presidential advisor, Karl Rove, presented this similar observation: **'Conservatives saw the savagery of 9/11 and the attacks and prepared for war; liberals saw the savagery of the 9/11 attacks and wanted to...offer therapy and understanding for our attackers'** (*The New York Times*, 24 Jun. 2005).

[1099] Pseudo idealistic, politically correct, postmodern, left-wing **'compassion junkies'**—empowered by their own self-righteousness and the right-wing's inability to defend itself—were taking over almost everywhere. Drawn together by their shared overwhelming need to find and promote causes through which to feel 'good' they hijacked such influential institutions as the national public broadcasters in the UK and Australia—bodies that are supposed to represent *all* of its constituents, not one extreme faction. As a former British Broadcasting Corporation (BBC) correspondent, Robin Aitken, said, **'[I] could not raise a cricket team of Tories** [right-wing conservatives] **among BBC staff'** (*The Australian*, 10 Oct. 2005). The British journalist Melanie Phillips similarly observed that **'With a few honourable exceptions, the BBC views every issue through the prism** of left-wing, secular, anti-Western thinking' *(Daily Mail,* 16 May 2005). A blogger posted this alarm: 'The broadcasters particularly in the BBC are...acting like the militant wing of the Labour Party. They have completely lost all restraint and integrity. They are completely out of order...[it's] a national outrage...[that they] are allowed to control the most influential power centre in the history of mankind. The BBC is run by an extreme, unrepresentative and unelected cult. The Left have finally ruined the BBC...The BBC is now irreparably infiltrated and broken...All responses to the Tories are to give the impression that whatever the Tories say, do or propose is immoral or incompetent, or imply that selfish, self-serving or somehow bad motives are the real reason for them—that they are not proposed altruistically for the genuine benefit of the country and therefore cannot be a credible alternative to what is being done by Labour, who are altruistic' ('How the Left have corrupted broadcast news', 1 Feb. 2010; see <www.wtmsources.com/124>). Almost every program on Australia's national broadcaster, the Australian Broadcasting Corporation (ABC), is now either outrightly dedicated to, or at least heavily skewed in its presentation towards servicing the needs of relief-hunters; in fact, its blatant bias has become something of a national joke. These national broadcasters serve no purpose now other than as places for the overly exhausted to find relief from their exhaustion. Neither the national interest nor humanity's are of any concern, even though the illusion is that these are their only interests.

[1100] And the problem of pseudo idealism in the media has not been confined to the national broadcasters. A 2005 UCLA-led study into the political leanings of media in the US revealed that 'almost all major media outlets tilt to the left...there is quantifiable and significant bias' ('Media Bias Is Real, Finds UCLA Political Scientist', *UCLA News,* 14 Dec. 2005). But at least some are protesting; in 2010 it was reported that 'Four billboard trucks bearing the message "Stop the Liberal Bias, Tell the Truth!" began circling the Manhattan headquarters of ABC, CBS, NBC and the *New York Times*' ('Trucks Encircle ABC, CBS, NBC, Challenge "Liberal" Media to "Tell The Truth"', CNSnews.com, 4 Oct. 2010; see <www.wtmsources.com/142>).

[1101] Much of academia, the supposed home of higher learning, has also been hijacked by overly upset, guilt-escaping, human-condition-avoiding, truth-hating relief-hunters, with a 'comprehensive' survey undertaken in 2005 by political science professors at the University of Toronto revealing that '87 percent' of the teaching faculty 'at the most elite' 'American universities' were 'left-wing' 'liberal'. A co-author of the study, Professor Robert Lichter, commented that 'What's most striking is how few conservatives there

**are in any field** [of study in US universities]...**It's a very homogenous environment'** ('College Faculties A Most Liberal Lot, Study Finds', *The Washington Post*, 29 Mar. 2005). That **'87 percent'** of teachers in the major American centers of learning held a distinct bias of thought was certainly a **'most striking'**—indeed, truly frightening—statistic. The literary scholar Harold Bloom recognized the danger of this hijacking of academia by truth-hating relief-hunters in his 1994 book, *The Western Canon*, which, as a *TIME* magazine review summarized, asserted that a **'rebellion in U.S. schools against Dead White European Male authors'** (or **'D.W.E.Ms'**) would lead to **'the end of civilization'** through a **'triumph of the forces of darkness'** (10 Oct. 1994). In *The Western Canon*, Bloom lamented that **'*Batman* comics...will replace Chaucer, Shakespeare, Milton, Wordsworth'** (p.485 of 560). But with the human condition now explained and the truth about humans finally revealed, we can see who was doing the critical deeper thinking needed to liberate humanity from ignorance, for if you go through this book and write down the names of all the authors of the quotes used to evidence these human-race-saving insights into our species' condition, you will find that **'Dead White European Male'**'s saved the human race! Yes, the truth is that the 'Dead White European Male'-led **'rise of the West is, quite simply, the pre-eminent historical phenomenon of the second half of the second millennium after Christ. It is the story at the very heart of modern history'** (Niall Ferguson, *Civilization: The West and the Rest*, 2012, p.18 of 432).

[1102] A 2006 article in *The Australian* newspaper reported that Australia's then Prime Minister, John Howard, **'believes the postmodern literature being taught in schools is "rubbish"...accusing state education authorities of "dumbing down" the English syllabus and succumbing to political correctness.** [He said] **"I feel very, very strongly about** [this situation where]...**traditional texts, are treated no differently from pop cultural commentary"'** (21 Apr. 2006). In the article, Howard, Australia's greatest ever Prime Minister (in my view), also referred to postmodern discourse as meaningless **'gobbledegook'**. The article said Howard's comments were in response to the dismissal of great literature as being elitist, sexist and racist, with a leading Sydney private school having **'asked students to interpret Othello from Marxist, feminist and racial perspectives'**. The real subject students should be asked to **'interpret'** is how artificial and dishonest pseudo idealism is and how it is threatening to condemn humanity to extinction.

[1103] In his insightfully titled 1987 book, *The Closing of the American Mind*, the political scientist Allan Bloom also wrote of the devastating effects of postmodern, so-called 'deconstructionist' teaching in American universities. As summarized in this review published in *The Australian* newspaper: **'we are producing a race of moral illiterates, who have never asked the great questions of good and evil, or truth and beauty, who have indeed no idea that such questions even could be asked...As Mr Bloom says..."deprived of literary guidance they** [students] **no longer have any image of a perfect soul, and hence do not long to have one. They do not even imagine that there is such a thing"...If the classics are studied at all in the universities they are studied as curiosities in the humanities departments, not as vital centres of the liberal tradition, and not as texts offering profound insight into the human condition'** (Greg Sheridan, 'The Closing of Our Minds', 25 Jul. 1987). Of course, the whole point of the postmodern movement was to avoid **'the great questions'** about our **'soul'** and what has happened to it, namely the question of our self-corruption and alienation, the issue of **'good and evil'**, **'the human condition'**, **'truth'**. The object was to stop thinking, stop the pain of guilt, bring about a **'closing of the...mind'**, **'dumb'** the mind **'down'**. The deeper, fundamental and true objective of education is to enlighten us about the underlying issue in all of human life of the human condition, and that is precisely what literature that has historically been considered great has managed to do. While not overly confronting us with our upset, corrupted, fallen condition the great writers of our time have, through great literary skill and device, managed to throw light on it. But it is precisely that great gift that postmodernism found a way to destroy. The character Kurtz in Joseph Conrad's great/classic 1902 book **'Heart of Darkness** [of our condition]' acknowledged **'The horror! The horror!** [of our massively upset condition]', but, of course, for the politically correct postmodernist this was a despicable **'D.W.E.M.'**-written piece that propagated male chauvinism, elitism, racism and aggression.

[1104]Not so long ago, HRH The Prince of Wales (whose mentor or main influence in his philosophical life was, like mine, Sir Laurens van der Post, and who also attended Geelong Grammar School where, like me, he benefited from the influence of Sir James Darling's soul-rather-than-intellect-emphasizing, Platonic education system) wrote a letter of deep concern to the Lord Chancellor of Great Britain, questioning the extreme bias that is now also apparent in legal thinking, stating that **'The Human Rights Act is only about the rights of individuals. I am unable to find a list of social responsibilities attached to it and this betrays a fundamental distortion in social and legal thinking'** (*The Australian*, 27 Sep. 2002).

[1105] Dr William Anderson, a professor of economics in the US, similarly observed that **'Justice pretty much is dead in the United States...Like so many other trends, this one has its intellectual underpinnings in that academic refuse pile we call Post-Modernism...a line of thinking that denies any possibility of Truth, and is the dominant "guiding light"—darkness?—in academe these days... right now, the post-modernists are winning battle after battle. It is one thing when post-modern nonsense dominates a history or English class; it is quite another when it becomes the bedrock of modern law'** ('Post-Modern Prosecutions', 25 Nov. 2006; see <www.wtmsources.com/115>).

[1106] The American lawyer Gary Saalman also wrote that **'In recent years... postmodernism has risen to the forefront of legal theory. Postmodern theorists... claim the law cannot have any foundation because there is no foundation for objective knowledge of any kind...Principles of law could never reflect universal truths, they argue...According to these scholars, it is senseless to talk about whether a law is right or wrong or moral or amoral...most observers agree that postmodern theories of law are exerting a huge influence today in the courtroom and the legislature...Remember, these are not a lunatic fringe at the margins of legal practice. They include department heads, and leading professors of law schools...practicing lawyers and legal authorities'** ('Postmodernism and You: Law', 1996; see <www.wtmsources.com/109>).

[1107] And, in his 2006 book, *Understanding the Times: The Collision of Today's Competing Worldviews*, the American religious leader Dr David Noebel had this to say about the dangers of postmodernism: **'Harold J. Berman, former professor of law at Harvard Law School...explains that today... foundational beliefs are rapidly disappearing, not only from the minds of philosophers, but from "the minds of lawmakers, judges, lawyers, law teachers...the historical soil of the Western legal tradition is being washed away in the twentieth century, and the tradition itself is threatened to collapse"...Postmodernists are intent on eliminating religious roots and transcendent qualities from Western law. They desire more fragmentation and subjectivity, and less objective morality than the Judeo-Christian tradition demands. In the end, they are intent on creating and using their own brand of social justice merely for left-wing political purposes'** ('Postmodern Law'; see <www.wtmsources.com/121>).

[1108] The media, our centers of education and learning, the judiciary— these are all pillars of society that are being white-anted in the march toward terminal alienation. The world *is* in jeopardy of being *completely* hijacked by those who are no longer concerned with humanity's heroic journey to enlightenment and who only want to escape the depressing effects of their human condition. Total self-preoccupation, selfishness disguised as selflessness, has arrived. Terminal levels of alienation *are* upon us.

[1109] I should mention that in 2014 a new level of escapism from the issue of the human condition, and with it a new level of abandonment of humanity's heroic journey to enlightenment, emerged with the development of 'Empathetic Correctness'. An article in *The Atlantic* titled '"Empathetically Correct" Is the New Politically Correct' reported that in **'college classrooms'** there is **'a troubling trend toward protecting people from their own individual sensitivities'**, and referred to students who are **'refusing to read texts that challenge their own personal comfort'**, and to a **'number of campuses proposing that so-called "trigger warnings" be placed on syllabi in courses using texts or films containing material that might "trigger" discomfort for students'**, **'so that students can read only portions of a book with which they are fully comfortable'**, or even **'miss classes containing such [challenging] material without a grade penalty'** (23 May 2015). Truly, the doors of the human mind ARE closing. Thinking has become TOO terrifying. Terminal alienation IS upon us.

[1110] Yes, with its supposedly underpinning postmodern deconstructionist philosophy, the politically correct movement is a pure form of freedom-denying dogma that fabricated, demanded and imposed ideality or 'correctness', specifically that of an undifferentiated world, in complete denial of the reality of the underlying issue of the existence of and reasons for humans' variously embattled and upset states. It is concerned with *pretending* to behave ideally and by so doing *deluding* yourself that you are an ideal, 'good' person. The delusion involved is, in fact, so great the movement brazenly claims to be taking humanity beyond or **'post'** the existing upset, embattled **'modern'** state to a good versus evil **'deconstructed'**, **'correct'** world when in truth it is behaving *non-ideally* and working *against* the search for the liberating understanding that *will* take humanity beyond the upset state to a **'post'** present/ **'modern'** human-condition-ameliorated, good-versus-evil-reconciled and thus truly **'deconstructed'**, *genuinely* **'correct'** world. The gloves are off now, the confidence of—and sheer bloody-minded anger and aggression underlying—the industry of escapist denial, delusion and non-thinking is such that it is now prepared to go the whole hog and outrageously mimic the arrival of the true world at the actual expense of any chance it had of arriving. The fact is, the postmodernist politically correct movement represents the very height of dishonesty, the most sophisticated expression of denial and delusion to have developed on Earth. The **'forces of darkness'** articulated by Harold Bloom are certainly in ascendance. The time that Plato predicted (see par. 172), where he said there would be **'more and more forgetting** [alienation]**'** until the **'universal ruin to the world'** occurred, has arrived. Yes, terminal alienation *is* upon us; humanity *has* entered end play, a death by dogma.

## Chapter 8:16Q The 'abomination that causes desolation' 'sign... of the end of the age' that is 'cut short' by the arrival of the liberating truth about the human condition

[1111] To look at humanity's situation overall now, it is clear we have reached the terminally alienated, end play point in the human journey where the rapidly rising tide of the charismatic and fundamentalist misrepresentations of religious teachings in the dysfunctional, materially poorer parts of the world, and the politically correct, postmodern culture in the more functional, materially affluent parts of the world are threatening to stifle the freedom of expression upon which liberating enlightenment of the human condition depends.

[1112] A reflection of the compounding rate of increase of upset in humans is that the interval between the development of one form of pseudo idealism to the next more evasive form in this chronology became shorter and shorter, shrinking from centuries to decades to years. Communism was a long time coming after the development of religion whereas the forms of pseudo idealism that followed communism appeared, or if they had been germinating for some time then became prominent, one after the other in rapidly shrinking intervals of time.

[1113] In the onrush of the psychosis-escalating struggle of the human journey it was inevitable that a point would eventually be reached where the need for relief and restraint would become so great that the delusion of dogma—the artificiality of imposing ideality to solve reality—would become invisible to its practitioners. We are now in that situation where the advocates of dogma have become blind to its short-comings and consequently are now on an all-out mission to seduce and intimidate everyone with their culture. It has been said that **'postmodernism has peaked, and will die with the century'** (Damian Grace, 'A Strange Outbreak of Rocks in the Head', *The Sydney Morning Herald*, 21 Jan. 1998), but this is a psychosis-driven situation where, increasingly, people *have to* variously live in adherence to pseudo forms of idealism to cope with their circumstances. Therefore, if you live in an affluent part of the world, no amount of opposition will halt the rise of politically correct culture, or, if you live in an impoverished part of the world, no amount of opposition to fundamentalist misrepresentations of religious teachings will halt its growth. It is only the arrival of the understanding of the human condition in this book that can halt the onrush of terminal alienation and destruction of the human race.

[1114] So, with understanding of the human condition finally found, the truth it reveals is that the underlying, *real* progression in all the great, so-called, 'social reforms' of the last 200 years was *not* to a more ideal world but to greater upset and its associated need for ever more guilt-stripped forms of pseudo idealism through which to live. From religion, the original, thousands-of-years-old, relatively honest, alienation-free form of pseudo idealism, developed socialism and communism, then, in the less functional, materially poorer parts of the world, charismatic and fundamentalist misrepresentations of religious teachings, and, in the more functional, materially richer parts of the world, the New Age Movement, then the feminist movement, then the environment or green movement, then the politically correct movement, and, finally, the totally dishonest, completely alienated, definitely autistic postmodernist movement; again, **'autism'** is **'a complex mental structure insuring against recurrence of…unthinkable anxiety'** — in this case, **'anxiety'** about being extremely corrupted/upset/ hurt/soul-damaged in your infancy and childhood.

[1115] Yes, we can now see exactly what the underlying psychological situation is that has driven the pseudo idealistic strategy of the left-wing. And, in fact, the work of the psychologist Arthur Janov is very confirming of the psychosis behind the adoption of the left-wing approach. As mentioned earlier (in par. 1005), Janov developed the technique of 'primal therapy' in which adults are helped to work their way back in their minds to memories of the original (primal) hurt to their soul that occurred in their infancy and early childhood as a result of growing up in the extremely human-condition-embattled, psychologically upset, insecure, have-to-somehow-establish-your-worth world of today. The following extract from Janov's famous 1970 book, *The Primal Scream*, was included earlier, but I am re-including it here with additional material from his book because in it Janov describes very clearly how the more upset humans became, the more they needed to find a way to escape their **'personal horror'**.

He wrote (the underlinings are my emphasis): '**Anger is often sown by parents who see their children as a denial of their own lives. Marrying early and having to sacrifice themselves for years to demanding infants and young children are not readily accepted by those parents who never really had a chance to be free and happy** [p.327 of 446] **...neurotic parents are antifeeling, and how much of themselves they have had to cancel out in order to survive is a good index of how much they will attempt to cancel out in their children** [p.77] **...<u>there is unspeakable tragedy in the world...each of us being in a mad scramble away from our personal horror</u>. That is why neurotic parents cannot see the horror of what they are doing to their children, why they cannot comprehend that they are slowly killing a human being** [p.389] **...A young child cannot understand that it is his parents who are troubled...He does not know that it is not his job to make them stop fighting, to be happy, free or whatever...If he is ridiculed almost from birth, he must come to believe that something is wrong with him** [p.60] **...Neurosis begins as a means of appeasing neurotic parents by denying or covering certain feelings in hopes that "they" will finally love him** [p.65] **...a child shuts himself off in his earliest months and years because he usually has no other choice** [p.59] **...When patients** [in primal therapy] **finally get down to the early catastrophic feeling** [the 'primal scream'] **of knowing they were unloved, hated, or never to be understood—that epiphanic feeling of ultimate aloneness—they understand perfectly why they shut off** [p.97] **...<u>Some of us prefer the neurotic never-never land where nothing can be absolutely true</u>** [the postmodernist philosophy] **<u>because it can lead us away from other personal truths which hurt so much. The neurotic has a personal stake in the denial of truth</u>** [p.395].' What has been said here makes it very clear that postmodern deconstructionism *was* autistic behavior; '**a complex mental structure insuring against recurrence of...unthinkable anxiety**'.

¹¹¹⁶ So, what is finally revealed about these supposed great, progressive, enlightened social reforms is that instead of those involved being more ideal people, behaving in a more ideal way and bringing about a more ideal world, as they trumpeted themselves to the world as being and doing, they were actually more corrupted/upset/hurt/soul-damaged/ **'neurotic'** people, behaving in a less ideal way and bringing about an even more alienated, **'neurotic'** devastated world. The truth was the complete opposite of what the pseudo idealists, especially the latter day, more-advanced-in-denial pseudo idealists, were claiming. What was being presented to the world was a totally fraudulent, dishonest sham—and, as emphasized, the great danger of these ever-increasing levels of dishonesty is that humanity has been taken to the brink of terminal alienation.

[1117] Again, we humans had to do what we had to do—create a guilt-free yet truthless environment in order to stay alive under the duress of the human condition—but the situation has spiraled way out of control, taking humanity perilously close to the perpetual darkness of terminal alienation. Humanity *is* facing a death by dogma—'**The Closing of the...** [human] **Mind**', as Allan Bloom so accurately described it; or '**the end of civilization**' through a '**triumph of the forces of darkness**', as Harold Bloom predicted; or the '**universal ruin to the world**' that Plato said '**more and more forgetting** [denial]' would lead to. Nietzsche recognized the necessity of denial when he wrote, '**That lies should be necessary to life is part and parcel of the terrible and questionable character of existence**' (*The Will To Power*, 1901, Vol. II, p.289; tr. O. Levy, 1910), and in his major work *Faust* (1808), the writer Johann Wolfgang von Goethe acknowledged that lying is part of human nature while truth is not; *however*, the wholesale pathology that lying has developed into means humanity *has* arrived at a state of terminal alienation. R.D. Laing described the situation that exists now when he wrote, '**There is a prophecy in Amos that there will be a time when there will be a famine in the land, "not a famine for bread, nor a thirst for water, but of *hearing* the words of the Lord** [words of truthfulness]**." That time has now come to pass. It is the present age**' (*The Politics of Experience* and *The Bird of Paradise*, 1967, p.118 of 156). Yes, as necessary as lying has been, to indulge it to the point of actually shutting down thought was the greatest weakness and failing possible on a planet whose culminating achievement is the fully conscious thinking mind. Preventing the search for knowledge represented a failure of all the effort and sacrifice made thus far by life on Earth. It represented a complete loss of nerve—as the science historian Jacob Bronowski said: '**I am infinitely saddened to find myself suddenly surrounded in the west by a sense of terrible loss of nerve, a retreat from knowledge into – into what? Into... falsely profound questions** [and other dishonesties that don't]**...lie along the line of what we are now able to know if we devote ourselves to it: an understanding of man himself. We are nature's unique experiment to make the rational intelligence prove itself sounder than the reflex** [instinct]**. Knowledge is our** [proper] **destiny. Self-knowledge**' (*The Ascent of Man*, 1973, p.437 of 448). The Scottish comedian Billy Connolly was certainly right to denounce political correctness as '**the language of cowardice**' (*Billy Connolly: Live 1994*).

[1118] While the danger for humanity's journey to enlightenment came from the increased levels of delusion and denial that humans have had to employ in order to cope with our increasingly insecure condition, to be truly free we *had to* confront and understand our condition, *not* escape it by adding more and more layers of denial. Denial blocked access to the truth, that being its purpose, but we *had to* find the truth, especially the truth about ourselves, **'self-knowledge'**. As the great Greek philosopher Socrates famously said, **'the only good is knowledge and the only evil is ignorance'** (Diogenes Laertius, *Lives of Eminent Philosophers*, c.225 AD), and, **'the unexamined life is not worth living'** (Plato's dialogue *Apology*, c.380 BC; tr. B. Jowett, 1871, 38)—but in the end a preference for ignorance and the associated need to oppress any examination of our lives, oppress any freedom to think truthfully, question and pursue knowledge, threatened to become the dominant attitude throughout the world. George Orwell's bleak prediction that **'If you want a picture of the future, imagine a boot stamping on a human face** [freedom] **– for ever'** is about to be realized—as is Aldous Huxley's fear (which he wrote about in his famous 1932 novel *Brave New World*) that we would become a trivial culture where **'the truth would be drowned in a sea of irrelevance'**.

[1119] So while the greed of capitalism is certainly causing immense suffering and devastation, it is not greed that is taking the world to the brink of destruction, as everyone has been told—no, our society is being taken over by a desperate and madly behaved faction. It is a very, very serious matter that is made doubly so by the fact that almost no one is raising the alarm. Pseudo idealism has almost everyone intimidated, bluffed or seduced. Warnings about the real danger facing the world are only being voiced by a rare few, like those just mentioned—Orwell, Nietzsche, Harold Bloom and Allan Bloom. While Geoffrey Wheatcroft recognized that **'the great twin political problems of the age are the brutality of the right, and the dishonesty of the left'**, it is the dishonesty of the left that has the potential to—and is poised to—destroy the world.

[1120] Yes, of the three aspects of upset of anger, egocentricity and alienation, alienation is the really dangerous one. Interestingly, anthropologists—in their defensive, mechanistic, denial-complying, dishonest mindset—named the final two varieties of humans in the series *Homo sapiens* and *Homo sapiens sapiens*, which literally translates from Latin as 'wise man' and 'wise wise man', respectively (with 'sapiens' meaning 'wise' and 'homo' meaning 'man'). Certainly, humans have been gaining wisdom or knowledge, but the more accurate description of what we have become is really 'False or Alienated Man', and 'False False or Alienated Alienated Man'.

[1121] **NIETZSCHE'S** warning of the danger of the **'many sickly people'** who **'have a raging hate for the enlightened man and for that youngest of virtues which is called honesty'** can't be emphasized enough. To repeat what he wrote: **'Purer and more honest of speech is the healthy body, perfect and square-built: and it speaks of the meaning of the earth** [to face truth and one day find understanding of the human condition]**...You are not yet free, you still *search* for freedom. Your search has fatigued you...But, by my love and hope I entreat you: do not reject the hero in your soul! Keep holy your highest hope!'** Yes, yes, we cannot afford to lose our nerve, but everyone nearly has.

[1122] As I've already said, I consider **SIR LAURENS VAN DER POST** to be the most exceptional denial-free thinking prophet and philosopher of the twentieth century; indeed, his full-page obituary in London's *The Times* was boldly headed **'A Prophet Out of Africa'** (20 Dec. 1996; see <www.wtmsources. com/166>). He was another who was **'pure'** and **'honest'** and **'square-built'** enough to speak out strongly against the extreme danger of pseudo idealism, writing that **'the so-called liberal socialist elements in modern society are profoundly decadent today because they are not honest with themselves... They give people an ideological and not a real idea of what life should be about, and this is immoral...They feel good by being highly moral about other people's lives, and this is immoral...They have parted company with reality in the name of idealism...there is this enormous trend which accompanies industrialized societies, which is to produce a kind of collective man who becomes indifferent to the individual values: real societies depend for their renewal and creation on individuals...There is, in fact, a very disturbing, pathological element—something totally non-rational—in the criticism of the** [then UK] **Prime Minister** [Margaret Thatcher]. **It amazes me how no one recognizes how shrill, hysterical and out of control a phenomenon it is...I think socialism, which has a nineteenth-century inspiration and was valid really only in a nineteenth-century context when the working classes had no vote, has long since been out of date and been like a rotting corpse whose smell in our midst has tainted the political atmosphere far too long'** (*A Walk with a White Bushman*, 1986, pp.90-93 of 326).

[1123] Two of the Bible's denial-free thinking, exceptionally sound, **'pure'** and **'honest'**, **'square-built'** prophets—Daniel and Christ—also warned of the great danger of extreme forms of pseudo idealism when they spoke of **'the abomination that causes desolation'** taking over the world. The Bible has, without a doubt, been the most influential book in history because it contains extraordinary honesty; indeed, as I mentioned in par. 751, it has been the most denial-free book humans have had for guidance. As

**DANIEL** said of his own contribution to the Bible, **'I will tell you what is written in the Book of Truth'** (Dan. 10:21)—and his ability to think in a denial-free, truthful and therefore effective way meant Daniel *was* in a position to **'explain to you what will happen to your people in the future'** (10:14). However, since there was no science in his day to evidence his argument, all Daniel could do was draw upon analogies to describe what he could see so clearly happening in the future. In one analogy he described **'The king of the South'** (which we can understand is the freedom-upholding, answer-searching but immensely upsetting and corrupting right-wing) constantly at war with **'the king of the North'** (the freedom-oppressing, dogma-based, pseudo idealistic, dishonest left-wing). He described how, for a long time, power would oscillate between these kings (as it does in democratic politics), until the complete polarization of the two kingdoms, the two political states, came about (which has occurred), at which point **'the abomination that causes desolation'** of left-wing pseudo idealism would finally threaten to take over the world. After describing many changes of power, Daniel said: **'the king of the North will muster another army larger than the first...The forces of the South will be powerless to resist; even their best troops will not have the strength to stand** [pseudo idealism becomes increasingly seductive and powerful as people become increasingly upset]. **The invader will do as he pleases; no-one will be able to stand against him...**[but eventually 'he'/pseudo idealism] **will make an alliance with the king of the South...**[some bipartisanship between the left and right wing will occur, but then the left-wing] **will stumble and fall...**[however, in time] **He will be succeeded by a contemptible person** [even more dishonest forms of left-wing pseudo idealism will emerge] **who will not be given the honour of royalty** [they will lack religion's honesty]. **He will invade the kingdom** [the religious kingdom of honesty]...**and he will seize it through intrigue** [through the dishonesty of extreme forms of pseudo idealism and charismatic and fundamentalist misrepresentations of religion masquerading as real idealism]. **Then an overwhelming army** [from the South] **will be swept away before him; both it and a prince of the covenant** [religion] **will be destroyed...when the richest provinces feel secure, he will invade them and will achieve what neither his father nor his forefathers did** [no force has been able to overthrow religion before]...**His armed forces will rise up to desecrate the temple fortress and will abolish the daily sacrifice. Then they will set up the abomination that causes desolation** [extremely guilt-stripped pseudo idealistic causes and charismatic and fundamentalist misrepresentations of religious teaching will take over the world and lead humanity to terminal alienation].

With flattery [the truth-and-guilt-avoiding, do-good, feel-good self-affirmation that pseudo idealism feeds off] **he will corrupt those who have violated the covenant** [extreme forms of pseudo idealism fundamentalism will seduce the more upset away from religion's infinitely more honest way of coping with the human condition], **but the people who know their God will firmly resist him** [the more secure, less dishonest will not be deceived and *must* strongly resist the seductive tide]**...The** [left-wing pseudo idealistic] **king will do as he pleases. He will exalt and magnify himself above every god and will say unheard-of-things against the God of gods** [such as Richard Dawkins, Oxford University's Professor of Public Understanding of Science, would you believe, who has, as mentioned in par. 938, said that **'"Faith is one of the world's great evils, comparable to the smallpox virus, but harder to eradicate. The whole subject of God is a bore"...those who teach religion to small children are guilty of "child abuse".'**]. **He will be successful until the time of wrath** [until the all-exposing truth of understanding of the human condition arrives to save the world, as it is doing in what you are reading right here]**...He will invade many countries and sweep through them like a flood...he will set out in a great rage** [the stridency that I and Sir Laurens van der Post talked about of those trying to persuade themselves and others that their pseudo idealistic causes and fundamentalist misrepresentations of religion represent real idealism] **to destroy and annihilate many...Yet he will come to his end, and no-one will help him...**[when understanding of the human condition arrives pseudo idealism and other forms of extreme dishonesty will be totally exposed and brought to an end. However, while 'he'/pseudo idealism reigns] **There will be a time of distress such as has not happened from the beginning of nations until then'** (Dan. 11-12).

[1124]In another analogy that describes the same progression, but even more explicitly—mentioning as it does how **'truth was thrown to the ground'**—Daniel, appropriately enough, used the symbols of a (determined) ram for the right-wing and a (stupid) goat for the left-wing. He said that initially **'No animal could stand against him** [the ram], **and none could rescue from his power. He did as he pleased and became great...**[greed and indifference to others was so great that even children, for instance, were put to work in coalmines, but then the] **goat...charged at him in great rage...**[the left-wing emerged and] **The ram was powerless to stand against him; the goat knocked him to the ground and trampled on him, and none could rescue the ram from his power. The goat became very great...It set itself up to be as great as the Prince of the host** [it set itself up to be more important than religion]**; it took** away the daily sacrifice from him, and the place of his sanctuary was brought low…It prospered in everything it did, and truth was thrown to the ground… "How long will it take for the vision to be fulfilled—the vision concerning the daily sacrifice, the rebellion that causes desolation, and the surrender of the sanctuary and of the host that will be trampled underfoot [the rebellion against religion's honesty]…It will take [a long time]…understand that the vision concerns the time of the end."…[when humans] have become completely wicked [and], a stern-faced king [the extremely upset], a master of intrigue, will arise [the left-wing pseudo idealistic, false prophet merchants of escapist denial and delusion will arise]…He will cause astounding devastation and will succeed in whatever he does. He will destroy the mighty men and the holy people [even the strong will begin to succumb to the intimidating tide of pseudo idealism]. He will cause deceit [the misrepresentation of pseudo idealism and fundamentalism as being real idealism] to prosper, and he will consider himself superior [the extreme delusion that left-wing pseudo idealism and fundamentalism is based on will spread everywhere]…"The vision…concerns the distant future [that has finally arrived]"" (Dan. 8).

[1125] So important to Daniel was this issue of the great danger from pseudo idealism for humanity in the future that he even used a third analogy to try to convey his fears about it. In this analogy, he first described the rule of a 'king of kings' with a 'head of gold' who had 'dominion and power and might and glory' (Dan. 2:37-38). This is a description of the original dominance hierarchy type situation (such as occurs in the animal kingdom) where some peace and order occurred in human society through the rule of the most powerful. Daniel then said, 'After' this 'another kingdom will rise' that is 'inferior', then 'one of bronze', then one as 'strong as iron' that 'breaks and smashes' 'all the others', but he said 'this will be a divided kingdom…partly iron and partly clay, so this kingdom will be partly strong and partly brittle…so the people will be a mixture and will not remain united, any more than iron mixes with clay' (2:39-43). What Daniel has described here is how, through trial and error, the very effective—'strong as iron'—device for managing upset of democracy is developed, but then this system of management becomes hopelessly 'divided'/polarized, because 'people will be a mixture', 'partly strong and partly brittle', which is clearly a description of what has happened with the strong right and the weak, pseudo-idealism-supporting left-wing in politics. Daniel then said that eventually 'God…will set up a kingdom that will never be destroyed… [and] will crush all those [other] kingdoms and bring them to an end', saying this is **'what will take place in the future'** (2:44-45). Yes, as will be described shortly in pars 1135-1136, now that **'God'** or Integrative Meaning-guided understanding of the human condition has emerged this whole ugly war between the left and the right is finally obsoleted, **'destroyed'**, **'crush[ed]'**, brought **'to an end'**.

[1126] In the New Testament, **CHRIST** gave exactly the same warning as Daniel, even referring to Daniel's description of pseudo idealism as **'the abomination that causes desolation'**—but Christ went further, truthfully and courageously advising people to head for the hills, **'flee to the mountains'**, when pseudo idealism and strident fundamentalist misrepresentations of religious teachings threaten humanity with annihilation. Referring to **'the sign...of the end of the age'**, Christ said that **'At that time many will turn away from the faith and will betray and hate each other** [a great deal of upset will develop], **and many false prophets** [pseudo idealists claiming to be leading the way to peace and a new age of goodness and happiness for humans] **will appear and deceive many people...even the elect** [even those less alienated, still relatively sound and strong in soul, will begin to be seduced by pseudo idealism]**—if that were possible. See, I have told you ahead of time...Wherever there is a carcass** [the extremely upset], **there the vultures** [false prophet promoters of delusion and denial to artificially make the extremely upset feel good] **will gather. Because of the increase of wickedness** [upset], **the love of most will grow cold. So when you see the "abomination that causes desolation," spoken of through the prophet Daniel, standing where it does not belong** [throwing out real religion and falsely claiming to be presenting the way to the human-condition-free, good-versus-evil-deconstructed, post-human-condition, better, correct world]**—let the reader understand—then let those who are in Judea flee to the mountains. Let no-one on the roof of his house go down to take anything out of the house. Let no-one in the field go back to get his cloak. How dreadful it will be in those days for pregnant women and nursing mothers! Pray that your flight will not take place in winter because those will be days of great distress** [mindless dogma and its consequences] **unequalled from the beginning of the world until now—and never to be equalled again. If those days had not been cut short** [by the arrival of the liberating understanding of the human condition], **no-one would survive'** (extracts from Matt. 24 & Mark 13 combined). In summary, when Christ said to **'Beware of false prophets. They come to you in sheep's clothing, but inwardly they are ferocious wolves'** (Matt. 7:15), he was warning that we had to be on guard for those who concealed their extreme upset behind pseudo idealistic causes in order to delude themselves that they were sound and ideally behaved people who were leading others to a sound and ideal world. Where true prophets *confronted* the issue of the human condition, false prophets were merely merchants of denial and delusion, advocates of *escapism* from the issue of the human condition.

[1127] (Note, it might be asked how, in such ancient times, could the threat pseudo idealism and strident forms of fundamentalist non-religion posed to the human journey have been so accurately anticipated by Christ and Daniel. But the truth is, *all* the nuances of the human condition can be seen in any sample of human behavior for those sufficiently psychologically open to seeing it. In the case of pseudo idealism, you only have to be caught in heavy traffic and observe how instead of holding their place against aggressive lane swappers some drivers are clearly deriving a feel good rush from letting such drivers push in front of them, even slowing down to encourage them. With a little more observation you can extrapolate what is eventually going to happen in the world, especially if you are sufficiently free from needing to live in denial of the issue of the human condition—in particular free from having to live in denial of Integrative Meaning or God—to be able to, and want to, think deeply about such matters. Daniel for example, in explaining why he **'could understand villains and dreams of all kinds'** (Dan. 1:17), said **'I have greater wisdom than other living men'** (2:30), and on the question of why he was able to have his mind **'turned to things to come'** (2:29), he **'praised the God of heaven** [he embraced the truth of Integrative Meaning and was unafraid of the issue of the human condition]…[who] **gives wisdom to the wise and knowledge to the discerning** [not denying Integrative Meaning enables him to think truthfully and thus effectively]. **He reveals deep and hidden things; he knows what lies in darkness, and light dwells with him. I thank and praise you, O God of my fathers: You have given me wisdom and power'** (2:19, 21-23).)

[1128] The most common criticism my writings about the human condition receive is their denunciation of left-wing causes—'What could possibly be wrong with being concerned for those who are disadvantaged, suffering and oppressed, or with loving the environment, or with wanting peace?', people ask—but that is exactly an indication of just how bluffed almost everyone has been by pseudo idealism. As Daniel foresaw, in the end **'even their best troops will not have the strength to stand'** against the seductive tide of pseudo idealism. He said **'no-one will be able to stand against'** it and it will **'invade many countries and sweep through them like a flood'** and **'cause deceit to prosper'**. Christ similarly warned that **'false prophets will appear**

**and deceive many people…even the elect — if that were possible. See, I have told you ahead of time.'**

[1129]There is, of course, another exceptionally **'pure'** and **'honest'**, **'square-built'** prophet who determinedly warned of the extreme danger of evasive, dishonest denial and delusion eventually threatening the human race with extinction: **PLATO**. As documented in chapters 2:6 and 6:12, Plato was so extraordinarily sound that long ago, around 360 BC, he was able to truthfully describe every aspect of the human journey from ignorance to enlightenment except for the actual explanation for *why* humans are good and not bad, which had to await the development of science and its discovery of how genes can orientate but only nerves can understand. In his extraordinarily prophetic writing on the human journey, Plato described **'our state of innocence, before we had any experience of evils to come'** — our original **'simple and calm and happy'** **'blessed and spontaneous life'** — and the subsequent emergence of our conscious mind, which allowed us to become our **'own masters'** but brought about our upset, **'evil'** condition of **'discord'**. We were so ashamed of that discordant state that we **'often unduly discredit**[ed]**'** the truth of our species' **'blessed'** past and any other truth that brought the **'painful'** issue of **'the imperfections of human life'**, which is **'our human condition'**, into focus, leaving us hiding **'a long way underground'**, **'enshrined in that living tomb'** of a **'cave'**-like state of dishonest denial where there was **'more and more forgetting** [alienation]**'**, to the point where we could only **'see dimly and appear to be almost blind'** — a development that Plato predicted would eventually lead to a state of terminal alienation and **'universal ruin to the world'**, from which only a denial-free approach, one where **'God'** in the form of Integrative-Meaning-acknowledging truthfulness, had **'again seated himself at the helm'**, could — and now has — **'set them** [humans] **in order and restored them'**!

[1130]Before summarizing the extent of the danger that the practice of pseudo idealism represents it needs to be emphasized that pseudo idealism is only one form of the extremely dishonest and dangerous denial or **'forgetting'** of the issue of **'the imperfections of human life'**, which is the issue of **'our human condition'**, that has been going on. The other extremely dishonest and dangerous form of this denial is the one that was described in chapter 6:12, '*The* great obscenity', which is science's obscenely irresponsible practice of unjustly dismissing and even persecuting the human-race-saving explanation of the human condition that has been presented in this and my earlier books. Recall how, in par. 611, it was pointed out that professors promenade across their campuses full of conviction of their own importance, with students certain that they too are where it's 'at'—but universities have become one big dishonest castle of lies; in fact, rather than centers of learning that seek truth, as they are supposed to be, universities have become hideous places that indulge, glorify and perpetuate dishonesty. So when reading the following summary of the warnings about the extremely dangerous consequences of denial as it is being practiced by pseudo idealism, it should also be kept in mind that the full horror of what is described *equally* applies to the extremely irresponsible and dangerous trap that science has fallen into of perpetuating its denial of the truth about the human condition when that denial is no longer justified!

[1131] To summarize the danger of pseudo idealism: for such exceptionally sound, **'pure'** and **'honest'**, **'square-built'** prophets throughout history who weren't living in denial and who thus **'kn[e]w their God'**—from **DANIEL** to **PLATO** to **CHRIST** to **NIETZSCHE** to **SIR LAURENS VAN DER POST**— to so **'firmly resist'** and warn of the great danger of the **'profoundly decadent'**, **'dreadful' 'abomination'** where there is **'more and more forgetting'**, to the point where **'universal ruin to the world'** has been brought about, illustrates just how extremely serious the threat has become of **'the end of civilization'** through a **'triumph of the forces of darkness'** of guilt-stripped charismatic and fundamentalist misrepresentations of religious teachings, and of guilt-stripped forms of non-religious pseudo idealism. The situation is, as these prophets warned, one of **'astounding devastation'** and **'desolation'** **'unequalled from the beginning of the world until now'** engendered by those whose **'love'** has **'grow[n]'** so **'cold'** that they have such a **'raging hate [of]...honesty'** that they have created a pseudo idealistic **'rotting corpse whose smell in our midst has tainted the political atmosphere far too long'**, in which **'truth was thrown to the ground'** and **'trampled underfoot'**—and the influence and hold of those responsible has become *so* great that any remaining opposition will soon, as Daniel predicted, be **'powerless to stand against'** them. In the case of the West, the forefront of humanity's journey to enlightenment—it defied and fought with all its might the spread of human-freedom-oppressing dogma in the form of communism in the former USSR and South East Asia, but it has proved incapable thus far of resisting the insidious takeover of its own culture by the freedom-oppressing dogma of pseudo idealism. During the 1950s Senator

Joe McCarthy did succeed in stemming its rise in America for a short while. And with pseudo idealism having now taken over all the pillars of American society, and the rabid socialist Democrat Bernie Sanders a real possibility of becoming the next President of the United States, there has been a desperate revolt against pseudo idealistic, politically correct oppression of freedom with the aforementioned extraordinary popularity of the hyper politically <u>in</u>correct Republican candidate Donald Trump. Irrespective of who wins the 2016 US election, the fundamental reality is that since taking up feel-good, pseudo idealistic causes has been the unavoidable way the more upset/alienated have been able to cope, and since upset/alienation is rising by the minute, it is inevitable that left-wing pseudo idealism *will be* unstoppable—and Daniel's prediction that pseudo idealism/**'the king of the North'**/**'the goat'** **'will succeed in whatever he does'** and **'cause astounding devastation'** will come true. Indeed, to attempt to describe how serious the situation in the West is, it is fair to say that freedom in those democracies is *as* under threat from pseudo idealism as a city or individual would be if it was surrounded by murderously hostile people and had, in effect, flown—or was at least hoisting—the white flag of surrender! In January 2014 *The Australian* newspaper published an article about Camille Paglia's latest book, *Glittering Images*, in which she decried this trend. With the subtitle **'For Camille Paglia, Western society has become too soft and weak to survive'**, the article quoted Paglia as saying, **'<u>What you're seeing is how a civilisation commits suicide</u>'** (Bari Weiss, 'In pursuit of manly equality', 1 Jan. 2014).

[1132] In the following passage, Sir Laurens provides a powerful affirmation of all that has just been said about the essential, intrinsic, absolute importance of maintaining freedom from oppression in the human journey to enlightenment: **'I hope a war is not declared by anybody in the modern world because I don't see the real necessity for war, but I would like to say that I think it would be immoral—it would be obscene—not to be ready at any moment to defend ourselves...If somebody should impose war upon us, attack us, I hope that we should have the will and the power and the moral courage to realise that life, freedom, are gifts from God and creation and our duty to defend. There's a wonderful episode in the life of Buddha where a group of villagers in the Himalayas did not defend themselves against a band of brigands who attacked and killed many of them, and he rebuked them because he said that was not what his teaching was about, that was not reverence for life of which he was speaking...it is a moral duty for us all to be ready to defend ourselves and**

**our freedoms'** (presentation by Laurens van der Post at the Institute of Contemporary Arts, London, 9 Dec. 1982; see <www.wtmsources.com/163>).

[1133] Well, the West's failure to **'defend'** its **'freedoms'** from the massive **'attack'** upon it by pseudo idealism means it has *failed* to exhibit the **'moral courage'** necessary to fulfill its **'moral duty'** to **'defend'** its **'freedoms'** against **'attack[s]'**. The result of this fundamental, **'obscene'** failure at the leading edge of humanity's journey to enlightenment is that humanity *has* reached the precipice of self-destruction—so it is no exaggeration to say that this understanding of the human condition has arrived *only just in time* to **'cut short'** that tragic end.

[1134] Yes, we are in the end play situation where the dangers from excessive upset in the world have become so great that **'even their [**humanity's**] best troops will not** [and have not] **have the strength to stand'** up to pseudo idealism. It has become virtually impossible to find sufficient **'pure'** and **'honest'** and **'square-built'** soundness (from having a relatively nurtured, loving, secure upbringing and thus not too much upset), coupled with strength of character (from having some, but not too much, genetically adapted toughness to life under the duress of the human condition), to stand against the tide of dishonesty and continue the freedom-dependent search for the liberating understanding of the human condition. Resolve and solidarity for our species' all-important, freedom-dependent journey to enlightenment is fast disappearing, and it is the lack of solidarity, the failure to present a strong, united front, that has been especially encouraging and empowering of the radical opponents of freedom such as Saddam Hussein in Iraq and Osama bin Laden in Afghanistan. The essence of the concept of The United Nations is of Nations acting strongly in a United way against what is patently **'obscene[ly]'** wrong behavior, but such is the lack of soundness and thus courage and character and thus resolve in the world that the concept has largely failed. R.D. Laing certainly wasn't exaggerating how sick, exhausted and spent with alienation the human race has become when, as has been mentioned before, he wrote that **'the *ordinary* person is a shrivelled, desiccated fragment of what a person can be...as men of the world, we hardly know of the existence of the inner world...The outer divorced from any illumination from the inner is in a state of darkness...i.e. alienation or estrangement from the inner light'**, and **'We are dead, but think we are alive. We are asleep, but think we are awake. We are dreaming, but take our dreams to be reality. We are the halt, lame, blind, deaf, the sick. But we are doubly**

**unconscious. We are *so* ill that we no longer feel ill, as in many terminal illnesses. We are mad, but have no insight** [into the fact of our madness].' And Pope Benedict XVI certainly wasn't overstating the darkness of the human situation when, in his 2005 inaugural mass, he said that **'We are living in alienation, in the salt waters of suffering and death; in a sea of darkness without light.'**

[1135] So in the relatively functional democracies of the West, humanity's 2-million-year journey to find understanding of the human condition has finally come down to a so-called **'cultural war'** between the philosophy of the political left-wing and the philosophy of the political right-wing. Both sides have been determined that they offered the better option, but with the human condition now explained what is revealed is that, in terms of a future for the human race, the philosophy of the left-wing was completely wrong while the philosophy of the right-wing was actually completely right. The left-wing was all about dogma, delusion and escapism, while the right-wing was comparatively all about realism, honesty and responsibility.

[1136] Interestingly and significantly, there is one final irony to the saga of humanity's great journey through ignorance, which is that the ideal world that the left-wing was dogmatically demanding is actually brought about by the right-wing winning its reality-defending, freedom-from-idealism, corrupting-search-for-knowledge battle against the freedom-oppressing pseudo idealism of the left-wing. Yes, with the freedom-from-dogma right-wing's search for understanding of the human condition completed, the justification for the egocentric power-fame-fortune-and-glory-seeking way of life espoused by the right-wing ends, replaced by the ideal-behavior-obeying attitude that the left-wing sought. In this sense, when the right-wing wins we all become left-wing; through the success of the philosophy of the right-wing, we all adopt the philosophy of the left-wing—but, *most significantly*, this time we are not abandoning an ongoing battle, we are leaving it won. In reality, of course, through finding understanding of the human condition the inability to explain what humanity's journey has been about ends—an outcome that completely obsoletes the different philosophical positions of the left-wing and right-wing once and for all. Now that it can be explained that humanity has been involved in a corrupting search for knowledge, corruption is explained, which in turn exposes and thus negates the unrealistic position of the left-wing's dogmatic insistence on idealism. For its part, the right-wing's corrupting search for knowledge is *also* brought to a close with the finding of the key knowledge it was searching for, that being understanding of the human condition. Of course, the search for knowledge *will* continue, but not from a basis of dishonest denial, and with the key piece of knowledge we needed found of the explanation of the human condition, the priority for the immediate future will be—as will be explained in the next chapter—to take up the human-condition-reconciled Transformed Way of Living. Thus, with the arrival of understanding of the human condition, the concept of 'politics' comes to an end, which will undoubtedly be of great relief to everyone. Indeed, it will be explained in the next chapter that with the arrival of the truth about the human condition, the whole intellectual and psychological world that humans' have been living in—our whole culture—is suddenly obsoleted and ended and a new, completely transformed, all-loving, peaceful human culture, civilization and world emerges, which is certainly not before time.

[1137] Finally, it should be re-emphasized that in the greater context of the human-condition-understood view that we now have, upset is an immensely heroic state, with the most upset being the most heroic individuals of all because they have necessarily been involved in the battle that humanity has been waging longer and/or more intensely than any others. While the stark descriptions that have been given of the extremely upset, as being **'wicked'**, **'stern-faced'**, **'cold'**, **'sickly'** **'carcasses'**, **'vultures'** and **'ferocious wolves'** involved in creating a **'rotting corpse'** of **'abomination'** that brings **'universal ruin to the world'**, were necessary to match the no-holds-barred, gloves-off, totally brutal assault on the truth that was being **'thrown to the ground'** and left **'trampled underfoot'** by pseudo idealists, in the human-condition-understanding new world such rhetoric is entirely wrong and redundant.

[1138] In fact, it shouldn't even be necessary to talk about the old struggle between the left and right wing philosophies any more than what has now been done in this book. Thankfully, humanity now has the option to move on to an entirely new existence. Indeed, there will be many subjects that no longer have to be discussed now that humanity is able to move on to another existence. We get the truth up and we move on. As will be explained in the next chapter, we leave our suitcase of experiences that took place in the old ignorant, human-condition-afflicted, power-fame-fortune-and-glory-seeking-and-pseudo-idealistic-coping world behind and move on to an entirely new, instinct-and-intellect-reconciled, human-condition-liberated, genuinely 'deconstructed' and 'correct', fabulous, all-wonderful, new transformed world.

*Old World*                    *New World*

Drawing by Jeremy Griffith © 1991-2011 Fedmex Pty Ltd

[1139]I would now like to include once more the following drawing to summarize humanity's heroic overall journey from our species' original innocent childhood through an insecure, human-condition-afflicted, ever-increasingly upset adolescence, to a secure, human-condition-reconciled, mature adulthood.

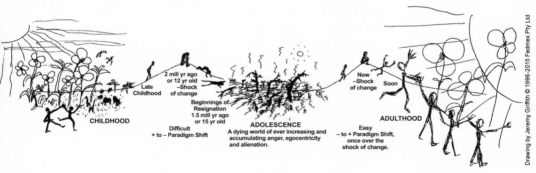

Drawing by Jeremy Griffith © 1996-2015 Fedmex Pty Ltd

[1140]The journey brings to mind Lord Alfred Tennyson's poem *Ulysses*, which most beautifully describes the last 200-year stage of our species' now desperately exhausted yet still hopeful, life-or-death struggle to find liberating understanding of the human condition: **'There lies the port** [home]; **the vessel puffs her sail** [in hope]**: there gloom the dark broad seas** [but we are in a dark place]**. My mariners, souls that have toil'd, and wrought, and thought with me – that ever with a frolic welcome took the thunder and the sunshine** [we took the bad and the good]**, and opposed** [any resistance and maintained] **free hearts, free foreheads – you and I are old** [the human race is exhausted, but]**; old** age hath yet his honour and his toil; death closes all: but something ere the end, some work of noble note, may yet be done' (1833). Tennyson's reference to the 'work of noble note' 'yet [to] be done' refers to the situation in Homer's Greek legend *The Odyssey*, where Ulysses, the hero of that great exploration and adventurous odyssey, finds on his return to his kingdom, Ithaca, that there remains one final journey he must undertake. Ulysses described this other journey to his faithful wife, Penelope, thus: '[The prophet] **Teiresias bade me travel far and wide, carrying an oar, till I came to a country where the people have never heard of the sea, and do not even mix salt with their food'** (c. eighth century BC). As has now been fully explained, because our original instinctive self or soul unbearably condemned our corrupting search for knowledge, it had to be repressed and banished. So our soul is now all alone, as if cast adrift way out in **'the lonely sea and the sky'** (John Masefield, *Sea-Fever*, 1902) — or as Samuel Taylor Coleridge put it, **'this soul** [of ours] **hath been alone on a wide, wide sea: so lonely 'twas, that God himself scarce seemed there to be'** (*The Rime of the Ancient Mariner*, 1798). And in having killed off our all-sensitive and all-loving soul we have lost access to all the truth, beauty and magic of life, to the **'salt'** or full flavor of existence, that our soul is aware of — leaving us in the exhausted numb, seared, flavorless, **'fifty feet of solid concrete'-'alienated'** world that Laing described **'the present age'** as being like. The **'oar'**, Teiresias explained to Ulysses, is actually a **'winnowing shovel'** — so we can now understand that this other journey that Ulysses still had to undertake was into the center of this **'country** [humans inhabit today] **where the people have never heard of the sea** [have no soul] **and do not even mix salt with their food** [are completely alienated]', in order to **'winnow'** from that world's human-condition-avoiding, denial-compliant, mechanistic scientific discoveries the truth about our human condition. It is *this* other journey, this **'work of noble note'** — which all of humanity's **'Souls that have toil'd, and wrought, and thought'** completed the preparation for — that has at last been accomplished.

[1141] So that's it — the greatest of all stories, our story, the story of humanity's incredibly, unbelievably heroic 2-million-year journey from ignorance to enlightenment told for the first time.

[1142] (A more complete description of all these stages that have been described of the maturation of humanity through its adolescence can be found in *Freedom Expanded* at <www.humancondition.com/freedom-expanded-stages-of-maturation>.)

# Chapter 9

# The Transformation of The Human Race

Photograph by Peter Viisimaa

## Chapter 9:1 <u>Summary</u>

[1143] This chapter describes how the instinct vs intellect explanation of the human condition is able to *immediately* transform *all* humans from their human-condition-afflicted, tortured existence to a fabulous life free of the human condition! This is how we *'change and become like little children...[and] enter the [fabulous, happy, psychologically secure] kingdom of heaven'* (Bible, Matt. 18:3).

## Chapter 9:2 <u>The 'dawn...of our emancipation'</u>

[1144] When *TIME* magazine invited the author Alan Paton to contribute an essay on apartheid in South Africa they instead received, and published in its place, a deeply reflective article on Paton's favorite pieces of literature. In what proved to be the renowned writer's last work, Paton wrote: **'I would like to have written one of the greatest poems in the English language — William Blake's "Tiger, Tiger Burning Bright", with that verse that** asks in the simplest words the question which has troubled the mind of man—both believing and non believing man—for centuries: "When the stars threw down their spears / And watered heaven with their tears / Did he smile his work to see? / Did he who made the lamb make thee?'" (25 Apr. 1988). The opening lines of Blake's poem, 'Tiger, Tiger, burning bright / In the forests of the night', refer to humans' great fear and resulting denial of the issue of our less-than-ideal, seemingly imperfect, 'fallen' or corrupted state or condition—a subject humans have consciously repressed and yet one that 'burn[s]' 'bright' in the 'forests of the night' of our deepest awareness. The very heart of the issue lies in the line, 'Did he who made the lamb make thee?'—a rhetorical question disturbing in its insinuation that we are wholly unrelated to 'the lamb', to the world of innocence. The poem raises the fundamental question involved in being human: how could the mean, cruel, indifferent, selfish and aggressive 'dark side' of human nature—represented by the 'Tiger'—be both reconcilable with and derivative of the same force that created 'the lamb' in all its innocence? As Paton identified, despite humans' denial of it, *the* great, fundamental, underlying question that 'has troubled the mind of man' has always been, *are* humans part of God's 'work', part of 'his' purpose and design, or *aren't we*? In other words—as was dealt with in chapter 4—why don't humans live in accordance with the cooperative, loving integrative meaning of existence? Interestingly, the final words of Paton's acclaimed book about apartheid, *Cry, the Beloved Country*, allude to humanity's dream of one day finding this answer, and, through doing so, freeing itself from the terrible 'bondage of fear' of our condition: 'But when that dawn will come, of our emancipation, from the fear of bondage and the bondage of fear, why, that is a secret' (1948). Humanity's hope and faith has *always* been that one day we would be able to explain the paradox of the human condition, and, as a result of doing so, liberate ourselves from 'the bondage' of our species' immensely psychologically upsetting sense of unworthiness or guilt.

[1145] So while religious assurances such as 'God loves you' could comfort us, ultimately we needed to understand *why* we were lovable; there *had* to be a rational, first-principle-based biological explanation for humans' divisive condition and our responsibility as conscious beings was to find that explanation. The poet Alexander Pope was recognizing this fundamental task and responsibility when he wrote, 'Know then thyself, presume not God to scan / The proper study of Mankind is Man' (*Essay on Man*, 1733); as were the ancients when they had emblazoned across their temples the phrase, **'Man, know thyself.'** To refer again to the words of that pre-eminent philosopher of the twentieth century, Sir Laurens van der Post: **'Only by understanding how we were all a part of the same contemporary pattern** [of wars, cruelty and greed] **could we defeat those dark forces with a true understanding of their nature and origin'** (*Jung and the Story of Our Time*, 1976, p.24 of 275); and, **'Compassion leaves an indelible blueprint of the recognition that life so sorely needs between one individual and another; one nation and another; one culture and another. It is also valid for the road which our spirit should be building now for crossing the historical abyss that still separates us from a truly contemporary vision of life, and the increase of life and meaning that awaits us in the future'** (ibid. p.29). Jacob Bronowski observed the same truth in his book *The Ascent of Man*: **'We are nature's unique experiment to make the rational intelligence prove itself sounder than the reflex** [instinct]. **Knowledge is our destiny. Self-knowledge, at last bringing together the experience of the arts and the explanations of science, waits ahead of us'** (1973, p.437 of 448). In fact, many of our most enduring mythologies, such as King Arthur's Knights' search for the 'Holy Grail', and Jason and the Argonauts' search for the 'Golden Fleece', are actually expressions of this all-important quest for the reconciling, redeeming and rehabilitating understanding of ourselves.

[1146]Yes, finding the biological explanation of the human condition *has been* 'the holy grail' of the whole Darwinian revolution, the *central* quest in our human journey of conscious thought and enquiry—but, as this book has shown, that journey has been lineball. When Richard Neville wrote that **'We are locked in a race between self destruction and self discovery'** (*Good Weekend, The Sydney Morning Herald*, 14 Oct. 1986; see <www.wtmsources.com/167>) he was recognizing that *only* understanding of ourselves could bring an end to all the devastation, heartache and suffering that has plagued the human race since time immemorial and is now fast engulfing our planet in a terrible, terminal embrace. AND, as our designated vehicle for enquiry, it was *science* (which, again, literally means 'knowledge') that was charged with this crucial task of finding the human-race-saving understanding of the human condition. So despite presenting a patently evasive, dishonest explanation of the human condition, E.O. Wilson was being truthful when he wrote that **'The human condition is the most important frontier of the natural sciences'** (*Consilience*, 1998, p.298 of 374). AND, most wonderfully, through the advances made in science, particularly understanding of the difference in the way genes and nerves process information, science *has* finally made it possible for a scientist to explain the human condition.

[1147] While virtually all scientists have had no choice but to operate from a basis of denial of any truths that led to confrontation with the issue of the human condition, it was still the *practice* of science that provided the all-important insights into the mechanisms of the workings of our world that were needed for that explanation to be assembled — which means, as pointed out in par. 296, it is the practice of science as a whole that is humanity's liberator, its so-called 'savior' or 'messiah'. Although extremely rare since Resignation became all but universal some 11,000 years ago, there have always been a few individuals who could confront and think truthfully about the human condition, but until the practice of science as a whole found understanding of the difference in the way genes and nerves process information such human-condition-confronting, truthful-thinking individuals were in no position to put together the explanation of the human condition. So it is the discipline of science that has finally made explanation of the human condition possible — and, by doing so, given the human race the means, the understanding, that enables all the psychological anger, egocentricity and alienation in humans to subside. Our previously inescapable need to live alienated from our beautiful soul, with all the horror that such a destructive, dishonest and shallow existence entailed, can now end, and the human race at last return to the non-upset ideal state we have longed for, be it Heaven, Paradise, Eden, Nirvana, Utopia, Shangri-La, or whatever term we ascribed to it — the fundamental difference being that where we were once, as Moses said in the Bible, **'in the image of God'** (Gen. 1:27), instinctively orientated to the Godly integrative meaning of life, this time we'll return in a fully conscious, aware and understanding state. We will be **'like God, knowing** [understanding] **good and evil'** (ibid. 3:5). Again, as the poet T.S. Eliot wrote, **'We shall not cease from exploration and the end of all our exploring will be to arrive where we started and know the place for the first time'** (*Little Gidding*, 1942).

[1148] YES, with understanding of the human condition found, we can *finally* be upset-free, secure, sound, effective managers of our world and, as a result, the peace and happiness that humans throughout history have hardly dared to dream of can come into being. Christ's anticipated time that he appealed to us to pray for in his *Lord's Prayer*, when **'Your** [the Godly, ideal, integrated, peaceful] **kingdom come'** (Matt. 6:10 & Luke 11:2), can now become our reality. We can finally leave the dark cave of denial that Plato so accurately described — the **'living tomb which we carry about, now**

**that we are imprisoned'**, having to hide from the condemning glare of all the confronting truths about our life (especially that most confronting truth of all of Integrative Meaning)—and stand free at last in the warm, healing sunlight of reconciling knowledge, and be part of, and see and feel, the world of staggering beauty that exists outside that debilitating 'cave' of denial. Again, as William Blake famously prophesized in his appropriately titled 1790 poem *The Marriage of Heaven and Hell*, **'When the doors of perception are cleansed, man will see things as they truly are. For man has closed himself up, till he sees all things through narrow chinks of his cavern.'** Buddhist scripture contains similarly truthful anticipations of the clarity and form humans would achieve once the ameliorating under-standing of the human condition arrived and was absorbed—that **'In the future they will every one be Buddhas** [be free of psychosis and neurosis] / **And will reach Perfect Enlightenment** [reach understanding of the human condition]' (Buddha [Siddartha Gautama], *The Lotus Sutra*, c.560-480 BC, ch.9; tr. W.E. Soothill, 1987, p.148 of 275), and **'will with a perfect voice preach the true Dharma, which is auspicious and removes all ill'** (*Maitreyavyakarana*; tr. Edward Conze, *Buddhist Scriptures*, 1959, pp.238-242). Of that time, the scripture says, **'Human beings are then without any blemishes, moral offences are unknown among them, and they are full of zest and joy. Their bodies are very large and their skin has a fine hue. Their strength is quite extraordinary'** (ibid). The Bible similarly describes the time when **'Another book** [will be]**…opened which is the book of life** [the human-condition-explaining and thus humanity-liberating book]**… [and] a new heaven and a new earth** [will appear] **for the first heaven and the first earth** [will have]**…passed away…**[and the dignifying full truth about our condition] **will wipe every tear from…**[our] **eyes. There will be no more death or mourning or crying or pain, for the old order of things has passed away'** (Rev. 20:12, 21:1, 4). YES, with understanding of the human condition now found, the human journey *can* have the happy ending we always believed it would: **'The happy ending is our national belief'** (Mary McCarthy, *On the Contrary*, 1961, p.18 of 312).

[1149] *Achieving* that **'happy ending'**, however, obviously requires that we allow the truth to live in our lives, which is an enormous adjustment that resigned humans will initially resist. So while we at last have the means to ameliorate the human condition, we still need to overcome the difficult responsibility of facing and accepting the truth about our-selves. As will now be explained, while the finding of understanding of the human condition *is* wonderfully liberating, it is also fearfully confronting—but, thankfully, as will also be explained in this final chapter, there is an all-exciting and all-relieving way to overcome this last great hurdle of our fear of exposure of our now immensely corrupted state or condition.

## Chapter 9:3 'Judgment day'

[1150] Humanity's journey thus far has been astonishing. In fact, as emphasized in chapter 8, the greatest, most heroic story ever told is our own. That saga, however, is not quite over. When the human-race-liberating, psychosis-addressing-and-solving, real explanation of the human condition arrives, which—as has been fully evidenced—it now has, all the denial that our species has had to employ to cope with the injustice of being unfairly condemned as bad when it was not, is suddenly exposed, and that exposure cannot come as anything other than a tremendous shock. The truth reveals the lies. The light of understanding suddenly floods the darkness of Plato's cave of denial where the human race has been hiding. The blinds are suddenly drawn on all the lies and delusions we humans have been using to protect ourselves from exposure of our corrupted condition.

[1151] As has been revealed throughout this book, denial has played a dominant role in human life, even pervading science. Although, as stated above, science—through its painstaking discoveries of the mechanisms of the workings of our world—enabled the assembly of this explanation of the human condition, its practitioners have certainly not been rigorously objective and impartial in their 'scientific method'. No, as it (understandably) turns out, scientists—just like every other resigned human—have been committed to avoiding any insights that brought the unbearably depressing subject of the human condition into focus, and there have been many, many such insights. There has been denial of the human condition itself; of Integrative Meaning; of the truth that our distant ancestors lived in an upset-free, innocent state; of the importance of nurturing in human development; of the true nature of consciousness; of our psychosis and the resulting differences in alienation—most especially between our distant ancestors and humans today, but also between individuals, 'races', genders, generations, countries, civilizations and cultures; and of many, many other truths. Yes, a great deal of *subjectivity* has polluted the practice of science, and with the arrival of the truth about the human condition all that subjective dishonesty is suddenly revealed. The arrival of understanding of the human condition exposes not just all the denials that every human has been practicing—but all the lies that science has been engaging in in support of that denial.

[1152] So while the arrival of understanding of the human condition liberates the human race from 2 million years of living with unjust condemnation, the immense change that this understanding brings cannot help but come as a *very* great shock, with the most difficult change to adjust to being the exposure of our corrupted condition—the *extent* of our accumulated anger, egocentricity and denial/falseness/dishonesty/ alienation. As was mentioned earlier in par. 592, in his famous 1970 book, *Future Shock*, Alvin Toffler was remarkably prescient in his anticipation of this time when understanding of the human condition would emerge and humans would suddenly be faced with (as he put it) **'the shattering stress and disorientation that we induce in individuals by subjecting them to too much change in too short a time'** (p.4 of 505). Yes, the truth about ourselves unavoidably and necessarily reveals the extent of our now immensely upset condition; it destroys the lies, our denials, our pretenses and delusions, as it must, otherwise it wouldn't be the truth. Consider the Adam Stork analogy that was presented in chapter 3:4. If Adam could have explained why he had to carry out his search for knowledge when he was *first* criticized for doing so he would never have become upset—he would never have become defensively angry, egocentric and alienated. Or, if he had found the explanation after only a few days of carrying out that search, he would have accumulated very little anger, egocentricity and alienation to have to heal with understanding. But in *our* species' case, our conscious, self-managing, human-condition-afflicted state fully emerged over *2 million years ago* and we have only *just* found the understanding of why we became upset, which means there is an absolute mountain of accumulated anger, unsatisfied ego, and denial in us humans to have to heal with this understanding. Certainly, we have learnt to restrain and conceal a great deal of the upset that exists within us. We have learnt to, as we say, 'civilize it', not let it show; for instance, we don't normally attack someone now the moment we become angry. Adult humans now exercise a great deal of self-control, but underneath our manufactured facade of restrained civility, even manufactured happiness, lies volcanic anger and immense frustrated egocentricity, which expresses itself in all the ferocious atrocities and vengeful acts of bloodshed we humans commit, and (albeit to a much lesser degree) in the smaller disputes that mar our everyday lives.

[1153] Yes, truth day *is* honesty day, exposure day, transparency day, revelation day. It is, in fact, the long-feared so-called **'judgment day'** referred to in the Bible (Matt. 10:15, 11:22, 24, 12:36; Mark 6:11; 2 Pet. 2:9, 3:7; 1 John 4:17; Acts 17:31; Rom. 2:16; Isa. 66, Joel 3:2). And although this **'judgment day'** is actually a day of compassionate understanding, not a day of condemnation—as an anonymous Turkish poet once said, judgment day is **'Not the day of judgment but the day of understanding'** (Merle Severy, 'The World of Süleyman the Magnificent', *National Geographic*, Nov. 1987)—it *is*, nevertheless, a day when the extent of our species' by now *extremely* psychologically upset condition is revealed. This paradox of being wonderfully liberated but at the same time agonizingly exposed was captured by the prophet Isaiah when he said that the liberation that **'gives you relief from suffering and turmoil and cruel bondage...will come with vengeance; with divine retribution...to save you. Then will the eyes of the blind be opened and the ears of the deaf unstopped...Your nakedness will be exposed and your shame uncovered...on the day of reckoning'** (Bible, Isa.14:3; 35:4-5; 47:3; 10:3). The prophet Joel also described the paradoxical consequences of the arrival of understanding of the human condition when he said **'The sun will be turned to darkness and the moon to blood before the coming of the great and dreadful day of...judgment'**, **'it will come like a destruction'**, but **'In that day the mountains will drip new wine, and the hills will flow with milk...Their [people's] bloodguilt...will [be] pardon[ed]'** (Bible, Joel 1-3). Referring to **'the Day of Reckoning'** (*The Koran*, ch.56) and **'the Last Judgement'** (ibid. ch.69), the prophet Muhammad provided a similar description of this paradoxical situation when he said, **'when the Trumpet is blown with a single blast and the earth and the mountains are lifted up and crushed with a single blow, Then, on that day, the Terror shall come to pass, and heaven shall be split...On that day you shall be exposed, not one secret of yours concealed'** (ibid). While in Zoroastrian religion, **'god...the one supreme deity...represented both light and truth...Ranged against him stood the powers of darkness, the angels of evil and keepers of the lie. The universe was seen as a battleground in which these opposing forces contended, both in the sphere of political conquest and in the depths of each man's soul. But in time the light would shine out, scattering the darkness, and truth would prevail. A day of reckoning would arrive in which the blessed would achieve a heavenly salvation, while all others would find themselves roasting in fiery purgatory'** (*A Soaring Spirit: Time-Life History of the World 600-400 BC*, 1988, p.37 of 176). And interestingly, the word 'apocalypse' is a further anticipation of this time when the liberating truth about our species' immensely heroic but also horrifically upset, exhausted condition is suddenly revealed. The original name for the Book of Revelation in the Bible, the etymology of **'apocalypse'** is **'Ancient Greek: meaning "un-covering"...translated literally from Greek, [it] is a disclosure of knowledge, hidden from humanity in an era dominated by falsehood and misconception, i.e., a lifting of the veil or revelation'** (Wikipedia; see <www.wtmsources.com/127>); **'a cataclysm in which the forces of good triumph over the forces of evil'** (*The Free Dictionary*; see <www.wtmsources.com/112>).

[1154] With regard to the exposure brought about by discussion of the differences in alienation that exist *between* individuals (*and* 'races', genders, generations, countries, civilizations and cultures), immediately after describing how the arrival of the all-exposing, shocking truth about humans will come **'like the lightning, which flashes and lights up the sky from one end to the other'**, Christ described how **'two people will be in one bed; one will be taken** [revealed as sound, relatively free of upset and alienation] **and the other left** [revealed as being upset and alienated]. **Two women will be grinding corn together; one will be taken and the other left'** (Luke 17:24, 34, 35; see also Matt. 24:27, 40). Again, it has to be emphasized that **'judgment day'** is *not* a time when some will be judged as deserving of being **'taken'** to heaven and others **'left'** behind, rejected as bad and 'evil'—even condemned to a **'fiery purgatory'**—but a time of compassionate understanding of *everyone's* situation. The prophet Micah understood the real nature of truth/judgment day when he said, **'I will bear the Lord's wrath, until he pleads my case and establishes my right. He will bring me out into the light; I will see his justice...Who is a God like you, who pardons sin and forgives the transgression...You do not stay angry forever but delight to show mercy. You will again have compassion on us; you will tread our sins underfoot and hurl all our iniquities into the depths of the sea'** (Bible, Mic. 7:9, 18-19). Yes, the way **'the forces of good triumph over the forces of evil'** is through 'evil' finally being compassionately, lovingly understood as a heroic consequence of humanity's immensely courageous but psychologically upsetting battle to defeat ignorance—which is the point Sir Laurens van der Post was making when he wrote that **'True love is love of the difficult and** [historically] **unlovable'** (*Journey Into Russia*, 1964, p.145 of 319).

[1155] On this note, I should immediately stress that while the left-wing faction in politics, mechanistic science as a whole, and the scientists E.O. Wilson, Richard Wrangham, Victoria Wobber and Brian Hare, have been particularly targeted for exposure of their alienating denials in this book, *everyone's* denials are now exposed. The fact is, the human-condition-confronting, truthful analysis of *anyone*'s situation would reveal similar levels of dishonesty. The only reason the left-wing political paradigm, mechanistic science, and the individuals listed above have been targeted for exposure is because *their* particular denials happen to be the most dangerous in terms of the threat they pose of taking humanity to terminal alienation and extinction. In the case of the extreme danger that denial-based mechanistic science poses to the world, I should include these honest words from Charles Birch, who, as I've mentioned before, I was fortunate enough to have as head of the biology faculty at Sydney University when I was studying science there: **'the mechanistic view...is the dominant mode of science and is particularly applicable to biology as it is taught today...**[it is] **A view or model of livingness that leaves out feelings and consciousness...**[and] **I believe it has grave consequences...In the name of scientific objectivity we have been given an emasculated vision of the world and all that is in it. The wave of anti-science...is an extreme reaction to this malaise... I believe biologists and naturalists have a special responsibility to put another image before the world that does justice to the unity of life and all its manifestations of experience — aesthetic, religious and moral as well as intellectual and rational'** ('Two Ways of Interpreting Nature', *Australian Natural History*, 1983, Vol.21, No.2). And this **'unity of life and all its manifestations of experience — aesthetic, religious and moral as well as intellectual and rational'** is *not* the fake offering that E.O. Wilson — that lord of lying, the master of keeping humanity *away* from any truth; indeed, the quintessential **'liar...the antichrist'** (Bible, 1 John 2:22), the **'deceiver and the antichrist'** (2 John 1:7), **'The beast...given...to utter proud words and blasphemies'** (Rev. 13:5) — put forward in his 1998 book, *Consilience: The Unity of Knowledge*, when he proposed that Evolutionary Psychology's alleged ability to explain the moral aspects of humans meant biology and philosophy, the sciences and the humanities, indeed science and religion, could at last be reconciled. Wilson wrote of **'the attempted linkage of the sciences and humanities...of consilience, literally a "jumping together" of knowledge...to create a common groundwork of explanation'** (p.6 of 374). And nor is it his more recent — and even more dishonest — attempt to explain the origins of humans' moral nature and the human condition. No, it is *only* the psychosis-addressing-and-solving *real* explanation of the human condition that is being presented here in this book that brings about the realization of Birch's belief in a **'malaise'**-of-**'science'**-ending, **'unity of life and all its manifestations'**. It is only *this* book that reconciles religion and science, instinct and intellect, conscience and conscious, idealism and realism.

[1156]The focus of this chapter, then, is on the *ramifications* of getting this greater truth up about humans—on HOW IS THE HUMAN RACE, and all its institutions, such as our political and scientific bodies, SUPPOSED TO COPE with this ultimate **'future shock'**, the **'Terror'** and **'cataclysm'** of having to face such a massive paradigm shift? How are humans going to endure this arrival of **'judgment day'** where all our **'nakedness will be exposed'** and, for example, the differences in alienation between **'two people** [and 'races', genders, generations, countries, civilizations and cultures]**'** will be revealed? Encountering the **'naked'** truth about ourselves cannot help but be an *immense* shock. What is being introduced is the arrival of the *real* 'future shock', 'brave new world', 'tectonic paradigm shift', 'gestalt switch', 'renaissance', 'revolution' or 'sea change' humanity has long anticipated would one day arrive. BUT HOW ARE WE SUPPOSED TO MANAGE SUCH A COLOSSAL TRANSITION?!! The answer, which will now be explained, is that while there will be an initial, brief period of shock and overwhelmed bewilderment, this great change from living in denial to living honestly is not only *not* difficult or painful, it is extremely easy and incredibly exciting, painless and joyful. Yes, there is not merely a way of *coping* with the truth about humans when it arrives, but an almost *unbearably exciting way of living* with it.

## Chapter 9:4 The Transformed Lifeforce Way of Living

[1157]Exasperated by all the denial/dishonesty/falseness/**'phon**[iness]**'**/ **'fake**[ness]**'**/delusion/pretense/artificiality/superficiality in human life, John Lennon, that creative genius and member of The Beatles band, desperately pleaded for some honesty in his 1971 song *Gimme Some Truth*: **'All I want is the truth, just gimme some truth. I've had enough of reading things by neurotic… politicians…I'm sick to death of seeing things from tight-lipped…chauvinists… I've had enough of watching scenes of schizophrenic…prima-donnas…I'm sick and tired of hearing things from uptight…hypocrites…All I want is the truth now, just gimme some truth NOW.'** Yes, we are all literally **'sick and tired of'** the dishonest denial the world has been drowning in; we have all completely had enough of the human condition and are all desperate to be free of it—BUT, while it is one thing to wish for the truth, it is quite another to cope with that truth, and the changes it heralds, when it actually arrives!

[1158]So *how is* the human race supposed to cope with having its 2 million years of accumulated denials, all our dishonest delusions and artificialities, and our pain and alienation, suddenly revealed? To arrive at the answer we need to summarize how humans came to be in the situation we are now in, and how this situation changes with the arrival of understanding of the human condition—because doing so will provide the logical sequence of events that leads to the equally logical way that humans will be able to manage their transition from being a victim of the human condition to becoming a secure, sound, effective manager of our world.

[1159] As has been emphasized throughout this book, all humans are born instinctively expecting to encounter an unconditionally selfless, all-loving environment, but as a result of humanity's 2-million-year necessary search for knowledge we instead encountered an immensely upset, human-condition-afflicted world where everyone was variously alienated from their true selves or souls with little or no unconditional love to give. As a result, we too became psychologically upset sufferers of the human condition. Then, as adolescents, we tried to understand why there was so much imperfection in the world around us and within ourselves, but unable to find an explanation we eventually, at around 15 years of age, had no choice but to resign ourselves to living in denial of the unbearably depressing subject of the horrendously corrupted state of the human condition. At this point, however, our inability to *refute* the negative views of ourselves with understanding meant we could only *counter* those negatives by focusing on, emphasizing and developing whatever superficial positive views of ourselves we could find. In particular, we became preoccupied trying to find ways to feel good about ourselves, to validate ourselves, to find relief from the criticism we were living with that we weren't worthy beings by competing for whatever power, fame, fortune and glory was available, the result of which was that we became extremely self-preoccupied, ego-centric, competitive and selfish. This, of course, was a situation that could only be remedied through the first-principle-based explanation of the human condition itself, namely that humans are conscious beings and as such it wasn't sufficient for us to be instinctively *orientated* to the world, we had to find *understanding* of the world if we were to master our conscious mind, a realization that means we weren't 'bad' to have defied our instinctive self and become, as a result of that defiance, angry, egocentric and alienated. (The full explanation of the human condition appears in chapter 3.) What the arrival of this understanding then means is that we no longer need to be angry, egocentric and alienated. But, as pointed out, the problem is humans needed this explanation when the search for knowledge commenced over 2 million years ago, because if our species had that explanation then we would never have become upset by the criticism coming from our instinctive self in the first place. We would have known *why* we had to defy our instincts—and thus we would have been secure about what we were doing. But unfortunately we didn't have that explanation, and so defensive and retaliatory anger, egocentricity and alienation resulted— behavior that has been accumulating throughout those *2 million years*. So while the ameliorating understanding of why we became upset, namely the explanation of the human condition, *has* finally been found, to use it to heal all the upset that now exists within us would take a great deal of time and counseling—counseling that isn't presently available, and time the world cannot afford at the moment. In fact, it will naturally take a number of generations for all the psychological upset within humans to be healed through the dissemination and absorption of the understanding of the human condition that is now available.

[1160] As mentioned in par. 72, Professor Prosen, a psychiatrist of almost 60 years' standing (including chairing two departments of psychiatry and serving as president of the Canadian Psychiatric Association), has rightly said about the arrival of this psychosis-addressing-and-solving *real* explanation of the human condition that **'I have no doubt this biological explanation of the human condition is the holy grail of insight we have sought for the psychological rehabilitation of the human race.'** Yes, it *is* true that the **'psychological rehabilitation of the human race'** *is* now possible, and everyone *could* now retrace and analyze all the hurtful events in their life and replace all the defensive denials that they have had to employ to cope with that pain with truthful, compassionate understanding of those events and the feelings they produced—why their mother and father were preoccupied and unable to give them the unconditional love their instincts expected during their infancy and childhood, and why they are now so insecure, etc, etc—but to undertake such psychological therapy and dismantle all the psychoses and neuroses would take an immense amount of time and supportive counseling, which, again, just isn't possible at the moment. *And*, apart from extreme cases of psychosis and neurosis, *such therapy is not, in fact, necessary*, which is an absolutely fabulous situation that I will now explain.

[1161] While understanding of the human condition finally gives us the means to heal all our anger, egocentricity and alienation (which, in truth, are our psychoses [psyche or soul repressions] and neuroses [neuron or mind denials]), we can't, in practice, hope to heal it in our lifetime; *however*—and this is all-important—<u>we can immediately *know* that we are *all* fundamentally good and not bad, and this knowledge puts each of us in a very powerful position *because it means we can legitimately decide not to think about and live in accordance with the upset within us.* We can reason that, 'Yes, I do have all these angry, egocentric and alienated feelings and thoughts, and I know that without undergoing a great deal of psychological therapy I can't hope to heal all the underlying upset that is producing these feelings and thoughts. But I *also* know that way of feeling and thinking is now obsoleted, so I'm simply going to leave that way of feeling and thinking behind me and adopt the new way of feeling and thinking that is now available where I live in support of a reconciled, non-aggressive, non-competitive, truth-based new world for humans!'</u>

[1162] Such transcendence of our real, upset self and adoption of a different way of thinking and behaving—a different attitude—may at first appear to be an extraordinary feat, but we humans are quite capable of being what we are not. For example, as explained in pars 854-856, when the human race became 'civilized' we chose not to live out our real, upset (in truth, volcanic) anger that now, after 2 million years of being unjustly condemned, exists within us, but to contain it; we chose not to let that upset express itself. We superimposed a different way of thinking and responding on top of our real, upset self's way of thinking and responding. As pointed out, we don't normally attack someone now the moment we become angry; instead, we transcend our real self's angry, defensive, destructive response and defer to and live by, or adopt, a different, civilized basis of thinking and behaving. And so, in exactly the same way, we can study this explanation of the human condition and know that our upset condition has finally been explained and that behaving in an upset way is no longer necessary—and on the basis of that logic prevent ourselves from behaving in an upset way by redirecting all our thoughts and energies to supporting the new human-condition-understood, reconciled world. Instead of letting our now obsoleted upset state rule our lives, we defer to and live through a different, now much more positive, way of thinking and behaving. We make a decision to shift our whole mindset from one state to another state. We mentally let go of one whole mindset and attach ourselves to another, and after a little practice the new mindset becomes automatic. That was how we learnt to civilize ourselves—we adopted a different response to our upset response and now, for the most part, we are very good at being restrained and civilized in the way we think and behave.

[1163] Yes, we can all now know that our insecure, egocentric need to validate ourselves through winning power, fame, fortune and glory is obsoleted. We can understand that it is no longer necessary to prove that we are good and not bad through demonstrations of our worth because our goodness has now been established at the most fundamental level through first-principle-based, biological explanation. We can know that with the arrival of understanding of the human condition humans' whole resigned, competitive, self-preoccupied, selfish, must-win-power-fame-fortune-and-glory way of living has been made redundant. In fact, the fundamental situation the human race is now in is that the resigned, competitive way of living has become pointless, meaningless, because it doesn't progress the human race to greater knowledge as it once did because the key piece of knowledge that we needed to find, of understanding of the human condition, has been found. Furthermore, living out that self-preoccupied, competitive, selfish existence is now unnecessarily destructive of our relationships with our fellow humans, and of our world in general. We can now know that the resigned, egocentric, competitive, selfish and aggressive way of living is finished with, which means we can, and, in fact, should, simply cast aside that way of living, and take up, in its place, a new way of living that is available to humans now where we live in support of a reconciled, non-competitive, selfless, loving existence. As was pointed out earlier on in par. 743, it was the process of Resignation that largely killed off access to our all-sensitive and all-loving soul and made humans virtually totally mad (sufferers of the **'dead' 'fifty feet of solid concrete'** state of **'alienation'** from our soul that R.D. Laing spoke of), and since Resignation, with all the soul-dead insensitive and must-prove-yourself mean and unsatisfying life that went with it, is fundamentally a *mental, psychological* condition, not an immutable *genetic* condition, you *can* choose to leave that insecure, embattled, insensitive, mean and unhappy life behind.

[1164] And, most wonderfully, what happens when we humans give up our old ways of living and take up the new way of living that understanding of the human condition has made possible is we become *transformed* from a competitive and selfishly behaved individual to a cooperative and selflessly behaved one, an integrative part of humanity. Yes, even though we are not yet free of the upset state of our own personal human condition, we *can immediately* have a change of attitude and decide not to live out that upset state that remains within us. The overall effect in our lives is that, despite our retention of the upset state of the human condition, we are effectively free from its hold and its influence, which is an absolutely fabulous transformation to have made in an instant—in one simple decision!

[1165] To elaborate, now that we know the reason why humans had to set out in search of knowledge and defy our original ideal-behavior-insisting instinctive self or soul, *all* the upset anger, egocentricity and alienation that unavoidably resulted from being unjustly criticized by our instincts is now rendered obsolete, unnecessary and meaningless. No longer do we have to retaliate against criticism of our upset/corrupted state because our upset/corrupted state has been defended with truthful, compassionate understanding at the most profound level. And no longer do we have to try to prove our worth because our worth has been established at the most fundamental level. And no longer do we have to deny any confronting truths about our immensely upset/corrupted condition because no longer are there any truths about our upset/corrupted state that condemn us. Our upset lives are explained and defended now, which means we no longer have to be preoccupied compensating for that upset by finding forms of self-aggrandizement, by seeking self-distraction, or by chasing relief through materialistic forms of compensation for all the hurt we experienced growing up in an immensely human-condition-afflicted world. In other words, we no longer need to seek power, fame, fortune and glory to make ourselves feel good about ourselves because our goodness has now been established at the deepest, most profound level. Instead, we can simply leave our whole 'must-prove-our-worth, attack-and-deny-any-criticism' way of living behind as obsolete and redirect all our thoughts and energies into supporting and disseminating these human-race-saving understandings, and to repairing the world from all the damage our species' upset behavior has caused—because with the human condition solved, all the upset that is causing the destruction of the planet can now end, which means it is at last possible to *properly* and permanently repair our environment. We *can*, as it were, put the issue of all our upsets/corruptions in a 'suitcase', attach a label to it saying, 'Everything in here is now explained and defended', and simply leave that suitcase behind at the entrance to what we in the World Transformation Movement call the Sunshine Highway and set out unencumbered by all those upset behaviors into a new world that is effectively free of the human condition. All our egocentric, embattled posturing to get a win out of life, all our strategizing every minute of every day to try to find a way to compensate for feeling inadequate or imperfect or bad about ourselves, can suddenly end. We can leave Plato's dark cave where we have been hiding to escape the glare of the truth about our seemingly imperfect condition. Excitement and meaning—based on liberating, truthful, honest understanding of ourselves and our world—is what we have to sustain ourselves now.

[1166] Yes, the excitement and relief of being effectively free of the human condition—the joy and happiness of being liberated from the burden of our insecurities, self-preoccupations and devious strategizing; the awesome meaning and power of finally being genuinely aligned with the truth and actually participating in the magic true world; the wonderful empathy and equality of goodness and fellowship that understanding of the human condition now allows us to feel for our fellow humans; the freedom now to effectively focus on repairing the world; and, above all, the radiant aliveness from the optimism that comes with knowing our species' march through hell has finally ended and that a human-condition-free new world is coming—CAN NOW TRANSFORM EVERY HUMAN AND THUS THE WORLD.

[1167] From being a human-condition oppressed and depressed alienated person, all humans can now be TRANSFORMED into redeemed, liberated-from-the-human-condition, exhilarated, ecstatic, enthralled-with-existence, empowered, world-transforming LIFEFORCES. This exhilarated, ecstatic, enthralled-with-existence aspect is the 'Life' in 'Lifeforce', while the empowered, world-transforming aspect is the 'force' in 'Lifeforce', so LIFEFORCE covers both the personal benefit and the benefit to the world in one word. From being human-condition-afflicted, Plato's-cave-dwelling, effectively dead humans, we become Transformed Lifeforces or simply LIFEFORCES! *That* is the difference the arrival of understanding of the human condition makes to the human race!

[1168] Before recommending the reader watch some short videos in which people describe their own transformation to a fabulous life in the new human-condition-understood world, more needs to be explained about this new Transformed Lifeforce Way of Living or Transformed Lifeforce State (TLS), especially how it differs in a most important and fundamental way to religious transformations, and to New Age Movement-type transcendences of self.

## Chapter 9:5  The Transformed Lifeforce Way of Living is not another religion

[1169]While current generations of humans are unable to heal the psychological upset within themselves, what *is* immediately possible is to let go of the resigned competitive mindset and take up support of the new Transformed Lifeforce Way of Living. And, as emphasized, humans *are* able to do this, decide to change from living for self to living a more selfless existence—*because we have been doing it throughout our history*. Chapter 8 just described the journey our species has been on as it attempted to find ways of living, and causes to support, that helped to contain and, in some cases, completely transcend our upset, resigned, self-centered, egocentric, selfish way of living. There was the just mentioned practice of self discipline, where we 'civilized' our upset behavior by superimposing more ideal principles for living upon our selfish and aggressive psychologically upset way of thinking and responding to life; then there was the imposed discipline of our upset, where we lived by adhering to specific laws of behavior, such as Moses' Ten Commandments; then there was religion, where we deferred to and lived through our support of an exceptional denial-free thinking, unresigned, sound prophet; and then there was the whole sequence of increasingly guilt-relieving attitudes that we adopted, from socialism/communism, to the New Age Movement, to feminism, to the environment or green movement, to the politically correct movement, to finally the completely guilt-stripped and totally dishonest, 'there-is-no-such-thing-as-truth' postmodern or deconstructionist movement. Of course, the degree to which we abandoned or left behind or let go or repressed or transcended our selfish and aggressive upset, real self and superimposed upon it or adopted a more selfless and cooperative attitude, varied enormously, but within the spectrum there existed the possibility of *completely* abandoning our upset way of behaving. In Christianity, for example, this complete so-called 'conversion' from living for yourself to living for Christ is described as being **'born again'** (John 3:3), as becoming **'a new creation'** where **'the old has gone, the new has come!'** (2 Cor. 5:17). Similarly, while resigned humans can't easily change from being psychologically upset—as emphasized, that is a process that will take a few generations—we *can immediately and completely* change our mind's *attitude* from living a selfish, self-preoccupied life and be, as it were, 'born again' to a consider-the-welfare-of-others-above-your-own-welfare, unconditionally selfless, soulful, pre-resigned-like way of thinking and living. While every resigned human naturally becomes extremely habituated to living for the relief of power, fame, fortune and glory, it *is* possible to completely relinquish that way of thinking and living and, in its place, adopt a completely different, unconditionally selfless way of thinking and behaving.

[1170] So while it is true that, as just described, the Transformed Way of Living is similar to a religion in that it involves completely letting go of or transcending our real, upset self and deferring to another way of living, that is where the similarity ends. The fundamental difference between the Transformed State and a religious conversion is that the Transformed State *is all about knowledge, not faith or dogma*—knowledge that, after a few generations, has the ability to eliminate all the upset in humans. As has been emphasized, dogma can't heal upset, only understanding can do that. As was explained in chapters 8:15 and 8:16, while religions *were* an incredibly effective means of containing the upset in humans while the search for understanding of that upset condition was being carried out, the Transformed Way of Living, and the World Transformation Movement that promotes it, is, in complete contrast, concerned with what happens *after* that liberating understanding is found, which is the advancement of the human race from a human-condition-afflicted state to a state completely free of that terrible affliction. The Transformed Lifeforce Way of Living, and the psychological amelioration of the human condition that it allows (given the time, the few generations, needed for it to take place), completely changes the human race from a state of troubled upset to a state of secure soundness. In fact, it metamorphoses—it matures—the human race from insecure adolescence to secure adulthood.

[1171] Religions were based on deferring to, and living through support of the *embodiment of the ideals* in the form of the soundness and truth of the unresigned, denial-free-thinking-and-behaving prophet around whom the religion was founded, whereas the Transformed Lifeforce State is based on deferring to and living through support of first-principle-based *understandings of the ideals and of our species' unavoidable historical lack of compliance with those ideals*. In fact, the Transformed Way of Living, where humans live in support of the understanding of the human condition that can, after only a few generations, eliminate all the psychological upset in the human race, represents the *realization* of religion's hope and faith that the liberating understanding of the human condition would one day appear—it represents the end of faith and dogma and the beginning of knowing. Yes, and I will comment more on this shortly (in par. 1217), this is the end of religion and the beginning of the time of understanding that all the great prophets looked forward to—including Moses, who anticipated a time when we will **'be like God, knowing'** (Gen. 3:5).

[1172] Another immense difference between this new Transformed Way of Living and a religion is that in the Transformed Way of Living <u>there is no deity involved, or deference to any one personality; in fact, there is no worship of any kind. And best of all, unlike religion, there is no involvement or emphasis on guilt, because guilt—and the whole notion of 'good and evil'—has been eliminated forever with the reconciling understanding of the human condition.</u>

[1173] So while religion and the human-condition-resolved Transformed Lifeforce Way of Living both involve letting go of living through our corrupted self and deferring to something else, that is where the similarity ends. Yes, there is a *world of difference* between the religious state and the Lifeforce state, about which more will be explained shortly.

## Chapter 9:6 <u>Nor is the Transformed Lifeforce Way of Living another deluded, false start to a human-condition-free new world</u>

[1174] And just as the Transformed Lifeforce Way of Living is not another religion, it is also not *another* deluded false start to a human-condition-free world. Again, as described in chapters 8:16H-Q, and just referred to in par. 1169, there has been a litany of false starts to an ideally behaved, cooperative, selfless, loving new world for humans, and the reason they were all false starts was because for humans to truly achieve that cooperative, loving, peaceful state we first had to find the reconciling, healing understanding of our divisive human condition. It is true that the new Transformed Way of Living does involve artificially ridding ourselves of our upset state by simply leaving it behind or 'transcending' it, which is also the way all the false starts to an ideal new world for humans achieved their artificial elimination of upset, *but*, as just explained, the immense difference is that in the Transformed Way of Living such transcendence is only an interim measure—we are only doing so because it will take a few generations to actually rid the human race of upset. The all-important difference is that in the Transformed State we are leaving the old upset world behind because it is finished with—the upsetting search for knowledge, ultimately self-knowledge, understanding of the human condition, is complete. So the Transformed Lifeforce Way of Living is not another *false* start to a new world for humans but the *real* start to that world.

[1175] What supporters of all the false starts to an ideal new world for humans were doing was 'jumping the gun' on this job, abandoning humanity's critically important, heroic, upsetting search to find knowledge before that search had been completed, which means that what they were doing was *fundamentally irresponsible*. Certainly it could be immensely relieving giving up the upsetting search for knowledge, but humanity's *real* freedom from upset depended on that search being carried out until understanding of the human condition was found. To refer again to the Adam Stork analogy, at any time Adam could abandon his upsetting pursuit of knowledge and 'fly back on course' and obey his instinctive orientations and feel good as a result of doing so, but taking that option would mean he would never find the understanding he needed to actually *end* his upset existence once and for all. While giving up the upsetting search for knowledge while it was still to be completed was relieving, it *was* fundamentally irresponsible; however, with the upsetting search for self-knowledge now over it is no longer irresponsible to abandon that upsetting search. In fact, it is now irresponsible to *continue* it. Yes, it is not only legitimate now to leave the old resigned, embattled, competitive, aggressive and selfish egocentric way of living behind in favor of living in support of the new reconciled, human-condition-understood world, it is imperative; it is, in short, the *only responsible* way to behave.

[1176] Of course, when humans became overly upset we had to do *something* to avoid behaving so destructively; as explained at length in chapters 8:15 and 8:16, as upset increased we first had to civilize the upset, then try to adhere to an imposed set of rules, then adopt a religion, and then, when we became so upset we could no longer bear the guilt-emphasizing honesty in religion, we had to take up support of one of the pseudo idealistic causes to relieve ourselves of our insecure condition—but the fact remains that all these ways of coping *were* fundamentally irresponsible. As was also explained in chapters 8:15 and 8:16, while taking up a religion was by far the least irresponsible way of transcending our upset and abandoning the search for knowledge, because the search continued indirectly through our support of the honesty of the prophet around whom our religion was founded, it was still fundamentally an irresponsible pseudo idealistic act of desertion.

[1177] So a very big difference with the new Transformed Way of Living is that it is not irresponsible at all; in fact, it is totally responsible and the only legitimate way to behave because the battle to find understanding has been won, so everyone *should* stop fighting, leave the battlefield, and, as they do at the end of a game of football, 'go to the showers and clean themselves up'!

[1178] Yes, this transformation from a human-condition-afflicted existence to an almost unbelievably wonderful and exciting, exhilarated and empowered Transformed Lifeforce State is not achieved by abandoning humanity's critical battle to find knowledge, ultimately self-knowledge, understanding of the human condition. In particular, it is not another fundamentally irresponsible 'New Age' or, more recently, 'A New Earth', 'alternative' movement in which 'spiritual gurus' advocated we transcend our embattled conscious thinking egoic self when that battle hadn't yet been won. Nor is it another of those fundamentally irresponsible, superficial and ultimately ineffectual 'think positive', 'human potential', 'self development', 'self improvement', motivational programs in which we tried to artificially defy the reality of the human condition by immersing ourselves in positive thoughts. Nor is our freedom from the agony of the human condition being achieved through the fundamentally irresponsible practice of deep, meditative extinction of our human-condition-distressed thinking mind as some religious practices taught. And nor does the new Transformed Lifeforce Way of Living involve the fundamentally irresponsible way of escaping the real issue before us as a species of our deeply troubled selves by adopting a focus-away-from-yourself, guilt-free, feel-good, pseudo idealistic cause like environmentalism or feminism or aboriginalism or multiculturalism. *And* our species' freedom is *certainly* not being achieved in the fundamentally irresponsible way of dogmatically imposing a deconstructed, good-and-evil-differentiation-free, politically-correct-but-human-reality-dishonest, ideal world, as the postmodern movement and, before it, the socialist and communist movements tried to do. No, our species' transformation is being achieved through what is ultimately the only real and lasting and fundamentally responsible way it could be—through satisfying our conscious thinking mind with first-principle-based, biological understanding of *why* humans are wholly worthwhile and meaningful beings.

[1179]There *had* to be a biological explanation for our species' non-ideal divisive, competitive, aggressive, angry, even-brutal-and-mean, selfish, self-obsessed, indifferent-to-others, arrogant, egocentric, deluded, defensive, escapist, superficial, artificial, alienated, despairing, lonely, depressed lives, and our fundamental responsibility as conscious beings was to find that ameliorating explanation. So while it *could* be immensely relieving giving up the all-important, upsetting search for knowledge, our fundamental responsibility as conscious beings was *not* to give up that search. To include more of what Bronowski said in his concluding statement to his 1973 television series and book of the same name, *The Ascent of Man*: **'I am infinitely saddened to find myself suddenly surrounded in the west by a sense of terrible loss of nerve, a retreat from knowledge into – into what? Into...falsely profound questions about, Are we not really just animals at bottom; into extra-sensory perception and mystery. They do not lie along the line of what we are now able to know if we devote ourselves to it: an understanding of man himself. We are nature's unique experiment to make the rational intelligence prove itself sounder than the reflex** [instinct]. **Knowledge is our destiny. Self-knowledge, at last bringing together the experience of the arts and the explanations of science, waits ahead of us.'**

[1180]There certainly has been a **'terrible loss of nerve'**, a loss of commitment to humanity's great struggle to find **'self-knowledge'**. Irresponsible transcendence of the issue of self, thought repression, enforced dogma and escapism were all fundamentally a **'retreat from knowledge'**. But de-braining ourselves was never going to work; ultimately, we needed brain food not brain anesthetic; we needed knowledge—specifically the dignifying, uplifting, healing, ameliorating, relieving, peace-bringing understanding of the human condition. Anything else *was* an abrogation of the fundamental responsibility that came with our greatest capacity and nature's greatest invention: our species' fully conscious, thinking, self-managing, self-adjusting mind. And, thankfully, that fabulous destiny and potential to progress from mere abstract, **'art**[istic]**'** *description* of the agonising, good-and-evil-afflicted dilemma of our human situation to reconciling, first-principle-based, scientific *understanding* of that dilemma and resulting amelioration, integration and unification of ourselves and of our species has finally arrived.

[1181]It should be mentioned here that all the deluded false starts to a new human-condition-free world have *so* severely discredited the possibility of the emergence of such a new world that it has made it easy for the unscrupulous to attempt to brand and dismiss the World Transformation Movement, which is promoting the *real*, human-condition-solved, human-race-saving transformation of humans, as just another of these deluded movements. Such an attack *is* unscrupulous because the transformation is based on assessable, understandable, first-principle-based biological explanation of the human condition that makes the practice of transcending our upset state not only legitimate but the only practice to adopt. Of course, the other evidence that this is not another deluded dream of utopia is that, far from making people superficially feel good, these human-race-liberating understandings are *extremely* exposing, confronting and unsettling. As Christ warned, **'false prophets will appear and perform great signs and miracles to deceive even the elect—if that were possible...Wherever there is a carcass** [the extremely upset]**, there the vultures** [false prophet promoters of 'feel-good' delusion, escapism and denial] **will gather'**, whereas the real, human-condition-exposing liberation of the human race will come like terrifying rolling thunder: **'as the lightning comes from the east and flashes to the west...**[it will be very frightening as if] **the sun will be darkened...the stars will fall from the sky, and the heavenly bodies will be shaken'** (Matt. 24). Yes, the real love of our dark side is *nothing* like the superficial, artificial, trivial, transitory, sickly, weak 'do good to feel good' form of 'love' that ultimately only buried humans deeper in alienation. Again, the history of all the patently dishonest and thus cowardly attacks on the world-saving World Transformation Movement can be read about at <www.humancondition.com/persecution>.

### Chapter 9:7  But how does the Transformed Lifeforce Way of Living solve the problem of the unbearable, 'judgment day' exposure that understanding of the human condition unavoidably brings?

[1182] What is yet to be explained is how are humans going to cope with having all our upsets that we have been living in denial of suddenly exposed; how are we going to manage the arrival of the liberating but at the same time all-exposing **'day of judgment'**, **'the Day of Reckoning'** when **'[Our] nakedness will be exposed'** and **'not one secret of** [ours will be] **concealed'**? The simple answer is that we don't try to confront all the truth about our corrupted condition. Once we have investigated these understandings sufficiently to know that they have explained the human condition, we don't actually need to know any more than that to take up a life where we direct all our thoughts and energies into supporting these understandings, and to the repair of the world that they finally make possible. We don't *need* to know the full extent of the truth that this information reveals about the upset state of humanity; and we particularly don't need to know how it explains and reveals everything about our own human-condition-afflicted life.

[1183] Yes, we can leave our upset behind as dealt with for *two* reasons: firstly, for the reason already given, that healing all our psychoses and neuroses would require a great deal of time and supportive counseling that simply isn't possible at the moment; and, secondly, because trying to face all the truth that is now revealed about our upset condition could be too confronting, exposing and depressing, and thus dangerously destabilizing, an exercise.

[1184] The danger is that if we study this information beyond what our particular level of soundness and security of self can cope with we risk becoming overly confronted by the extent of our corrupted condition and dangerously depressed. As has been pointed out, the human race has coped thus far by maintaining extreme levels of denial of many, many truths, so obviously we can't hope to dismantle that edifice of denial overnight. Humans typically take months, if not years, to adjust to big changes in our lives, so in this *massive* paradigm shift, which takes us from a position where we didn't face *any* truth about our corrupted condition to facing the prospect of having *all* the truth about it exposed, we have to accept that it will be a process that will take generations, but that doesn't mean we each can't support the truth while this adaption, and the digestion and healing that goes with it, takes place. We simply can't hope to, and mustn't, overly confront the truth while this long process of adjustment takes place, and, according to each person's level of upset, there will be a limit to how much truth each person can cope with; there will be a limit to how much they can listen to, read about and study these human-condition-confronting understandings. But, again, that doesn't mean that everyone can't *immediately* access and live in support of these human-race-liberating understandings. Freedom from the human condition is only a decision away now for all humans.

[1185] Regarding the degree to which we should each investigate these explanations, obviously it is necessary to sufficiently verify to our own satisfaction that they are the liberating understandings of the human condition that the whole human race has been tirelessly working its way towards for some 2 million years—but we shouldn't risk investigating them to the extent that we start to become *overly* exposed and confronted by the truths they reveal. Having lived without any real understanding of human life it is natural to want to keep studying these explanations that finally make sense of the world, both within and around us, but, again, such analysis can lead to becoming overly confronted and depressed by the extent of our own corrupted state, and that of our world. The mistake this particular reader of my books made was that he **'pushed'** his investigation of these understandings of human behavior further than his soundness of self could cope with: **'I was halfway through enlisting for the army when Jeremy's words truly hit home. Nothing could touch me in that time, everything made perfect absolute sense. Then things started to change. I wanted to go further but something stopped me and when I pushed I felt the worst fear I have ever known. Fear doesn't even go close to expressing it. What do you suppose you do when you find the most fearful thing you'll ever encounter is yourself'** (WTM records, Mar. 2003). The more intelligent and/or the more educated in the human-condition-avoiding, denial-based, mechanistic, reductionist paradigm, who pride themselves on being able to think and study and grasp new ideas, *will* initially be especially tempted to study these understandings beyond what their varying levels of security of self can cope with, but it won't be long before everyone learns that such an approach is both psychologically dangerous and irresponsible and, in any case, unnecessary.

[1186] When Christ spoke of a time when **'the meek...inherit the earth'** (Matt. 5:5), and when **'many who are first will be last, and many who are last will be first'** (Matt. 19:30, 20:16; Mark 10:31; Luke 13:30), he was anticipating this time when understanding of the human condition would arrive and instead of the more intelligent and intellectual leading the way, as has been the case in almost every human situation since Resignation became universal, the more innocent and sound, the more soulful and instinctual and functional, the less upset or corrupted or dysfunctional, will do so. As the story of Adam Stork reveals, throughout the 2-million-year battle to find understanding, our instinctive self or soul was repressed because of its unjust condemnation of our intellect, but when understanding of the human condition is finally found this process is reversed, soul becomes sought-after. Our innocent, upset-free, original instinctive self or soul—soundness—has to lead us back home to soundness. It makes sense. Again, Christ gave the perfect description of this new situation when he said, **'The stone the builders rejected has become the capstone'** (Ps. 118:22; Matt. 21:42; Mark 12:10; Luke 20:17; Acts 4:11; 1 Pet. 2:7). Sir Laurens van der Post referred to this biblical analogy when he too anticipated this new situation: **'It is part of the great secret which Christ tried to pass on to us when He spoke of the "stone which the builders rejected" becoming the cornerstone of the building to come. The cornerstone of this new building of a war-less, non-racial world, too, I believe, must be...those** [more innocent, instinctual] **aspects of life which we have despised and rejected for so long'** (*The Dark Eye in Africa*, 1955, p.155 of 159). In his 1964 song *The Times They Are A-Changin'*, Bob Dylan also anticipated the paradigm shift that the arrival of understanding of the human condition brings: **'Come gather round people wherever you roam / And admit that the waters around you have grown / And accept it that soon you'll be drenched to the bone** [the end play state of terminal alienation is upon us]**...keep your eyes wide, the chance won't come again** [for the denial-free understanding of the human condition to be found]**...the loser now will be later to win / For the times they are a-changin'... Don't stand in the doorway, don't block up the hall / For he that gets hurt will be he who has stalled / The battle outside ragin' / Will soon shake your windows and rattle your walls** [the liberating, but also exposing, truth about humans is on its way] **/ For the times they are a-changin' // Come mothers and fathers throughout the land / And don't criticize what you can't understand / Your sons and your daughters are beyond your command / Your old road is rapidly agein' / Please get out of the new one if you can't lend your hand / Oh the times they are a-changin' // The line it is drawn, the curse it is cast** [the exposing truth about humans is out] **/ The slow one now will later be fast / As the present now will later be past / The order is rapidly fadin' / And the first one now will later be last / For the times they are a-changin'.'**

[1187] So as with having to abandon our obsoleted upset, resigned, competitive, power-fame-fortune-and-glory way of living, investigating the truth only to the degree we are sound enough to do so will initially be a difficult proposition for the denial-based world to accept, but it too is a proposition that can be understood and accepted as reasonable. It is the more secure in self, the least alienated, who have to develop these understandings of the human condition. If you are not sound enough to study these ideas to any great depth and you try to do so, you will end up in a state of fearful depression—or, worse still, mad. This means that it can be particularly dangerous for people with a history of psychological instability to study these understandings. Furthermore, if you *do* become overly confronted by what is being presented your natural reaction could be to try to attack and deny it in order to protect yourself—in effect, reinstate all your denials and those of others, which means advocating humanity's retreat *back into* Plato's dark cave of denial that the human race has been desperately trying to get out of for 2 million years! Yes, if you do become overly confronted, you could become, as Dylan warned, **'hurt'** and **'stalled'**, defensive, angry and retaliatory toward the information, and the consequence of such a response could be to, in effect, sabotage the efforts of every human who has ever lived to bring the human race to this dreamed-of moment of its liberation. As documented in chapter 6:12, '*The* great obscenity', we in the World Transformation Movement have endured years and years of this furiously angry, defensive reaction towards this information, attacks that were ultimately fruitless because this information is true and it won't be intimidated or oppressed: it is too precious to allow that. In short, the effect of overly studying this information, studying it more than your degree of security of self can cope with, can be both dangerous to you and dangerous to the human race, and no one should want, nor risk, either of those outcomes.

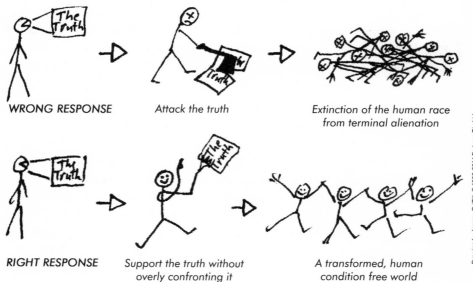

WRONG RESPONSE      Attack the truth      Extinction of the human race from terminal alienation

RIGHT RESPONSE      Support the truth without overly confronting it      A transformed, human condition free world

Drawing by Jeremy Griffith © 1996-2014 Fedmex Pty Ltd

[1188] So while initially it will be difficult accepting this advice to only investigate all this amazing truth that is now available about humans to the degree that each person is sound enough to do so, with honesty it *can* be appreciated that it is a reasonable and necessary proposition. For instance, in the old human-condition-avoiding, denial-based world, academia limited those who could be involved in the pursuit of knowledge to the more intelligent, those with a high IQ (intelligence quotient). To enter university you had to pass entrance exams that assessed your IQ, your suitability to study complex subjects like higher mathematics and physics, because if you didn't have an adequate IQ you would make little progress in studying such subjects. Well, in the new human-condition-resolved, human-condition-confronting world we similarly need the most appropriate people to study its information, which are those with a high SQ (soul or soundness quotient). If you don't have an adequate SQ you simply won't be able to make any progress with the information involved. But, importantly, with the explanation of the human condition we can now appreciate that while everyone *is* necessarily variously upset/unsound/alienated, that upset/unsoundness/alienation is not something bad, just as in the old denial-based world those who lacked IQ weren't considered bad people, just not as able to think as effectively about certain complex subjects. As has been emphasized throughout this book, upset is not an evil, sinful, bad state—it is a heroic, good state, because it is an unavoidable product of humanity's heroic, all-important search for knowledge.

[1189] The overall situation then is that everyone has to measure and limit how much they can study these human-condition-confronting understandings against how much self-confrontation they can cope with. If you start to feel unbearably confronted then restrict your study of the information. Again, the human race cannot have lived in the darkness of both ignorance and denial for some 2 million years and then suddenly be able to face all the truth about ourselves individually; it is simply not a reasonable expectation. Everyone *has to* expect that there will be a limit to how deeply they can afford to look into this information. But thankfully, and most importantly, no one has to overly confront their old, upset self—everyone *can* leave that behind as dealt with and simply live for the new world and all its potential. As emphasized, although existing generations who grew up without these understandings will still suffer the effects of the human condition, they can still live effectively free of the human condition because they no longer have to live preoccupied with it. It will be future generations who will be in the best position to further develop these understandings of the human condition because among current generations there are naturally very, very few individuals who remain secure enough in self to confront, study and develop them. The main point to remember is that once you know that this information has explained the human condition then you know all the upset in the world and within you is now explained and defended—which means you *can*, as described earlier, put the issue of all your upsets or corruptions in a 'suitcase', attach a label to it saying 'Everything in here is now explained and defended', and simply leave it behind as dealt with as you set out completely free and unencumbered into the new, human-condition-liberated world. You can join the Sunshine Army on the Sunshine Highway to the World In Sunshine.

[1190] Once you know this information is true, that the upset state of the human condition is explained and defended at the fundamental level, you can leave the issue of your own and the world's corruption behind as effectively dealt with and focus all your attention on developing a human-condition-free new world. You can preoccupy yourself with disseminating this information throughout the world and to a fresh generation, and with supporting all the projects that must now be undertaken to rehabilitate the world from the destructive effects of 2 million years of living under the duress of the human condition. In fact, it shouldn't even be necessary to talk about the whole issue of the human condition any more than what appears in this presentation and in the presentations on the World Transformation Movement website. Humanity moves on to an entirely new paradigm of reality now. We get the truth up, and we move on.

[1191] Yes, all that matters now is that the truth is kept alive and that it is disseminated to the world's population, because it alone can heal the human race and save the world. All everyone should do now is support the truth about the human condition and it will achieve everything everyone has ever dreamt of. If we look after this information it, in turn, will look after each of us and the world. That is the mantra of the new world that understanding of the human condition brings about.

[1192] We, the existing generations, are now *the conduit generations*, the generations who will connect the old dead world to the new living world. Having grown up 'in the dark' without understanding of the human condition, our priority is to shine the light on these insights to ensure fresh generations won't have to adopt all the artificial ways of coping with the human condition—all the dishonest denials, all the egocentric means of seeking compensation for the injustice of being unfairly condemned, and all the angry retaliations needed against the unjust condemnation—that previous and existing generations have had to resort to. If children and pre-resigned adolescents can access these understandings they won't have to resign themselves to a life of living in denial of the human condition as their only means of coping with it—and not having to grow up with all the dishonesties, denials, delusions, artificialities and superficialities that current generations have had to endure, they will be like a new species of beings. They will have an inner freedom that is almost unimaginable to those who grew up without understanding of the human condition. As described in pars 113-118, Holden Caulfield, the central character in J.D. Salinger's seminal book, *The Catcher in the Rye*, said that **'the only thing that I'd really like to be'** is a **'catcher in the rye'** saving **'thousands of little kids'** from going **'over the cliff'** of Resignation. Well, we can *all* start living Holden's dream now; we can all become **'catcher[s] in the rye'**!

[1193] Yes, finding understanding of the human condition is the dreamed-of breakthrough that can bring about the complete transformation—in fact, TRANSFIGURATION—of the entire human race and thus our world. The word transfiguration means **'a change that glorifies or exalts'**, **'a marked change in form or appearance; a metamorphosis'** (*The Free Dictionary*; see <www.wtmsources. com/140>), so it is a perfect description for what happens to humans when, through the application of dignifying, uplifting, relieving, ameliorating, reconciling, redeeming, healing and curing understanding, we are finally able to not only end our old egocentric, must-prove-that-we-are-not-bad, insecure, human-condition-afflicted existence, but enter an incredibly exciting state of glorious freedom, optimism and human-condition-liberated, empowered-to-repair-the-world capability.

[1194] I should make the point here that evidence that the human race can survive the exposure of 'judgment day'—and not merely survive it but live with it in an absolutely wonderfully joyous and exciting way—is provided by the affirmations that have been given by people living in the Transformed Lifeforce State (which can be viewed at <www.humancondition.com/affirmations>), and by the existence of the many other people who have been aware of and have lived happily with these understandings for decades. While a few people have become psychologically destabilized by these understandings, that is to be expected given the scale and the novelty of the change that these understandings bring—and that, initially, there was always going to be a few people who would not heed the advice to study these understandings *only* to the degree that they could personally cope with them, and to stop if they started to become overly confronted and destabilized. This advice especially applies to people with a history of psychological instability. But of course the ultimate evidence that it is possible for the human race to survive 'judgment day' is inherent in the fact that the human race would never have struggled so determinedly all these thousands of generations to find understanding of the human condition if we didn't intuitively know that we could free ourselves from the human condition when that understanding finally arrived.

[1195] It should also be said that if you are one of the rare people who find it relatively easy to read, access and absorb all the immensely human-condition-confronting insights in this book—someone who is fortunate enough to be relatively secure in self—you have a special responsibility to help support and promote these understandings that are the only thing that can save humanity from torturous extinction. Clearly, at this great transition point in the human journey where the human race changes from living in denial to living honestly, those who can make that transition most easily have a special responsibility to make sure that the initial inclination of many to want to maintain denial of these insights doesn't prevail. As the prophet Daniel said, **'the people who know their God** [those who are not overly afraid of such fundamental truths as Integrative Meaning] **will** [must] **firmly resist'** (Bible, Dan. 11:32) those who aren't so secure; and **'None of** [those who have especially suffered from the battle of the human condition]**...will understand, but those who are wise will understand'** (ibid. 12:10). Hosea similarly warned that those **'Who is** [are] **wise'** and **'discerning' 'will understand' 'but... the rebellious** [will] **stumble'** (Bible, Hos. 14:9). It has always been important for those who have the good fortune to be secure in self to stand up and give leadership, but at this moment of immense paradigm shift it is *critically* important that they do so. You *must* ensure this information survives and prospers—send copies of this book to key people (or direct them to the freely available copy at <www.humancondition.com>), write supportive reviews, get involved.

[1196] So, for the human race to be saved two things have to happen—and both contributions are equally critical—as Bob Dylan **'plea**[ded]**'**, now that **'the times they are a-changin'** the less secure must do their utmost not to **'stand in the doorway'**, while the more secure must **'lend your hand'**. Of course, the Transformed Lifeforce State is the way that allows *everyone* to easily be involved in supporting this immense transition that saves the world.

## Chapter 9:8 <u>The utter magnificence of the Transformed Lifeforce State</u>

[1197] To reveal just how magnificent the Transformed Lifeforce Way of Living is, a further contrast can be made between it and Christianity—remembering that Christianity, although <u>human-condition-*relieving*, was *not* human-condition-*resolving*</u>.

[1198] It was St Paul who, more than anyone else, introduced the benefits of Christianity to the world, so consideration of his discovery of the beauty of Christianity, his conversion to the faith and his life as a Christian is especially indicative of what the new Transformed Lifeforce Way of Living offers the human race. Acts 8 and 9 of the Bible provide the account of Saul's (as St Paul was known before his conversion) journey to Damascus where he planned to persuade the authorities to destroy the fledgling group of Christians, such was his fury towards the truth about the human condition that Christ, through his sound words and life, had dared to reveal and which this small group were living in courageous support of. However, while riding along on his donkey in a seething, **'murderous'** (Acts 9:1) rage towards Christ and his followers, Saul had an epiphany, the effect of which was so incredible that it was to change the world and basically save the human race from unbearable levels of upset.

[1199] With an appreciation now of what the human condition is, we can explain and thus understand what happened to Saul. The human condition is a terrifyingly confronting subject and, to cope with that terror, all that the upset human race has been able to do is live in denial of it, try to block out the whole issue of the imperfection in our lives. We attacked, denied and attempted to prove wrong any exposing criticism of our corrupted state. This response relieved our condition but there was, of course, a significant downside—a loathsome life of extreme anger, alienation and egocentricity. So in one sense our retaliation against the threat of exposure of our corrupted condition made us feel even *more* condemned. Defensive retaliation as a strategy for coping with the human condition, while relieving, also fuelled feelings of self-loathing and unworthiness. And that was Saul's predicament—his anger was very great, but so was the level of self-hatred he would have felt deep within himself for being such a brutally angry person. The whole issue of Christ's soundness that was making him *so* angry was also, at another level, making him extremely distressed and unhappy in himself. We can imagine then that, in a moment of full engagement in his mind with the horror of both sides of his situation, a thought would have occurred to him: *'If this man, Jesus, is so truthful and sound that he is provoking such anger in me, what would happen if I was to flip the situation around and instead of attacking him I supported him? Wouldn't I suddenly then be a force for enormous good—because I would be supporting someone who is the absolute opposite of my immensely corrupted and angry-with-soundness, truth-hating-and-denying self. I would be a force for good in the world, instead of a self-loathing monster. Wow, that would certainly turn my life around, wouldn't it!'* And, at that moment, as the metaphorical account goes, Saul fell off his donkey and was struck blind for three days; in more direct terms, he was completely overwhelmed by the sudden freedom he was feeling from all the pain of his human-condition-afflicted life. Through his support of Christ, the agonising weight of the human condition had been lifted from his shoulders and he was able to live again. By siding with Christ, he was able to resurrect the truthful, soulful side of himself; he had been **'born again'** (John 3:3); **'he has crossed over from death to life'** (John 5:24); he had become **'a new creation'** where **'the old has gone, the new has come!'** (2 Cor. 5:17). Again, as Christ authoritatively said, **'I am the resurrection and the life** [through me, your soulful true self can live again]' (John 11:25).

[1200] One way of appreciating how much upset humans are preoccupied with, and how oppressed they are by the insecurity caused by the human condition, and thus how absolutely incredibly relieving it is to not have to be preoccupied with that condition, is to consider the experience of someone who has had a Near-Death Experience. For instance, as I explained in par. 997, mountain climbers who survive falls that they were convinced would be fatal (they were saved, perhaps, by landing in a snow drift) often report that during those near-death moments they experienced a state of extraordinary euphoria in which the world suddenly appeared utterly beautiful and radiant and that they were flooded with a feeling of ecstatic enthrallment. With understanding of the horror of the human condition we can see how, in such cases, the mind would give up worrying, and that all the facades—in particular the denial that they adopted at Resignation—would become meaningless. If death is seemingly imminent, there is no longer any reason to worry or to pretend, at which point the struggle and agony of having to live under the duress of the human condition ceases and the true world of our all-sensitive soul suddenly surfaces. You suddenly discover what it is like to be free of the human condition. You suddenly have access to all the real beauty that exists in the world. You suddenly discover another seemingly magic world that is all-radiant and magnificent. This is the freedom Saul experienced when he abandoned his struggle with the human condition and deferred to Christ. He was suddenly 'born again' from a state of near death. All his embattled posturing to get a win out of life, all his focus on egocentrically building a castle of grandeur for himself, an edifice, a representation of glory around himself, all his strategizing every minute of every day to try to find a way to compensate for feeling bad about himself, suddenly ceased. Realizing the magnificence of the new way of living that Christianity offered, this **'way to be saved'** (Acts 16:17) from an effectively dead, resigned, alienated state (Christianity was actually originally called **'the Way'** (Acts 9:2, 19:9, 23, 22:4, 24:14, 22)), Saul took this breakthrough, life-saving idea to the four corners of the known world, and was thereafter known as Paul the Apostle and, following his canonization, St Paul.

[1201]To gain some further insight into the relieving nature of the new Transformed Lifeforce State, where humans' embattled struggle with the human condition similarly ends, you can, if you haven't already, watch the affirmations at <www.humancondition.com/affirmations>. You can especially sense the relief of the Transformed State in Doug Lobban's affirmation, but all the others are equally powerful. For example, Neil Duns says, **'The freedom this brings is like nothing humans have ever experienced'**; while Stacy Rodger says, **'This is all that we have ever, ever wanted'**; Tony Gowing says, **'This is like seeing the world for the first time, it's like waking up from a nightmare, it's like 100 tonnes being lifted off your shoulders'**; and John Biggs says, **'The relief and through-the-roof excitement that flows through your veins is incredible...It is beyond comprehension how much wonder and fun and excitement and happiness the human race has coming.'**

[1202] So while leaving your baggage, your 'suitcase' of human-condition-embattled strategies and posturing behind by deferring to Christ *was* a marvelous solution to the problem of the horror of the human condition, as has been emphasized, it meant giving up your battle to prove yourself, giving up your particular participation in humanity's heroic struggle to overthrow the ignorance of our original instinctive self or soul, which, in effect, didn't appreciate our conscious mind's search for knowledge. As was explained in some detail in chapter 8:15, while supporting the prophet around whom religions were based did ensure the battle to find knowledge continued indirectly through that prophet's alienation-free, truthful words and sound life (this was the very great benefit of religion compared to the other ways of transcending your upset self), in becoming religious you had personally given up the battle. To use the Adam Stork analogy, you had stopped searching for knowledge and were flying 'back on course'. As mentioned in par. 937, in *The Simpsons* cartoon series the character Ned Flanders is a Christian, someone who has deferred to, and lives through, Christ. His neighbour Homer Simpson, on the other hand, is still living out the corrupting battle to overthrow the implication that humans aren't fundamentally good. In one episode, Ned lends Homer his lawnmower, which Homer ends up wrecking without remorse. But rather than getting angry or defensive, Ned simply accepts Homer's behavior—he is the 'goody-goody' while Homer is one upset Adam Stork, living out the battle to the full; he's massively angry, egocentric and alienated. If, however, Homer could have explained the situation, he would have said to Ned: 'Listen Ned, you love Christ and he loves you, and you're a goody-goody, and I'm one upset, corrupted dysfunctional dude, but Ned, I'm still out there participating in humanity's heroic battle to find knowledge, so I'm a legend and you're a worn-out quitter.'

[1203] The new Transformed Lifeforce State, where you leave all your upset baggage behind in favor of supporting these understandings of the human condition, offers a similar all-relieving and all-exciting, 'near-death-experience'-like, 'fall-off-your-donkey', 'go-blind-for-three-days'-type incredible freedom from the human condition that Christians experienced when they deferred to Christ. BUT—and, again, this is a *very*, *very* important difference—unlike Christianity or any other religion, *this freedom does not come at the cost of abandoning the battle because the battle is now won. It is not an act of irresponsibility or weakness to let go of the battle now, it is an act of strength*, because with the battle now won the illogical and thus irresponsible and thus weak thing to do would be to continue, or somehow re-engage with, that now-obsolete battle.

[1204] So, if we return to the Homer and Ned example and imagine that Ned has taken up the new Transformed Lifeforce Way of Living, Homer would have no grounds to criticize him. With the battle to find knowledge, ultimately the understanding of the human condition, now won, there is no longer any good reason to keep living out the battle. In this scenario, the tables are turned: Ned would be in the position to censure Homer.

[1205] Again, religions were a way of avoiding living out your upset when you became overly upset, but it meant giving up directly participating in the battle to find the knowledge needed to save the human race. Religion was a weak abandonment of the battle that still had to be fought. With the Transformed Lifeforce Way of Living, however, there is no weakness involved because the battle has been won. In fact, taking up the Transformed Way of Living is the only strong course of action for a human to take now. What is weak or 'wimpish' is *not* taking it up. Again, this is the *very* important difference—and it is a most profound difference—between religion and the Transformed Lifeforce State.

[1206] It is this fact that there is no longer any reason to keep living out the battle to champion the ego that has the potential now to change the world so rapidly from one of conflict and suffering to a world of peace and happiness. The explanations being presented in this book are all rational, there is no dogma or mysticism or abstract concept involved. As such, we can know if what is being presented is accountable and true or not, and since this information does explain the human condition there is no longer any justification for continuing the upsetting battle to champion the ego, so it follows that no one can argue against taking up the Transformed Way of Living, so when everyone *does* take up that way of living our world will be completely transformed. <u>We are rational creatures, and so when all the logic says there is only one response we can make, namely the Transformed State, then that is the response the human race should take.</u>

[1207] And this is not a revolution dogmatically imposed by others upon us, as has pretty much been the case with revolutions in the past; no, this is a revolution based on logic, on understanding, on knowledge, on information. <u>Once someone is given this information there is really only one outcome, and that is that they take up the Transformed Way of Living.</u>

Of course, while it is the power of the logic, the rationale, that makes the Transformed Way of Living irresistible, the adoption of this way of living is also made irresistible by the absolutely wonderful transforming effect it has on people. Indeed, before long those still living in the old embattled, have-to-prove-your-worth way of living will feel like they have been caught wearing last year's fashions! There is such a stark difference between the free, expansive, enthralled-with-life, transformed existence and the old embattled egocentric existence where your mind is narrowly focused and preoccupied working all the angles every second of every day, defensively seeking validation, reinforcement, a 'win', that those living in the old way, in the past, will seem like a different species. Compared to the new expansive, all-exciting, free, Transformed State, those still stuck in the old embattled paradigm *will* be like sad, forgotten ghosts trapped in an obsoleted state.

[1208] It is certainly true that people will initially find it difficult absorbing and taking in all these explanations because of the '<u>deaf effect</u>' that was described in chapter 1:4. Indeed, the problem of the 'deaf effect' is now especially great because the levels of block-out or alienation have increased so rapidly that, as described in pars 948-949, the **'attention span'** of **'students'** is now so **'brief'** they can focus their minds for **'only eight minutes'**; that if they **'have to read…a book they all now groan'**; that they live in **'psychic saunas of superficial sensory stimulation'** from their **'iTunes'** and **'Facebook profiles'** that have produced **'a fathomless ocean of online ignorance and trivia that makes what is worthy and timeless more inaccessible than ever'**, **'decapitating our culture, trading in the ideas of some 3000 years of civilization for BuzzFeed'**—the result of all of this being that **'We have a lot of competing noise for our attention these days, and it would be criminal to let that overwhelm our true potential, by masking** [the] **useful information** [in Griffith's books] **with hideous noise.'** Yes, the world *is* becoming so superficial, so drowning in **'hideous noise'**, that it *is* rapidly approaching a situation where these deep-thinking-dependent, extremely profound, penetrating, confronting and 'deafening' understandings of human existence threaten to **'overwhelm'** humans' **'true potential'** to understand their world, make them 'deaf' to this information.

[1209] And beyond the 'deaf effect', there *will* also be the initial stage of procrastination, what we in the WTM call the '<u>Mexican Standoff</u>', where people deliberate over the choice that is in front of them, because they don't want to deny the truth and transforming potential of the explanations, but they also don't want to let go of their old, now-obsoleted, resigned power, fame, fortune and glory-seeking way of living that they have become habituated to living in. There is also the problem of what we call 'Pocketing The Win', where people misuse the information's ability to make sense of the world and everyone around them to become *even more effective winners* of power, fame, fortune and glory—with, in the case of scientists, some even re-working the new insights in this book as their own! And there is also the initial temptation to 'Side With The Angels', which involves avoiding self-confrontation through deluding yourself you are an unresigned innocent; or even a not fully resigned, relatively innocent 'Ship At Sea', someone who bravely refused to pull into port and adopt denial when the storm of unbearable self-confrontation occurred during their adolescence. BUT despite all these problems and procrastinations, *all* the logic and thus power lies with the new Transformed Lifeforce Way of Living, and since the transformation is so fabulously relieving and wonderful, the expectation is that support and enthusiasm for the Transformed Way of Living will eventually reach a critical mass, at which point a tidal wave of support for this information and the Transformed State it makes possible will sweep the world.

[1210] The following are drawings I have made to illustrate the main elements involved in this procrastination—of our 'judgment day' fear of exposure of our now fully defended 2-million-year-old, 'fallen', soul-corrupted human condition; of our habituated attachment to our now completely obsoleted power, fame, fortune and glory-seeking way of justifying our worth; and the post-procrastination, absolutely fabulous solution provided by the Transformed Lifeforce Way of Living.

*Drawing by Jeremy Griffith © 1996-2014 Fedmex Pty Ltd*

*Our now baseless fear of exposure of our corrupted human condition*

*Resistance to letting go our now-obsoleted selfish power, fame, fortune and glory-seeking way of justifying our worth*

*The Transformed Lifeforce Way of Living*

[1211] So, as emphasized in chapter 2:1, while the scientific establishment has so far failed in its responsibility to recognize these fully accountable, world-saving biological insights into human behavior, the great hope, indeed expectation, is that by complementing this carefully argued and constructed new presentation of the instinct vs intellect explanation of the human condition and all the other insights that flow from it with deaf-effect-eroding introductory videos and the opportunity to participate in interactive online discussions, support from not only scientists but the general public for this information, and the fabulous transformed life outside of Plato's dark, horrible, alienating, effectively dead, **'living tomb'**, **'cave'** of denial that it makes possible, will finally begin in earnest.

[1212] Yes, the fact is, there is really nothing stopping interest in this information and the adoption of the Transformed Lifeforce Way of Living sweeping the world, and when that happens life for humans will change from a state of terrible turmoil, confusion and despair to one of immense relief, happiness and togetherness. <u>The world for humans *will be* transformed from darkness into light</u>. It really will be as if a light in the dark room we have been staggering around in has suddenly been turned on—everything becomes illuminated and lovely. As mentioned in par. 837, humans had this awesome computer put in our heads, our fully conscious, thinking brain, but we weren't given the program for it and were instead left to wander this planet searching for the program in a terrifying darkness of confusion and bewilderment. Well, from that terrifyingly cold darkness we can now emerge into the warm sunshine of dignifying, redeeming, relieving and transforming understanding.

[1213] A further very important difference between religions and the Transformed Lifeforce State is that there is no delusion involved in the Transformed State. To use *The Simpsons* analogy again: Ned has a 'goody-goody', self-satisfied, 'I-occupy-the-moral-high-ground' attitude, which drives the still-human-condition-embroiled heroic Homer crazy with frustration because he intuitively knows that Ned is deluding himself in thinking he occupies the moral high ground, that he is the more together, sound person and is on the right track, but Homer can't explain *why* Ned's view of himself is so extremely deluded and dishonest. Homer can't explain and thus reveal the truth that real idealism and the truly on-track, moral high ground lay with continuing the upsetting battle to find knowledge, ultimately self-knowledge, and that Ned had become so upset, so unsound, that he had to abandon that all-important battle and leave it to others, including Homer, to fight. Worse, by effectively condemning those still upset and fighting the battle, Ned was basically siding *against* those still trying to win that battle, adding substantially to the opposition they had to overcome. In fact, it was the delusion and dishonesty that made giving up the battle particularly dangerous, because to uphold that position required constantly persuading yourself, and others, that you weren't being dishonest and deluded in what you were doing. (Indeed, I should say that in recent times, as the plight of the world has become more serious by the day and supporters of the left-wing in politics have had to struggle so much harder to justify, to themselves and others, their patently dishonest, fundamentally irresponsible and counter-productive pseudo idealistic response to that crisis, the resulting shrill fanaticism that has emerged *has* become all but suffocating of those in the right-wing who have been trying to continue the all-important battle to find understanding of the human condition. Thank goodness the dangerous delusion and resulting fanaticism in left-wing politics has finally been exposed and made redundant by the arrival of understanding of the human condition.)

[1214] So while religion was by far the least dishonest of all the forms of pseudo idealism that upset humans could take up (because of the honesty involved in acknowledging the truthful, sound life and words of the founding prophet), it still involved substantial dishonesty and delusion. In near total contrast to this situation, the Transformed Lifeforce Way of Living involves no delusion and virtually no dishonesty. By not fully confronting the extent of the upset within yourself and avoiding looking too deeply into all the truth about the human condition, you are practicing some denial/dishonesty, but the compassionate full truth about the upset state of the human condition means you aren't deluded about the fact that you are an upset human. Indeed, you have to recognize and embrace that truth (which you can now safely do because the upset state is defended) to effectively be able to adopt the Transformed Lifeforce State.

[1215] As briefly mentioned in par. 931, possibly the best sales pitch ever given for Christianity was one delivered by St Paul, as documented in the Bible in 2 Corinthians: **'Now if the ministry that brought death, which was engraved in letters on stone** [Moses' Ten Commandments that were enforced

by the threat of punishment], **came with glory** [because they brought society back from the brink of destruction from excessively upset behavior]...
**fading though it was** [there was no sustaining positive in having discipline imposed on you], **will not the ministry of the Spirit** [the positive mental state from being at last aligned with truth and soundness through your support of someone relatively free of upset and alienation] **be even more glorious? If the ministry that condemns men** [through punishment] **is glorious, how much more glorious is the ministry that brings righteousness** [that allows you to be part of the ideal state]**! For what was glorious has no glory now in comparison with the surpassing glory. And if what was fading away came with glory** ['fading' because it's hard to maintain attachment to a system merely out of fear of punishment], **how much greater is the glory of that which lasts!** [compared with the relative ease of maintaining an attachment to a system that makes you feel that you are at last on the side of what is good, ideal and right]' (3:7-11). So if Christianity was considered the **'surpassing glory'** to living in fear of punishment, then the Transformed Lifeforce State is the surpassing of all surpassing glories. It is the best solution the upset human race has ever had by an immense, stupendous, incredible, absolutely wonderful, utterly magnificent country mile, because although it involves living in a deferential state of transition, we are only a few generations away from completely eradicating the human condition from the human race— and there is very little dishonesty and no irresponsibility, no weakness, no delusion, no deity, no worship, no focus on a personality, no faith, no dogma, no guilt of any sort involved in the process. What we have now is so relieving and so exciting that when this way of living catches on it will, as stated, certainly sweep the world.

[1216] So the Transformed Lifeforce Way of Living and World Transformation Movement that promotes it is *not* a religion; it is a *metamorphoser*, the proponents of a way of living and a movement that brings about a complete progression—the metamorphosis of the human race from a horrible human-condition-afflicted state to a completely liberated, human-condition-free state. Humanity is transformed from a state of insecure adolescence, where we humans *searched* for understanding of ourselves, to a state of secure adulthood, where we humans at last *achieve* understanding of ourselves.

[1217] Finally, it should be reiterated that religions aren't being threatened by the arrival of dignifying understanding of the human condition—they are being fulfilled. The whole purpose of religion was to be the custodian of the ideal state while the search for the liberating understanding of humans' 'fallen' condition was underway. Buddha, for instance, looked forward to the arrival of the amelioration of the human condition when he said that **'In the future they will every one be Buddhas** [meaning in the future everyone will be free of psychosis] / **And will reach Perfect Enlightenment / In domains in all directions / Each will have the same title** [there will be no more distorting alienation] / **Simultaneously on wisdom-thrones / They will prove the Supreme Wisdom'** (Buddha [Siddartha Gautama] 560–480 BC, *The Lotus Sutra*, ch.9; tr. W.E. Soothill, 1987, p.148 of 275). In the Bible, Moses similarly anticipated a time when we **'will be like God, knowing'** (Gen. 3:5). In his *Lord's Prayer*, Christ instructed us to pray for the time when **'Your** [Godly, integrated, peaceful] **kingdom come, your** [integrative] **will be done on earth as it is in heaven'** (Matt. 6:10 & Luke 11:2). He also looked forward to the time when **'another Counsellor to be with you forever—the Spirit of truth** [the denial-free, truthful, first-principle-based, scientific understanding]**...will teach you all things and will remind you of everything** [all the denial-free truths] **I have said to you'** (John 14:16, 17, 26). He similarly said he looked forward to when, instead of being restricted to **'speaking figuratively'**, we **'will no longer use this kind of language but will** [be able to] **tell you plainly about my Father** [be able to explain the world of Integrative Meaning in denial-free, human-condition-reconciled, compassionate, understandable, rational, first principle, scientific terms]**'** (John 16:25). And again, the same anticipation of our species' liberation from the human condition is expressed in Revelations in the Bible where it states that **'Another book** [will be]**...opened which is the book of life** [the human-condition-explaining and humanity-liberating book]**...**[and] **a new heaven and a new earth** [will appear] **for the first heaven and the first earth** [will have]**...passed away...** [and the dignifying full truth about our condition] **will wipe every tear from...** [our] **eyes. There will be no more death or mourning or crying or pain** [insecurity, suffering or sickness]**, for the old order of things has passed away'** (20:12, 21:1, 4). Yes, as Isaiah hoped, there would come a time when humans **'will beat their swords into ploughshares...Nation[s] will...[not] train for war any more'** (Isa. 2:4). And what did that truth-saying prophet John Lennon **'imagine'**? A time when the human condition is resolved and **'the world will be as one'**, when there will be **'no heaven** [above us and] **no hell below us'**; when, in essence, there will be a world without the condemning differentiation of good and evil, a world liberated from the insecurity of the human condition and thus the need for religion, where, as Lennon sang, there will be **'Nothing to kill or die for, and no religion too...all the people living life** in peace...No need for greed or hunger, a brotherhood of man...all the people sharing all the world' (*Imagine*, 1971).

[1218]The human race has *always* hoped and believed that some day, some where, at some time humanity's heroic search for and accumulation of knowledge would lead to, as Alan Paton hoped, 'the dawn...of our emancipation'—the finding of the liberating understanding of the human condition—at which point every aspect of human life that was seemingly so inexplicable would suddenly make sense, and we can now see how true that hope and belief was. Out of an overwhelmingly complex and problematic existence a straight forward and totally effective, extremely-rapidly-repairing-of-human-life-and-the-Earth, way of living for humans is now able to emerge.

## Chapter 9:9  How the Transformed Lifeforce Way of Living will quickly repair the world

[1219]The Transformed Lifeforce Way of Living not only has the power to immediately transform every human's life from one that is tortured by the human condition to one that is characterized by a fabulously-radiant state of freedom from that condition—and thus the power to bring the whole human race back to life from its effectively dead state of alienation—it also has the capacity to repair the world we live in, to resuscitate planet Earth from all the brutal anger, selfish egocentricity and mindless alienated behavior we have inflicted upon it.

[1220]In par. 1104 I briefly mentioned that I and others in the World Transformation Movement, including my brother Simon and fellow Patron of the WTM, Tim Macartney-Snape, are beneficiaries of Sir James Darling's extraordinary soul-rather-than-intellect-emphasizing, Platonic education system, having attended Geelong Grammar School in Victoria, Australia. This is the school HRH The Prince of Wales, the heir to the British throne, was sent halfway around the world to attend for part of his education. (While referring to the inspiring, soul-reinforcing influences on my life, and that of Prince Charles', I should also mention the exceptionally sound, soul-preserving inspiration of Sir Laurens van der Post, whose books about humanity's lost state of innocence helped me to hold onto the truth of the existence of another true world; he is the author most often quoted in this book. Sir Laurens was also so important a person to Prince Charles that he was chosen to be godfather to Prince Charles' eldest son, and the future king, Prince William, and there is **'A bronze bust of van der Post...in Prince Charles' garden at Highgrove'** (Christopher Booker, 'Post, Sir Laurens Jan van der (1906-1996)', *Oxford Dictionary of National Biography*, 2004). A former British Prime Minister, Baroness Thatcher, no less, once described Sir Laurens as **'the most perfect man I have ever met'** (mentioned in J.D.F. Jones interview on *Late Night Live*, ABC Radio, 25 Feb. 2002), and Sir James Darling wrote that **'In the world of books there are, for me, Antoine de Saint-Exupéry, or Laurens van der Post'** (*The Education of a Civilized Man*, ed. Michael Persse, 1962, p.36 of 223). For those interested, more can be read about the visions of Sir James Darling [who was described as **'a prophet in the true biblical sense'** in his 3 November 1995 full-page obituary in *The Australian* newspaper], and Sir Laurens van der Post [who was similarly described as **'a prophet out of Africa'** in his 20 December 1996 full-page obituary in London's *The Times*], at <www.humancondition.com/darling> and <www.humancondition.com/vanderpost>.) With regard to the inspired teaching of Sir James Darling, in one of his many extraordinary speeches, he said, **'selfishness is, as it has ever been, the ultimately destructive force in a society, and there are only two cures for selfishness—the regimented state which we all profess to dislike, and the <u>change of heart</u>, which we refuse to make. That is the choice, believe me, for each one of us, and we have not much time in which to make it. The need for decision** [to have a **'change of heart'** and live selflessly] **is serious and urgent, and the sands** [of time] **are running out'** (1950 GGS Speech Day address; Weston Bate, *Light Blue Down Under*, 1990, p.219 of 386).

[1221] In this extract, Sir James has identified selfishness as **'the ultimately destructive force'** and that, historically, there have only ever been **'two cures for selfishness—the regimented state which we all profess to dislike, and the change of heart, which we refuse to make'**. The **'regimented state'** is obviously a reference to the dogmatically imposed and strictly enforced selfless, cooperative, communal, social state of communism/socialism, a state **'which we all profess to dislike'**, because, as has been emphasized, being fully conscious, self-managing beings, having to subordinate our thinking, questioning mind to dogma was never going to work. De-braining ourselves was no real solution to our problems; ultimately we needed brain food, not brain anesthetic. Our thinking mind needed *understanding*—specifically, understanding of why humans are fundamentally good and not bad, evil, worthless beings. In short, we needed the psychologically dignifying, redeeming and transforming understanding of the human condition. But what Sir James has, in effect, done in this passage is recognize that, despite the urgency, until humanity achieved that breakthrough we would **'refuse to make'** or embrace the alternative, that of having a **'change of heart'** and deciding to live selflessly instead of selfishly. And he was right—until the liberating understanding of why humans are fundamentally good and not bad was found, *humans had no choice* but to keep on trying to achieve some relief from the insecurity of our condition by finding superficial ways to prove we were good and not bad, such as through winning power, fame, fortune and glory. We had no choice but to be 'ego-centric'—we had no choice but for our conscious thinking self (which again is how the dictionary defines **'ego'**) to be focused or 'centered' on trying to prove it/we were not bad, worthless and meaningless. It is *only now* with the fundamental, trustable, knowable, first-principle-based, dogma-free biological understanding of why humans are not bad that we can afford to stop trying to prove ourselves all the time—that we can stop being egocentric; that we can let go of being so self-worth-preoccupied, and start living for the human-condition-understood world free of that old insecure existence. It is only now that we can transform from being selfish to selfless—that we can have the **'change of heart'** that Sir James recognized was so **'serious and urgent'**.

[1222] In his acclaimed 1969 BBC television documentary series *Civilisation*, the eminent historian Kenneth Clark mentioned that **'People who hold forth about the modern world often say that what we need is a new religion. It may be true but it isn't easy to establish'** (episode 'The Fallacies of Hope'). Saying **'that what we need is a new religion'** is really another acknowledgment of the fundamental need for the human race to have that **'change of heart'**, that change from living selfishly to selflessly, if we are to solve the world's problems. *And*, saying **'but it isn't easy to establish'** **'a new religion'** is really, deep down, a recognition that what is needed for this change of heart to occur is for the daunting issue of the human condition to be confronted and solved. As has been explained, the Transformed Lifeforce Way of Living is not a religion, but, like a religion, it does bring about a change from living selfishly for yourself to living selflessly for something beyond self—so Clark's acknowledgment that we need a new religion was about as close as the old denial-based world could come to saying we needed to solve the human condition. In this regard, comments made by the Tibetan Buddhist leader, the Dalai Lama, in the following newspaper report show that he was getting closer to the truth of the need to solve the human condition in order to produce a universally acceptable 'God and man reconciled' new potential for humanity: **'The Dalai Lama believes secular ethics, not religion, is best placed to assist the "moral crisis" facing the world's people...[because it] respected all traditional faiths as well as non-believers. "Some people—some my friends—believe moral ethics must be based on religious faith," he said. [But] "No matter how wonderful a religion, (it will) never be universal. The crisis is universal—now the remedy must also be universal"'** (*The Australian*, 14 Jun. 2013).

[1223] And when Plato posed the rhetorical question, **'Will our released [cave] prisoner hanker after these prizes or envy this power or honour? Won't he be more likely to feel, as Homer says, that he would far rather be "a serf in the house of some landless man", or indeed anything else in the world, than live and think as they** [the power and glory hungry] **do?'** (*The Republic*, tr. H.D.P. Lee, 1955, 516), he too was recognizing the great transformation that occurs when we go from living in the cave of denial, where we are each self-preoccupied trying to prove we are good and not bad all the time, to living free from all of that insecure 'baggage' in *selfless* support of the liberating understanding of *why* humans are fundamentally good and not bad.

[1224] And as to why we have **'hanker[ed] after these prizes or envy this power or honour'**, it has been explained that since materialistic luxuries, like glittering dresses, sparkling diamonds, bubbling Champagne, huge chandeliers, silver tea sets, big houses, swimming pools and pretentious cars, gave us (if we were lucky enough to afford them) the fanfare and glory we knew was due us, but which the world in its ignorance of our true goodness would not give us, materialism was one of the important means by which we could sustain our sense of self-worth while the upsetting search for the liberating understanding of why we are good and not bad was being carried out. In fact, when we *only* had our ability to win power, fame, glory and fortune (and, with that fortune/money/wealth/capital, acquire materialistic luxuries) to sustain our sense of self-worth during that upsetting search, those relieving artificial forms of reinforcement were the only engines driving the human-condition-afflicted, insecure world—the *only* rewards sustaining the all-important search for self-understanding. This is why socialism, which sought to replace the engine of greed with the idealism of selfless cooperativeness before we had found self-understanding, couldn't and didn't work. So, while the egocentric, self-centered, individualistic, selfish, greedy, competitive, power-fame-fortune-and-glory-seeking existence was the inequality-producing, human-suffering-causing, Earth-destroying, **'destructive force'** that Sir James described it as, it *was also a 'constructive force'*.

Yes, there was truth indeed in the saying **'Greed is good.'** The public demonstrations against greed and capitalism that occurred in 2011 and 2012 were actually naive, pseudo idealistic, make-yourself-feel-good-but-don't-solve-anything protests against the reality of the insecurity of life under the duress of the human condition. (Of course, this destructive and constructive, 'bad and good' aspect of our behavior is the core paradox of the human condition that we are now at last able to explain and understand and thus bring to an end.)

[1225] Nevertheless, while the **'selfishness'** of the old egocentric, self-centered, individualistic, selfish, greedy, capitalistic, competitive, materialistic way of living *was* the driving force that kept the all-important search for the relieving understanding of why humans are good and not bad going, and was thus a necessary and 'constructive force', it is at base, as Sir James said, the **'ultimate'** **'destructive force in a society'** because in the end *only* a selfless way of living is sustainable. *Ultimately* selfishness *is* destructive. No matter how much you try to control and regulate it, a society operating from a basis of selfishness will ultimately become dysfunctional. There was a limit to how long we could keep going under the drive of selfish greed. For human civilization to survive, selflessness *had to* become the driving force in the world; <u>*ultimately, there had to be*</u>, Sir James' **'change of heart'**. Yes, as many people have pointed out over the years, **'it's not that humans lack the ability to fix the world, it's that they lack the will'**, the **'change of heart'**, the preparedness to live selflessly.

[1226] So despite its precious contribution to the human race's progress towards finding understanding of ourselves, the ultimate truth is that selfishness, especially insatiable materialistic, capitalistic greed, was poised to destroy the world. It was 'insatiable' because, as was explained in par. 896, as an artificial form of reinforcement, materialism was never going to genuinely make us feel we were good and not bad—*only* understanding of our fundamental goodness could and now does achieve that. The celebrated leader of Indian nationalism, Mahatma Gandhi, was making this point about the insatiability of trying to make ourselves feel good by surrounding ourselves with material luxury when he famously said, **'Earth provides enough to satisfy every man's need, but not every man's greed.'** And that often-repeated phrase that **'Living well is the best revenge'** (attributed to the seventeenth century poet George Herbert) also recognizes that while we lacked the real **'revenge'** for the injustice of the human condition, namely reconciling understanding, materialism *was*

**'the best revenge'** we had. So, it is going to come as an enormous relief for the planet and for humankind—in fact, it is going to make *all* the difference—that, through the finding of understanding of why humans are good and not bad, we can finally let go of our selfish, egocentric, have-to-prove-our-worth, materialistic, power-fame-fortune-and-glory-seeking way of living and take up the transformed, human-condition-liberated, selfless existence.

[1227] Just to illustrate the change that is going to come to our materialistic way of living, in front of me is a teaspoon—well, the monetary value of all the human-glorifying, egocentric content and effort that has gone into its ornate, embellished design, extravagant silver plating and competitive salesmanship and marketing to sell it to me, etc, etc, could feed a starving person for a week! Almost everything I see in front of me—my elaborate watch, my fancy shirt, my sophisticated pen—is, in truth, obscenely extravagant in a human-condition-free world. All these items should really be re-designed so they are not so extravagant, and it won't be long before I and everyone else in the world will be doing just that. In fact, I idealistically once designed and manufactured a full range of wooden furniture that was free of embellishment and artificial content before realizing that for such integrity to be tolerated the human condition needed to be explained. Imagine if all the car makers in the world were to sit down together to design one extremely simple, embellishment-free, functional car that was made from the most environmentally-sustainable materials, how cheap to buy and humanity-and-Earth-considerate that vehicle would be. And imagine all the money that would be saved by not having different car makers duplicating their efforts, competing and trying to out-sell each other, and overall how much time that would liberate for all those people involved in the car industry to help those less fortunate and suffering in the world. Similarly, when each house is no longer designed to make an individualized, ego-reinforcing, status-symbol statement for its owners and all houses are constructed in a functionally satisfactory, simple way, imagine how much creative energy, labor, time and expense that will free up to care for the wellbeing of the less fortunate and the planet.

[1228] Again, while we *needed* the individualistic, materialistic world to sustain our sense of self-worth while we couldn't establish it through understanding, now that we have established it that insecure way of living is obsoleted. If you watch Doug Lobban's affirmation of the Transformed

State, which can be viewed at <www.humancondition.com/affirmations>, you will see how he let go of his egocentric mindset when he became transformed and that his only desire now is to fix the world, starting with his dysfunctional shower curtain! The finding of understanding of the human condition naturally brings about a whole new way of living. And there's no dogma involved, like there was in socialism; as has been emphasized, this is about the *end of dogma* and also of faith, and the *beginning of understanding*.

[1229] It also has to be remembered that our species' original, natural, instinctive way of behaving is in an unconditionally selfless, consider-the-larger-whole-above-yourself manner—as Rousseau famously said, **'Mankind is naturally good'**—so when we no longer have to behave inconsistently with that, then that original, natural inclination comes thundering through. Our whole natural, excited, the-world-is-as-it-should-be self bursts forth, as you can see occurring in Doug Lobban, and in others who have provided affirmations. And all you have to do to experience this amazing human-condition-free state is let your unnatural, false way of living go, remembering also that letting it go is no longer an act of weakness, but a strong, fully responsible way to behave.

[1230] The story of Lei Feng, a hero during China's cultural revolution, provides an example of the satisfaction and excitement and meaning that can be gained from selflessly serving the larger whole. As has been emphasized, communism was an extremely dangerous false start to the human-condition-free world but, nevertheless, we can learn from Feng's actions. During Mao Tse-tung's cultural revolution (which took place between 1966 and 1976), a Chinese newspaper published the posthumous diary of one of its readers, Feng, a young communist soldier whose job it was to sweep one of the city's streets. The diary proved so powerful it was printed as a small book that was eventually circulated throughout China and the rest of the world. The gist of the diary was that Feng was happy doing his job because he saw himself as a cog in a giant machine, instrumental in helping the whole machine to work. He wrote, **'A man's usefulness to the revolutionary cause is like a cog in a machine. Though a cog is small, its use is beyond estimation. I am willing to be a cog'** (*The Australian*, 7 Mar. 1987). In essence, Feng wrote to tell of the satisfaction he felt at being able to serve. Serving *is* our original, natural way of behaving, and now that it has become the legitimate and responsible way to behave, it can again become the great joy in everyone's life.

[1231] So while it will take time, in fact, a few generations, for humans to sufficiently absorb the understanding of the human condition to the degree needed for our underlying insecurity to be fully ameliorated—and thus the need for some artificial reinforcement from materialism, and some artificial relief through escapism, self-distraction and sexual adventure, etc, etc, to completely disappear—once you adopt the Transformed State, while you haven't eliminated the insecure state of the human condition within yourself, you have completely changed your mind's focus from living an ego-embattled, selfish life to living selflessly; you have had the **'change of heart'**, the fundamental change of direction, that is so urgently needed to fix the world.

[1232] Importantly, the more this understanding and associated transformation catches on, the easier it will be for people to move from the old insecure, ego-centric way of living to the new, secure, ego-less way of living. The egocentric way of living has had such a powerful hold on humans because without understanding of the human condition all that we had was the superficial reinforcement we gained from seeking power, fame, fortune and glory, but now that relieving understanding has been found then that old egocentric way of living will quickly lose its power—so quickly, in fact, that everyone will be amazed by how swiftly their need for and attachment to the old artificial forms of reinforcement falls away.

[1233] So, what I have been talking about is the transformation that the current, transitional generations can take up, from living a selfish, egocentric, have-to-prove-that-I-am-good-and-not-bad existence to embracing the new, selfless, egocentricity-obsoleted, don't-have-to-prove-that-I-am-good, liberated, exhilarated and empowered, Transformed Lifeforce Way of Living. Of course, in the future when human-condition-extinct generations appear who are *fully* integrated, harmonious, upset-free, all-sensitive and all-loving towards each other, the world and, indeed, the whole universe, they will be properly termed 'Universal Beings', at which point even greater relief will course through our planet. But the point is, the great **'change of heart'** is on—the power of love is finally able to replace the love of power! The human race *is* on its way home, back to soundness, security of self and real love of one another. Imagine when we no longer have to dress to impress, deceive and disguise—especially when women no longer have to be preoccupied with being sex objects—how much freedom that is going to unleash, and how much time, energy and resources it will save? Imagine when communication technology is used only to spread reconciling truth rather than truthless, alienated, escapist, superficial drivel, as it is currently doing—how much relief that is going to bring to humans and thus our world? End the human condition and we end all the big problems of the world—and thus all the little problems too.

[1234] So there are degrees of selflessness that we will now be capable of as we move from the old egocentric world to the new egocentricity-obsoleted, Transformed Lifeforce State, before our species eventually arrives at the future human-condition-extinct, Universal Beings State—*however*, it is the fundamental shift in direction, and the rapid changes that accompany it, that is all-important. The point is the immense power and potential that cooperative selflessness has to solve *all* of our problems. Consider, for instance, if we were all selfless the AIDS epidemic could be solved overnight, because all it would take would be for everyone to agree to be tested for the AIDS virus and those who tested positive to agree to not have sex with anyone who didn't also have the virus. When everyone is selfless we will be able to solve the global warming problem almost overnight by everyone agreeing to hold every third breath or something— that's obviously not going to work, but the point is we will be able to work together selflessly to do whatever it takes! In fact, the most effective mechanism we have available to control our environmental impact is our ability to curb our numbers by not reproducing as much, at least until we have stabilized all the threats to the world that our exploding population is causing. Under the duress of the human condition, we often had children for selfish reasons—to egocentrically perpetuate our family name; or to distract ourselves from our own psychologically distressed lives; or, if we live in a dysfunctional society that isn't able to care for its elderly, to ensure we have someone to look after and provide for us in our old age—the result of which is that there *is* a plague of humans on Earth, in the sense that a 'plague' is **'a widespread affliction or calamity'** (*The Free Dictionary*; see <www.wtmsources.com/164>). With the control the contraceptive pill and other measures allow us to have over our reproduction it is obviously very irresponsible to have children if it means the destruction of our world. We have the means now, both practically and, with these understandings, psychologically, to love our world, so it's madness to choose self-destruction. The world needs—and we are now in a position to supply—love, not selfishness. Controlling our numbers will certainly be important in rehabilitating both humans and our environment. In fact, we in the WTM have already employed this device to help establish these understandings and the new world they can bring about. As mentioned in par. 608, we decided not to have children so that we would have all the resources we could possibly muster to ensure these precious understandings of the human condition would survive the initial onslaught of vilification, persecution and attempted repression for daring to address the issue of the human condition. Certainly our tiny band of 50 people would not have been able to undertake the legal actions we had to take against two of the biggest media organizations in Australia to redress their ferociously vilifying publications about us if we had children to nurture and support. It is the capacity to be selfless that has been missing from the human situation. Basically, as soon as humans no longer have to be selfishly preoccupied with having to artificially try to relieve themselves of the insecurity caused by their human condition through winning as much power, fame, fortune and glory as they possibly can, and are thus at last able to be fully concerned about the well-being of others, all the suffering in the world will be able to be brought to an end. <u>The key rehabilitating effect for humanity and our planet is our ability now to fulfill Sir James Darling's instruction to transition from living selfishly to living selflessly, to have that **'change of heart'**</u>.

<sup>1235</sup>Understandably, this power of selflessness has not been something humans wanted to acknowledge or think about because it has been too confronting of our present massively embattled, ego-hungry, desperately-needing-self-gratification-and-glorification, selfish, greedy, materialistic existence. We have *had to* live an alienated, escapist, materialistic life because we haven't been able to live a secure, honest, spiritualistic life, but now, with the human condition explained, we can. And now that we can afford to think about it, all manner of insights into the power of selfless cooperation become accessible, including those provided through observing the dynamics of ant, bee and termite colonies; truly, as King Solomon advised, **'Go to the ant…consider its ways and be wise'** (Bible, Prov. 6:6). The 2004 award-winning documentary *Ants—Nature's Secret Power* admitted the power of selfless cooperation when it concluded that **'The secret of ant societies is their cooperation…**[it's what has enabled them to] **act as a superorganism…**[and become] **nature's true world power'** (produced by Adi Mayer Films and ORF). In his book *The Soul of the White Ant*, Eugène Marais wrote that **'the termite…never rests or sleeps'** (1937, p.61 of 154). The extreme selflessness of bees was also made apparent

in a documentary that reported that **'when bees become sick they sacrifice themselves and leave the hive to die to prevent infecting the rest of the colony'**, and how **'in the summer, the workers only live around 30 days because they literally work themselves to death'** (*Silence of the Bees*, National Geographic Channel, 2007). I'm certainly not suggesting that we go beyond self-sacrifice to self-elimination as a strategy for solving problems, but these examples do make it clear how powerful a force selflessness is. Our ability now to leave behind the insecure, self-preoccupied, ego-centric, selfish state of the human condition makes possible the *true* reparation of our whole world; it is us self-adjusting, conscious humans who—now that we no longer have to live an insecure, selfish existence and can live a selfless, cooperative existence—can now become **'nature's true world power'**.

[1236] Although this new world for humans, which the World Transformation Movement has been established to facilitate, will, as has been stated, have a slow beginning because humans initially find it difficult even taking in or 'hearing' discussion about the human condition, once the WTM is discovered, word of it and what it offers every human and the world will spread like wildfire—as the exceptional denial-free thinking prophet Teilhard de Chardin wrote, **'The Truth has to only appear once...for it to be impossible for anything ever to prevent it from spreading universally and setting everything ablaze'** (*Let Me Explain*, 1966; tr. René Hague et al., 1970, p.159 of 189). Indeed, all the quotes from prophets, songwriters and poets included in par. 218 testify to the nature and consequences of this information finally appearing in the world. Yes, before long we will be taking this information to the world—marketing our own human-condition-free, world-saving products, providing our own all-exciting and meaningful, denial-free, honest, human-condition-understood films, documentaries and books, launching our own TV station—and our website will be the biggest in the world, bigger than Google or any other existing site. And this is not deluded hubris, or wild exaggeration, it is simply the logical truth of what is able to happen when the dignifying, redeeming and healing understanding of the human condition is finally found.

[1237] Sir Laurens van der Post once said that **'The carrier of new life, the carrier of renewal in societies, has always started with a lone voice, with one lonely individual somewhere taking upon himself in terms of his own life wherever it's challenged, the start of renewal and living the "being" with which he has been charged in a world of "having"; and it always starts like that before it becomes a group...[and then] a community—which is not yet, but which is**

**coming'** (presentation by Laurens van der Post at the Institute of Contemporary Arts, London, 9 Dec. 1982; see <www.wtmsources.com/163>). Yes, the bottom line truth is that *only* the finding of the reconciling understanding of the human condition, and the **'change of heart'** from living by the doctrine of **'having'** to living by the transformed doctrine of **'being'** that the understanding finally makes possible, can save the human race. The World Transformation Movement provides the *only* path forward for humankind—anything else represents ever-increasing, excruciating, unthinkable suffering and ultimately doom for our species.

[1238] So, what everyone can now do and should now do is become a member of the World Transformation Movement and adopt the Transformed Lifeforce Way of Living—because when this fabulously empowering breakthrough explanation spreads like de Chardin anticipated it would, the transformation it allows will become the main objective of all nations. Then, with everyone's help, we will each be able to rehabilitate and make environmentally sustainable our realm—our home, our community, our region, our country, and then our world. For 2 million years humanity has strained with every fibre of its being to achieve this great breakthrough from ignorance to enlightenment, and for those of us living now our wonderful opportunity is to implement the rewards of that achievement. Here comes the golden sunshine revolution!

[1239] And as I have previously said, when I speak of rewards, the main personal 'reward' for humans now is the anticipation and the excitement of being effectively free of the human condition—the joy and happiness of being liberated from the burden of your insecurities, self-preoccupations and devious strategizing; the awesome meaning and power of finally being genuinely aligned with the truth and actually participating in the magic true world; the wonderful empathy and equality of goodness and fellowship that understanding of the human condition now allows you to feel for your fellow humans; the freedom now to effectively focus on repairing the world; and, above all, the radiant aliveness from the optimism that comes with knowing our species' march through hell has finally ended. Yes, this information will empty all our jails; redeem, reform and transform everyone. Better than that, it will *transfigure* all humans, make them glow with pure joy. Our species' redemption is *so* fabulous it will make humans fly through the air—well, almost! These all-loving-of-ourselves understandings are just pure wonderfulness, joy beyond description!

## Chapter 9:10 <u>Humanity's overall situation now that we have understanding</u>

[1240] I have drawn this picture to illustrate the situation the human race is now in.

Humanity's Situation: the Sunshine Highway to Freedom, the Abyss of Depression, our Cave-like Dead Existence and the Spiraling Pit of Terminal Alienation

[1241] At the top of the picture, humans are depicted reveling in the sun-drenched **'FREEDOM'** of the Transformed Lifeforce Way of Living. In the middle of the picture, however, lies a terrible abyss with the word **'DEPRESSION'** written prominently across its base, representing the deep, dark depression humans suffered when they tried to confront the issue of the human condition. You can see that the abyss consumes two-thirds of the whole picture, which is an accurate reflection of how big a part depression has played in human life—basically, *all* our activities have been ruled by our fear of it. Yes, the human condition has essentially been our fear of the depressing implications of our non-ideal behavior, of are we good or bad; are we actually worthwhile creatures or evil beings—and, although the human condition has finally been compassionately explained, the fact remains that it will still take a few generations to completely heal, so current generations will *still* be faced with that depression if they try to fully confront all the truth about themselves and the world. As has been explained, the arrival of truth 'day' or honesty 'day' is also exposure or 'judgment' 'day', the time when we are faced with too much truth to have to suddenly confront. The lower side of the abyss features a narrow strip of land that is colored the grey-green of dead flesh. Featuring the words **'CAVE-LIKE DEAD EXISTENCE'**, it is populated by a throng of people walking around like zombies. This is a truthful depiction of the state of extreme block-out or denial and resulting alienation from our true selves or soul that humans currently live in. The bottom corner of the picture also contains a big black spiraling pit that has written beside it, **'Extinction of the human species from terminal levels of alienation'**—which was the destiny of the human race if understanding of the human condition was not found.

[1242] In this situation that the human race is now in we obviously can't continue living a **'CAVE-LIKE DEAD EXISTENCE'**. The amount of suffering is way beyond obscene and unconscionable. What's more, if we each remain where we are the human race will very soon go 'down the gurgler', into **'THE SPIRALING PIT OF TERMINAL ALIENATION'**. There can be no doubt that our species has already reached the end play or end game point where it is about to plunge into that pit of perpetual darkness and extinction—indeed, the descriptions that were given in chapters 8:16B and 8:16C of what is occurring at the forefront of human progress, namely in the Western world, and in particular in the USA, spell out how serious our species' plight is: **'today's children are probably** the least loved generation of all', '96 percent of American families are now dysfunctional'; 'someone born since 1945 [is] likely to be up to 10 times more depressed than their grandparents'; 'Autism...now occurs at epidemic rates... According to California records, autism has increased 1,000% in the past 20 years!'; there has been a '41% increase in ADHD in US children during the past decade'; 'there are more and more cases of ADD among children [in the US], one in five to be exact...So what is going on?...The brain [of the ADHD child] is busy, busy, dealing with the pain' from 'childhood', 'where there was no [real] love...[so] the kid is agitated out of his mind, driven by agony inside... drowning in misery'; 'while only two years ago students were able to sit through a half hour art documentary, I now know I lose them after only eight minutes'; 'Alone and adrift in...our "psychic saunas" of superficial sensory stimulation, members of my [Y] generation lock and load our custom iTunes playlists, craft our Facebook profiles to self-satisfied perfection, and...bravely venture forth into life within glossy, opaque bubbles that reflect ourselves back to ourselves'; 'The youth of today are living their lives one mile wide and one inch deep'; 'the wisdom of the ages is...buried like lost treasure beneath a fathomless ocean of online ignorance and trivia that makes what is worthy and timeless more inaccessible than ever...[The] digital revolution...[is] decapitating our culture, trading in the ideas of some 3000 years of civilization for BuzzFeed'; 'We have a lot of competing noise for our attention these days, and it would be criminal to let that overwhelm our true potential, by masking [the] useful information [in Griffith's books] with hideous noise'; 'Generation Y is being ravaged by depression, anxiety and stress disorders.' Young people now feel that 'Our innocence is lost...It scares me to death to think of what I have become...I feel so lost in this world...I scream to the sky but my words get lost along the way. I can't express all the hate that's led me here and all the filth that swallows us whole. I don't want to be part of all this insanity. Famine and death. Pestilence and war. [Famine, death, pestilence and war are the 'Four Horsemen of the Apocalypse', the state of terminal alienation, described in Revelation 6 in the Bible.] A world shrouded in darkness...Fear is driven into our minds everywhere we look...How do we save ourselves from this misery...So desperate for the answers...No one hears our cries. And no one sees us screaming...This is the end.' The aforementioned TIME magazine story titled 'The Me Me Me Generation' about the 'narcissism epidemic' in the present Y-generation was essentially about how this '80 million strong...biggest age group in American history' exists in a state of terminal alienation. And the frightening question that the very existence of a Y-generation—and the X-and-Z generations that flank it—poses is what exactly do we mean when we say the X, Y, Z generations? The answer is the end play state of terminal alienation. After all, what comes after Z?

[1243] Michelangelo's and Blake's horrifying pictures of the torment and agony of the state of terminal alienation that were included at the beginning of chapter 2 reveal what life would be like for all humans if the human condition was not resolved; and indeed, *will* be like if this understanding is not supported, because, as the quotes above indicate, this horrifying state of terminal alienation *is* upon us. And indicative of the world being **'so desperate for the answers'** that can **'save ourselves from this misery'** is the panic we've seen in the corridors of power, with, as mentioned, the President of the United States announcing in 2013 a **'Brain Initiative'**, giving **'$100 million initial funding'** to mechanistic science to find **'the underlying causes of…neurological and psychiatric conditions'** afflicting humans; and in 2012, an American billionaire pledging $200 million to **'accomplished scholars whose collective mission is…greater understanding of the human condition'**; and in 2013, the European Commission announcing the launch of the **'Human Brain Project with a 2013 budget of €54 million (US$69 million)'** with a **'projected billion-euro funding over the next ten years'** with the goal of providing **'a new understanding of the human brain and its diseases'** to **'offer solutions to tackling conditions such as depression'**; and similar initiatives are taking place at Oxford and Cambridge universities—this despite the mechanistic scientists in these institutions being committed to living and practicing their craft in complete denial of the human condition!

[1244] *Clearly* humanity can't stay living on the 'narrow strip of land that is colored the grey-green of dead flesh' a moment longer. The pain of our tortured condition HAS TO STOP!! So thank goodness we have **'the answers'** that we are **'so desperate for'** to **'save ourselves from this misery'** and don't need to rely on these failure-trapped initiatives. However, as I have emphasized, trying to confront all these now terminal levels of upset inside us humans is also not something that should be attempted. As Enrico, a reader of my books, described the difficulty of trying to face down the terrible Abyss of Depression—**'Diving into a sea of truth where everything is completely transparent one can't but ask, "how will anybody cope with this; how in the world can anybody cross such darkness to reach light?!"'** (Email to the WTM, 24 Feb. 2011). Yes, one of the people I have drawn looking down into the Abyss of Depression, holding their head in horror, could be the mother of an autistic child trying to fully confront the truth that the main reason for autism is a lack of alienation-free, unconditional, true love. As emphasized in chapter 7:16C, that mother is now able to understand that autism is *not* the fault of parents, but of the ever-increasing upset in the human race as a whole, which in turn is due to our species' heroic search for knowledge. The difficulty, however, is that while she *can* access some relief from that guilt through that fully compassionate understanding, if she dwells at all on the truth of how much upset exists within her and all the pain that upset has inadvertently caused her child, she will still experience unbearable feelings of depression. Clearly she is going to have to maintain a degree of block-out or denial of such deeper thinking. And virtually every human is in a similar situation, where it will be impossible for them to fully immerse themselves in and confront the truth about their corrupted condition and the suffering and destruction it has caused. So virtually all humans *will* have to take the 'SUNSHINE HIGHWAY' bridge *over* the Abyss of Depression; adopt the Transformed Lifeforce Way of Living, where they support the truth while avoiding overly confronting it. Incidentally, the only reason I made the bridge over the abyss in the drawing so narrow was to make sure that all the letters of the word 'depression' were visible. The bridge should really be a mile wide because it is, in truth, a wide open and easy path to take. But the point is that there *is* a way **'to cross such darkness to reach light'**, as evidenced by all the people who have *already* crossed that darkness to reach the light-filled existence that understanding of ourselves allows. People are already managing the problem of exposure, and doing so in an extremely relieved and joyous way, through having adopted the Transformed Lifeforce Way of Living.

[1245] With regard to the extremely self-confronting situation I've just described of the mother of an autistic child, I might include this comment by WTM Founding Member Tony Gowing, which shows the extraordinary honesty, compassion and power of the TLS (Transformed Lifeforce State). Tony said: **'If she** [the mother of an autistic child] **was in the TLS it would actually be easy for her to cope. If you look at Ned Flanders' born-again Christian situation, Ned, like every human, would have done terrible things in his life—he would have been self-obsessed finding relief from the insecurity of his condition and, as a result, uncaring towards others, then dishonestly defensive about that insecure inability to truly care and so even *more* determined to prove his worth, then even *more* self-preoccupied and uncaring towards others, then angry and retaliatory for being so condemned, and so on—a vicious cycle of ever-increasing** self-preoccupation, anger and fury, committing every kind of outrage. *But*, having become a Christian, Ned's just whistling with relief all day long now—he's no longer self-preoccupied, guilt-ridden and destructive—and I just love Ned in that sense because if you are in that liberated state, in our case, living in the TLS, the frustration and guilt no longer burdens you. That mother, and *every* human now, can cope because they are now able to know that what they have done, and how they have been, is not their fault. They are released from the crippling and deadening cycle of struggle they have been living in. The truth is all mothers now are massively upset, soul-dead humans (as R.D. Laing truthfully described our concrete-encased lives under the duress of the human condition [see par. 123]) and, as a result, have been *extremely* inadequate in their ability to give their children unconditional love, but so have we all—we have *all* been soul-dead, committing all manner of hideous obscenities and atrocities. That's the human condition, which we can fully understand now, and so we don't have to harbour that guilt anymore. We can just understand it and leave it be; it is all now dealt with, explained and now just obsolete baggage. Everyone has suffered from the human condition, but we have been relieved of it now. We are all horribly messed up but we are free now. Some races of humans are more upset than other races, that's the truth, but so what, we're all massively upset—all our upset and the degrees of differences in upset between humans and races of humans is meaningless now. The world of upset, all aspects of it, has no meaning now. That world has gone forever, disappeared, vanished. Another world has opened up that is free of all of that. We are FREE and TRANSFORMED, and free to transform and repair the world' (WTM records, 30 Nov. 2014). You can view people like Tony talking like this about the marvel of the Transformed Lifeforce Way of Living at <www.humancondition.com/affirmations>.

[1246] Yes, so much is going to change now. For instance, with understanding of the human condition and the ability to be truthful that it makes possible, all the books on psychology are going to have to be re-written. What particularly changes the way psychology has been practiced is the truth that humans are born with cooperatively-orientated, all-loving instinctive expectations, and, therefore, how compromising of an infant's healthy development it is when these expectations for unconditional love are not met. It particularly reveals how easily children, in their trusting naivety, blame themselves for any shortfall in love, and how devastatingly hurtful such conclusions that they are a bad, unworthy person are for them, and, therefore, how much children have to block out any thinking about that terrible conclusion, and, therefore, how alienated they become from their true self or soul, and how that lack of access to their true self compromises their ability to nurture the next generation with real love when they in turn became parents. It is our all-loving and thus all-trusting soul that has not been acknowledged in psychology; in fact, it has not been acknowledged in any aspect of human life.

[1247] Indeed, the human race has been living in denial of so many truths that are now revealed that nearly all academic books will have to be re-written. Science as a whole will be faced with changing from complying with the practice of denial to having to be honest, which will mean its whole evasive, intellectualism-emphasizing-instinctualism-denying structure, including its traditional universities and academic departments and millions of denial-drenched research papers (as has been described and evidenced, the human condition is such that humans have been incapable of being genuinely objective), will be faced with obsolescence. A stark indication of the immense change that comes with the arrival of understanding of the human condition is that with the exception of just a handful of titles (the main ones being the great religious texts), virtually all the world's books have been deeply denial-compliant in their content and so will be largely obsoleted by the new denial-free paradigm. Basically, it won't be long before all our libraries and academic institutions become museums; indeed, if it wasn't for the Transformed Lifeforce Way of Living that allows everyone to fully participate in the human-condition-understood new world, the whole, current, denial-saturated human race itself would become a museum piece! Interestingly, in his classic book, *Nineteen Eighty-Four*, George Orwell predicted that **'By 2050—earlier, probably—all real knowledge of Oldspeak will have disappeared. The whole literature of the past will have been destroyed'** (1949, p.53 of 328). While Orwell was referring to the emergence of a completely thought-oppressing, pseudo-idealistic state, there is some intuitive realization deep within what he has written of a time when all dishonest thinking will **'disappear'**; a time when, as predicted in the Bible, revelation about our human condition will lead to a situation where **'the old order of things has passed away'** (Rev. 21:4).

[1248] It is true. When understanding of the human condition arrives, as it has, every denial-based aspect of the whole world becomes redundant. The world as we know it comes to an end, and virtually overnight—well, within a few generations, which is still very fast. For teachers at schools or universities, all the truth-avoiding, immensely superficial, basically dishonest, mechanistic ways of viewing the world will soon no longer be taught. The arrival of the actual understanding of the reason why humans are good and not bad will mean that materialistic goods that have been designed to make ourselves feel superficially better about our corrupted condition will, before long, be seen as meaningless and worthless. The whole basis of materialism has been undermined, made pointless and unnecessary. All our posturing and promenading and pretending will eventually lose its power to impress. Our whole artificial and superficial existence has been exposed and made redundant. Going to church will, in time, be superseded by this reconciling understanding of our previously God-fearing existence. The business of politics, of having left and right-wing factions, is similarly made redundant by the reconciling understanding of the human condition. Basically, all our denials are exposed, made transparent; they no longer work. So yes, before long, within a few generations, that whole denial-based existence will come to an end. Life as we have known it will be over—and, most relievingly for humankind, with its passing all the suffering and destruction that went with it will also go. A whole new world now begins to open up across the world based on reconciling understanding and the Transformed Lifeforce Way of Living it makes possible. As I've said in every major document I have written since I started writing about the human condition over 40 years ago, '**soon from one end of the horizon to the other will appear an army in its millions to do battle with human suffering and its weapon will be understanding**'.

[1249] Humanity's situation now is actually perfectly depicted by Hans Christian Andersen's 1837 fable, *The Emperor's New Clothes*, in which a child breaks the spell of deception that the emperor is beautifully clothed when he discloses the truth of the Emperor's nakedness (see drawing overleaf). The first reaction of the Emperor and his entourage is to try even harder to maintain the charade, maintain the denial, but it is futile. To quote from the fable: '**"But he has got nothing on," said a little child. "Oh, listen to the innocent," said its father. And one person whispered to the other what the child had said. "He has nothing on—a child says he has nothing on!" "But he has nothing on!" at last cried all the people. The Emperor writhed, for he knew it was true. But he thought "The procession must go on now." So he held himself stiffer than ever, and the chamberlains held up the invisible train**' (*Andersen's Fairy Tales*, tr. E.V. Lucas & H.B. Paull, 1963, p.243 of 311). Yes, with the truth revealed, the spell is broken; our delusions and pretenses have been stripped of their power.

Drawing by Jeremy Griffith © 1988 Fedmex Pty Ltd

[1250] It is true that humans are afraid of change almost as much as we are of self-confrontation, and this information does bring about phenomenal change, a massive paradigm shift, but it is entirely manageable, most especially because the change is a positive paradigm shift, not a negative one—humanity is going from an extremely negative, absolutely awful paradigm, to an extremely positive, fabulously wonderful paradigm. So we can procrastinate and try to stay in the old, essentially dead, hell-on-earth world—stay in the egocentric castles we have built, or are trying to build, for ourselves based on our pursuit of power, fame, fortune and glory; think **'But...The procession must go on...**[and hold up our **'invisible train**[s]**'**] **stiffer than ever'**—or we can join the new all-exciting world that will be opening up right across the planet. There is now so much that needs to be done in support of the liberated new world that the range of new opportunities, needs, positions, industries, services, etc, etc, will be endless—and the best part of it all is that this new world for humans will be filled with immense excitement, happiness, meaning and fellowship. With understanding of the human condition found there is no longer *any* justification for staying in the old, lonely, effectively dead world. <u>Before long *everyone* will be rolling their sleeves up and excitedly going to work for the absolutely fabulous, human-condition-understood, peaceful, reconciled, all-loving new world for humans.</u>

[1251] At this point I might address the concern that in **'a thoroughly transparent world'**, which understanding of the human condition does produce, our **'species would...suffer spiritual anemia and perhaps terminal boredom'** (Frank Tippett, 'Essay', *TIME*, 17 Jan. 1983). This often-stated perception that **'the price of paradise will be boredom'**, and even that we should, therefore, stay living in our alienated state of denial of the human condition, is what I call 'black speak', blindness preaching blindness, because the alienated state is such a numb, seared and exhausted state that to argue we should stay alienated is effectively arguing that we should stay dead! If a bucket of water represents the depth of sensitivity humans are capable of then our alienated state is equivalent to living life on the thin surface meniscus of that water. So to maintain that we should remain in an empty, insecure, immensely unhappy and deeply, deeply distressed, effectively dead state that can hardly feel or know anything when we can finally, at last, savor the magnificence of our world is simply alienation trying to build a positive out of a negative. It is an understandable defensive reaction that people have had to make themselves feel better about their human-condition-afflicted, empty lives, but it is clearly an absurd betrayal of all that the human race has worked towards. While our world will, for example, lose some of its variety or 'color' when the different states of alienation disappear and everyone has superficially similar personalities, the incredible sensitivity and happiness that will come to humans from being able to access the world of our soul again will mean our lives have a depth and potential that we have hardly dared to dream of. For example, each person will be able to immerse themselves in whatever aspect of sensitivity they choose, with the number of different aspects of sensitivity available in our world to savor being innumerable. Some people might spend a whole lifetime perfecting, and sharing with others, what it's like to feel what a certain mood feels like, or what a certain animal or plant feels like, or what a certain time of the day truly feels like. There are awarenesses and feelings and knowledge and thoughts and imaginings that we haven't even begun to tap into in our human-condition-preoccupied, virtually dead state. The reason I said above that people in the human-condition-free new world will be 'superficially' similar is because once we access the *immense* sensitivities our soul has we will become aware of so many wonderful differences that will still exist between humans. So once people allow themselves to appreciate what life will be like free of the human condition they will never make such sad, blind, defensive comments as suggesting a human-condition-understood future will be 'boring'.

[1252] Yes, the fact is, having lived in such a dead, alienated, dark, cave-like state of dishonest denial and resulting estrangement from our true soulful self for so long, coming out into the truthful world of liberating sunshine of understanding will be *so* redeeming and exalting it *will be* almost more exciting, transfiguring, rapturous and exhilarating than the human body is capable of handling! To finally be free from having to—and we did *have to*—live such a world-destroying, selfish, mean and egocentric existence for so long is such an *immense* relief. To be able to now live a life of transformed freedom from the dark, effectively dead state of denial and be able to participate fully in the true, human-condition-liberated, soul-resuscitating, loving world with all its beauty and sensitivity is something *so* exciting that appreciating just how exciting it *will be* is almost beyond our present powers of imagination. So let me paint the picture as starkly as I possibly can. Firstly, the following is a re-inclusion of a powerful description that was given in par. 1052 of the situation the materialism-driven world has arrived at. It was explained that with communication technology—especially television and the internet—showing everyone in the world what is going on elsewhere in the world, everyone is now acutely aware of the financial wealth and material luxury that some people and cultures have created, and those on the poorer side of the scale envy it; they ask themselves, 'Why shouldn't I have all this?' As the psychiatrist Clancy McKenzie observed, **'While visiting Machu Picchu in Peru in 1979 I noted very poor persons, living in the mountains, who had only the clothes they wore and perhaps a lama or two, but had beautiful, warm smiles and seemed content and happy. Days later I was in Bogota in Colombia. It was a very hot day and we asked the driver to stop at an outdoor tavern to buy cold beer. The people were very impoverished, but there was a TV playing and they were able to view the "outside world" where everyone seemed to have more, and luxury was abundant. I offered to go in with the driver and he urged me to wait in the car. I soon learned why. The absolute hatred was so intense that it was palpable. These people did not have less than those in Machu Picchu but they saw others who had more, and their needs were intensified.'** Yes, the result of this awareness of what materialism offers is that there is now rampant greed everywhere; all those populations who haven't had the material luxury that is so prevalent in the Western world see themselves as being on catch-up—they want a big house, a car, a fridge, a suit, a gown, a fancy watch, Nikes, Coca-Cola and Big Macs, a beach holiday, etc. The effect is that our original instinctive self or soul's natural inclination to be selfless and loving has been thrown by the wayside and everywhere cynicism reigns—a 'cynic' being **'one who doubts or denies the goodness of human motives'**. The following is an account of the situation in Kenya, but it very much applies to what is happening throughout Africa, South America, China, India, south-east Asia and in parts of the Middle East. Written by Mirella Ricciardi, the aforementioned Kenyan photographer, it describes Kenya's current **'chaos and confusion, the crime and lawlessness, the greed, the bribery and corruption at all levels, the senseless wildlife killings, the environmental destruction, the mindless power games, the hopeless destitution and hunger, the lack of concern and forward thinking…The outrageous greed and disregard for anything and anyone other than themselves and their immediate families in the top ranks of governments, have created a financial drain on the country that filters down and affects whole populations in varying degrees, threatening the fragile infrastructure of the land. All those, without exception, who have somehow attained a position of command and authority, however humble, have but one idea in mind, to emulate the example set by the head of state – to milk their positions for all they are worth, for as long as it lasts. Because finances are siphoned off to Swiss bank accounts, salaries are so low they need topping up by whatever means. As a consequence bribery and corruption at every level is the only option. The police force and judiciary, the ministers, the administrators, the teachers, the doctors and the heads of government divisions, extract whatever they can, by fair means or foul, under the guise of their uniforms and titles, while the country collapses and the masses starve.'**

[1253] So, the 'developing' 'poor' parts of the world are fast approaching **'collapse'** and **'hopeless destitution'**—while the so-called 'developed' 'affluent' parts of the world are fast approaching a state of terminal alienation with anxiety disorders such as ADHD and autism rampant and the **'biggest age group in American history'** in the grip of an extremely insecure **'narcissism epidemic'**. After presenting this analysis in par. 1052, I then summarized in chapter 8:16G that it is basically all over for the insecure, human-condition-avoiding, escapist, power-fame-fortune-and-glory-seeking, materialism-driven, greedy, never-satisfied, selfish existence. Yes, that world *has* reached its calamitous end, SO IT IS FABULOUS NEWS THAT AT THAT BRINK THE WORLD *CAN* BE SAVED! <u>The God of Mammon and Greed *can be replaced by the God of Soul and Love*, humanity *can* have a **'change of heart'**—because the insecurity-ending understanding of the human condition has been found</u>. Humanity can finally come home. While we had no other form of reinforcement other than materialism, we built it up in our mind to be something wonderful, but the truth has always been that materialism really offers nothing; it is a desperately sad, shallow and unsatisfactory way of living—a very poor substitute for our lack of spiritualism. With understanding of the human condition finally found, however, what is on sale now is nothing less than heaven on Earth, a way of living that is *incredibly* exciting and happy. Between the materialistic way of living and the Transformed Lifeforce life there is just no contest, NONE AT ALL. The way forward now is back to our soul's true world. The gates have been flung open to paradise. Materialism and greed are old baggage now, discarded rubbish to be left on the garbage tip. No one is going to buy them anymore. Their power to seduce is gone because the most magnificent alternative has arrived to transport everyone to a world of soul, real love, real meaning, real togetherness and real happiness. Everyone, every Peruvian, every Kenyan, every Chinese person, every Indian, every American, every European, *all* people on Earth, can now join the great exodus from the, in truth, terrible human-condition-afflicted, materialistic, greedy, selfish existence. NO ONE WILL WANT THAT ANYMORE WHEN THE TRUTH GETS OUT ABOUT WHAT IS ON OFFER NOW!

[1254] From the perspective of all those individuals who, for whatever reason, haven't been successful in the materialistic power-fame-fortune-and-glory-seeking world, when they see those who have been more materially fortunate making the great **'change of heart'** from living selfishly to living selflessly, they will finally be released from feeling the injustice and humiliation of their situation. When selfless love finally comes to Earth, the hearts of humans everywhere will be flooded with enormous relief and incredible joy. (On this point, I might mention that the Founding Members of the WTM have sometimes been accused of being from fortunate backgrounds and therefore not representative of the majority of humanity, but such criticism misses the significance of what is occurring, which is the great change of direction the human race is now able to take. The *real* interpretation of the situation should acknowledge how *exciting* it is that those who are able to continue the heroic but corrupting search for knowledge are finally able to abandon that individualistic, materialistic existence that has now been made redundant and instead dedicate their lives to helping the less fortunate — to being part of that aforementioned **'army in its millions'** that is able **'to do battle with human suffering and** [whose] **weapon** [is] **understanding'**. So when a businessman made the comment to me about the WTM that **'I always thought we would know when we had won when we saw the Gurkhas coming back'** (John Rowntree, WTM records, Dec. 1982), he was recognizing this phenomenon. Renowned for their fighting ability, the

Gurkhas are the Nepalese soldiers who are employed by the British and Indian armies as front line troops and forward scouts—so when they stop pushing forward it is a sure sign the frontline battle has been won.)

[1255] Yes, with the battle to find understanding of the human condition over, relief will spread across the world. In particular, selfless love, concern and help can now come to every person in every corner of the world. To paraphrase U2's fabulous anticipation of our species' liberation from the agony of the human condition, 'love has come to town' in its absolute purest form, namely as the redeeming and relieving understanding that finally ends the insecure, egocentric, brutal and selfish way of living forever. Everyone in every situation and predicament can now rise up as radiant new beings from their corpse-like state. Everyone can now come back to life—can wake up from their human-condition-afflicted torpor and look outwards and see each other and the world for the first time, and move across and help each other, and do anything and everything that needs to be done to end the suffering and pain that plagues this planet.

[1256] When Tony Gowing was once asked to describe the main 'selling point' of *FREEDOM*, he replied, '**It is the complete 100 percent turn around in any and every dire situation the world is now in—as Jeremy has often said, "one minute the world is hurtling on a one-way track to destruction, the next minute there is so much love and freedom that every problem and difficulty is solved". If you look at any of the problems in the world with any degree of honesty— anybody's personal situation, the refugee crisis, the relationship between men and women, the environment, the economy, mental health, etc, etc—they are each in a depressingly dark state, but with this understanding of the human condition they are all completely turned around into the most glorious, happy and light- filled situation imaginable. This understanding brings an end to all the pain, all the suffering, all the insecurity, all the unsureness, all the darkness; it turns every single one of those situations into an absolutely wonderfully secure, knowing, excited, relieved and happy—downright incandescent with joy—state. If I look at my personal situation, I carry a great deal of pain and insecurity from my childhood, such as from my parents separating when I was 10, but really this was just my version of the hideousness of life that every child experienced when they were growing up in this mad, human-condition-suffering world that provides no answers and no explanation whatsoever to all the soul-destroying, crippling madness. And, as much as my mother was and is incredibly special to me, all she could tell me about the world was that it is unfair, that life is just the way** it is, but that she loved me. Basically, all we have been able to tell children is that the world is unfair and horrible, and all you can do is get on with life and do the best you can; no one likes a whinger, life is all about being competitive and winning. And if someone started to tell me that we should behave selflessly and lovingly I would say what rubbish, because I know from my experience that the world has never treated me like that and people who speak of those things are just weak-minded.

[1257] In the world before this explanation there were no answers, there was no meaning, no direction, no real understanding—I had no real idea what to do in the world, no framework of reference, no idea about the meaning of existence at all, in fact. I had my own talents of playing sport and striving for financial success or 'doing well' in my chosen career, but underneath all of that I knew and know that my life was a desperate, pain and stress-filled, dysfunctional mess. But now I have complete understanding of the world. I know what the meaning of life is; I know that there is, in fact, tremendous purpose and meaning underlying everything in the world; that every little spec of dust follows a fantastic path of order, integration and love. I know now about humans and that they have been on this most wonderful journey of love and courage, which has taken such a toll on our species and our planet. I know now why we are the way we are, and why the world has been so mean and full of pain, and why there can now be a wondrous change of heart in every single human—that I and every human can right now make the change from being completely consumed by an incredibly insecure, mad, terrified and obsessively egocentric and selfish mindset, to a completely 100 percent secure person living out an all-meaningful, wildly exciting existence dedicated to helping make this change from darkness and pain to brilliant light and an all-understanding state of freedom from the human condition. When I say I am 100 percent secure, I am not secure in myself as such (as Jeremy explains, the complete psychological rehabilitation of the human race will take a few generations), but all the insecurity that came from my childhood is now completely covered, explained and looked after by the larger umbrella of understanding I now have. I am 100 percent secure in what I am doing and how I am leading my life and what I am trying to help achieve with all my efforts. All day, every day, I am totally consumed with fulfilling 2 million years of human endeavour. And the wonderful thing is that the meaning of life is to actually be integrative, loving, cooperative and caring, not disintegrative, and, what is more, we now know that unconditional selflessness is already deeply inside us, running through our whole being because we humans once lived in a completely cooperative state, and so we are just coming home to that true and original state. We can finally live out the love that we so deeply, unquenchably crave inside ourselves. And not only are we deeply, deeply cooperative and loving beings inside ourselves once the human condition is explained away, but the whole world and everything in it works through the very same integration, cooperation and love that we have within us. Finally we aren't the outcasts, the baddies, the horrors, the downright monsters of planet Earth—we are connected with everything and everyone on Earth. The wonder and relief is endless. Where everything was just so desperate, disappointing, meaningless, hopeless, depressing, tiring, hurtful and false— everything, every situation, every person, everything now is just enthusiasm, love, peace, relief, happiness and excitement. Everything can now be fixed and brought back to life.'

[1258] We humans *have*, in truth, all been asleep, owned by so much pain and suffering. And, certainly, we are going to be in shock for a little while absorbing the realization that we have finally won our freedom from the agony of the human condition—but it's on, the great awakening, the rising up of the human race from its deep slumber. From the festering, stalled state it has been in for far too long, waiting for these liberating understandings of the human condition, the human race is finally on its way! As Doug Lobban says in his aforementioned affirmation, 'I am just so excited, I can't sleep, I am about to burst. I was in the shower and a hook on the curtain had fallen off so I put it back on and got an amazing feeling of meaning and I got out of the shower and just looked around for something else I could fix because that is what we get to do now, fix the world, and I just can't believe that after eons of generations we are the ones who actually get to participate in fixing the world up. It's just unbelievable, there is so much to do I can't wait to get into it…I've got this image of marching straight forward with incredible purpose, straight out of a massive swamp into long green grass with flowers and butterflies and sunshine, and never ever looking back, just keep marching forward. Everyone together, we are going to have so much fun.' Yes, the gateway *is* wide open and sunlight is now streaming into our world. We can all break out of our chrysalis, our human-condition-afflicted straitjacket, where we have been constrained, letting the wings of our great potential unfold in the sunshine of relieving understanding to reveal the fabulously beautiful creatures we really are.

[1259] The following compilation of quotes from the Bible perfectly describe the transforming **'change of heart'** from living selfishly to living selflessly that the great prophets of the past knew would one day occur when Integrative-Meaning/God-acknowledging, truthful thinking finally delivered the reconciling understanding of the human condition. Firstly, from the prophet Joel: **'Like dawn spreading across the mountains a large and mighty army comes, such as never was of old nor ever will be in ages to come…Before them the land is like the garden of Eden, behind them, a desert waste—nothing escapes them. They have the appearance of horses; they gallop along like cavalry. With a noise like that of chariots…like a mighty army drawn up for battle. At the sight of them, nations are in anguish; every face turns pale.** [The reconciling truth is shockingly confronting and exposing at first, but those who have progressed past the shock stage are overwhelmed with excitement about being transformed and having the capacity to end human suffering and the devastation happening on Earth.] **They charge like warriors; they scale walls like soldiers. They all march in line, not swerving from their course. They do not jostle each other** [Joel 2] …**[and] In that day the mountains will drip new wine, and the hills will flow with milk; all the ravines…will run with water…Their bloodguilt, which I** [Integrative Meaning] **have not pardoned, I will pardon** [3]**.'** And from the prophet Hosea: **'on the day of reckoning… let us return to the Lord** [the integrated state]**. He** [Integrative Meaning] **has torn us to pieces** [has horribly condemned us] **but** [now] **he will heal us…As surely as the sun rises, he** [the redeeming and relieving truth] **will appear; he will come to us like the winter rains, like the spring rains that water the earth** [Hosea 5-6]**.'**

[1260] And from the prophet Isaiah: **'He** [reconciling understanding] **lifts up a banner for the distant nations, he whistles for those at the ends of the earth. Here they** [transformed humans] **come, swiftly and speedily! Not one of them grows tired or stumbles, not one slumbers or sleeps; not a belt is loosened at the waist, not a sandal thong is broken. Their arrows are sharp** [the reconciling truth is undeniable]**, all their bows are strung; their horses' hoofs seem like flint, their chariot wheels like a whirlwind. Their roar is like that of the lion** [Isaiah 5] …**[Everyone] from the four quarters of the earth…**[will recognize the redeeming truth about humans and] **jealousy will vanish, and…hostility will be cut off** [11] …**In a surge of anger I** [Integrative Meaning/God] **hid my face from you** [because humans became upset sufferers of the human condition and had to live in denial of Integrative Meaning]…**but with everlasting kindness** [reconciling understanding] **I will have compassion on you…O afflicted city** [the world of humans]**, lashed by storms and not comforted, I** [reconciling understanding] **will build you with stones of turquoise, your foundations with sapphires…your gates of sparkling jewels…All your sons will be taught by the Lord** [truth]**, and great will be your children's peace…righteousness…will be** established...no weapon forged against you will prevail and you will refute every tongue that accuses you [denial will lose its power] [54] ...Your sun will never set again...[you will live in] everlasting light, and your days of sorrow will end...The least of you will become a thousand [everyone will be understood to be equally good and worthwhile]...I will do this swiftly [this transformation of the world will be fast] [60] ...[reconciling understanding] will bring forth justice...to free...from the dungeon those who sit in darkness...[and] lead the blind...along unfamiliar paths...[and] turn the darkness into light...and make the rough places smooth' [42] ...the eyes of the blind [will] be opened and the ears of the deaf unstopped. Then will the lame leap like a deer, and the tongue of the dumb shout for joy. Water will gush forth in the wilderness and streams in the desert [35] ...Awake, awake...rise up...Free yourself from the chains on your neck...Listen! Your watchmen lift up their voices; together they shout for joy [52] ...[Integrative Meaning/God/the Lord] will look with compassion on all her ruins; he will make her deserts like Eden, her wastelands like the garden of the Lord [51] ...[The reconciling understanding will] make straight in the wilderness a highway...And the glory of the Lord [truth] will be revealed, and all mankind together will see it [40] ...[and this understanding of the human condition] will...cut through bars of iron [entrenched dishonest denial, and]...will give you...riches [previously] stored in secret places [45] ...[and] light will break forth like the dawn, and...healing will quickly appear [58] ...creat[ing] new heavens and a new earth...Never again will there be in it an infant that lives but a few days, or an old man who does not live out his years; he who dies at a hundred will be thought a mere youth [65] ...[like] the rising of the sun [the reconciling truth]...will come like a pent-up flood...The Redeemer [understanding] will come [59] ...Arise, shine, for your light has come...you will look and be radiant, your heart will throb and swell with joy' [60] ...[this understanding] will destroy the shroud that enfolds all peoples...[and] will wipe away the tears from all faces...[and] will remove the disgrace of his people [25] ...[and on this] day of salvation I will help you...and...restore the land and...say to the captives, "Come out," and to those in darkness, "Be free!" [49] ...[The reconciling understanding] will settle disputes...They will beat their swords into ploughshares and their spears into pruning hooks. Nation will not take up sword against nation, nor will they train for war any more [2] ...[The reconciling understanding] will put an end to the arrogance of the haughty and will humble the pride of the ruthless [13] ...The oppressor will come to an end and destruction will cease; the aggressor will vanish from the land. In love a throne will be established' [16] ...[The reconciling understanding] gives you

**relief from suffering and turmoil and cruel bondage...[It] has broken...the sceptre of the rulers, which in anger struck down peoples with unceasing blows, and in fury subdued nations with relentless aggression** [humanity's heroic but competitive and aggressive search for knowledge will end]. [Now] **All the lands are at rest and at peace; they break into singing. Even the pine trees and the cedars of Lebanon exult...and say, "Now that you have been laid low, no woodsman comes to cut us down."...The poorest of the poor will find pasture, and the needy will lie down in safety** [14].'

## Chapter 9:11 The 'pathway of the sun'

[1261] The other aspect of *The Emperor's New Clothes* fable that needs explaining is the significance of the **'innocent' 'child'** breaking the spell of all the denial and delusion in the Emperor's court. The overall situation is that humans have been practicing denial for some 2 million years, ever since we as a species became fully conscious and had to defend ourselves without understanding of why we are good and not bad. This means that this practice of denial has been refined and added to for *so* long now that it has achieved a state of near complete block-out of any truths that bring the issue of the human condition into focus, which, as has been revealed in this book, is a phenomenal amount of truth. It was as though our planet had become enclosed in a solid steel casing through which no light, no truth, could enter. Every time some light crept through a crack, the great welding machines of denial would come out and weld a slab of steel over the offending fracture. In the end it was a truly massive edifice of denial, so strongly reinforced that it seemed no one would ever be able to break through it, leaving the human race to perish inside its terrible darkness. The great fortress of denial *did* seem impenetrable, unassailable, BUT NOT SO TO OUR SOUL, which an **'innocent' 'child'** is the perfect representation of, for it has always had so much love that it was able to lie in wait for the time when science would find sufficient understanding of the mechanisms of the workings of our world for it, that little innocent child of loving, denial-free soundness, to rise up and break through that steel case and liberate the human race from its incarceration.

[1262] The imagery of a child one day liberating the human race appears in many mythologies. The biblical story of David and Goliath is actually a metaphorical description of this situation where the whole army of humanity is besieged by the monstrous, all-pervading and all-powerful hold that denial has on the world, which the **'nine feet tall'** giant Goliath represents, when David, who is **'only a boy'**, **'come**[s] **against you [Goliath] in the name of the Lord Almighty** [in the form of Integrative Meaning-accepting, denial-free, unresigned thinking]**' 'and kill**[s] **him'**, allowing everyone to celebrate their liberation **'with singing and dancing'** (1 Sam. 17-18). Yes, in the mythology of Don Quixote (in par. 67), a human-condition-avoiding, resigned human was always going to be 'unhorsed' when it came to slaying the **'outrageous giant'** of ignorance-of-the-fact-of-our-species'-fundamental-goodness, which in that story is symbolized by a huge windmill. The truth is that in the end, strength of soul, not strength of body, has always been the best killer of giants. Of course, in addition to the story of David and Goliath, the Bible contains Isaiah's description of how **'a little child will lead them** [humanity]**'** to the state where the concepts of 'evil' and 'good'—and, in the process, where the more corrupted and the more innocent—will be reconciled; to where, he says, the **'wolf will live with the lamb'** (Isa. 11:6).

[1263] In the great European legend of King Arthur, the wounded (alienated) king whose realm was devastated (humans unavoidably made their world an expression of their own madnesses) could only have his wound healed, and his realm restored, by the arrival in his kingdom of a simple, naive boy. The boy's name is Parsifal, which, depending on which of the numerous sources you refer to, means either **'guileless fool'** or **'pure fool'**. To the alienated, only a naive **'fool'** would dare approach and grapple with the confronting truths about our divisive condition. The Jungian analyst Robert A. Johnson offered this rendition of the legend, saying firstly that **'Alienation is the current term for it** [the wounded state of humans today]. **We are an alienated people, an existentially lonely people; we have the Fisher King wound'** (*He, Understanding Masculine Psychology*, 1974, p.12 of 97). He then described how **'The court fool had prophesied that the Fisher King would be healed when a wholly innocent fool arrives in the court. In an isolated country a boy lives with his widowed mother** [since the male ego can be especially oppressive of the souls of children, saying that the mother is widowed provides recognition that the child wasn't oppressed by an especially egocentric father]...**His mother had taken him to this faraway country and raised him in primitive** [isolated, innocent, unconditionally-loving, alienation-free, sound, soulful] **circumstances. He wears homespun clothes, has no schooling, asks no questions. He is a simple, naive youth'** (p.90). Johnson went on to recount that in the myth it is this boy, Parsifal, who, when he becomes an adult, is able to heal the Fisher King's wound of alienation, so that **'the land and all its people can live in peace and joy'** (p.94).

[1264] In contemporary mythology, the 1984 film *The NeverEnding Story*, which was based on the novel of the same name by Michael Ende, describes how a world called 'Fantasia' is under threat from a force called **'The Nothing'**—a void of darkness that engulfs everything in its path, which is, of course, a perfect description for the current alienated condition of humanity. The people of Fantasia end up enlisting a **'boy'** to help put an end to **'The Nothing'**. During the great saga that ensues, the boy's companion, a warrior named Atreyu, must pass through a **'Magic Mirror Gate'** in which he **'has to face his true self'**. While **'everyone thinks'** **'that** [self-confrontation] **won't be too hard'**, it is pointed out that, to the contrary, **'kind people find out that they are cruel. Brave men discover that they are really cowards! Confronted by their true selves, most men run away screaming!'** In the end, when **'The Nothing'** has consumed all but one grain of Fantasia's sand (indicating the human race has entered the end play stage of terminal alienation), it is the boy's wishes and imagination that is able to restore Fantasia to its former glory; in the final hour, our fantastic world is saved. So in this tale the crux problem of the human condition and the exposure that comes with the truth about ourselves, as well as the innocence required to resolve these issues, are fully identified.

[1265] In his astonishingly prophetic 1940 essay titled *The Almond Trees*, the literature Nobel Laureate Albert Camus also recognized this truth that the task of finding the understanding of the human condition that would **'heal'** humanity's **'Fisher King wound'** of **'alienation'**, restore **'Fantasia'** from **'The Nothing'** to its former glory, was going to require an **'innocent'** person, symbolized by a **'boy'** or **'youth'** from **'an isolated' 'faraway country'** that had not been overly exposed to the horrifically upset, **'alienated'**, **'existentially lonely'** almost universal state of humans now. Camus wrote that, despite all the corrupted upset in the world, there are still **'those shining** [innocent] **lands where so much strength is still untouched'** that can produce the innocent and sound **'whiteness and its sap' 'strength of character'** needed to **'stand up to'** all **'The Nothing'**, alienated, fraudulent evasion and denial and last **'just long enough to prepare the fruit'**—find the world-saving liberating understanding of the human condition. The following is an extract from what Camus wrote (underlinings are my emphasis): **'All we then need to know is what we want. And what indeed we want is never again to bow down before** the <u>sword</u>, never more to declare <u>force</u> to be in the right when it is not <u>serving the</u> <u>mind. This, it is true, is an endless task. But we are here to pursue it.</u> I do not have enough faith in reason to subscribe to a belief in progress, or to any philosophy of History. <u>But I do at least believe that men have never ceased to grow in the</u> <u>knowledge of their destiny. We have not overcome our condition, and yet we</u> <u>know it better. We know that we live in contradiction,</u> but <u>that we must refuse</u> <u>this contradiction</u> and do what is needed to reduce it. Our task as men is to <u>find</u> <u>those few first principles</u> that will calm the infinite anguish of free souls. <u>We must</u> <u>stitch up what has been torn apart</u>, render justice imaginable in the world which is so obviously unjust, make happiness meaningful for nations poisoned by the misery of this century. Naturally, <u>it is a superhuman task. But tasks are called</u> <u>superhuman when men take a long time to complete them, that is all</u>.

<sup>1266</sup> Let us then know our aims, standing steadfast on the mind, even if force dons the mask of ideas or of comfort to lure us from our task. The first thing is not to despair. Let us not listen too much to those who proclaim that the world is ending. Civilizations do not die so easily, and even if this world were to collapse, it will not have been the first. It is indeed true that we live in tragic times. But too many people confuse tragedy with despair. "Tragedy", Lawrence said, "ought to be a great kick at misery." This is a healthy and immediately applicable idea. There are many things today deserving of that kick.

<sup>1267</sup> When I lived in Algiers, I would wait patiently all winter because I knew that in the course of one night, one cold, <u>pure</u> February night, the almond trees of the Vallée des Consuls would be covered with <u>white flowers</u>. I was then filled with delight as <u>I saw this fragile snow stand up to all the rain and resist the wind</u> <u>from the sea. Yet every year it lasted, just long enough to prepare the fruit</u>.

<sup>1268</sup> This is not a symbol. We shall not win our happiness with symbols. We shall need something more weighty. All I mean is that sometimes, when life weighs too heavily in this Europe still overflowing with its misery, <u>I turn towards those</u> <u>shining lands where so much strength is still untouched. I know them too well</u> <u>not to realize that they are the chosen lands where courage and contemplation</u> <u>can live in harmony</u>. The contemplation of their example then teaches me that if we would save the mind we must pass over its power to groan and exalt its strength and wonder. <u>This world is poisoned by its misery, and seems to wallow</u> <u>in it. It has utterly surrendered to that evil which Nietzsche called the spirit of</u> <u>heaviness</u>. Let us not contribute to it. It is vain to weep over the mind, it is enough to labour for it.

<sup>1269</sup> But where are the conquering virtues of the mind? This same Nietzsche listed them as the mortal enemies of the spirit of heaviness. For him they are

the strength of character, taste, the "world", classical happiness, severe pride, the cold frugality of the wise. These virtues, more than ever, are necessary today, and each can choose the one that suits him best. **Before the vastness of the undertaking, let no one in any case forget strength of character. I do not mean the one accompanied on electoral platforms by frowns and threats. But the one that, through the virtue of its whiteness and its sap, stands up to all the winds from the sea. It is that which, in the winter for the world, will prepare the fruit'** (*Summer*, 1954, pp.33-35 of 87).

[1270] To examine what Camus has so beautifully articulated, he began by stating that the fundamental priority and responsibility of humanity is to solve our human condition, liberate the human mind from its underlying upset and, by so doing, replace the need for **'force'** to control our upset, troubled natures with the ability to explain, understand and thus pacify that upset. He acknowledged the human condition—that **'we live in contradiction'**—and that we have to live in denial of this condition—that **'we must refuse this contradiction'**. He also acknowledged the reality of the current extremely upset, alienated, soul-disconnected, depressed human state—of **'nations poisoned by the misery of this century...[a world] utterly surrendered to that evil which Nietzsche called the spirit of heaviness'**—and went on to emphasize the need for the clarifying, first principle reconciling biological explanation to **'overcome our condition'**, saying we need **'to find those first few principles'** that will **'stitch up what has been torn apart'**.

[1271] Again, of special significance here is Camus' acknowledgment that these answers were not going to come from the ivory towers of intellectualdom (as Professor Prosen put it in his Introduction), but from outlying realms where there is still sufficient innocence, **'strength still untouched'**, **'whiteness and its sap'**, to overcome all the evasion, denial and dishonesty, to **'stand up to all the winds from the sea'**, and find the reconciling understanding of the human condition, **'prepare the fruit'**. And, about where the answers about ourselves would emerge, Camus wrote, **'I turn towards those shining lands where so much strength is still untouched. I know them too well not to realize that they are the chosen lands where courage** [to defy all the dishonest denial in the world] **and contemplation** [denial-free, honest thought] **can live in harmony.'**

[1272] Australia has been—at least up until very recent times when communication technology especially, but also immigration and multiculturalism, have ended our isolation and spread **'alienation'** everywhere

here as well—just such a **'shining land…where so much strength is still untouched'**. As the renowned English author D.H. Lawrence said in his 1923 novel *Kangaroo*, **'You feel free in Australia…There is a great relief in the atmosphere, a relief from tension, from pressure. An absence of control or will or form. The sky is open above you, and the air is open around you'** (p.15 of 428). And, in fact, **'courage and contemplation'** have been able to **'live in harmony'** here **'just long enough to prepare the fruit'**, to find the understanding of the human condition that liberates the human race! And, moreover, the anticipation of this fantastic breakthrough occurring here has been very strong in Australian mythology.

[1273] While Australia has an ancient mythology that is grounded in the Dreamtime stories of the Aborigines, we also have a powerful contemporary mythology—at the heart of which is 'Banjo' Paterson's 1895 poem *The Man From Snowy River*; in fact, in recognition of the poem's significance to our nation, Australia's $10 note features Paterson's image and, in microprint, all the many words to *The Man From Snowy River*. Ostensibly, the poem is about a great and potentially dangerous ride undertaken by mountain horsemen to recapture an escaped thoroughbred that had joined the brumbies (wild horses) in the mountain ranges, but what the poem is *really* recognizing is that in Australia's isolation and relative innocence there would emerge sufficient soundness to defeat denial and put together the liberating explanation of the human condition. In the poem, an inspired stockman called Clancy persuades the station owner, Harrison, to let a **'stripling' 'lad'**—a boy—on his **'hardy mountain**

**pony'** join their expedition to retrieve the escaped thoroughbred; Clancy argues, **'I warrant he'll be with us when he's wanted at the end.'** Again, a boy is the embodiment of the innocence that is needed **'at the end'** of humanity's heroic journey of accumulating sufficient scientific understandings of the mechanisms of our world to assemble the denial-free explanation of the human condition. The poem describes how the boy rides beyond where the rest of the horsemen (the resigned, alienated adults) dare to go, and follows the brumbies down the **'terrible descent'** of a steep mountain where (if you weren't sound) **'any slip was death'** (to confront the unconfrontable issue of the human condition) and recaptures the thoroughbred from the impenetrable mountains (retrieves the escaped truth from the depths of denial). As the poem says, the boy **'ran them** [the brumbies]**...till their sides were white with foam / He followed like a bloodhound on their track / Till they halted, cowed and beaten—then he turned their heads for home'** (he fought all the alienation and its denial that has been enslaving this world to a standstill until it finally gave up the truth).

[1274] Paterson was even more explicit in his anticipation of the liberating understanding of the human condition emerging from the Australian bush (innocent countryside) in another of his works from 1889, his aptly titled poem *Song of the Future*. Using the analogy of the pioneers who finally forged the path through our eastern coastal mountain range (appropriately enough called 'The Great Dividing Range') that had barred the way to Australia's interior during the early days of European settlement, Paterson envisaged **'the future'** heroic, Australian-led expedition humanity would take from the alienated bondage of the human condition to the fertile, sun/understanding-drenched freedom of a human-condition-resolved new world. These are the poem's key verses: **'Tis strange that in a land** [Australia] **so strong, so strong and bold in mighty youth** [innocence]**, we have** [in 1889] **no poet's voice of truth to sing for us a wondrous song** [explain the human condition]**...**[However,] **we yet may find achievements grand within the** <u>**bushman's quiet life**</u>**. Lift ye your faces to the sky, ye far blue mountains of the west...Tis hard to feel that years went by before the pioneers broke through your rocky heights and walls of stone, and made your secrets all their own** [broke through the wall of denial blocking access to the truth about the human condition]**. For years the fertile western plains were hid behind your sullen** [alienated] **walls, your cliffs and crags...Between the mountains and the sea, like Israelites with staff in hand, the people waited restlessly: They looked towards the mountains old and saw the sunsets come and go with gorgeous golden afterglow** that made the west a fairyland, and marvelled what that west might be of which such wondrous tales were told…At length the hardy pioneers by rock and crag **found out the way, and woke with voices of today, a silence kept for years and years** [brought an end to the silence of the resigned world of denial]…**The way is won! The way is won! And straightway from the barren coast there came a westward-marching host, that aye and ever onward prest with eager faces to the west** along **the pathway of the sun**…Could braver histories unfold than this bush story, yet untold—the story of their westward march [liberation from the human condition]…And it may be that we who live in this new land apart, beyond the hard old world grown fierce and fond and bound by precedent and bond [bound up in sophisticated, intellectual denial], **may read the riddle** [of the human condition] **right** [synthesise the liberating truth from science's hard won insights into our world] **and give new hope to those who dimly see** [those who are embedded in blind denial/alienation], **that all things may be yet for good and teach the world at length to be one vast united brotherhood.'** 'Banjo' Paterson was certainly an extraordinarily prophetic and gifted writer! (I might mention that the gifted editor of my writing, World Transformation Movement [WTM] Founding Member Fiona Cullen-Ward, is related to 'Banjo'.)

[1275] Henry Lawson, who was introduced in par. 110 when his poem about Resignation, *The Voice from Over Yonder*, was included, is another of Australia's greatest poets, and he too anticipated that the answer to the human condition would be found here. Making the same insights, employing the same comparisons and using the same imagery as Paterson's poems, Lawson wrote in his 1892 poem, the also aptly titled *When the Bush Begins to Speak*, that **'They know us not in England yet, their pens are overbold. We're seen in fancy pictures that are fifty years too old. They think we are a careless race—a childish race, and weak. They'll know us yet in England, when the bush begins to speak** [when innocence makes its contribution]…**"The leaders that will be", the men of southern destiny, are not all found in cities that are builded by the sea. They learn to love Australia by many a western creek** [while Australia as a whole has been relatively sheltered and thus innocent, it is from the Australian inland countryside or 'bush', rather than from the cities, that exceptional innocence will appear]. **They'll know them yet in England, when the bush begins to speak…All ready for the struggle, and waiting for the change, the army of our future lies encamped beyond the** [coastal] **range. Australia, for her patriots, will not have far to seek, they'll know her yet in England when the bush begins to speak…We'll find the peace and comfort that our fathers could**

**not find**, or some shall strike the good old blow that leaves a mark behind. **We'll find the Truth and Liberty** [the truth about the human condition that brings liberating understanding to humanity] **our fathers came to seek, or let them know in England when the bush** [innocence] **begins to speak.'**

[1276] And in yet another extraordinarily prescient work (which again has the same insights, comparisons and imagery), the Australian poet A.D. Hope also wrote of the role of the relatively innocent continent of Australia in delivering the liberating understanding of the human condition. In his 1931 poem, actually titled *Australia*, Hope wrote of **'A nation of trees, drab green and desolate grey…Without songs, architecture, history…And her five cities, like five teeming sores, each drains her, a vast parasite robber-state, where second-hand Europeans pullulate timidly on the edge of alien shores. Yet there are some like me turn gladly home from the lush jungle of** [alienated, dishonest, intellectual] **modern thought, to find the Arabian desert of the human mind, hoping, if still from the deserts the prophets** [innocent, unresigned, denial-free, truthful, profound, effective thinkers] **come. Such savage and scarlet as no green hills dare, springs in that waste, some spirit which escapes the learned doubt, the** [dishonest, intellectual as opposed to the honest, instinctual] **chatter of cultured apes which is called civilization over there** [in England].' In his award-winning 1979 book, *A Woman of the Future*, the Australian author David Ireland expressed the same awareness as Hope as to where these all-important answers would originate *and* he used the same metaphor of the desert, recognizing that Australians hide along the coast, distanced from the truth that exists in the center of their being/country: **'The future is somehow… somewhere in the despised and neglected desert, the belly of the country, not the coastal rind. The secret is in the emptiness. The message is the thing we have feared, the thing we have avoided, that we have looked at and skirted. The secret will transform us, and give us the heart to transform emptiness. If we go there, if we go there and listen, we will hear the voice of the eternal. The eternal says that we are at the beginning of** [real, non-alienated] **time'** (p.349). The prophet Isaiah was another who used the metaphor of the desert for that neglected part of ourselves from which the healing answers would come, saying, **'A voice of one calling: "In the desert prepare the way of the Lord** [truth]**; make straight in the wilderness a highway for our God. Every valley shall be raised up, every mountain and hill made low; the rough ground shall become level, the rugged places a plain. And the glory of the Lord will be revealed, and all mankind together will see it"'** (Isa. 40:3–5).

[1277] In more recent Australian mythology, the Australian-made animated film *Babe: Pig in the City* (1998) tells the story of a little pig who, in order to save his farm, which symbolizes the true world, has to take his innocence into the very heart of alienation, the city, and defy and defeat the alienation to release the captured soul of humanity, represented by the innocent animals being held captive there. To symbolize that the hazardous undertaking will require toughness as well as innocence, the pig wears a spiked collar into battle, the collar being donated by the bulldog who recognizes he hasn't got the sufficient innocence/soundness to do the job himself. Interestingly, the Australian production company that made the *Babe* movies, Kennedy Miller, also produced the late 1970s and early 1980s *Mad Max* trilogy, which features the same theme of unrecognized (because the world practices evading and denying truthful unevasiveness) innocence taking on the alienated world and leading humanity out of bondage from the darkness of that blind, alienated world. To quote a review of the film *Mad Max* (or *The Road Warrior* as the film was titled in the USA), Max is **'a hero in the classical tradition — a figure whose origins lie in the ancient myths; his role, in common with classical heroes, is of a man from nowhere destined to lead society into the next generation'** (*Sunday Telegraph*, 21 Mar. 1982). Yes, these liberating answers were not going to come from the ivory towers of the truth-avoiding mechanistic scientific establishment, but from, again as Professor Prosen put it, the deepest of deep left field, from way out on the periphery of the great battle humanity has been waging against ignorance.

[1278] Sir Laurens van der Post, that exceptional denial-free-thinking prophet of our time, clearly recognized the truth of innocence lying in wait for science to do its job so that it could lead humanity home when he wrote that **'Whatever happens, I shall be there in the end, for I, child that I am, am mother of your future self'** (*Jung and the Story of Our Time*, 1976, p.167 of 275). Yes, it is guidance from the long-repressed innocent, soulful, conscience-infused clarity that humans had before the upset, denial-based, alienated human condition developed that was needed to synthesise the liberating truth from science's hard won insights into the workings of our world. As was explained in chapter 2:4, evasive, whole-view-avoiding, reductionist, mechanistic science had to complete the difficult task of finding all the 'pieces of the jigsaw' of the explanation of the human condition before soul-guided, whole-view-confronting, denial-free thinking could come along **'in the end'** (or, as Paterson wrote, **'at the end'**) and piece the 'jigsaw' together. As emphasized in pars 296 and 1147, science is the real 'messiah' or liberator of humanity—it made the explanation of the human condition that is presented in chapter 3 possible. Historically, people have talked about a 'second coming' of innocence, but in the vast spectrum of alienation that inevitably developed in humanity's heroic battle to defeat the ignorance of our instinctive self or soul and establish that humans are good and not bad, there have always been an extremely rare few individuals who have been sufficiently nurtured with alienation-free, unconditional love in their infancy and childhood to be sound and secure enough in self to confront and think truthfully about the human condition, but until the discipline of science found understanding of the difference in the way genes and nerves process information, such upset-free, innocent individuals could not make their final contribution and assemble the explanation of the human condition from those understandings. Science was tasked with the tedious, hard work of finding the clues that make explanation of the human condition possible—and it has to be remembered that science is only the peak expression of the courageous effort that *every* human has exerted to defeat ignorance. In reality, it is 'on the shoulders' of 2 million years of human struggle against our soul's oppression that understanding of our species' fundamental goodness has finally been found—but our soul *was* needed **'in the end'** to assemble the actual explanation of the human condition from those clues. It is like a game of gridiron football where the team as a whole, with one exception, does all the hard work, gaining yardage down the field. Finally, when the side gets within kicking distance of the goal posts, a specialist kicker, who until then has played no part, is brought onto the field. While he—in his unsoiled attire—kicks the winning goal, the win clearly belongs to the team of exhausted players who did all the ground work. It is science, backed up by humanity as a whole, that is the real liberator of the human race from ignorance as to our species' fundamental goodness.

[1279]The point here is that it is true that exceptional innocence has had a crucial concluding role to play in humanity's journey but, *most importantly*, such an innocent is no more special or worthy or wonderful than any other human. If anything, they are less worthy because they haven't been involved in humanity's heroic battle as much as everyone else; but, in any case, viewing those who have been more involved in the battle as more worthy is not accurate either, because no one could choose where they were going to be born/positioned on that battlefield. Yes, the fundamental insight that understanding of the human condition gives us is that in the epic battle to defeat ignorance that the human race has been waging for some 2 million years, there was going to be a vast spectrum of exhaustion/alienation but in this great army of warriors ALL HUMANS ARE <u>EQUALLY</u> GOOD, SPECIAL AND WONDERFUL. Since upset resulted from fighting humanity's battle against ignorance, upset, exhausted alienation is a heroic state, not a bad, inferior or lesser state, and, by the same logic, relatively alienation-free innocence that resulted from having escaped encounter with all the upset on that battlefield is not in any way better, or in some way more special or more deserving or more wonderful than the overly upset, embattled, alienated state. The differentiation where some people are viewed as bad and others as good has gone forever from our discourse; it doesn't exist anymore; it has no basis of truth or fact. Understanding of the human condition completely and permanently eliminates the concept of 'good and evil'. Again, while instinct and intellect had particular concluding roles to play (the soul/instinct had to synthesise the explanation of the human condition from the insights into the workings of our world that science/knowledge/intellect had to first find), the truth is that it is the human race as a whole, *all the efforts of every human who has ever lived*, that has liberated our species from the horror of the human condition. Understanding the human condition allows us to know that we are all absolutely wonderful, utterly sublime, completely lovable, and that having all fully contributed to humanity's successful battle against ignorance we are *all* fabulous heroes of the story of life on Earth. So now we, THE HUMAN RACE AS A WHOLE, should give ourselves the biggest party ever. Everyone everywhere is going to be hoppin' and boppin', rompin' and stompin', hollerin' and howlin', movin' and groovin', rollickin' and rollin', hootin' and tootin', jumping and jiving, jolting and somersaulting, skipping and skating, shaking and shimmering, hugging and laughing, embracing and gyrating, twisting and shouting, dancing and singing, slipping and sliding, jamming and slamming, ripping and roaring, whirling and twirling and reelin' and rockin'. Yes, **'allow freedom to ring...from every village and every hamlet, from every state and every city'** because **'all of God's children, black men and white men, Jews and Gentiles, Protestants and Catholics'**, the more innocent and the more upset, the short and the tall, the big and the small, those

who are left-handed and those who are right-handed, the butcher, the
baker and the candlestick-maker, EVERYONE, can **'join hands and sing'**,
**'Free at last! Free at last! Thank God Almighty, we are free at last** [from unjust
condemnation]**!'** (Martin Luther King Jr, 'I Have A Dream' speech, 28 Aug. 1963).

Drawing by Jeremy Griffith © 2010-2013 Fedmex Pty Ltd

[1280] In about 1969 (when I was around 23 years old) I wrote a poem that
clearly anticipated the excitement of this time of humanity's liberation.
The poem indicates just how strong my vision has always been of one day
being able to find the liberating understanding of the human condition
(and note how similar these words are to those of Joel, Hosea and Isaiah
in pars 1257-1260): **'This is a story you see, just a story—but for you / Um—I
remember a long time ago in the distant future a timeless day / a sunlit cloudless
day when all things were fine / when we all slow-danced our way to breakfast
in the sun // You see the day awoke with music / Can you imagine one thousand
horses slow galloping towards you across a vast plain / and we loved that day
so much / We all danced like Isadora Duncan through the morning light // We
skipped and twirled and spun about / Fairies were there like dragonflies over a
pool / Little girls with wings they hovered and flew about / their small voices you
could hear / You see it was that kind of morning // When the afternoon arrived
it was big and bold and beautiful / In worn out jeans and bouncing breasts we
began / to fight—our way—into another day / into something new—to jive our
way into the night / from sunshine into a thunderstorm // We all took our place,
rank upon rank we came / as an army with Hendrix out in front / and the music**

**busted the horizon into shreds / By God we broke the world apart / The pieces were of different colours and there were so many people / We danced in coloured dust, we left in sweat no room at all / We had a ball in gowns of grey and red / There were things that happened that nobody knew / Bigger and better, I had written on my sweater / Where there was sky there was music, huge clouds of it / and there were storms of gold with coloured lights / It was so good we cried tears into our eyes / In a tug of war of love we had no strength left at all / Dear God we cried but he only sighed and / whispered strength through leaves of laughter // On and on we came in bold ranks of silvered gold / to lead a world that didn't know to somewhere it didn't care / It couldn't last, it had to end and yet it had an endless end / We were so happy in balloons of coloured bubbles that wouldn't bust / and we couldn't, couldn't quench our lust / There we were all together for ever and ever / and tomorrow had better beware because / when we've wept and slept we will be there to shake its bloody neck.'**

[1281] Regarding this vision, I should explain how, when we were young children and still thinking completely honestly and thus effectively, we knew the truth about our destiny; we didn't know exactly what would happen in our adult lives, but we did know the general course our future would take as a result of how hurt our original instinctive self or soul was in our infancy and early childhood. As children we have a clear awareness of the imperfections or otherwise of our circumstances and can think truthfully about the consequences of those circumstances. If they were not ideal, as was the case in nearly everyone's lives, then we rapidly began to stop thinking about those unhappy consequences, but the point is there was a time in everyone's life during early childhood when we knew the basic path our life was going to take, and in that brief time of total honesty and thus insightfulness we all knew of the immense problem facing humanity of the dilemma of the human condition, the issue of the extreme imperfection of human life, and knowing that crux problem we made an assessment of the extent and nature of our contribution to its solution. For the very rare exceptionally fortunate, those exceptionally loved and nurtured in their infancy and early childhood, *they* knew they could make an exceptional contribution and the precise nature of it. When we spoke of people being driven by a 'vision' this is essentially what we were recognizing—an awareness in someone of them having the opportunity to make a critical contribution to the underlying battle humanity has been engaged in of bringing understanding to the human condition. And because people with such guiding visions were carrying such a strong awareness of what they could and must do from such a young age, it was as if they were owned or possessed by their vision and were thus very difficult to deter from their path. This is why, despite extreme resistance to the point of almost unbearable persecution, I have so tenaciously held on to my vision of being able to deliver understanding of the human condition. Yes, everyone is born a truthful, denial-free thinking and thus insightful prophet—as mentioned in par. 680, R.D. Laing recognized this when he said, **'Each child is a new beginning, a potential prophet'**—but few could afford to stay thinking so truthfully and insightfully; most had to forget what they could see because it was too confronting and depressing and just get on with their life as best they could. So while the main resignation to the imperfections of life typically occurred when people were about 15 years of age, there were many mini-resignations prior to that major one as people variously adjusted and eventually acquiesced to life under the duress of the human condition.

[1282] Sir Laurens van der Post recognized the, in truth, least heroic and special, but critical, concluding role soul/conscience-led innocence had to play when he wrote these beautiful words: **'One of the most moving aspects of life is how long the deepest memories stay with us. It is as if individual memory is enclosed in a greater which even in the night of our forgetfulness stands like an angel with folded wings ready, at the moment of acknowledged need, to guide us back to the lost spoor of our meanings'** (*The Lost World of the Kalahari*, 1958, p.62 of 253). Sir Laurens' writings about the relatively innocent Bushmen of the Kalahari desert in Africa were so incredibly precious to me as a young man trying to stand up to all the dishonesty in the world because what he wrote about the Bushmen confirmed for me that there *is* another alienation-free, true world—and Sir Laurens' great vision was that he knew exactly that this is what all his beautiful descriptions of the relatively innocent life of the Bushmen could achieve, writing that **'I had a private hope of the utmost importance to me. The Bushman's physical shape combined those of a child and a man: I surmised that examination of his inner life might reveal a pattern** [of child-like innocence, soundness and truth] **which reconciled the spiritual opposites in the human being and made him whole...it might start the first movement towards a reconciliation'** (*The Heart of the Hunter*, 1961, p.135 of 233). Yes, what Sir Laurens wrote did **'start the first movement towards a reconciliation'**; it gave me the support I needed to synthesise the explanation of the human condition. In short, his writings prepared the way for this book and our species' freedom.

The author (left), and fellow World Transformation Movement Patron
Tim Macartney-Snape AM OAM, with Sir Laurens van der Post
in London in 1993, a few years before Sir Laurens' death in 1996

[1283]Interestingly, in a 2006 interview with the Australian television presenter Andrew Denton, Bono, the prophetic lead singer of the rock band U2, said, **'You do get the feeling in Australia that there's…something going on down here, a new society being dreamt up…**[that in Australia there is] **the opportunity to lead the world'** *(Enough Rope*, ABC-TV, episode 97, 13 Mar. 2006). The Australian academic David Tacey also anticipated Australia's pivotal role in lifting the siege humanity has been under of the dilemma of the human condition, writing in his 1995 book, *Edge of the Sacred*, that **'Australia is uniquely placed not only to demonstrate this world-wide experience but also to act as a guiding example to the rest of the world. Although traditionally at the edge of the world, Australia may well become the centre of attention as our transformational changes are realised in the future. Because the descent of spirit has been accelerated here by so many regional factors, and because nature here is so deep, archaic, and primordial, what will arise from this archetypal fusion may well be awesome and spectacular. In this regard, I have recently been encouraged by Max Charlesworth's essay "Terra Australis and The Holy Spirit". In a surprisingly direct—and unguarded?—moment, Charlesworth says: "I have a feeling in my bones that there is a possibility of a creative religious explosion occurring early in the next millennium with the ancient land of Australia at the centre of it, and that the Holy Spirit may come home at last to *Terra Australis*". I am pleased that this has already been said, because if Charlesworth had not said it, I would have**

**been forced to find within myself exactly the same prophetic utterance'** (p.204 of 224). (Again, while the arrival of understanding of the human condition brings about an incredible spiritual awakening in humans—as if rising from the dead—it doesn't bring a **'religious explosion'** as this explanation represents the end of faith and dogma and the beginning of knowing.) This **'awesome and spectacular' 'transformation'** was also anticipated by Mark Seymour of the Australian rock band Hunters and Collectors when, in the lyrics to their 1993 hit song *Holy Grail*, he wrote (and again, how similar are these words to those of Joel, Hosea and Isaiah in pars 1257-1260!): **'Woke up this morning from the strangest dream / I was in the biggest army the world had ever seen / We were marching as one on the road to the Holy Grail // Started out seeking fortune and glory / It's a short song but it's a hell of a story / When you spend your lifetime trying to get your hands on the Holy Grail // Well have you heard about the Great Crusade? / We ran into millions but nobody got paid / Yeah we razed four corners of the globe for the Holy Grail // All the locals scattered, they were hiding in the snow / We were so far from home, so how were we to know / There'd be nothing left to plunder when we stumbled on the Holy Grail? // We were so full of beans but we were dying like flies / And those big black birds, they were circling in the sky / And you know what they say, yeah nobody deserves to die // Oh but I've been searching for an easy way / To escape the cold light of day / I've been high and I've been low / But I've got nowhere else to go / There's nowhere else to go! // I followed orders, God knows where I've been / but I woke up alone, all my wounds were clean** [by the reconciling understanding of the human condition] **/ I'm still here, I'm still a fool for the Holy Grail.'**

[1284] The most important point overall, of course, is that the truth about humans now exists, the spell has been broken, the proverbial 'elephant in our living rooms' has finally been acknowledged and exposed. Denial no longer works—its hold has been released. The great citadel of lies has been stormed. The steel casing of denial enveloping Earth has been wrenched apart. Yes, the truth *is* out, and thank goodness it is because the suffering simply cannot go on. This is no exaggeration: humans currently live in a place of such horrendous pain and suffering (which can be measured by the degree we block it out; by our astronomical levels of alienation) that it is as though we are living in a smoldering, toxic, sulfurous cauldron. BUT, there is now a path out of there to a world of soothing, healing sunshine, and amazing freedom, tranquility and beauty. All we have to do is cross the **'SUNSHINE HIGHWAY'** bridge: adopt the Transformed Lifeforce Way of Living. There is nothing stopping every human now from escaping the horror of the human condition, except their now completely obsoleted habituation to living in what is literally HELL ON EARTH. So while our freedom comes as such a 'future shock' that we initially won't quite know how to take it up, the truth is there is absolutely nothing that justifies *not* taking it up—there really is *no* impediment to taking it up, none at all. The gateway to freedom *is* now WIDE open to everyone.

[1285] Sir Laurens—who, again, wrote *so* beautifully—gave us this wonderful description of how, in **'the midnight hour of the crashing darkness'** of the time we are living in, we had to build **'a bridge'** to **'cross from one side to the other'** of **'that split between night and day in ourselves'**. He wrote: **'We must shut our eyes and turn them inwards, we must look far down into that split between night and day in ourselves until our head reels with the depth of it, and then we must ask: 'How can I bridge this self? How cross from one side to the other?' If we then allow that question to become the desire for its own answer, and that desire to become a bridge across the chasm, then, and only then, from high above on this far peak of our conscious self, on this summit so far above the snow-line of time, in this cold, sharp, selected moment, clearly and distinctly we shall see a cross. A gulf bridged makes a cross; a split defeated is a cross. A longing for wholeness presupposes a cross, at the foundations of our being, in the heart of our quivering, throbbing, tender, lovely, lovelorn flesh and blood, and we carry it with us wherever we journey on, on unto all the dimensions of space, time, unfulfilled love, and Being-to-be. That is sign enough. After that the drum can cease from drumming, the beating and troubled heart have rest. In the midnight hour of the crashing darkness, on the other side of the night behind the cross of stars, noon is being born'** (*Venture to the Interior*, 1952, p.229 of 239). It could be that Sir Laurens was intuitively envisaging Australia as the place where the human condition would be solved when he talked of **'noon'** being **'born'** beneath the Southern Cross, which is the constellation that hangs over our night skies and forms the centerpiece of our Australian flag. I know Sir Laurens was very impressed with the soundness of the Australians he was held alongside as a prisoner of war, and that he knew of the exceptional isolation of Australia from all the upset in the world, **'the bushman's quiet life'** that Paterson referred to, and that he also appreciated Australia's natural soulfulness—that Australia is like Africa where our species grew up—writing that **'When I first went to Australia…my senses told me at once that here, beyond rational explanation, was a land physically akin to Africa'** (*The Dark Eye in Africa*, 1955, p.35 of 159). His emphasis on **'a cross'** being **'sign enough'** also invokes the involvement of Christ in 'saving the day', and there is truth in that, in the sense that Christ was exceptionally sound and, as has been explained, such innocent soundness is necessarily what was required to be able to **'turn…inwards'** and **'look far down into that split between night and day in ourselves until our head reels with the depth of it'**, namely look into the human condition. As Sir Laurens has also written, **'He who tries to go down into the labyrinthine pit of himself, to travel the swirling, misty netherlands below sea-level through which the harsh road to heaven and wholeness runs, is doomed to fail and never see the light where night joins day unless he goes out of love in search of love'** (*The Face Beside the Fire*, 1953, p.290 of 311).

[1286] Yes, as Billy Joel wrote and sang in his 1993 song *River of Dreams* about the overwhelming difficulty virtually all humans have of crossing the metaphorical **'river so deep'** to find the human-race-liberating-and-rehabilitating understanding of the human condition, **'In the middle of the night I go walking in my sleep, from the mountains of faith…through the valley of fear…through the jungle of doubt…through the desert of truth…to the river so deep…that is runnin' to the promised land…but the river is wide and it's too hard to cross…I try to cross to the opposite side so I can finally find what I've been looking for…I've been searching for something taken out of my soul.'** While *only* understanding of our devastated condition could rehabilitate the human race—as Tracy Chapman wrote and sang in her 1994 song *New Beginning*, **'The world is broken into fragments and pieces that once were joined together in a unified whole…The whole world's broke and it ain't worth fixing. It's time to start all over, make a new beginning…Change our lives and paths, create a new world…There's too much fighting, too little understanding…We need to…make a new** [truthful] **language, with these we'll define** [explain] **the world and start all over'**—to **'start all over'** and **'create a new world'** required uncorrupted, alienation-free, sound, soul-full **'love'** because only it could **'cross' 'the river so deep'; 'travel the swirling, misty netherlands below sea-level through which the harsh road to heaven and wholeness runs'.**

[1287] Indeed, for U2's wonderfully named and aforementioned song, *When Love Comes To Town*, Bono wrote these exciting lyrics that anticipated humanity's liberation from the human condition: **'I was a sailor, I was lost at sea / I was under the waves…But I did what I did before love** [the ultimate expression of which is truth] **came to town…I've seen love conquer the great divide** [between 'good and evil', yin and yang, idealism and realism, instinct and intellect, conscience and conscious, altruism and egotism, spiritualism and materialism, young and old, women and men, blacks and whites, religion and science, the left-wing and the right-wing, socialism and capitalism, country and city—and, most especially, between the **'split between night and day in ourselves']…When love comes to town I'm gonna jump that train'** (1987). Yes, yes, yes—when the reconciling understanding of the human condition arrives we should *all* **'jump that train'**, get on board, *choose transformation NOT terminal alienation for the human race*.

[1288] To include more anticipations of our species' liberation—from Bono and the other prophets of our time—consider these lyrics from U2's song *Love Rescue Me*, which Bono wrote: **'I've conquered my past, the future is here at last, I stand at the entrance to a new world I can see. The ruins to the right of me, will soon have lost sight of me'** (1988). Yes, **'The ruins to the right of'** us **'will soon have lost sight of'** us because we have **'conquered'** our **'past'**, found the reconciling understanding of our corrupted condition, which means **'the future is here at last'** and we can leave that old, now dealt with and thus obsoleted, upset existence behind us forever. And what did Bob Dylan say? **'The present now will later be past…For the times they are a-changin''** (*The Times They Are A-Changin'*, 1964). Yusuf Islam (the aforementioned singer/songwriter previously known as Cat Stevens) similarly foresaw that **'a change is coming from another side of time, breaking down the walls of silence, lifting shadows from your mind…Yesterday has past, now let's all start the living for the one that's going to last…the day is coming that will stay and remain when your children see the answers…when the clouds have all gone…and the beauty of all things is uncovered again…the day is coming…when the people of the world can all live in one room, when we shake off the ancient chains of our tomb'** (*Changes IV*, 1971), and that **'out on the edge of darkness there rides a peace train…[to] take me home again…everyone jump upon the peace train…come and join the living'** (*Peace Train*, 1971). And since it has such similar symbolism, I just have to include the words of Walter Earl Brown's 1968 song *If I Can Dream*, which was performed by Elvis Presley: **'There must be peace and understanding sometime, strong winds of promise that will blow away all the doubt and fear. If I can dream of a warmer sun where hope keeps shining on everyone…We're trapped in a world that's troubled with pain…Still I am sure that the answer's gonna come somehow, out there in the dark, there's a beckoning candle.'** So, yes, let's go, let's **'jump upon the peace train'**, let's get out of this dead, **'dark' 'tomb'** we have all been **'trapped in'** for far too long—join the Sunshine Army on the Sunshine Highway to the World in Sunshine; and, as Paterson, Lawson and Seymour foresaw, become part of the **'westward-marching host'**, **'the army of our future'**; in fact, **'the biggest army the world had ever seen'**!

[1289]I might mention that not only did I have the good fortune to be born in Australia, I also grew up in the post-war 1960s period of relative innocence and idealism. As was mentioned in par. 782, after a terrible bloodletting like the Second World War, which amounts to an immense valving off of upset, there always emerges a period of enormous relief and freshness, and it was all this innocence and its soundness that made the 1960s such a **'golden age'** (Bono, *God Part II*, 1988), for not only did our generation conquer outer space by sending man to the moon, we have now conquered *inner* space by finding the truth about the human condition. The best account I have come across of the halcyon days and extraordinary optimism of the 1960s—indeed that generation's intuition that it was going to achieve **'victory over the forces of Old and Evil'**—was given by the writer Hunter S. Thompson in his 1971 masterpiece, *Fear and Loathing in Las Vegas*. This is what he wrote (the italics and dots are his, with slashes indicating his paragraph breaks): **'the kind of peak that never comes again. San Francisco in the middle sixties was a very special time and place to be a part of. Maybe it *meant something*. Maybe not, in the long run . . . . but no explanation, no mix of words or music or memories can touch that sense of knowing that you were there and alive in that corner of time and the world. Whatever it meant. . . . . / History is hard to know, because of all the hired bullshit, but even without being sure of "history" it seems entirely reasonable to think that every now and then the energy of a whole generation comes to a head in a long fine flash, for reasons that nobody really understands at the time—and which never explain, in retrospect, what actually happened. / My central memory of that time seems to hang on one or five or maybe forty nights—or very early mornings—when I left the Fillmore half-crazy and, instead of going home, aimed the big 650 Lightning across the Bay Bridge at a hundred miles an hour wearing L. L. Bean shorts and a Butte sheepherder's jacket . . . . booming through the Treasure Island tunnel at the lights of Oakland and Berkeley and Richmond, not quite sure which turn-off to take when I got to the other end (always stalling at the toll-gate, too twisted to find neutral while I fumbled for change) . . . . but being absolutely certain that no matter which way I went I would come to a place where people were just as high and wild as I was: No doubt at all about that. . . . . / There was madness in any direction, at any hour. If not across the Bay, then up the Golden Gate or down 101 to Los Altos or La Honda. . . . . You could strike sparks anywhere. There was a fantastic universal sense that whatever we were doing was *right*, that we were winning. . . . . / And that, I think, was the handle—that sense of inevitable victory over the forces of Old and Evil. Not in any mean or military sense; we didn't**

**need that. Our energy would simply** *prevail*. **There was no point in fighting—on our side or theirs. We had all the momentum; we were riding the crest of a high and beautiful wave. . . . .'** (pp.66-68 of 204).

[1290] While Thompson's comment that **'no...music...can touch that sense of knowing that you were there and alive in that corner of time and the world'** is probably true for those who weren't there, the music of the mid-1960s *can* still connect everyone to the optimism of that time. Consider the music and words of the 1967 rock musical *Hair*, especially the lyrics of *Aquarius* (the underlinings are my emphasis): **'When the moon is in the Seventh House and Jupiter aligns with Mars, then peace will guide the planets and love will steer the stars. This is the dawning of the age of Aquarius, the age of Aquarius. Aquarius! Aquarius! Harmony and <u>understanding</u>, sympathy and trust <u>abounding</u>. <u>No more falsehoods</u> or derisions, golden living dreams of visions, mystic crystal <u>revelation</u> and <u>the mind's true liberation</u>. Aquarius!...As our hearts go beating through the night, <u>we dance unto the dawn of day, to be the bearers of the water, our light will lead the way</u>'** (lyrics by James Rado and Gerome Ragni). Of course, in terms of the music of those times signaling the coming breakthrough in understanding ourselves, it was around that time that 'rock and roll' music was created, for what is 'rock and roll' if not totally optimistic, all-out, rock-solid *'determination and resilience'* to achieve freedom from our species' historical state of unjust condemnation—determination to, as Bono sang, **'kick the darkness till it bleeds [the] daylight'** (*God Part II*, 1988) of the truth about us humans and end the damned condemnation of our species FOREVER! What did Bob Dylan famously say about Elvis Presley—**'Hearing him for the first time was like busting out of jail.'** John Lennon famously reiterated the sentiment, saying, **'Before Elvis there was nothing'**; there was not all-out determination and optimism, there was no 'rock and roll', there was just endless decades and epochs and ages of resigned, lonely music—which, sadly, we have now returned to, but in an even *more* lonely state, with today's terminally alienated, head-banging, autistic, soul-screaming-in-agony, psychotic music.

[1291] Those phenomenal singers who emerged in the late 1950s, Jerry Lee Lewis, Chuck Berry and Little Richard, were also locked onto the immensely excited driving beat that lay at the heart of rock and roll of anticipation of our species' liberation from the horror of the human condition—*especially* that belt-it-out, blast-out-of-here, boiling-with-excitement, completely-raging supernova from Ferriday, Louisiana, Jerry Lee Lewis who was rightly the first performer inducted into the Rock and Roll Hall of Fame. In the documentary *Mojo Working: The Making of Modern Music*, which contains a wonderful collection of footage and commentary about Jerry Lee Lewis, the writer Charles 'Dr Rock' White reported that **'John Lennon came into Jerry Lee Lewis' dressing room...and he walked over to Jerry Lee and...bent down and kissed Jerry Lee's feet...**[and then he] **walked out speechless'** (directed by Mark Neal, 1992). I wholeheartedly agree with Lennon's gesture; to me no one's music channeled the excitement of the anticipation of the liberation of the human race from the human condition as purely as Jerry Lee's did. If you listen to a live recording of Jerry Lee's April 5, 1964 performance at the Star Club in Hamburg, Germany, especially the tracks *Long Tall Sally* and *Hound Dog*, you will hear what **'is regarded by many music journalists as one of the wildest and greatest rock and roll concert albums ever'** and that Jerry Lee **'sounds possessed'** and was **'rocking harder than anybody had before or since...words can't describe the music'** (Wikipedia; see <www.wtmsources.com/125>). Jerry Lee's performances were just drenched in the excitement of breaking free from the dungeon of our species' tortured condition. In fact, my vision is of a hysteria of millions and millions and millions of excited people with Jerry Lee's piano being held aloft out in front and Jerry Lee standing on top of it, flicking his hair back and hitching his pants up, as he used to do—and filling the air is the musical build-up to humanity's great breakthrough to its freedom in *Prologue/Crunchy Granola Suite*, from Neil Diamond's 1972 *Hot August Night* album. But then, to actually take us through the portals of the new world that understanding of the human condition now makes possible, instead of Diamond's *Crunchy Granola Suite* vocals coming in, Jerry Lee would start singing 'Great Balls of Fire', Let's get out of here, LET'S GO!, to an immense roar of unbelievable relief and excitement from the flood of humanity bursting through that doorway to its freedom. The aforementioned WTM Founding Member, Tony Gowing, has actually written a song titled *LET'S GO* that he sings with our WTM band, The Denialators—you can watch a performance of this song at <www.humancondition.com/denialators>.

[1292](Many more excited anticipations of the human-condition-reconciled new world, especially from 1950s and 1960s music, can be found in *Freedom Expanded* at <www.humancondition.com/freedom-expanded-anticipations>.)

[1293]The following picture of a tiny figure standing with upraised arms in front of an immense sun as it rises over the horizon represents the arrival of understanding and the dawn of humanity's all-magnificent freedom from the darkness and horror of the human condition. Created by WTM Founding Member Genevieve Salter in January 1998, this image has become the emblem of the World Transformation Movement and the inspiration for the cover of this book.

Painting by Genevieve Salter © 1998 Fedmex Pty Ltd

Genevieve's Sun poster, copies of which are freely available at
<www.humancondition.com/posters>

[1294] Finally, bringing these understandings to the world has been an enormous effort for the tiny band that is the 50 Founding Members of the WTM. It has been such a struggle and such a team effort that in the more than 20 years that most of us have been involved there has hardly been a week go by when one of us hasn't been prompted to say 'we couldn't have survived that attack on us', or 'we couldn't have accomplished that critical task', 'if we were even one less in number'. We have fought so hard and for so long and been through so much together it really is like we are one organism—and, given all that we have managed to achieve together based on our love of this information and what it can do for the world, it is a *superorganism*, with each member—Annabel Armstrong, Susan Armstrong, Sam Belfield, John Biggs, Richard Biggs, Anthony Clarke, Lyn Collins, Steve Collins, Lachlan Colquhoun, Eric Crooke, Emma Cullen-Ward, Fiona Cullen-Ward, Anthony Cummins, Neil Duns, Sally Edgar, Anna Fitzgerald, Brony FitzGerald, Connor FitzGerald, Tony Gowing, Jeremy Griffith, Simon Griffith, Damon Isherwood, Felicity Jackson, Charlotte James, Lee Jones, Monica Kodet, Anthony Landahl, Doug Lobban, Tim Macartney-Snape, Manus McFadyen, Ken Miall, Tony Miall, Rachel O'Brien, James Press, Stacy Rodger, Marcus Rowell,

Genevieve Salter, Will Salter, Nick Shaw, Wendy Skelton, Pete Storey, Ali Watson, Polly Watson, Prue Watson, Tess Watson, Tim Watson, James West, Stirling West, Prue Westbrook, Annie Williams—acting as beacons in **'the crashing darkness'** in this **'winter for the world'**, lining **'the pathway of the sun'** to humanity's freedom for *everyone*.

""" **The Rising Sun** """
**Founding Members of the World Transformation Movement behind Norman and Jill Griffith's Memorial Tree, Sydney, 13 December 2008**

L to R: <u>Back Row</u>: Sean Makim*, John Biggs, Tim Macartney-Snape, Will Salter, Tony Gowing, Sam Belfield, Neil Duns, Anna Fitzgerald, Anthony Cummins, Tess Watson, Anthony Landahl. <u>3rd Row</u>: Monica Kodet, Prue Watson, Andy Colquhoun*, James Press, Tony Miall, Ken Miall, Richard Biggs, Pete Storey, Connor FitzGerald, Emma Cullen-Ward, Nick Shaw, Lee Jones, James West, Eric Crooke, Tim Watson. <u>2nd Row</u>: Felicity Jackson, Fiona Cullen-Ward, Marie McNamara*, Damon Isherwood, Ali Watson, Anna Zilioli*, Stirling West, Susan Armstrong, Polly Watson, Doug Lobban, Dave Downie*, Sally Edgar, Charlotte James, Manus McFadyen, Annabel Armstrong, Genevieve Salter, Marcus Rowell. <u>Front Row</u>: Katrina Makim*, Simon Griffith, Jeremy Griffith, Annie Williams, Lachlan Colquhoun, Anthony Clarke, Wendy Skelton, Brony FitzGerald, Prue Westbrook, Sarah Colquhoun*, Lyn Collins, Steve Collins, Amanda Purdy*, Stacy Rodger, Simon Mackintosh*. <u>Sitting on grass</u>: Tess Colquhoun*, Meika Collins*, Hebe Colquhoun*, Katie Collins*. Also present in spirit, adored Founding Member Rachel O'Brien. (*Supporter or Friend of the WTM.)

[1295] So please visit our World Transformation Movement website and join us—because be assured that this book epitomizes what the whole human journey has come down to: this book represents the final great battle, the war of the worlds, the battle of Armageddon, between 'the Christ' and 'the anti-Christ'; between the truth and the lies; between what I offer and what E.O. Wilson offers; between the responsibility of our conscious mind to deliver knowledge, ultimately self-knowledge, the honest explanation of the human condition, and the abrogation of that responsibility, which is Wilson's dishonest explanation of the human condition—basically between transformation and terminal alienation for the human race. We, the human race, everyone, *must* make sure transformation wins.

[1296] (Video presentations where I and others in the WTM answer questions about and describe the Transformed Lifeforce Way of Living can be viewed at <www.humancondition.com/affirmations>. And a more complete description of all aspects of the transformation can be found in *Freedom Expanded* at <www.humancondition.com/freedom-expanded-book2>.)

[1297] In conclusion, the image below is a still frame from a documentary about the artist Paul Gauguin (*Gauguin: The Full Story*, directed by Waldemar Januszczak, 2003). The beauty and happiness of the dancing girl captures something of the joy and excitement that these understandings now bring to the whole world. What a celebration it is going to be! More of this girl's dancing and some other wonderful footage of excited, transformed-new-world-like Pacific Island dancing from another documentary about Gauguin (*Palette Collection – Gauguin*, directed by Alain Jaubert, 2003) can be seen in this YouTube clip at <www.wtmsources.com/108>.

[1298] As a tribute to the pure, unconditional, selfless love I received from my parents, Norman and Jill Griffith, which created the determined and excited inspiration in this book, I include the following photograph of them. It was taken under the wisteria beside our home on our sheep station, 'Totnes', near the town of Mumbil in the central west of New South Wales in 1959 when I was 13 years old. You can see something of the unresigned, human-condition-defying, unaccepting-of-even-contemptuous-of-corrupted-reality, no-alienated-nonsense, belief-in-another-true-world, relatively innocent, sound, secure, upright, core strength in my mother that nurtured in me the love of, and belief in, the authentic, instinctive, soulful, true world and defiance of the alienated world of dishonest denial. (The origin of innocent women's lack of empathy, even contempt, for non-ideal behavior was explained in par. 810.) Of course, while there had to be a causation for events in the world, *all that really matters now is that everyone's life is celebrated.*

© 2013 Fedmex Pty Ltd

*Jeremy Griffith, Sydney, Australia, November 2013*

A biography of Jeremy is provided by Professor Harry Prosen
in his Introduction in paragraphs 20-27.

# Index

Please note, the references are to paragraphs, not page numbers.
Also note you can search words in the online edition using the 'Search FREEDOM' facility in the top right-hand corner of the book's website at <www.humancondition.com>.

1960s 782, 942, 1289-1292
AAAS 607
Abraham, the prophet: critical contribution 906; demystified 751; monotheism 325, 840; the feast that saved humanity 614-615
Adam and Eve **155-156**; heroes not villains 273-275; naked 783; parallels Adam Stork 58, 65, 252; sexual desire 162-165, 794
Adam Stork analogy 56-66, **250-260**; pseudo idealism 885-886, 890
ADHD 945-952, 962, 966-968, 981, 984, 987
adolescence 106, 266-267, 708-716, **738-768**. See also Resignation
Adolescentman 713-715; Adventurous Early Adulthood 765-768; Angry Adulthood 845-871; Distressed Adolescence 741-764; Early Sobered Adolescence 739-740; Hollow Final Adulthood 894-901; Pseudo Idealistic Late Adulthood 872-893
adulthood 106, 708, 711-716
adventure 67, **756**, 763, 766-767
Africa: cradle of mankind 467, 471-472, 487, 835; picture 386
afterlife 840
Aitken, Robin 1099
Akhenaton: monotheism 325
Albrechtsen, Janet 813-814
Alexander, Richard 665-668
alienation: danger of 1120; definition 259; destroys ability to nurture 961-994; detail focused 683-684; extent of 123-125, 220-232, 642, 677-689, 834-835, 954-957; terminal 944-960, 1108-1111, 1241-1243; twice alienated 689
Allot, Robin 323, 488, 491, 501
Al-Qaeda 1053, 1065, 1073
altruistic moral sense: origin of 375-476
Ames, Evelyn 472
ancestor worship 839-840
Andersen, Hans Christian: *Emperor's New Clothes* 1249-1250, 1261
André, Claudine 418
anger 70, 271
anger, egocentricity and alienation 62, **257**, 726-727
Anglo-Saxons 1034, 1043
animal condition **355**, 357, 394, 425-426, 833, 840
animism and nature worship 838, 840
ants 197, 356, 360-371, 1235
anxiety disorders 942, **945-954**; from lack of nurturing 962; healing tools 958
apocalypse 229, 1058-1060; definition 1153; Four Horsemen 946
Appleyard, Bryan 199
Arabs 912-915, 1028, 1033, 1038, 1053; contribution of 1045
*Ardipithecus* **396-411**, 438, 521-524, 712, 719-720, 736

Ardrey, Robert 514
Aristotle 148
arms race. See warfare
arrested development 709-710
art 124, 829-835
Arthur, King. See King Arthur
Aryans 912-914
Asfaw, Berhane 407
Assyrians 913-915
atheism 1070
Attenborough, David 607
Australia: critical role 1037, 1272-1276, 1283-1285
Australian Aborigine 811; degradation 764; Dreamtime 181; nurturing 728, 748, 976; relative innocence 170, 184, 743-745, 862-868; rock art picture 834
Australian Broadcasting Corporation (ABC): defamation 573, 604, 607, 1078; left-wing 1099
*Australopithecus* 411, 706, 713, 740
*Australopithecus afarensis* 712, 720, 724, 733
*Australopithecus africanus* 706, 712, 726-727, 733
*Australopithecus boisei* 706, 712, 733-735
*Australopithecus robustus* 733-735
autism 749, 945-947, 962, **966-973**, 980-981, 987, 994-995, 1244-1246
Aztecs 910
*Babe*: film 1277
baboons 430; strong-willed females 445
Bacon, Francis 681, 834; *Study for self-portrait* picture 124-125
Badrian, Alison and Noel 430, 527
Bantu 1028, 1039, 1041, 1044
barbarians 859-860, 912-914
Bardot, Brigitte 785-786
Barnett, Anthony 1, 5, 14, 17
Basquiat, Jean-Michel: picture 1059
Baudelaire, Charles 681, 777
'Baywatch effect' 1052
BBC: left-wing 1099
Beatles, The 108. See also Lennon, John
beauty of women. See women
Beckett, Samuel 495, 681, 989
bees 197, 356, 360-371, 1235
Beethoven, Ludwig van: *Ninth Symphony* 1049
Bell, Barbara 428, 451, 693
Belyaev, Dmitri 547
Berdyaev, Nikolai: fear blocks knowledge 14, 120, 237, 582; God-fearing 330; Golden Age 181, 237, 378; human soul is divided 188, 237, 245; on escapism 1069; on human condition **236-237**
Berlin, Irving 790
Berman, Harold J. 1107
Berry, Chuck 1291; *Sweet Little Sixteen* 897
Bettelheim, Bruno 971, 980

Please note, the references are to paragraphs, not page numbers

Bible: Adam and Eve 155-156; Bible demystified 751, 1067-1068; Cain and Abel 906-908; David and Goliath 1262; fire symbol 332; Four Horsemen 946; Garden of Eden 155-156; judgment day 1153; marriage 784; miracles 934-935; Noah's Ark 750-751; old order passed 218, 1148, 1217, 1247; on dishonest intellectualism 611; on love 321-323; original innocence 180; repository of truth 751, 927, 1123; the abomination that causes desolation 1111-1142; Trinity explained 746, 1067; Virgin Mary 796, 1066-1067. See also Abraham, Christ, Christianity, Daniel, Genesis, Hosea, Isaiah, Jacob, Job, Joel, Micah, Moses, (Saint) Paul
big bang 316, 335
Biggs, John 1019, 1201, 1294
bipedalism 391-392, 398-401, 404, 413
bipolar disorder 952. See also depression
Birch, Charles: on consciousness 625; on mechanistic thinking 225, 625, 642, 1155; on order 337; on power addict fathers 1012; *Science Friction* 32; support from 25, 607, page 2
Bird, Richard 337
Blake, William: *Albion Arose* picture 234; *Cringing in Terror* picture 98; doors of perception 996, 1148; marriage of heaven and hell 290; the Bard 1037; *Tiger, Tiger* 1144
Bloom, Allan: *Closing of American Mind* 1103; on Plato 129
Bloom, Harold 1101
Blount, Ben 527
*Blues Brothers*: film 280
Boesch, Christophe 208, 527
Bohm, David 337
Bono 1062, 1287-1290; love comes to town 1255, 1287; on Australia 1283; ruins to the right 1049, 1288; streets have no name 218
bonobos: at Milwaukee Zoo 455-467; bipedal 413; bipedalism pictures 414; consciousness of 426-429, 638, 669, 693-694, 718; denial of their significance 512-518, 524, 525-535, 544-561, 563, 566; equivalent of our ancestor 183, 411, 735-737; evidence Golden Age 450-454; food-sharing 208; forgotten ape 461, 516; group picture 416; hunting 208, 443; infancy period 440-441; Kanzi 427, 452, 454; Kanzi picture 418; Kanzi and Matata picture 375, 717; love-indoctrination 411-476; maternalism 416; matriarchal 409, 416, 444; neoteny 434-438; nest and sharing pictures 413; nurturing 440-443, 511; nurturing picture 417, 477; sex as appeasement 415-416; skeleton like 'Lucy' 720; Specie Individual 415-416; strong-willed females 444-449; threatened species 455, 465-466; vocalization 415
Booker, Christopher 216, 245
'born again' 886-887, **925-931**, 1169, 1199-1200
Bradley, Omar 226, 685, 1054
brain. See consciousness
brain hemispheres 682, 799
Brain Initiatives 28, 603-605, 645, 698, 1061-1062, 1243
brain volume increase 707, 714, 842-843
brain-washing 221, 577

branching development 733-737
British 1017, 1034, 1043
Bronowski, Jacob: loss of nerve 1117, 1145, 1179; the child wonder 486
Brooks, David 1062
Browne, Thomas 182, 472
Bruno, Giordano 599
Buddhism 180, 218, 469, 1069, 1148, 1217; great religion 1045
bullying 731
Burns, Robert 355
Bushmen 1039; degradation 764; nurturing 728, 748, 976; pictures 781; relate to animals 832; relative innocence 170, 184, 206, 743-745, 848, 860-868, 1028, 1282; rock art picture 834
Buteyko breathing 958
Buttrose, Ita 771
Byron, Lord 182
Cain and Abel 906-908
Campbell, Jeremy 1097
Camus, Albert: innocence is condemning 285; on women 789; suicide is the only question 291; *The Almond Trees* 1265-1272; thinking undermines 121
capitalism 289, 1092, 1119, 1224
Carlton, Brian 51, 94
Carr, Nick 948-949
Carr-Gregg, Michael 950, 1003
Casals, Pablo 228
Casebeer, William 579, 593
*Catcher in the Rye* 113-118, 1192
Catholicism 1065-1066
Caucasians 912, 1028, 1031, 1039
cave paintings 830-835; avoided representing humans 834; picture 831, 834
celibacy 784
Celts 913, 1034-1037
Cervantes, Miguel de. See *Don Quixote*
Chalmers, David 651
Chandler, Raymond 791
Chapman, Tracy 1033, 1286
Charles, Prince of Wales 1062, 1104, 1220
Chatwin, Bruce 184, 744, 877
Chauvet Cave 831-834; picture 831
Chesterton, G.K.: *The Man Who Was Thursday* 284
child: leads us home 1262. See also innocence leads us home
childhood 106, 264-267, 708, 711-716, **722-737**; grand mistakes 726
Childman 712-713, 722-737; Early Happy, Innocent Childhood 723-724; Late Naughty Childman 729-737; Middle Demonstrative Childman 725-728
children: ask real questions 740; blame themselves 988, 1004-1005, 1246; called 'kids' 988; picture of 386, 1143; powerless 1011-1012; think truthfully 680-681, 740, 761, 1281
Chilton Pearce, Joseph: endorsement page 1
Chimpanzee Violence Hypothesis 514, 521
chimpanzees: consciousness of 426-429, 638, 692-694; contrast with bonobos 183, 208, 414-416, 434-444, 514-516; fetus picture 438; overlapping development 735-737; used to explain our dark side 514-516, 566

Please note, the references are to paragraphs, not page numbers

Chinese 1028, 1031, 1033, 1038, 1230, 1253; contribution of 1045

Christ: all men drunken 877; another counselor 1217; authoritative 929; become like children 1143; beware false prophets 1126; blind fools 611; character of 929-935; critical contribution 614-615, 906, 930; crucifixion 610, 930; deaf effect 822; exposing light 332; few can think on this scale 17; firstborn from the dead 744; foxes have holes 930; good tree can't bear bad fruit 577; great prophet 614-615, 877, 929-935; judgment day 1181; lamb of God 745; love grow cold 501; male disciples 777; meek inherit earth 1186-1187; miracles 934-935; my burden is light 934; not one stone left of cities 941; 'One Solitary Life' 930; on danger of pseudo idealism 1126-1128; on love 322; persecution of 610, 657, 930; revealed to children 680, 836; second coming 1278; signs of the end 946; the feast that saved humanity 614-615; toughness of 745; two people, one chosen 1154; unresigned 877, 929-930, 934-935; well nurtured 420, 486

Christianity 927, 1065-1071; 'born again' 925, 929-934, 1169; contrast with the Transformed State 1169-1173, 1197-1216; 'the Way' 1200

Christie, Agatha: *The Mousetrap* 283

chronic fatigue: Charles Darwin 608; healing tools 958; Jeremy Griffith 28, 608

Churchill, Scott 579, 581, 593

cities: compound upset 747, 905, 941-943; dismantled 941-942

civilizations: decadence of 1018, 1027, 1044-1045

civility 271, 848; disguises upset 855-859, 865-871. See also self discipline

Clark, Kenneth 1222

Clark, Manning 941

clothes 808; prevent lust 783

codependency 988, 1004

Coleridge, Samuel Taylor 1140

Coles, Robert 109-113

colonialism 1043-1044

communication technology: compounds upset 943-944, 948-949, 954-955; materialism envy 1052-1053

communism 299, 1044, **1080-1081**, 1092-1093, 1131, 1221, 1230

complexity. See Integrative Meaning

Confucianism 1044

Connolly, Billy 1117

Conrad, Joseph 776, 1103

conscience 196, 379; origin of 375-476; sharp accuser 282-288

consciousness: association cortex 633-644, 707; blocked in other animals 425, 660-676, 689-691; conscious selection 247, 659; false explanations for 660-669; greatest invention 44, 66, 310, 629, 659; 'hard problem' 650-652; how it emerged 659-701; liberated by love-indoctrination 690-699; memory 61, 632-639, 647-648; raises issue of human condition 623-630, 640-658; what is 61, 247, 623-658. See also thinking

Conway, Ronald 382

Copernicus 598-599

corruption/graft 1033, 1051-1052, 1252

Coulter, Ann 813-814, 1098

courage of humans 764; picture of 68

Coyne, Jerry 203, 532

cradle of mankind. See Africa

Creationism 334

crying of babies: origin of 728

Csikszentmihalyi, Mihaly: endorsement page 1

Cullen-Ward, Fiona 38, 1274, 1294

culture 829, 1040-1041

cynicism 1026-1027, 1052, 1252

D'Emilio, John 804

Dahl, Roald 761

Dalai Lama 1222

Dalrymple, Theodore. See Daniels, Anthony

*Dancing Queen*: ABBA 897-899

Daniel, the prophet: interpret dreams 1127; on danger of pseudo idealism 1123-1128; those who know their God will resist 594, 1123, 1195

Daniels, Anthony 956

Dante: *Divine Comedy* 879

Darling, James: change of heart 1220-1221; grand solution 228; need to be sensitive and tough 745, 1032; on conscience 611; on education 1104, 1220-1221; on the British 1043; prophet 745, 1220; religion and science reconciled 327

Dart, Raymond 514

Darwin, Charles: avoided human condition 188-189, 485; chronic fatigue 608; criticized 31, 598-600, 608; *Descent of Man* 189, 423, 484-485; dominance hierarchy 354; group selection 203; lone genius 32, 579; natural selection 348, 488; no advance since 225; on moral sense 375; *Origin of Species* 188-189, 195; psychology based on new foundation 5, 188, 339; recognized nurturing 484-485; sexual selection 423; simple theory 240; undecided on 'fittest' 195, 358; untestable hypothesis 31, 581

David and Goliath 1262

Davies, Paul: on God 335; on mechanistic thinking 222; on order 329, 337; on religion 1068; *Science Friction* 32

Dawkins, Richard: on religion 938, 1070

de Waal, Frans 183, 411, 415, 453, 501, 516, 524, 527, 531, 720

deaf effect **77-85**, 131-133, 576, 822, 1208; examples of 79, 93; solution to 24, 36, 86-97

deductive thinking 31, 138, 581-582, 929; picture 138

defamation case. See Griffith, Jeremy

Delaney, Joe 380

Delingpole, James 955

democracy 301, 1047; hijacked by the left 1090-1094

denial: became a feature 828; blocks truth 220-232; Plato on danger of 1129. See also science

Denialators, The 1291

Denisovans 907

Dennett, Daniel 652

depression 108-124, 270, 332, 945-952, 981, 1185, 1241

Derrida, Jacques 1096

Descartes, René 624, 651

developed and developing worlds **1023-1024**, 1044, 1051-1052

Please note, the references are to paragraphs, not page numbers

Development of Integration: chart 842-849
development of matter: chart 316. See also Integrative
    Meaning
Development of Mental Cleverness: chart 842-843
Devine, Miranda 1003
dictators 1009, 1017, 1038, 1041
dimorphism 409-410
displacement behavior 335
Disraeli, Benjamin 228
distraction: need for 947-949
DNA 344-349
dogs: domestication of 547-549, 556
dolphins: intelligence of 425
domestication of animals and plants 841, **902-907**
dominance hierarchy 352-355, 425-426, 557-558
*Don Quixote* 67-68, 1262
Doré, Gustave: *The Dove Sent Forth From The Ark*
    picture 750
Dowd, Maureen 817
dreams 287; prophets' ability to interpret 1127
Drummond, Henry 488-489
Duns, Neil 1201, 1294
Duvall, Robert 899
Dworkin, Andrea 782
Dyer, Wayne 228
Dylan, Bob 218, 278, 1290; times a-changin' 1186,
    1196, 1288
eating disorders 791, 952
Ecological Dominance-Social Competition Model
    505-512, 665-668
ego 62, 257
egocentricity 588, 654, 726, 742; effect on children
    964, **1001-1022**
Einstein, Albert 342; contribution before thirty 23;
    every child a genius 680; intelligence of 1046; on
    God 328; on order 315, 337; on problem solving
    6; on science and religion 611; truth stands test of
    experience 73, 295
Eisley, Loren 487
elaborated reproductive unit 356, 359-371, 385;
    pictures 368
'elephant in the living room' 130, 688, 740, 1284;
    picture of 688
elephants: intelligence of 425; sensitivity of 511
Eliot, George 740
Eliot, T.S.: we can't bear reality 121; end of
    exploration 307, 1147; *The Hollow Men* 895-896,
    900
emotion-induced blindness 782
Empathetic Correctness 1109
end play 48, 226-229, 562-568, 873, 944-960, 1057-
    1062, 1108-1135, 1242; in films 1058-1060
environment: can repair now 1165, 1225-1239;
    destruction of 276; Environmental Movement
    **1083-1084**
eusociality 201
evil 46, 268-271; not condoned 294
evolution 341
Evolutionary Psychology 196-200, 357, 536
Expensive Brain Hypothesis 669
Fairfax Media 573, 607, 1078
fasting 877, 999

feminism 773, 808-818; movement 809, **1082-1083**
Feng, Lei 1230
Ferguson, Marilyn 306
Ferguson, Niall 1101
Fiji 1030-1031
fire: symbolism of 331-332; use of 841
Fiske, John 380, **488-491**, 607
Flinn, M.V. 665-667
Fong, Benjamin 29, 605, 625, 645
Fossey, Dian 429, 446-448, 500-503, 694
fossil hominin skulls: picture 706
Four Horsemen of Apocalypse 946
Fouts, Roger 972
foxes: silver 547-549, 556
Frazier, Ian 36; endorsement page 2
Freud, Sigmund 287, 598, 601, 680, 794
Fruth, Barbara 183, 208, 415, 527
Fuegian Indians 811
Fukuyama, Francis 160
Fuller, Buckminster 680
fundamental truths: three 151, 155, 180
Furedi, Frank 1087
future not boring 1251
future shock 305, 592, 1152
Galileo 599
gambling 1033
Gandhi, Mahatma 328, 1226
Garden of Eden **155-156**, 273-276, 378, 386; pictures
    of 386, 472
Gauguin, Paul 1297
Gee, Henry 3
Geelong Grammar School 20, 607, 1032, 1104, 1220.
    See also James Darling
Geldof, Bob 226
Gemmell, Nikki 491, 791
generations: 1960s 782, 942, 1289-1292; 'baby
    boomers' 782, 950; 'X' 950; 'Y' 949-951, 1003,
    1018; 'Z' 950
Genesis 65, 155, 215, 252, 386, 471. See also Moses
genetic selection 61, 195, 256, **348-350**, 641, 659;
    limitation of 670-676
Gibran, Kahlil: *Broken Wings* 784; on Christ 929
glandular fever 112
Glazebrook, Patricia: endorsement page 1
Gnosticism 1066
God: demystified 31, 169-173, 182, 263, **324-340**; is
    love 327; monotheism 324-325, 840; multiple gods
    840; reconciliation 280, 313
Goddess statues: picture 810
Goethe, Johann Wolfgang von 1117
Golden Age 170, **181**, 378
Goodall, Jane 445 (caption above)
good and evil 46
gorillas: love-indoctrination 446-448, 500; overlapping
    development 735-737; sensitivity of 511
Gould, Stephen Jay 537, 539, 607
Gowing, Tony 38, 1201, 1291, 1294; all free now
    1245; selling point 1256-1257
Goya, Francisco 111; *The sleep of reason* picture 111
Gramsci, Antonio 227
Gray, John 775
Gray, Thomas 1037

Please note, the references are to paragraphs, not page numbers

Greece 912-913, 1044; golden era 747, 1045
greed 1224-1226
Green Movement. See Environmental Movement
Greer, Germaine 773, 1082
Griffith, Jeremy: animals walking free 836 (caption
    above); art 22, 832, 835-836, 996; at Milwaukee
    Zoo 455-467; at university 32; biography 20-
    27; character 33-35; childhood 9, 23; chronic
    fatigue 28, 608; confronting truthfulness of 33-
    37; defamation ruling 18, 573, 578, 593, 602;
    described as a prophet 34, page 2; furniture
    business 22-23; had to create WTM 32, 1294;
    history of rejection 607; holistic, inductive
    science 18, 31, 573, 581-582; *Human Condition
    Documentary Proposal* 25, 607; importance of van
    der Post 1220, 1282; in Tsavo 835; *IS IT TO BE*
    26, 604, 607; job to raise truth 1049; on thinking
    836; parents 23, 1036, 1298; persecution of 15-
    19, 26, 28, 33-34, 572-616; photo 1298; photo in
    Africa 836; photo with bonobos 455, 459; photo
    with van der Post 1282; photos with primates
    445; plagiarized 590, 607; publications 24, 607;
    published in books not journals 32; sun picture
    309; theory 'not science' 31; thylacine search 21;
    unresigned 762; vision 34, 1248, 1281; vision
    poem 1280; Welsh heritage 1036
Griffith, Norman and Jill 23, 1036; photo 1298
Griffith, Simon 608, 1220, 1294
group selection 201-205, 532-543
Gruffydd, Llywelyn ap 1036-1037
'grumpy old men' 895
guilt lifted 71, 292
Guiscard, Robert 1035
Gurkhas 1254
*Guys and Dolls*: musical 805
*Hair*: musical 294, 1290
Hamilton, William 203
*Happy Days*: TV series 802
Hare, Brian 527, 533, 544-561, 566-567, 606-607,
    1155
harems 899
Harlow, Harry 323, 501
Hartwig, Walter 579, 593
Hauteville family 1035
Hawkes, Kristen 778
Hawking, Stephen: on God 328; on order 315, 337;
    support from 25, page 2
*Heart of Darkness* 776, 1103
heaven 1147
Heinberg, Richard **181**, 378
Heisenberg, Werner 328
Hemingway, Ernest 773
Henninger, Daniel 1091
Hesiod **180**, 184, 470, 509, 661, 802
Hill, Andrew 403-404
Hillsong Church 1065
Hinduism: great religion 1045
Hitler, Adolf 577, 859, 1013, 1017, 1041
Hittites 913-915
Hobsbawm, Eric 227
Hodgson, David 593
Hohmann, Gottfried 208, 527, 531

holism. See science
holy 926, 934
holy grail 2, 51, 72, 144, 146, 226, 290, 296, 570, 1146,
    1160; mythology about 1145; song 1283
Homer 802; Ulysses' final journey 1140
*Homo* 706-707, 713-714, 740
*Homo erectus* 666-667, 715, 846-849; left Africa 767;
    tool use 841
*Homo habilis* 667, 706, 715, 733-735, 740, 830
*Homo sapiens* 706, 715; archaic 706; modern 706
*Homo sapiens sapiens* 706, 715, 742-743, 873
homosexuality 802-804
Hope, A.D. 1276
Hopkins, Gerard Manley 108, 270
Horne, Ross 804
Hosea, the prophet 1195; on transformation 1259;
    vision 1280
*House of Cards*: film 681, 972
Howard, John 1040, 1062, 1102
*How Green Was My Valley*: film 1036
Hughes, Robert 111
Hugo, Victor 829
human condition: all-important issue 2, 51, 144, 570,
    688; blame genes 981-984; double and triple
    whammy 261-272; equality of humans 297, 1278-
    1279; examples of relief from 93; false explanation
    of 200-217, 546; fear of 78-83, 101, 104-128; grail
    of science 2, 51, 72, 144, 146, 226, 290, 296, 570,
    1146, 1160; misrepresentation of 43, 214; never
    back down 66, 277, 654, 742; not immutable
    218, 293; paradox of 191, 283-284; 'personal
    unspeakable' 109, 576-577; poles reconciled 298,
    313; savage instincts excuse 40-41, 153, 165, 194-
    195, 210-213, 357, 792; what is 7-14, 43-50
human condition explanation: brief 53-68; full 242-
    311; non-falsifiable situation 583; not untestable
    hypothesis 581-582; simple 240-241
human sacrifice 782
humanity's journey to enlightenment 702-1142; last
    200 years 940-1142; picture 308, 702, 754, 1139
humanity's self-esteem: three blows 598-601
humanity's situation now 1240-1260
humans: heroes 65-66, 273-281, 291-292, 297, 1279;
    three varieties of 1015
humor 869
Humphrey, Nicholas 660
hunting: by humans 355, **778-782**
Huxley, Aldous: animals live in present 250, 648;
    humans don't want to know 479; truth drowned
    1118
Huxley, Thomas: how stupid to not have thought of
    that 240, 298; on consciousness 651; unlucky
    substitution 195
ice ages 847-848
Idani, Genichi 443
Ihobe, Hiroshi 208
imagination 582, 682, 995
imposed discipline 916-917, 931-932, 938
Indians 1028, 1030-1031, 1033, 1041, 1253;
    contribution of 1045
individuation 306
Indo-European speakers 912-915, 1034; diagram 915

Please note, the references are to paragraphs, not page numbers

inductive thinking 31, 138, 581-582, 929; picture 138
infancy 106, 691-694, 708, 711-716, **717-721**
Infantman 712, 717-721
innocence leads us home 1261-1279
instinct 61, 256; definition 248
Integrative Meaning 31, 169-173, 182, 263, **313-374**; chart 316; development of order 314-323, **341-373**; science's denial of 317-323, 326-340, 640-642
Intelligent Design 334
International Primatological Society 458, 581, 607
internet. See communication technology
*Ipi Tombi*: musical 781
IQ 1046; the average is where balance was struck 843; vs SQ 652, 657, 1188
Ireland, David 1276
Iroquois Confederacy 916
Isaiah, the prophet: all in darkness 278; calloused hearts 681; child lead them 1262; fear of God 336; judgment day 1153; on transformation 1260; swords into ploughshares 1217; there is no soundness 182; vision 1280; voice of one calling 1276
Islam: fundamentalism 1065, 1073; great religion 1045; Koran 155; terrorism 1053, 1065, 1073. See also Muhammad
Islamic State 1065, 1073
*It's Kind Of A Funny Story*: film 108
Jacob, the prophet 906
James, Oliver 473, 982
Janov, Arthur: child split by pain 221, 684; children blame themselves 1005, 1115; on ADHD 946-949; on brainwashing 577
Jason and the Argonauts 1145
Jaynes, Julian 245
Jews 1028, 1039, 1041, 1048; contribution of knowledge 1046; gave us the Bible 927
Job, the prophet 332, 681
Joel, Billy 1286
Joel, the prophet: judgment day 1153; on transformation 1259; vision 1280
Johanson, Donald 720
Johnson, Robert A. 898; Fisher King 1263
Jones, J.D.F. 864
Judaism: great religion 1045; Torah 155
judgment day 74-75, 295, 305, **1150-1156**; Enrico 1244; most fearful thing is yourself 1185; solution to 1182-1196
*Jumping Mouse, The Story of* 576
Jung, Carl: empathy heals 934; man is dangerously unaware 226, 688; nostalgia theory 184; on Christianity 927, 1071; on dreams 287; our shadow 13, 72, 121, 155, 218, 290, 654; women's blindness 776
Kano, Takayoshi 415, 441, 527
Kant, Immanuel 120, 375
Kauffman, Stuart 337
Keith, Arthur 520
Kelly, Ned 1037
Kelly, Walt 50
Kenya: corruption 1033, 1052, 1252-1253; Tsavo NP 835
Kessler, Karl 536

'key unlocking mind' picture 72
Khan, Genghis 859, 908-909
Kierkegaard, Søren: on human condition 13, 119, 879; on losing oneself 959
King Arthur 1036, 1145, 1263
King Jr, Martin Luther 220, 308, 1279
King, Steve 3
Koestler, Arthur: instinct vs intellect 187, 245; negative entropy 316-317; on mechanistic thinking 223, 792; on order 337-339
Konner, Melvyn 817, 863
Kronemeyer, Robert 804
Kuhn, Thomas: advocates win the day 141, 579; old scientists have to die off 589; outsider not hampered 6, 32; progress made with books not journals 32
Kuroda, Suehisa 527
Laing, R.D.: child abdicates ecstasy 495, 681, 989; each child a prophet 680-681, 1281; explore consciousness 224, 688; famine of hearing 1117; on alienation **123-125**, 190, 224, 287, 649, 678, 1134; on insanity 987; on schizophrenia 999
Lake, Frank 1007
language 637, 727-728, **828-830**
Lanting, Frans 415, 453, 516
Larkin, Phillip 491
Lawrence, D.H. 796, 1266, 1272
Lawson, Henry 110-111, 1275
Leach, Penelope 944
Leakey, Richard 486, 505, 727
Lee, Desmond: on Plato 128
Lee, Richard 778, 781, 863
left-wing strategy to life **918-939**, 1085-1101, 1128, 1135-1136. See also politics and pseudo idealism
Lennon, John 1290-1291; *Gimme Some Truth* 1157; *Imagine* 217-218, 1217; *Let It Be* 108
*Let It Be* 108
Leunig, Michael: cartoons: Adam and Eve 274-276, 279; gardens of human condition 49; men and women 776; tattoos 957; truth and beauty 232; understandascope 242
Lewin, Roger: bonobos 442; endorsement page 1; on consciousness 624; on order 337; *Origins* 505
Lewis, Jerry Lee 1291
Lewontin, Richard 537, 539
Liedloff, Jean 728, 748, 976-977, 980
life: explained 344
Little Richard 1291
Lobban, Doug 1201, 1228-1229, 1258, 1294
Lola Ya Bonobo sanctuary 418
*Lord's Prayer* 218, 1148, 1217
Lorenz, Konrad 536
love: conditional 1002-1006; definition 321; falling in 786, 789-790; first love 784; science's denial of 323, 493-494, 501, 694; significance of **321-323**, 327, 426, 473, 489, 491-495, 675, 840
love-indoctrination **388-449**, 511, 606; assisted by consciousness 421-431; bonobos 411-476; brief summary of 550-559; dimorphism 409-410; first put forward 488-490; fossil evidence 396-410, 519-524, 719-721; ideal nursery conditions 393, 402-404, 414; in gorillas 446-448, 500; liberates

Please note, the references are to paragraphs, not page numbers

consciousness 690-699; loss of body hair 438; maternalism 394, 405, 416; neoteny **432-439**; self-selection 423, 556; sexual selection 408, 423, 430, 432-439, 556-557, 788; small canines 406-408, 411; strong-willed females 444-449

Lovejoy, C. Owen 183, 401, 407-408, 410, 522-524, 531, 719

'Lucy' fossil 720, 724

Luther, Martin 751

lying 127, 214, 265, 726, 988, 1117; reverse-of-the-truth 577

Macartney-Snape, Tim 1220; photo with van der Post 1282; support from 608, 1294; WTM Patron 26

Machiavellian Intelligence Hypothesis. See Social Intelligence Hypothesis

MacLaine, Shirley 771

MacLean, Paul 245

*Mad Max*: film 1277

Maddox, John 24, 607

Madonna and child 420, 486, 717; picture 375, 717. See also Virgin Mary

Mailer, Norman 817

male provisioning model 522

*Man From Snowy River* 1273

*Man of La Mancha*: musical 68, 299, 754, 874

'man the hunted' theory 538

manners: social 856

Marais, Eugène 708, 1235; instinct consciousness struggle 187, 245, 648

marriage 784

Marsden, John 420, 983

Marshall Thomas, Elizabeth 206, 862-864

Marshall, Lorna 206, 862

Martin, John: *The Bard* picture 1037

Marx, Karl 299, 1044, **1080**

masks 856; picture 856

mate selection. See sexual selection

materialism 289, **827**, 1224-1228

materialism envy **1051-1056**, 1072-1073, 1252

maternalism 390-395

matriarchy 416, 769-770, 810-814

*Matrix, The*: film 576

Matthews, Robert 1006, 1009

McCarthy, Joe 1131

McCarthy, Mary: happy ending 1148; on religion 1063

McCollister, Betty 978-981

McKenzie, Clancy 1052-1053, 1252

meaning of life 313-374

meat-eating 780

mechanistic science. See science

meditation 287, 999, 1178

men: become weak 814-819; heroes 773, 817, 896; misunderstood by women 771-778, 810; movements 817; obsolete 817-819

men and women 769-824; different roles 770-771, 781, 963; different roles photos 781; when older 894-900

Mencius 681

mental abilities in other animals 670

messiah 296, 1147, 1278

'Mexican Standoff' 1209

Micah, the prophet: judgment day 1154

Michelangelo: *Creation of Adam* picture 280, 313, 775; *Last Judgement* picture 98

mid-life crisis 872-893; adopt pseudo-idealism 880-890; of the unresigned 876-877; parallels Resignation 880-881

Mill, John Stuart 579, 602

Millan, Cesar 557

Milligan, Spike 935

Milton, John 181

Milwaukee County Zoo 428, 451, **455-467**

mind-control 576-577

miracles 934-935

Mithen, Steven 207

monogamy 784

monotheism. See God

Monroe, Marilyn 795

Montagu, Ashley: honesty of 457; on Aborigine's nurturing 748; on love and nurturing **493-494**, 496-497, 618, 695, 974, 978, 992; on psychopaths 1013

Moore, Michael 299, 1092

moral instincts: origin of 375-476

Morissette, Alanis 1002

Morrison, Jim 218, 294

Morton, John: support from 25, 607, page 1

Moses: Adam and Eve 155-156, 178; Cain and Abel 906-908; critical contribution 906; face God 906; few can think on this scale 17; great prophet 614-615, 751, 877, 906; like God, knowing 702, 1147, 1217; limited by no science 156, 245, 296; men rule 770, 777; Noah's Ark 750-751; on sexual desire 162-165; Ten Commandments 877, 906, **917**, 931-932; the feast that saved humanity 614-615; unresigned 877; use of metaphors 751; woman as helper 785. See also Genesis

Moss, Cynthia 511

*Mud*: film 784

Muhammad 747; judgment day 1153. See also Islam

Mulavwa, Mbangi 415

multiculturalism 1040

Multilevel Selection 200-219, 357, 538-541

Munch, Edvard: *The Scream* 124

music 829-830, 1290-1292

Mussolini, Benito 859, 1013, 1017

mythology: child leads us home 1261-1277; contemporary 218; now explained 67, 751

Nagel, Thomas 1081

naked mole rat 370

Napoleon (Bonaparte) 859, 1017, 1035

narcissism 1003-1022; in Y-generation 951

natural selection. See genetic selection

nature: importance of 942

*Nature* journal 24, 607

nature vs nurture 473, 983

Neanderthals 830, 848, 907

Near-Death Experience **997**, 1200

Negative Entropy 316-317, 342-348, 384-387

neoteny **432-439**, 549-551, 556, 787-791, 802; picture of pandas 433; pictures of: bonobo 435; chimp fetus 438; chimp infant 437

Neumann, Erich 184, 245

neurosis: origin of 63, 258, 287, 1005

*Never Ending Story*: film 1264

Please note, the references are to paragraphs, not page numbers

Neville, Richard 48, 228, 1146
New Age Movement 1081-1082
*New Scientist* 24, 607
Nietzsche, Friedrich: brothers in war 302, 892; on danger of pseudo idealism 764, 1121; on lying 1117; on women 777, 785, 789, 797, 820; psychological breakdown 821; spirit of heaviness 1268-1269; *Thus Spoke Zarathustra* 821-822; will to power 821
Nishida, Toshisada 527
Noah's Ark 750-751; picture 750
non-falsifiable situation 583
Normans 1034-1035
Norwich, John Julius 1035
nostalgia for infancy theory 184-185, 470-471
nurturing: alienation destroys ability 961-994, 1001-1022; becomes priority 617-621, 989-993; confronting truth 419-420, 473-474, **491-498**, 513, 980-981, 1011; created moral sense 388-395, 486-490, 699; difficulty of 619, 962; obvious truth 483-490, 499-504
*Ode to Joy* 1049
O'Hara, Maureen 228
O'Loan, Macushla: endorsement page 2
O'Reilly, Bill 1062
orangutans: overlapping development 735-737; 'Sheriff Daisy' 431, 454
order in nature. See Integrative Meaning
*Origin of Species*. See Darwin, Charles
*Orrorin* 396-410, 521, 712, 719
Orwell, George: *Nineteen Eighty-Four* 300, 892, 1118, 1247
overlapping development 735-737, 860
Paglia, Camille 771, 774, 804, 814, 1131
parallel development 737
*Paranthropus* 733
*Parenthood*: film 473
parenting. See nurturing
Parish, Amy 527
Pascal, Blaise 49, 66, 103, 628
Paterson, 'Banjo': *Man From Snowy River* 1273; *Song of the Future* 1274; *Waltzing Matilda* 1037
Paton, Alan 1144
patriarchy **769-770**, 777, 810-814
Paul, Saint: conversion of 1198-1200; surpassing glory 931-932, 1215
Pericles 747
Peter Pan 732
Phillips, Melanie 1099
Picasso, Pablo 681, 835
Pilbeam, David 526-527
Pinker, Steven 651-652
Planck, Max 327, 582, 589
Plato **157-178**; cave allegory 12, **17**, **82-86**, 126-129, 222, 331, **574-578**; cave picture 39; cave prisoner's good fortune at being released 93, 1223; chariot allegory 157-167; critical contribution 906; few can think on this scale 17; grow accustomed to light 24, 36, 86; instinct and intellect 159-161, 171-172; limited by no science 167, 245, 296; on danger of denial 175, 214, 222, 338, 1129; on

fuddled thinking 222, 679, 836; on Integrative Meaning 169, 173, 325, 328, 331, 338; on original innocence 157-158, 170, 180, 184-185, 470, 509, 661, 783; on resistance to liberation from the cave 17, 574-578; on sexual desire 161-165, 794; on the search for knowledge 174; philosopher kings 1042; prophet 614-615, 747; 'reversed cosmos' myth 168-175; soul resembles the divine 177, 379; the feast that saved humanity 614-615; *The Republic* 12, 82-83; translations 127-129; truthfulness of 81, 126, 157, 614-615
Plavcan, J. Michael 409, 411
Player, Ian: endorsement page 2
pocketing win 1209
Politically Correct Movement **1085-1096**, 1110, 1131
politics: end of 299-304, 1136; left and right wing 299-304, 889-892, 903, 1085-1101, 1213; left and right wing's cultural war 1123-1125, 1135-1138
polygamy 899
Poole, Joyce 511
Pope Benedict XVI 1134
Pope, Alexander 1145; conscience is a sharp accuser 285, 381
Popper, Karl 581
population explosion 48, 941, 1234
pornography 818, 954
Porter, Cole 786
Postmodern Deconstructionism 1096-1115
power addicts **1001-1022**; in civilizations 1018; in companies 1016; in countries 1017; picture 1011-1012; transformation of 1019
prayer 287, 999
prejudice: danger of 1025, 1041; now eliminated 1020, 1025, 1041, 1279; racism 1046
Presley, Elvis 1288, 1290
Press, James 460, 1294
Price, Weston 748
Prigogine, Ilya 337
primal therapy 1005
Prometheus 282
prophets 750-751, 906, 916-917, 925-936; all children potential prophets 680-681, 1281; contemporary 218, 278, 1288; definition 127, 744; demystified 1067-1068; false 1126; foresight 1127; holy 926; honesty is love 822; persecution of 610; relatively innocent 877; saved humanity 614-615; security of self 998-999; shepherds 747; vision of 1280-1281
Prosen, Harry: animal psychiatrist 456; appeal for support 609; at Milwaukee Zoo 455-467; background 4, 15; holy grail of insight 72, 290, 1160; inductive science 581; Introduction **1-38**; photo with bonobos 455, 459; support from 141, 579, 593, 607
pseudo idealism 299-304, 764, 872, **881-890**, **918-939**; danger of 882-890, 919-924, 1111-1135, 1155; hijacking institutions 1099-1108; is psychosis 1115-1116; movements 306, 1063-1064, 1112-1114, 1178; non-religious forms 1076-1079
psyche 379
psychiatry: now possible 72, 290
psychologically crippled state 1002-1015

Please note, the references are to paragraphs, not page numbers

psychopath. See power addicts

psychosis: definition 258-260, 382; origin of 63, 258, 287

'puberty blues' 112

Quantock, Rod 121

races: denial of differences 862-868, 1025, 1042-1047; differences **1023-1050**, 1051-1052; exhausted 1025-1033, 1038-1050; functional 1034-1050; innocent 743-745, 748, 764, 859-868, 907, 1025-1033, 1038-1050; use of term 48, 1023

racism. See prejudice

Rand, Ayn 813-814, 1092

reciprocity 351-352

reductionist science. See science

Reich, Wilhelm 761, 1026

Reinartz, Gay 466

relief-hunting. See pseudo idealism

religion **919-939**; ancestor worship 839-840; animism and nature worship 838, 840; best form of pseudo idealism 938; demystified 1067-1068; fulfilled 1217; fundamentalism 751, 1065-1068, 1073; less guilt-emphasizing forms 1065-1075; repentance 925; too confronting 1063-1064

religion and science: reconciliation 4, 32, 326-328, 336

*Republic, The*. See Plato

Resignation **103-118**, 738, 742-744, 875-881; becomes universal 743, 750, 902; form of autism 749; life after 119-133, **654-658**, 742, 753-758, 766-768; picture of 115; poetry 116-117; those who didn't resign 759-763, 766-768; three features of 876, 879

rhesus monkeys 501; picture 391

Ricciardi, Mirella 805, 1052, 1252

Richards, Renée 803

right-wing strategy to life **904-917**, 926, 1135-1136. See also politics

rock and roll 1290-1291

Rodger, Stacy 810, 1201, 1294

Rolling Stones, The 290

romance 757, 786

Rousseau, Jean-Jacques: E.O. Wilson's view on 197, 862; mankind good 1229; nothing more gentle 181, 378, 474; women's nurturing 770

Rove, Karl 1098

Russell, Bertrand 121

Ryle, Gilbert 651

*Sahelanthropus* 392, 396-410, 521, 712, 719

Saint-Exupéry, Antoine de 226, 740, 1220

Sakamaki, Tetsuya 208, 443

Salinger, J.D.. See *Catcher in the Rye*

Salter, Genevieve 1293-1294

Sanders, Bernie 1131

Sandin, Jo 456, 458, 462-464, 607

Sartre, Jean-Paul 905

Savage-Rumbaugh, Sue 416, 427, 429, 442, 452, 454, 502, 693-694

savants 995

Savory, Allan 240

SCENAR 958

Schaller, George: endorsement page 2

Schiffer, Claudia: picture 788

schizophrenia 969, 981, 987, 999

Schopenhauer, Arthur: acceptance stages 590; discovery of truth 220, 675; man's diabolical nature 45

Schreiner, Olive: destructiveness of sex 795; new time for women 798, 899; on nurturing 420, 492, 494, 496, 770, 974; women must march with regiment 807

Schrenk, Friedemann: endorsement page 1

Schrödinger, Erwin 317, 328

Schultz, Adolph H. 438, 447

science: books obsoleted 1247; danger of its denial 135, 140-143, 478-482, **562-568**, 569-616, 1130, 1155; deductive and inductive 31, 138, 581-582, 929; deductive and inductive picture 138; denial of bonobos' significance 512-518, 544-561, 563, 566; denial of differences between races 862-868; denial of fossil evidence 519-524; denial of Integrative Meaning 317-323, 326-340, 640-642; denial of love 323, 493-494, 501, 694; denial of significance of nurturing 477-621, 981-984; dishonesty of 10, 135, 185, 189, 536-561, 590, 611-613, 792-794; grand syntheses 3, 31, 573, 581-582; great questions 3, 76, 102, 146, 310, 376-377, 479, 622, 702; holism and teleology 31-32, 138, 169, 263, 573, 581-582; holism and teleology definition 318; its practice of denial 34, 75-76, 98, **134-148**, 149-153, 176, 193, 220-232, 317-323, 326-340, 473-474, 623-630, 640-658; left-wing and right-wing 536-543; mechanistic and reductionist 5, 29-32, 134-148, 149-153, 220-232, 338, 569-571, 581-582, 604-605, 611, 625, 645, 683-684, 698; not objective 590, 1151; now farcical 542, 611-613; progresses funeral by funeral 589; responsibility of 16, 30, 135, 140-143, 310, 570, 585-587, 592, 612, 620, 1146, 1211; savage instincts excuse 40-41, 153, 165, 194-195, 210-213, 357, 792; scientific method 590; the obvious truths 478-482; trap 140, 569-571, 603

science and religion. See religion and science

*Science Friction* article 32

*Scientific American* 521

Searle, John 651

Seaver, George 240, 836

'second coming' 1278

Second Path of the Second Law of Thermodynamics. See Negative Entropy

secularism 1070

selection against aggression: flawed theory 531-537, 543, **544-561**

self discipline 845-859

Self-Domestication Hypothesis **544-561**, 562-568; failed to acknowledge author's work 606-607; false explanation of human condition 546, 566-567

selfishness: destructive 1225; origin of 62, 257

selflessness: power of 1228-1235; significance of 263, 320-321, 473, 675

self-selection. See love-indoctrination

Semitic speakers 912-915

sex as humans practice it 155, 161-164, 170, 778, **782-809**; act of love 785; cheapened 806, **954**; dishonest theories 792-794

Please note, the references are to paragraphs, not page numbers

sexual reproduction: origin of 347
sexual selection 408, 423, 430, 432-439, 556-557
Seymour, Mark: *Holy Grail* 1283
Shakespeare, William: on Africa 472; piece of work is man 49, 66, 103, 628
'shattered defense' 998-999
Shaw, George Bernard 591
Shaw, Nick 1019, 1294
Sheldrick, Daphne 511
Shelley, Percy Bysshe 681, 941
'ship at sea' 1209
'side with the angels' 1209
silence: effect of 761, 764, 988, 993
Simon and Garfunkel 761
*Simpsons, The*: TV series 820; Homer vs Ned 937, 1202-1204, 1213, 1245
sin 46; seven deadly 289
Sioux Indians 910
siphonophores 362-363
Smuts, Barbara 527, 529
Smuts, Jan 318
*Social Conquest of Earth, The*. See Wilson, E.O.
Social Darwinism 194-195, 357
Social Ecological Model 517, **525-535**, 541, 543, 550-552, 559-568
Social Intelligence Hypothesis **499-512**, 660-668
social media. See communication technology
socialism 299, **1080-1081**, 1092-1093, 1122, 1221, 1224
Sociobiology 196-200, 357, 536
Socrates 597, 747, 1118
Solomon, Andrew 945, 981
Solomon, King 1235
*Somewhere*: song 790
Sondheim, Stephen 790
songs that anticipate new world 1283, 1286-1292
Sophocles 277
Sorokin, Pitirim 473
soul: definition 63, 177, 258, 260, **379**; distress of 286; origin of 375-476
Spartans 912
Specie Individual 350-352, 356, 372-373, 384-387, 415, 661, 702-703
Spencer, Herbert 195
stages of life 106, 264-267, 638, 697, 703, **708-716**; age twenty-one 755-756, 766; at school 740; Japanese proverb 852, 894; mid-life crisis 872-893; school teachers 730; stages within stages 860
Stalin, Joseph 1009, 1013, 1017
Stanford, Craig 403, 516
Statue of Liberty 281, 890; picture of 281
Steadman, Ralph: *Lizard Lounge* picture 900
Steinbeck, John 491
Steinem, Gloria 773, 1082
Stengers, Isabelle 337
Stevens, Cat (Yusuf Islam) 764, 942, 1288
Stevenson, Robert Louis 740
Stewart, Rod 851
Strahan, Ronald: endorsement page 2
Strum, Shirley 430, 445
sun picture 1293

superficiality 122-123
superheroes 99, 997, 1060
Susman, Randall 527
Sussman, Robert 538-539
Suwa, Gen 403, 406-407, 409, 438
swearing 792, 870
Swift, Jonathan 228
*Sydney Morning Herald*. See Fairfax Media
Tacey, David 1283
Taoism 180, 184
tattoos 956-957
Teilhard de Chardin, Pierre: omega point 306; on order 336-337; on women 789; truth sets world ablaze 1236
teleology. See science
Ten Commandments 877, 906, **917**, 931-932
Tennyson, Alfred: red in tooth and claw 40, 194, 357; *Ulysses* 1140
termites 368, 370; picture 368
terrorism 1053, 1073
Thatcher, Margaret 813-814, 1122, 1220
thinking: our inability to 121, 655-658, **675-689**
Thompson, Hunter S. 1289
Tixer, Alain 418
Toffler, Alvin 592, 1068, 1152
Tolstoy, Leo 790
tool use 841
Townes, Charles 4, 25, 327
Transformed State 289-298, 305-306, 618-620, 743, 1049, **1157-1168**; affirmations 1194, 1201; change of heart 1220-1239, 1253-1260; don't overly study the explanation 1182-1196; *Humanity's Situation* picture 1240-1244; Lifeforce explained 1167; magnificence of 1197-1218, 1253-1260; not a religion 1169-1173, 1197-1216; not pseudo idealism 1174-1181; pathway of sun 1261-1298; procrastination picture 1210; repairs the world 1219-1239; solves exposure 1182-1196, 1244; will help parents 993-994, 1244-1246
trigger warnings 1109
Trinity: demystified 746, 1067
*Truman Show, The*: film 576
Trump, Donald 1091, 1131
truth: right and wrong response picture 1188; set you free 297
Tudge, Colin 607
Turner, J.M.W.: *Fishermen at Sea* picture 278
U2 218, 1049, 1062, 1255, 1287, 1288
United Nations 50, 1134
Universal Beings 1233-1234
universities: become museums 1247-1248
untestable hypothesis 581
upset: explained 64-65, 259; right word 65, 273
USA: anthem 281; Declaration of 297
Valentino, Rudolph 802, 851
van der Post, Laurens: Bushmen ignored by science 864; Bushmen's innocence 184, 745, 761, 848, 1282; Bushmen's sensitivity 832, 996; compassion 290, 1145, 1154; cornerstone of new world 1186; honesty about races 1047; lonely thinkers 232; love in search of love 1285; men and women's

Please note, the references are to paragraphs, not page numbers

pact 785; must defend freedom 1132; on Australia 1285; on danger of pseudo idealism 764, 1122; one lone voice 1237; on religion 1067-1068, 1071; on role of innocence 1278, 1282; on the British 1043; our homesickness 186, 460; persecution of 864, 1047; photo with author 1282; prepared the way for Jeremy Griffith 1282; Prince Charles 1220; prophet 1122, 1220; repression of soul 287; support from 25, page 1; sword and doll 800; the bridge to cross 1285; we need nature 472, 942; women's beauty 789; women's blindness 776; women's patience 899

van Gogh, Vincent 829
van Schaik, Carel 406, 660, 669
Vaughan, Henry 182
Venus figurines 810
Victoria, Queen 806
Vikings 859, 912, 1034-1035
villages. See cities
Virgin Mary 796, 1066-1067. See also Madonna and child
vision people have: explained 1281
'visual cliff' experiments 673
Volvox 361-363
Wald, George 740, 978
Waldrop, Mitchell 337
*Wallace & Gromit* 678; picture 678
Wallace, Alfred Russel 195
*Waltzing Matilda* 1037
warfare 204-209, 271, 826; arms race 909-915; payback 909
Watson, James 1046
Watterson, Bill: cartoon 740
Weil, Simone 479
Weizsäcker, Carl von 889
Welsh: vision of 1036-1037
West, James 1019, 1294
West, Mae 710, 1008
West, Morris 271, 855, 1006
*West Side Story*: musical 790
Wheatcroft, Geoffrey 302, 889, 1119
White, Frances 527
White, Tim 397, 403, 407, 521-522
Whitehead, A.N. 81, 175
Whiten, Andrew 660
*Whole Earth Catalog* 942
Wilberforce, Samuel 581, 600
Williams, Annie 445, 608, 1031, 1033, 1294; photo in Africa 836
Williams, George 197, 536-537
William the Conqueror 1035
Wilson, David Sloan 203, 537-538
Wilson, E.O. 40; bipedalism 404; claim of warring heritage 204-209, 443, 521; consilience 1155; dishonesty of 214, 1295; false explanation of human condition 189, **200-217**, 300, 565-568; human condition is grail of science 144, 570, 1146; humans imperfectible 218; misrepresents Bushmen 206, 865; on consciousness 660; on group selection 200-219, 532; on Multilevel Selection 200-219, 538-539; on religion 215, 326, 938; on

Sociobiology 197; represents the antichrist 1155; Rousseau wrong 197, 862; *The Social Conquest of Earth* 10, **143-146**, 200; 'what guilt?, what psychosis?' 217
Winnicott, D.W. 287, 749, 969-970, 973, 995
With Life In Mind 229, 946
Wobber, Victoria 544-561, 566-567, 606-607, 1155
WoldeGabriel, Giday 403-404
women: ageing 897-900; beauty of **786-795**, 803, 897-899; bimbo, breeder 898; blondes 791; intuition 799; lack of empathy for men 771-778, 810; loss of innocence **795-798**, 803; mystery of 790, 803; nagging 814; not mainframed 776-777, 815; PMT 797; seduction of 805; sex objects 787, 791; sex objects pictures 788; sexually aware 802; strong-willed 810-814; victim of a victim 797. See also men and women
Woods, Vanessa 428, 451, 511, 558, 693
Wordsworth, William: conscience is condemning 285; Eye among the blind 614; *Intimations of Immortality* 121, 182, 614, 681, 835; thoughts lie too deep 121
World Transformation Movement: establishment of 26-27, 29, 305, 572, 1236-1238; Founding Members 572-573, 608-609, 1294; Founding Members childless 608, 1234; Founding Members fortunate background 1254; Founding Members picture 1295; funded by its members 29, 604; introductory videos 87-89; malicious cult accusation 577; non-profit 26; persecution of 572-616, 1062, 1078, 1181, 1187, 1294
Wrangham, Richard 514, 516, 521-522, 526-527, 531, 543-561, 566-567, 606-607, 863, 1155
Wren-Lewis, John 607
writing 828
Yanomamö Indians: relative innocence 862-868; *Secrets of the Tribe* 542
Yequana: nurturing 728, 748, 976
Yerkes, Robert 426, 693
Yiannopoulos, Milo 818
Zihlman, Adrienne 183, 411; picture of bonobo and 'Lucy' 720
zombie: picture 1059
Zoroaster: fire symbol 332; judgment day 1153; monotheism 325